Advanced Thermoelectrics
Materials, Contacts, Devices, and Systems

Series in Materials Science and Engineering

Advanced Thermoelectrics

Materials, Contacts, Devices, and Systems

Edited by
Zhifeng Ren
Yucheng Lan
Qinyong Zhang

CRC Press
Taylor & Francis Group
Boca Raton London New York

CRC Press is an imprint of the
Taylor & Francis Group, an **informa** business

CRC Press
Taylor & Francis Group
6000 Broken Sound Parkway NW, Suite 300
Boca Raton, FL 33487-2742

First issued in paperback 2019

ISBN-13: 978-1-4987-6572-5 (hbk)
ISBN-13: 978-0-367-87797-2 (pbk)

Library of Congress Cataloging-in-Publication Data

Names: Ren, Zhifeng, editor. | Lan, Yucheng, editor. | Zhang, Qinyong, editor.
Title: Advanced thermoelectrics : materials, contacts, devices, and systems /
edited by Zhifeng Ren, Yucheng Lan, Qinyong Zhang.
Other titles: Series in materials science and engineering.
Description: Boca Raton, FL : CRC Press, Taylor & Francis Group, [2017] |
Series: Series in materials science and engineering | Includes
bibliographical references.
Identifiers: LCCN 2017010759| ISBN 9781498765725 (hardback ; alk. paper) |
ISBN 1498765726 (hardback ; alk. paper)
Subjects: LCSH: Thermoelectric materials. | Nanostructured materials. |
Thermal conductivity.
Classification: LCC TK2950 .A38 2017 | DDC 620.1/1297--dc23
LC record available at https://lccn.loc.gov/2017010759

Visit the Taylor & Francis Web site at
http://www.taylorandfrancis.com

and the CRC Press Web site at
http://www.crcpress.com

This book is dedicated to Dr. Jui H. Wang, the late Einstein Chair Professor at the State University of New York at Buffalo, who retired in 2006 at the age of 85 and passed away March 2016 at the age of 95.

In 1990, I came to know Professor Jui H. Wang, who changed my life forever and kept giving me hope and encouragement to pursue persistently until successful. He took me in as a postdoctoral researcher to work with him on high-performance superconducting materials. Through his exemplary hard work, smart decisions, and devotion, we made breakthroughs in the areas of bulk thallium-based superconductors, long-length superconducting tapes, and superconducting films. Our joint work resulted in one paper in *Science* (vol. 271, pp. 329–332 [1996]) and two in *Nature* (vol. 387, pp. 481–483 [1997] and vol. 395, pp. 360–362 [1998]), in addition to over 60 in other journals and three major research grants from the National Science Foundation, Department of Energy, and Army Research Office all in 1998. In addition, our research branched into other areas such as carbon nanotubes, especially aligned carbon nanotubes, for which we jointly published a milestone paper in *Science* (vol. 282, pp. 1105–1107 [1998]), which opened a brand-new field for many people worldwide to follow on.

Building from that success, I moved to the Department of Physics at Boston College as a tenure-track associate professor in 1999. In 2003, I teamed up with Professor Gang Chen at the Massachusetts Institute of Technology to study thermoelectric materials; we published our first paper in *Science* (vol. 320, pp. 634–638 [2008]) and have subsequently published more than 150 papers on thermoelectric materials, contacts, devices, and systems, thereby making it possible for me to edit this book. Since 2013, Ren has been with the University of Houston continuing the research on thermoelectrics and also branched into other fields such as high thermal conductivity materials (boron arsenide), efficient catalysts for hydrogen and oxygen evolution reactions (HER and OER) and water splitting, amphiphilic surfactant for enhanced oil recovery, etc.

Zhifeng Ren
Houston, Texas

Contents

PART I   Thermoelectric Materials

PART II   *Thermoelectric Contacts*

PART III   *Thermoelectric Modules and Systems*

Series Preface

The series publishes cutting-edge monographs and foundational textbooks for interdisciplinary materials science and engineering.

Its purpose is to address the connections among properties, structure, synthesis, processing, characterization, and performance of materials. The subject matter of individual volumes spans fundamental theory, computational modeling, and experimental methods used for design, modeling, and practical applications. The series encompasses thin films, surfaces, and interfaces, and the full spectrum of material types, including biomaterials, energy materials, metals, semiconductors, optoelectronic materials, ceramics, magnetic materials, superconductors, nanomaterials, composites, and polymers.

It is aimed at undergraduate and graduate level students, as well as practicing scientists and engineers.

Proposals for new volumes in the series may be directed to Lu Han, senior publishing editor at CRC Press, Taylor & Francis Group (lu.han@ taylorandfrancis.com).

Preface

Thermoelectric devices were designed and fabricated during the 1960s and 1970s. These devices showed low efficiency because of insufficient properties of thermoelectric materials at that time. Since the 1990s, with the development of nanotechnology and nanoengineering, thermoelectric materials with improved properties have been discovered and fabricated. Such novel thermoelectric nanomaterials possess better thermoelectric properties and can be commercialized to harvest solar energy and waste heat. Thermoelectric devices have been developed from these thermoelectric nanomaterials and tested in laboratories. This book covers important advances in the development of thermoelectric devices made of nanocomposites. Our hope is that it will benefit all the students, scientists, and engineers who are working on thermoelectric devices today and will push forward the commercialization of thermoelectric devices.

The contents emphasize approaches to improving the electrical and thermal conductivity of contacts, stability issues, and challenges to large-scale demonstration, with particular attention to vertical integration. After an introductory section, the next chapters cover traditional materials, followed by new nanomaterials, then a few chapters on contacts and properties of devices, and finally device applications. The book offers readers the state of the art on the synthesis of thermoelectric nanocomposites, techniques for good contacts and stability of devices made of the nanocomposites, and potential applications, especially for emerging waste heat recovery and solar energy conversion.

Editors

Professor Zhifeng Ren is an M. D. Anderson Professor in the Department of Physics and the Texas Center for Superconductivity at the University of Houston, Texas. Professor Ren has been a pioneer in the field of nanostructured thermoelectric materials and devices. His group published an article in *Science* in 2008 establishing the ball milling/hot pressing method as the way to produce thermoelectric nanocomposites with significantly enhanced thermoelectric properties in a bismuth–telluride system. This work pioneered the field of nanostructured thermoelectric materials and has been cited over 2000 times. His laboratory has since successfully enhanced the thermoelectric properties of various thermoelectric nanomaterials such as $YbAgCu_4$, PbTe/PbSe, skutterudites, half-Heuslers, and SiGe alloys, covering 20–1300 K. These thermoelectric nanomaterials were fabricated into thermoelectric devices to harvest waste heat and convert solar energy (published in the journal *Nature Materials* in 2011). His group has published more than 400 papers in peer-reviewed journals, including *Nature, Science, Physical Review Letters, Journal of the American Chemical Society, Nano Letters, Advanced Materials, Small, Advanced Functional Materials, Advanced Energy Materials, Energy and Environmental Science, ACS Nano, Advances in Physics,* and *Proceedings of the National Academy of Sciences,* with over 30,000 total citations.

Professor Yucheng Lan is an associate professor at Morgan State University, Baltimore, Maryland, and was a research assistant professor in the Department of Physics and the Texas Center for Superconductivity at the University of Houston, Texas. Professor Lan has worked on microstructures of thermoelectric nanomaterials and devices. He is the author of 130 peer-reviewed papers published in journals, including *Nature Nanotechnology, Science, Physical Review Letters, Advances in Physics, Nano Letters, Advanced Materials, Advanced Functional Materials, Advanced Energy Materials,* and *Proceedings of the National Academy of Sciences.* His work has been cited over 9000 times with an *h*-index of 34 based on Google Scholars.

Professor Qinyong Zhang is a professor in the Center for Advanced Materials and Energy at the Xihua University of China, Chengdu, China. Professor Zhang has worked on the synthesis of thermoelectric nanomaterials and design of waste heat recovery systems, including the technology of large-scale thermoelectric nanomaterial production and module manufacturing. He is the author of 20 peer-reviewed papers published in several journals, including *Nano Letters, Energy and Environmental Science, Nano Energy,* and *Journal of Applied Physics.*

Contributors

Ernst Bauer
Institute of Solid State Physics
Vienna University of Technology
and
Christian Doppler Laboratory for
    Thermoelectricity
Vienna, Austria

Stephen R. Boona
Center for Electron Microscopy and
    Analysis
Ohio State University
Columbus, Ohio

David Broido
Department of Physics
Boston College
Chestnut Hill, Massachusetts

Cheng Chang
School of Materials Science
    and Engineering
Beihang University
Beijing, China

Gang Chen
Mechanical Engineering Department
Massachusetts Institute of Technology
Cambridge, Massachusetts

Lidong Chen
State Key Laboratory of High
    Performance Ceramics and Superfine
    Microstructure and Shanghai
    Institute of Ceramics
Chinese Academy of Sciences
Shanghai, China

Renkun Chen
Materials Science and Engineering
    Program and Department of
    Mechanical and Aerospace
    Engineering
University of California, San Diego
San Diego, California

Shuo Chen
Department of Physics
and
Texas Center for Superconductivity
University of Houston
Houston, Texas

Eyob Chere
Department of Physics
and
Texas Center for Superconductivity
University of Houston
Houston, Texas

Johannes de Boor
Institute of Materials Research
German Aerospace Center
Koeln, Germany

Sonika Gahlawat
Department of Mechanical Engineering
University of Houston
Houston, Texas

Andriy Grytsiv
Institute of Materials Chemistry
and Research
University of Vienna
and
Institute of Solid State Physics
Vienna University of Technology
and
Christian Doppler Laboratory for
Thermoelectricity
Vienna, Austria

Li Han
Department of Energy Conversion
and Storage
Technical University of Denmark
Roskilde, Denmark

Ran He
Department of Physics and Texas
Center for Superconductivity
University of Houston
Houston, Texas

Matt Heine
Department of Physics
Boston College
Chestnut Hill, Massachusetts

Joseph P. Heremans
Department of Mechanical and
Aerospace Engineering, Department
of Physics, and Department of
Materials Science and Engineering
Ohio State University
Columbus, Ohio

Naomi Hirayama
Department of Materials Science
and Technology
Tokyo University of Science
Kagurazaka, Japan

Lihong Huang
Center for Advanced Materials and
Energy
Xihua University
Chengdu, China

Le Thanh Hung
Department of Energy Conversion
and Storage
Technical University of Denmark
Roskilde, Denmark

Tsutomu Iida
Department of Materials Science
and Technology
Tokyo University of Science
Kagurazaka, Japan

Qing Jie
Department of Physics and Texas
Center for Superconductivity
University of Houston
Houston, Texas

Hyungyu Jin
Department of Mechanical Engineering
Stanford University
Stanford, California

Giri Raj Joshi
Sheetak Inc.
Austin, Texas

C. Robert Kao
Department of Materials Science
and Engineering
National Taiwan University
Taipei, Taiwan

Hee Seok Kim
Department of Physics
and
Texas Center for Superconductivity
University of Houston
Houston, Texas

and

Department of Mechanical Engineering
University of South Alabama
Mobile, Alabama

Yasuo Kogo
Department of Materials Science
and Technology
Tokyo University of Science
Kagurazaka, Japan

Machhindra Koirala
Department of Physics
and
Texas Center for Superconductivity
University of Houston
Houston, Texas

Yucheng Lan
Department of Physics and Engineering
Physics
Morgan State University
Baltimore, Maryland

E. M. Levin
Division of Materials Sciences
and Engineering
US Department of Energy Ames
Laboratory
and
Department of Physics and Astronomy
Iowa State University
Ames, Iowa

Cheng-Chieh Li
Department of Materials Science
and Engineering
Northwestern University
Evanston, Illinois

Jing Li
Department of Materials Science
and Engineering
Shenzhen Graduate School
Harbin Institute of Technology
Shenzhen, China

Weishu Liu
Department of Materials Science and
Engineering
Southern University of Science and
Technology
Shenzhen, China

Zihang Liu
Department of Physics and
Texas Center for Superconductivity
University of Houston
Houston, Texas

and

School of Materials Science and
Engineering
Harbin Institute of Technology
Harbin, China

Jun Mao
Department of Physics
and
Texas Center for Superconductivity
University of Houston
Houston, Texas

Natalio Mingo
Laboratoire d'Innovation pour les
Technologies des Energies Nouvelles
Commissariat à l'énergie
atomique-Grenoble
Grenoble, France

Austin J. Minnich
Division of Engineering and Applied
Science
California Institute of Technology
Pasadena, California

Takao Mori
International Center for Materials
    Nanoarchitectonics
National Institute for Materials Science
and
Graduate School of Pure and Applied
    Sciences
University of Tsukuba
Tsukuba, Japan

Pham Hoang Ngan
Department of Energy Conversion
    and Storage
Technical University of Denmark
Roskilde, Denmark

Keishi Nishio
Department of Materials Science
    and Technology
Tokyo University of Science
Kagurazaka, Japan

Ngo Van Nong
Department of Energy Conversion
    and Storage
Technical University of Denmark
Roskilde, Denmark

Nini Pryds
Department of Energy Conversion
    and Storage
Technical University of Denmark
Roskilde, Denmark

Pengfei Qiu
State Key Laboratory of High
    Performance Ceramics and Superfine
    Microstructure and Shanghai
    Institute of Ceramics
Chinese Academy of Sciences
Shanghai, China

Zhensong Ren
Department of Physics
and
Texas Center for Superconductivity
University of Houston
Houston, Texas

Gerda Rogl
Institute of Materials Chemistry
    and Research
University of Vienna
and
Institute of Solid State Physics
Vienna University of Technology
and
Christian Doppler Laboratory
    for Thermoelectricity
Vienna, Austria

Peter Rogl
Institute of Materials Chemistry
    and Research
University of Vienna
and
Christian Doppler Laboratory
    for Thermoelectricity
Vienna, Austria

Udara Saparamadu
Department of Physics
and
Texas Center for Superconductivity
University of Houston
Houston, Texas

Xun Shi
State Key Laboratory of High
    Performance Ceramics and Superfine
    Microstructure and Shanghai
    Institute of Ceramics
Chinese Academy of Sciences
Shanghai, China

Sunmi Shin
Materials Science and Engineering
    Program
University of California, San Diego
San Diego, California

Jing Shuai
Department of Physics
and
Texas Center for Superconductivity
University of Houston
Houston, Texas

Jiehe Sui
School of Materials and Engineering
Harbin Institute of Technology
Harbin, China

Koen Vandaele
Department of Inorganic and Physical
    Chemistry
Ghent University
Ghent, Belgium

Bo Wang
Key Laboratory of Fluid and Power
    Machinery of Ministry of Education
Center for Advanced Materials and
    Energy
Xihua University
Chengdu, China

Dezhi Wang
Department of Physics
and
Texas Center for Superconductivity
University of Houston
Houston, Texas

Kenneth White
Department of Mechanical Engineering
University of Houston
Houston, Texas

Bartlomiej Wiendlocha
Faculty of Physics and Applied
    Computer Science
Akademia Górniczo-Hutnicza im
    University of Science and Technology
Kraków, Poland

Yu Xiao
School of Materials Science
    and Engineering
Beihang University
Beijing, China

Atsuo Yasumori
Department of Materials Science
    and Technology
Tokyo University of Science
Kagurazaka, Japan

Guanting Yu
School of Materials Science
    and Engineering
Zhejiang University
Hangzhou, China

Mona Zebarjadi
Department of Electrical and Computer
    Engineering
University of Virginia
Charlottesville, Virginia

Hao Zhang
Department of Physics
and
Department of Chemistry
University of Houston
Houston, Texas

Qian Zhang
Department of Materials Science and
    Engineering
Harbin Institute of Technology
Shenzhen, China

Qinyong Zhang
Key Laboratory of Fluid and Power
  Machinery of Ministry of Education
Center for Advanced Materials and
  Energy
Xihua University
Chengdu, China

Xiao Zhang
School of Materials Science
  and Engineering
Beihang University
Beijing, China

Huaizhou Zhao
Department of Physics
and
Texas Center for Superconductivity
University of Houston
Houston, Texas

Li-Dong Zhao
School of Materials Science
  and Engineering
Beihang University
Beijing, China

Xinbing Zhao
School of Materials Science
  and Engineering
Zhejiang University
Hangzhou, China

Yiming Zhou
School of Materials Science
  and Engineering
Beihang University
Beijing, China

Tiejun Zhu
School of Materials Science
  and Engineering
Zhejiang University
Hangzhou, China

PART I
THERMOELECTRIC MATERIALS

CHAPTER 1

Introduction to Thermoelectric Materials, Contacts, Devices, and Systems

Zhifeng Ren, Yucheng Lan, and Qinyong Zhang

Thermoelectric technology that converts heat into electrical energy or vice versa is so powerful and important because it does not involve any moving parts nor does it release any by-products into the environment. Therefore, it has a long lifetime and is environmentally friendly. A thermoelectric device can have three functions: electrical power generation when heat is provided and cooling or heating when electrical power is provided. Then why is it not being used everywhere for power generation or cooling or heating? It is simply because of the low conversion efficiency η for power generation, defined as

$$\eta = \frac{T_h - T_c}{T_h} \left[\frac{\sqrt{1 + (ZT)_{ave}} - 1}{\sqrt{1 + (ZT)_{ave}} + T_c/T_h} \right],$$

from which we see that the efficiency is always less than the Carnot efficiency $[(T_h - T_c)/T_h]$ since the factor depending on the average thermoelectric figure of merit $(ZT)_{ave}$ is less than 1. In addition, this efficiency formula was derived by assuming that the Seebeck coefficient S, electrical conductivity σ, and thermal conductivity κ are all temperature independent, with the thermoelectric figure of merit ZT defined as $S^2\sigma/\kappa$. Obviously the higher the $(ZT)_{ave}$, the higher the efficiency. Therefore, in the past almost 200 years since the discovery of voltage generation from heat by Seebeck in 1821, scientists have been trying to improve the ZT of known materials or searching for new materials with higher ZT. The early searching was mainly focused on elements and, many years later, on compounds when it is realized that phonons can be significantly scattered by point defects to reduce the thermal conductivity κ without affecting the power factor $PF = S^2\sigma$ too much. The best-known example is Bi_2Te_3, in which the atomic mass and size difference between Bi and Te makes the lattice thermal conductivity very low (~1.5 W/m K) and the power factor pretty high (~40 µW/cm K²).

The improvement of ZT has been proved extremely challenging since the three parameters S, σ, and κ are all interrelated; changing one will normally affect the other two, which is why ZT has not been increased to much higher than 1.0 until the last 10 years, when it was realized that the only way to separate the influence on charge carries and phonons is to have nanostructures with grain size larger than the electron mean free path but smaller than the phonon mean free path so that electron transport is not affected but phonon scattering is significantly enhanced. Under such a guiding principle, thermal conductivity has been significantly reduced in almost all the existing thermoelectric materials and newly discovered materials by creating nanostructures such as nanoprecipitates/inclusions in matrix, nanograins with many grain boundaries, and defects with a wide distribution of sizes, which has resulted in ZT improvement by at least 30% in many known materials, materials covering temperatures as low as 10 K such as $FeSb_2$ and as high as 1500 K such as borides and carbides, which are discussed in detail in Chapters 2 through 10 of this book.

When the reduction of thermal conductivity reaches a certain level, further improvement in ZT has to come from the improvement in the power factor (PF). In the past few years, new methods have been extensively reported to improve power factors, such as the creation of resonant states, band convergence, modulation doping, and simple electrical conductivity improvement without affecting the Seebeck coefficient. Clearly, when the lattice thermal conductivity is reduced to its lowest level, the amorphous limit, further improvement on ZT has to come from the power factor, which is the direction for future research. These topics are discussed in detail in Chapter 11.

Materials research heavily involves trial and error, which is very time and money consuming, even though it very often produces great discoveries and surprises, but it would be much more time and money efficient if properties can be first theoretically predicted to guide and minimize the number of experimental tries. Therefore, simulation is indispensable in research, especially in any materials research. Chapter 12 discusses some of the most recent findings in simulation and modeling in thermoelectric materials.

Seeking higher and higher ZT is absolutely the ultimate goal in thermoelectric materials research. However, there has been too much emphasis on peak ZT, not $(ZT)_{ave}$, which is not completely correct since a higher peak ZT does not translate into higher efficiency. In fact, a higher $(ZT)_{ave}$ is much more important than a higher peak ZT to the effective conversion efficiency. Therefore, enhancement on $(ZT)_{ave}$ has to be emphasized. However, how to evaluate the $(ZT)_{ave}$ is not a settled issue in the community among the few available methods. Furthermore, the conversion efficiency formula shown earlier is not reliable in predicting the efficiency either since it is assumed that all the Seebeck coefficient S, electrical conductivity σ, and thermal conductivity κ are temperature independent across a large temperature difference for especially very large temperature differences between the hot and cold sides. Fortunately, this situation has recently been improved by redefining engineering ZT, $(ZT)_{\text{eng}}$, as follows:

$$(ZT)_{\text{eng}} = \frac{\left[\int_{T_c}^{T_h} S(T)\mathrm{d}T\right]^2}{\int_{T_c}^{T_h} \rho(T)\mathrm{d}T \int_{T_c}^{T_h} \kappa(T)\mathrm{d}T}(T_h - T_c),$$

which considers the accumulative effect of the Seebeck coefficient S, electrical conductivity σ, and thermal conductivity κ between the cold side T_c and hot side T_h, instead of ZT at a particular temperature T. Based on $(ZT)_{\text{eng}}$, a new efficiency formula η_{CTD} (efficiency based on cumulative temperature-dependent properties) is derived as

$$\eta_{\text{CTD}} = \eta_c \frac{\sqrt{1+(ZT)_{\text{eng}}a_1\eta_c^{-1}} - 1}{a_0\sqrt{1+(ZT)_{\text{eng}}a_1\eta_c^{-1}} + a_2},$$

where a_i is defined as the following by taking the Thomson and Joule heat:

$$a_i = \frac{(T_h - T_c)S(T_h)}{\int_{T_c}^{T_h} S(T)\mathrm{d}T} - \frac{\int_{T_c}^{T_h} \tau(T)\mathrm{d}T}{\int_{T_c}^{T_h} S(T)\mathrm{d}T}W_T\eta_c - iW_J\eta_c \quad (i = 0, 1, 2).$$

Using $(ZT)_{\text{eng}}$, we can now accurately predict/calculate the efficiency very reliably without any ambiguity since $(ZT)_{\text{eng}}$ has a monotonic relationship with the efficiency. Because of this monotonicity, $(ZT)_{\text{eng}}$ instead of $(ZT)_{\text{ave}}$ should be used as the practical barometer of any thermoelectric materials to reliably predict the conversion efficiency.

The efficiency for a power generator is very important, but only half of the importance. Output power density is the other half of the importance because the end users also pay attention to how much they pay for the electrical power, meaning \$$/W$. Output power density is determined by power factor PF and device geometry as

$$P_d^{\text{CTD}} = \frac{m}{(m+1)^2}\left(\frac{T_h - T_c}{L}\right)(PF)_{\text{eng}},$$

where *m* is related to the internal and external resistance and $(PF)_{\text{eng}}$ is defined as

$$(PF)_{\text{eng}} = \frac{\left[\displaystyle\int_{T_c}^{T_h} S(T)\,dT\right]^2}{\left[\displaystyle\int_{T_c}^{T_h} \rho(T)\,dT\right]},$$

which is different from the traditional power factor. The details dealing with $(ZT)_{\text{eng}}$, efficiency, $(PF)_{\text{eng}}$, and output power density are presented in Chapter 13.

The thermoelectric properties have been the focus ever since the discovery by Seebeck, which is the right thing to do. However, when the thermoelectric properties are good enough for some applications, mechanical properties have to be considered for device liability. Mechanical properties have not been paid enough attention since the beginning. Now, it is time to seriously consider them since many materials have pretty good *ZT*s, but very weak mechanical properties, which prevents us from making useful modules using these materials. The details on the study on mechanical properties are described in Chapter 14.

With better and better materials, contact layers (also called diffusion barriers) have to come into play for making any useful modules. The studies on contacts are much less intensive than those on materials, but the importance of contacts is not emphasized, and the challenges are huge. The requirements for contacts are as follows:

1. Zero electrical contact resistance
2. Zero thermal contact resistance
3. Strong mechanical strength
4. Matching of coefficient of thermal expansion with the thermoelectric materials
5. Long-term stability at high temperature

Up to now, contact layers using Ni by electrocoating or electroless coating are commercialized for large crystals Bi_2Te_3 only, not even for nanosized Bi_2Te_3. Reports on successful contact layers have only been seen in p-type $Bi_{0.4}Sb_{1.6}Te_3$. Most of the good thermoelectric materials have not been studied for good contact layers. Without good contacts, good modules cannot be realized. Chapters 15 through 17 are dedicated to studies on contacts.

With good contact layers, the next step is to make modules. Module assembling is not simple and is basically an art and has been kept as secret. There are very few reports on module assembling since it involves much more secret engineering. Nevertheless, we can still see some reports on module assembling efforts. Chapters 18 through 20 summarize the reported efforts on module assembling. Modules based on oxides are discussed in Chapter 21 due to its late submission.

Assuming that good modules are made, then the final step is to assemble them into an integrated system to generate electrical power or provide cooling to end users. With each step further toward real application, the secrecy becomes more. Up to now, there has not been much success on large generators or cooling units. Most of the efforts were focused on generators using solar power or waste heat in vehicles. Chapter 22 presents the preliminary work on this topic.

Finally, we hope that this book can serve as a starting material for young people and professionals to stimulate their unlimited imaginations to make the thermoelectric field more successful.

CHAPTER 2

Low-Temperature Thermoelectric Materials

Koen Vandaele, Joseph P. Heremans, Yiming Zhou, Li-Dong Zhao, Huaizhou Zhao, Zhifeng Ren, Machhindra Koirala, and Stephen R. Boona

Contents

2.1 Bismuth–Antimony Alloys

Koen Vandaele and Joseph P. Heremans

2.1.1 Introduction

Almost two centuries ago, Thomas Seebeck[1] reported that the solid with the highest negative Seebeck coefficient was elemental bismuth and that the one with the second highest positive Seebeck coefficient was elemental antimony. Both elemental Bi and Sb are semimetals, meaning that they have an equal amount of electrons and holes at the Fermi surface. As a result, the partial thermopower (S, for Seebeck coefficient) of the electrons, which is negative, counteracts the partial thermopower of the holes, which is positive, and the total thermopower is quite low (for Bi at 300 K, on the order of $S_\perp \approx -50$ µV/K in the direction perpendicular to the trigonal axis,[2] and $S_\parallel \approx -105$ µV/K parallel to the trigonal axis[2]). As a result, the thermoelectric figure of merit (ZT) of elemental Bi at 300 K is limited to $ZT_\parallel = 0.38$ from 250 to 300 K along the trigonal axis and $ZT_\perp = 0.07$ perpendicular to it, too low to be useful. Elemental Sb has many more charge carriers than elemental Bi, and its thermoelectric performance is irrelevant.

Theory promises much better performance. If, by some yet-unknown process, one could lift the overlap between the electron and hole bands of elemental Bi, then the optimal ZT that could be achieved in n-type material would be $ZT_\parallel = 1.8$ and even $ZT_\perp = 1.0$.[2] The optimum carrier concentration at which these numbers would be realized is calculated by Gallo et al.[2] to be on the order of $n_{OPT} = 1 \times 10^{18}$ cm^{-3}, about one-third the concentration in elemental Bi. Nobody has managed to make such a crystal to date, but two ideas have been tried that are the objective of this review: alloying with Sb and making nanowires. A second promising theory was developed for p-type Bi by Thonhauser et al.[3]: if the material could be doped p-type sufficiently to reach the heavy hole band at the h-point of the Brillouin zone (explained hereunder), which is the equivalent of doping PbTe so as to reach the Σ-band, the material could achieve $ZT_\perp$ = 1.44. Note that while the n-type material is optimal with the heat and current fluxes along the trigonal axis, the p-type material is optimal with the fluxes in plane. This idea was attempted by Jin et al.,[4] but a sufficiently high doping level at 300 K could not be reached.

Overall, the best ZT that has been obtained in Bi–Sb alloys at zero magnetic field is achieved in a single crystal of composition $Bi_{94.5}Sb_{5.5}$ doped with potassium as a resonant level. It has a maximum n-type of $ZT_\parallel$ = 0.7 between 100 and 150 K and $ZT_\parallel$ > 0.5 from 50 to 300 K. This development is described in Section 11.3, which is dedicated to resonant levels.

2.1.2 Crystal Lattice and Electronic Properties

Bismuth and antimony crystallize in the rhombohedral structure $(R\bar{3}m)$, with a trigonal axis (denoted 3), three two-fold binary axes (1), each normal to the trigonal axes and a mirror plane ("binary plane"), and a center of inversion. The bisectrix axis (2) completes a set of orthonormal axes. The lattice constants of pure Bi and Sb and for $Bi_{1-x}Sb_x$ alloys in the range x = 0–30 at.% are given in Table 2.1 at various temperatures. The lattice constants a_H and c_H of single-crystal Bi–Sb solid solutions linearly vary with the Sb content following Vegard's law.[5]

The semimetal behavior of bismuth originates from the small band overlap of the fifth and sixth bands, which have essentially the orbital character of the three 6p electrons. The 6s electrons form a band deep in the valence block. Figure 2.1a displays the location of the Fermi surfaces of bismuth on the Brillouin zone. The two half-ellipsoidal hole pockets of bismuth are located at the T points along the trigonal axis (3), which can be described by the parabolic model $E = \hbar^2 k_1^2 / 2m_1 + \hbar^2 k_2^2 / 2m_2 + \hbar^2 k_3^2 / 2m_3$. The mass tensor is expressed in terms of the crystallographic axes, which are also the principal axes of the ellipsoids in Figure 2.1. The values are reported in Table 2.2.[6]

TABLE 2.1 Unit Cell Parameters for Bi, Sb, and BiSb Alloys

	T (K)	a_H (Å)	c_H (Å)
Bi	4.2	4.5330	11.797
	78	4.5350	11.814
	298 ± 3	4.5460	11.862
Sb	4.2	4.3007	11.2220
	78	4.3012	11.2320
	298 ± 3	4.3084	11.2740
$Bi_{1-x}Sb_x$	4.2		$11.803-40.75 \times 10^{-4}x$
	78	$4.534-21.92 \times 10^{-4}x$	$11.814-48.75 \times 10^{-4}x$
	298 ± 3	$4.546-23.84 \times 10^{-4}x$	$11.863-51.66 \times 10^{-4}x$

Source: Cucka, P., and Barrett, C. S.: The crystal structure of Bi and of solid solutions of Pb, Sn, Sb and Te in Bi. *Acta Crystallographica.* 1962. 15. 865–872. Copyright Wiley-VCH Verlag GmbH & Co. KGaA. Reproduced with permission; Schiferl, D., and Barrett, C. S.: The crystal structure of arsenic at 4.2, 78 and 299°K. *Journal of Applied Crystallography* 2, 30, 1969.

Note: x in atomic percent (at.%).

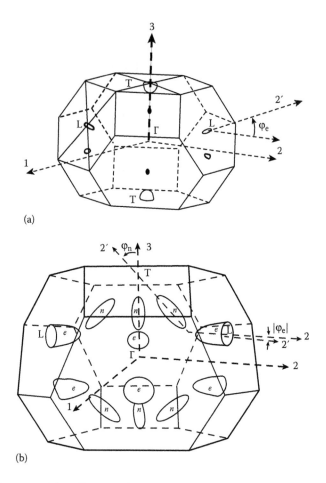

(a)

(b)

FIGURE 2.1   Electron and hole Fermi surfaces of elemental (a) Bi and (b) Sb. Both semimetals have electrons in ellipsoidal pockets centered at the L-points of their Brillouin zone and inclined vis-à-vis the bisectrix axis φ_e. The masses reported in the tables are given in the principal axes of the ellipsoids. The holes in Bi are in ellipsoidal pockets centered at T-point with their principal axes aligned with the crystallographic axes, as shown. The holes in Sb are of a different nature from those in Bi and are in more elongated pockets centered around the h-points internal to the Brillouin zone. They are of the same nature as the heavy holes in PbTe located in tubes at the Σ-points of the face-centered cubic (fcc) crystal structure.

In the notation we use, m_i' represents the masses along the principal axes of the ellipsoidal Fermi surfaces; m_i, the mass tensor elements in the crystallographic axes. For the T-point holes of Bi, we have $m_i = m_i'$. These hole masses are, in a first approximation, independent of energy.

The electrons are located in six half-quasi-ellipsoidal pockets centered at the L-points of the Brillouin zone (Figure 2.1a), which are strongly elongated along a direction near the bisectrix (2) axis. The electron ellipsoids are tilted out of the binary–bisectrix plane by an angle $\varphi_{e_{2\to 2'}}$,[7] also reported in Table 2.2. The highly anisotropic shape is responsible for the small effective masses along the two directions (1, 2) in the basal plane, yielding a very high electron mobility. Bismuth has a direct bandgap of E_g = 13.6 meV at

TABLE 2.2 Band Parameters for Carriers in Bismuth and Antimony at 4.2 K

Carriers		$m'_1 (m'_e)$	$m'_2 (m'_e)$	$m'_3 (m'_e)$	$\varphi_{e_2 \to z'}$ (°)	E_F (meV)	$n = p$ (cm^{-3})	N_V
Bi	L-electron	$m'_{1,0} = 0.00119$	$m'_{2,0} = 0.266$	$m'_{3,0} = 0.00228$	6	27.2	2.7×10^{17}	3
	T-hole	0.064	0.064	0.69	0	10.8	2.7×10^{17}	1
Sb	L-electron	0.0093	1.14	0.088	−4	93.1	3.74×10^{19}	3
	H-hole	0.068	0.92	0.050	53	84.4	3.74×10^{19}	6

Source: Journal of Physics and Chemistry of Solids, S32, Dresselhaus, M. S., Electronic properties of the group V semimetals, 3–33, Copyright (1971), with permission from Elsevier.

Note: m'_1, m'_2, and m'_3 are the effective masses of the charge carriers in fraction of the free electron mass m_e; E_F is the Fermi energy of the electrons or holes, specified with respect to the energy extrema; $n = p$ is the carrier concentration; and N_V is the number of valleys in the Fermi surfaces. The masses are reported along the principal axes of the ellipsoidal Fermi surfaces. In the case of the electrons at the L-point of the Brillouin zone of Bi, the masses are those at the bottom of the conduction band.

0 K between the conduction band and the "light" hole band at the L-point; E_g is temperature dependent.[8] Due to very strong spin–orbit interactions, both L-point bands display a highly nonparabolic dispersion relationship for which a Dirac-type two-band model is used[6]:

$$\gamma(E) = E\left(1 + \frac{E}{E_g}\right) = \frac{\hbar^2}{2m_0}\vec{k} \cdot \overline{\overline{\alpha}} \cdot \vec{k}, \quad \text{with } \overline{\overline{\alpha}} = \begin{pmatrix} m'_{1,0} & 0 & 0 \\ 0 & m'_{2,0} & 0 \\ 0 & 0 & m'_{3,0} \end{pmatrix}^{-1}. \quad (2.1)$$

Here, the values for the effective masses $m'_{i,0}$ are given in Table 2.2 in the coordinates that correspond to the principal axes of the ellipsoids shown in Figure 2.1. To calculate the dispersion relations and masses along the crystallographic axes, a tensor-tilting procedure must be followed with a rotation matrix for the angle $\varphi_{e_{2 \to 2'}}$. On top of that, in Dirac-type bands, the effective mass has an energy dependence that can be derived from Equation 2.1: at any given energy $m'_{i,E}(E) = m'_{i,0}\left(1 + 2E/E_g\right)$, where $m'_{i,0}$ is the bottom of the band mass given in Table 2.2. The effective masses of the electrons at the L-point of Bi are also temperature dependent above approximately the liquid nitrogen temperatures, as is the direct energy gap E_g, so that the band structure of Bi at 300 K looks quite different from that at 4 K.[9] In particular, the energy overlap has increased so much that the electron and hole concentrations increase[2] from ~3 × 10^{17} cm^{-3} at $T \leq 77$ K to ~3 × 10^{18} cm^{-3} at 300 K.

The Brillouin zone of antimony is depicted in Figure 2.1b. Fermi surface hole pockets are located at six equivalent h-points, while the six electron pockets are at the L-points. The deviation of the simple cubic structure is more pronounced in antimony than in bismuth, which results in a higher degree of overlap between the valence and conduction bands ($E_g \approx 180$ meV, an order of magnitude larger than in Bi) and a carrier density in antimony about two orders of magnitude larger than in Bi (see Table 2.2). The effective masses at the bottom of the bands are given in Table 2.2.[6] The effective masses of electron and holes in Sb are assumed independent of energy and temperature, and the dispersion is assumed parabolic. Again, the masses in Table 2.2 are given in the principal axes of the ellipsoids, and the tilt procedure (the angles φ are in Table 2.2) must be followed to calculate the dispersions in the crystallographic axes system.

Solid solutions of Bi and Sb form throughout the composition range. Bi-rich Bi$_{1-x}$Sb$_x$ alloys become semiconductors for ~9 at.% < x < ~22% (see Figure 2.2a).[10] This can be understood from the fact that the hole bands in Bi and in Sb are at different points in the Brillouin zone and move differently with x. For x < 9%, the alloys have a Bi-like valence band; for ~9 at.% < x < ~15%, they are direct gap semiconductors with electron and hole bands symmetric and located at the L-points of the Brillouin zone; and for ~15 at.% < x < ~22%, the valence band becomes Sb-like[4] (not shown). Above x = 22%,

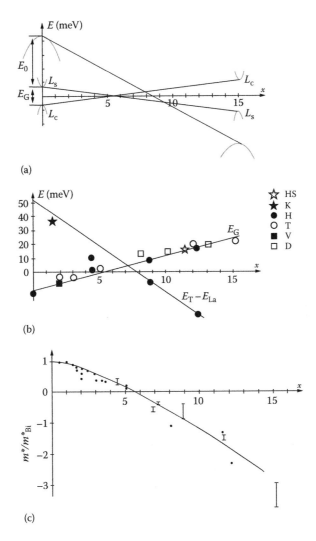

FIGURE 2.2 (a) Band structure of $Bi_{1-x}Sb_x$ alloy at approximately 0 K. (After Heremans, J. P., and Michenaud, J.-P., *Journal of Physics C: Solid State Physics*, 18, 6033, 1985.) (b) Value of the direct band gap E_G at L-point and the energy difference of between the band extrema $E_T - E_{La}$ at the T- and L-points (for $x < 5\%$, $E_T - E_{La} = E_G + E$, where E_0 is the overlap). The letters refer to the authors of the studies (see text). (c) Effective mass of the L-point carriers m^*, normalized to that of pure bismuth (Table 2.2).

the alloys are Sb-like semimetals. There is a Dirac point: the direct band-gap at the L-point decreases in the range of $0 < x < 5.5\%$, until the bonding band L_s and antibonding band L_a cross and a zero gap state occurs at $x \approx 5.5$ at.%. The alloy with the record ZT is K-doped $Bi_{94.5}Sb_{5.5}$ at the composition of the Dirac point where the electrical conductivity is optimal. Solids with the Fermi energy exactly at the Dirac point would have zero thermopower, but this sample is doped with the resonant impurity K and maintains a high thermopower. Beyond the composition of the Dirac point, the bands invert and the bandgap opens with increasing antimony content; this material is a topological insulator.[11] When the antimony concentration exceeds $x \approx 7\%$ or 8%, the overlap between the valence band at the T-point and the conduction band and the L-point vanishes and the semimetal to semiconductor transition occurs. In Figure 2.2b, we summarize the literature values for the variation of the L- and T-point band edge energies with respect to each other[10]: the letters refer to the authors of the studies: D to Dugue,[12] H to Hiruma et al.,[13] HS to Hebel and Smith,[14] K to Kao et al.,[15] T to Tichovolsky and Mavroides,[16] and V to Vecchi et al.[17] A maximum bandgap of approximately 30 meV is reached around $x \approx 15$–17%. The variation of the effective masses of the L-point electron and hole tensor elements is given in Figure 2.2c, as a fraction normalized to the value reported for Bi L-electrons in Table 2.2. It is a fair assumption to consider that the T-holes change little in effective mass for $x < 10$ or 15 at.% and that their tensor elements are the same as in Table. 2.1. We are not aware of a study of the temperature dependence of the electron band parameters for $x \neq 0$, but we suspect that it is strong. It is possible to reduce the carrier concentration of the semiconducting alloys at a low temperature to 10^{14} cm^{-3}, but they never freeze out.

The transport properties of Bi and Sb were reviewed by Issi.[18] Both Bi and the BiSb alloys can be doped n- or p-type at low temperatures: Se[19] and Te[20,21] are donors and Pb[19,22,23] and Sn[19,21,24] are acceptors. Potassium forms a resonant level while interstitial Li[25] is a donor. Since the 6s electrons of Bi form a band deep in the valence block, and only the three 6p electrons form states near the Fermi energy, Bi is essentially trivalent in the solid; yet the group III elements In, Ga, and, in principle, Al are acceptors.[26] While all these dopants work below about 150 K, it has not been possible to alter the carrier concentration of Bi much at 300 K, presumably because of the temperature of the band structure.[27]

2.1.3 Thermoelectrics

$Bi_{1-x}Sb_x$ alloys possess much better thermoelectric properties below 250 K than Bi_2Te_3,[28] particularly in the composition range of 5% < x < 22%, so that they are the most suitable alloys for cryogenic Peltier[28–30] or Nernst–Ettingshausen cooling.[31] In conventional Peltier modules, one needs both n-type and p-type materials, while only n-type BiSb shows good ZT values. Below 150 or 120 K, a possible Peltier cooler geometry that would require no p-type element can use a very thin layer or very narrow-diameter wire of a high-temperature superconductor to sustain the return current between the hot and the cold plates while being too thin to conduct much heat: this could use only an n-type BiSb alloy. The thermoelectric performance of BiSb alloys has been reviewed before.[28–30,32] Their figure of merit is reported as a function of temperature in Figure 2.3a, and at 77 K, a function of x in Figure 2.3b,[32] where we have added the performance of the newly developed K-doped $Bi_{94.5}Sb_{5.5}$. Jandl and Birkholz[33] describe how single crystals of the same alloy at the composition corresponding to the formation of a Dirac point ($Bi_{95}Sb_5$), but doped with 145 ppm Sn, can reach $ZT_{\parallel} \approx 1.4$ at 280 K in an external magnetic field $H = 1.2$ T along the bisectrix axis. Jandl and

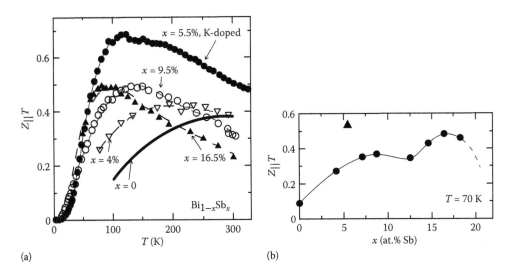

FIGURE 2.3 (a) Thermoelectric ZT of $Bi_{1-x}Sb_x$ along the trigonal axis as a function of temperature. The undoped samples are from the literature, as compiled by Lenoir et al.[32] The K-doped sample with $x = 5.5\%$ is from Section 11.3, where K is a resonant impurity in Bi. (b) Temperature for various values of the Sb content x as at 70 K; the undoped samples are circles, and the K-doped sample is a triangle.

Birkholz[33] reported on that sample even with $ZT_{\parallel} \approx 1.2$ (280 K) in a moderate field of 0.3 T achievable with a permanent magnet; we have not been able to reproduce those results.

The maximum temperature gradient ΔT_{max} that can be obtained with Peltier coolers is approximately $\Delta T_{max} = \frac{1}{2}ZT_c^2$, with T_c (K) as the temperature of the cold side of the module. Cryogenic Peltier coolers, even with the materials with $ZT_{\parallel} \approx 1$, therefore require multiple stages. This limitation can be circumvented in Nernst–Ettingshausen coolers,[31] for which BiSb alloys are eminently suitable. The Nernst and Ettingshausen effects are high in materials with bipolar conduction, which is the case for $Bi_{1-x}Sb_x$ alloys at low values of x. In an Ettingshausen cooling device, the current is passed along the trigonal axes of the $Bi_{1-x}Sb_x$ single crystal, while the external field is applied along a bisectrix axis. The performance of an Ettingshausen refrigerator is characterized by the product of the Nernst coefficient and the applied magnetic field, also called the isothermal thermomagnetic power, and the thermal and electrical conductivities. The isothermal thermomagnetic figure of merit, for the current along the trigonal, the heat flow along the binary, and the magnetic field along the bisectrix axes is[33] $Z_{31} = (N_{31}B_2)^2/\kappa_{11}\rho_{33}$, where N_{31} is the Nernst coefficient in that geometry. Optimal values for $Z_{31}T$ were obtained[33] again in $Bi_{95}Sb_5$-doped with 145 ppm Sn: $Z_{31}T \approx 0.36$ (180 K; 145 ppm Sn; 1.2 T), $Z_{31}T \approx 0.32$ (180 K; 145 ppm Sn; 0.3 T). Generally, values above $Z_{31}T > 0.2$ are possible[33] from 130 to 200 K with $H < 0.6$ T. An Ettingshausen cooler based on $Bi_{97}Sb_3$ has been able to reach[34] a $\Delta T \approx 42$ K starting from 160 K, at 0.75 T, proving that all-solid-state cooling to 118 K is possible using BiSb alloys.

2.1.4 Nanowires

The second possible way to make semiconductors out of Bi is via size quantization: Hicks and Dresselhaus[35,36] predicted that low-dimensional structures could lead to an enhancement of the thermoelectric properties through a reduction of the thermal conductivity and an enhancement of the Seebeck coefficient. First, the lattice thermal conductivity can be reduced without too much loss in carrier mobility if the phonon mean free path is of the same length scale as the wire diameter or longer, while the electron mean free path is much smaller. Second, the Seebeck coefficient in a low-dimensional system is expected to be enhanced at a certain carrier concentrations through size effects,[37] which give spikes in the density of states (DOS) and works via the same mechanism as is described for resonant levels in Section 11.3.

Experimental work on Bi nanowires was undertaken jointly by Dresselhaus and Heremans et al.,[38] and a strong enhancement of the thermopower in Bi nanowire composites was identified that corresponded well with the prediction of size quantization effects. The temperature dependence of the resistivity of bismuth nanowires with different diameters are shown in Figure 2.4.[39] In order to eliminate the influence of the cross section of the nanowire arrays, the resistivity was normalized by its value at 300 K. Although the electron and hole density of semimetallic Bi decreases with about one order of magnitude upon reducing the temperature from 300 to 70 K, bulk Bi exhibits a positive value for dR/dT over the entire temperature range. This is because the phonon-limited mobility decreases with temperature as almost a T^{-4} law.[37] The 70 nm diameter wires have a $dR/dT > 0$ below 100 K, which indicates that they still behave as semimetals in that region. Bismuth nano wires with a diameter below 49 nm have a negative temperature-dependent dR/dT over the entire range, indicating that the carrier density has become temperature dependent at all temperatures. This proves the existence of the size-effect-induced metal-to-semiconductor transition for nanowires with a diameter below about 50 nm. The thermopower, on the other hand, drastically increases as the wire diameter drops from 200 to 9 nm and decreases again for 4 nm wires. Wires (200 nm) exhibit a similar behavior as bulk bismuth, namely, they have a linear relationship between the temperature and Seebeck coefficient below about 90 K and then flatten as in bulk bismuth. The temperature dependence of the 9 nm wires shows a $1/T$ relationship at high temperatures, which is typical for nearly intrinsic semiconductors, but the Seebeck coefficient drops at low temperatures. A similar trend is observed for 4 nm nanowires. The behavior of the very narrow wires, with a diameter below 10 nm, is characteristic of weak localization, not of classical band transport in a semiconductor.

A similar study was undertaken with antimony nanowires.[40] Although the effective masses for the holes of Sb are of same magnitude as for bismuth, the electron effective masses are much higher. The energy overlap and carrier concentrations (see Table 2.2) are also orders of magnitude larger. All this results in a totally different behavior of the temperature dependence of

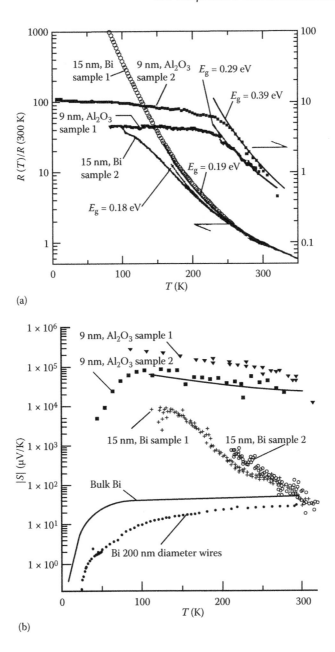

FIGURE 2.4 (a) Resistivity ratio and (b) thermopower of Bi nanowire composites with wire diameters indicated. The values for E_g indicated are obtained by fitting an Arrhenius function with activation energy $2E_g$ to the temperature dependence of the resistivity. (After Heremans, J. P. et al., *Physical Review Letters*, 88, 21, 2002.)

the resistivity compared to Bi: basically, the semiconductor regime is not achieved for Sb nanowires before localization effects take over.

2.1.5 Summary

Bi, Sb, and the Bi–Sb alloys form a complete class of thermoelectric semiconductors that are particularly suited to thermoelectric applications below room temperature. There are reports in the literature where a *ZT* on the order of 0.7 is reached at some temperature below 300 K and several that measure and predict values in excess of 1, perhaps as high as 1.4 theoretically. Given that these materials contain only earth-abundant elements and that Bi is regularly ingested by humans in medication or applied to the skin in cosmetics without documented deleterious effects, further research might provide a practical material for thermoelectric cooling. One potential drawback needs to be mentioned: the highest *ZT*s are obtained in single crystals with the fluxes oriented perpendicular to the natural cleavage plane. While single crystals are easy to prepare, this property might create difficulties with the mechanical stability of Peltier or Ettingshausen modules.

2.2 $CsBi_4Te_6$

Yiming Zhou and Li-Dong Zhao

$CsBi_4Te_6$ was first reported by Chung et al.[41] When they were investigating the corresponding Te analogs of $K_2Bi_8S_{13}$ and β-$K_2Bi_8Se_{13}$, they found the unexpected structure of $CsBi_4Te_6$. While Cs was used as the alkali metal in these analogs, instead of $Cs_2Bi_8Te_{13}$, researchers found a new compound of $CsBi_4Te_6$. This new compound shows an excellent intrinsic low-temperature thermoelectric property that is similar to Bi_2Te_3.[42]

2.2.1 Crystal Structure

$CsBi_4Te_6$ crystallizes in the space group *C2/m* and has a layered anisotropic structure. It is composed of anionic $[Bi_4Te_6]^-$ slabs and Cs^+ ions residing in the interlayer space. The $[Bi_4Te_6]^-$ slabs can be regarded as a fragment out of a NaCl lattice. This fragment is two Bi-octahedrals thick and four Bi-octahedrals wide (12 × 23 Å²) on the *ac*-plane. All the fragments pile up along the *b* axis and form an infinite long rod. The $[Bi_4Te_6]^-$ slabs are, in essence, one dimensional, and this structure is responsible for the characteristic needlelike morphology of the crystal, as shown in Figure 2.5.[42–44]

Given that $CsBi_4Te_6$ can be regarded as a reduced form of Bi_2Te_3, the added one electron from Cs per two equivalents of Bi_2Te_3 does not give a formal intercalation compound. However, the added electrons result in a dramatic restructuring of the Bi_2Te_3 framework and localize on a Bi atom, giving formally Bi^{2+} (rare in Bi chemistry) and Bi–Bi bonds of 3.2383(10) Å. The infinite long rods of the $[Bi_4Te_6]^-$ slabs arranged side by side and are connected via these bonds. The Bi–Bi distance in $CsBi_4Te_6$ is comparable to a

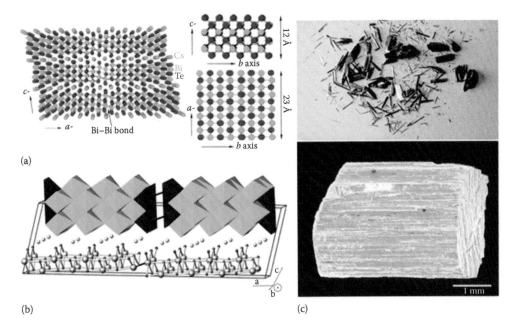

FIGURE 2.5 (a) Structure of $CsBi_4Te_6$; (b) ball-and-stick models and octahedral models of $CsBi_4Te_6$; (c) crystals (*top*) and oriented ingot (*bottom*) of $CsBi_4Te_6$. (Reprinted with permission from Chung, D. Y. et al., *Journal of the American Chemical Society*, 126, 6414–6428. Copyright 2004 American Chemical Society; from Chung, D. Y. et al. *Science* 287(5455):1024–1027, 2000. Reprinted with permission of AAAS; reprinted with permission from Lykke, L. et al., *Physical Review B*, 73, 6, 2006. Copyright 2006 by the American Physical Society.)

Bi–Bi distance of 3.267(6) Å in BiTe, a metallic compound possessing a layer of Bi atoms inserted between Bi_2Te_3 layers.

The anionic $[Bi_4Te_6]^-$ slabs are separated by Cs^+ layers, which are in two different coordination environments. The Cs (1) ions are located at the corner of each $[Bi_4Te_6]^-$ slab and coordinated by 10 peripherally distributed Te atoms. The Cs (2) ions are located between the $[Bi_4Te_6]^-$ slabs and coordinated by 9 terminal Te atoms. The average Cs–Te distance is 4.025 Å. Further crystal information is summarized in Table 2.3.

2.2.2 Electronic Band Structure

The band structure for $CsBi_4Te_6$ was performed with and without a spin-orbit coupling (Figure 2.6).[43,45] The spin–orbit coupling shifts the conduction band toward the valence band, resulting in an ultranarrow gap of 0.04 eV for $CsBi_4Te_6$. This value is consistent with the energy gap range of 0.04–0.08 eV calculated from the formula $E_g \approx 2S_{max}T_{max}$, in which S_{max} means maximum Seebeck coefficient and T_{max} means the temperature at maximum Seebeck coefficient.

According to the carrier excitation formula, the wider the bandgap, the higher the temperature of maximum achievable ZT. The bandgap of $CsBi_4Te_6$ is the narrowest reported for thermoelectric compounds, so there is no surprise that $CsBi_4Te_6$ performs best at low temperature. Generally, when alkali

TABLE 2.3   Summary of Crystallographic Data and Structural Analysis for CsBi₄Te₆

Formula	$CsBi_4Te_6$
Formula weight	1734.43
Crystal habit	Silvery white needle
Crystal size (mm³)	0.028 × 0.051 × 0.307
Space group	$C2/m$ (No. 12)
a (Å)	51.9205(8)
b (Å)	4.40250(10)
c (Å)	14.5118(3)
β (°)	101.4800(10)
Z, V (Å³)	8, 3250.75(11)
D_{calc} (g/cm³)	7.088
Temperature (K)	293(2)
λ (Mo Kα) (Å)	0.71069
Absorption coefficient (mm)	55.899
$F(000)$	5592
$\theta_{min} - \theta_{max}$ (°)	1.43–28.17
Index range	$-68 \leq h \leq 68, -5 \leq k \leq 5, -19 \leq l \leq 18$
Total reflections collected	18,450
Independent reflections	4373 [R(int) = 0.0767]
Refinement method	Full-matrix least-squares on F^2
Data/restraints/parameters	4373/0/134
Final R index [$I > 2\sigma(I)$]^{a,b}	$R_1 = 0.0434, wR_2 = 0.1017$
R indices (all data)^{a,b}	$R_1 = 0.0585, wR_2 = 0.1127$
Extinction coefficient	0.000076(6)
Largest differential peak and hole (e/A³)	4.344 and −2.490
GOF on F^2	1.05

Source: Chung, D. Y. et al., *Journal of the American Chemical Society*, 126, 6414–6428. Copyright 2004 American Chemical Society. With permission.
Abbreviation: GOF goodness of fit.
^{a} $R_1 = \Sigma ||F_o| - |F_c|| / \Sigma |F_o|$.
^{b} $wR_2 = \left\{ \Sigma \left[w \left(F_o^2 - F_c^2 \right)^2 \right] \middle/ \Sigma \left[w \left(F_o^2 \right)^2 \right] \right\}^{1/2}$.

metals are introduced into a binary semiconductor, the bandgap should be expanded. However, $CsBi_4Te_6$ shows a contrary trend. The reason lies in the presence of the Bi–Bi bond. This kind of homoatomic bond in semiconductors seems to have a band-narrowing effect.

2.2.3 Synthesis Method

All premanipulations were carried out in a vacuum under a dry nitrogen atmosphere in a Dri-Lab glove box and in a Schlenk line.[3] Cs_2Te was obtained by stoichiometric reactions of elemental cesium and tellurium in liquid NH_3. It was dried and ground to give a fine homogeneous powder prior to use. Several methods were used to obtain $CsBi_4Te_6$.

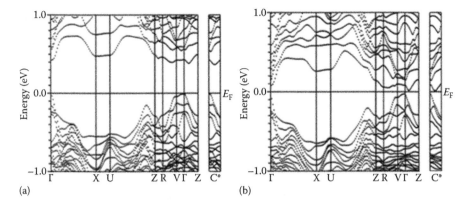

FIGURE 2.6 Electronic band structure of $CsBi_4Te_6$ (a) before adding spin–orbit interaction and (b) after adding spin–orbit interaction. (Reprinted with permission from Chung, D. Y. et al., *Journal of the American Chemical Society*, 126, 6414–6428. Copyright 2004 American Chemical Society; reprinted with permission from Larson, P. et al., *Physical Review B*, 65, 5, 2002. Copyright 2002 by the American Physical Society.)

2.2.3.1 Method A

Cs_2Te and Bi_2Te_3 were thoroughly mixed and loaded into an alumina thimble, which was plugged with a graphite lid and subsequently sealed inside a silica tube at a residual pressure of < 10^{-4} Torr. The mixture was heated to 573 K over 24 h followed by heating to 973 K at a rate of ~3 K/h. It was isothermed there for 2.5 days and then slowly cooled to 573 K at a rate of 4 K/h followed by cooling to 323 K in 12 h. The product was isolated by dissolving away the residual Cs_2Te with several portions of degassed dimethylformamide under a nitrogen atmosphere until the solvent remained clear. After being washed with ether and dried, shiny, long, silvery needles of $CsBi_4Te_6$ were obtained in quantitative yield.

2.2.3.2 Method B

A mixture of Cs metal and Bi_2Te_3 was loaded in an alumina thimble. An exothermic reaction took place during mixing. The mixture was sealed by the same procedure as that mentioned earlier and heated to 873 K at a rate of 5 K/h. The mixture was isothermed there for 1 day followed by slow cooling at a rate of 4 K/h to 423 K. The alumina thimble containing the product was immersed overnight in dried and degassed methanol under a nitrogen atmosphere. The relatively large crystals (>5 mm long) of $CsBi_4Te_6$ were taken out of the thimble by carefully scratching with a spatula and were washed with methanol.

2.2.3.3 Method C

A mixture of Cs metal and Bi_2Te_3 was loaded and sealed as mentioned earlier. The mixture was heated at 523 K for 24 h to complete the reaction of Cs metal. The resulting black material mixed with unreacted Bi_2Te_3 was slowly melted in a flame torch for a minute and then quenched in air. This method also gives a quantitative yield and is suitable for a large-scale synthesis.

2.2.3.4 Method D

Cs metal and Bi_2Te_3 were loaded separately in the two ends of an H-shaped silica tube. The other two ends were sealed under vacuum ($< 10^{-4}$ Torr) while keeping the Cs-containing end in liquid N_2 temperature to avoid the evaporation of the metal. The tube was heated to 523 K over 24 h and isothermed there for 1 day followed by heating to 853 K over 24 h. After 2 days at 853 K, it was cooled to 323 K for 12 h. The product obtained at 100% yield was washed several times with degassed methanol under a nitrogen atmosphere.

2.2.3.5 Doping Method

All doped materials were synthesized using doped Bi_2Te_3 as a starting material. To obtain doped Bi_2Te_3, each dopant was mixed with it and melted at 1073 K in a rocking furnace and quenched in air. In the cases of Bi and Sb doping, stoichiometrically doped $Bi_{2+x}Te_{3-x}$ and $Bi_2Sb_xTe_{3-x}$ ($x < 0.2$) were used as starting materials. The reaction of doped Bi_2Te_3 with Cs was performed by method D. After the synthesis of the doped $CsBi_4Te_6$, crystal growth was carried out by a Bridgman technique to obtain well-oriented ingots. The ingots were then annealed at 523 K for 2 days before measurements of charge-transport properties. A doping level in units of molar percentage (mol%) was applied in this doping study.

2.2.4 Thermoelectric Transport Properties

2.2.4.1 As-Prepared CsBi₄Te₆

As prepared directly from the synthesis, the room-temperature conductivity and Seebeck coefficient of the single crystals of $CsBi_4Te_6$ are in the range of 900–450 S/cm and 90–150 μV/K. The properties distribute in a range due to the impurities. So all these kinds of materials are called *as-prepared* rather than *undoped*. All those as-prepared samples are invariably p-type, which is indicated by the positive Seebeck coefficient. The weak temperature dependence of conductivity as well as the large value of the Seebeck coefficient and its positive temperature dependence correspond to the behaviors of a degenerate narrow-gap semiconductor.

The room-temperature thermal conductivity measurement on oriented ingots along the *b* axis shows a value of about 1.25 W/mK for as-prepared samples. The low thermal conductivity originates from several factors. The rather complex and anisotropic layer structure of $CsBi_4Te_6$ along with relatively large cell volume and heavy atoms leads to low lattice thermal conductivity. This is because heavy atoms cause low acoustic phonon frequencies, and a large unit cell means extra optical vibrational modes to minimize acoustic vibrational-mode heat capability. Meanwhile, the thermal-displacement parameters of Cs atoms are about 1.6 times larger than those of Bi and Te atoms, suggesting that Cs atoms may play the role of rattlers. While Cs atoms rattle dynamically in a cage, $CsBi_4Te_6$ is similar to the typical phonon-glass, electron-crystal material skutterudites in which rattling atoms can scatter heat-carrying acoustic phonons.[46]

The thermoelectric transport properties values mentioned earlier can give a reasonable estimation of ZT for $CsBi_4Te_6$. The room-temperature ZT value for as-prepared $CsBi_4Te_6$ is about 0.2. The ZT indicates that this material is an excellent candidate for further optimization.

2.2.4.2 p-Type CsBi₄Te₆

To enhance the thermoelectric ZT, various chemical impurities such as SbI_3, BiI_3, CuCl, SnTe, In_2Te_3, Sb, Bi, Sn, Zn, Mg, Te, Ge, Pt, and Pr are used to modify the charge transport of $CsBi_4Te_6$, as shown in Figure 2.7. Most of these dopants are chosen to achieve a change in carrier concentration or the type of major carrier. To evaluate the effectiveness of dopants, the power factor (PF) ($S^2\sigma$) was calculated as a function of temperature (Figure 2.8).[43,45,47]

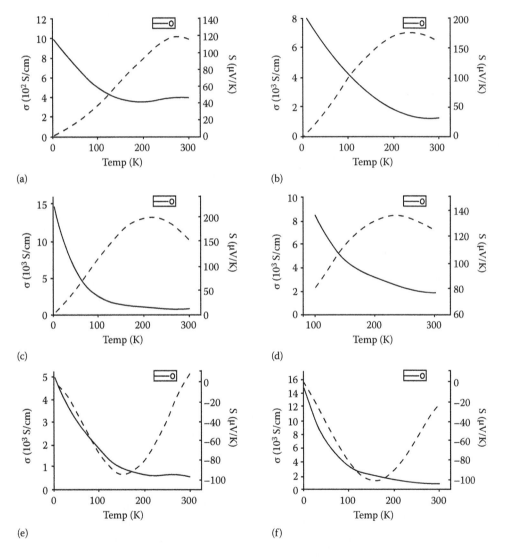

FIGURE 2.7 Variable-temperature electrical conductivity and Seebeck coefficient for single crystals of (a) as-prepared, (b) 0.05% SbI_3-doped, (c) 0.1% Bi-doped, (d) 0.06% Sb-doped, (e) 3.0% In_2Te_3-doped, and (f) 1.0% Te-doped $CsBi_4Te_6$. (Reprinted with permission from Chung, D. Y. et al., *Journal of the American Chemical Society*, 126, 6414–6428. Copyright 2004 American Chemical Society.)

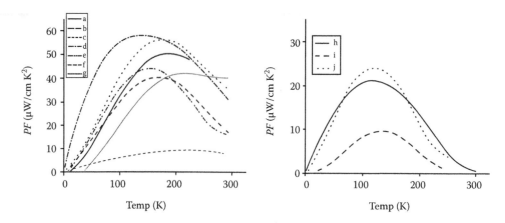

FIGURE 2.8 PFs as a function of temperature for a $CsBi_4Te_6$ sample doped with 0.05% SbI_3 (a), 0.1% Bi (b), 0.3% Sb (c), 0.06% Sb (d), 0.3% BiI_3 (e), 3.0% In_2Te_3 (i), 1.0% Te (j), and 0.5% Sn (h). For comparison, the PF data of an as-prepared $CsBi_4Te_6$ sample (f) and a commercial p-type-doped $Bi_{2-x}Sb_xTe_3$ (g) are shown. (Reprinted with permission from Chung, D. Y. et al., *Journal of the American Chemical Society*, 126, 6414–6428. Copyright 2004 American Chemical Society; reprinted with permission from Greanya, V. A. et al., *Physical Review B* 65(20):6. Copyright 2002, American Institute of Physics; reprinted with permission from Larson, P. et al., *Physical Review B* 65(4):5. Copyright 2002, American Institute of Physics.)

SnTe was applied to substitute a Bi atom with a Sn atom. Due to the lack of an electron, the conductivity of $CsBi_4Te_6$ should be promoted and the Seebeck coefficient decreased. Compared to as-prepared $CsBi_4Te_6$, doping in SnTe is responsible for the increased conductivity from 899 to 2584 S/cm and the decreased Seebeck coefficient from 103 to 54 µV/K at room temperature.[43]

SbI_3 and BiI_3 were chosen as dopants because the iodine atom can substitute at the Te site. The measurement of the Seebeck coefficient, however, shows a totally different result from what was implied before. The samples display as p-type, rather than n-type, over the range of doping. The most promising material was 0.05 mol% SbI_3-doped $CsBi_4Te_6$ having a maximum PF of ~51.5 µW/cm·K² at 184 K, while the conductivity achieves ~1927 S/cm and the Seebeck coefficient reaches ~163 µV/K. As for the BiI_3-doped sample, it reaches the maximum PF of ~45.0 µW/cm·K² at 147 K, which is lower than that of SbI_3-doped $CsBi_4Te_6$ and is the record low temperature for a p-type thermoelectric compound. The maximum property temperature of the p-type $CsBi_4Te_6$ is about 70–100 K lower than that of the Bi_2Te_3 system, ensuring it as a bright prospect in low-temperature cooling applications.

The unexpected result from SbI_3 and BiI_3 doping proves that the Te site was replaced by either the Bi atom or the Sb atom. Each replacement introduces a hole for $CsBi_4Te_6$. The fate of iodine is not clear, and no trace was found. However, based on these results, iodine is not the necessary element in the doping. Sb and Bi were used directly as the dopants, achieving a similar PF at about 40–60 µW/cm·K² to SbI_3- and BiI_3-doped ones.

2.2.4.3 n-Type CsBi₄Te₆

Because iodine dopants such as SbI_3 and BiI_3 do not introduce an extra electron on the Te site, many dopants have been investigated in order to synthesize n-type $CsBi_4Te_6$. Sn doping above 0.5% turned the material to n-type by changing a major carrier to electrons. This corresponds with the result that the Seebeck coefficient has a negative dependence on the SnTe-doping level. The n-type behavior appears when the number of electrons exceeds the number of holes. The Sn atoms may substitute for Bi atoms in the structure. The properties of Sn-doped $CsBi_4Te_6$ do not show a systematic dependence on doping level. The room-temperature conductivity is in the range from 560 S/cm for 1.0% Sn doping to 1655 S/cm for 0.5% Sn doping.

In_2Te_3 can serve as a weak donor when doped in $CsBi_4Te_6$; thus, a high concentration of In_2Te_3 leads to n-type behavior. A maximum Seebeck coefficient of –95 μV/K was observed at 160 K from a sample doped to 3.0%. Both an excess and a lack of Te can lead to n-type behavior, although the method varies. When $CsBi_4Te_6$ was prepared with a slight excess of Te, the excess Te atoms may occupy Bi sites, introducing electrons into this structure. A maximum Seebeck coefficient of –95 μV/K was observed at 160 K from a 1.0% Te-doped sample. While an as-prepared p-type $CsBi_4Te_6$ was annealed at 523 K for over 2 h under vacuum, it turned out to be an n-type material. A maximum Seebeck coefficient of –93 μV/K was observed at 220 K. The explanation for this p–n conversion is perhaps due to the fact that Te atoms escape from the $[Bi_4Te_6]^-$ framework, providing extra electrons.

The elemental dopants Zn and Mg provided n-type conductivity probably because of their small size, permitting them to insert in interstitial spaces of the structure and releasing two electrons per atom. The room-temperature thermal conductivity measurement on oriented ingots along the *b* axis shows values between 1.25 and 1.85 W/mK for a lightly and heavily doped sample. The slight shift of thermal conductivity indicates that the lattice thermal conductivity has little dependence on doping level. While the conductivity was promoted due to the doping, the electronic contribution to thermal conductivity was increased following the Wiedemann–Franz (W-F) law. The p-type $CsBi_4Te_6$ *ZT* calculated is shown in Figure 2.9.

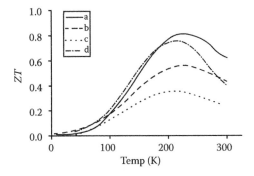

FIGURE 2.9 Variable temperature *ZT* for a 0.05% SbI_3 (a), 0.06% Sb (b), 0.3% BiI_3 (c), and 0.1% Bi-doped sample (d). (Reprinted with permission from Chung, D. Y. et al., *Journal of the American Chemical Society*, 126, 6414–6428. Copyright 2004 American Chemical Society.)

2.2.5 Concluding Remark

$CsBi_4Te_6$ is a promising low-temperature thermoelectric compound with layered anisotropic structure. The unique Bi–Bi bonds lead to an ultranarrow bandgap and make the compound perform excellently at low temperatures. Many materials were used as dopants to modify the thermoelectric properties of $CsBi_4Te_6$. SbI_3-, BiI_3-, SnTe-, Sb-, and Bi-doped samples display a p-type behavior. Doping with In_2Te_3, Te, Zn, Mg, and Sn and annealing of p-type $CsBi_4Te_6$ yields n-type behavior. The thermoelectric properties of $CsBi_4Te_6$ are sensitive to the type and doping level. Low doping levels seem to affect $CsBi_4Te_6$ charge transport properties dramatically. The maximum ZT value of 0.82 at 225 K was obtained by doping with 0.05% SbI_3. The wide range of working temperature ensures the various applications of this thermoelectric compound.

2.3 FeSb$_2$

Huaizhou Zhao and Zhifeng Ren

Cryogenic cooling (≤77 K) using thermoelectric materials remains challenging so far. The strongly correlated electron systems (including Kondo insulator and heavy fermions), such as CeB_6,[48] $YbAl_3$,[49] $FeSi$,[50] $FeSb_2$,[51–59] and $CrSb_2$,[60] have been investigated for cryogenic cooling applications. However, the ZT, which determines the cooling efficiency, is much lower than those thermoelectric materials working at or above 300 K. It is known that $ZT = (S^2\sigma/\kappa)T$, where S, σ, κ, and T are the Seebeck coefficient, electrical conductivity, thermal conductivity, and absolute temperature, respectively. Because of the large Seebeck coefficient S and high electrical conductivity σ observed in the strongly correlated electron systems, large PFs have been reported. For example, some $FeSb_2$ single crystals recently were reported[58] to have a PF of around 78×10^{-4} W/mK. That is about twice the highest known PF when compared to other regular systems, although it decreases to 5.5×10^{-4} W/m K^2 for polycrystal samples.[59]

In this context, we would trace this particularly large, thermoelectric PF back to the origins of crystal structure and electrical band structure in $FeSb_2$. As shown in Figure 2.10a, $FeSb_2$ crystallizes to the orthorhombic structure with $a = 5.83$ Å, $b = 6.54$ Å, and $c = 3.20$ Å, and the structure belongs to space group $Pnnm$.[51] The unit cell contains two formula units. The Sb atoms are tetrahedrally coordinated by three Fe atoms and one Sb atom. The Fe atoms are octahedrally coordinated by six Sb atoms. The Fe–Sb and Sb–Sb dimer bonds are 2.58, 2.60, and 2.88 Å, respectively. There are no direct Fe–Fe bonds, and the shortest Fe–Fe distance is 3.20 Å.

The density functional calculations for $FeSb_2$ have been performed by Lukoyanov et al.,[61] and the calculated DOS is shown in Figure 2.10b. As shown by Lukoyanov et al.,[61] the Fermi energy lies in a small bandgap, the DOS broken down in Fe $3d(t_{2g})$ and Sb $4p$ states is displayed. Narrow bands

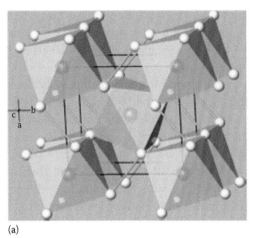

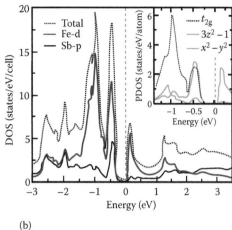

(a)   (b)

FIGURE 2.10   (a) Structure of $FeSb_2$. Red (dark) atoms are Fe and white atoms are Sb. (Reprinted with permission from Bentien, S. et al., *Physical Review B*, 74, 205105, 2006. Copyright 2006 by the American Physical Society.) (b) Total and partial DOSs for $FeSb_2$ from the local density approximation (LDA) calculation. Inset shows partial t_{2g}-DOS and $3z^2 - r^2$ and $x^2 - y^2$ orbital DOS of Fe $3d$ states; the Fermi energy corresponds to zero. (With kind permission from Springer Science+Business Media: *European Physical Journal B*, 53, 2006, 205–207, Lukoyanov, A. V. et al.)

appear above and below the bandgap with predominantly Fe $3d$ character. It appears that the Sb $4p$ states contribute only to the valence band peak just below the Fermi energy. A small gap (≈ 0.3 eV) is observed between relatively flat bands with $3d$ character in the DOS. The sharp features of the DOS in the vicinity of the Fermi level favor a large Seebeck coefficient in $FeSb_2$, as expected from the Mahan–Sofo theory in which good TE materials have a delta-like DOS.

As a Kondo insulator, the enhancement of thermopower in $FeSb_2$ can be accomplished by introducing electron–electron correlations. Here, the narrow bandgaps in $FeSb_2$ play a vital role, and the resistivity of $FeSb_2$ single crystal follows semiconductor-like temperature dependence. By applying an Arrhenius law $\rho = \rho_o \exp(E_g/2k_B T)$, two transport gaps were estimated to be $E_{g1} = 4$–10 meV and $E_{g2} = 26$–36 meV for the single-crystal sample.[54] The two transport gaps are believed to be intrinsic to $FeSb_2$. The nature of the extended $3d$ bands of iron might lead to hybridization with the conduction band, accounting for the opening of the narrow gaps and the formation of the narrow bands at the gap edges. Especially for the small gap E_{g1}, it has been detected by thermopower, Nernst effect, optical conductivity, and the electronic specific heat experiments. A substantial increase of the thermopower and the PF in $FeSb_2$ has been observed in the temperature range between 10 and 30 K, even though the Seebeck coefficient measured by two major individual groups varies enormously.[48,54] More consistently, Sun et al.[54] observed that upon cooling, $\rho(T)$ exhibits a plateau followed by an increase due to the opening of the small transport gap E_{g1} (4–10 meV). Concomitant to the opening of E_{g1}, huge charge-carrier mobility has been detected. This approved that the opening of this small transport gap is mainly responsible for the fascinating thermoelectric properties of $FeSb_2$.

It is noted that a huge thermopower in the low-temperature range has been reported for many clean semiconductors and typically is ascribed to the phonon-drag effect.[62] In crystals with only small numbers of defects and charge carriers, the phonon–electron interaction gives rise to a directional momentum transfer from heat-carrying phonons to conduction electrons along the temperature gradient. Thus, a net charge flow is induced, which adds to the diffusion thermopower. However, a comparison of the thermoelectric properties of $FeSb_2$ with those of the reference system $RuSb_2$ indicates that the phonon drag plays only a minor role for $FeSb_2$ in terms of the huge difference in the thermopower at around 10 K for $FeSb_2$ and $RuSb_2$. The Jonker plot in Figure 2.11 lends evidence against a dominating phonon drag effect in $FeSb_2$.[54] As shown in Figure 2.11, assuming that the $|S_{max}|$ of $FeSb_2$ and $FeSb_{2-x}As_x$ roughly follows the log n function (dashed line) as theoretically predicted, the slope is found to be enhanced by a prefactor A of larger than 30 relative to the calculated one based on the free-electron model (solid line). The enhancement factor A suggests that the strong electron–electron correlations are the major cause of the thermoelectric enhancement. The electron correlations assume a heavy charge-carrier effective mass and large nonparabolicity of the bands, which might result in an enhancement of the thermopower. The As-doped $FeSb_{2-x}As_x$ later was revealed to turn into heavy fermions with its effective mass being estimated to be around 14 m_0, indicative of the likely heavy charge-carrier effective mass in $FeSb_2$.

Interestingly, not only are the electrical transport properties such as the charge-carrier mobility and the Seebeck coefficient very sensitive to the structural defects and carrier concentration in single-crystal $FeSb_2$, but

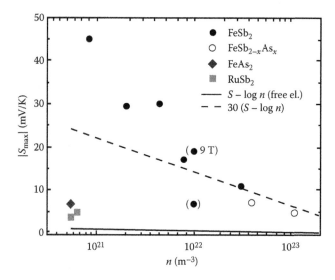

FIGURE 2.11 Dependence of the charge-carrier concentration n on the absolute thermopower maximum $|S_{max}|$ for $FeSb_2$, $FeSb_{2-x}As_x$, and the reference systems $RuSb_2$ and $FeAs_2$. Note that the temperature of $|S_{max}|$ in $FeSb_{2-x}As_x$ is slightly higher (~15 K) than that of the others (~10 K). (Sun, P. et al., *Dalton Transactions*, 39, 1012, 2010; reproduced by permission of The Royal Society of Chemistry.)

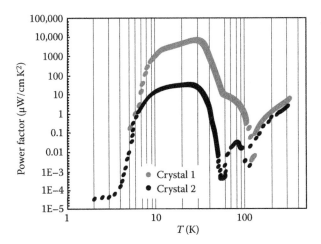

FIGURE 2.12 Comparison of PF along the *c* axis for two crystals with and without MIT. The low resistivity around MIT leads to a record high thermoelectric power factor (TPF). Crystal 1 has two orders of magnitude higher TPF between 8 and 100 K. (Reprinted with permission from Jie, Q. et al, *Physical Review B*, 86, 115121, 2012. Copyright 2012 by the American Physical Society.)

they also exhibit greatly enhanced electrical conductivity while maintaining thermopower at relatively high levels. This has been accomplished by subtle structural-difference-induced metal-insulator transition (MIT) for electronic transport along the orthorhombic *c* axis of $FeSb_2$ single crystals. By this means, as shown in Figure 2.12, the thermoelectric PF is enhanced to a record high of $S^2\sigma \sim 8000$ μW/cm K^2 at 28 K. It is still found that the large thermopower in $FeSb_2$ can be rationalized within the correlated electron model, which is consistent with Steglich group's work.[52] This result suggests that MIT in $FeSb_2$ crystals is governed by subtle structural differences and can be tunable by synthesis procedures.

In order to enhance the *ZT* in $FeSb_2$ for real cryogenic cooling applications, the most important strategy would be to suppress gigantic lattice thermal conductivity in $FeSb_2$. The lower limit of the lattice thermal conductivity in $FeSb_2$ has been calculated to be as low as 0.3 W/m K at 50 K through the model proposed by Zhu et al.[63] The nanostructure approach has proven to be a very efficient way to reduce the lattice contribution to the thermal conductivity in many thermoelectric material systems.[64,65] Nanostructured $FeSb_2$ first was synthesized by ingot formation through melting and solidification, then by ball milling and hot pressing with different processing parameters. Scanning electron microscopy (SEM) images in Figure 2.13 show how the grain size changes as a function of ball milling time and hot pressing temperature. Figure 2.14a shows the temperature dependence of thermal conductivity for all samples and the single crystals grown from vapor transport and self-flux methods.[66] For all samples throughout the temperature range, thermal conductivity substantially decreased with decreasing grain size. When compared with single-crystal $FeSb_2$, there is a reduction by more than three orders of magnitude in the thermal conductivity from 500 W/m K down to around 0.1 W/m K at 20 K in the nanostructured sample S15 hr-200°C.

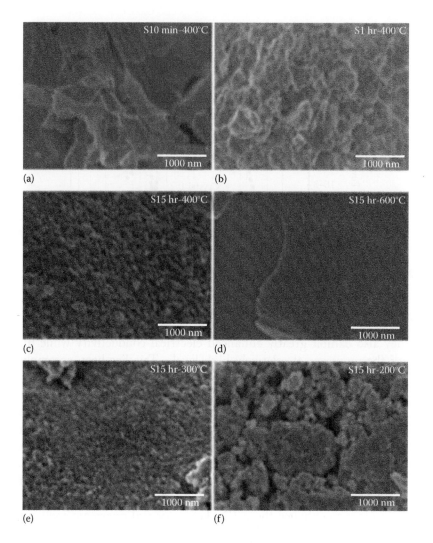

FIGURE 2.13 SEM images for nanostructured samples that were prepared with different conditions. (a) Hot pressed at 400°C by using powders ball milled for 10 min, (b) hot pressed at 400°C by using powders ball milled for 1 h, (c) hot pressed at 400°C by using powders ball milled for 15 h, (d) hot pressed at 600°C by using powders ball milled for 15 h, (e) hot pressed at 300°C by using powders ball milled for 15 h, and (f) hot pressed at 200°C by using powders ball milled for 15 h. (Reprinted with permission from Zhao, H. Z. et al. 2011. *Applied Physics Letters*, 99, 163101; Copyright 2011, American Institute of Physics.)

Such large thermal conductivity suppression by nanostructuring at low temperature is much more significant than any other nanostructured thermoelectric materials at high temperatures.

Figure 2.14b shows the temperature dependence of electrical resistivity for all the polycrystal samples. The data were fit using Arrhenius law to find approximate energy gaps. Sample S15 hr-600°C has two gaps of 28.2 and 4.2 meV. When the pressing temperature is lowered further, e.g., sample S15 hr-300°C, only one gap appears with a value of 21 meV. The change in the bandgaps corresponds to the increase of crystal defects that are probably due to

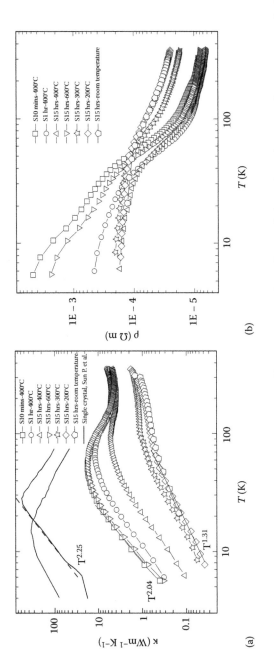

FIGURE 2.14 Thermoelectric properties for nanostructured samples. (a) Temperature dependence of thermal conductivity; fittings were applied to sample S10 min–400°C and S15 hr–200°C. Two solid curves correspond to thermal conductivity from single crystal samples. (b) Temperature dependence of resistivity.

(Continued)

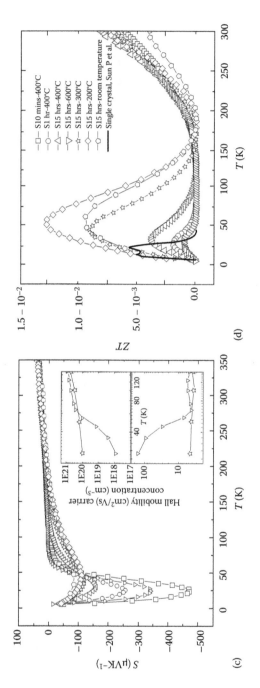

FIGURE 2.14 (CONTINUED) Thermoelectric properties for nanostructured samples. (c) Temperature dependence of Seebeck coefficient; the insets indicate the temperature-dependent carrier concentration and Hall mobility for S15 hr-600°C and S15 hr-300°C, respectively. (d) Temperature dependence of *ZT*. (Reprinted with permission from Zhao, H. Z. et al. 2011. *Applied Physics Letters*, 99, 163101; Copyright 2011, American Institute of Physics.)

the decreased grain size and increased carrier concentration. It appears that the smaller energy gap located in the temperature range of 7–20 K was suppressed, and the larger bandgap was decreased, as can be seen in the sample pressed at room temperature whose bandgap is reduced to 18 meV. Measurements of the carrier concentration, inset of Figure 2.14c, increased as well, confirming the narrowing of the energy gaps. As has been discussed above, the opening of bandgaps E_{g1} and E_{g2}, as well as the carrier concentration, are vital parameters in order to maintain the large Seebeck coefficient in $FeSb_2$.

The temperature-dependent Seebeck coefficients (S) are shown in Figure 2.14c. It shows that S decreases as grain size is decreased, which could mean that carriers are generated. Defects typically increase carrier concentration, which decreases the S. For the S15 hr-300°C sample, the carrier concentration at 25 K is 9.75×10^{19} cm^{-3}, which is higher than the carrier concentration for S15 hr-600°C at 25 K, 8.36×10^{17} cm^{-3}. On the other hand, the mobility for S15 hr-300°C is lower at 4.52 cm^2/V s than that of the S15 hr-600°C sample at 160 cm^2/V s. It is quite likely that an increase in the Seebeck coefficient can be realized by tuning carrier concentration through doping or composition adjustment, providing the potential for much future work. The optimized ZT in nanostructured $FeSb_2$ was found to be 0.013, compared to 0.005 for single-crystal $FeSb_2$, an increase of 160%.

Based on our early study, it appeared that the higher carrier concentration in nanostructured $FeSb_2$ mostly originated from its high defect density as compared to single-crystal or microsized samples.[59] As a result of the high defect density, the electrical conductivity of the nanosized $FeSb_2$ is significantly higher than the microsized polycrystal samples in the low temperature range (roughly ≤200 K), but the Seebeck coefficient is also significantly lower, which makes the PF much lower. Increasing the electrical conductivity in nanostructured $FeSb_2$ without degrading the Seebeck coefficient is challenging. We realized that localized Cu nanoparticles (NPs) can provide a large number of free electrons for higher electrical conductivity without changing the band structure of $FeSb_2$ for high Seebeck coefficient due to modulation doping, similar to what was observed in SiGe alloy system.[67] The schematic band alignment between $FeSb_2$ and Cu and the structure of $FeSb_2Cu_y$ nanocomposite are shown in Figure 2.15a and b, respectively. Since $FeSb_2$ is a n-type semiconductor with a bandgap of ~28 meV,[11] with its Fermi level located at the conduction band edge, the difference between the conduction band edge and vacuum level can be regarded as the same as the work functions for $FeSb_2$ at different crystal orientations. The work functions of $FeSb_2$ have been calculated to be 4.514 eV for the (001) plane, 4.852 eV for the (010) plane, and 4.723 eV for the (100) plane.[68] According to the alignment of Fermi levels, the band bends for the (001) and (010) planes, leaving an energy barrier in the range of 0.15–0.2 eV at their interfaces. However, due to the similar work functions between the (100) plane and Cu, which is 4.7 eV, the electron transfer between them would be much easier. It is reasonable to expect that Cu NPs can donate electrons from their conduction bands to $FeSb_2$, which will increase the carrier concentration in

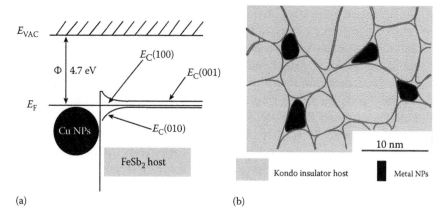

FIGURE 2.15 (a) Schematic of the band alignment between $FeSb_2$ and Cu; (b) distribution of Cu NPs in the nanocomposite; scale bar indicates that the grains of $FeSb_2$ are around 50 nm on average and ~5 nm for Cu NPs. (Koirala, M. et al. 2013. *Applied Physics Letters*, 102:213111. Copyright 2013, American Institute of Physics.)

the $FeSb_2$ host. As expected, the TE properties of $FeSb_2$ nanocomposite were enhanced significantly upon Cu NP modulation doping.[69]

The measured TE properties of $FeSb_2Cu_y$ composites are shown in Figure 2.16. Figure 2.16a shows the temperature dependence of thermal conductivity for all $FeSb_2Cu_y$ composites, as well as the pure nanostructured $FeSb_2$; reduced thermal conductivity was observed for most samples. It is seen that with the increase in Cu content (inset of Figure 2.16a), thermal conductivity at 60 K slowly decreased to a minimum of 0.39 W/m K for $FeSb_2Cu_{0.045}$ and from 0.44 to 0.39 W/m K for the pure nanostructured $FeSb_2$. We believe that this is due to phonon scattering at the interface of $FeSb_2$ and Cu. Seebeck coefficient results are shown in Figure 2.16b. Interestingly, the Seebeck coefficient only slightly decreases from –102 to –93 µV/K, even though the electrical conductivity is increased by a factor of ~2 as shown is Figure 2.16c for the $FeSb_2Cu_{0.045}$ sample with Cu NPs compared to pure $FeSb_2$. The relatively high Seebeck coefficients at different concentrations of Cu NPs can be understood as the result of modulation doping since the $FeSb_2$ matrix is not affected significantly. Regarding the electrical conductivity shown in Figure 2.16c, we have achieved significant improvement by incorporating Cu NPs to the nanostructured $FeSb_2$. Finally, a significant PF improvement below 200 K was observed for all Cu NP-incorporated nanocomposites compared to pure $FeSb_2$. As can be seen from Figure 2.16d, a maximum PF ~ 1.64×10^{-4} W/m K^2 at 60 K was obtained for $FeSb_2Cu_{0.0225}$ before decreasing to 1.39×10^{-4} W/m K^2 for $FeSb_2Cu_{0.045}$. Such a large improvement (~90%) over the pure $FeSb_2$ is comparable with the results achieved by other approaches such as resonant doping,[70,71] band engineering,[72,73] and modulation doping.[67] Combined with the slight decrease of thermal conductivity shown in Figure 2.16a, the ZT of ~0.027 (Figure 2.16e) has been achieved, that is ~110% enhancement over ~0.013 achieved in the nanostructured pure $FeSb_2$ at 60 K. These results suggest that

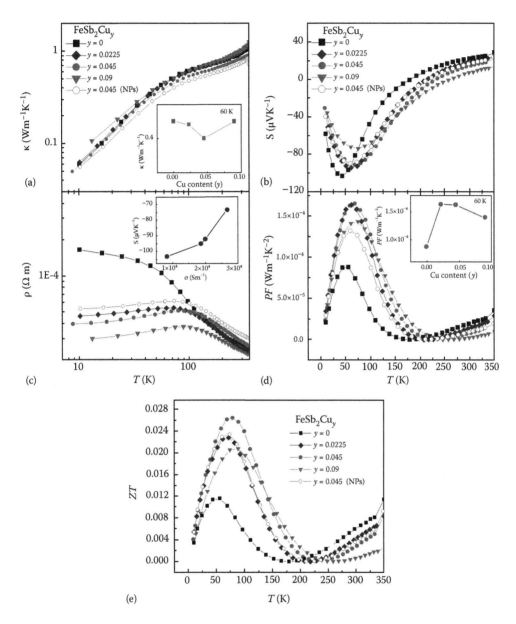

FIGURE 2.16 Thermoelectric properties of $FeSb_2Cu_y$ (y = 0, 0.0225, 0.045, 0.09) samples. (a) Temperature dependence of thermal conductivity; inset shows the measured thermal conductivity versus Cu content at 60 K. (b) Temperature dependence of Seebeck coefficients. (c) Temperature dependence of electrical resistivity; inset shows the peak Seebeck coefficient versus electrical conductivity at corresponding temperatures. (d) Temperature dependence of PF; inset shows the peak value of PF at 60 K versus Cu content. (e) Temperature dependence of ZT for $FeSb_2$ and $FeSb_2Cu_y$ nanocomposites. (Koirala, M. et al. 2013. *Applied Physics Letters*, 102:213111. Copyright 2013, American Institute of Physics.)

a similar strategy could be extended to other Kondo insulators to enhance their TE properties if the modulation dopant does not severely react with the matrix.

2.4 YbAgCu$_4$ as a Cryogenic Temperature Thermoelectric Material

Machhindra Koirala and Zhifeng Ren

2.4.1 Motivation

For the space program, infrared sensors on satellites are used for detecting and tracking objects and missiles. Infrared sensors, such as MgCdTe and Si (bolometer), work only if they are cooled to a certain temperature in the cryogenic temperature range.[74] By using the Stirling thermal cycle, these detectors can be cooled to a cryogenic temperature, but these cooling mechanisms have many complications related to heavy weight, mechanical vibrations, as well as mechanical failures.[75] In this scenario, the design of reliable and space-compatible devices is always the first choice. The use of solid-state cryogenic thermoelectric cooler could be the best replacement for the Stirling cooling systems. On these circumstances, the development of lightweight and shape- and size-compatible cryogenic thermoelectric cooling system attracts much attention from the thermoelectric community.[76] There has not been enough progress made toward developing cryogenic thermoelectric cooler due to the lack of research on thermoelectric materials in the low temperature range. In this section, we discuss the current status of research on low-temperature thermoelectric materials and the thermoelectric properties of some rare earth (RE) alloys that we think will be a good direction for research to bring thermoelectric cryogenic cooling into a reality.

2.4.2 Thermoelectrics in Low Temperature

The efficiency of power generation and coefficient of performance for cooling of a thermoelectric device is related to thermoelectric ZT [$ZT = (S^2\sigma/\kappa) T$, where S is Seebeck coefficient, σ is electrical temperature, κ is thermal conductivity, and T is temperature]. The materials with higher ZT will have higher efficiency for power generation as well as higher coefficient of performance for cooling.[77] So, the focus of thermoelectric material research is the synthesis of thermoelectric materials with high ZT. Heavily doped semiconductors are considered to be one of the best families of thermoelectric materials. Thermoelectrics at room and high temperatures are well-developed fields, and many materials were synthesized with high ZT. At cryogenic temperatures, the development of highly efficient thermoelectric materials is more challenging. As we know that ZT is the product of Z (or $S^2\sigma/\kappa$) and working temperature T, the value of Z of these materials should be very high to obtain high ZT at low temperature.[78] In the temperature range above 150 K, Bi_xSb_{1-x} is one of the best thermoelectric materials having a peak ZT of 0.5 at 150 K.[79] Below 150 K, most of the well-known thermoelectric materials do not have a good Seebeck coefficient. More research into other materials may be required, rather than focusing only on heavily doped semiconductor thermoelectric materials. The other research directions points toward Kondo materials. Researchers are attracted toward Kondo insulators, such as $FeSb_2$ and $FeSi$,[52,80] and heavy fermions[81,82] because of their high Seebeck coefficients at low temperatures.

2.4.3 Rare Earth Compounds

RE alloys are well known for their high Seebeck coefficient at cryogenic temperatures.[83] The high Seebeck coefficient of these materials is due to hybridization between the localized *f*-band of Yb or Ce with the delocalized conduction band. The band hybridization changes the overall DOS of electrons near the Fermi level. In some circumstances, there is a formation of a peak (bump) on the electron DOSs near the Fermi level.[84] The presence of a bump in DOS near the Fermi level increases the Seebeck coefficient, which is much higher than the value of Seebeck coefficient suggested by Mott's relation. For Yb-based materials, the bump on the DOS is just below the Fermi level and has a negative Seebeck coefficient. Similarly for Ce-based materials, the DOS has a bump just above the Fermi level, and it has positive Seebeck coefficients.[85] In some materials, there is a crossover of the Seebeck coefficient from negative to positive and vice versa, which could be due to the presence of crystal-field splitting.[86]

In the presence of the *f*-band, the DOS of the Fermi level can be expressed as[87,88]

$$N_f(\varepsilon_F) = \frac{W(T)}{T_o^2 + [W(T)]^2}, \qquad (2.2)$$

where $W(T) = T_f\, e^{-T_f/T}$ is the width of *f*-band at temperature T. T_f is the temperature-dependent parameter related to quasi-elastic line width of the neutron spectra. $k_\beta T_o$ is the energy difference between the central position of *f*-band from the Fermi level and defined as $T_o = (\varepsilon_F - \varepsilon_f)/k_\beta$, where ε_F is the Fermi energy, ε_f is the central position of the *f*-band, and k_β is the Boltzmann constant. The Seebeck coefficient of a system is energy derivative DOS at the Fermi level and can be written as

$$S = c_1 T + c_2 T \left[\frac{\partial \log N_f(\varepsilon)}{\partial \varepsilon} \right]_{\varepsilon = \varepsilon_F}, \qquad (2.3)$$

where the first term gives the Seebeck coefficient for isostructural materials without the presence of *f*-band near the Fermi level, while the second term gives the contribution of *f*-band to the Seebeck coefficient. The coefficients c_1 and c_2 give the strength of nonmagnetic and magnetic contributions to the Seebeck coefficient, respectively. Using the expression for DOS as discussed earlier, the Seebeck coefficient can be simplified as

$$S = c_1 T + c_2 \frac{T T_0}{T_0^2 + [W(T)]^2}. \tag{2.4}$$

At very low temperatures, the width of the f-band is small and all the magnetic moments are screened by conduction electrons, and these materials do not have a good Seebeck coefficient. As the temperature rises, there is a good hopping between the conduction band and f-band, thereby making the drastic change in the electron DOS at the Fermi level. When there is a sharp peak in the DOS at the Fermi level, there is a significant increase in the value of the Seebeck coefficient. At a very high temperature, the lattice contribution (nonmagnetic contribution) to the Seebeck coefficient plays a dominant role compared to the magnetic contribution. Therefore, the materials behave like the general metal without having the f-band near the Fermi level. For these kinds of f-electron metallic system, there is a good chance to have a large peak in the Seebeck coefficient at certain temperature range. Mahan[83] listed a set of Ce- and Yb-based thermoelectric materials based on published data.

2.4.4 YbAgCu₄ Introduction

YbAgCu$_4$ was discovered by Rossel et al.,[89] and it has a fcc structure. YbAgCu$_4$ is a heavy fermions with a Sommerfield constant of 250 mJ/mol K^2.[90] Most of the transport properties of YbAgCu$_4$ are governed by the hybridization of the f-band of Yb with the conduction band. The Yb atom in YbAgCu$_4$ experiences small, cubic, crystalline field splitting, but its effect mostly is masked by the Kondo effect.[91] Mahan[83] listed YbAgCu$_4$ as one of the high-PF materials on the basis of previously published results. It has been reported[85] that the PF of YbAgCu$_4$ can have a value of 235 μW/c K^2. One of the important characteristics of YbAgCu$_4$ is that it has mysteriously low thermal conductivity compared to other metallic Kondo systems.[92] The low thermal conductivity of this material makes it more favorable for thermoelectric research and applications compared to other Kondo systems that have high thermal conductivities.

2.4.5 YbAgCu₄ Results

YbAgCu$_4$ can be prepared by melting Yb, Ag, and Cu in a vacuum or argon environment. Different methods such as tantalum vacuum tube sealing and arc melting methods[93,94] were implemented to synthesize a YbAgCu$_4$ ingot. One of the key aspects of designing thermoelectric materials is the design of a sample with low thermal conductivity. The low-thermal-conductivity samples can be prepared by making nanostructured samples.[95] For our synthesis approach, we have prepared a YbAgCu$_4$ ingot using the arc-melting method. Yb, Ag, and Cu granulars were melted in an argon environment by using the arc-melting method. Extra Yb (5%) was used during the melting process to compensate for the volatile loss of Yb. The ingot was ball milled in the argon environment using a ball milling machine, and the obtained powder was hot pressed at different conditions to obtain samples with different grain sizes. Each of the sample disks were cut into a sample bar with dimensions of 3 mm ×

3 mm × 5 mm, and electrical and thermal transport measurements were carried out using a quantum design physical properties measurement system.

Figure 2.17a shows the electrical resistivity of the samples hot pressed at different temperatures. The sample hot pressed at a higher temperature has a smaller electrical resistivity than the sample hot pressed at a lower temperature. During the sample synthesis process, the sample hot pressed at a higher temperature has a larger grain size than the sample hot pressed at a lower temperature. Hence, it is reasonable to have higher electrical conductivity for the sample hot pressed at a higher temperature than that for the sample

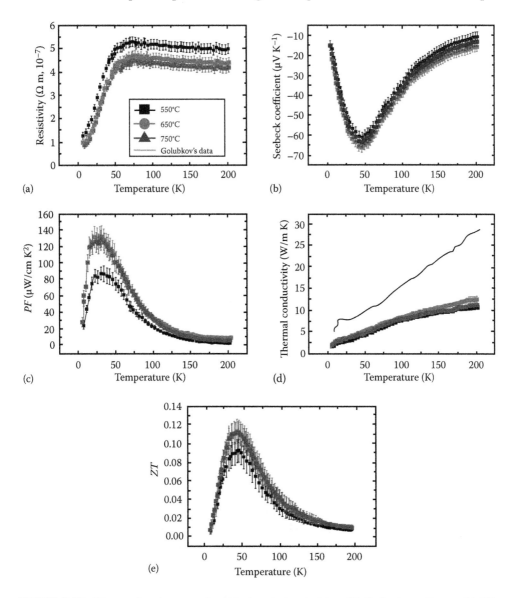

FIGURE 2.17 Thermoelectric properties: (a) electrical resistivity; (b) Seebeck coefficient; (c) PF; (d) thermal conductivity; (e) ZT of nanostructured $YbAgCu_4$ hot pressed at 550°C, 650°C, and 750°C. (From Koirala M. et al., *Nano Lett.*, 14(9), 5016–5020. Copyright 2014 American Chemical Society.)

hot pressed at a lower temperature. The sample with a small grain size may have a larger number of grain boundaries and defects, which increase the scattering mechanisms of electrons. Figure 2.17b shows the Seebeck coefficients of the samples. The data are similar to the resistivity trends, as with the electrical conductivities. Further, the Seebeck coefficients are inversely related to each other. The sample hot pressed at 700°C has a peak Seebeck coefficient of 66 μV/K at 42 K.[96] Although the value of the Seebeck coefficient is not high compared to general thermoelectric materials that have a Seebeck coefficient of 200–300 μV/K, the value of the Seebeck coefficient here is high compared to the other RE materials that are being studied in the cryogenic temperature range. The Seebeck coefficient minima (maxima on the absolute value of Seebeck coefficient) can be explained on the basis of the modification of the DOS of electrons at the Fermi level due to the presence of the *f*-band very close to it. Figure 2.17c shows the PF for nanostructured $YbAgCu_4$ samples. Despite the relatively poor Seebeck coefficient compared to other thermoelectric materials, the PF of $YbAgCu_4$ is much higher than the PF of general thermoelectric materials. This high PF is due to high electrical conductivity and a moderate Seebeck coefficient. The peak PF that was calculated by Mahan[83] was 235 μW/cm K^2 on the basis of earlier published data. For the nanostructured sample, the value of 131 μW/cm K^2 seems to be a reliable value. The grain size of the samples affects the PF, and it could be small for nanostructured samples. These PF values are much higher than the PF of general thermoelectric materials. The materials with high PF are capable of handling a large amount of heat and can convert heat into electricity and vice versa. One of the important characteristics of $YbAgCu_4$ is that the thermal conductivity of this material is much smaller compared to the value suggested by any theoretical models. Thermal conductivity further decreases for nanostructured samples than single-crystal or large-grain-sized samples. Figure 2.17d shows the thermal conductivity of the hot-pressed samples, which are 30–50% smaller than the samples prepared by other methods.[92] Figure 2.17e shows the ZT of nanostructured $YbAgCu_4$ samples. Because of the high PF and relatively low thermal conductivity, the peak ZT of nanostructured $YbAgCu_4$ is 0.11 at 45 K.

Although the peak ZT of 0.11 is too low to be desirable, given the cryogenic temperature range, obtaining a high ZT is challenging with ZT a product of Z and T. For example, PbTe-based thermoelectric materials have a peak ZT of 2 at 1000 K. To obtain the same ZT at 50 K, the value of Z should be 20 times higher compared to the Z of the sample having a ZT value of 2 at 1000 K, which is almost impossible with our current knowledge about materials and their research. While there are multiple ways to enhance ZT at low temperatures, the best way is to nullify the lattice thermal conductivity in these materials. In that situation, the Wiedemann–Franz Law can be implemented for thermal conductivity, and the ZT value can be approximated as S^2/L. Considering the value of $L = 2.4 \times 10^{-8}$ W Ω/K^2, the Seebeck coefficient of the thermoelectric materials must be ~150 μV/K to obtain the peak ZT of ~1. However, it is hard to nullify the lattice part of thermal conductivity, but it can be reduced using different approaches.

To reduce the lattice part of thermal conductivity, different approaches could be implemented such as point defects, nanostructuring, and mass fluctuation.[97,98] Because of the large phonon mean free path and small electron mean free path, the method of point-defect scattering does not help much for decreasing the thermal conductivity for heavy fermions and Kondo systems.[99] From the experimental results, it had been obtained that the thermal conductivity of the doped system is equal to the thermal conductivity of the undoped system.[99] The best way to reduce thermal conductivity in that situation is the method of nanostructuring where the grain boundary scattering can scatter the long-energy phonon. The electron mean free path of heavy fermion materials is much smaller than the phonon mean free path; hence, it will be much easier to find the certain grain size window where the phonon can be scattered well without affecting the electrical transport properties.

2.4.6 YbAg$_x$Cu$_{5-x}$

YbAgCu$_4$ has peak ZT of 0.11 at 45 K. By changing the ratio of Ag and Cu on YbAg$_x$Cu$_{5-x}$, there is a change in the lattice parameter for the crystal.[100] The change in lattice parameter changes the chemical pressure of the system. The chemical pressure of the crystal changes the electrical transport properties. When the concentration of Ag is reduced, there is a shift in the peak Seebeck position toward a lower temperature.[101] The shift of the peak Seebeck coefficient position also changes the peak ZT position. Figure 2.18 shows the peak ZT for YbAg$_x$Cu$_{5-x}$ for x = 0.5, 1, and 1.5. If the peak position of the ZT is shifted by changing the concentration of Ag and Cu, a segmented leg can be prepared with variation in the concentration of Cu and Ag, which will increase the average ZT of the thermoelectric leg, which ultimately increases the coefficient of performance of the devices.

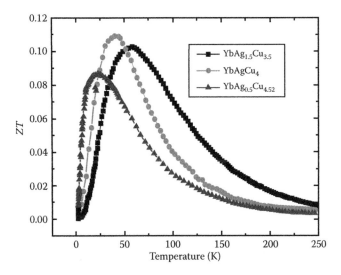

FIGURE 2.18 ZT of YbAg$_x$Cu$_{5-x}$ with x = 0.5, 1.0, and 1.5 hot pressed at 650°C.

2.4.7 Other Materials

The study on $YbAg_xCu_{5-x}$ has been focused on the low temperature range (20–60 K). However, there are some other Yb-based RE alloys such as $YbAl_3$ and $YbCu_2Si_2$ with peak ZTs of 0.23 and 0.14 at 300 and 125 K, respectively.[102,103] The thermoelectric properties of these materials have also been optimized by doping and chemical pressure.[102,103] Either through making segmented legs or with more study on these materials, we can find many RE alloys that can be used as thermoelectric materials in the temperature range from 20 to 300 K.

As mentioned earlier, these Yb-based RE alloys are n-type thermoelectric materials with a working temperature range from 20 to 300 K. Ce-based RE alloys are also being investigated as p-type thermoelectric materials. To summarize the result of p-type thermoelectric materials, $CeAl_3$ has a peak ZT of 0.016 at 55 K,[104] $CeCu_6$ has peak ZT of 0.024 at 60 K,[105] and $CePd_3$ has a peak ZT of 0.23 at 200 K.[106]

2.4.8 Conclusion

In conclusion, the thermoelectric figure of merit (ZT) of low-temperature thermoelectric materials is very low for application purposes. The investigation of RE alloys could be a good direction to bring thermoelectric cryogenic cooling toward reality. These materials need further study and optimization to make them practical for applications.

2.5 Thermoelectric Properties of Ce-Based Intermetallic Alloys and Compounds

Stephen R. Boona

2.5.1 Introduction

There are significant challenges inherent to optimizing the thermoelectric properties of conventional semiconductor materials, especially at low temperatures. These challenges invite us to consider alternate types of material systems that hold potential for achieving favorable thermoelectric performance. Among these are intermetallic materials containing rare earth (RE) elements, which often display interesting transport properties due to strong electronic correlations. These correlations usually correspond with interactions between narrow *f*-electronic levels and delocalized charge carriers, sometimes leading to unusual combinations of electronic and thermal properties that are favorable for thermoelectric applications. Although these narrow *f* levels are highly sensitive to perturbations, the properties of some RE intermetallic materials can be tuned gradually through systematic changes in the chemical composition. Such modification strategies have been deployed successfully in certain instances to achieve values of the thermoelectric figure of merit $ZT \geq$ 0.3, which is remarkably high for a metal.

This section primarily focuses on a subset of such materials, namely, Ce-based intermetallic alloys and compounds. Particular emphasis will be given here to the thermoelectric properties of materials derived from $CePd_3$, which is the canonical Ce-based intermetallic material for thermoelectric applications. A review of the developments in this field is provided, along with a discussion of how further study may lead to additional enhancements in *ZT*.

2.5.2 Thermoelectric Metals

The general motivation for exploring thermoelectric effects in RE intermetallic systems stems directly from the Wiedemann-Franz (W-F) law, which is the single most significant hindrance preventing the development of thermoelectric materials for use at low temperatures. This law quantitatively describes the amount of heat carried by electronic currents in solids, which is an unavoidable consequence of applying a temperature gradient to any electrical conductor. The formula describing W-F is given by $\kappa_e = LT/\rho$, where κ_e is the electronic thermal conductivity, L is the Lorenz number, T is the absolute temperature, and ρ is the electrical resistivity. The value of L for a free electron gas is $L_0 = 2.44 \times 10^{-8}$ W Ω/K^2, which is applicable to a surprisingly large assortment of materials within about a factor of 2.[107,108]

Usually, the only other relevant contribution to thermal conductivity in solids comes from phonons (κ_p), such that $\kappa_{total} = \kappa_p + \kappa_e$. Although κ_e becomes immeasurably small in electrical insulators, there exists a so-called "amorphous limit" for κ_p that corresponds with the minimum possible phonon conductivity. This occurs when the phonon scattering length is on the order of the interatomic spacing,[109] and typical values of $\kappa_{p,min}$ in real materials are on the order of 0.1 W/m K. This value is approximately constant at high temperatures, while at low temperatures, it is roughly proportional to the heat capacity.

The thermoelectric figure of merit *ZT* is described by the relation $ZT = S^2T/\rho(\kappa_e + \kappa_p)$, where S is the Seebeck coefficient. By substituting in the W-F law $\kappa_e = LT/\rho$, we arrive at an alternate general expression $ZT = S^2/L(1 + \kappa')$, where κ' is the ratio of κ_p/κ_e. If we assume ideal conditions, such that a material displays temperature-independent $\kappa_p = \kappa_{p,min} = 0.1$ W/m K and $L = L_0$, we can produce a plot like that in Figure 2.19. This figure shows the Seebeck coefficient necessary at a given temperature to achieve $ZT = 1$ for each of the labeled values of ρ. The solid horizontal line reflects the limiting case of a good metal, where $\kappa_e \gg \kappa_p$, in which case we can simplify the *ZT* equation even further to $ZT \approx (S/156)^2$. In any case, Figure 2.19 makes it abundantly clear that unless a material *substantially* violates the W-F law, the only hope for achieving large *ZT* values at low temperatures is to seek out good electrical conductors that also have large S. Since S and ρ are intrinsically linked in conventional materials through the carrier density n, good conductors almost always display low S. There are only a few exceptions to this, where certain physical mechanisms in bulk materials allow for partially decoupling these parameters. Among these are resonant levels in narrow gap semiconductors,[110] magnon–electron drag in magnetic conductors,[111] and strong electronic correlations in RE intermetallics.[112] The latter are

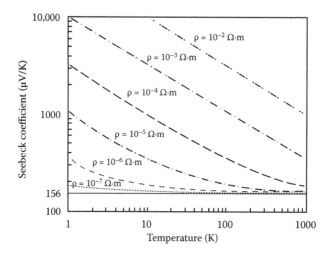

FIGURE 2.19 Seebeck coefficient versus temperature needed to achieve $ZT = 1$ for the given value of electrical resistivity, assuming a constant phonon thermal conductivity of 0.1 W/m K and $L = L_0$. From this plot, we see that high ZT can be achieved below room temperature only in materials with simultaneously low electrical resistivity and large Seebeck coefficients.

the focus of this section, in particular those with RE = Ce, which are almost always p-type.

Exploiting strong correlations to achieve favorable thermoelectric properties is by no means a new idea; it was first famously proposed in the early 1990s[113] that one way to partially decouple S and ρ (thereby enabling large ZT) is to design materials with sharp features in their electronic DOS. This design concept naturally points to RE intermetallics, since Mahan[113] described $4f$ levels as "nature's closest approximation to a delta function." However, after a few decades of concerted effort on this front, recent considerations[114] suggest that there may be practical limits to the maximum possible thermoelectric performance of these materials. This is due primarily to the fact that $4f$ DOS peaks have finite widths in real materials, as well as unavoidable background contributions from s- and p-electrons that short circuit the intended effect.

In spite of this, RE intermetallics still are recognized as one of the only successful classes of materials in which reasonably high ZT values can be realized at low temperatures. For example, studies of Ce-based alloys have reported ZT values as high as 0.125 in $CeNi_{1.2}Cu_{0.8}Al_3$ at 100 K[115] and 0.3 in $CePd_2Pt$ above 250 K.[116] However, to date, no RE intermetallic material has been discovered with simultaneous $S > 130$ μV/K and $\rho < 10^{-5}$ Ω m, effectively capping the maximum potential ZT of these materials well below unity (vis-à-vis Figure 2.19).

Many of the largest S values observed within this class of materials are found in solid solutions of homogeneously intermediate valence (IV) compounds (e.g., $CePd_3$[117] and $Ce(Sn_{1-x}In_x)_3$[118]), in which the valence state of each Ce atom rapidly fluctuates in time. The origin of these anomalously large S values is associated with free carriers scattering in and out of the Ce $4f$ states, which nearly overlap with the Fermi level E_F. This overlap results in an enormous DOS over a small energy range for free carriers, which generally corresponds with anomalously large electronic heat capacity (and therefore entropy) per carrier in these systems,[119] along with large charge-carrier effective mass ("heavy fermions"). The complexity of these electron–electron interactions make it almost impossible to search reliably for promising new RE intermetallic materials using ab initio computational approaches. In some materials, s–d and s–f hybridization can cause E_F to fall within a semiconductor-like energy gap. This results in the formation of a Kondo insulator[120] state that behaves more like a conventional semiconductor at low temperatures, as seen in $Ce_3Pt_3Sb_4$[121] and $CeRu_4P_{12}$.[122] While these materials can display S values significantly larger than 130 μV/K, they do not display metallic ρ.

The apparent upper limit for conducting IV materials of $S_{max} \approx 130$ μV/K is found in $CeRu_4Sb_{12}$ within a narrow range of temperatures close to 100 K.[123] In contrast to this, $CePd_3$ displays a similar $S_{max} \approx 115$ μV/K that results in larger thermoelectric PF S^2/ρ over a much broader temperature range,[118] making it far more appealing for energy conversion. Given its status as the most promising and most studied archetype Ce-based intermetallic alloy for thermoelectric applications, $CePd_3$ is the primary focus of this section.

2.5.3 $CePd_3$

$CePd_3$ is a well-known IV material sometimes referred to as a "lightly doped Kondo insulator"[124] with an effective carrier mass of about 40 times the free electron value.[125] The IV effect is manifested prominently in this material through a significant deviation of the lattice parameter from the lanthanide contraction trend seen in the other $REPd_3$ compounds,[126] which occurs due to the significant size difference between the effective atomic radii of the Ce atoms in their different valence states.

Although RE intermetallics can often deviate from $L = L_0$,[127] a rough projection in the limiting case of $\kappa' \ll 1$ indicates that $CePd_3$ has a maximum potential of $ZT \approx (115/156)^2 \approx 0.5$. Increasing the ZT in this material then becomes an exercise in reducing κ' and/or enhancing (or at least not destroying) the strong electronic correlations responsible for large S. There have been numerous successful attempts at reducing the thermal conductivity of this material, for example, by introducing point defects through solid-solution mixing and interstitial site filling,[128] but these nominal improvements sometimes mask the fact that there is often actually no significant change in κ'. On top of that, almost every modification strategy destroys the IV state and turns the material into a conventional metal (i.e., a poor thermoelectric). The following sections explain some of these modification strategies and discuss their efficacy and shortcomings, as well as some ways in which the challenges they present might be overcome.

2.5.4 Ce Site Substitutions

Solid solutions of the type $Ce_{1-x}R_xPd_3$ have been reported for a wide variety of R elements, including Gd,[129] Y,[129–132] Nd,[132,133] La,[132,134] Th,[133] Sc,[135] U,[136] and Mn.[137] The nominal benefit of this approach is that it offers a way to create mass mismatches within the material, which should lead to phonon scattering. However, it does so without altering the local bonding environment of the Ce atoms, which should preserve the favorable electronic properties. One main caveat is that the differences in the ionic radii of the substituted elements typically produce strain, and thus, the R atoms exert "chemical pressure" that can have a significant impact on the Ce ions.

It is no surprise then that available reports on the thermoelectric properties of these alloys typically show a substantial decrease in S as x is increased for all R elements. Very few of these reports include thermal conductivity data, so the effectiveness of Ce substitutions with regard to phonon scattering has not been explored in great detail. $Ce_{1-x}Sc_xPd_3$[138] represents one of the only alloys for which a complete thermoelectric dataset is available. In this series, a complete, solid solution exists[135] with an apparent collapse of the IV state for $x > 0.3$. Among all possible 3+ and 4+ elements that can be substituted for Ce, Sc represents the maximum possible mass mismatch, as well as one of the largest possible size differences. Both of these factors should result in significant phonon scattering, and therefore, one might at least naively anticipate that any impact on the thermal conductivity should be maximized in this series.

The results of Ref. 141 indeed show a nominal ~25% decrease in lattice thermal conductivity for $x \geq 0.1$, although there is also a simultaneous drop in S and increase in ρ. Together, these effects mean that no value of x produces enhanced ZT. Based on these results, in conjunction with the broader body of work on Ce site substitutions in $CePd_3$, it appears highly unlikely that this particular approach is capable of producing improved thermoelectric performance.

2.5.5 Pd Site Substitutions

Within the phase space of $CePd_{3-x}T_x$ (T = transition or noble metal), reports on $CePd_{3-x}Rh_x$ and $CePd_{3-x}Ag_x$ solid solutions by Lackner et al.[139,140] show that substitutions on the Pd site can result in ~50% reductions in κ_p. However, similar to alloying on the Ce site, these substitutions also have a significant impact on S and ρ. Although the net changes tend to be detrimental overall, Lackner[139,140] did report a slight increase in ZT for one sample, $CePd_{2.7}Rh_{0.3}$, which reaches values of 0.15 at 100 K and 0.25 at 300 K. These correspond with increases of approximately 50 and 10%, respectively, over the base compound. A similar study by Sales[141] briefly mentions an increase in ZT for $CePd_{2.655}Pt_{0.3}$ up to approximately 0.15 at 100 K and 0.23 at 175 K. No other transport data for that or any other compositions in the series were included in Ref. 141, although the ρ of $CePd_{3-x}Pt_x$ had been previously published.[142]

These studies led to further work by Boona and Morelli[116] on the $CePd_{3-x}Pt_x$ system, which has a solubility limit at $x = 1$.[143] Since Pt and Pd have almost identical valence configurations, ionic radii, and electronegativities, the substitution of one for the other does not significantly change the nearest neighbor environment of the Ce atoms, leading to minimal impact on S. Meanwhile, the substantial mass difference between Pt (195.08 amu) and Pd (106.42 amu) sharply decreases κ_p. When combined, these effects result in enhanced ZT at all temperatures for all x, reaching a maximum of $ZT = 0.3$ above 250 K in $CePd_2Pt$,[116] which is a 35% improvement over $CePd_3$.

In fact, a slight enhancement in S actually occurs at 300 K as x is increased, which coincides with a more gradual slope of $S(T)$ above $T \approx 150$ K. Since Pt substitution does not alter the IV state of the CE atoms significantly, this modest increase in S can be explained best as a result of the chemical pressure introduced by the slightly smaller Pt atoms, which cause a small contraction of the unit cell. This affects S similar to mechanical pressure, which has been shown to reduce the lattice parameter[144] and increase S by 30% when the material is subjected to several gigapascals of external pressure.[145,146] In that regard, Pt substitution and mechanical pressure are the only mechanisms known to increase S at 300 K in $CePd_3$.

2.5.6 Interstitial Filling: $CePd_3M_x$

Perhaps the most well-studied modification strategy in $CePd_3$ is partially filling the central 1b interstitial site of the cubic Cu_3Au structure with various elements. This results in the formation of a pseudoperovskite phase with composition $CePd_3M_x$, where M can be almost any type of element; reports exist for M = Na,[147] Be,[148,149] Mg,[147] Ca,[147] B,[147,150] Al,[147,150–152] Ga,[147,150,153] In,[147,150,153] Si,[147,150] Ge,[147,150,154] Sn,[147,150,153] Pb,[150,153] Sb,[153] Bi,[153] Te,[147,150] Mn,[155] and Zn.[150] For certain M elements, mostly those from rows 3–5 of the periodic table, an ordered $(CePd_3)_8M$ superstructure phase forms; this occurs spontaneously for elements from rows 4 and 5, while it appears to happen only after annealing for Al and elements from row 3.

Lackner et al.[128] first demonstrated that point-defect phonon scattering causes $CePd_3B_x$ to display lower κ_p compared to $CePd_3$, and this also was observed by Gumeniuk et al.[148] in $CePd_3Be_x$. Boona and Morelli[147] studied the effect of M atoms on κ_p in a wide array of materials and found two empirical trends. First, for fixed x, κ_p decreases as the mass of M is increased up to ~30 amu, above which no further reduction in κ_p is observed. Second, for fixed M mass, κ_p decreases as x is increased up to ~0.05, above which a similar saturation effect is observed. Additional parameters such as lattice strain and the onset of superstructure formation may also have an effect on κ_p, but these mechanisms have not yet been studied systematically in any detail.

While adding interstitial atoms to the material lowers κ_p, it also significantly affects the materials' electronic and magnetic properties, including S and ρ. Intuitively, one might expect that inserting M atoms into the interstitial site results in negative chemical pressure that forces the lattice to expand to accommodate these atoms. However, systematic studies have shown that the changes in S, ρ, and lattice parameter that occur due to M insertion are independent of the size of the M atoms and instead primarily depend on the M atoms' electronic properties.[147,150,156,157] A similar observation has been made in the related compound $EuPd_3B_x$.[158] In fact, it is actually possible to predict to first order the structural, electronic, and magnetic properties of any $CePd_3M_x$ material by considering only the M atoms' valence electron count N_V and filling fraction x, which can be combined to define an effective valence $V_{eff} = (x)(N_V)$. This concept, first developed by Boona and Morelli,[147] is almost surprisingly accurate for predicting the lattice parameter and S of the entire $CePd_3Be_x$ series[148] published in 2016. To demonstrate this, Figure 2.20 has been included to show the data originally published in Ref. 147 (filled symbols) along with the new data in Ref. 149 (empty symbols).

The results of the study in Ref. 147 elucidate the dependence of κ_p on the mass of M, as well as an inverse relationship between S and V_{eff}. Based on these results, interstitial site filling should produce enhanced ZT if the M atoms are relatively heavy and have few valence electrons. This inspired the effort in Ref. 147 to integrate sulfur on the M site in the $CePd_3$ structure as an electron acceptor, but detailed follow-up inspection shows that this approach ultimately failed.[159] Ref. 159 also discusses several unsuccessful attempts to incorporate M = Sr, Ba, and Ag into the structure, but the resulting ingots were coated with obvious secondary phases. X-ray powder diffraction and transport property analysis of the ingots' interiors closely resemble that of unmodified $CePd_3$, suggesting that these M elements are not easily incorporated into the 1b interstitial site. Nonetheless, as demonstrated in Ref. 147, the insertion of small amounts of Mg results in modest ZT enhancement at 80 K from 0.07 in $CePd_3$ to 0.11 in $CePd_3Mg_{0.05}$.

Based on these overall results, it does not appear that interstitial site filling by itself can lead to any substantial improvements in the thermoelectric performance of $CePd_3$. This approach is effective for lowering κ_p, but the

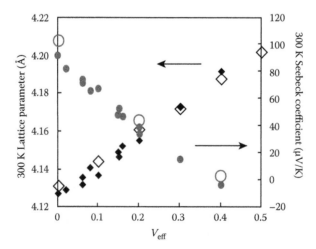

FIGURE 2.20 Lattice parameter (*diamonds*) and Seebeck coefficient (*circles*) at ~300 K for a variety of $CePd_3M_x$ materials. The data are plotted as a function of the effective valence V_{eff} of the interstitial site, which is determined by the valence electron count of M multiplied by the filling fraction x. Solid symbols represent a variety of materials from Ref. 147, while empty symbols represent $CePd_3Be_x$ from Ref. 149.

concomitant impacts on S and ρ are too detrimental to result in ZT enhancement except for small improvements over narrow temperature ranges. However, if an approach can be found that generates electron acceptors in the material, this could be combined with interstitial filling to cancel the competing doping effects and perhaps significantly enhance ZT.

2.5.7 Nanostructured CePd₃

Nanostructuring is a popular approach for improving the thermoelectric properties of various materials.[160] At very low temperatures, thermal conductivity is dominated by phonons with long mean free paths, which tend to be scattered primarily by grain boundaries.[161] Nanostructuring therefore tends to be most effective for reducing thermal conductivity at cryogenic temperatures. Its relevance to RE intermetallic materials is discussed here briefly.

One interesting consequence of high surface-area-to-volume ratios in RE intermetallics is that surface states begin to play a more significant role in determining a material's overall properties. One example is elemental Sm, for which angle-resolved X-ray photoelectron spectroscopy (ARXPS) experiments have shown[162,163] that a relatively large concentration of Sm^{2+} ions are found at the sample surface, while the bulk consists of predominantly Sm^{3+}. These ARXPS results are supported by calculations of surface energy differences between Sm^{2+} and Sm^{3+} ions that reveal the former to exist in a lower energy state when exposed at a surface.[162] Similar phenomena have been studied as a function of grain size in numerous other RE compounds observed through various experimental techniques.[164–172] This includes CePd₃, for which the following two reports suggest that nanostructuring may alter the electronic properties of the material.

The first is a series of spectroscopic and diffraction experiments performed by Lin et al.[173] on <10 nm particles created through laser ablation. In addition to a shift in lattice parameter, the authors observe a systematic increase in the low-temperature magnetic susceptibility as the particle size is reduced, indicating that approximately 25% of the Ce atoms shift toward the 3+ state. Seemingly contradictory to this, they also report X-ray absorption data that show a shift in the average valence of the Ce atoms toward the 4+ configuration. Together, these results suggest that different behavior occurs in the bulk and at the surface. The second report on nanostructured CePd₃ is a study of magnetization, nuclear magnetic resonance (NMR), and computational analysis of 30, 60, and 150 nm CePd₃ particles created via ball milling.[174] These samples do not show any shift in lattice parameter, but they do show systematic increases in magnetic susceptibility, in addition to a change in the NMR Knight shift that indicates an increase in the average $4f$ occupancy as the particle size is reduced. The authors of both studies caution that defects and/or surface oxidation may be responsible at least partially for the observed changes, but overall, their results appear consistent with the idea that particle size reduction can have a direct impact on the IV state.

One possible interpretation of these combined results is that Ce^{3+} is the preferred configuration for Ce atoms exposed at the surface. If so, this has an important consequence for transport properties. The average Ce valence in bulk $CePd_3$ is approximately +3.45 at room temperature,[175,176] so each unique unit cell exposed at a grain surface requires "capturing," on average, approximately 0.45 additional electrons from the conduction band to transform the Ce atoms into the 3+ state. This suggests that nanostructuring $CePd_3$ could be a clever way of pinning electrons at the particle surfaces, thereby increasing the effective hole concentration in the bulk without modifying the chemical composition or structure. No published articles exist describing the thermoelectric properties of nanostructured $CePd_3$, although this idea is discussed in Ref. 159.

2.5.8 Universal Trends and Combined Approaches

A careful comparison of the properties observed in the different series discussed earlier leads to additional insights regarding the apparently universal nature of these trends.[177]

One such universal trend that emerges across both the $CePd_3M_x$ and $CePd_{3-x}Pt_x$ series is the relationship between S and lattice parameter. In both series, it was observed that an approximately linear correlation exists between these properties, which results from their shared dependence on the average valence configuration of the Ce atoms. This analysis holds even in stoichiometric $CePd_3$ when comparing how S and the lattice parameter[178] both change as a function of temperature. The results of all three series are included in Figure 2.21. The universality of this behavior is somewhat surprising, since different mechanisms determine the properties of each series. The $CePd_3M_x$ system is dominated by the effective valence of the interstitial site, indicating that the properties of this system are driven entirely by changes in electronic structure; $CePd_{3-x}Pt_x$ is controlled by chemical

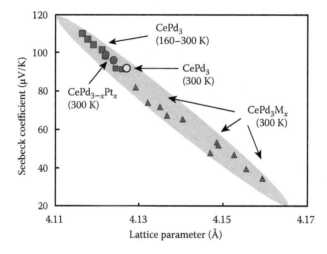

FIGURE 2.21 Seebeck coefficient as a function of lattice parameter for the labeled materials. A universal correlation exists between these properties for all three series, even though the behavior of each series is dominated by a different physical mechanism: thermal expansion in stoichiometric $CePd_3$ (*squares*), chemical pressure in $CePd_{3-x}Pt_x$ (*circles*), and electron count in $CePd_3M_x$ (*triangles*).[177]

pressure from the smaller Pt atoms, and in stoichiometric $CePd_3$, we see simply the effects of thermal expansion.

Empirical insights such as this are useful for screening ways of increasing ZT. Although each of the mechanisms discussed earlier has a limited effectiveness on its own for improving the thermoelectric performance of the material, it may be possible to combine them together in such a way that the properties are improved beyond what each approach can achieve separately. This basic idea has been attempted before in studies inspired by the general correlation between S and the lattice parameter of $CePd_3$, which has been observed by others as the material is modified in various ways. These efforts typically have focused on combining mechanisms that result in either a decrease or zero net change in the lattice parameter, such that κ_p is reduced while S is enhanced or, at least, remains unchanged.

The first such attempt was reported by Ijiri and DiSalvo,[132] where they explored the effects of dual substitutions on the Ce site. In their study, they examined solid solutions of $Ce_{1-x}Y_xPd_3$ and $Ce_{1-x}Nd_xPd_3$, as well as a series of samples where Ce was partially replaced in equal parts by large La and small Y atoms. Their results very clearly indicated that even if the average interatomic spacing was held constant, the thermoelectric properties of the material were changed substantially in ways not expected based on the simple combination of the Mattheissen and Nordheim–Gorter rules for solid solutions. The $Ce_{1-x}(La_{0.5}Y_{0.5})_xPd_3$ samples showed larger changes in the transport properties relative to those seen in $Ce_{1-x}Nd_xPd_3$, even though these samples also had a very similar average lattice parameter to $CePd_3$. This suggests that in addition to doping effects, local strains and distortions in the lattice can have significant effects on the IV state. These results demonstrated that simply maintaining the average lattice parameter at a constant value does not maintain the favorable thermoelectric properties.

The second attempt at combining approaches was reported by Lackner et al.[139], where they tried to balance the lattice parameter reduction due to Rh_{Pd} substitution by adding B and/or substituting Ag_{Pd} to expand the lattice back toward its original size. Although this approach did increase S in some $Ce(Pd_{1-x}Rh_x)_3B_{0.05}$ samples relative to their singly modified parent compounds, none of these combinations resulted in $ZT > 0.2$. The results of this second study nonetheless serve as proof of principle that it is possible in certain instances to combine modification strategies in such a way that κ_p is reduced while S is at least partially restored to its original value, resulting in ZT values not otherwise attainable through each approach alone. Further investigation of combined approaches potentially may result in enhanced ZT.

2.5.9 Related Compounds

Given the sensitivity of the IV state in $CePd_3$ to even small changes in composition and structure, it is generally very difficult to find ways of decreasing κ_p without also affecting S and ρ. For this reason, it is desirable to identify other base compounds that are not already optimally doped so that the thermoelectric properties may be improved simultaneously.

Several other strongly correlated materials are known to display intermediate valence, mixed valence, and/or Kondo-type behavior that may give rise to favorable thermoelectric properties. These compounds exist within a wide range of stoichiometry ratios that can be described by the general chemical formula $R_xT_yM_z$, where R is a lanthanide or actinide element with $1 < x < 3$, T is a transition metal or noble metal element with $1 < y < 3$, and M is a p-block element with $0 < z < 4$. These compounds form in various crystal structures that also span a wide range of complexity, including relatively simple perovskite-like RT_3M_z (Cu$_3$Au-type; e.g., CePd$_3$B$_x$), complex cubic $R_3T_3M_4$ (Y$_3$Au$_3$Sb$_4$-type; e.g., Ce$_3$Pt$_3$Sb$_4$), and hexagonal RT_2M_3 (PrNi$_2$Al$_3$-type; e.g., CeNi$_2$Al$_3$). The available data on thermoelectric properties of these compounds suggest that some of them may be close to the ranges necessary for high *ZT*.

For example, the size and complexity of the Ce$_3$Pt$_3$Sb$_4$ unit cell is favorable for intrinsically low κ_p. Indeed, typical κ_p values for materials with this structure are on the order of a few Watts per meter-Kelvin, including as low as 1.5 W/m-K in Sm$_3$Au$_3$Sb$_4$.[179] Generally, it is possible to observe a wide variety of behavior in the electronic properties of these materials. Many of them can be loosely classified as narrow-gap semiconductors, and accordingly, they tend to display relatively large *S* and low ρ values typical for such compounds. Some Ce$_3$T$_3$M$_4$ materials can be classified as Kondo insulators,[180] while others in this family behave more like metallic Kondo lattices, similar to CePd$_3$. Given the large number of isostructural compounds with similar lattice parameters in this family, an enormous number of possible variations exist that can be formed through single or double substitutions on the R, T, and/or M sites. Such solid solution mixing can be expected to result in a robust range of electronic properties while simultaneously reducing κ_p.

A demonstration of this composition-dependent variety can be seen in the ρ of RCu$_3$Sb$_{4-x}$Sn$_x$ alloys, where R = La, Ce, Pr, Nd, and Sm and $x = 0.05$ and 0.10.[181] While all the compounds without Ce show simple metallic behavior, the samples containing Ce show a dramatically different temperature dependence that corresponds with the ln(T) behavior typical for Kondo lattices.

Two of the few complete thermoelectric studies available on Ce$_3$Pt$_3$Sb$_4$ are those in Refs. 180 and 182, which explore the properties of Ce$_3$Pt$_{3-x}$Cu$_x$Sb$_4$ and Ce$_{3-x}$Nd$_x$Pt$_3$Sb$_4$, respectively. In these studies, the material can be doped systematically through substitutions on both the Ce and Pt sites. Cu substitution on the Pt site, for example, reduces ρ at 100 K by three orders of magnitude from 10^{-2} Ωm in Ce$_3$Pt$_3$Sb$_4$ to 10^{-5} Ωm in Ce$_3$Cu$_3$Sb$_4$. The value of *S* also changes dramatically, from a peak at 100 K of 300 μV/K in Ce$_3$Pt$_3$Sb$_4$, to –175 μV/K in Ce$_3$Pt$_{2.5}$Cu$_{0.5}$Sb$_4$, to approximately 0 in Ce$_3$Cu$_3$Sb$_4$. The sensitivity of these systems to compositional variation is apparent even in stoichiometric Ce$_3$Pt$_3$Sb$_4$ control samples,[182] which display an order of magnitude variation in ρ at 300 K between two nominally identical samples, along with peak *S* values that range from approximately 230 to 350 μV/K. The only difference between the samples is that one was measured directly after arc melting and the other after powder processing. Since these changes in the transport properties are comparable to those resulting from compositional modifications, it is imperative for any further studies on these materials to focus on improved stoichiometric control in order to distinguish better between the effects of off-stoichiometry and intentional doping.

Based on reports from Sun et al.,[115,183] $CeNi_2Al_3$ can be considered another candidate system for low-temperature thermoelectric applications, this time without any expensive noble metals. Sun[115] reports that the substitution of Cu onto the Ni site results in the simultaneous optimization of all three thermoelectric properties in $x = 0.8$. At 100 K, S reaches ~90 µV/K, ρ decreases to 10^{-6} Ω m, and κ_p drops under 5 W/m K, resulting in ZT ~ 0.125. As seen in Figure 2.22, this is the largest ZT value ever reported in any Ce-based intermetallic system below 80 K. Above 80 K, this record is set by $CePd_2Pt$, which is also included in Figure 2.22. The other lines represent the only other successful enhancements of ZT in $CePd_3$.

Although Ref. 115 reports S values in $CeNi_{1.2}Cu_{0.8}Al_3$ as high as 90 µV/K at 80 K, the largest value observed in Ref. 173 is 65 µV/K at the same temperature for $x = 0.7$. Similar samples studied in Ref. 159 also indicate lower peak values for S, consistent with the report in Ref. 183. While 65 µV/K is still substantially higher than what typically is observed in any conventional metal, it remains approximately three times too small to achieve ZT values that are technologically relevant. For this reason, this *specific* alloy system may not be a viable candidate for low-temperature thermoelectric applications, but the richness of this phase space suggests that other substitutions on the Ce, Ni, and/or Al sites may be more promising. Since this type of compositional modification does not appear to cause any appreciable change in the magnitude of ρ, emphasis should be placed on exploring how chemical pressure and doping can maximize S.

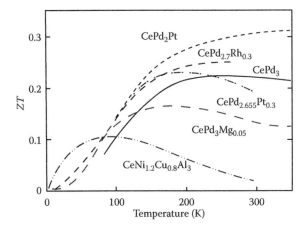

FIGURE 2.22 ZT versus temperature for $CeNi_{1.2}Cu_{0.8}Al_3$,[115] $CePd_2Pt$,[116] $CePd_{2.7}Rh_{0.3}$,[140] $CePd_{2.655}$ $Pt_{0.3}$,[141] and $CePd_3Mg_{0.05}$.[147] Together, these data represent the largest ZT values ever observed in any Ce-based intermetallic materials, including the only four known instances where modifications of $CePd_3$ have led to increased ZT.

2.5.10 Summary

Due to the W-F law, the only hope for achieving high ZT at low temperatures is to find metallic materials with large Seebeck coefficients. Strong electronic correlations in Ce-based intermetallic alloys represent one of the only known ways to achieve this in p-type conductors. The difficulty of accurate first-principles calculations in these systems makes it almost impossible to identify promising alloy systems from first principles alone. Empirical exploration can lead to fruitful insight, but this approach often is hampered by synthesis challenges combined with the sensitivity of electronic correlations to even small changes in chemical composition.

Nonetheless, ZT values as high as 0.3 have been achieved in $CePd_3$-based alloys through careful selection of substitution elements that impede phonon conduction without significantly altering the electronic structure. Further improvements in ZT may be realized in other alloy systems where the electronic structure of the base compound is not optimized already, such that chemical substitution results in simultaneous improvement of thermoelectric properties.

References

1. Seebeck, T. J. 1825. Magnetische polarisation der metalle und erze durch temperatur-differenz. *Abh. preuss. Akad. Wiss.* 1822–1823:265–373.
2. Gallo, C. F., Chandrasekhar, B. S., and Sutter, P. H. 1963. Transport properties of bismuth single crystals. *J. Appl. Phys.* 34:144.
3. Thonhauser, T., Scheidemantel, T. J., and Sofo, J. O. 2004. Improved thermoelectric devices using bismuth alloys. *Appl. Phys. Lett.* 85:588–590.
4. Jin, H., Jaworski, C. M., and Heremans, J. P. 2012. Enhancement in the figure of merit of p-type $Bi_{100-x}Sb_x$ alloys through multiple valence-band doping. *Appl. Phys. Lett.* 101:053904.
5. Cucka, P., and Barrett, C. S. 1962. The crystal structure of Bi and of solid solutions of Pb, Sn, Sb and Te in Bi. *Acta Crystallogr.* 15:865–872.
6. Dresselhaus, M. S. 1971. Electronic properties of the group V semimetals. *J. Phys. Chem. Solids* S32:3–33.
7. Chao, P. W., Chu, H. T., and Kao, Y. H. 1974. Nonlinear band-parameter variations in dilute bismuth-antimony alloys. *Phys. Rev. B* 9:4030–4034.
8. Vecchi, M. P., and Dresselhaus, M. S. 1974. Temperature-dependence of band parameters of bisumuth. *Phys. Rev. B* 10:771–774.
9. Heremans, J., and Hansen, O. P. 1979. Influence of non-parabolicity on intra-valley electron–phonon scattering: The case of bismuth. *J. Phys. C Solid State Phys.* 12:3483.
10. Heremans, J. P., and Michenaud, J.-P. 1985. Electronic magnetostriction of $Bi_{1-x}Sb_x$ alloys. *J. Phys. C Solid State Phys.* 18:6033.
11. Hsieh, D., Qian, D., Wray, L., Xia, Y., Hor, Y. S., Cava, R. J., and Hasan, M. Z. 2008. A topological Dirac insulator in a quantum spin Hall phase. *Nature* 452:970.
12. Dugue, M. 1965. Propriétés électriques des solutions solides bismuth-antimoine. *Phys. Status Solidi* 11:149.

13. Hiruma, K., Kido, G., and Miura, N. 1981. Far-infrared magnetoreflection in bismuth and bismuth-antimony alloys in high magnetic fields. *Solid State Commun.* 38:859–863.
14. Hebel, L. C., and Smith, G. E. 1964. Interband transitions and band structure of a BiSb alloy. *Phys. Lett.* 10:273.
15. Kao, Y. H., Brown, R. D., and Hartmann, R. 1964. Shubnikov–de Haas effect and cyclotron resonance in a dilute BiSb alloy. *Phys. Rev.* 136:A358.
16. Tichovolsky, E. J., and Mavroides, J. G. 1969. Magnetoreflection studies on the band structure of bismuth-antimony alloys. *Solid State Commun.* 7:927–931.
17. Vecchi, M. P., Pereira, J. R., and Dresselhaus, M. S. 1976. Anomalies in the magnetoreflection spectrum of bismuth in the low-quantum-number limit. *Phys. Rev. B* 14:298–317.
18. Issi, J.-P. 1979. Low-temperature transport properties of the group V semimetals. *Aust. J. Phys.* 32:585.
19. Thompson, N. 1936. Electrical resistance of bismuth alloys. *Proc. R. Soc. A.* 155:111–123.
20. Li, S. S., and Rabson, T. A. 1970. The Nernst and the Seebeck effects in Te-doped Bi–Sb alloys. *Solid-State Electron.* 13:153–160.
21. Noothoven van Goor, J. M. 1971. Donors and acceptors in bismuth. *Philips Res. Rep. Suppl.* 4:1–91.
22. Noguchi, H., Kitagawa, H., Kiyabu, T., Hasezaki, K., and Noda, Y. 2007. Low temperature thermoelectric properties of Pb- or Sn-doped Bi-Sb alloys. *J. Phys. Chem. Solids* 68:91–95.
23. Thompson, N. 1936. Electrical resistance of bismuth alloys. *Proc. Royal Soc. Lond. Math. Phys. Sci.* 155:111–123.
24. Jin, H., Jaworski, C. M., and Heremans, J. P. 2012. Enhancement in the figure of merit of p-type $Bi_{100-x}Sb_x$ alloys through multiple valence-band doping. *Appl. Phys. Lett.* 101:053904.
25. Orovets, C. M., Chamoire, A. M., Jin, H., Wiendlocha, B., and Heremans, J. P. 2012. Lithium as an interstitial donor in bismuth and bismuth-antimony alloys. *J. Electron. Mater.* 41:1648.
26. Jin, H., Wiendlocha, B., and Heremans, J. P. 2015. P-type doping of elemental bismuth with indium, gallium, and tin: A novel doping mechanism in solids. *Energ. Environ. Sci.* 8:2027–2040.
27. Heremans, J., and Hansen, O. P. 1983. Temperature dependence of excess carrier density and thermopower in tin-doped bismuth: The pseudo-parabolic model. *J. Phys. C Solid State* 16:46234636.
28. Smith, G. E., and Wolfe, R. 1962. Thermoelectric properties of bismuth-antimony alloys. *J. Appl. Phys.* 33:841.
29. Yim, W. M. and Amith, A. 1972. Bi–Sb alloys for magneto-thermoelectric and thermomagnetic cooling. *Solid-State Electron.* 15:1141.
30. Lenoir, B., Dauscher, A., Cassart, M., Ravich, Yu. I., and Scherrer, H. 1998. Effect of antimony content on the thermoelectric figure of merit of $Bi_{1-x}Sb_x$ alloys. *J. Phys. Chem. Solids* 59:129.
31. Goldsmid, H. J. 2010. *Introduction to Thermoelectricity.* Berlin: Springer Verlag.
32. Lenoir, B., Dauscher, A., Devaux, X., Martin-Lopez, R., Ravich, Yu. I., Scherrer, H., and Scherrer, S. 1996. Bi-Sb Alloys: An update. In *Proceedings of the Fifteenth Institute of Electrical and Electronics Engineers International Conference on Thermoelectrics.* Pasadena, CA: Institute of Electrical and Electronics Engineers.

33. Jandl, P., and Birkholz, U. 1994. Thermogalvanomagnetic properties of Sn-doped $Bi_{95}Sb_5$ and its applications for solid-state cooling. *J. Appl. Phys* 76:7351–7366.

34. Scholz, K., Jandl, P., Birkholz, U., and Dashevskii, Z. M. 1994. Infinite stage Ettingshausen cooling in Bi-Sb alloys. *J. Appl. Phys.* 75:5406.

35. Hicks, L. D. and Dresselhaus, M. S. 1993. Thermoelectric figure of merit of a one-dimensional conductor. *Phys. Rev. B* 47:16631–16634.

36. Hicks, L. D., and Dresselhaus, M. S. 1993. Effect of quantum-well structures on the thermoelectric figure of merit. *Phys. Rev. B* 47:12727–12731.

37. Heremans, J. P. 2005. Low-dimensional thermoelectricity. *Acta Phys. Pol.* 108:609–634.

38. Heremans, J., Thrush, C. M., Lin, Y. M., Cronin, S., Zhang, Z., Dresselhaus, M. S., and Mansfield, J. F. 2000. Bismuth nanowire arrays: Synthesis and galvanomagnetic properties. *Phys. Rev. B* 61:2921–2930.

39. Heremans, J. P., Thrush, C. M., Morelli, D. T., and Wu, M. C. 2002. Thermoelectric power of bismuth nanocomposites. *Phys. Rev. Lett.* 88:21.

40. Heremans, J. P., Thrush, C. M., Lin, Y.-M., Cronin, S. B., and Dresselhaus, M. S., 2001. Transport properties of antimony nanowires. *Phys. Rev. B* 63:085406.

41. Chung, D.-Y. et al. 1997. High thermopower and low thermal conductivity in semiconducting ternary K-Bi-Se compounds: Synthesis and properties of β-$K_2Bi_8Se_{13}$ and $K_{2.5}Bi_{8.5}Se_{14}$ and their Sb analogues. *Chem. Mater.* 9(12):3060–3071.

42. Chung, D. Y. et al. 2000. $CsBi_4Te_6$: A high-performance thermoelectric material for low-temperature applications. *Science* 287(5455):1024–1027.

43. Chung, D.Y. et al. 2004. A new thermoelectric material: $CsBi_4Te_6$. *J. Am. Chem. Soc.* 126(20): 6414–6428.

44. Lykke, L., Iversen, B. B., and Madsen, G. K. H. 2006. Electronic structure and transport in the low-temperature thermoelectric $CsBi_4Te_6$: Semiclassical transport equations. *Phys. Rev. B* 73(19):195121.

45. Larson, P. et al., 2002. Electronic structure of $CsBi_4Te_6$: A high-performance thermoelectric at low temperatures. *Phys. Rev. B* 65(4):045205.

46. Sales, B. et al. 1997. Filled skutterudite antimonides: Electron crystals and phonon glasses. *Phys. Rev. B* 56(23):15081.

47. Greanya, V. A. et al. 2002. Angle-resolved photoemission study of the high-performance low-temperature thermoelectric material $CsBi_4Te_6$. *Phys. Rev. B* 65(20):205123.

48. Harutyunyan, S. R., Vardanyan, V. H., Kuzanyan, A. S., Nikoghosyan, V. R., Kunii, S., Wood, K. S., and Gulian, A. M. 2003. Thermoelectric cooling at cryogenic temperatures. *Appl. Phys. Lett.* 83(11):2142–2144.

49. Rowe, D. M., Gao, M., and Kuznestsov, V. L. 1998. Electrical resistivity and Seebeck coefficient of hot-pressed $YbAl_3$ over the temperature range 150±700 K. *Philos. Mag. Lett.* 77(2):105–108.

50. Sales, B. C., Delaire, O., McGuire, M. A., and May, A. F. 2011. Thermoelectric properties of Co-, Ir-, and Os-doped FeSi alloys: Evidence for strong electron-phonon coupling. *Phys. Rev. B* 83:125209.

51. Bentien, S. Madsen, G. K. H., Johnson, S., and Iversen, B. B. 2006. Experimental and theoretical investigations of strongly correlated $FeSb_{2-x}Sn_x$. *Phys. Rev. B* 74(20):205105.

52. Bentien, S., Johnsen, S., Madsen, G. K. H., Iversen, B. B., and Steglich, F. 2007. Colossal Seebeck coefficient in strongly correlated semiconductor $FeSb_2$. *Europhys. Lett.* 80:17008.

53. Sun, P., Oeschler, N., Johnsen, S., Iversen, B. B., and Steglich, F. 2009. $FeSb_2$: Prototype of huge electron-diffusion thermoelectricity. *Phys. Rev. B* 79(15):153308.

54. Sun, P., Oeschler, N., Johnsen, S., Iversen, B. B., and Steglich, F. 2010. Narrow band gap and enhanced thermoelectricity in $FeSb_2$. *Dalton Trans.* 39:1012–1019.

55. Sun, P., Søndergaard, M., Sun, Y., Johnsen, S., Iversen, B. B., and Steglich, F. 2011. Unchanged thermopower enhancement at the semiconductor-metal transition in correlated $FeSb_{2-x}Te_x$. *Appl. Phys. Lett.* 98:072105.

56. Diakhate, M. S., Hermann, R. P., Möchel, A., Sergueev, I., Søndergaard, M., Christensen, M., and Verstraete, M. J. 2011. Thermodynamic, thermoelectric, and magnetic properties of $FeSb_2$: A combined first-principles and experimental study. *Phys. Rev. B* 84(12):125210.

57. Mani, J. Janaki, A., Satya, T., Kumary, T. G., and Bharathi, A. 2012. The pressure induced insulator to metal transition in $FeSb_2$. *J. Phys.: Condens. Matter* 24:075601.

58. Jie, Q., Hu, R., Bozin, E., Llobet, A., Zaliznyak, I., Petrovic, C., and Li. Q. 2012. Electronic thermoelectric power factor and metal-insulator transition in $FeSb_2$. *Phys. Rev. B* 86(11):115121.

59. Zhao, H. Z., Pokharel, M., Zhu, G. H., Chen, S., Lukas, K., Jie, Q., Opeil, C., Chen, G., and Ren, Z. F. 2011. Dramatic thermal conductivity reduction by nanostructures for large increase in thermoelectric figure-of-merit of $FeSb_2$. *Appl. Phys. Lett.* 99:163101.

60. Sales, C., May, A. F., McGuire, M. A., Stone, M. B., Singh, D. J., and Mandrus, D. 2012. Transport, thermal, and magnetic properties of the narrow-gap semiconductor $CrSb_2$. *Phys. Rev. B* 86:235136.

61. Lukoyanov, A. V., Mazurenko, V. V., Anisimov, V. I., Sigrist, M., and Rice, T. M. 2006. The semiconductor-to-ferromagnetic-metal transition in $FeSb_2$. *Eur. Phys. J. B* 53:205–207.

62. Weber, L., and Gmelin, E. 1991. Transport properties of silicon. *Appl. Phys. A Mater. Sci. Process* 53:136–140.

63. Zhu, S., Xie, W., Thompson, D., Holgate, T., Zhou, M, Yan, Y., and Tritt, T. M. 2011. Tuning the thermoelectric properties of polycrystalline $FeSb_2$ by the in situ formation of Sb/InSb nanoinclusions. *J. Mater. Res.* 26(15):1894–1899.

64. Poudel, Q. et al. 2008. High-thermoelectric performance of nanostructured bismuth antimony telluride bulk alloys. *Science* 320(5876):634–638.

65. Wang, Z., Alaniz, J. E., Jang, W., Garay, J. E., and Dames, C. 2011. Thermal conductivity of nanocrystalline silicon: Importance of grain size and frequency-dependent mean free paths. *Nano Lett.* 11(6):2206–2213.

66. Sun, P., Oeschler, N., Johnsen, S., Iversen, B. B., and Steglich, F. 2009. Huge thermoelectric power factor: $FeSb_2$ versus $FeAs_2$ and $RuSb_2$. *J. Phys. Conf. Ser.* 150:012049.

67. Yu, B., Zebarjadi, M., Wang, H., Lukas, K., Wang, H. Z., Wang, D. Z., Opeil, C., Dresselhaus, M. S., Chen, G., and Ren, Z. F. 2012. Enhancement of thermoelectric properties by modulation-doping in silicon germanium alloy nanocomposites. *Nano Lett.* 12:2077–2082.

68. Zhao, H. Z., Pokharel, M., Chen, S., Liao, B., Lukas, K., Opeil, C., Chen, G., and Ren, Z. F. 2012. Figure-of-merit enhancement in nanostructured $FeSb_{2-x}Ag_x$ with $Ag_{1-y}Sb_y$ nanoinclusions. *Nanotechnology* 23:505402.

69. Koirala, M., Zhao, H. Z., Pokharel, M., Chen, S., Dahal, T., Opeil, C., Chen, G., and Ren, Z. F. 2013. Thermoelectric property enhancement by Cu nanoparticles in nanostructured $FeSb_2$. *Appl. Phys. Lett.* 102:213111.

70. Heremans, J. P., Jovovic, V., Toberer, E. S., Saramat, A., Kurosaki, K., Charoenphakdee, A., Yamanaka, S., and Snyder, G. J. 2008. Enhancement of thermoelectric efficiency in PbTe by distortion of the electronic density of states. *Science* 321:554–558.

71. Zhang, Q. Y. et al. 2012. Enhancement of thermoelectric figure-of-merit by resonant states of aluminium doping in lead selenide. *Energy Environ. Sci.* 5:5246–5251.

72. Pei, Y. Z., Shi, X. Y., LaLonde, A., Wang, H., Chen, L. D., and Snyder, G. J. 2011. Convergence of electronic bands for high performance bulk thermoelectric. *Nature* 473:66–69.

73. Zhang, Q., Cao, F., Liu, W. S., Lukas. K., Yu, B., Chen, S., Opeil, C., Broido, D., Chen, G., and Ren, Z. F. 2012. Heavy doping and band engineering by potassium to improve the thermoelectric figure of merit in p-type PbTe, PbSe, and PbTe$_{1-y}$Se$_y$. *J. Am. Chem. Soc.* 134:10031–10038.

74. Rogalski, A. 2002. Infrared detectors: An overview. *Infrared Phys. Technol.* 43:187–210.

75. Vermeulen, H. 2013. Cryogenic circulators: The solution for cooling problems. *Cold Facts* 29:46–48.

76. Taylor, R. A., and Solbrekken, G. L. 2008. Comprehensive system-level optimization of thermoelectric devices for electronic cooling applications. *IEEE Trans. Compon. Packag. Technol.* 31:23–31.

77. Angrist, S. W. 1977. *Direct Energy Conversion*. Boston, MA: Allyn and Bacon.

78. Vendernikov, M. V., and Iordanishvili, E. K. 1998. A. F. Ioffe and origin of modern semiconductor thermoelectric energy conversion. *17th Int. Conf. Thermoelectr.* 1:37–42.

79. Smith, G. E., and Wolfe, R. 1962. Thermoelectric properties of bismuth antimony alloys. *J. Appl. Phys.* 33:841–846.

80. Mahan, G. D., and Sofo, A. M. 1996. The best thermoelectric. *Proc. Nat. Acad. Sci. USA* 93:7436–7439.

81. Schweitzer, H., and Czycholl, G. 1991. Resistivity and thermopower of heavy-fermion systems. *Phys. Rev. Lett.* 67:3724–3727.

82. Ignatov, M. I., Bogach, A. V., Burkhanov, G. S., Glushkov, V. V., Demishev, S. V., Kuznetsov, A. V., Chistyakov, O. D., Shitsevalova, N. Y., and Sluchanko, N. E. 2007. Anomalous thermpower in heavy-fermion compounds CeB$_6$, CeAl$_3$, and CeCu$_{6-x}$Au$_x$. *J. Exp. Theor. Phys.* 105:58–61.

83. Mahan, G. D. 1998. *Good Thermoelectr. Solid State Phys.* (Academic press, New York) 51:81–157.

84. Aynajian, P., da Silva Neto, E. H., Gyenis, A., Baumbach, R. E., Thompson, J. D., Fisk, Z., Bauer, E. D., and Yazdani, A. 2012. Visualizing heavy fermions emerging in a quantum critical Kondo lattice. *Nature* 486:201–206.

85. Rowe, D. M. 1995. *CRC Handbook of Thermoelectrics*. Boca Raton, FL: CRC Press.

86. Levy, P. M., and Zhang, S. 1989. Crystal-field splitting in Kondo systems. *Phys. Rev. Lett.* 62:78–81.

87. Freimuth, A. 1987. Correlation between transport properties and quasielastic linewidths of Ce and Yb compounds with unstable 4f-shell. *J. Magn. Magn. Mater.* 68:28–38.

88. Grade, C. S., and Ray. J. 1995. Thermopower and resistivity behavior in Ce-based Kondo-lattice systems: A phenomenological approach. *Phys. Rev. B* 51:2960–2965.

89. Rossel, C., Yang, K. N., Maple, M. B., Fisk, Z., Zirngiebl, E., and Thompson, J. D. 1987. Strong electric correlation in a new class of Yb-based compounds: $YbXCu_4$ (x=Ag, Au, Pd). *Phys. Rev. B* 35:1914–1918.

90. Graf, T., Lawrence, J. M., Hundley, M. F., Thompson, J. D., Lacerda, A., Haanappel, E., Torikachvili, M. S., Fisk, Z., and Canfield, P. C. 1995. Resistivity, magnetization, and specific heat of $YbAgCu_4$ in high magnetic fields. *Phys. Rev. B* 51:15053–15061.

91. Schlottmann, P. 1993. The heavy fermion compound $YbAgCu_4$. *J. Appl. Phys.* 73:5412–5414.

92. Golubkov, A. V., Parfen'eva, L. S., Smirnov, I. A., Misiorek, H., Mucha, J., and Jezowski, A. 2001. Thermal conductivity of the "light" heavy-fermion compound $YbIn_{0.7}Ag_{0.3}Cu_4$. *Phys. Solid State* 43:1811–1815.

93. Graf, T., Movshovich, R., Thompson, J. D., Fisk, Z., and Canfield, P. C. 1995. Properties of $YbAgCu_4$ at a high pressure and magnetic field. *Phys. Rev. B* 52:3009–3107.

94. Casanova, R., Jaccard, D., Marcenat, C., Hamdaiui, C., and Besnus, M. J. 1990. Thermoelectric power of $YbMCu_4$ (M=Ag, Au, and Pd) and $YbPdSi_2$. *J. Magn. Magn. Mater.* 90–91:587–588.

95. Poudel, B. et al. 2008. High-thermoelectric performance of nanostructured bismuth antimony telluride bulk alloys. *Science* 320:634–638.

96. Koirala, M., Wang, H., Pokharel, M., Lan, Y., Guo, C., Opeil, C., and Ren, G. F. 2014. Nanostructured $YbAgCu_4$ for potentially cryogenic thermoelectric cooling. *Nano Lett.* 14:5016–5020.

97. Levander, A. X. et al. 2011. Effect of point defects on thermal and thermoelectric properties of InN. *Appl. Phys. Lett.* 98:012108/1–0121108/3.

98. Nolas, G. S., Sharp, J., and Goldsmid, H. J. 2001. *Thermoelectrics: Basic Principles and New Materials Developments.* Berlin: Springer-Verlag.

99. Zhang, Y., Dresselhaus, M., Shi, Y., Ren, Z. F., and Chen, G. 2011. High thermoelectric figure-of-merit in Kondo insulator nanowires at low temperatures. *Nano Lett.* 11:1166–1170.

100. Ruzitschka, R., Hauser, R., Bauer, E., Soldevilla, J. G., Gomez Sal, J. C., Yoshimura, K., Tsujii, N., and Kosuge, K. 1997. Volume dependence of the physical behavior of $YbCu_{5-x}Ag_x$. *Physics B* 230–232:279–281.

101. Tsujii, N., He, J., Yoshimura, K., Kosuge, K., Michor, H., Kreiner, K., and Hilscher, G. 1997. Heavy-fermion behavior in $YbCu_{5-x}Ag_x$. *Phys. Rev. B* 55: 1032–1039.

102. Lehr, G. J., and Morelli, D. T. 2013. Thermoelectric properties of $Yb_{1-x}(Er,Lu)_x Al_3$ solid solutions. *J. Electron. Mater.* 42:1697–1701.

103. Lehr, G. J., Morelli, D. T., Jin, H., and Heremans, J. P. 2015. $YbCu_2Si_2$-$LaCu_2Si_2$ solid solutions with enhanced thermoelectric power factors. *J. Electron. Mater.* 44:1663–1667.

104. Pokharel, M., Dahal, T., Ren, Z., Czajka, P., Wilson, S., Ren, Z. F., and Opiel, C. 2014. Thermoelectric properties of $CeAl_3$ prepared by hot-press method. *Energy Convers. Manag.* 87:584–588.

105. Pokharel, M., Dahal, T., Ren, Z. F., and Opiel, C. 2014. Thermoelectric properties of nanocomposite heavy fermion $CeCu_6$. *J. Alloys Compd.* 609:228–232.

106. Mahan, G. D., Sales, B. C., and Sharp, J. 1997. Thermoelectric materials: New approaches to an old problem. *Phys. Today* 50(3):42–47.

107. Kumar, G. S., Prasad, G., and Pohl, R. O. 1993. Experimental determinations of the Lorenz number. *J. Mater. Sci.* 28:4261–4272.

108. Minnich, A. J., Dresselhaus, M. S., Ren, Z. F., and G. Chen. 2009. Bulk nanostructured thermoelectric mateirlas: Current research and future prospects. *Energ. Environ. Sci.* 2:466–479.

109. Cahill, D. G., Watson, S. K., and Pohl, R. O. 1992. Lower limit to the thermal conductivity of disordered crystals. *Phys. Rev. B* 46:6131.

110. Heremans, J. P., Wiendlocha, B., and Chamoire, A. M. 2012. Resonant levels in bulk thermoelectric semiconductors. *Energ. Environ. Sci.* 5:5510–5530.

111. Watzman, S. J., Duine, R. A., Tserkovnyak, Y., Jin, H., Prakash, A., Zheng, Y., and Heremans, J. P. 2016. Magnon-drag thermopower and Nernst coefficient in Fe, Co, and Ni. arXiv:1603.03736.

112. Goncalves, A. P., and Godart, C. 2014. New promising bulk thermoelectrics: Intermetallics, pnictides, and chalcogenides. *Eur. Phys. J. B* 87:42.

113. Mahan, G. D., and Sofo, J. O. 1996. The best thermoelectric. *Proc. Natl. Acad. Sci.* 93:7436–7439.

114. Sales, B. 2014. Electronic correlations and thermoelectric performance. *B. Am. Phys. Soc.* http://meetings.aps.org/link/BAPS.2014.MAR.Q25.1

115. Sun, P., Ikeno, T., Mizushima, T., and Isikawa, Y. 2009. Simultaneously optimizing the interdependent thermoelectric parameters in $Ce(Ni_{1-x}Cu_x)_2Al_3$. *Phys. Rev. B* 80:193105.

116. Boona, S. R., and Morelli, D. T. 2012. Enhanced thermoelectric properties of $CePd_{3-x}Pt_x$. *Appl. Phys. Lett.* 101(10):101909-4.

117. Gambino, R. J., Grobman, W. D., and Troxen, A. M. 1973. Anomalously large thermoelectric cooling figure of merit in the Kondo systems CePd3 and $CeIn_3$. *Appl. Phys. Lett.* 22:506.

118. Sakurai, J., Ohyama, T., and Komura, Y. 1985. Thermoelectric power and electrical resistivity of $Ce(In_{1-x}Snx)_3$ and $(Ce_{1-x}Lax)In_3$. *J. Magn. Magn. Mater.* 52:320–322.

119. Behnia, K., Jaccard, D., and Flouquet, J. 2004. On the thermoelectricity of correlated electrons in the zero-temperature limit. *J. Phys.: Condens. Matter.* 16:5187–5198.

120. Schlesinger, Z., Fisk, Z., Zhang, H.-T., and Maple, M. B. 1997. Is FeSi a Kondo insulator? *Physica B* 237–238, 460–462.

121. Jones, C. D. W., Regan, K. A., and DiSalvo, F. J. 1998. Thermoelectric properties of the doped Kondo insulator: $Nd_xCe_{3-x}Pt_3Sb_4$. *Phys. Rev. B* 58:16057.

122. Sato, H., Abe, Y., Okada, H., Matsuda, T. D., Abe, K., Sugawara, H., and Aoki, Y. 2000. Anomalous transport properties of RFe_4P_{12} (R = La, Ce, Pr, and Nd). *Phys. Rev. B* 62(22):125–130.

123. Abe, K., Sato, H., Matsuda, T. D., Namiki, T., Sugawara, H., and Aoki, Y. 2002. Transport properties in the filled-skutterudite compounds $RERu_4Sb_{12}$ (Re = La, Ce, Pr, and Nd); an exotic heavy fermion semimetal $CeRu_4Sb_{12}$. *J. Phys.: Condens. Matter.* 14:11757–11768.

124. Bucher, B., Schlesinger, Z., Mandus, D., Fisk, Z., Sarrao, J., DiTusa, J. F., Oglesby, C., Aeppli, G., and Bucher, E. 1996. Charge dynamics of Ce-based compounds: Connection between the mixed valent and Konto-insulator states. *Phys Rev. B* 53:2948–2951.

125. Beyermann, W. P., Gruner, G., Dalichaough, Y., and Maple, M. B. 1988. Relaxation-time enhancement in the heavy-fermion system $CePd_3$. *Phys. Rev. Lett.* 60:216.

126. Dhar, S. K., Malik, S. K., and Vijayaraghavan, R. 1981. *Mat.* Boron addition to RPd_3 compounds (R = rare earth). *Res. Bull.* 16(12):1557–1560.

127. Costa, T. A., Hewson, A. C., and Zlatic, V. 1994. Transport coefficients of the Anderson model via the numerical renormalization group. *J. Phys. Condens. Matt.* 6(13):2519.

128. Lackner, R., Bauer, E., and Rogl, P. 2006. Study of the thermoelectric properties of $CePd_3B_x$. *Physica B* 378:835.

129. Schneider, H., and Wohlleben, D. 1981. Electrical and thermal conductivity of $CePd_3$, YPd_3, $GdPd_3$ and some dilute alloys of $CePd_3$ with Y and Gd. *Z. Phys.* B 44:193.

130. Kappler, J. P., Krill, G., Besnus, M. S., Ravet, M. F., Hamdaoui, N., and Meyer, A. 1982. Electronic structure of cerium in $Ce_{1-x}Y_xPd_3$ and $Ce_{1-x}M_xAl_2$ (M = Sc, Y). *J. Appl. Phys.* 53(3):2152.

131. Jaccard, D., Besnus M. J., and Kappler, J. P. 1987. Thermoelectric power investigation of the (Ce,Y)Pd_3 system. *J. Magnet. Magnet. Mater.* 63:572.

132. Iijiri, Y., and DiSalvo, F. J. 1997. Thermoelectric properties of $RxCe_{1-x}Pd_3$ (R = Y, $La_{0.5}Y_{0.5}$, Nd). *Phys. Rev. B* 55:1283.

133. Proctor, K. J., Jones, C. D. W., and DiSalvo, F. J. 1999. Modification of the thermoelectric properties of $CePd_3$ by the substitution of neodymium and thorium. *J. Phys. Chem. Sol.* 60(5):663.

134. Scoboria, P., Crow, J. E., and Mihalisin, T. 1979. Resistive behavior and intermediate valence effects in $La_{1-x}Ce_xPd_3$ and $CePd_{3-x}Rh_x$. *J. Appl. Phys.* 50:1895.

135. Gambke, T., Elschner, B., and Schaafhausen, J. 1980. Intermediate valence of Ce in $Ce_{1-x}Sc_xPd_3$. *Phys. Lett. A* 78:413.

136. Mishra, S. N. 2003. Local magnetism and Kondo behavior of Ce in U-substituted $CePd_3$. *J. Phys. Condens. Matt.* 15:5333.

137. Schaeffer, H., and Elschner, B. 1985. ESR and hyperfine structure of Mn impurities in the metallic host $CePd_3$. *Sol. State Commun.* 53:611.

138. Boona, S. R., and Morelli, D. T. 2012. Thermoelectric properties of $Ce_{1-x}Sc_xPd_3$. *J. Electron. Mater.* 41(6):1199–1204.

139. Lackner, R., Bauer, E., and Rogl, P. 2007. The influence of substitution and doping on the thermoelectric properties of $CePd_3$. *Proc. 26th Intl. Conf. Thermoelectrics* 386–389.

140. Lackner, R. 2007. The effect of substitution and doping on the thermoelectric properties of $CePd_3$. Doctoral Dissertation, Institute of Solid State Physics, Vienna University of Technology, Austria.

141. Sales, B. C. 1997. Novel thermoelectric materials. *Curr. Opin. Solid St. M.* 2:284–289.

142. Veenhuizen, P. A., Yang, F. M., van Nassou, H., and de Boer, F. R. 1987. Magnetic properties and electrical resistivity of (Ce,Y)Pd_3 and Ce(Pd,Pt)$_3$. *J. Magn. Magn. Mater.* 63:567–571.

143. Rambabu, D., Dhar, S. K., Malik, S. K., and Vijayaraghavan, R. 1982. Crystal structure and valence state of cerium in $CePd_{3-x}Pt_x$ alloys. *Phys. Lett. A* 87:294.

144. Oomi, G., Onuki, Y., and Komatsubara, T. 1990. Effect of pressure on the electrical resistivity and lattice constant of $CePd_3$. *Phys. B Condens. Matt.* 163:405–408.

145. Chandra Shekar, N. V., Rajagopalan, M., Meng, J. F., Polvani, D. A., and Badding, J. V. 2005. Electronic structure and thermoelectric power of cerium compounds at high pressure. *J. Alloys Comp.* 388(2): 215–220.

146. Pedrazzini, P., Jaccard, D., Deppe, M., Geibel, C., and Sereni, J. G. 2009. Multiprobe high-pressure experiments in $CePd_{0.6}Rh_{0.4}$ and $CePd_3$. *Physica B Condens. Matter* 404(19):2898–2903.

147. Boona, S. R., and Morelli, D. T. 2012. Relationship between structure, magnetism, and thermoelectricity in $CePd_3M_x$ alloys. *J. Appl. Phys.* 112:063709.

148. Qachaou, A., Panissod, P., Nejjar, M., El Harfaoui, M., Faris, M., and Meziane Mtalsi, D. 1998. Alloying effect on the Cerium valence in the series $CePd_3B_x(Be_x)$. *Moroccan J. Condens. Matter* 1(1):1–5.

149. Gumeniuk, R., Schnelle, W., Kvashnina, K. O., and Leithe-Jasper, A. 2016. Kondo effect and thermoelectric transport in $CePd_3Be_x$. *J. Phys.: Condens. Matter* 28:165603.

150. Jones, C. D. W., Gordon, R. A., Cho, B. K., DiSalvo, F. J., Kim, J. S., and Stewart, G. R. 1999. Comparisons of electrical, magnetic, and low temperature specific heat properties in group 13 and group 14 $Ce_8Pd_{24}M$ compounds (M = B, Al, Ga, In and Si, Ge, Sn Pb). *Physica B Condens. Matt.* 262:284–295.

151. Tchoula Tchokonte, M. B., Du Plessis, P. D. V., and Kaczorowski, D. 2004. Magnetic and electrical properties of $Ce_8Pd_{24}M_x$ alloys (M = Al, Ga, Ge). *J. Magnet. Magnet. Mater. Proc. Inter. Conf. Magnet.* 2003:272–276.

152. Tchokonte, T., Du Plessis, P. D. V., Strydom, A. M., and Kaczorowski, D. 2006. Magnetic and electrical transport studies of $Ce_8Pd_{24}Al_x$. *Physica B Condens. Matt.* 378:849–850.

153. Cho, B. K., Gordon, R. A., Jones, C. D. W., DiSalvo, F. J., Kim, J. S., and Stewart, G. R. 1998. Specific heat and heavy-fermionix behavior in $Ce_8Pd_{24}M$ (M = Ga, In, Sn, Sb, Pb, and Bi). *Phys. Rev. B* 57:15191.

154. Mitra, C., Dhar, S. K., and Ramakrishnan, S. 1999. Dense-Kondo antiferromagnetism in $(CePd_3)_8Ge$. *Solid State Commun.* 110(12): 701–705.

155. Singh, S., and Dhar, S. K. 2003. Magnetic ordering of the Mn sublattice: Dense Kondo-lattice behavior of Ce in $(RPd_3)_8Mn$ (R = La, Ce). *Phys. Rev. B* 68(14):144433.

156. Malik, S. K., Vijayaraghavan, R., Boltich, E. B., Craig, R. S., Wallace, W. E., and Dhar, S. K. 1982. Effect of Si addition on the valence state of Ce in $CePd_3$. *Solid State Commun.* 43(4):243–245.

157. Beaurepaire, E., Panissod, P., and Kappler, J. P. 1985. Local environment effects on the IV→3⁺ transition in $CePd_3B_x$. *J. Magn. Magn. Mater.* 47:108–110.

158. Schmitt, M., Gumeniuk, R., Trapananti, A., Aquilanti, G., Strohm, C., Meier, K., Schwarz, U., Hanfland, M., Schnelle, W., Leithe-Jasper, A., Rosner, H. 2012. Tuning the Eu valence in $EuPd_3B_x$: Pressure versus valence electron count–A combined computational and experimental study. arXiv:1203.1865.

159. Boona, S. R. 2013. Structure-processing-property relationships and thermoelectricity in strongly correlated $CePd_3$ and related compounds. Doctoral Dissertation, Michigan State University.

160. Heremans, J. P., Dresselhaus, M. S., Bell, L. E., and Morelli, D. T. 2013. When thermoelectrics reached the nanoscale. *Nature Nano.* 8(7):471–473.

161. Berman, R. 2002. *Defects and Goemetry in Condensed Matter Physics*. New York: Cambridge University Press.

162. Johansson, B. 1979. Valence state at the surface of rare-earth metals. *Phys. Rev. B* 19:6615–6619.

163. Rosengren, A., and Johansson, B. 1982. Valence instability of the samarium metal surface. *Phys. Rev. B* 26:3068–3078.

164. Chen, Y. Y., Yao, Y. D., Wang, C. R., Li, W. H., Chang, C. L., Lee, T. K., Hong, T. M., Ho, J. C., and Pan, S. F. 2000. Size-induced transition from magnetic ordering to Kondo behavior in (Ce,Al) compounds. *Phys. Rev. Lett.* 84(21):4990–4993.

165. Han, S. W., Booth, C. H., Bauer, E. D., Huang, P. H., Chen, Y. Y., and Lawrence, J. M. 2006. Lattice disoarder and size-induced Kondo behavior in $CeAl_2$ and $CePt_{2+x}$. *Phys. Rev. Lett.* 97(9):097204.

166. Mukherjee, K., Iyer, K. K., and Sampathkumaran, E. V. 2012. Contrasting magnetic behavior of fine particles of some Kondo lattices. *Solid State Commun.* 152(7):606–611.

167. Chen, Y. Y., Huang, P. H., Ou, M. N., Wang, C. R., Yao, Y. D., Lee, T. K., Ho, M. Y., Lawrence, J. M., and Booth, C. H. 2007. Kondo interactions and magnetic correlations in $CePt_2$ nanocrystals. *Phys. Rev. Lett.* 98(15):157206.

168. Rojas, D. P., Barquin, L. F., Fernandez, J. R., Espeso, J. I., and Sal, J. C. G. 2010. Magnetization and specific heat of nanocrystalline rare-earth $TbAl_2$, $TbCu_2$, and $GdAl_2$ alloys. *J. Phys. Conf. Series* 200:072080.

169. Zhou, G. F., and Bakker, H. 1995. Mechanically induced structural and magnetic changes in the $GdAl_2$ Laves phase. *Phys. Rev. B* 52(13):9437.

170. Rojas, D. P., Barquin, L. F., Espeso, J. I., Fernandez, J. R., and Chaboy, J. 2008. Reduction of the Yb valence in $YbAl_3$ nanoparticles. *Phys. Rev. B* 78(9):094412.

171. Kim, J. S., Stewart, G. R., and Samwer, K. 2009. Evolution of physical properties with decreasing size in $Ce(Ru_{0.4}Rh_{0.6})_2Si_2$. *Phys. Rev. B* 79(16):165119.

172. Sampathkumaran, E. V., Mukherjee, K., Iyer, K. K., Mohaptra, N., and Das, S. D. 2011. Magnetism of fine particles of Kondo lattices obtained by high-energy ball-milling. *J. Phys. Condens. Matt.* 23(9):094209.

173. Lin, Y. H., Wang, C. R., Dong, C. L., Ou, M. N., and Chen, Y. Y. 2011. Size effects on mixed valence $CePd_3$. *J. Phys.: Conf. Series* 273(1):012041.

174. Mohanta, S. K., Mishra, S. N., Iyer, K. K., and Sampathkumaran, E. V. 2013. Microscopic evidence for 4f localization with reduced particle size in correlated electron system $CePd_3$. *Phys. Rev. B* 87(12):125125.

175. Holland-Moritz, E., Loewenhaupt, M., Schmatz, W., and Wohlleben, D. K. 1977. Spontaneous relaxation of the local 4f magnetization in $CePd_3$. *Phys. Rev. Lett.* 38(17):983–986.

176. Harris, I. R., Norman, M., and Gardner, W. E. 1972. The electronic state of cerium in some $CeRh_{3-x}Pd_x$ alloys. *J. Less Comm. Metals* 29(3):299–309.

177. Boona, S. R., and Morelli, D. T. 2013. Structural, magnetic, and thermoelectric properties of some $CePd_3$-based compounds. *J. Electron. Mater.* 42(7):1592–1596.

178. Razafimandimby, H. A., and Erdos, P. 1982. Z. Intermediate valence theory of CePd3 and $UNi_{5-x}Cu_x$. *Phys. B Condens. Matt.* 46:193–197.

179. Young, D., Mastronardi, K., Khalifah, P., Wang, C. C., Cava, R. J., and Ramirez, A. P. 1999. $Ln_3Au_3Sb_4$: Thermoelectrics with low thermal conductivity. *Appl. Phys. Lett.* 74(26):3999–4001.

180. Jones, C. D. W., Regan, K. A., and DiSalvo, F. J. 1999. $Ce_3Cu_xPt_{3-x}Sb_4$: Modifying the properties of a Kondo insulator by substitutional doping. *Phys. Rev. B* 60(8):5282.

181. Horyn, A., Romaka, V., and Gorelenko, Y. 2007. Crystal structure and electric transport properties of solid solutions of substitution $R_3Cu_3Sb_{4-x}Snx$ (R = La, Ce, Pr, Nd, Sm). *Ser. Khim.* 48(1):223–228.

182. Jones, C. D. W., Regan, K. A., and DiSalvo, F. J. 1998. Thermoelectric properties of the doped Kondo insulator: $Nd_xCe_{3-x}Pt_3Sb_4$. *Phys. Rev. B* 58(24):16057.

183. Lu, Q., Sun, P., Huo, D., Mizushima, T., Isikawa, Y., and Sakurai, J. 2004. Thermal and transport properties of $Ce(Ni_{1-x}Cu_x)2Al_3$: The dominant role of electronic change. *J. Phys. Soc. Jpn.* 73:681–686.

CHAPTER 3

Materials for Near-Room Temperatures

Weishu Liu, Yucheng Lan, Jiehe Sui, Zihang Liu, and Zhifeng Ren

Contents

3.1 Bismuth Chalcogenides

Weishu Liu, Yucheng Lan, and Zhifeng Ren

3.1.1 Introduction

The history of V_2–VI_3 compounds for thermoelectric application near-room temperature started from Bi_2Te_3 and later moved to its alloys with element Sb at Bi site and Se at Te site.[1-3] The early works mainly focused on the well-grown single crystals or casted ingots, which were already well reviewed by many authors over the past decades.[4-6] However, the Bi_2Te_3-based ingots are mechanically brittle due to the intrinsic laminar crystalline structure characterized with van der Waals bonds. In contrast, their polycrystalline counterparts were proved to be more mechanically robust because the grain boundaries work well as a block to the propagation of fracture cracks.[7,8] The decreased lattice thermal conductivity was widely observed in both the p-type $(Bi, Sb)_2Te_3$ and n-type $Bi_2(Te, Se)_3$ polycrystalline bulks made by various nanoapproaches. However, the thermoelectric figure of merit (ZT) values of these nanostructured materials were not inevitably enhanced due to various reasons. One of the challenges was the controllable charge carrier concentration, which was sensitive to the native defects, grain boundaries, and protection atmosphere. In this chapter, we will focus our discussion on the latest advances in the performance enhancement and puzzle understandings for the p-type $(Bi, Sb)_2Te_3$ and n-type $Bi_2(Te, Se)_3$ polycrystalline bulks.

3.1.2 p-Type $(Bi, Sb)_2Te_3$

The classic compositions of the p-type $Bi_xSb_{2-x}Te_3$ ranged from $x = 0.3$–0.6, in which the antisite defects were identified as the dominant charge carrier provider. Generally, $Bi_xSb_{2-x}Te_3$ with more Bi has less positive charge carriers because of the decreased concentration.[9] Here, the Kröger–Vink notations were used to describe the electric charge and lattice position for point defect species in crystals. The prime (′) represents the negative charge, the dot (˙) represents the positive charge, and the cross (×) for the neutral charge. A recent theoretical calculation for Sb_2Te_3 suggests that the dominant native defect is antisite defect in Sb-rich case while vacancy in a Te-rich case,[10] as shown in Figure 3.1.[10] The native acceptor behavior makes a well explanation for the reason why Sb_2Te_3-rich $Bi_xSb_{2-x}Te_3$ is usually a p-type semiconductor.

3.1.2.1 Early Works

The early efforts in the polycrystalline p-type $Bi_xSb_{2-x}Te_3$ achieved notable enhancements in mechanical strength rather than thermoelectric performance compared with its state-of-the-art single crystal counterpart.[11-17] Tokiai et al.[7] fabricated a polycrystalline p-type $Bi_xSb_{2-x}Te_3$ with significantly enhanced mechanical strength by using ball milling (BM) (dry process + wet process) with hot pressing (HP) (cold isostatic press + hot isostatic press) with elemental powders as the raw materials. Nevertheless, some oxide impurities, such as Bi_2O_3, TeO_3, and Sb_2O_3, were observed in their p-type polycrystalline

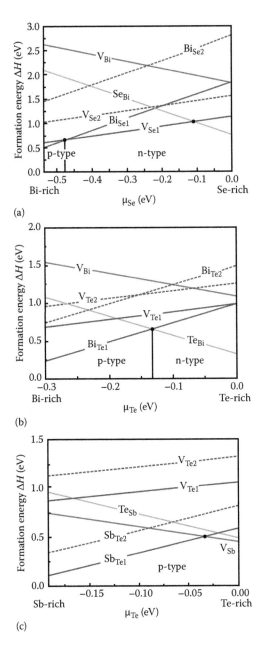

FIGURE 3.1 The formation energy of ΔH as a function of anion chemical potential for all the possible defects in (a) Bi_2Se_3, (b) Bi_2Te_3, (c) Sb_2Te_3. V_{Bi}, V_{Sb}, V_{Se}, and V_{Te} stand for the bismuth vacancy, antimony vacancy, selenium vacancy, and tellurium vacancy, respectively, while Bi_{Se}, Bi_{Te}, Sb_{Te}, Se_{Bi}, and Te_{Bi} are the antisite defects. 1 and 2 are labeled to distinguish Se(Te) in different layers. X-rich (X for Bi, Sb, Se, or Te) indicates the extreme growth conditions. Vertically lines highlight the boundary of the carrier types. (Reprinted with permission from Zhang, M. et al., *Physical Review B*, 88, 235131, 2013. Copyright 2013 by the American Physical Society.)

$Bi_xSb_{2-x}Te_3$ which could have resulted from the usage of wet BM process. Furthermore, the Seebeck coefficient of $Bi_{0.5}Sb_{1.5}Te_3$ continuously increased from 50 to 280 µV/K, together with an increased electrical resistivity from 10 to 41 µΩm, when the BM time increased from 1 to 10 h. The maximum power factor (*PF*) of this polycrystalline $Bi_{0.5}Sb_{1.5}Te_3$ (*PF* = 19 µW/cm K² for the 10 h BM sample) was much lower than that of the single crystal counterpart (*PF* = 32–46 µW/cm K²),[18] which therefore led to a low *ZT* value of 0.48 at 25°C. By adding extra Sb in $Bi_xSb_{2-x}Te_3$ and using less Bi content, the enhanced *PF* of ~40 µW/cm K² and *ZT* of 0.78 at 25°C were reported in the $Bi_{0.45}Sb_{1.55}Te_3$ by the same group.[13] Even so, the *ZT* value of this polycrystalline $Bi_{0.45}Sb_{1.55}Te_3$ was still inferior compared with its single crystalline counterpart (*ZT* = 0.96 in $Bi_{0.45}Sb_{1.55}Te_3$).[18] Navratil et al.[14] synthesized a polycrystalline $Bi_{0.45}Sb_{1.55}Te_3$ with a *ZT* value of 0.93 at 27°C through the fabrication route of grinding, cold pressing, and annealing by using corresponding ingot as the raw material.[14] The grinding $Bi_{0.45}Sb_{1.55}Te_3$ powders were sieved into two size scales. The sintered $Bi_xSb_{2-x}Te_3$ sample, by using the finer grinding powders (grain size <130 µm), displayed higher Seebeck coefficient and electrical resistivity than that of the one by using coarser grinding powders (grain size 130–300 µm), which suggested that a donor-like effect of grinding and pressing process. This donor effect even changed the p-type $Bi_xSb_{2-x}Te_3$ into n-type.[11] One of the possible reasons was that Te vacancies generated from the mechanical deformation in the grinding process annihilated the antisite defects and, hence, reduced the hole concentration.[14] Oh et al.[16] conducted a systematic investigation in the $Bi_xSb_{2-x}Te_3$ system and got optimized *ZT* values of 0.92 and 0.88 at room temperature in polycrystalline $Bi_{0.4}Sb_{1.6}Te_3$ by using ingots and elemental powders as raw materials, respectively.[16] Additionally, a donor-like effect of oxygen was confirmed in polycrystalline $Bi_{0.4}Sb_{1.6}Te_3$ by Oh et al.[16] Seo et al.[17] reported that an enhanced *ZT* value from 0.81 to 0.88 was obtained in polycrystalline $Bi_{0.45}Sb_{1.55}Te_3$ by a thermal extrusion process. It was believed that the texture developed by the thermal extrusion process increased carrier mobility. These early efforts failed to get superior *ZT* values in polycrystalline $Bi_xSb_{2-x}Te_3$ bulks compared with their single crystal counterpart. However, they outlined the most important factors that govern the thermoelectric performance of a p-type $Bi_xSb_{2-x}Te_3$ polycrystalline bulk: (1) Bi/Sb ratio, (2) extra Te or Sb, (3) BM conditions, (4) HP conditions, (5) protection atmosphere, and (6) morphology of grain boundaries. Systematically considering all these factors is a big challenge to make the polycrystalline p-type $Bi_xSb_{2-x}Te_3$ with a better thermoelectric performance over their single crystalline counterpart.

3.1.2.2 BM-HP Route

In 2008, high peak *ZT* values of 1.3 and 1.4 near 100°C were reported in p-type $Bi_xSb_{2-x}Te_3$ nanostructured bulks by using elemental chucks and high-quality BiSbTe ingot as raw materials, respectively.[19,20] These nanostructured $Bi_xSb_{2-x}Te_3$ bulks were made by a joint route of high-energy BM and direct current (dc) HP. Here, the dc HP is a current assistant HP technique utilizing the self-heating effect of joule heat for heating when current passes through the sample within a graphite die, which was also called spark plasma sintering (SPS) or plasma activate sintering technique by different groups. The importance of the fast-speed HP to obtain the nanostructured or nanocrystalline thermoelectric bulk materials has been explained in our previous work.[21] Shortly, the fast-sintering technology is the key to ensure that the nanostructures of the ball milled powders remain in the final sintered bulks because of the limited grain growth. Figure 3.2 shows the temperature-dependent thermoelectric properties of the nanostructured $Bi_xSb_{2-x}Te_3$ bulks and their ingot counterpart. Because of well protection atmosphere and systematic fabricate condition optimization, the room-temperature PF of both nanostructured bulks is comparable to their ingot counterpart (~46 µW/cm K^2). Since the carrier mobility of the polycrystalline bulks is lower than that of their single crystal counterpart, a higher carrier concentration is necessary for the polycrystalline materials to achieve the comparable PF. The carrier concentration of the polycrystalline $Bi_xSb_{2-x}Te_3$ was adjusted by the BM and HP conditions and other dopants. The combination of high-energy BM and dc HP (BM-HP) endowed the polycrystalline $Bi_xSb_{2-x}Te_3$ bulks with finer nanostructures and, hence, lower lattice thermal conductivity. Figure 3.3 shows some general features of the nanostructured $Bi_xSb_{2-x}Te_3$.[22] Firstly, the bulks consisted of widely size-distributed grains ranging from 10 nm to 3 µm, as shown in Figure 3.3a and b. At the fine grain end of the distribution, there were 12% of the grains with a diameter less than 20 nm and 5% with a diameter in the 20−40 nm range in one bulk made from BiSbTe ingot. Secondly, there were many nanoinclusions either embedded in the grains or located at the grain boundaries with size from 10–100 nm, as shown in Figure 3.3c and d. In the nanostructured $Bi_xSb_{2-x}Te_3$ synthesized from ingot, some of interesting nanoinclusions were identified: one was the Te-poor $Bi_xSb_{2-x}Te_{3-\delta}$ without clear boundary from the matrix (coherent inclusion) and another was the pure Te with a high-angle boundary from the surrounding materials (incoherent inclusion). The simultaneous appearance of these two types of nanoinclusions suggested that the BM-HP process could create some kind of local compositional fluctuation. Furthermore, the microstructure of the nanostructured $Bi_xSb_{2-x}Te_3$ made from the elemental chunks is quite similar to the one made from ingot, having many Te-poor coherent $Bi_xSb_{2-x}Te_{3-\delta}$ nanoinclusions. However, the difference is that Sb nanoprecipitations rather than Te nanoprecipitations were found in the nanostructured $Bi_xSb_{2-x}Te_3$ bulk made from the elemental chunks. Thirdly, some atomic level defects, such as point defects and threading dislocations were identified, as shown in Figure 3.3e through f. For example, the dislocation concentration of nanocrystalline $Bi_xSb_{2-x}Te_3$ was ~10^{11} cm^{-2}, at least 10 times higher than the value of ~5×10^9 cm^{-2} found in the single crystalline ingot. However, it is too challenging to numerically consider these scale defects to clarify the reason why the nanostructured $Bi_xSb_{2-x}Te_3$ synthesized from ingot has ~10% lower thermal conductivity. Recently, similar BM-HP routes were used for reinvestigating the polycrystalline $Bi_xSb_{2-x}Te_3$ by several other groups.[23–26] However,

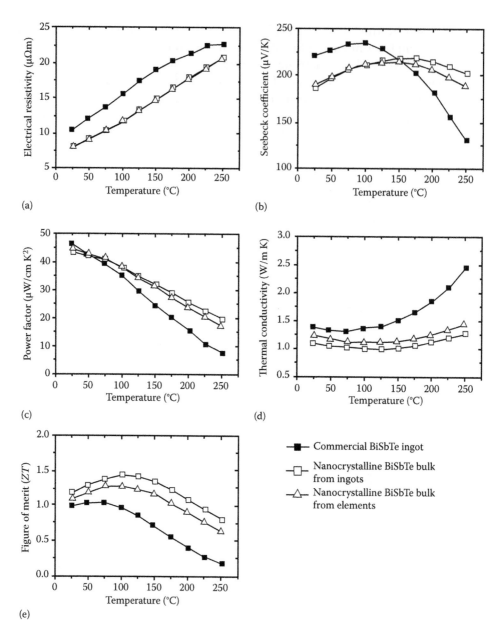

FIGURE 3.2 Temperature-dependent thermoelectric properties of nanocrystalline $Bi_xSb_{2-x}Te_3$ bulks made by high-energy BM and dc HP: (a) electrical resistivity, (b) Seebeck coefficient, (c) PF, (d) thermal conductivity, and (e) figure of merit. The data of the commercial BiSbTe ingot and nanocrystalline BiSbTe bulk made from ingots are from Poudel et al.,[19] while those for nanocrystalline BiSbTe bulk made from elements are from Ma et al.[20] (Reprinted with permission from Ma Y. et al., Enhanced thermoelectric figure-of-merit in p-type nanostructured bismuth antimony telluride alloys made from elemental chunks. Nano Lettets 8, 2580. Copyright 2008. American Chemistry Society.)

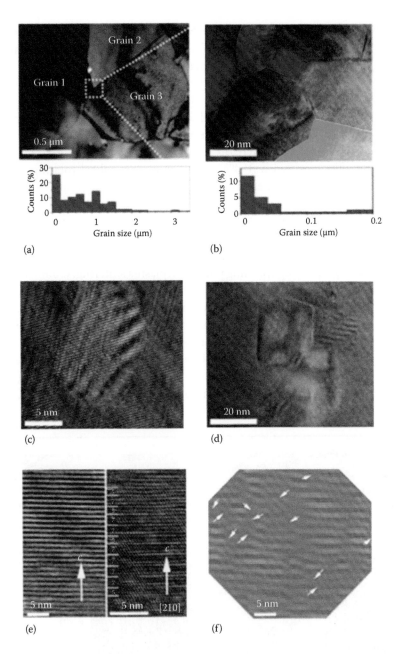

FIGURE 3.3 Multisize scale crystalline structure in the $Bi_xSb_{2-x}Te_3$ bulks made by high-energy BM and dc HP: (a) TEM image of large grains; (b) HRTEM image of small grains; (c) HRTEM image of a tellurium-poor nanoinclusion without a boundary; (d) HRTEM image of a tellurium nanoinclusion with a high-angle boundary; (e) HRTEM images of dislocation in grains with excited {003} reflections (*left*), along the [210] zone (*right*) (numbers 5 and 7 represent five- and seven-layer lamellae, respectively); and (f) inverse FFT (IFFT) of dislocation in grains with excited {105} reflections (Wiener-filtered) (white arrows indicate threading dislocations). Grain size distribution histogram in two size scales are present in the lower panel of a and b, respectively. (Reprinted with permission from Lan Y. C. et al., Structure study of bulk nanograined thermoelectric bismuth antimony telluride. *Nano Letters*, 9, 1419–1422. Copyright 2009. American Chemical Society.)

only few of the nanostructured $Bi_xSb_{2-x}Te_3$ with a Seebeck coefficient closed to 190–210 $\mu V/K^{-1}$ had a high $ZT > 1.2$,[24,26] which suggested the importance of synergistically tuning the thermal and electrical properties.

3.1.2.3 MS-HP Route

Besides the high energy BM, rapid solidification through melt spinning (MS) was another good way to get the nanostructured precursors before sintering. Kim et al.[27] fabricated $Bi_{0.5}Sb_{1.5}Te_3$ ribbons early by melt spinning the $Bi_{0.5}Sb_{1.5}Te_3$ ingot with 3–4% extra Te on a Cu wheel rotating at a surface speed of 47 m/s, and then fast-speed (5 min) hot pressed into the polycrystalline bulks at different temperatures. Microscale compositional inhomogeneity was identified in both casted ingot and the final hot pressed bulks. Due to the low PF (~33 $\mu W/cm\ K^2$), only a ZT of 0.92 at room temperature was obtained. Later, Xie et al.[28,29] reported a nanocrystalline $Bi_{0.52}Sb_{1.48}Te_3$ bulk made by using similar melting spinning plus spark plasma sintering. Compared with the BM-ed powders, the MS-ed ribbons has wider size scale structures, including amorphous phase, nanocrystalline phase, and microcrystalline phase due to the different cooling rate from the contact surface with cooper roller to the free surface with air.[30] Furthermore, slight compositional variation was identified along the cross section of ribbons from contact surface to contact surface (Te = 57.7 at.%; Sb = ~ 28.0 at.%; Bi = 14.4 at.%) to free surface (Te = 61.6 at.%; Sb = ~ 29.5 at.%; Bi = 9.2 at.%),[29] as shown in Figure 3.4. Compared

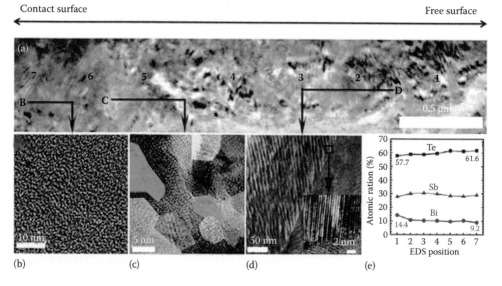

FIGURE 3.4 Position-related evolution of microstructure and composition along the cross section of melt-spun ribbon from the contact surface (with Cu roller) to free surface (with air). (a) TEM image of the cross section of ribbon. Positions 7 and 1 are near the contact surface and free surface, respectively. (b) HRTEM image of amorphous phase. (c) HRTEM image of nanocrystalline grains. (d) TEM image of microcrystalline grains. (e) Energy-dispersive spectroscopy (EDS) results of related positions in (a). (Reprinted with permission from Xie, W. J. et al., Identifying the specific nanostructures responsible for the high thermoelectric performance of (Bi, Sb)$_2$Te$_3$ nanocomposites. *Nano Letters*, 10, 3283–3289. Copyright 2010. American Chemical Society.)

with Kim's sample, Xie's nanostructured sample has a comparable PF (~35 μW/cm K²) but much lower thermal conductivity and, hence, a superior *ZT* value of 1.56 at 300 K. They later clarified that the measurement direction is vertical to the SPS pressure direction for electrical properties while parallel to the SPS pressure direction for the thermal conductivity.[30] The real thermal conductivity vertical to the SPS pressure direction is slightly higher than that parallel to the SPS pressure direction. Recently, the same group conducted a more systematic investigation of the MS-SPS effect has on the $Bi_{0.5}Sb_{1.5}Te_3$,[31] as shown in Figure 3.5. Compared to the ingot counterpart, an enhanced *ZT* value of 1.22 at 340 K was achieved in MS-SPS-ed $Bi_{0.5}Sb_{1.5}Te_3$ with a significantly reduced lattice thermal conductivity of 0.55 W/m K and a mildly reduced PF (~34 μW/cm K²). Multisize scale structural features were also identified in the MS-SPS-ed bulk, including fine grains, nanoinclusions, and dislocations, which is generally similar to the one made by the BM-HP route. Compared with their ingot counterpart, the MS-SPS-ed bulk is more

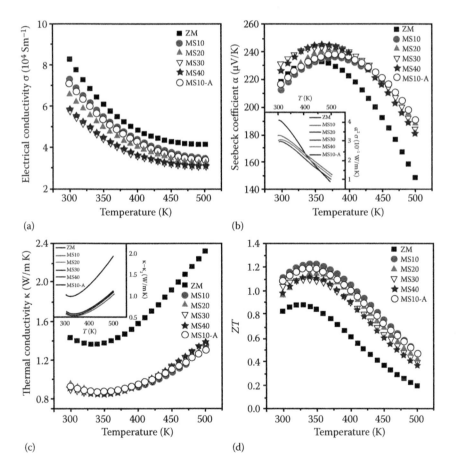

FIGURE 3.5 Temperature-dependent thermoelectric properties of nanocrystalline $Bi_{0.5}Sb_{1.5}Te_3$ made by the MS and plasma activate sintering: (a) electrical conductivity, (b) Seebeck coefficient, (c) thermal conductivity, (d) figure of merit. *(Continued)*

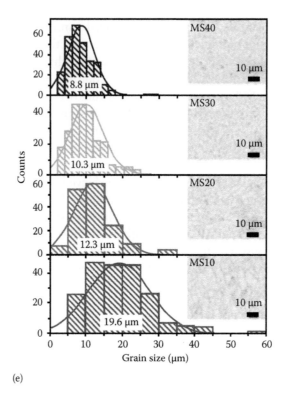

(e)

FIGURE 3.5 (CONTINUED) Temperature-dependent thermoelectric properties of nanocrystalline $Bi_{0.5}Sb_{1.5}Te_3$ made by the MS and plasma activate sintering: (e) grain size distribution. ZM means the zone melted ingot. MSX (X = 10, 20, 30, 40) presents that the nanocrystalline $Bi_{0.5}Sb_{1.5}Te_3$ bulk was made from the ribbon by spinning on Cu roller with linear speed of X = 10, 20, 30, 40 m/s. ZM10A was the ZM10 sample after further annealing. (Reproduced with permission from Zheng Y. et al.: Mechanically robust BiSbTe alloys with superior thermoelectric performance: A case study of stable hierarchical nanostructured thermoelectric materials. *Advanced Energy Materials*. 5, 1401391–1401411. Copyright 2015 Wiley-VCH Verlag GmbH & Co. KGaA.)

mechanically robust with a 26–40% enhanced fracture toughness of K_{IC}. It is noted that the PF of the MS-SPS-ed nanocrystalline bulks is lower than that of the one fabricated from the BM-HP process. Ivanova et al.[32] suggested that an anneal in Ar or H_2 process could improve the PF of the MS-HP-ed $Bi_{0.5}Sb_{1.5}Te_3$, as a result, an enhanced ZT of 1.3 near-room temperature. In addition to BM and melting spinning to obtain the nanostructured precursors, there are also other methods to synthesize the $Bi_xSb_{2-x}Te_3$ nanopowders such as bottom-up chemical synthetic methods.[33,34] However, it is too challenging for these chemical metallurgy routes to get a high PF comparable with the MS-HP-ed and MA-HP-ed nanocrystalline $Bi_xSb_{2-x}Te_3$. Recently, Nguyen et al.[35] reported a new spark erosion (SE) technique for producing high-quality p-type $Bi_{0.5}Sb_{1.5}Te_3$ nanoparticles at a production rate as high as 135 g/h and very low energy consumption (<2.0 kWh/kg). The nanoparticles subjected to SE were observed to be well defined and round with an average size of 20–50 nm. The SE-SPSed nanocrystalline $Bi_{0.5}Sb_{1.5}Te_3$ shows a comparable PF and reduced thermal conductivity and, hence, a significantly enhanced ZT value of 1.36 at 87°C compared with the starting ingot ($ZT = 1$ at 50°C).

3.1.2.4 GB Engineering

In addition to tuning the size and shape of the precursor particles or ribbons, the final nanostructures can be further refined by dispersing extra nanoparticles or modifying particle surfaces. A notable suppression of the grain growth was observed by using the extra nanoparticles in the $Bi_xSb_{2-x}Te_3$, such as SiC (100 nm),[36,37] C_{60},[38] and B_4C.[39] The enhanced phonon scattering comes from the increased concentration of grain boundaries and newly formed phase boundaries. However, the carrier mobility of $Bi_xSb_{2-x}Te_3$ was also reduced by these nanoparticles. SiC has been considered as a stable and inert secondary phase to $Bi_xSb_{2-x}Te_3$. However, the carrier concentration of $Bi_{0.3}Sb_{1.7}Te_3 + x$ (vol%) SiC showed a notable increase from 1.8×10^{19} cm^{-3} to 3.5×10^{19} cm^{-3} when the content of SiC increases from $x = 0$ to $x = 0.6$,[37] as shown in Table 3.1. There are two possible reasons: (1) Some of SiC nanoparticles decomposed during high-energy BM, and then the Si or C got into the $Bi_{0.3}Sb_{1.7}Te_3$ lattice as a new acceptor. (2) Some of the native defects at the phase boundaries (SiC/$Bi_{0.3}Sb_{1.7}Te_3$) behaved as new acceptor. Zhang et al.[40] reported a more effective way to suppress the grain growth of p-type $Bi_{0.4}Sb_{1.6}Te_3$ by using the some oleic acid (OA) in the MA-HP route. It was found that the OA significantly reduced the particle agglomeration during BM and, hence, suppressed grain growth during HP process. In contrast to having a flake-like grains, $Bi_{0.4}Sb_{1.6}Te_3$ with OA had more round grains with the random distribution, creating less anisotropy between the directions vertical/parallel the pressure direction. Furthermore, slight increases in both the Seebeck coefficient and electrical resistivity were observed. A combination of small grains (200–500 nm) and nanopores leads to a decrease in lattice thermal conductivity, therefore reaching an optimized ZT of 1.3 at 100°C for the sample with 2.0 wt% OA.

Recently, a group of researchers from South Korea reported that a large amount of extra Te liquid phase (25 wt.%) could significantly modify the grain boundary of nanocrystalline $Bi_{0.5}Sb_{1.5}Te_3$ by MS-SPS route.[41] Dense dislocation

TABLE 3.1 Seebeck Coefficient, Carrier Concentration, and Mobility of the Canocomposite $Bi_{0.3}Sb_{1.7}Te_3 + X$ vol% SiC ($X = 0. 0.1, 0.4,$ and 0.6) at 323 K

Sample	Seedbeck Coefficient ($\mu V/K$)	Carrier Concentration (10^{19} cm^{-3})	Carrier Mobility (cm^2/V s)	m_x^*/m_0^*
$x = 0$	187.94	1.81	365.4	1
$x = 0.1$	186.14	2.48	284.2	1.222
$x = 0.4$	196.18	3.39	212.1	1.586
$x = 0.6$	187.52	3.51	164.1	1.552

Source: Li, J. H. et al.: BiSbTe-based nanocomposites with high ZT: The effect of SiC nanodispersion on thermoelectric properties. *Advanced Functional Materials.* 2013. 23. 4317–4323. Copyright Wiley-VCH Verlag GmbH & Co. KGaA. Reproduced with permission.

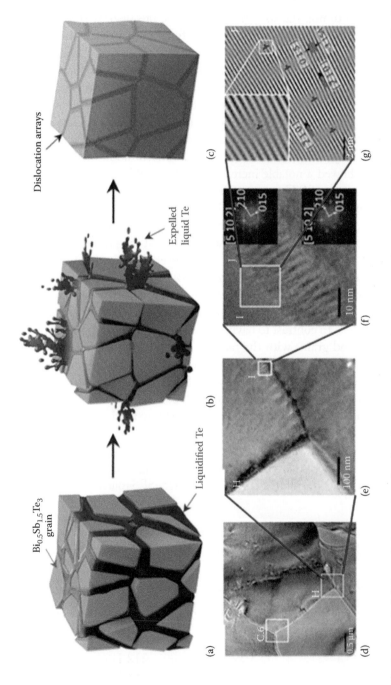

FIGURE 3.6 (a through c) Schematic illustration showing the generation of dislocation arrays during the liquid phase compacting process. The Te liquid (red) between the $Bi_{0.5}Sb_{1.5}Te_3$ grains flows out during the compacting process and facilitates the formation of dislocation arrays embedded at the grain boundaries. (d through g) TEM images of dislocation array from low resolution to high resolution. (Reprinted with permission from Kim, S. I. et al., *Science* 348, 109–114, 2015. Copyright 2015. AAAS.)

arrays were formed at the low-energy grain boundary when the extra Te liquid was expelled out during spark plasma sintering process, as shown in Figure 3.6. The novel nanostructures were claimed for the very low lattice thermal conductivity (0.33–0.34 W/m K) and, hence, the new record ZT of 1.67 (∥, parallel to the pressure direction) and 1.85 (⊥ vertical to the pressure direction). Anisotropic electrical conductivity σ was observed in their Te-MS-ed $Bi_{0.5}Sb_{1.5}Te_3$ in which $σ_⊥$ is 16.5% higher than $σ_∥$. In contrast, only 6.5% difference in thermal conductivity was detected. The real role of the dense dislocation arrays is still unclear. The extremely low lattice thermal conductivity desires more experimental confirmations and theoretical understandings. Besides the low thermal conductivity, the Te-MS-ed $Bi_{0.5}Sb_{1.5}Te_3$ showed a clear enhanced PF of ($PF_⊥$ ~40 µW/ cm K^2) compared with previous MS-HP-ed $Bi_{0.5}Sb_{1.5}Te_3$ bulks. Grain boundary engineering was used to tailor not only the transport of the phonons, but also the transport of electrons. Ko et al.[42] observed a significantly increased Seebeck coefficient in the Pt/Sb_2Te_3 nanocomposite from 115.6 to 151.6 µV/K. Enhanced PF in the $Bi_{0.5}Sb_{1.5}Te_3$ was reported by embedding a few selected metal (Ag, Co, Cu, Mn, Mo, Ni, Pd, Tb, Zn) nanoparticles, in which the $Bi_{0.5}Sb_{1.5}Te_3$/ Ag nanocomposite showed the largest incensement of 25%.[43,44] The Hall coefficient measurement suggested an increased effective carrier mass from $m^* = 0.8m_0$ to $m^* = 1.2m_0$. At the same time, a decreased lattice thermal conductivity was observed due to the increased phonon scattering by the phase boundary. As a result, a high ZT of 1.35 was obtained in the $Bi_{0.5}Sb_{1.5}Te_3$/Ag nanocomposite. A similar increased effective carrier mass m^* was also observed in the $Bi_{0.5}Sb_{1.5}Te_3$-expanded graphene.[45] Additionally, Zhang et al.[46] observed a significant reduction in the lattice thermal conductivity in the Ag nanoparticle-dispersed (60 nm) Bi_2Te_3 made from a chemical metallurgy route.

3.1.3 n-Type $Bi_2(Te, Se)_3$

The classic compositions of the n-type $Bi_2Te_{3-x}Se_x$ ranged from $x = 0.2$ to 0.8. A recent theoretical calculation for Bi_2Se_3 suggests that the dominant native defect is vacancy $V_{Se}^{··}$ in Bi-rich case while anti-site defect Se_{Bi}' in a Se-rich case, which makes Bi_2Se_3 a very strong n-type. Parker and Singh[47] theoretically suggested that a p-type Bi_2Se_3 could have a potential high ZT of 1.5. However, it is too challenging to experimentally achieve a p-type Bi_2Se_3 with high hole concentrations of ~10^{19} cm^{-3}. On the other hand, the main charge carriers type of Bi_2Te_3 was stoichiometric ratio dependent.[10,48] It suggests that the dominant native defect is Bi_{Te}' in Bi-rich case (p-type), while $Te_{Bi}^·$ in a Te-rich case (n-type), as shown in Figure 3.1. Furthermore, in an alloy of $Bi_2Te_{3-x}Se_x$, the real dominant defect is more complicated. In the Bi rich case, the donor-like $V_{Se}^{··}$ and acceptor-like Bi_{Te}' coexisted. Besides stoichiometric ratio (Te, Se)/Bi and alloying ratio Te/Se, the defect related to the plastic deformation also needs to be taken into account.

3.1.3.1 Equivalent Doping Effect of BM-HP

Similar to the p-type $Bi_xSb_{2-x}Te_3$, the early works to prepare fine-grained n-type $Bi_2Te_{2-y}Se_y$ materials were also motivated by the expectation to improve the mechanical strength for more robust and reduced thermal conductivity for higher thermoelectric ZT values.[15,49,50] However, it was even more challenging to achieve an enhanced ZT in the fine-grained n-type $Bi_2Te_{2-y}Se_y$ materials than in the p-type $Bi_xSb_{2-x}Te_3$ materials. One of the challenges is controllable carrier concentration. The BM-HP process was not a pure donor-like effect in the n-type $Bi_2Te_{2-y}Se_y$ materials. For example, the BM time and HP temperature were most widely investigated; however, there were still many conflicting experimental results that puzzle the readers,[49–62] as shown in Table 3.2. Some results suggested that BM and HP have a donor-like effect on the n-type $Bi_2Te_{2-y}Se_y$ which corresponds to decreased Seebeck coefficient and increased carrier concentration, while others just suggested an opposite acceptor-like effect. Kanatzia et al.[50] observed an acceptor-like behavior for the BM time (from 1 to 15 h) based on the cold pressed bulks, while a donor-like behavior for the BM time based on hot pressed bulks. Unfortunately, they did not give enough explanations to clarify this puzzle. However, we would also like to give the credits to these early works. Besides the main factors that determined the thermoelectric properties, similar to what was summarized for the p-type $Sb_xBi_{2-x}Te_3$, they also additionally indicated the main challenges for n-type $Bi_2Te_{2.7}Se_{0.3}$ polycrystalline or nanostructured materials: (1) carrier concentration heavily depends on the plastic deformation and thermal history and (2) a notable side effect from grain boundary to the transport of electrons.

First, we will uncover these puzzles by understanding the evaluation of the dominant atomic defects due the plastic deformation and thermal history. Mechanical BM does not only reduce the particle size but also results into heavy plastic deformation within each particle. Heavy plastic deformation produces many line defects and point defects. Schultz et al.[63] described a simple experiment by sharply bending a single crystal Bi_2Te_3 about an axis parallel to the basal plane to introduce the dislocations; however, these dislocations did not make the concentration change of the charge carrier. The plastic deformation usually produces vacancy–interstitial pair together due to the dislocation intersections. In Bi_2Te_3, the nonbasal slip gives the average vacancy–interstitial pair of Te and Bi in a ratio of 3/2, while a slip along the basal plane only produces the Te vacancy–interstitial pair. Due to the van de Waals bonds, the dislocation slip along the basal plane is much easier than that along nonbasal plane. It was therefore believed that the concentration of the $V_{Te}^{\bullet\bullet}$–Te_i pair was more than that of $V_{Bi}^{'''}$–Bi_i vacancies. According to the net increase in the negative charge carrier after subjecting a pure plastic deformation, Schultz et al.[63] further suggested that the interstitials were undoubtedly present but apparently not as electrically active as the vacancies. This dominant donor effect of $V_{Te}^{\bullet\bullet}$ was therefore widely accepted for the Bi_2Te_3 ingot changed from p-type to n-type after being directly subjected to hot extrusion,[63] or BM and then HP.[50,51] The thermal treatment process, such as HP or annealing, has two major effects: (1) loss of Te making Bi_2Te_3 more Bi-rich and (2) annihilation of the high-energy state defect pair $V_{Te}^{\bullet\bullet}$–Te_i. The loss of Te or Se of $Bi_2Te_{3-x}Se_x$ during the HP and annealing process was widely confirmed.[58,63,64] It is noted that the loss

TABLE 3.2 Effects of BM Time and HP or Annealing Temperature on the Seebeck Coefficient or Carrier Concentration in n-Type $Bi_2Te_{3-y}Se_y$ Fine-Grained Materials

Start Material before BM-HP	BM Time (h)	HP Temperature °C	S (μV/K)	N (10^{19} cm^{-3})	Equal Effect	Reference
BM Time						
Bi_2Te_3 ingot	1 → 15	CP	−191 → −236		Acceptor	Kanatzia et al.[50]
	1 → 15	400	−168 → −150		Donor	
Bi_2Te_3 mix	3 → 9	300		7 → 13	Donor	Kuo et al.[52]
Bi_2Te_3 mix	1-80	400		6.2 → 1.6	Acceptor	Zhao et al.[53]
$Bi_{1.94}Sb_{0.01}Te_{2.76}Se_{0.29}$ ingot	0.5 → 10	420	−120 → −208	9.5 → 4.2	Acceptor	Oh et al.[48]
$Bi_2Te_{2.7}Se_{0.3}$ mix	12 → 18	500	−154 → −133		Donor	Liu[54]
$Bi_2Te_{2.7}Se_{0.3}S_{0.01}$ mix	10 → 18	500	−208 → −224		Acceptor	Liu[55]
$Bi_2Se_{0.21}Te_{2.79}$ ingot	1 → 8	CP		9 → 12	Donor	Lin et al.[58]
HP or Annealing						
Bi_2Te_3 mix	3	350 → 450	−112 → −127	6.9 → 6	Acceptor	Zhao et al.[59]
$Bi_2Te_{2.85}Se_{0.15}$ mix	6	320 → 440	−147 → −135		Donor	Fan et al.[61]
$Bi_2Te_{2.79}Se_{0.21}$ ingot	15	250 → 350	−155 → −110	4.5 → 6	Donor	Lu and Liao[62]
$Bi_2Te_{2.7}Se_{0.3}$ mix	12	300 → 500	−214 → −294		Acceptor	Liu[56]

Note: CP: cold press; ingot: melting ingot; mix: elemental mixtures.

of Te and Se in the conventional $Bi_2Te_{3-x}Se_x$ could generate both donor $V_X^{\cdot\cdot}$ (X = Te, Se) and acceptor Bi'_{Te} in the conventional $Bi_2Te_{3-x}Se_x$ depending on the content of Se. At the same thermal history, the loss of the Te and Se in a BM- or grinding-derived power compact would be much higher than that in the casted ingot or well-grown single crystals because the former one has more escape channels than the latter one. Generally, BM and HP would creative extra $V_{Te}^{\cdot\cdot}$ and $V_{Se}^{\cdot\cdot}$ in the final polycrystalline bulks. The chalcogen vacancies are the well-known donors; however, they could further react with some native antisite defects and result in the opposite effect, as shown in Figure 3.7. In a highly Te-rich case, the chalcogen vacancies could cancel the dominant defect $Te_{Bi}^{\cdot}$, leading to a decreased n-type carrier concentration. This was the reason why the increasing BM time resulted into a Seebeck coefficient increase in the chalcogen-rich $Bi_{1.94}Sb_{0.01}Te_{2.76}Se_{0.29}$[48] and $Bi_2Te_{2.7}Se_{0.3}S_{0.01}$.[55] Besides artificially starting with chalcogen-rich samples, oxygen is another source for turning a stoichiometric $Bi_2Te_{3-x}Se_x$ into a chalcogen-rich case. In the case without well protection, oxygen was easily absorbed on the grinding particles and left in the final sintered pucks as a strong n-type donor.[16,63,65,66] It was suggested that oxygen resolved into Bi_2Te_3 and took up the place of Te(2) position with the smallest electronegativity, consequently forming $Bi_2Te_{3-x}O_x$ solid solution.[65] The oxygen could make an original stoichiometric $Bi_2Te_{3-x}Se_x$ become the chalcogen-rich and high

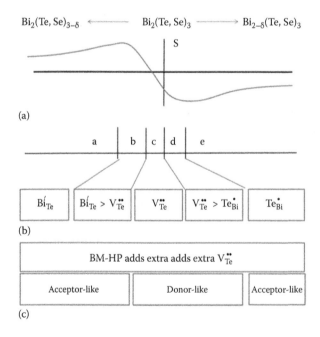

FIGURE 3.7 Schedule picture for the effect of BM time and HP temperature for the Bi_2Te_3. Shown in (a) is the plot of Seebeck coefficient versus charge carrier concentration. Shown in (b) is the composition-dependent domains. Domain a is p-type, Te-rich range with dominant defect of V_{Bi}''; domain b is p-type depletion range with coexisting V_{Bi}'' and $Te_{Bi}^{\cdot}$; domain c is n-type depletion range with coexisting V_{Bi}'' and $Te_{Bi}^{\cdot}$; domain d is n-type depletion range with coexisting V_{Bi}'' and $V_{Te}^{\cdot\cdot}$; and domain e n-type, Bi-rich range with dominant defect of $V_{Te}^{\cdot\cdot}$.

n-type carrier concentration. This explains why Zhao et al.[53] had a Bi_2Te_3 with a high carrier concentration and a low Seebeck coefficient and observed acceptor-like BM (1–80) and HP processes (from 350°C to 450°C).[59]

3.1.3.2 Reproducibility from Batch to Batch

The second challenge for the n-type $Bi_2Te_{2.7}Se_{0.3}$ connects with the reproducibility of the thermoelectric properties from batch to batch. This is a question that has not yet received enough attentions. Figure 3.8a and b shows the temperature-dependent electrical resistivity and Seebeck coefficient of 10 batches of $Bi_2Te_{2.7}Se_{0.3}$ made by BM and HP methods under the same fabrication conditions, indicating a poor reproducibility from batch to batch.[61] The coefficient of deviation for the electrical resistivity of 10 batches was larger than 13%, suggesting an uncontrollable carrier concentration. However, the application of the same fabrication route to the p-type $Bi_{0.4}Sb_{1.6}Te_3$ achieved a good reproducibility

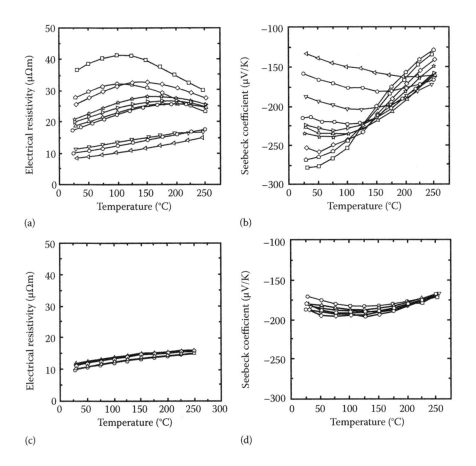

(a)   (b)

(c)   (d)

FIGURE 3.8 (a and b) Temperature dependence of electrical resistivity and Seebeck coefficient of 10 batches of $Bi_2Te_{2.7}Se_{0.3}$ samples made by BM plus dc-HP method. (c and d) Temperature dependence of electrical resistivity and Seebeck coefficient of 8 batches of $Cu_{0.01}Bi_2Te_{2.7}Se_{0.3}$ samples made by the same method and conditions as those shown in a and b. (Liu, W. S. et al.: Thermoelectric property studies on Cu doped n-type $Cu_xBi_2Te_{2.7}Se_{0.3}$ nanocomposites. *Advanced Energy Materials*. 2011. 1. 577–587. Copyright Wiley-VCH Verlag GmbH & Co. KGaA. Reproduced with permission.)

with coefficient of deviation less than 3%. The puzzle is that both the high-energy BM and the fast-speed HP processes are a nonthermodynamic process. Generally, ball milled powders are homogenous in macroscale, but inhomogeneous in microscale, characterized with a broad particle size distribution, nanoinclusions, and atomic defects. The later fast-speed HP does not give enough time to recover all the charge active defects because we intentionally expect that such defects or nanofeatures could be left as the effective phonon scattering centers. Furthermore, the evaporation of Te or Se was another important factor to cause the scattering from batch to batch. Liu et al.[60] gave an explanation for the reproducibility difference puzzle between n-type $Bi_2Te_{3-y}Se_y$ and p-type $Bi_xSb_{2-x}Te_3$ by considering the dominant defect under a metal rich case due to loss of the chalcogen elements as follows:

$$Bi_2Te_3 = 2Bi_{Bi}^{\times} + (3-x)Te_{Te}^{\times} + xTe(g)\uparrow + xV_{Te}^{\cdot\cdot} + 2xe, \qquad (3.1)$$

$$Bi_2Te_3 = \left(2 - \frac{2}{5}x\right)Bi_{Bi}^{\times} + (3-x)Te_{Te}^{\times} + xTe(g)\uparrow$$
$$+ \left(\frac{2}{5}xV_{Bi}''' + \frac{3}{5}xV_{Te}^{\cdot\cdot}\right)^{\times} + \frac{2}{5}xBi_{Te}' + \frac{2}{5}xh. \qquad (3.2)$$

For the n-type $Bi_2Te_{3-y}Se_y$, the loss of Te or Se will be favorable to form the donor-like $V_{Te}^{\cdot\cdot}$ as shown in Equation 3.1. However, the loss of Te will be favorable to form the acceptor-like Bi_{Te}' as shown in Equation 3.2. Considering that both the n-type $Bi_2Te_{3-y}Se_y$ and p-type $Bi_xSb_{2-x}Te_3$ have the same amount of chalcogen element loss at the same BM and HP conditions, it is found that the free carrier concentration with Te vacancy dominant is five times higher than that when Bi antisite is dominant. Hence, a little random fluctuation of missing Te generates more serious irreproducible electrical resistivity and Seebeck coefficient in n-type Bi_2Te_3 rather than p-type Bi_2Te_3. It is worth pointing out that both the situations described by Equation 3.1 for n-type and Equation 3.2 for p-type are the ideal cases. In the real case, there are minor acceptor-like Bi_{Te}' or Bi_{Se}' antisite defects besides the major donor-like $V_{Te}^{\cdot\cdot}$ and $V_{Se}^{\cdot\cdot}$ vacancies in n-type Bi_2Te_3 alloying, while there is also minor donor-like $V_{Te}^{\cdot\cdot}$ besides the major acceptor-like Sb_{Te}' or Bi_{Te}' antisite defects in p-type Bi_2Te_3. The alloy with Sb usually increases the concentration of antisite defect on Te-site $\left(Sb_{Te}'\right)$ and hence gives more holes due to the smaller electronegativity difference between Sb (2.05) and Te (2.10) than that between Bi (2.02) and Te (2.10). The alloy with Se usually increases the concentration of vacancy on the Te site $\left(V_{Se}^{\cdot\cdot}\right)$ and, hence, gives more electrons because Se has a lower energy of evaporation (37.70 kJ/mol) than Te (52.55 kJ/mol). The concentration of vacancy $\left(V_{Te}^{\cdot\cdot}\text{ and }V_{Se}^{\cdot\cdot}\right)$ in n-type $Bi_2Te_{2-x}Se_x$ will be higher than that in p-type $Sb_{2-x}Bi_xTe_3$, which is another reason why we have a more serious reproducibility problem in n-type $Bi_2Te_{3-x}Se_x$ rather than in p-type $Sb_{2-x}Bi_xTe_3$.

The key to improve the reproducibility of n-type $Bi_2Te_{2.7}Se_{0.3}$ was to suppress the generation of Te vacancy, including both whole V_{Te} and fractional V_{Te}.[60] Reducing the energy of the BM process, such as decreasing BM rotation speed and adjusting the BM medium filling parameter, would decrease the mechanical deformation and thus, reduce, the generation of Te vacancy. However, we desired the high BM energy to achieve finer grains and high hot press temperature to optimize grain boundaries. One of the alternative choices was to raise the formation energy of the $V_{Te}^{\cdot\cdot}$ and $V_{Se}^{\cdot\cdot}$ by doping. Figure 3.8c and d presents the temperature dependence of electrical resistivity and Seebeck coefficient for eight batches of $Cu_{0.01}Bi_2Te_{2.7}Se_{0.3}$. The reproducibility of the Cu-added n-type $Cu_{0.01}Bi_2Te_{2.7}Se_{0.3}$ is obviously improved. The coefficient of variation of electrical resistivity, Seebeck coefficient, and PF for $Cu_{0.01}Bi_2Te_{2.7}Se_{0.3}$ were 1.92%, 1.00% and 0.98%, respectively, which is much lower than 13.23%, 6.50%, and 3.47% for $Bi_2Te_{2.7}Se_{0.3}$, respectively. Here, the Cu was believed to raise the formation energy of chalcogen vacancy and suppress escaping Te and Se. However, it is worthy to point out that Cu ion is mobile which could lead to some thermal stability problem under a thermal gradient. Instead, slight extra Se or S was used to suppress the appearance of Te vacancy and, hence, minimize the performance fluctuation from batch to batch.

3.1.3.3 Texture for the Enhanced ZT

Besides the challenges mentioned earlier, the simple BM-HP route is still not enough to achieve an enhanced *ZT* in the n-type $Bi_2Te_{3-x}Se_x$ material compared with the single crystalline counterpart. The significantly decreased lattice thermal conductivity is offset by the reduced PF due to the decreased carrier mobility. Here, we use the $Bi_2Te_{3-x}Se_x$ single crystals grown by Scherrer and Sherrer[4,5] as a reference. The optimized PF of $Bi_2Te_{3-x}Se_x$ single crystals is around 50 μW/cm K² for x = 0.075, 46 μW/cm K² for x = 0.15, and 42 μW/cm K² for x = 0.45 when the Seebeck coefficient is around −200 μV/K. However, many of the early efforts on fine-grained or nanostructured $Bi_2Te_{3-x}Se_x$ by the BM-HP route only resulted into a PF less than 20 W/cm K².[48,49,53,61–64] The deteriorated effect of the oxygen could be one of the reasons for the low PF.[65–68] However, even with a well protection, the nanocrystalline $Bi_2Te_{3-x}Se_x$ with randomly distributed grains only reach a power of 24–28 μW/cm K².[54–57,69] The Cu-doped nanocrystalline $Bi_2Te_{2.7}Se_{0.3}$ was reported to have slightly superior PF of 31 μW/cm K².[60] However, it is still ~30% lower than the PF of Scherrer an Sherrer's[4,5] single crystals. It is worthy to point out that the similar BM-HP could make a nanocrystalline $Bi_{0.4}Sb_{1.6}Te_3$ having a PF of 45 μW/cm K² comparable with its single crystal counterpart.[20] It was well known that the n-type $Bi_2Te_{2.7}Se_{0.3}$ has higher anisotropic electrical transport properties between the in-plane and out-of-plane directions than the p-type $Bi_{0.4}Sb_{1.6}Te_3$ (use the basal plane containing the van der Waals bonds).[4,5] This fact leads to the n-type $Bi_2Te_{2.7}Se_{0.3}$ suffering more from the randomly distributed grains. An effective way to reconstruct the transport channel of the electrons is to further align the randomly distributed grains. Two general approaches were reported to achieve the texture in the Bi_2Te_3-based materials: hot extrusion and hot forging. The hot extrusion route is characterized when the material flowing direction is parallel to the pressure. Figure 3.9 illustrates two main hot extrusion routes: one is the unequal channel linear extrusion (Figure. 3.9a); another is equal channel angular extrusion (Figure 3.9b). For the unequal channel linear extrusion, the deformation is mainly determined by the channel size reduction ratio (or extrusion ratio). Seo et al.[70] fabricated n-type $Bi_2Te_{2.85}Se_{0.15}$ by hot extrusion in the temperature range of 300–440°C under an extrusion ratio of 20:1. The hot extrusion gave rise to a slightly preferred orientation of grains with an orientation factor $F(0\ 0\ l)$ of 0.16. Here, the orientation factor was calculated by the Lotgering method using the following[71]:

$$F = \frac{P - P_0}{1 - P_0},\qquad(3.3)$$

where $P = I(0\ 0\ l)/\Sigma I(h\ k\ l)$, $P_0 = I_0(0\ 0\ l)/\Sigma I_0(h\ k\ l)$. P and P_0 are the ratios of the integrated X-ray intensities of all $(0\ 0\ l)$ planes to those of all $(h\ k\ l)$ planes for the preferentially oriented and randomly oriented samples, respectively. At a higher temperature of 450°C, Hong et al.[72] obtained stronger texture with an orientation factor of 0.17–0.2 in the n-type $Bi_2Te_{2.85}Se_{0.15}$ at an extrusion ratio of 16:1. By using an even higher temperature of 480°C and high extrude ratio of 25:1, Hong and Chun[73] got a higher orientation factor of 0.23, which also resulted in a marked PF of 53 µW/cm K^2. Besides the extrusion ratio and temperature, Yang et al.[74] showed that the extrusion angle also played an important role in plastic deformation. The die with lower extrusion angle had longer or larger regions of deformation, resulting in lower deformation and, thence, less micropores.[74] It was found that hot extruded material had a ringlike texture in which the normal to the basal plane orients to the radial direction and [110] orients to the extrusion direction.[75] Equal channel

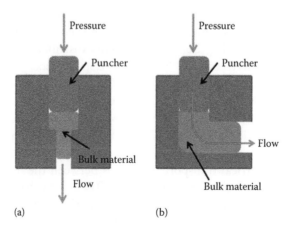

FIGURE 3.9 Schematic figure of the hot extrusion: (a) unequal channel linear extrusion and (b) equal channel angular extrusion.

angular extrusion (ECAE) is another widely used thermal extrusion process.[76,77] For example, Fan et al.[76] reported an orientation factor of 0.26–0.28 in n-type $Bi_2Te_{2.85}Se_{0.15}$ by performing ECAE under a ram speed of 2.5 mm/min in the temperature range 380–480°C in argon atmosphere. The stainless steel die was used for the extrusion process, and graphite powders were coated with the extrusion die to reduce the friction force. Hayashi et al.[77] observed continuously improved carrier mobility from 140 cm²/V s (without ECAE) to 240 cm²/V s (six times of ECAE) by applying multitime ECAE. The orientation imaging microscopy confirmed the formation of textured grains and twin boundaries, which also explains the high PF of 40 μW/cm K and an improved ZT of 0.9 near-room temperature in the sample after six times of ECAE. The thermal extrusion has also been used in the p-type $Bi_{2-x}Sb_xTe_3$ materials.[70,78]

Compared with hot extrusion, hot forging involves slightly simple and easy processing and also has two common modes: radial flow hot forging and uniaxial flow hot forging as shown in Figure 3.10. The (00l) texture has been early identified in the hot pressed Bi_2Te_3-based bulks. It was found that the hot pressed bulk from the spherical particles (by gas atomizing method) showed an isotropic thermoelectric properties, while the hot pressed bulk from the flake-like particles showed anisotropic behavior.[79] Jiang et al.[80] obtained a strong texture ($F = 0.8$) in the n-type $Bi_2Te_{0.79}Se_{0.21}$ by using big grinding

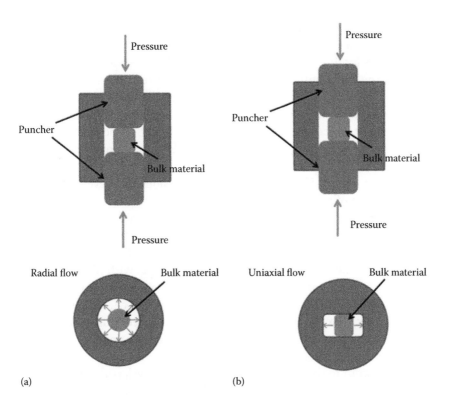

(a)   (b)

FIGURE 3.10 Schematic figure of the hot forging: (a) radial flow hot forging and (b) uniaxial flow hot forging.

particles (180–380 μm). However no notable enhancement in the ZT was observed due to the comparable thermal conductivity and reduced PF compared with as-grown ingot. The orientation factor was almost independent of sintering temperature in the range of 400–440°C. Furthermore, the orientation factor was significantly reduced from 0.85 to 0.4 when the particles size reduced from 180–380 to 90–120 μm, which suggested that these textures are controlled by the size and morphology of the starting particles. Besides the plastic flow, similar compacting texture was also observed in a p-type $Bi_{0.4}Sb_{1.6}Te_3$ case (100–175 μm) with strong morphological anisotropy.[81] When the particle size went down in nanoscale, the texture due to the particle morphology become much weaker.[30,69] In order to get a high orientation in fine-grained Bi_2Te_3, a flowing assistant hot forging is necessary. Usually, a hot pressed bulk was used for the hot forging to develop the strong texture. Zhao et al.[82] reported a highly textured Bi_2Te_3 with an orientation factor of $F = 0.67$ by radial flowing hot forging from ϕ15 mm rod into ϕ20 mm, which corresponds to a PF of 33 μW/cm K², with 70% enhancement compared the originally pressed bulk. Hu et al.[83] reported that multitime radial flow multitime hot-forging could continuously improve the grain alignment of the $Bi_2Te_2Se_1$. They also named their process a repetitive hot deformation. In their experiment, casted ingots were ground into fine powders and then hot pressed into a ϕ10 mm rod at 400°C for 30 min. Subsequently, the HP bulk was placed in the center of a larger die with an inner diameter of 16 mm and re-pressed at 550°C (HD1). Then, a disk of ϕ10 mm was cut from the ϕ16 mm HD1 bulk, and the ϕ 16 mm die was put back for the second time, hot forging into ϕ 16 mm bulk (HD2). These processes were repeated to get the HD3 bulk, which shows a higher orientation factor of 0.48 compared with $F = 0.2$ for the HP bulk. However, a further hot forging cannot get notable higher orientation factor ($F = 0.46$ for the HD4). Figure 3.11 shows the temperature-dependent thermoelectric properties of the multitime radial flowing hot forged $Bi_2Te_2Se_1$. The PF was continuously increased from 13 μW/cm K² (HP bulk) to 29 μW/cm K² (HD4 bulk), and the peak ZT was improved from 0.57 (HP) to 1 (HD4). Besides the developed texture, the same group of authors also suggested that the re-pressed process might further introduce some new lattice defect resulting in decreased lattice thermal conductivity because of a recrystallization process.[84,85] It was noted that they might overestimate the electronic thermal conductivity by using an empirical Lorenz number 2.0×10^{-8} V²/K². The sample with higher texture would have an overestimated electrical conductivity and an overestimated κ_{carr} ($\kappa_{carr} = L\sigma T$); as a result, it could lead to a underestimated κ_{lat}, from the relationship, i.e., $\kappa_{lat} = \kappa_{tot} - \kappa_{carr}$. The new developed lattice defect might cause a decreased thermal conductivity for some special case, such as a zone melt ingot as a start rod for the hot deformation.[85]

In contrast to the radial flow hot forging, we developed the uniaxial flow hot forging (W. S. Liu and Z. F. Ren, unpublished, 2012), as shown in Figure 3.12a. Firstly, BM-ed nanopowders were hot pressed into a rod (ϕ = 0.5″ and length = 0.5″) at 500°C, which was referred to as as-pressed (AP) bulk. Then half of the

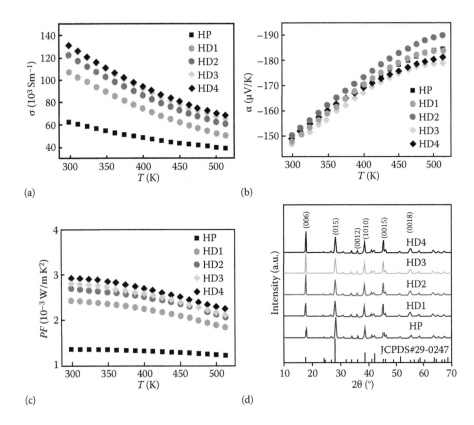

FIGURE 3.11 Thermoelectric properties and (00l) texture of the multitime radial flow hot forged $Bi_2Te_2Se_1$: (a) electrical conductivity, (b) Seebeck coefficient, and (c) PF. (d) XRD patterns of textured $Bi_2Te_2Se_1$. (Reprinted from *Acta Materialia*, 60, Hu, L. P. et al., Improving thermoelectric properties of n-type bismuth-telluride-based alloys by deformation induced lattice defect and texture enhancement, 4431–4437, Copyright (2012), with permission from Elsevier.)

rod was placed in a big die and was uniaxial repressed into a plate (length = 1″ and width = 0.5″) at 525°C, which was identified as first re-pressed (RP1) bulk. Then, the plate was polished, cleaned, and then cut into two identical pieces and stacked together for re-pressing into a plate (length= 1″ and width = 0.5″) again, which was referred as second re-pressed (RP2) bulk. Then, the RP2 plate was repetitively polished, cleaned, cut, stacked, and re-pressed into the third re-pressed (RP3) bulk. The advantage of uniaxial flow hot forging is that it is easy to execute the multitime repressing. Figure 3.12b through e shows the thermoelectric properties of multitime repressed $Cu_{0.005}Bi_2Te_{2.7}Se_{0.3}$ and their X-ray powder diffraction (XRD) patterns. The significant increased peak PF from 28.7 μW/cm K² (AP) to 42.2 μW/cm K² (RP3) was consistent with the developed (00l)-texture by the multitime repressing. The (00l) orientation factor from the XRD pattern was 0.22 for RP1 bulk, 0.30 for RP2 bulk, and 0.38 for RP3 bulk. In our case, the lattice thermal conductivity was slightly increased with the improved texture, finally reaching a highest peak ZT of 1.1 at 100°C in RP3 $Cu_{0.005}Bi_2Te_{2.7}Se_{0.3}$, which is near 30% higher than that of the AP one (0.85 at 100°C).

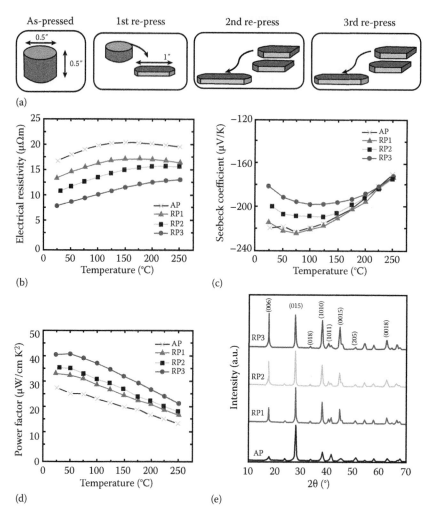

FIGURE 3.12 Thermoelectric properties and (001) texture of the multitime uniaxial flow hot forged $Cu_{0.005}Bi_2Te_{2.7}Se_{0.3}$: (a) schematic picture of the multi-times uniaxial flow hot forging process, (b) electrical resistivity, (c) Seebeck coefficient, and (d) PF. (e) XRD patterns of (001)-textured $Cu_{0.005}Bi_2Te_{2.7}Se_{0.3}$. (Courtesy of W. S. Liu and Z. F. Ren, 2012.)

3.1.4 Outlook

In this section, we reviewed the most recent advances in Bi_2Te_3-based materials by using the powder metallurgy routes, such as BM-HP and MS-SPS. We did not have enough space to give enough praise for the efforts on the chemical metallurgy routes for the nanostructured Bi_2Te_3-based thermoelectric materials. We also discussed some challenges met in the n-type nanocrystalline $Bi_2Te_{3-x}Se_x$, such as the equivalent doping effect of BM-HP, reproducibility from batch to batch, and texture for the enhanced ZT. It is noted that the highest temperature usage for Bi_2Te_3 is around 250°C due to poor thermoelectric performance and thermal stability at higher temperature. Some new V_2- to VI_3-type compounds, such as Bi_2Te_2S, and Bi_2SeS_2, have shown to be promising for overcoming the usage temperature limit, which desires more some attentions.

3.2 MgAgSb Low-Temperature Thermoelectric Materials

Jiehe Sui, Zihang Liu, and Zhifeng Ren

3.2.1 Introduction

For low-temperature applications, p-type $Bi_xSb_{2-x}Te_3$ thermoelectric materials have received dominant attentions since 1950s.[86,87] The highest ZT could surpass the benchmark of ~1 by nanostructuring or hot deformation.[88,89] Nevertheless, the strong bipolar effect above 373 K severely degrades its performance,[90] which restricts the utilization merely around room temperature. Even worse, due to the fact that tellurium (Te) is an extremely scarce element on the earth's crust, this further limits the widespread application.[91] Therefore, searching for a better alternative without Te that can work at higher temperature is necessary and critical in thermoelectric field.

Recently, MgAgSb-based materials have been developed as a promising p-type candidate for power generation below 550 K,[92–96] possessing an exciting conversion efficiency of about 8.5% for T_c = 293 K and T_h = 518 K.[97] The heavy valence band and intrinsically weak electron–phonon scattering contribute to the relatively high PF; meanwhile, the distorted crystal structure and multiscale microstructure result in low thermal conductivity,[96] which will be discussed in detail later.

3.2.2 Crystal Structure and Band Structure

3.2.2.1 Crystal Structure

As first reported by Kirkham et al.,[98] pristine MgAgSb has complicated phase transitions from room-temperature α phase to intermediate-temperature β phase and finally to high-temperature γ phase. α-MgAgSb has a tetragonal structure with space group $I\bar{4}C2$, as shown in Figure 3.13a. Sb atoms form a face-centered cubic closely packing sublattice, with Mg and Ag atoms lying in the centers of all the Sb octahedrons and half of the Sb tetrahedrons, respectively. The filled sites form one-dimensional chains, running in all three primary directions. But the Mg–Sb rock salt lattice is a distorted structure, rotated by 45° about the c axis. β-MgAgSb phase is also tetragonal

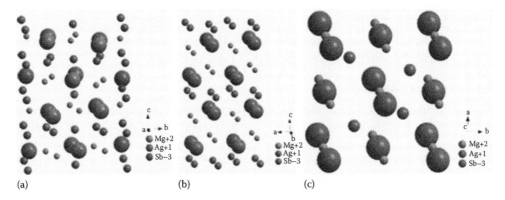

FIGURE 3.13 Crystal structure of MgAgSb: (a) α phase, (b) β phase, and (c) γ phase.

with space group *P4/nmm* (129) shown in Figure 3.13b. Mg and Sb atoms form a distorted rock salt lattice, as in the low-temperature phase, and Ag atoms fill half of the distorted Mg–Sb pseudocubes. The structure is similar to the Cu_2Sb structure with Mg and Ag each replacing one of the two Cu sites. γ-MgAgSb is cubic with space group $F\bar{4}3m$ (216), as shown in Figure 3.13c. In fact, this crystal structure is identical to half-Heusler structure. Mg and Sb atoms form the rock salt sublattice, and Ag atoms fill half of the tetrahedral interstices, alternating between filled and empty sites.

3.2.2.2 Band Structure

Since α-MgAgSb has an excellent thermoelectric performance, the following discussions mainly focus on its structure, properties, and their relationship. Ying et al.[99] first calculated the band structure within the density functional theory (DFT) using the projector augmented wave method. The result showed that α-MgAgSb was a narrow bandgap semiconductor with an indirect gap energy of ~0.1 eV. The top of the valence band was at the X-point, while the bottom of the conduction band was located at the G-point. Moreover, the DOSs of the valence band edge was much higher than that of the conduction band edge. Similarly, Li et al.[100] performed the band structure calculation of α-MgAgSb based on DFT plus Generalized Gradient Approximation-Perdew Bruke Ernzerhof (GGA-PBE) function, which had an indirect bandgap of ~0.063 eV. Sheng et al.[101] also predicted that α-MgAgSb was a semiconductor using the Becke–Johnson (mBJ) and Heyd–Scuseria–Ernzerhof (HSE) methods, while it was a semimetal using PBE method, as shown in Figure 3.14.

However, Miao et al.[102] presented a comprehensive theoretical study of the structural, electronic, and thermoelectric properties of MgAgSb by combining first-principles calculations and Boltzmann transport theory. They claimed that the α phase was better classified as a semimetal with highly dispersive conduction and valence bands overlapping slightly and crossing the Fermi level at distinct k-points using different calculation methods, including GGA-PBE, local-density approximation (LDA), and GGA-PBEsol. They assumed that previous investigations restricted their calculations to a limited path within the primitive Brillouin zone that did not cross the hole pockets. Moreover, a semimetal–to-semiconductor transition in α-MgAgSb had been predicted with raising the pressure.[103] The true band structure needed to be clarified by further optical properties measurement.

3.2.3 Synthesis Method

Kirkham et al.[98] first tried to prepare the pristine MgAgSb using a normal method, including vacuum melting, long-time annealing, and hot pressing procedures. However, there were high contents of secondary phases based on the XRD pattern analysis, namely, Sb and Ag_3Sb. Then, Ying et al.[99] used a similar controlled fabrication process and finally obtained a much lower content of impurity phases and the mass fraction of around 2%. Zhao et al.[92] first employed a simple and novel two-step BM and quick hot press method,

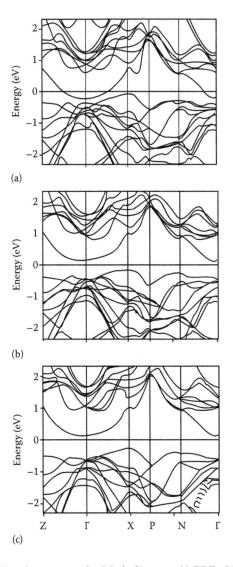

FIGURE 3.14 Calculated band structure of α-MgAgSb using (a) PBE, (b) mBJ, and (c) HSE functional. (Reprinted with permission from Sheng, C. et al., Predicting the optimized thermoelectric performance of MgAgSb. *Journal of Applied Physics*, 119, 19, 195101, Copyright 2016, American Institute of Physics.)

finely tuned the composition from MgAgSb to $MgAg_{0.97}Sb_{0.99}$, and obtained the pure phase of nanostructured MgAgSb within the detection limit of the XRD analysis.

3.2.4 Microstructure

The following part mainly focuses on the microstructure of nanostructured MgAgSb. The SEM image and medium-magnification transmission electron microscopy (TEM) image showed the typical grain size of around 150 nm (Figure 3.15a). Moreover, some nanoinclusions ranging from 5 to 10 nm

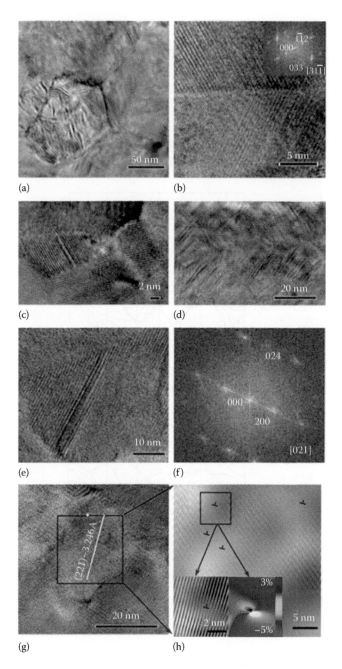

FIGURE 3.15 (a) Medium-magnification TEM image showing mesoscale grains; (b) high-magnification TEM image of the grain boundary showing good crystallization; the inset corresponds the FFT image along the *c* zone axis direction; (c) high-magnification TEM image showing nanoinclusions; (d) high-magnification TEM image showing high-density stacking faults; (e) high-magnification TEM image showing specific morphology of stacking fault; (f) FFT image along the [021] direction; (g) high-magnification TEM image showing a large quantity of dislocations inside the grain; (h) corresponding IFFT images showing a density of dislocation ~3.3 × 10^{11} cm^{-2} with the enlarged dislocations and GPA images as insets. (Reproduced with permission from Liu, Z. et al.: Lithium doping to enhance thermoelectric performance of MgAgSb with weak electron–phonon coupling. *Advanced Energy Materials*. 6. 1502269. Copyright Wiley-VCH Verlag GmbH & Co. KGaA.)

under high magnification can also be observed (Figure 3.15c). Moreover, a high number of defects could be seen in some grains, mainly including a high density of nanoscale stacking faults and dislocations. Careful investigations of the stacking faults showed that the typical morphology of the nanoscale stacking fault is about 20 nm in length (Figure 3.15d and e). The corresponding fast Fourier transformation (FFT) image further demonstrated the existence of stacking faults, playing a vital role in scattering those short to medium wavelength phonons. More importantly, a large quantity of dislocation could be clearly observed inside the grain, as shown in Figure 3.15g. To study the surrounding strain field around the dislocation cores, a high-quality high-resolution transmission electron microscopy (HRTEM) image was analyzed via geometric phase analysis (GPA), as shown in the inset in Figure 3.15h, both of which could act as effective scattering centers for short-wavelength phonons.

3.2.5 Thermoelectric Properties

3.2.5.1 Thermoelectric Properties of Pristine MgAgSb

Because there are some differences about thermoelectric properties between unheated and heated MgAgSb samples, it should be noted that all the thermoelectric properties discussed later are about the heated sample. As shown in Figure 3.16a, the electrical resistivity first increases from 3.6×10^{-5} Ω m at 300 K to 4.3×10^{-5} Ω m at 348 K and then sharply decreases to 2.1×10^{-5} Ω m due to the strong bipolar effect. Moreover, it shows a dependence of temperature (ρ versus $T^{3/2}$), which confirms a carrier transport mechanism dominated by acoustic phonon scattering before the intrinsic excitation. The relatively high electrical resistivity is ascribed to the low carrier concentration, e.g., 2.7×10^{19} cm^{-3} at 300 K. As expected, the Seebeck coefficients show the same tendency with the electrical resistivity. The large Seebeck coefficients, e.g., 255.8 µV/K at 300 K, are due to the typical heavy band feature, as shown in Figure 3.16b. The high DOS effective mass at the Fermi level (m^*) at 300 K can be roughly calculated assuming a single parabolic band model and an acoustic phonon scattering mechanism.[96] The obtained value is around $2.4m_e$ at 300 K, as shown in Figure 3.16c, which can be ascribed to the flat valence band ($m_b^* = 1.3m_0$) rather than a highly degenerated electronic band. The deformation potential E_{def} can be roughly estimated by the following equation when the acoustic phonon scattering is dominant[104,105]:

$$E_{def}^2 = \frac{(8\pi)^{1/2}(h/2\pi)^4 e\rho v_L^2}{3(m_b^*)^{5/2}(k_B T)^{3/2}\mu_H}, \tag{3.4}$$

where ρ is density (~6.2 g/cm³), v_L is the longitudinal velocity of sound (~3708 m/s), and μ_H is the drift mobility ($\mu_H = \mu/r_H$). The obtained deformation potential E_{def} is about 6.6 eV. Typically, the reported deformation potential E_{def} for thermoelectric materials ranges from 5 to 35 eV,[105,106] which means

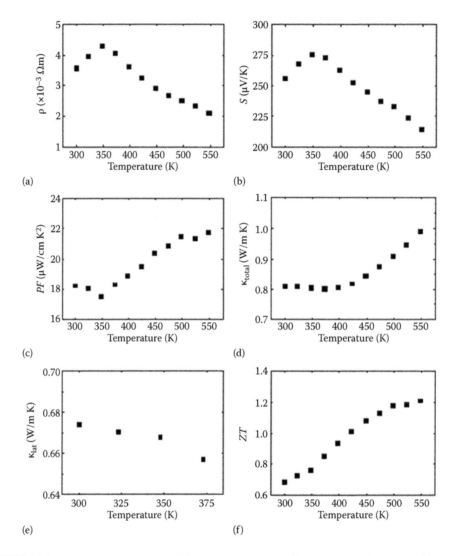

FIGURE 3.16 Temperature-dependent (a) electrical resistivity, (b) Seebeck coefficient, and (c) PF and (d) total thermal conductivity, (e) lattice thermal conductivity, and (f) ZT value for pristine nanostructured $MgAg_{0.97}Sb_{0.99}$.

that the deformation potential E_{def} of MgAgSb is at the lower end, demonstrating weak electron–phonon coupling. The calculated PF maintains a relatively high value at the whole measured temperature range, e.g., 18.2 µW/cm K^2 at 300 K and 21.8 µW/cm K^2 at 548 K. As shown in Figure 3.16d, the total thermal conductivity κ_{total} first decreases at a low-temperature range due to the Umklapp scattering process and then increases at a higher temperature because of the bipolar effect. The difference between the total thermal conductivity κ_{total} and the electronic thermal conductivity κ_{ele} before intrinsic excitation can be approximately estimated as the lattice thermal conductivity κ_{lat}, neglecting the contribution of bipolar effect. From the W-F relationship,

κ_{ele} can be calculated ($\kappa_{ele} = L\sigma T$), where L is the Lorenz number calculated based on the single parabolic model with an acoustic phonon scattering mechanism.[96] As shown in Figure 3.16e, room-temperature κ_{lat} is around 0.66 W/m K, much lower than that of the state–of-art thermoelectric materials. This intrinsically low κ_{lat} is due to the combination of weak bonding and intricate microstructure. Finally, the obtained ZT is shown in Figure 3.16f, with peak ZT and average ZT of around 1.2 and 1.0, respectively. Furthermore, it is a very promising candidate to fill the gap in the p-type ZT spectrum between low-temperature BiSbTe alloys and medium to high temperature CoSb$_3$-based and PbTe-based materials.

3.2.5.2 Optimizing the Thermoelectric Properties upon Doping

Due to the relatively low carrier concentration of pristine MgAgSb, effective doping aims to increase the carrier concentration and optimize the electrical transport properties. As shown in Figure 3.17, Li doping is more

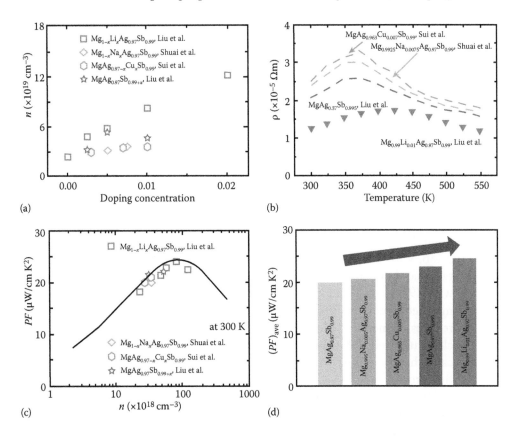

FIGURE 3.17 (a) Carrier concentration as a function of doping concentration, (b) comparison of temperature-dependent electrical resistivity of the optimized compositions, (c) carrier concentration-dependent PF. The solid line in c is calculated assuming a single parabolic band model and an acoustic phonon scattering mechanism; (d) comparison of average PF of the optimized compositions for MgAgSb system, including Li doping, Na doping, Cu doping, and Sb content tuning. (Reproduced with permission from Liu, Z. et al.: Lithium doping to enhance thermoelectric performance of MgAgSb with weak electron–phonon coupling. *Advanced Energy Materials.* 6. 1502269. Copyright 2016 Wiley-VCH Verlag GmbH & Co. KGaA.)

effective to supply holes as an acceptor dopant in the MgAgSb system, compared to Na and Cu dopings or Sb content tuning. The highest carrier concentration has already exceeded 10^{20} cm^{-3} with 2% Li doping concentration or more. That is why Li doping can lead to the lowest electrical resistivity and highest PF in the MgAgSb system. Assuming the single parabolic band model and an acoustic phonon scattering mechanism,[97] the calculated carrier concentration dependence of PF at room temperature is shown in Figure 3.17. The optimum carrier concentration for PF is around 9×10^{19} cm^{-3}, in good agreement with the measured Li doping data. Furthermore, the average PF from 300 to 548 K somewhat increases upon doping, e.g., 19.6 μW/cm K^2 for pristine MgAgSb and 24.6 μW/cm K^2 for 1% Li doping.

Although the electrical transport properties show a regular change upon extrinsic doping, the thermal transport properties seem to be abnormal. For example, doping with Ni or Cu observably suppresses the lattice thermal conductivity compared with other methods,[93,95] especially in the low-temperature range. The underlying mechanism for this phenomenon is still unknown that needs further investigations through first principles calculations or neutron diffraction. Due to the point defect scattering, doping slightly decreases the lattice thermal conductivity except for Ni or Cu doping.[93,95] Since weak chemical bonding and intricate microstructure scatter the heat-carrying phonons from atomic scale to microscale, extrinsic doping would not significantly affect the phonon transport in general.

3.2.6 Contact and Conversion Efficiency

Silver pad has been demonstrated as a proper electrical contact with low electrical resistance for MgAgSb using a simple one-step hot press technique,[96] as shown in Figure 3.18a. The obtained sample is given in Figure 3.18b, with a size of $3 \times 3 \times 5$ mm^3. The reason for choosing silver contact mainly includes two parts: First, the coefficients of thermal expansion with 19.5×10^{-6}/°C for silver and 20×10^{-6}/°C for the MgAgSb are well matched; second, silver is one of the component elements of MgAgSb that reduces elemental diffusion due to the smaller concentration gradient. A clean and well-defined interface between MgAgSb and silver contact indicates that there is only minimal elemental interdiffusion during the fabrication process.

Due to the combination of high average ZT and low electrical resistance, the measured single thermoelectric leg device efficiency of MgAgSb compound with hot pressed silver contacts is around 8.5% operating between 293 and 538 K using home-designed system (Figure 3.18c). The high efficiency value substantially outperforms the previous reports of other thermoelectric modules in the same temperature range, including commercial p-type Bi_2Te_3-based material shown in Figure 3.18d.

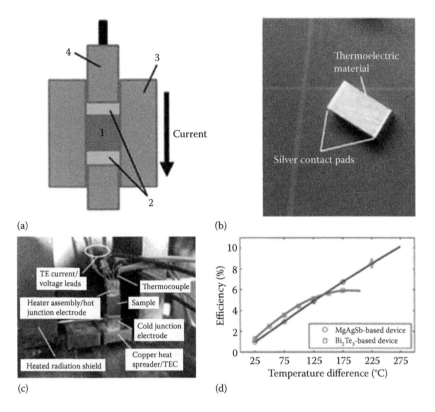

FIGURE 3.18 (a) Thermoelectric material powder (1) is sandwiched between electrical contact pad material (2) inside the graphite die (3) of the hot press. While the graphite piston (4) applies pressure, the current flowing through piston, die, and materials increases the hot press temperature. (b) p-Type MgAgSb sample as fabricated with a three-step process (BM, HP, cutting) with silver contact pads. (c) Fabricated single thermoelectric leg device with the sample soldered to the copper heater assembly (also acting as the hot junction electrode) and the copper cold junction electrode. The device is soldered onto a thermoelectric cooler, which is mounted onto a liquid-cooled cold plate and surrounded by a heated radiation shield. (d) Efficiency results of a single thermoelectric leg device based on the MgAgSb compound (*circles*) compared to a device based on a commercial p-type doped bismuth telluride sample (*squares*). Solid lines correspond to simulation results obtained with the iterative method based on temperature-dependent material properties extracted from average material properties measurements.

3.2.7 Conclusion

Pristine α-MgAgSb exhibits typical semiconductor properties and possesses excellent thermoelectric performance at the low-temperature range. Proper doping increases the carrier concentration and thus optimizes the electrical transport properties. In addition, the combination of weak chemical bonding and multiscale microstructure scatters the heat-carrying phonons. Therefore, a high peak ZT and average ZT can be finally achieved. More importantly, a record high conversion efficiency of 8.5% for T_c = 293 K and T_h = 518 K has been experimentally demonstrated for hot pressed MgAgSb with Ag contact, which highlights the realistic potential of MgAgSb for power generation in the low-temperature range.

References

1. H. J. Goldsmid and R. W. Douglas, The use of semiconductors in thermoelectric refrigeration, *British J. Appl. Phys.* 5, 386, 1954.

2. H. J. Goldsmid, Recent study of bismuth tellurite and its alloys. *J. Appl. Phys.* 32, 2198, 1961.

3. M. Imamuddin and A. Dupre, Thermoelectric properties of p-type Bi_2Te_3-Sb_2Te_3-Sb_2Se_3 alloys and n-type Bi_2Te_3-Bi_2Se_3 alloys in the temperature range 300 to 600 K. *Phys. Stat. Solidi A* 10, 415–424, 1972.

4. H. Scherrer and S. Sherrer, Bismuth telluride, antimony telluride, and their solid solution. In *CRC Handbook of Thermoelectrics*, edited by D. M. Rowe. CRC Press, Boca Raton, FL, 1995.

5. H. Scherrer and S. Sherrer, Thermoelectric properties of bismuth antimony telluride solid solutions. In *Thermoelectrics Handbook: Macro to Nano*, edited by D. M. Rowe. CRC Press: Boca Raton, FL, 2005.

6. V. A. Kutasov, L. N. Lukyanova, and M. V. Vedernikov, Shifting the maximum figure-of-merit $(Bi, Sb)_2(Te,Se)_3$. In *Thermoelectrics Handbook: Macro to Nano*, edited by D. M. Rowe. CRC Press: Boca Raton, FL, 2005.

7. T. Tokiai, T. Uesugi, and K. Koumoto, Relationship between thermoelectric properties and microstructure on p-type $(Bi_2Te_3)_{0.25}(Sb_2Te_3)_{0.75}$ system using milled powder. *J. Ceram. Soc. Jap.* 103, 1182–1187, 1995.

8. Y. Pan, T. R. Wei, Q. Cao, and J. F. Li, Mechanical enhanced p and n-type Bi_2Te_3-based thermoelectric materials reprocessed from commercial ingots by ball milling and spark plasma sintering. *Mater. Sci. Eng. B* 197, 75–81, 2015.

9. Z. Stary, J. Horak, M. Stordeur, and M. Stolzer, Antisite defects in $Bi_{2-x}Sb_{1-x}Te_3$ mixed crystals. *J. Phys. Chem. Solid.* 49, 29–34, 1998.

10. J. M. Zhang, W. M. Ming, Z. G. Huang, G. B. Liu, X. F. Kou, Y. B. Fan, K. L. Wang, and Y. G. Yao, Stability, electronic, and magnetic properties of the magnetically doped topological insulators Bi_2Se_3, Bi_2Te_3, and Sb_2Te_3. *Phys. Rev. B* 88, 235131, 2013.

11. Br. W. R. George, R. Sharples, and J. E. Thompson, The sintered bismuth telluride. *Proc. Phys. Soc.* 74, 768–770, 1959.

12. F. A. A. Amin, A. S. S. Al-Ghaffari, M. A. A. Issa, and A. M. Hassib, Thermoelectric properties of fine grained (75%Sb_2Te_3-25% Te_2Se_3) p-type and (90%Bi_2Te_3-5% Bi_2Se_3-5% Sb_2Se_3) n-type alloys. *J. Mater. Sci.* 27, 1250–1254, 1992.

13. T. Tokiai, T. Uesugi, and K. Koumoto, Microtructure and thermoelectric properties of p-type bismuth telluride prepared by HIP-ing of the metal powder mixture. *J. Ceramic Soc. Japan* 103, 797–803, 1995.

14. J. Navratil, Z. Stary, and T. Plechacek, Thermoelectric properties of p-type antimony bismuth telluride alloys prepared by cold pressing. *Mater. Res. Bull.* 31, 1559–1566, 1996.

15. S. Sugihara, S. Tomita, K. Asakawa, and H. Suda, High performance properties of sintered Bi_2Te_3–based thermoelectric materials. *Fifteenth Int. Conf. on Thermoelectrics, Pasadena, CA*, 46–50, 1996.

16. T. S. Oh, D. B. Hyun, and N. V. Kolomoets, Thermoelectric properties of the hot pressed $(Bi,Sb)_2(Te,Se)_3$ alloys. *Scripta Mater.* 42, 849–854, 2000.

17. J. Seo, K. Park, D. Lee, and C. Lee, Microstructure and thermoelectric properties of p-type $Bi_{0.5}Sb_{0.5}Te_3$ compounds fabricated by hot pressing and hot extrusion. *Scripta Mater.* 38, 477–484, 1998.

18. T. Calliat, L. Cailliard, H. Scherrer, and S. Scherrer, Transport properties analysis of single crystals $(Bi_xSb_{1-x})_2Te_3$ grown by the traveling heater method. *J. Phys. Chem. Solids* 54, 575–581, 1993.

19. B. Poudel, Q. Hao, Y. Ma, Y. C. Lan, A. Minnich, B. Yu, X. Yan et al., High-thermoelectric performance of nanostructured bismuth antimony telluride bulk alloys. *Science* 320, 634–638, 2008.

20. Y. Ma, Q. Hao, B. Poudel, Y. C. Lan, B. Yu, D. Z. Wang, G. Chen, and Z. F. Ren, Enhanced thermoelectric figure-of-merit in p-type nanostructured bismuth antimony telluride alloys made from elemental chunks. *Nano Lett.* 8, 2580–2584, 2008.

21. W. S. Liu, Z. F. Ren, and G. Chen, Nanostructure thermoelectric materials. In *Thermoelectric Nanomaterials*, edited by K. Koumoto and T. Mori. Springer-Verlag, Berlin, 2013.

22. Y. C. Lan, B. Poudel, Y. Ma, D. Z. Wang, M. S. Dresselhaus, G. Chen, and Z. F. Ren, Structure study of bulk nanograined thermoelectric bismuth antimony telluride. *Nano Lett.* 9, 1419–1422, 2009.

23. H. Y. Li, H. Y. Jing, Y. D. Han, G. Q. Lu, and L. Y. Xu, Effect of mechanical alloying process and sintering methods on the microstructure and thermoelectric properties of bulk $Bi_{0.5}Sb_{1.5}Te_3$ alloy. *Intermetallics* 43, 16–23, 2013.

24. L. P. Bulat, L. A. Drabkin, V. V. Karatayev, V. B. Osvenskii, Yu. N. Parkhomenko, M. G. Lavrentev, A. I. Sorokin et al., Structure and transport properties of bulk nanothermoelectrics based $Bi_xSb_{2-x}Te_3$ fabricated by SPS. *J. Electron. Mater.* 42, 2110–2113, 2013.

25. G. E. Lee, I. H. Kim, S. M. Choi, Y. S. Lim, W. S. Seo, J. S. Park, and S. H. Yang, Process controls for Bi_2Te_3-Sb_2Te_3 prepared by mechanical alloying and hot press. *J. Korean Phys. Soc.* 65, 2066–2070, 2014.

26. S. Jimenez, J. G. Perez, T. M. Tritt, S. Zhu, J. L. Sosa-Sanchez, J. Martinez-Juarez, and O. Lopez, Synthesis and thermoelectric performance of a p-type $Bi_{0.4}Sb_{1.6}Te_3$ materials developed via mechanical alloying. *Energy Convers. Manag.* 87, 868–873, 2014.

27. T. S. Kim, I. S. Kim, T. K. Kim, S. J. Hong, and B. S. Chun, Thermoelectric properties of p-type 25%Bi_2Te_3-75%Sb_2Te_3 alloys manufactured by rapid solidification and hot pressing. *Mater. Sci. Eng. B* 90, 42–46, 2002.

28. W. J. Xie, X. F. Tang, Y. G. Yan, Q. J. Zhang, and T. M Tritt, Unique nanostructures and enhanced thermoelectric performance of melt-spun BiSbTe alloys. *Appl. Phys. Lett.* 94, 102111/1–102111/3, 2009.

29. W. J. Xie, J. He, H. J. Kang, X. F. Tang, S. Zhu, M. Laver, S. Y. Wang et al., Identifying the specific nanostructures responsible for the high thermoelectric performance of $(Bi, Sb)_2Te_3$ nanocomposites. *Nano Lett.* 10, 3283–3289, 2010.

30. W. J. Xie, J. He, S. Zhou, T. Holgate, S. Y. Wang, X. F. Tang, Q. J. Zhang, and T. M. Tritt, Investigation of the sintering pressure and thermal conductivity anisotropy of melt-spun spark-plasma-sintered $(Bi,Sb)_2Te_3$. *J. Mater. Res.* 26, 1791–1799, 2011.

31. Y. Zheng, Q. Zhang, X. L. Su, H. Y. Xie, S. C. Shu, T. L. Chen, G. J. Tan et al., Mechanically robust BiSbTe alloys with superior thermoelectric performance: A case study of stable hierarchical nanostructured thermoelectric materials. *Adv. Energy Mater.* 5, 1401391, 2015.

32. L. D. Ivanova, L. I. Petrova, Yu. V. Granatkina, V. G. Leontyev, A. S. Ivanov, A. S. Varlamov, Yu. P. Prilepo, A. M. Sychev, A. G. Chuiko, and I. V. Bashkov, Thermoelectric and mechanical properties of the $Bi_{0.5}Sb_{1.5}Te_3$ solid state solution prepared by melt spinning. *Inorganic Mater.* 49, 120–126, 2013.

33. M. Scheele, N. Oeschler, I. Veremchuk, K. G. Reinsberg, A. M. Kreuziger, A. Kornwski, J. Broekaert, C. Klinke, and H. Weller, ZT enhancement in solution-grown $Sb_{2-x}Bi_xTe_3$ nanoplate. *ACS Nano* 4, 4283–4291, 2010.

34. R. J. Mehta, Y. L. Zhang, C. Karthik, B. Singh, R. W. Siegel, T. Borca-Tasciuc, and G. Ramanath, A new class of doped nanobulk high-figure-of-merit thermoelectric by scalable bottom-up assembly. *Nat. Mater.* 11, 233–240, 2012.

35. P. K. Nguyen, K. H. Lee, J. Moon, S. I. Kim, K. A. Ahn, L. H. Chen, S. M. Lee, R. K. Chen, S. Jin, and A. E. Berkowitz, Spark erosion: A high production rate method for producing $Bi_{0.5}Sb_{1.5}Te_3$ nanoparticles with enhanced thermoelectric performance. *Nanotechnology* 23, 415604–415607, 2012.

36. D. W. Liu, J. F. Li, C. Chen, and B. P. Zhang, Effect of SiC nanodispersion on the thermoelectric properties of p-type and n-type Bi_2Te_3-based alloys. *J. Electron. Mater.* 40, 992–998, 2011.

37. J. H. Li, Q. Tan, J. F. Li, D. W. Liu, F. Li, Z. Y. Li, M. M. Zou, and K. Wang, BiSbTe-based nanocomposites with high *ZT*: The effect of SiC nanodispersion on thermoelectric properties. *Adv. Funct. Mater.* 23, 4317–4323, 2013.

38. V. D. Blank, S. G. Buga, V. A. Kulbachinskii, V. G. Kytin, V. V. Medvedev, M. Y. Popov, P. B. Stepanov, and V. F. Stok, Thermoelectric properties of $Bi_{0.5}Sb_{1.5}Te_3/C_{60}$ nanocomposite. *Phys. Rev. B* 86, 075426, 2012.

39. H. R. Williams, R. M. Ambrosi, K. Chen, U. Friedman, H. Ning, M. J. Reece, M. C. Robbins, K. Simpson, and K. Stephenson, Spark plasma sintered bismuth telluride based thermoelectric materials incorporating dispersed born carbide. *J. Alloys Compd.* 626, 368–374, 2015.

40. Q. Zhang, Q. Y. Zhang, S. Chen, W. S. Liu, K. Lukas, X. Yan, H. Z. Wang et al., Suppression of grain growth by additive in nanostructured p-type bismuth antimony telluride. *Nano Energy* 1, 183–189, 2012.

41. S. I. Kim, K. H. Lee, H. A. Mun, H. S. Kim, S. W. Hwang, J. W. Roh, D. J. Yang et al., Dense dislocation arrays embedded in grain boundaries for high performance bulk thermoelectrics. *Science* 348, 109–114, 2015.

42. D. K. Ko, Y. J. Kang, and C. B. Murray, Enhanced thermopower via carrier energy filtering in solution-processable Pt–Sb_2Te_3 nanocomposites. *Nano Lett.* 11, 2841–2844, 2011.

43. Y. H. Lee, H. S. Kim, S. I. Kim, E. S. Lee, S. M. Lee, J. S. Rhee, J. Y. Jung, I. H. Kim, Y. F. Wang, and K. Koumoto, Enhancement of thermoelectric figure of merit for $Bi_{0.5}Sb_{1.5}Te_3$ by metal nanoparticle decoration. *J. Electron. Mater.* 41, 1165–1169, 2012.

44. S. Hwang, S. I. Kim, K. Ahn, J. W. Roh, D. J. Yang, S. M. Lee, and K. H. Lee, Enhancing the thermoelectric properties of p-type Bulk Bi-Sb-Te nanocomposite via solution-based metal nanoparticle decoration. *J. Electron. Mater.* 42, 1411–1416, 2013.

45. D. Suh, S. Lee, H. Mun, S. H. Park, K. H. Lee, S. W. Kim, J. Y. Choi, and S. Baik, Enhanced thermoelectric performance of $Bi_{0.5}Sb_{1.5}Te_3$ expanded graphene composite by simultaneous modulation of electronic and thermal carrier transport. *Nano Energy* 13, 67–76, 2015.

46. Q. H. Zhang, X. Ai, L. J. Wang, Y. X. Chang, W. Luo, W. Jiang, and L. D. Chen, Improved thermoelectric performance of silver nanoparticles-dispersed Bi_2Te_3 composite deriving from hierarchical two-phased heterostructure. *Adv. Funct. Mater.* 25, 966–976, 2015.

47. D. Parker and D. J. Singh, Potential thermoelectric performance from optimization of hole-doped Bi_2Se_3. *Phys. Rev. X* 1, 021005, 2011.

48. M. W. Oh, J. H. Son, B. S. Kim, S. D. Park, B. K. Min, and H. W. Lee, Antisite defects in n-type Bi_2 $(Te,Se)_3$: Experimental and theoretical studies. *J. Appl. Phys.* 115, 133706, 2014.

49. T. Tokiai, T. Uesugi, and K. Koumoto, Relationship between thermoelectric properties and microstructure on n-type $(Bi_2Te_3)_{0.95}(Bi_2Se_3)_{0.05}$ ceramics prepared from mixed powders. *J. Ceramic Soc. Japan* 103, 917–922, 1995.

50. A. Kanatzia, Ch. Papageorgiou, Ch. Lioutas, and Th. Kyratsi, Design of ball milling experiments on Bi_2Te_3 thermoelectric materials. *J. Elec. Mater.* 42, 1652–1660, 2013.

51. R. Ionescu, J. Jaklovszky, N. Nistor, and A. Chiculita, Grain size effects on thermoelectrical properties of sintered solid solutions based on Bi_2Te_3. *Phys. Stat. Sol. (a)* 27, 27–34, 1975.

52. C. H. Kuo, C. S. Hwang, M. S. Jeng, W. S. Su, Y. W. Chou, and J. R. Ku, Thermoelectric transport properties of bismuth telluride bulk materials fabricated by ball milling and spark plasma sintering. *J. Alloys Compd.* 496, 687–690, 2010.

53. L. D. Zhao, B. P. Zhang, W. S. Liu, and J. F. Li, Effect of mixed grain sizes on thermoelectric performance of Bi_2Te_3 compound. *J. Appl. Phys.* 105, 023704, 2009.

54. W. S. Liu, Sample ID: WS-BC-025, $Bi_2Te_{2.7}Se_{0.3}$ bulk balled for 12h, 15h, 18h and hot press at 500°C. 2010.

55. W. S. Liu, Sample ID: WS-NBT-05, $Bi_2Te_{2.7}Se_{0.3}S_{0.01}$ bulk ball milled for 10h, 14h, 15h, and hot pressed at 500°C. 2015.

56. W. S. Liu, Sample ID: WS-BC-029, $Bi_2Te_{2.7}Se_{0.3}$ bulk balled for 12h and hot press at 300°C, 400°C, 500°C. 2010.

57. W. S. Liu, Sample ID: WS-BC-041, $Cu_{0.02}Bi_2Te_{2.7}Se_{0.3}$ bulk balled for 20h and hot press at 300°C, 400°C, 500°C. 2010.

58. S. S. Lin and C. N. Liao, Effect of ball milling and post treat on crystal defects and transport properties of $Bi_2(Se,Te)_3$ compounds. *J. Appl. Phys.* 110, 093707, 2011.

59. L. D. Zhao, B. P. Zhang, J. F. Li, M. Zhou, and W. S. Liu, Effects of process parameter on electrical properties of n-type Bi_2Te_3 prepared by mechanical alloying and spark plasma sintering. *Physica B* 400, 11–15, 2007.

60. W. S. Liu, Q. Y. Zhang, Y. C. Lan, S. Chen, X. Yan, Q. Zhang, H. Wang, D. Z. Wang, G. Chen, and Z. F. Ren, Thermoelectric property studies on Cu doped n-type $Cu_xBi_2Te_{2.7}Se_{0.3}$ nanocomposites. *Adv. Energy Mater.* 1, 577–587, 2011.

61. X. A. Fan, J. Y. Yang, W. Zhu, H. S. Yun, R. G. Chen, S. Q. Bao, and X. K. Duan, Microstructure and thermoelectric properties of n-type $Bi_2Te_{2.85}Se_{0.15}$ prepared by mechanical alloying and plasma activated sintering. *J. Alloys Compd.* 420, 256–259, 2006.

62. M. P. Lu and C. N. Liao, Mechanical and thermal processing effects on crystal defects and thermoelectric transport properties of $Bi_2(Se,Te)_3$ compounds. *J. Alloys Compd.* 571, 178–182, 2013.

63. J. M. Schultz, J. P. McHugh, and W. A. Tiller, Effects of heavy deformation and annealing on the electrical properties of Bi_2Te_3. *J. Appl. Phys.* 33, 2443–2450, 1962.

64. Y. Eum, I. H. Kim, S. M. Choi, Y. S. Lim, W. S. Seo, J. S. Park, and S. H. Yang, Transport and thermoelectric properties of $Bi_2Te_{2.7}Se_{0.3}$ prepared by mechanical alloying and hot pressing. *J. Korean Phys. Soc.* 66, 1726–1731, 2015.

65. K. Takayanagi, N. Takezaki, and A. Negishi, Effects of powder oxidation on sintered thermoelements. *J. Japan. Inst. Met.Mater.* 30, 527–533, 1966.

66. D. C. Cho, C. H. Lim, D. M. Lee, S. Y. Shin, and C. H. Lee, Effect of oxygen content on thermoelectric properties of sintered Bi-Te based compounds. *Mater. Sci. Forum* 449–452, 905–908, 2004.

67. C. H. Lim, D. C. Cho, Y. S. Lee, C. H. Lee, K. T. Kim, and D. M. Lee, Effect of hydrogen reduction on the thermoelectric properties of spark-plasma-sintered Bi_2Te_3 based compounds. *J. Korean Phys. Soc.* 46, 995–1000, 2005.

68. L. D. Zhao, B. P. Zhang, W. S. Liu, H. L. Zhang, and J. F. Li, Effect of annealing on electrical properties of n-type Bi_2Te_3 fabricated by mechanical alloying and spark plasma sintering. *J. Alloys Compd.* 467, 91–97, 2009.

69. X. Yan, B. Poudel, Y. Ma, W. S. Liu, G. Joshi, H. Wang, Y. C. Lan, D. Z Wang, G. Chen, and Z. F. Ren, Experimental studies on anostropic thermoelectric properties and structures of n-type $Bi_2Te_{2.7}Se_{0.3}$. *Nano Lett.* 10, 3373–3378, 2010.

70. J. Seo, C. Lee, and K. Park, Effect of extrusion temperature and dopant on thermoelectric properties for hot-extrude p-type Te-doped $Bi_{0.5}Sb_{1.5}Te_3$ n-type SbI_3-doped $Bi_2Te_{2.85}Se_{0.15}$. *Mater. Sci. Eng. B* 54, 135–140, 1998.

71. S. K. Lotgering, Topotactical reactions with ferrimagnetic oxides having hexagonal crystal structures. *J. Inorg. Nuclear Chem.* 9, 113–123, 1959.

72. S. J. Hong, Y. S. Lee, J. W. Byeon, and B. S. Chun, Optimum dopant content of n-type 95% Bi_2Te_3+5% Bi_2Se_3 compounds fabricated by gas atomization and extrusion process. *J. Alloys Compd.* 414, 146–151, 2006.

73. S. J. Hong and B. S. Chun, Microstructure and thermoelectric properties of n-type 95% Bi_2Te_3-5%Bi_2Se_3 alloy produced by rapid solidification and hot extrusion. *Mater. Res. Bull.* 38, 599–608, 2003.

74. J. Y. Yang, R. G. Chen, X. A. Fan, W. Zhu, S. Q. Bao, and X. K. Duan, Microstructure control and thermoelectric properties improvement to n-type bismuth telluride based materials by hot extrusion. *J. Alloys Compd.* 429, 156–162, 2007.

75. S. Miura, Y. Sato, K. Fukuda, K. Nishimura, and K. Ikeda, Texture and thermoelectric properties of hot-extruded Bi_2Te_3 compounds. *Mater. Sci. Eng. A* 277, 244–249, 2000.

76. X. A. Fan, J. Y. Yang, W. Zhu, S. Q. Bao, X. K. Duan, C. J. Xiao, and K. Li. Preferential orientation and thermoelectric properties of n-type $Bi_2Te_{2.85}Se_{0.15}$ alloys by mechanical alloying and equal channel angular extrusion. *J. Phys. D: Appl. Phys.* 40, 5727–5732, 2007.

77. T. Hayashi, Y. Horio, and H. Takizawa, Equal channel angular extrusion technique for controlling the texture of n-type Bi_2Te_3 based thermoelectric materials. *Mater. Trans.* 51, 1914–1918, 2010.

78. C. H. Lim, K. T. Kim, Y. H. Kim, C. H. Lee, and C. H. Lee, Equal channel angular extruded $Bi_{0.5}Sb_{1.5}Te_3$ thermoelectric compound. *Mater. Trans.* 49, 889–891, 2008.

79. D. H. Kim, C. Kim, S. H. Heo, and H. Y. Kim, Influence of powder morphology on thermoelectric anisotropy of spark-plasma-sintered Bi-Te based thermoelectric materials. *Acta Mater.* 59. 405–411, 2011.

80. J. Jiang, L. D. Chen, S. Q. Bai, Q. Yao, and Q. Wang, Fabrication and thermoelectric performance of textured n-type $Bi_2(Te, Se)_3$ by spark plasma sintering. *Mater. Sci. Eng. B* 117, 334–338, 2005.

81. O. Ben-Yehuda, R. Shuker, Y. Gelbstein, Z. Dashevsky, and M. P. Dariel, Highly textured Bi_2Te_3-based materials for thermoelectric energy conversion. *J. Appl. Phys.* 101, 113707, 2007.
82. L. D. Zhao, B. P. Zhang, J. F. Li, H. L. Zhang, and W. S. Liu, Enhanced thermoelectric and mechanical properties in textured Bi_2Te_3 prepared by spark plasma sintering. *Solid State Sci.* 10, 651–658, 2008.
83. L. P. Hu, X. H. Liu, H. H. Xie, J. J. Shen, T. J. Zhu, and X. B. Zhao, Improving thermoelectric properties of n-type bismuth-telluride-based alloys by deformation induced lattice defect and texture enhancement. *Acta Mater.* 60, 4431–4437, 2012.
84. J. J. Shen, T. J. Zhu, X. B. Zhao, S. N. Zhang, S. H. Yang, and Z. Z. Yin, Recrystallization induced in-situ nanostructures in the bulk bismuth antimony telluride: A simple top-down route and improved thermoelectric properties. *Energy Environ Sci.* 3, 1519–1523, 2010.
85. L. P. Hu, H. J. Wu, T. J. Zhu, G. G. Fu, J. Q. He, P. J. Ying, and X. B. Zhao, Tuning multiscale microstructures to enhanced thermoelectric performance of n-type Bismuth-Telluride-based solution. *Adv. Energy Mater.* 5, 1500411–1500413, 2015.
86. D. M. Rowe, *CRC Handbook of Thermoelectrics*. CRC Press, Boca Raton, FL, 1995.
87. H. J. Goldsmid, *Introduction to Thermoelectricity*. Springer Science & Business Media, Berlin, 2009.
88. B. Poudel, Q. Hao, Y. Ma, Y. Lan, A. Minnich, B. Yu, X. Yan et al., High-thermoelectric performance of nanostructured bismuth antimony telluride bulk alloys. *Science* 320, 5876, 634–638, 2008.
89. L. Hu, T. Zhu, X. Liu, and X. Zhao, Point defect engineering of high-performance bismuth-telluride-based thermoelectric materials. *Adv. Funct. Mater.* 24, 33, 5211–5218, 2014.
90. H. Goldsmid and R. Douglas, The use of semiconductors in thermoelectric refrigeration. *British J. Appl. Phys.* 5, 11, 386, 1954.
91. J. Emsley, *Natures Building Blocks: An AZ Guide to the Elements*. Oxford University Press, Oxford, 2011.
92. H. Zhao, J. Sui, Z. Tang, Y. Lan, Q. Jie, D. Kraemer, K. McEnaney, A. Guloy, G. Chen, and Z. Ren, High thermoelectric performance of MgAgSb-based materials. *Nano Energy* 7, 97–103, 2014.
93. J. Shuai, H. S. Kim, Y. Lan, S. Chen, Y. Liu, H. Zhao, J. Sui, and Z. Ren, Study on thermoelectric performance by Na doping in nanostructured $Mg_{1-x}Na_xAg_{0.97}Sb_{0.99}$. *Nano Energy* 11, 640–646, 2015.
94. J. Sui, J. Shuai, Y. Lan, Y. Liu, R. He, D. Wang, Q. Jie, and Z. Ren, Effect of Cu concentration on thermoelectric properties of nanostructured p-type $MgAg_{0.97-x}Cu_xSb_{0.99}$. *Acta Mater.* 87, 266–272, 2015.
95. Z. Liu, J. Shuai, J. Mao, Y. Wang, Z. Wang, W. Cai, J. Sui, and Z. Ren, Effects of antimony content in $MgAg_{0.97}Sb_x$ on output power and energy conversion efficiency. *Acta Mater.* 102, 17–23, 2016.
96. Z. Liu, Y. Wang, J. Mao, H. Geng, J. Shuai, Y. Wang, R. He, W. Cai, J. Sui, and Z. Ren, Lithium doping to enhance thermoelectric performance of MgAgSb with weak electron–phonon coupling. *Adv. Energy Mater.* 6, 7, 1502269, 2016.
97. D. Kraemer, J. Sui, K. McEnaney, H. Zhao, Q. Jie, Z. Ren, and G. Chen, High thermoelectric conversion efficiency of MgAgSb-based material with hot-pressed contacts. *Energy Environ Sci.* 8, 4, 1299–1308, 2015.

98. M. J. Kirkham, A. M. dos Santos, C. J. Rawn, E. Lara-Curzio, J. W. Sharp, and A. J. Thompson, Ab initio determination of crystal structures of the thermoelectric material MgAgSb. *Phys. Rev. B* 85, 14, 144120, 2012.

99. P. Ying, X. Liu, C. Fu, X. Yue, H. Xie, X. Zhao, W. Zhang, and T. Zhu, High performance α-MgAgSb thermoelectric materials for low Temperature power deneration. *Chem. Mater.* 27, 3, 909–913, 2015.

100. D. Li, H. Zhao, S. Li, B. Wei, J. Shuai, C. Shi, X. Xi, P. Sun, S. Meng, and L. Gu, Atomic disorders induced by silver and magnesium ion migrations favor high thermoelectric performance in α-MgAgSb-based materials. *Adv. Funct. Mater.* 25, 41, 6478–6488, 2015.

101. C. Sheng, H. Liu, D. Fan, L. Cheng, J. Zhang, J. Wei, J. Liang, P. Jiang, and J. Shi, Predicting the optimized thermoelectric performance of MgAgSb. *J. Appl. Phys.* 119, 19, 195101, 2016.

102. N. Miao and P. Ghosez, Optimization of thermoelectric properties of MgAgSb-based materials: A first-principles investigation. *J. Phys. Chem. C* 119, 25, 14017–14022, 2015.

103. N. Miao, J. Zhou, B. Sa, B. Xu, and Z. Sun, Pressure-induced semimetal-semiconductor transition and enhancement of thermoelectric performance in α-MgAgSb. *Appl. Phys. Lett.* 108, 21, 213902, 2016.

104. Y. Pei, H. Wang, and G. Snyder, Band engineering of thermoelectric materials. *Adv. Mater.* 24, 46, 6125–6135, 2012.

105. H. Wang, Y. Pei, A.D. LaLonde, and G. J. Snyder, Weak electron–phonon coupling contributing to high thermoelectric performance in n-type PbSe. *Proc. Nat. Acad. Sci.* 109, 25, 9705–9709, 2012.

106. H. Xie, H. Wang, Y. Pei, C. Fu, X. Liu, G. J. Snyder, X. Zhao, and T. Zhu, Beneficial contribution of alloy disorder to electron and phonon transport in half-Heusler thermoelectric materials. *Adv. Funct. Mater.* 23, 41, 5123–5130, 2013.

CHAPTER 4

IV–VI Compounds for Medium Temperatures

Qian Zhang, Yu Xiao, Li-Dong Zhao, Eyob Chere,
Zhifeng Ren, Xiao Zhang, Cheng Chang, and E. M. Levin

Contents

4.1 Ternary and Quaternary Pb Chalcogenides

Qian Zhang

4.1.1 Introduction

Lead chalcogenides are traditional thermoelectric materials with decent *ZT* values. Both PbTe and PbSe have intrinsic two valence bands contributing to high hole DOSs and high Seebeck coefficients.[1–5] Resonant states could be created in p-type PbTe by Tl doping[6] and n-type PbSe by Al doping[7] to increase the DOSs close to the Fermi levels and Seebeck coefficients. Compared with PbTe and PbSe, PbS is also an ideal candidate because of the cheaper and more abundant S despite the single valence band and absence of the resonant states.[8–10] Now, it has been easy to obtain high *ZT*s > 1 in all p-type and n-type PbTe, PbSe, and PbS.[1,2,6–15] Since PbTe, PbSe, and PbS have similar face-centered cubic rock salt structure, ternary and quaternary Pb chalcogenides are considered for further optimization. There are always related reports with high ZTs > 2 in the lead chalcogenide samples with different content of Te, Se, and S.[16–18] Both $PbTe_{1-x}Se_x$ and $PbSe_{1-x}S_x$ are complete solid solution in the whole composition range. PbTe–PbS and $PbTe_{1-x}Se_x$–PbS (or PbTe–$PbSe_{1-x}S_x$) show very limited solubility. There is a miscibility gap in the phase diagram. In this chapter, we will review the high-performance lead chalcogenides and analyze the reason for the high performance with the categories of $PbTe_{1-x}Se_x$, $PbSe_{1-x}S_x$, $PbTe_{1-x}S_x$, and $PbTe_{1-x-y}Se_xS_y$.

4.1.2 $PbTe_{1-x}Se_x$

The lattice thermal conductivities are normally lowered by $PbTe_{1-x}Se_x$ solid solutions due to the phonon scattering from point defect. Tian et al.[19] detailed the phonon transport properties of PbTe, PbSe, and $PbTe_{1-x}Se_x$ by first principles calculations. The optical phonons were proven to comprise over 20% of the lattice thermal conductivity and provide strong scattering channels for the acoustic phonons. Alloying has advantages over nanostructuring in reducing the lattice thermal conductivity in this materials system. By combining the scattering of phonons, and the resonant electronic levels induced by Tl doping, Jaworski et al.[20] obtained high thermoelectric figure of merits in $PbTe_{1-x}Se_x$ and $PbTe_{1-x}S_x$ alloys.[21]

Because both PbTe and PbSe have light (L) and heavy (Σ) valence bands, high hole doping can move the Fermi level close to the Σ band, helping increase the DOS around the Fermi level and the Seebeck coefficient. It has been concluded that the convergence of the electronic bands can provide more benefit for the enhancement of the Seebeck coefficient from valley degeneracy.[22,23] The valence band extremum in PbTe occurs at the L-point in the Brillouin zone, where the valley degeneracy is 4 (see Figure 4.1a).

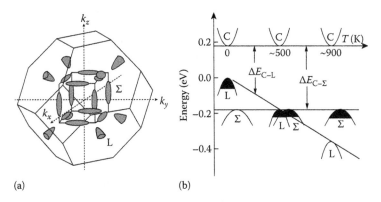

(a)   (b)

FIGURE 4.1 Valence band structure of $PbTe_{1-x}Se_x$.[23] (a) Brillouin zone showing the low degeneracy hole pockets (*half pocket*) centered at the L-point and the high degeneracy hole pockets (*whole pocket*) along the Σ line. The figure shows eight half pockets at the L-point so that the full number of valleys N_v is 4, while the valley degeneracy of the Σ band is N_v = 12. (b) Relative energy of the valence bands in $PbTe_{0.85}Se_{0.15}$. At 500 K, the two valence bands converge, resulting in transport contributions from both the L and Σ bands. C, CB; L, low degeneracy hole band; Σ, high degeneracy hole band. (Reprinted with permission from Pei, Y. Z. et al., *Nature*, 473, 66–69, 2011. Copyright 2011. Macmillan Publishers Ltd.)

The second valence band (Σ) has a valley degeneracy of 12. By producing the convergence of the L and Σ valence bands, we can obtain an increased valley degeneracy of 16. According to the temperature-dependent energy differences between the conduction (C) band edge and the L- and Σ-hole band edges of $PbTe_{1-x}Se_x$:

$$\Delta E_{C-L} = 0.18 + (4T/10{,}000) - 0.04x$$
$$\Delta E_{C-\Sigma} = 0.36 + 0.10x, \tag{4.1}$$

where x is the concentration of Se. The two bands start to converge when the temperature increases. However, the L band will move gradually below the Σ band at a relatively low temperature and depart from the convergence for PbTe (see Figure 4.1b). By alloying PbTe with PbSe, the convergence temperature will increase, which gives the most optimized Seebeck coefficient at a high temperature. By combining this effect with alloy scattering, high peak ZTs were obtained in Na-doped $PbTe_{1-x}Se_x$ (~1.8 in Te-rich samples of $Na_{0.02}Pb_{0.98}Te_{0.85}Se_{0.15}$ at 850 K)[23] and K-doped $PbTe_{1-x}Se_x$ (~1.6 in Te-rich samples of $K_{0.02}Pb_{0.98}Te_{0.75}Se_{0.25}$ at 773 K and ~1.7 in Se-rich samples of $K_{0.02}Pb_{0.98}Te_{0.15}Se_{0.85}$ at 873 K).[24] This temperature-induced band convergence has been studied only in rock salt IV–VI compounds and needs a deeper understanding.

Compared with Na-doped PbTe, there have been limited reports on K doping in PbTe due to the limited solubility of K in PbTe.[25,26] By alloying with PbSe, the balanced electronegativity resulted in the increased solubility and carrier concentration.[24] Figure 4.2a shows the room-temperature Pisarenko plots for K-doped PbTe, K-doped PbSe, and $K_{0.02}Pb_{0.98}Te_{1-x}Se_x$, in comparison with the Na-doped PbTe, Na-doped PbSe, and $Na_{0.02}Pb_{0.98}Te_{1-x}Se_x$. The Hall carrier concentration effectively increased from <6 × 10^{19} cm^{-3} in K-doped PbTe to the optimized concentration of 8–15 × 10^{19} cm^{-3} for K-doped $PbTe_{1-x}Se_x$, which is, however, still lower than Na-doped $PbTe_{1-x}Se_x$ in Te-rich compositions. For Se-rich compositions, more features come from K-doped PbSe. Owing to the smaller effective mass and larger energy difference between the heavy hole and light hole band edges in PbSe, the Seebeck coefficients are lower than those of Te-rich compositions. The pinning of the Fermi level by the heavy band happens only at the high temperature when the offset value of the two bands is small enough. Figure 4.2b presents the ZT values of best-optimized Na-doped and K-doped samples. The average ZT of the Te-rich compositions is higher even though Te is more expensive, so a trade-off between cost and performance needs to be considered for practical applications.[23,24]

N-Type $PbTe_{1-x}Se_x$ is studied mainly because of the lowered lattice thermal conductivity as shown in Figure 4.3a.[9,15,27–31] The lattice thermal conductivity dramatically decreased due to effective phonon scattering by point defects in Cr- and I-doped $PbTe_{1-x}Se_x$,[27–29] contributing to the increased ZT values (see Figure 4.3b). However, this enhancement is not as high as p-type

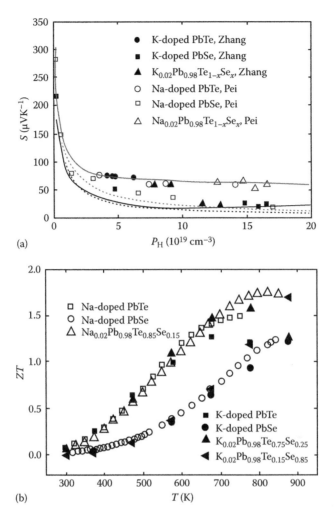

FIGURE 4.2 (a) Room-temperature Pisarenko plots for K-doped PbTe,[24] K-doped PbSe,[24] $K_{0.02}Pb_{0.98}Te_{1-x}Se_x$ (x = 0.15, 0.25, 0.75, 0.85, and 0.95),[24] Na-doped PbTe,[1] Na-doped PbSe,[2] and $Na_{0.02}Pb_{0.98}Te_{1-x}Se_x$ (x = 0, 0.05, 0.15, and 0.25).[23] Dashed gray curve is based on a single nonparabolic band model with the light hole effective mass of PbSe as m^*/m_e = 0.28. Solid black curve is based on two band models (light nonparabolic band and heavy parabolic band) with the heavy hole effective mass of PbSe as m^*/m_e = 2.5. Dashed gray curve is based on single nonparabolic band model with the light hole effective mass of PbTe of m^*/m_e = 0.36. Solid gray curve is based on two band models with the heavy hole effective mass of PbTe as m^*/m_e = 2. (b) Temperature dependence of ZTs for best optimized Na-doped PbTe,[1] Na-doped PbSe,[2] $Na_{0.02}Pb_{0.98}Te_{1-x}Se_x$,[23] K-doped PbTe,[24] K-doped PbSe,[24] and $K_{0.02}Pb_{0.98}Te_{1-x}Se_x$. (Reprinted with permission from Pei, Y. Z. et al., *Nature*, 473, 66–69, 2011. Copyright 2011. Macmillan Publishers Ltd; Zhang, Q. et al., *Journal of the American Chemical Society*, 134, 10031–10038, 2012. Copyright 2012. American Chemical Society.)

$PbTe_{1-x}Se_x$, because of the absence of the band convergence.[23,24] Both PbTe and PbSe have a single conduction band (CB) at the L-point of the first Brillouin zone. But we usually use a two-band Kane (TBK) model for the n-type PbTe, PbSe, and $PbTe_{1-x}Se_x$, including the nonparabolicity of the CB, which describes the conduction and light hole valence bands about the L-point.[7,12,13,15,27–31]

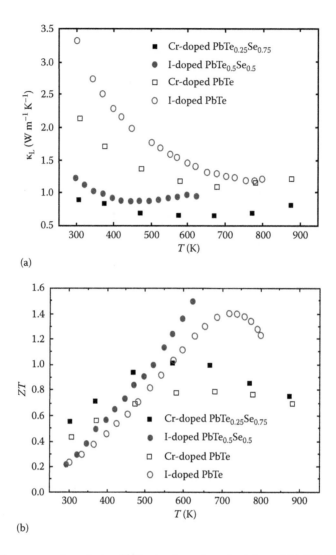

FIGURE 4.3 Temperature dependence of (a) lattice thermal conductivity and (b) *ZT*s for best-optimized Cr-doped PbTe,[15] I-doped PbTe,[9] Cr-doped PbTe$_{0.25}$Se$_{0.75}$,[29] and I-doped PbTe$_{0.5}$Se$_{0.5}$. (From Rawat, K. et al., *Nanotechnology*, 24, 215401 2013, Institute of Physics.)

Actually, forming a solid solution is really a good method to reduce the lattice thermal conductivity. However, the electron scattering from the same point defects results in the loss of the carrier mobility, making the net increase of *ZT* small.[32,33] Wang et al.[32] estimated that the thermoelectric quality factors *B* and *ZT* values are unchanged throughout the solid solution compositions in Br-doped PbTe$_{1-x}$Se$_x$ according to the following equation, where *K* is the bulk modulus; Ω is the average volume per atom; k_B is the Boltzmann constant; γ is the Grüneisen parameter; α is the Seebeck coefficient; *U* is the alloy scattering potential; Ξ is the deformation potential coefficient, ΔM and Δa are the difference in mass and lattice constants, respectively, between two constituents; *M* and α are the molar mass and

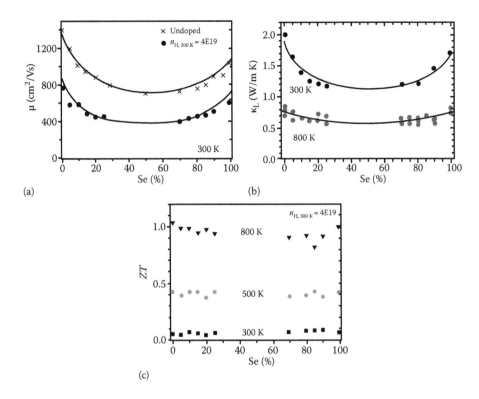

FIGURE 4.4 (a) Drift mobility at 300 K for $PbTe_{1-x}Se_x$ solid solutions with different doping levels. (b) Lattice thermal conductivity at 300 and 800 K as function of solid solution composition. The solid curve is calculated from the model described in the reference. (c) ZT at different temperatures for solid solutions with different compositions; doping level is the same for all samples. (H. Wang, A. D. LaLonde, Y. Z. Pei, and G. J. Snyder: The criteria for beneficial disorder in thermoelectric solid solutions. *Advanced Functional Materials*. 2011. 23. 1586–1696. Copyright Wiley-VCH Verlag GmbH & Co. KGaA. Reproduced with permission.)

lattice constant of the alloy, respectively; and ε is related to γ and elastic properties:

$$\frac{\mathrm{d}}{\mathrm{d}x}\frac{\Delta B}{B_{\mathrm{pure}}}\bigg|_{x=0} = \frac{\pi K\Omega}{4k_{\mathrm{B}}T}\left\{\frac{1}{4\gamma^2}\left[\left(\frac{\Delta M}{M}\right)^2 + \varepsilon\left(\frac{\Delta\alpha}{\alpha}\right)^2\right] - 10.6\left(\frac{U}{\Xi}\right)^2\right\}. \quad (4.2)$$

Figure 4.4 shows the drift mobility, lattice thermal conductivity, and ZT value as functions of the solid solution composition for Br-doped $PbTe_{1-x}Se_x$.[32] The decreased mobility compensated the decreased lattice thermal conductivity, resulting in the unchanged ZT values, consistent with the calculated results. This quantitative criterion was provided to determine the possible improvement by forming solid solutions for different material systems.

4.1.3 $PbSe_{1-x}S_x$

$PbSe_{1-x}S_x$ also forms a complete solid solution in the whole composition range.[34] Figure 4.5 presents the lattice thermal conductivity, Hall carrier mobility, thermoelectric quality factor, and ZT value for Cl-doped $PbSe_{1-x}S_x$.[35]

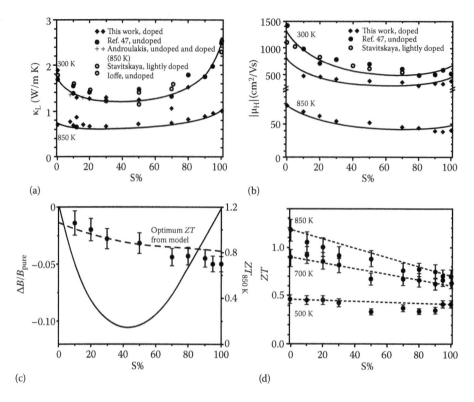

FIGURE 4.5 (a) Lattice thermal conductivity and (b) Hall mobility reduction at both 300 and 850 K. (c) Relative change of quality factor with composition (*line*) and change of ZT (*dashed line* and *squares*) at 850 K with composition. The dashed line is the calculated maximum ZT for each composition. (d) Measured ZT versus composition at different temperatures for $PbSe_{1-x}S_x$. The dashed lines are the linear average between ZT of the two binary compounds PbSe and PbS. The error bar represents a magnitude of 10% uncertainty on ZT. (H. Wang et al., *Journal of Materials Chemistry A*, 2, 3169, 2014. Reproduced by permission of The Royal Society of Chemistry.)

Similar with $PbTe_{1-x}Se_x$, we can rely on the alloy scattering to obtain low lattice thermal conductivity in $PbSe_{1-x}S_x$ (Figure 4.5a). But the carrier mobility is inevitably decreased (Figure 4.5b). It should be noted that the asymmetrical reduction of the mobility is due to the different effective masses in the alloys. So according to the criterion mentioned in the previous section, ΔB stays negative throughout the composition range (Figure 4.5c), showing no evident improvement of ZT over PbSe and PbS (Figure 4.5d). We also tried p-type $PbSe_{1-x}S_x$ with different dopings, which shows a similar trend on the composition dependence of ZT. However, considering the high natural abundance and inexpensiveness of Se and S, $PbSe_{1-x}S_x$ is still promising, especially when nanostructures are included. A high ZT value of ~1.2–1.3 at about 900 K was obtained with decreased lattice thermal conductivity and high carrier mobility as the pristine n-type PbSe in n-type Cl-doped PbSe containing as high as 16 atm.% PbS. The ~2–5 nm precipitates increase in density with increasing PbS concentration, showing the characteristic of inhomogeneity on the nanoscale in the solid solution, such as $PbTe–AgSbTe_2$ and $SnTe–AgSbTe_2$, which needs to be further studied.[36]

4.1.4 PbTe$_{1-x}$S$_x$

PbTe–PbS is a very important pseudobinary system (see Figure 4.6 for the schematic phase diagram between the PbTe and PbS phases), which can have a very low thermal conductivity ascribed to both point defects from alloying and nanostructures from phase separation.[37–41] Figure 4.7 presents the room-temperature lattice thermal conductivity for n-type In-doped PbTe$_{1-x}$S$_x$.[42] Different from the PbTe$_{1-x}$Se$_x$ (Figure 4.4b)[32] and PbSe$_{1-x}$S$_x$ (Figure 4.5a)[35] systems, which have the lowest lattice thermal conductivity in the middle composition, there are two inflection points in the PbTe$_{1-x}$S$_x$ system, corresponding to the XRD data and the phase diagram. The lattice thermal conductivity decreases in Pb$_{0.98}$In$_{0.02}$Te$_{0.95}$S$_{0.05}$, and Pb$_{0.98}$In$_{0.02}$Te$_{0.05}$S$_{0.95}$ is the result of alloying. The compositions within the immiscible gap spinodally decompose to Te-rich phase and S-rich phase without nucleation barrier, since the free energy curve has a negative curvature. Figure 4.7a through f shows the SEM images for the different compositions (x = 0.10, 0.20, 0.30, 0.50, 0.70, and 0.90 [filled circles]) with the same naturally formed striped structure, but different width of the strips. The lowest room-temperature κ_L is ~0.68 W/m K for Pb$_{0.98}$In$_{0.02}$Te$_{0.8}$S$_{0.2}$ with the thinnest strips (~30 nm in width) (Figure 4.7b). In Figure 4.8, the TEM images for this composition are presented. It is easy to find thinner strips (~2 nm) coming from nanoscale compositional fluctuation (Figure 4.8a), nanocrystals (~30 nm) due to the partial nucleation and growth features in some regions (Figure 4.8b), and many edge dislocations (Figure 4.8c). All these microstructures are believed to contribute to the low lattice thermal conductivity in this system. Clean boundaries and good crystallinity confirmed in Figure 4.8d are beneficial to the transport of charge carriers. Normally, a Te-rich composition has superior TE properties than a S-rich composition. However, considering the abundance of S, it is still promising to boost the TE properties by the addition of only a few percentage of PbTe into PbS.[43]

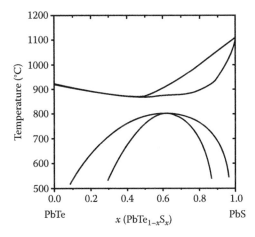

FIGURE 4.6   Schematic phase diagram of PbTe$_{1-x}$S$_x$.

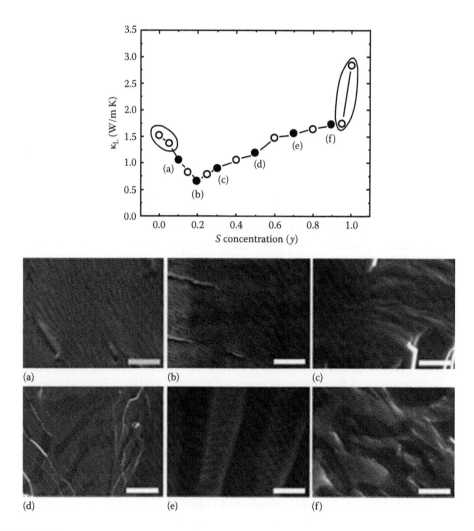

FIGURE 4.7 Room-temperature lattice thermal conductivity for In-doped $PbTe_{1-x}S_x$ ($0 \leq x \leq 1$). The SEM images for different compositions (*filled circles*) are shown in a through f, respectively. The scale bar is 200 nm for a, b, c, e, and f and 2 μm for d. (Reprinted from *Nano Energy*, 22, Q. Zhang, E. K. Chere, Y. M. Wang, H. S. Kim, R. He, F. Cao, K. Dahal, D. Broido, G. Chen, and Z. F. Ren, High hermoelectric performance of n-type $PbTe_{1-y}S_y$ due to deep lying states induced by indium doping and spinodal decomposition, 572–582, Copyright 2016, with permission from Elsevier.)

Due to the competition between alloying and phase separation, some unique cuboidal precipitates are observed in both p-type 3 atm.% Na-doped $PbTe_{0.8}S_{0.2}$ sample (Figure 4.9a)[18] and p-type 2.5 atm.% K-doped $PbTe_{0.7}S_{0.3}$ sample (Figure 4.9b)[17] with outstanding figure of merits. For 3 atm.% Na-doped $PbTe_{0.8}S_{0.2}$, a figure of merit ZT of ~2.3 was achieved at 923 K with the average size of the precipitates as ~118 nm. But the typical wavelength of phonons, which dominates the thermal conductivity, is only 1–10 nm.[18] The lower thermal conductivity is ascribed to the all-scale hierarchical architecturing. The superior high-power factor is because of the modulation in the hole concentration due to excessive Na diffusion and redissolution into the matrix grains. For 2.5 atm.% K-doped $PbTe_{0.7}S_{0.3}$, a high ZT value of

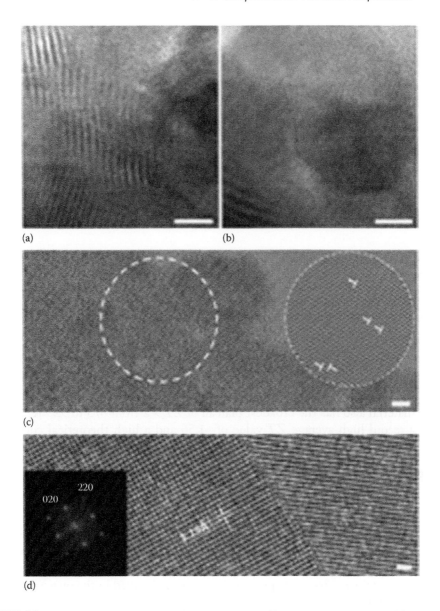

FIGURE 4.8 TEM images for $Pb_{0.98}In_{0.02}Te_{0.8}S_{0.2}$, showing (a) spinodal decomposition; (b) nanograins; (c) many edge dislocations and the extra half planes marked by ⊥ in magnified filtered image of the circular area; and (d) clean grain boundary. FFT of the left area of the boundary is presented in the inset to show the good crystallization. The scale bars are 10 nm for (a) and (b) and 2 nm for (c) and (d). (Reprinted from *Nano Energy*, 22, Q. Zhang, E. K. Chere, Y. M. Wang, H. S. Kim, R. He, F. Cao, K. Dahal, D. Broido, G. Chen, and Z. F. Ren, High hermoelectric performance of n-type $PbTe_{1-y}S_y$ due to deep lying states induced by indium doping and spinodal decomposition, 572–582, Copyright 2016, with permission from Elsevier.)

~2.2 at 923 K was obtained with the average size of the precipitates as ~4 nm.[17] These nanoscale regular-shaped precipitates, together with atomic scale defects and mesoscale grains, can effectively scatter phonons on multiple length scales, being considered to contribute to the low lattice thermal conductivity. The introduction of K in the phase-separated $PbTe_{0.7}S_{0.3}$

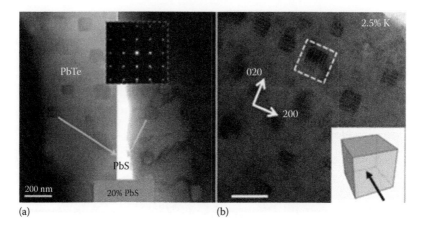

FIGURE 4.9   (a) Low-magnification TEM (*left*) and STEM-HAADF (*right*) images of the 3 atm.% Na-doped $PbTe_{0.8}S_{0.2}$ sample, which reveals cuboidal precipitates with inserted electron diffraction pattern along the 100 zone axis.[18] (b) Lattice image of 2.5 atm.% K-doped $PbTe_{0.7}S_{0.3}$ depicting cubic precipitates. Scale bar is 10 nm. (Reprinted with permission from Wu, H. J. et al., *Nature Communications*, 5, 4515, 2014. Copyright 2014. Macmillan Publishers Ltd.)

system enhances the carrier concentration for high electrical conductivity, suppresses the bipolar effect by pushing down the Fermi level away from the CB, and assists S alloying to tailor a more favorable band structure. This material also has the broadest plateau of $ZT > 2$ from 673 to 923 K, causing a record high average ZT value of ~1.56 and a high theoretical conversion efficiency of ~20.7% in a device implementation.

4.1.5 $PbTe_{1-x-y}Se_xS_y$

High ZTs > 2 have been achieved in ternary alloy $PbTe_{1-x}Se_x$ with band convergence and $PbTe_{1-x}S_x$ with nanostructures. Can we get an even higher ZT by combing the merits of $PbTe_{1-x}Se_x$ and $PbTe_{1-x}S_x$? Actually, this is a very long way if considering the complexity of the compositions. However, some primitive results have been obtained based on the analysis of some selected compositions. First, nanoprecipitates are also observed in the quaternary alloy as in Figure 4.10.[16,44,45] However, the experimental room-temperature lattice thermal conductivity almost fits in the theoretical curve for the $PbTe_{1-x-y}Se_xS_y$ alloy only assuming the point defect scattering involving multiple mass differences among Te, Se, and S (see Figure 4.11)[16]; Umklapp; and normal scattering processes. The currently obtained nanostructures are not beneficial for lattice thermal conductivity reduction. Much work is expected to control the size and the morphology of the nanoprecipitates. This system is an excellent platform to study phase competition between entropically driven atomic mixing (solid solution behavior) and enthalpy-driven phase separation.[16] Second, for p-type $PbTe_{1-x-y}Se_xS_y$, the alloying of PbS further modified the converged valence band of $PbTe_{1-x}Se_x$[16,44,46]; together with the strengthened point defect scattering by the inclusion of PbS, a high ZT value of ~2.0 was obtained in 2 atm.% Na-doped $(PbTe)_{0.84}(PbSe)_{0.07}(PbS)_{0.07}$.[16] Third,

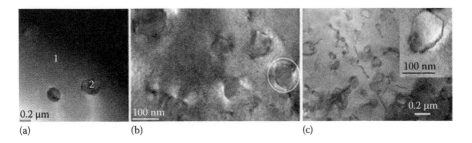

(a)   (b)   (c)

FIGURE 4.10 (a) Z-contrast image shows the existence of the large precipitate in $(PbTe)_{0.86}(PbSe)_{0.07}(PbS)_{0.07}$. (Reprinted with permission from Korkosz et al., *Journal of the American Chemical Society*, 136, 3225. Copyright 2014. American Chemical Society.) (b) Bright-field TEM micrograph of sulfur-rich precipitates in the tellurium-rich matrix in the Na-doped $(PbTe)_{0.75}(PbSe)_{0.1}(PbS)_{0.15}$ sample. Strong Moiré patterns are observed due to the double diffraction. (Reprinted from *Acta Materiala*, 80, S. Aminorroaya Yamini, H. Wang, Z. Gibbs, Y. Pei, D. Mitchel, S. X. Dou, and G. J. Snyder, Thermoelectric performance of tellurium-reduced quaternary p-type lead-chalcogenide composites, 365–372, Copyright 2014, with permission from Elsevier.) (c) Bright-field TEM micrograph of PbS-rich precipitates distributed within the PbTe-rich matrix of the sintered sample Cl-doped $(PbTe)_{0.75}(PbS)_{0.15}(PbSe)_{0.1}$. (Reprinted with permission from S. A. Yamini et al., *ACS Applied Materials and Interfaces*, 6, 11476–11483. Copyright 2014. American Chemical Society.) *Inset*: Detail of a precipitate showing Moiré fringes at oblique interfaces.

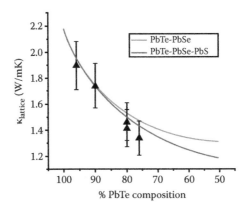

FIGURE 4.11 Comparison of the experimental lattice thermal conductivities and lattice thermal conductivities calculated based on point defect, Umklapp, and normal scattering processes. The experimental data points (*triangles*) fall very close, within 10% error, to the theoretical lattice thermal conductivity curve for the pseudoternary $(PbTe)_{1-2x}(PbSe)_x(PbS)_x$ pure solid solution. The calculated pseudobinary $(PbTe)_{1-x}(PbSe)_x$ lattice thermal conductivity (*top curve*) is similar to the calculated pseudoternary lattice thermal conductivity (*bottom curve*) at high PbTe compositions. (Reprinted with permission from Korkosz et al., *Journal of the American Chemical Society*, 136, 3225. Copyright 2014. American Chemical Society.)

for n-type $PbTe_{1-x-y}Se_xS_y$, similar with n-type $PbTe_{1-x}Se_x$, lowered thermal conductivity by point defect scattering is compensated by lowered carrier mobility. Therefore, a maximum figure of merit of ~1.1 at about 800 K was obtained in n-type $(PbTe)_{0.75}(PbSe)_{0.15}(PbS)_{0.1}$, similar to the performance of the single-phase alloys PbTe, PbSe, and $PbTe_{1-x}Se_x$.[45]

4.1.6 Conclusion

High-performance thermoelectric materials are always pursued for large-scale applications. Lead chalcogenides attract more and more attention due to the increasing ZT values. It seems easier to optimize this materials system with the suitable intrinsic characteristics, such as complex band structure, existing phase separation, and several possible resonant dopants. However, we must also notice the difficulties in this system, which need attention in future investigations. (1) Good TE materials should have good thermal stability. How can the long-term application with volatile elements Te, Se, and S be ensured? (2) How can the poor mechanical properties of this materials system be improved? (3) It is urgent to find a kind of matching contact material for the highly efficient TE device based on the lead chalcogenides. (4) Higher power factor is expected for the higher output power density. (5) The toxicity of lead will inevitably restrict some aspects of the application despite the high performance. Anyway, the optimization of the TE properties is always the first step to reach the application. Ternary and quaternary lead chalcogenides are promising TE materials worthy of being studied.

4.2 AgPb$_m$SbTe$_{2+m}$

Yu Xiao and Li-Dong Zhao

To enhance the thermoelectric performance, nanostructuring is an effective method to reduce lattice thermal conductivity since the lattice thermal conductivity is an independent parameter among all the interdependent thermoelectric parameters. There are two types of nanostructured thermoelectric materials; the first kind is a single phase, and it is defined by an assembly of nanosized particles or grains. The second kind is a system composed of a major bulk phase (matrix) containing a minor second phase embedded in the matrix, typically via precipitation from solid solution. In this section, we mainly introduce the AgPb$_m$SbTe$_{m+2}$ (LAST was named using the first letters of the composing elements lead antimony, silver, and tellurium) systems as typical thermoelectrics, which have been investigated for decades.

4.2.1 Crystal Structures

LAST compounds possess an average NaCl structure ($Fm\bar{3}m$ symmetry).[47] The atoms Ag, Pb, and Sb are statistically disordered in the structure on the Na sites, whereas the chalcogen atoms occupy the Cl sites (Figure 4.12). The formula is charge balanced because the average charge on the metal ions is 2⁺, and on the chalcogen ions, it is 2⁻.

The LAST system is characterized by a particular crystal structure instead of an explicit solid solution. LAST phases are inhomogeneous at the nanoscale with at least two coexisting sets of well-defined phases. The minority phase is richer in Ag and Sb in the nanosized length scale, and it is endotaxially embedded in the majority phase which is poorer in Ag and Sb. Moreover,

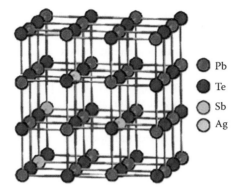

FIGURE 4.12   Crystal structure[1] of $AgPb_mSbTe_{m+2}$.

within each nanodomain extensive long-range ordering of Ag, Pb and Sb atoms can be observed. Indeed, the minority phase was successfully refined in space groups of lower symmetry than $Fm\bar{3}m$, including $P4/mmm$ and $R\bar{3}m$.[48] Various analogs of LAST compound can be described as the family of chalcogenide lead-based compounds including $Na_{1-x}Pb_mSb_yTe_{2+m}$ (SALT),[49] $Ag(Pb, Sn)_mSbTe_{2+m}$ (LASTT),[50] $AgPb_mMTe_{2+m}$ (m = Bi, Sb, La),[51,52] and $K_{1-x}Pb_mSb_yTe_{2+m}$ (PLAT).[53]

4.2.2 Synthesis Methods

- *Melting method*[47,49–51,53]: Ingots with nominal compositions of LAST system were prepared by melting and annealing, in quartz tubes under vacuum ($<10^{-4}$ Torr), mixtures of Ag, Pb, Sb, and Te at 1223 K for 4 h and cooling to 723 K for 40 h. Likewise, other analogs of LAST compound with different compositions can also be produced using melting method by optimizing the temperature and cooling speed. It is noticeable to point out that the inner walls of quartz tubes need carbon coating process to dispel side reactions when alkali or alkaline earth elements are incorporated in the system.

- *Mechanical alloying (MA)*[54]: The raw powders weighed according to appropriate ratio compositions are mixed in an MA form using a planetary mill with hardened steel jar and balls. After several hours of BM, the target compound with designed stoichiometry can be obtained. Generally, mechanical alloying can produce fine-grained powder, which plays a key role in enhancing dimensionless figure of merit (ZT), and bulk materials can be obtained by compressing the fine-grained powder using SP or HP method.

- *Hydrothermal synthesis*[55]: Te powder (5N) and other chemicals with analytical purity were used. The Te powder, $NaBH_4$, and NaOH were put into a Teflon-lined autoclave. Deionized water was poured into a beaker, placing the beaker on the plate of a magnetic force stirring device, and then $Pb(NO_3)_2$, $AgNO_3$, and $Sb(NO_3)_3$ were added, in that order, into the beaker, stirred with a magnetic bar until the materials were fully dissolved, and then the solution was gradually added into the autoclave, stirred with a glass bar. Additional deionized water was added into the autoclave until about 85% of the volume of the autoclave was filled, and then the autoclave was sealed. The autoclave was placed into an oven, heated up to 453 K, held for 20 h, and then naturally cooled to room temperature. Black precipitates were collected and washed with deionized water and absolute ethanol in a sequence for several times then filtered. The black product was dried in vacuum at 333 K for 6 h. The dried nanopowders with homogeneous grain size of about 40 nm can be compressed into bulks followed by SPS or HP method.[55]

4.2.3 Development in LAST System

The LAST system was first reported in 2004,[47] which exhibited a high thermoelectric figure of merit and attracted worldwide attentions. Over the past decade, the LAST system and its analogs have gotten great developments, as shown in Figure 4.13.

Kanatzidis's group first reported the cubic $AgPb_mSbTe_{m+2}$ system prepared via melting method in 2004. With $m = 18$ and doped appropriately, n-type semiconductors were produced and exhibited an extremely high ZT of ~2.2 at 800 K (see Figure 4.14), which arose from the particular microstructure generating a remarkably phonon scattering and keeping an ultrahigh power factor of ~28.0 μW/cm K^2 at 700 K in the $AgPb_{18}SbTe_{20}$ sample.[47] To understand the complex microstructure of the quaternary compound, powder/single-crystal XRD, electron diffraction and HRTEM were investigated. Extensive studies suggested that the LAST phases were inhomogeneous at the nanoscale with at least two coexisting sets of well-defined phases rather than a solid solution. Coherent nanocrystalline regions rich in Ag–Sb embedded in the matrix were observed, which can effectively depress thermal conductivity and simultaneously retain a high power factor (see Figure 4.14a and b).

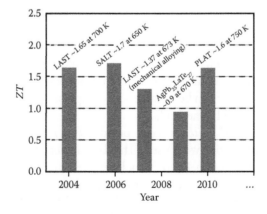

FIGURE 4.13 ZT of LAST system and its analogs as a function of year (LAST, SALT, and PLAT means $AgPb_mSbTe_{m+2}$, $Na_{1-x}Pb_mSb_yTe_{m+2}$ and $K_{1-x}Pb_mSb_yTe_{m+2}$, respectively).

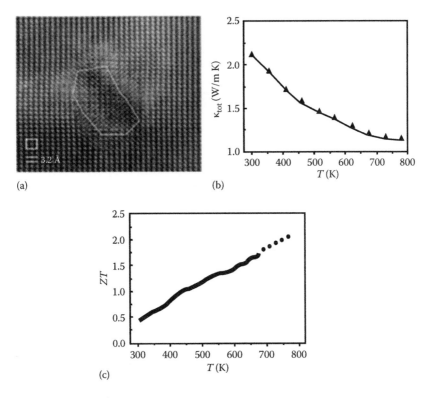

FIGURE 4.14 (a) TEM image of a $AgPb_{18}SbTe_{20}$ sample showing a nanosized region (a nanodot shown in the enclosed area) of the crystal structure that is Ag–Sb-rich in composition. (b) Total thermal conductivity (κ) in the range of 300–800 K. (c) Thermoelectric figure of merit ZT as a function of temperature. (From Hsu, F. et al., *Science*, 303, 818–821, 2004. Reprinted with permission of AAAS.)

The results provide experimental evidence for a conceptual basis that could be employed when designing high-performance thermoelectric materials.[48]

Kanatzidis's group conducted a comparative investigation of the $Ag_{1-x}Pb_m$ MTe_{m+2} (m = Bi, Sb) system to assess the roles of Sb and Bi on the thermoelectric properties. For comparable nominal compositions, the carrier concentrations were lower in the Sb analogs and the carrier mobilities were higher. The Seebeck coefficient dramatically decreased in going from Sb to Bi. Microstructure observations of both samples revealed that all systems contain compositional fluctuations at the nanoscopic level. Compared to PbTe, the lattice thermal conductivity of $AgPb_{18}BiTe_{20}$ was substantially reduced. However, the lattice thermal conductivity of the Bi analog was higher than the Sb analog, as shown in Figure 4.15. As a result, the dimensionless figure of merit ZT of $Ag_{1-x}Pb_{18}BiTe_{20}$, ZT of ~0.53 for x = 0 at 665 K, was found to be substantially smaller than that of $Ag_{1-x}Pb_{18}SbTe_{20}$, ZT of ~1.0 for x = 0.3 at 650 K.[51] Likewise, the new compound $AgPb_mLaTe_{m+2}$ system was investigated; a high electrical conductivity of ~1700 S/cm and a relatively low Seebeck coefficient of ~–60 µV/K were obtained at room temperature. A ZT of ~0.9 at ~670 K was achieved in the $AgPb_{25}LaTe_{27}$ sample.[52]

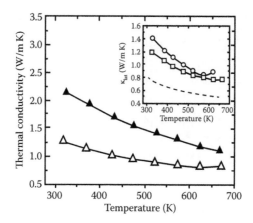

FIGURE 4.15 Temperature dependence of the total thermal conductivity (κ_{tot}) (*solid triangle*) and lattice thermal conductivity (κ_{lat}) (*open triangle*) of $AgPb_{18}BiTe_{20}$. The inset indicates the temperature dependence of the lattice thermal conductivity of $Ag_{1-x}Pb_{18}BiTe_{20}$ (x = 0, 0.3), compared with the lattice thermal conductivity (κ_{latt}) of $AgPb_{18}SbTe_{20}$ (*dashed line*). (Reprinted with permission from Han, M. K. et al., *Chemistry of Materials*, 20, 3512–3520, 2008. Copyright 2008. American Chemical Society.)

In further research, Na and K elements are chosen to substitute the Ag atoms in $AgPb_mSbTe_{m+2}$ (LAST-m) system and produced the new compounds $NaPb_mSbTe_{m+2}$ (SALT-m) and KPb_mSbTe_{m+2} (PLAT-m). Widely dispersed nanoscale particles were observed in both systems similar to the LAST-m system; however, all KPb_mSbTe_{m+2} samples show n-type semiconducting behavior as opposed to the p-type character of the $NaPb_mSbTe_{m+2}$ analogs. A maximum ZT of ~1.6 at 750 K was measured for an n-type PLAT system[7] and ~1.7 at 650 K for p-type SALT system.[49] Another successful example was conducted by Kanatzidis's group that Sn element partly substituted Pb sites in LAST-m samples. The existence of two chemical knobs, namely, the Pb/Sn ratio and the Ag and Sb concentration, allows for tuning of the electrical transport properties and therefore the identification of compositions with a high ZT of ~1.45 at 627 K.[50]

Jing-Feng Li's group prepared the $Ag_nPb_mSb_nTe_{m+2n}$ samples using a combined process of MA and SPS. Samples obtained by this method were considerably dense with fine grains. And the microstructure images (see Figure 4.16) indicated that the average grain size is about 1 μm. In TEM observation, many small dots, whose diameters are approximately 20 nm, are observed inside almost each grain. The present TEM observation together with these reported findings suggested that nanoscale inhomogeneities played an important role in the PbTe-based thermoelectric materials. The scattering of phonons was effectively and selectively intensified; thus, the thermal conductivity is significantly reduced while the electric properties largely remained the same.[54]

The ZT values for the composition $Ag_{0.8}Pb_{22}SbTe_{20}$ reached 1.37 at 673 K, and they found that the TE properties were much dependent on the content of Pb. Although this value is 20% lower than the ZT value of ~1.65 at the same temperature reported by Hsu et al.,[47] the study confirmed that high

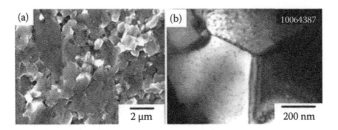

FIGURE 4.16 Cross-sectional (a) SEM image and (b) TEM image of $Ag_{0.8}Pb_{22}SbTe_{20}$ sample. (Reprinted with permission from Wang, H. et al., *Applied Physics Letters*, 88, 2006. Copyright 2006, American Institute of Physics.)

Seebeck coefficient can be obtained in the $Ag_nPb_mSbTe_{m+2n}$ bulk materials fabricated by another method much simpler than melting method.

Xinbing Zhao's group prepared $AgPb_{18}Sb_{1-x}Te_{20}$ compounds where $x = $ 0, 0.1, 0.3, and 0.5 by quenching the melts in liquid nitrogen to improve the homogeneity of these materials and expect a certain impact on their thermoelectric properties. Consequently, the quenched samples showed lower power factors compared to the slowly cooled compounds previously reported due to the absence of nanosized inhomogeneous regions in the liquid nitrogen-quenched compounds. It is believed that the preparation conditions, microstructure, and composition strongly affect the thermoelectric properties of this series of $AgPb_{18}Sb_{1-x}Te_{20}$.[56] Cai Kefeng Cai's group prepared $AgPb_mSbTe_{2+m}$ ($m = 10$–18) using hydrothermal synthesis method.[55] The samples showed a large positive Seebeck coefficient but low electrical conductivity, and no-uniform Seebeck coefficient was observed in the sample on a microscopic scale, shown in Figure 4.17.

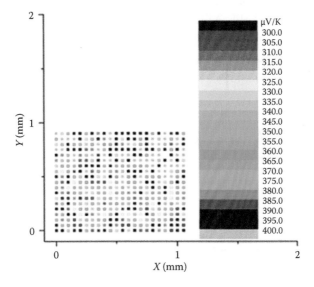

FIGURE 4.17 Seebeck coefficient scanning image in LAST-10 sample at room temperature. (Reprinted from *Journal of Alloys and Compounds*, 469, Cai, K. F. et al., 499–503, Copyright 2009, with permission from Elsevier.)

4.2.4 DFT Calculation of LAST System

Given the close structural relationship of $AgPb_mSbTe_{2+m}$ to PbTe, band structure calculations were examined near the bandgap region. Lower bandgap and resonant states were found near the top of the valence and the bottom of the CBs of bulk PbTe when Ag and Sb replaced Pb, as shown in Figure 4.18, and these states could be understood in terms of the modified Te–Ag (Sb) bounds. Further studies revealed that electronic structures near the gap were closely related to the microstructural arrangements of Ag–Sb atoms.[57] Band splitting was observed from the presence of the Ag–Sb pairs because the symmetry was lowered compared to the undoped PbTe, as shown in Figure 4.19. The valence band top was predominantly perturbed by Ag with an impurity-derived band predominantly formed out of Te p and Ag d states. This was the nearly flat band that formed the highest occupied band in Figure 4.19b, which was consistent with the resonant state near the valence band top in the single-particle DOSs.[51,57]

4.2.5 Summary and Outlook

The $AgPb_mSbTe_{2+m}$ (LAST) thermoelectric materials have been systematically investigated including crystal structure, microstructure, band structure, and thermoelectric performance. Some simple and new methods are introduced into synthesizing the LAST and its analogs system. The LAST materials exhibit an excellent thermoelectric performance in the temperature range from 300 to 800 K, mainly originating from the low thermal conductivity resulted from the compositional fluctuations in nanoscale and preserving a remarkably high power factor simultaneously. Meanwhile, the similar microstructures and disordering atom arrangement are observed in the analogs of LAST as well, including SALT, PLAT, $Ag(Pb, Sn)_mSbTe_{2+m}$ (LAST-T) and $AgPb_mMTe_{2+m}(m = Bi, La)$ systems. The extremely high ZT value makes the LAST system a potential thermoelectric material in power

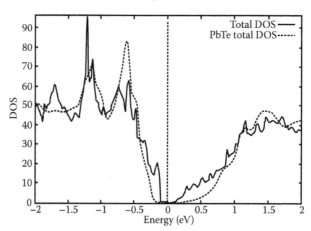

FIGURE 4.18 Total DOS of Ag–Sb pair with the Ag–Sb distance of ~11.19Å. (Reprinted with permission from Bilc, D. et al., *Physical Review Letters*, 93, 146403, 2004. Copyright 2004 by the American Physical Society.)

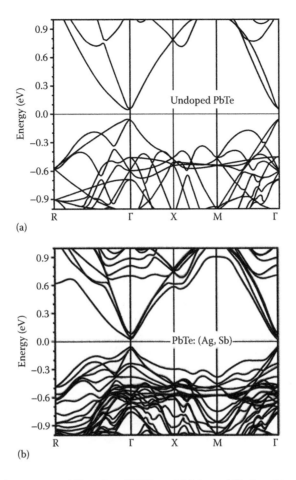

FIGURE 4.19 Band structures of (a) undoped PbTe and (b) Ag and Sb doped in a 64-atom supercell. (Reprinted with permission from Han, M. K. et al., *Chemistry of Materials*, 20, 3512–3520, 2008. Copyright 2008. American Chemical Society.)

generation application, shown in Figure 4.20. However, if more scattering mechanisms, such as ionized impurity or structure defects, are introduced in the LAST and its analogs systems, an even higher *ZT* value may be expected.

4.3 Improvement of Average *ZT* by Cr, In, etc. Doping

Eyob Chere and Zhifeng Ren

4.3.1 Introduction

Thermoelectric materials have superior advantages from the perspective of providing clean energy by direct conversion of heat to electricity without moving parts. However, their efficiency is insufficient compared to other heat engines due to their lower *ZT* values. Mathematically, the maximum thermal to electrical conversion efficiency η_{max} of a thermoelectric device is related to the average *ZT* by the following equation;

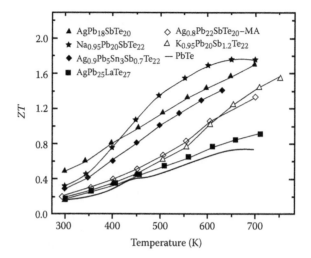

FIGURE 4.20 Temperature dependence of the *ZT* value in different samples. (From Hsu, F. et al., *Science*, 303, 818–821, 2004. Reprinted with permission of AAAS; Poudeu, F. et al.: High thermoelectric figure of merit and nanostructuring in bulk p-type Na$_{1-x}$Pb$_m$Sb$_y$Te$_{m+2}$. *Angewandte Chemie*, 2006, 45, 3835–3839. Copyright Wiley-VCH Verlag GmbH & Co. KGaA. Reproduced with permission; Androulakis, J. et al.: Nanostructuring and high thermoelectric efficiency in p-type Ag(Pb$_{1-y}$Sn$_y$)$_m$SbTe$_{2+m}$. *Advanced Materials*, 2006, 18, 1170–1173. Copyright Wiley-VCH Verlag GmbH & Co. KGaA. Reproduced with permission; reprinted with permission from Ahn, L. et al., *Chemistry of Materials*, 22, 876–882, 2010. Copyright 2010. American Chemical Society.)

$$\eta_{\max} = \left(1 - \frac{T_c}{T_h}\right) \frac{(1 + ZT_{\text{average}})^{1/2} - 1}{(1 + ZT_{\text{average}})^{1/2} + (T_c/T_h)}, \quad (4.3)$$

where T_h is the temperature at the hot junction and T_c is the temperature at the cold junction.[58–62] Equation 4.3 tells us that achieving an efficient thermoelectric device requires finding materials with high average *ZT* across the whole temperature range. One possible way of achieving this is by a segmented leg approach. In this approach, different thermoelectric materials of known high *ZT* values at different temperatures are combined to yield a high average *ZT* across the whole temperature range so as to boost the efficiency.[63–67] However, segmented legs suffer from the added complexity of bonding, interfacial mass diffusion, and thermal expansion mismatch that gradually lead to adverse device degradation and severe efficiency reduction. Therefore, it is more plausible to use a single material in a thermoelectric device to span the whole temperature range under which the device is operating.

Chalcogenide IV–VI compounds, such as lead chalcogenides and tin chalcogenides, are widely studied alloys for thermoelectric applications. Despite recent interests in tin chalcogenide thermoelectric materials, the practicality toward real device application is quite unpromising due to their lower *ZT* values in its polycrystalline form. On the other hand, lead chalcogenides have decent values of *ZT* at 573–873 K, making them promising candidates for medium-temperature power generation applications. However, the efficiency of lead chalcogenides is also not so high across the temperature range from 300–873 K due to their lower average ZT resulting from their corresponding lower ZT values near room temperature. In the last few decades, significant efforts have been put forward in improving the peak ZTs of Pb chalcogenides using various techniques such as alloying, doping, and band engineering with less emphasis given to improving the ZTs of these materials near room temperature. To mention a few, Tl acts as a resonant dopant in PbTe to enhance the ZT to ~1.5 by modifying the band structure[6]; a peak ZT value of ~1.7 at 873 K was achieved in $K_{0.02}Pb_{0.98}Te_{0.15}Se_{0.85}$ by potassium doping.[68] A ZT value of ~1.8 was obtained in p-type $Na_{0.02}Pb_{0.98}Te_{0.85}Se_{0.15}$ by band convergence.[69] Even though the peak ZTs of these materials at high temperatures are high, the average ZTs are low because of the very low ZTs below 400 K. Recently, significant studies have been forwarded to improve the thermal to electrical conversion efficiency of lead chalcogenides by improving the average ZT across the whole temperature range via systematic improvement of the ZT of these materials near room temperature.[70,71] Optimized doping to tune the carrier concentration to an optimum value on the order of 10^{18}–10^{19} cm^{-3} have proven to efficiently boost the ZT of Pb chalcogenides near room temperature. Consequently, the average ZT and efficiency of lead chalcogenide materials have been improved across the temperature range from 300 to 873 K. In the following sections, we will address the details of how optimized doping together with thermal conductivity reduction improved the thermoelectric performance of PbSe, PbTe–PbSe, and PbTe–PbS alloys.

4.3.2 Thermoelectric Performance of Cr-Doped PbSe

Transition metals such as Ti, V, Cr, Nb, and Mo were found to be effective dopants in increasing the Seebeck coefficient and power factor of n-type PbSe at temperatures below 500 K. From these dopants, Cr doping was found to yield the highest ZT of ~0.4 at room temperature with a peak ZT value of ~1 at 573 K. This high ZT value is attributed to the combined effect of high power factor and low thermal conductivity. The temperature dependence of the transport properties of Cr-doped PbSe ($Pb_{1-x}Cr_xSe$) (x = 0.0025, 0.005, 0.0075, and 0.01) compared with reported data on n-type $Bi_2Te_{2.7}Se_{0.3}$[72] is shown in Figure 4.21.

The electrical conductivity of all Cr-doped PbSe samples are higher than the electrical conductivity of the state-of-the-art $Bi_2Te_{2.7}Se_{0.3}$ at room temperature, and the Seebeck coefficient is comparable with the Seebeck coefficient of $Bi_2Te_{2.7}Se_{0.3}$ below 473 K. The maximum room temperature power factor reaches ~3.0 × 10^{-3} W/m K^2, higher than all the power factors of the reported doped PbSe and even the power factor of $Bi_2Te_{2.7}Se_{0.3}$.[27,72–78] PbSe has been shown to have lower lattice thermal conductivity than PbTe, which is attributed to the higher degree of anhamonicity of lattice vibrations.[73] Thus, by the simultaneous combination of high room-temperature power factor obtained via carrier concentration tuning with the relatively lower thermal conductivity, the room temperature ZT of PbSe boosted to ~0.4 (shown in

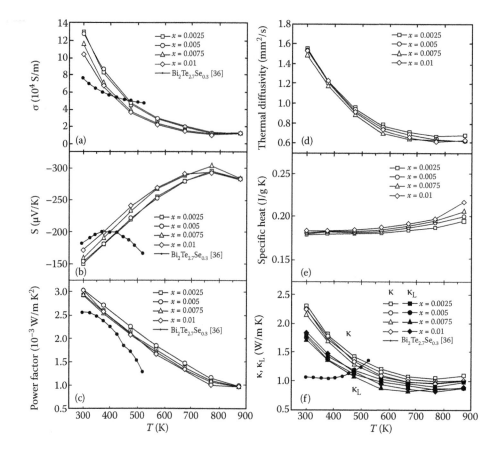

FIGURE 4.21 Temperature dependence of (a) electrical conductivity, (b) Seebeck coefficient, (c) power factor, (d) thermal diffusivity, (e) specific heat, and (f) total thermal conductivity and lattice thermal conductivity for $Pb_{1-x}Cr_xSe$ (x = 0.0025, 0.005, 0.0075, and 0.01) in comparison with reported data on n-type $Bi_2Te_{2.7}Se_{0.3}$ by Yan et al.[72] (*small solid circles*).

Figure 4.22), which is much higher than In-doped n-type PbSe,[74] although it is still lower than the room-temperature ZT of $Bi_2Te_{2.7}Se_{0.3}$.[72]

The ZT values continuously increase and reach ~1.0 at ~573 K for $Pb_{0.9925}Cr_{0.0075}Se$ and ~673 K for $Pb_{0.995}Cr_{0.005}Se$ and stay above 0.9 from 573 to 873 K, which strongly increases the average ZT of the PbSe-based materials. The higher room-temperature ZT is attributed to the higher room-temperature power factor ($PF = S^2n\mu q$) which is obtained by carrier concentration tuning in the range of ~10^{18}–10^{19} cm^{-3} and lower intrinsic thermal conductivity. When the carrier concentration is tuned in this range, both Seebeck coefficient (S) and Hall mobility (μ) maximized yielding to a higher power factor at room temperature.

4.3.2.1 Average ZT and Efficiency of Cr-Doped PbSe

Due to the improvement of the ZT of Cr-doped PbSe near room temperature, its average and device ZTs are significantly improved. Figure 4.23a and b shows the average and device ZTs of Cr-doped PbSe. The left panels in

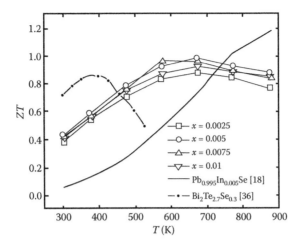

FIGURE 4.22 Temperature dependence of *ZT* for $Pb_{1-x}Cr_xSe$ (x = 0.0025, 0.005, 0.0075, and 0.01) in comparison with reported data on $Bi_2Te_{2.7}Se_{0.3}$ by Yan et al.[72] (*small solid circles*) and our previous data on In-doped PbSe[76] (*black line*).

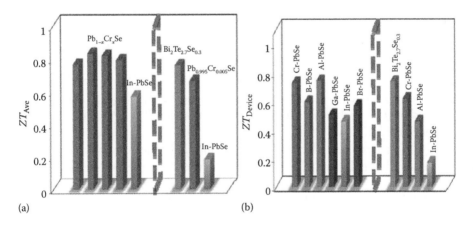

FIGURE 4.23 (a) Comparison of average *ZT* from 300 to 873 K between $Pb_{1-x}Cr_xSe$ (x = 0.0025, 0.005, 0.0075, and 0.01) and optimized In-doped PbSe (*left panel*).[74] Comparison of average *ZT* from 300 to 523 K for $Bi_2Te_{2.7}Se_{0.3}$,[72] 0.5 at.% Cr-doped PbSe and optimized In-doped PbSe (*right panel*).[74] (b) Comparison of device *ZT* operated between 300 and 873 K for 0.5 at.% Cr-doped PbSe (this work) and optimized B-,[74] Al-,[76] Ga-,[74] In-,[74] and Br-doped[77] PbSe (*left panel*). Comparison of device *ZT* operated between 300 and 523 K for $Bi_2Te_{2.7}Se_{0.3}$,[72] 0.5 at.% Cr-doped PbSe and optimized Al-[78] and In-doped PbSe,[74] (*right panel*).

the figures are for cold-side temperature of 300 K and hot-side temperature of 873 K, and the right panels in the figures are for cold-side temperature of 300 K and hot-side temperature of 523 K.

From Figure 4.23a and b, Cr-doped PbSe has both a preferred average *ZT* and device *ZT* compared to the other n-type PbSe, especially when working between 300 and 523 K, but not as good as $Bi_2T_{2.7}Se_{0.3}$. Furthermore, the device *ZT* of Cr-doped PbSe (between 300 and 873 K) is even higher than that of the reported n-type $PbTe:La/Ag_2Te$ (between 300 and 775 K) which has a peak *ZT* value of ~1.6 at 775 K.[79] Many of the best p-type PbTe

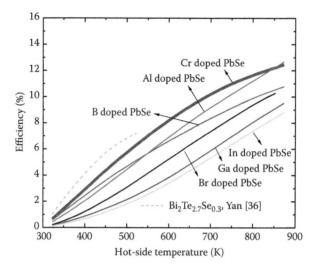

FIGURE 4.24 Temperature dependence of device efficiency for 0.5 at.% Cr-doped PbSe (this work) in comparison with the optimized B-doped PbSe,[73] Al-doped PbSe,[78] Ga-doped PbSe,[74] In-doped PbSe,[74] Br-doped PbSe,[77] and $Bi_2Te_{2.7}Se_{0.3}$,[72] with cold-side temperature of 300 K.

materials, such as PbTe:Na (between 300 and 750 K) with peak ZT value of ~1.4 at about 750 K,[80] PbTe:Tl (between 300 and 775 K) with peak ZT value of ~1.4 at about 775 K,[6] PbTe:Na/SrTe (between 300 K and 775 K) with peak ZT value of ~1.5 at about 775 K,[81] and $PbTe_{0.85}Se_{0.15}$:Na (between 300 and 800 K) with peak ZT value of ~1.8 at about 800 K,[69] could benefit from a flatter ZT curve.

The thermal to electrical conversion efficiencies of single leg Cr-doped materials are shown in Figure 4.24. Even though higher peak ZTs are achieved in other n-type PbSe materials (1.3 at 850 K for Al-doped PbSe,[78] 1.2 at 850 K for Br-doped PbSe,[77] and 1.2 at 873 K for Ga- and In-doped PbSe[74]), Cr-doped PbSe has the highest efficiency for a wide range of hot-side temperatures (350–873 K). At the lower temperature range, when the hot-side temperature is between 350 and 523 K, the device efficiency of Cr-doped PbSe is only slightly lower than that of the temperature-limited n-type $Bi_2Te_{2.7}Se_{0.3}$.[72] At high hot-side temperatures, the efficiency of Cr-doped PbSe is rivaled only by Al-doped PbSe, a material which has a much lower efficiency at lower hot-side temperatures.

These results emphasize the benefits of a flatter ZT curve: not only is the efficiency higher at high temperatures, but also the efficiency is higher at a wide range of hot-side temperatures. This improvement of thermoelectric performance is attributed to the improvement of thermoelectric properties near room temperature through carrier concentration tuning.

4.3.3 Thermoelectric Performance of Cr-Doped PbTe–PbSe Alloy

In the preceding section, we have seen that Cr doping in PbSe effectively improved the room temperature as well as the average ZT across a wide range of temperature yielding a promising boost in the thermal to electrical conversion efficiency. We may ask what would happen if Cr is doped into PbTe which is isostructural to PbSe. When Cr is doped into PbTe, it also significantly improved the room temperature power factor higher than that of Cr-doped PbSe and other n-type PbTe alloys.[71,82] However, the thermal conductivity of Cr-doped PbTe is higher compared to that of Cr-doped PbSe. The lower lattice thermal conductivity in PbSe due to its larger Grüneisen parameter values, and higher room-temperature power factor in Cr-doped PbTe, leaves a room for further optimization of the thermoelectric properties of these alloy systems by forming a Cr-doped $PbTe_{1-y}Se_y$ tertiary alloy to get high power factor and lower thermal conductivity. Improving the ZT of this alloy near room temperature and the average ZT across the whole temperature range from 300–873 K requires tuning the carrier concentration in the range of 10^{18}–10^{19} cm^{-3} by optimizing the Cr doping level and tuning the Te:Se ratio. It was found that Cr doping levels of 1 and 1.5 atm.% tailor the thermoelectric performance of PbTe–PbSe alloy more effectively. The temperature-dependent thermoelectric properties of 1 and 1.5 atm.% Cr-doped $PbTe_{1-y}Se_y$ are shown in Figures 4.25 and 4.26 at different Te:Se ratios.

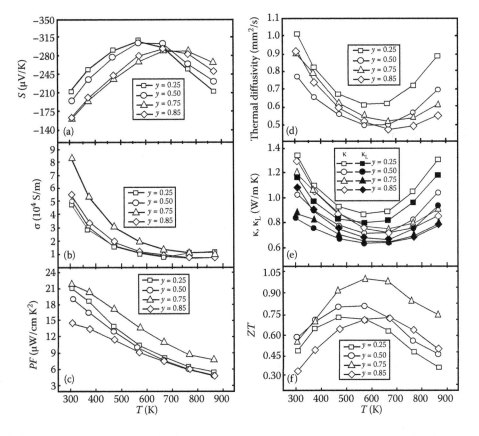

FIGURE 4.25 Temperature-dependent thermoelectric properties of $Cr_{0.01}Pb_{0.99}Te_{1-y}Se_y$ (y = 0.25, 0.5, 0.75, 0.85) with fixed Cr concentration of 1 atm.%. (a) Seebeck coefficient, (b) electrical conductivity, (c) power factor, (d) thermal diffusivity, (e) total thermal conductivity and lattice thermal conductivity, and (f) ZT.

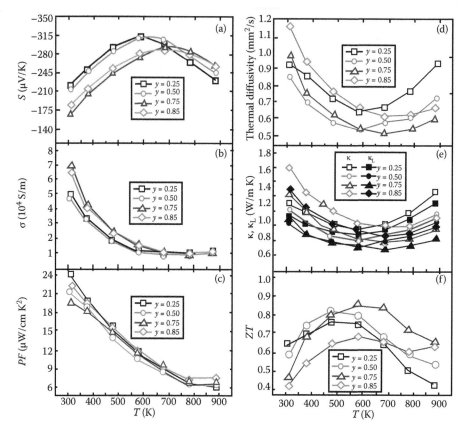

FIGURE 4.26 Temperature-dependent thermoelectric properties of $Cr_{0.015}Pb_{0.99}Te_{1-y}Se_y$ ($y = 0.25$, 0.5, 0.75, 0.85) with fixed Cr concentration of 1.5 atm.%. (a) Seebeck coefficient, (b) electrical conductivity, (c) power factor, (d) thermal diffusivity, (e) total thermal conductivity and lattice thermal conductivity, and (f) ZT.

At a fixed concentration of 1 atm.% Cr doping, the room-temperature Seebeck coefficient decreased from -211 to -157 µV/K with increasing Se concentration. The Seebeck coefficient of all the samples decreased at higher temperatures showing a bipolar transport nature, and it was found that the temperature at which the bipolar effect becomes important for Te-rich samples is lower than that of Se-rich samples. This is explained by the relatively higher carrier concentrations in Se rich samples compared to Te-rich ones. The electrical conductivity increased when the Se concentration increased to 75 atm.% and then decreased above this concentration showing a higher value at an optimum Se concentration of 75 atm.%. The high electrical conductivity manifested by the Se-rich sample $Cr_{0.01}Pb_{0.99}Se_{0.75}Te_{0.25}$ resulted in a higher power factor across the whole temperature range. This high electrical conductivity is attributed to the relatively higher carrier concentration ($\sim 8.12 \times 10^{18}$ cm^{-3}) as evidenced by the room-temperature Hall measurement. The thermal conductivity is heavily decreased compared to that of Cr-doped PbTe and PbSe due to phonon scattering by atomic mass disorder. Hence, by the combination of high power factor via carrier concentration optimization and reducing thermal conductivity by alloying, the room temperature ZT was improved to ~0.55 in Se-rich $Cr_{0.01}Pb_{0.99}Te_{0.25}Se_{0.75}$.

At a Cr doping concentration of 1.5 atm.%, as shown in Figure 4.26, the Seebeck coefficient follows a similar trend as in 1 atm.% Cr-doped samples (Figure 4.26) with increasing Se concentration.

The maximum room temperature Seebeck coefficient was –220 μV/K in Te-rich $Cr_{0.015}Pb_{0.985}Te_{0.75}Se_{0.25}$. The higher electrical conductivity of $Cr_{0.015}Pb_{0.985}Te_{0.75}Se_{0.25}$ at 1.5 atm.% Cr doping level is attributed to the higher carrier mobility of (~1120 cm²/V s) as confirmed by Hall coefficient measurement. This electrical conductivity in combination with the high Seebeck coefficient yields a higher room-temperature power factor of ~24 mW/cm K² (Figure 4.26c). The thermal conductivity is highly reduced due to the alloying effect. Thus, by the simultaneous reduction of the thermal conductivity and enhancement of the power factor, the room-temperature ZT is boosted to ~0.63 is in Te-rich $Cr_{0.015}Pb_{0.985}Te_{0.75}Se_{0.25}$. This ZT value is the highest value reported so far in both n-type and p-type lead chalcogenides at room temperature.

4.3.3.1 Average ZT and Efficiency of $Cr_xPb_{1-x}Te_{1-y}Se_y$

As already discussed in the preceding sections, the room-temperature ZT and average ZT of Pb chalcogenides are maximized at carrier concentrations in the range of ~10^{18}–10^{19} cm⁻³. Figure 4.27 shows the carrier concentration dependence of the room-temperature power factor, room temperature ZT, and average ZT of $Cr_{0.01}Pb_{0.99}Se_{0.75}Te_{0.25}$ and $Cr_{0.015}Pb_{0.985}Se_{0.25}Te_{0.75}$.

As seen in Figure 4.27, high power factor and ZT are obtained at room temperature at optimum carrier concentrations of 10^{18}–10^{19} cm⁻³. This yielded a high average ZT across the whole temperature range from 330 to 873 K.

Due to the improvement of the average ZT from 300 to 873 K, the thermal to electrical conversion efficiency of Cr-doped PbTe–PbSe alloy is significantly improved.

Figure 4.28 shows the efficiency of $Cr_{0.015}Pb_{0.985}Te_{0.75}Se_{0.25}$ and $Cr_{0.01}Pb_{0.99}Te_{0.25}Se_{0.75}$ at different hot-side temperatures.

The single leg efficiency of $Cr_{0.015}Pb_{0.985}Te_{0.75}Se_{0.25}$ and $Cr_{0.01}Pb_{0.99}Te_{0.25}Se_{0.75}$ is ~11 and ~13%, respectively, with a cold-side temperature of 300 K and hot-side temperature of 873 K. Despite having high ZTs exceeding 1.4 near 700 K for $I_{0.0012}PbTe_{0.9988}$, the efficiency is still lower than that of $Cr_{0.01}Pb_{0.99}Te_{0.25}Se_{0.75}$, especially below 700 K. The efficiency of $Cr_{0.01}Pb_{0.99}Te_{0.25}Se_{0.75}$ is also comparable with the efficiency of other thermoelectric materials with a cold-side temperature 300 K and hot-side temperature 773 K, for example, Bi_2Te_3–Bi_2Se_3–Bi_2S_3 (12.5%),[83] PbSe:Al (9.4%),[78,83] half Heuslers (8.4%),[83,84] filled skutterrudites (13.1%),[83,85] and PbTe: La (6.7%).[27,83] Thus, achieving high average ZT is more important to improve the efficiency.

4.3.4 Thermoelectric Performance of In-Doped PbTe–PbS Alloy

In the preceding sections, we have seen that carrier concentration optimization by Cr doping enhanced the thermoelectric performance of PbSe and PbTe–PbSe alloy. In a similar fashion, the carrier concentration

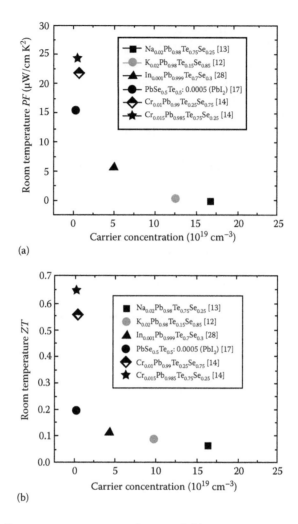

FIGURE 4.27 (a) Room-temperature power factor and (b) room-temperature ZT of $Cr_{0.01}Pb_{0.99}$ $Se_{0.75}Te_{0.25}$ and $Cr_{0.015}Pb_{0.985}Se_{0.25}Te_{0.75}$ as a function of carrier concentration compared with state-of-the-art PbTeSe compounds of high peak ZT values. *(Continued)*

of PbTe–PbS tertiary alloy could also be optimized to yield high room-temperature and average ZT. Unlike in PbSe and PbTe–PbSe alloy, Cr is not an efficient dopant to tailor the room-temperature thermoelectric property of PbTe–PbS. Therefore, we need to find another dopant that improves the thermoelectric performance of PbTe–PbS alloy near room temperature. It was found that indium forms deep lying states in $PbTe_{1-y}S_y$ and tunes the carrier concentration to an optimal level in the range of 10^{18}–10^{19} cm^{-3} to give the best thermoelectric properties near room temperature and simultaneously helps maintain a high thermoelectric performance at high temperatures by suppressing bipolar transport since minority carriers get trapped at the deep lying defect states. The thermal conductivity of this tertiary alloy

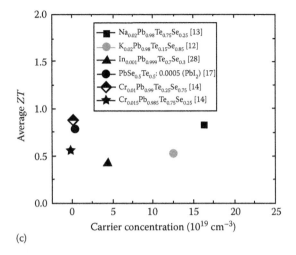

(c)

FIGURE 4.27 (CONTINUED) (c) Average ZTs of $Cr_{0.01}Pb_{0.99}Se_{0.75}Te_{0.25}$ and $Cr_{0.015}Pb_{0.985}Se_{0.25}Te_{0.75}$ as a function of carrier concentration compared with state-of-the-art PbTeSe compounds of high peak ZT values.

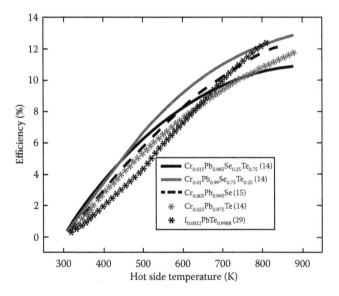

FIGURE 4.28 Temperature dependence of the calculated leg efficiencies of $Cr_{0.015}Pb_{0.985}Te_{0.75}Se_{0.25}$ (*solid black line*), $Cr_{0.01}Pb_{0.99}Te_{0.25}Se_{0.75}$ (*solid gray line*), $Cr_{0.005}Pb_{0.995}Se$ [72] (*dashed black line*), $Cr_{0.025}Pb_{0.975}Te$ (*star gray line*), and $I_{0.0012}PbTe_{0.9988}$ [78] (*star black line*) with a cold-side temperature at 300 K.

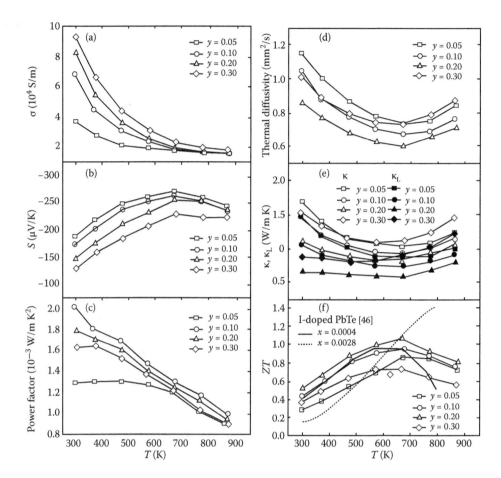

FIGURE 4.29 Temperature dependence of (a) electrical conductivity, (b) Seebeck coefficient, (c) power factor, (d) thermal diffusivity, (e) total thermal conductivity and lattice thermal conductivity, and (f) ZT for $Pb_{0.98}In_{0.02}Te_{1-y}S_y$ ($y = 0.05$, 0.10, 0.20, and 0.30). The ZT value of $PbTe_{1-x}I_x$ is plotted in f for comparison. (LaLonde, D. et al., *Energy and Environmental Science*, 4, 2090, 2011. Reproduced by permission of The Royal Society of Chemistry.)

is found to be lower due to alloy scattering and spinodal decomposition. The temperature-dependent thermoelectric properties of $PbTe_{1-y}S_y$ are shown in Figure 4.29.

Thus, by simultaneous carrier concentration tuning via In doping and thermal conductivity reduction by alloy scattering and spinodal decomposition, the room temperature ZT of $Pb_{0.98}In_{0.02}Te_{0.8}S_{0.2}$ is improved to as high as ~0.5.

4.3.4.1 Average ZT and Efficiency of In-Doped PbTe–PbS Alloy

In order to be considered for real device application, the engineering $(ZT)_{eng}$ (a more accurate estimate of the average ZT) needs to be maximized to get improved efficiency. The calculated engineering $(ZT)_{eng}$, and the efficiency of $Pb_{0.98}In_{0.02}Te_{0.8}S_{0.2}$, are higher or comparable with other reported n-type ternary and quaternary lead chalcogenides with even higher peak ZTs as shown in Figure 4.30c. $(ZT)_{eng}$ is the engineering figure of merit as a direct indicator, representing the overall performance of the material at given T_h and T_c where all the thermoelectric transport properties are temperature dependent.

An engineering $(ZT)_{eng}$ of ~0.7 and an efficiency of ~12% are achieved in $Pb_{0.98}In_{0.02}Te_{0.8}S_{0.2}$ with cold-side temperature of 323 K and hot-side temperature of 773 K by In deep lying doping. Despite high peak ZTs in I-doped $PbTe_{1-y}Se_y$ (~1.5 at about 600 K),[27] Cl-doped $PbSe_{1-y}S_y$ (~1.3 at about 900 K),[86] and Cl-doped $PbTe_{1-x-y}Se_xS_y$ (~1.1 at about 800 K)[87]

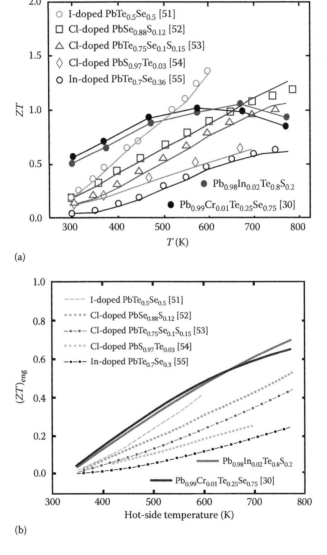

(a)

(b)

FIGURE 4.30 Temperature dependence of (a) ZT and (b) $(ZT)_{eng}$ for $Pb_{0.98}In_{0.02}Te_{0.8}S_{0.2}$ in comparison with the other reported n-type ternary and quaternary lead chalcogenides: Cr-doped $PbTe_{1-y}Se_y$,[71] I-doped $PbTe_{1-y}Se_y$,[88] Cl-doped $PbSe_{1-y}S_y$,[89] Cl-doped $PbTe_{1-x-y}Se_xS_y$,[90] Cl-doped $PbS_{1-y}Te_y$,[91] and In-doped $PbTe_{1-y}Se_y$,[92] with cold-side temperature of 323 K. *(Continued)*

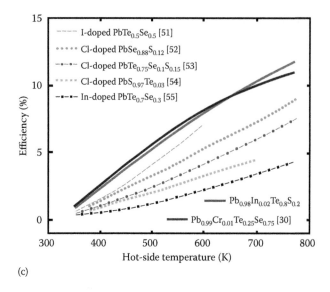

(c)

FIGURE 4.30 (CONTINUED) Temperature dependence of (c) device efficiency for $Pb_{0.98}In_{0.02}Te_{0.8}S_{0.2}$ in comparison with the other reported n-type ternary and quaternary lead chalcogenides: Cr-doped $PbTe_{1-y}Se_y,$[71] I-doped $PbTe_{1-y}Se_y,$[88] Cl-doped $PbSe_{1-y}S_y,$[89] Cl-doped $PbTe_{1-x-y}Se_xS_y,$[90] Cl-doped $PbS_{1-y}Te_y,$[91] and In-doped $PbTe_{1-y}Se_y,$[92] with cold-side temperature of 323 K.

(Figure 4.30a), the $(ZT)_{eng}$ (Figure 4.30b) and the efficiency (Figure 4.30c) of both $Pb_{0.99}Cr_{0.01}Te_{0.25}Se_{0.75}$[70] and $Pb_{0.98}In_{0.02}Te_{0.8}S_{0.2}$[89] are higher than others with T_h of up to 773 K. It clearly shows that the increase of $(ZT)_{eng}$ is more important than achieving high peak ZT.

4.3.5 Summary

IV–VI compounds, especially Sn and Pb chalcogenides, attracted significant interest for thermoelectric device applications. Pb chalcogenides in particular are widely studied for the last few decades for medium-temperature power generation applications due to their decent ZT values at high temperature (573–873 K). Despite their use for medium-temperature power generation applications, Pb chalcogenides still have lower thermal to electrical conversion efficiencies due to their lower average ZT over the temperature range of 300–873 K. This is due to their corresponding lower ZT values near room temperature. Recently, significant efforts have been made to improve the efficiency of Pb chalcogenides by improving the ZT near room temperature. The room-temperature thermoelectric performance of Pb chalcogenides was found to be improved by tuning the carrier concentration to an optimum value of $10^{18}–10^{19}$ cm^{-3} and deep lying doping. By a combination of systematic thermal conductivity reduction and power factor enhancement through carrier concentration tuning, room-temperature ZT as high as 0.63 and efficiency of up to ~13% have been achieved.

4.4 PbS

Xiao Zhang and Li-Dong Zhao

PbTe and its related alloys are one kind of the earliest studied middle-range temperature applications (600–900 K) of thermoelectric materials. But the content of Te element in the earth's crust is very scarce (0.01 ppm), even less than Pt (0.05 ppm) and Au (0.004 ppm). On the one hand, the price increases of Te elements rise rapidly; on the other hand, it is widely used in metallurgy, electronics, chemical industry, optoelectronics, and other fields; especially in recent years, the continuous production of CdTe optoelectronics thin film makes the scarce Te resources dwindling. In addition, the only commercial application of thermoelectric material Bi_2Te_3 also contains Te elements. Therefore, in order to have a sustainable development, and because of the need to find a material to replace the Te in PbTe-based thermoelectric materials, the attention of researchers began to transfer to PbSe compounds and PbS compounds. Selenium is 50 times more abundant than tellurium, while sulfur is among the 16 elements with the highest abundance in the earth's crust.[90] In this section, we will discuss the structure, the synthesis methods, the development trend of ZT figure for both n- and p-types, and the development prospects of PbS.

4.4.1 Crystal and Band Structures of PbS

Lead chalcogenides (PbS, PbSe, and PbTe) have a similar crystal structure, all like the rock-salt structure, shown in Figure 4.31a. Pb occupies the cation site, and S (Se, Te) occupies the anion position, and the chemical bond is a covalent bond. Their lattice constants are successively increased to 5.936, 6.124, and 6.462 Å, respectively, and the melting point decreases from 1114, 1065, to 917 K, respectively.[91–94]

Lead chalcogenides also have similar band structures, as illustrated in Figure 4.31b; the bandgap width for PbS, PbSe, and PbTe are 0.41, 0.27, and 0.32 eV at room temperature, respectively.[95] They belong to the narrow

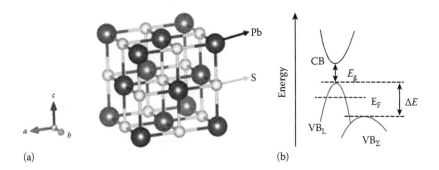

(a) (b)

FIGURE 4.31 Diagrammatic sketch of the crystal and band structure of lead chalcogenides (PbS, PbSe, and PbTe): (a) the rock-salt structure of lead chalcogenides (PbS, PbSe, and PbTe); (b) schematic showing the relative energy of the valence bands in lead chalcogenide (PbS, PbSe and PbTe) system.

bandgap semiconductor; the top of the valence band and the bottom of the CB are located at the L-point in the Brillouin zone. The small bandgap may lead to the occurrence of bipolar diffusion at high temperature. The energy separation between the two balance bands (L and Σ bands) influences the carrier mobility and the Seebeck coefficient. This double valence band structure provides an effective guarantee for structure engineering to enhance Seebeck coefficient and then the increase of the ZT value.[96]

4.4.2 Synthesis Methods of PbS

Due to the high vapor pressure of sulfur element, it is difficult to obtain highly dense ingot samples by furnace melting and cooling. While the Bridgman growth method could get dense PbS single crystals, however, single crystal growth is time consuming. Therefore, most researchers prepared polycrystalline PbS bulks, which involves several continuous steps, e.g., melting ingot, crushing into fine powders, followed by a sintering process. The details can be summarized as the starting materials are put inside fused silica tubes. The tubes are subsequently evacuated to a base pressure of ~10^{-4} Torr flame sealed and slowly heated to the programmed temperature for heated reaction, and then cooling according to the procedures has been set to cool in a high-temperature melting furnace. This method is conducive to reduce raw material volatilization and avoid the oxidation of the materials in the reaction process. In the high-temperature melting state, the diffusion coefficient of the reaction raw material is very large, so the raw material can be fully mixed and react, and the preparation time is short; the process is simple and easy to operate.

SPS, also known as field-assisted sintering technique or pulsed electric current sintering, is a new fast sintering densification method, which has been widely used in the preparation and development of ceramics, metals, semiconductors, and composite materials. The obtained ingots were crushed into powders and then densified by the SPS method (SPS-10-4, Thermal Technology). To prepare for SPS processing, the melt grown ingots were ground to a powder using a mortar and pestle to reduce the grain size to smaller particle size. These powders were then densified at a given temperature for several minutes in a graphite die under an axial compressive stress in an argon atmosphere. Highly dense disk-shaped pellets were obtained. In the case of air-sensitive starting materials, the manipulations and preparative steps and the grinding powders for SPS were carried out in a purified Ar atmosphere glove box, with total O_2 and H_2O level of <0.1 ppm.

4.4.3 Intrinsic Thermoelectric Transport Properties of PbS

Pei et al.[97] made a comparative study of the system for electrical and thermal transport properties of lead chalcogenides (PbTe, PbSe, and PbS), as shown in Figure 4.32. Compared with PbTe and PbSe, PbS has a larger bandgap and a higher lattice thermal conductivity; thus, in lower thermoelectric properties, the PbS compound has relatively high resistivity which increases

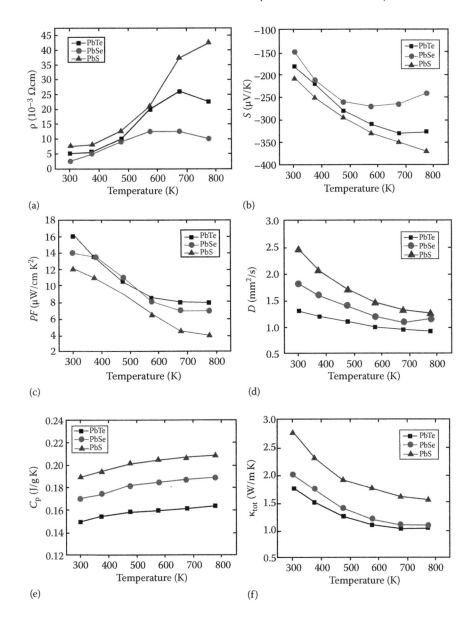

FIGURE 4.32 Electric transmission and thermal performance of PbTe, PbSe, and PbS: (a) electrical resistivity; (b) Seebeck coefficient; (c) power factor; (d) thermal diffusivity; (e) specific heat capacity; (f) total thermal conductivity as a function of temperature. (*Continued*)

with temperature, which is due to its greater ionic bonding component in PbS. And at high temperature, the optical wave significant scattering effect leads to decreased carrier mobility. Additionally, the contribution of bipolar thermal conductivity of PbS will undermine the final ZT.[98] Pei et al.'s[97] report indicates that inexpensive and earth-abundant PbS would be a robust alternative for PbTe and are very promising candidates for the thermoelectric power generation applications.

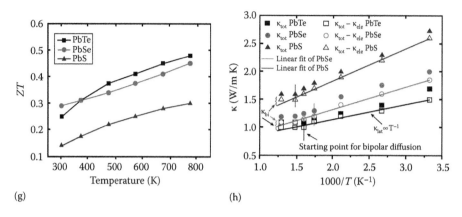

(g)   **(h)**

FIGURE 4.32 (CONTINUED) Electric transmission and thermal performance of PbTe, PbSe, and PbS: (g) the ZT figure as a function of temperature; (h) component details of thermal conductivity as a function of temperature; the solid line is linearly fitting to the lattice thermal conductivity at temperature ranges from room temperature to 723 K. (Reprinted from *Journal of Alloys and Compounds*, 514, Pei, Y. L., and Liu, Y., Electrical and thermal transport properties of Pb-based chalcogenides: PbTe, PbSe, and PbS, 40–44, Copyright 2012, with permission from Elsevier.)

As shown in Figure 4.32h, at high temperatures, κ_{bi} should be equal to $\kappa_{tot} - \kappa_{lat} - \kappa_{ele}$. Pei et al.[97] noted that the contribution of κ_{bi} is proportional to the width of the bandgap and occurs with the intrinsic excitation, resulting in an increase in total thermal conductivity at high temperature. Therefore, to enhance the ZT figures of the PbS-based materials to reach the level of PbTe-based materials, the total thermal conductivity and bipolar thermal conductivity should be supposed.

4.4.4 Promising n- and p-Type PbS

4.4.4.1 n-Type PbS

Johnsen et al.[99] investigated the n-type PbS by doping with $PbCl_2$, which is the first time for the nanostructuring paradigm to be extended to a bulk PbS-based system. $PbCl_2$ was used as a dopant, optimizing the carrier concentration. As PbTe is coherently and semicoherently embedded in a PbS matrix, nanostructured $PbS_{1-x}Te_x$ samples decrease the lattice thermal conductivity compared with pristine PbS without largely changing the electronic properties. Therefore, the composite sulfur-rich $PbS_{0.97}Te_{0.03}$ with 0.1% $PbCl_2$ achieves a maximum ZT of about 0.8 at 910 K. Zhao et al.[100] reported that the n-type thermoelectric performance improved PbS via doping and nanostructuring. Similarly, $PbCl_2$ doping optimizes the electrical transport properties (power factor). Selected metal sulfides Bi_2S_3, Sb_2S_3, SrS, and CaS acted as the second phases for scattering in the PbS matrix to lower the thermal conductivity. The thermal conductivity at 723 K for pristine PbS was reduced by about 50, 52, 30, and 42% by introducing 5.0 mol% of Bi_2S_3, Sb_2S_3, SrS, and CaS as second phase additions, respectively.[100] Finally, the PbS doping with 1.0 mol% $PbCl_2$ and incorporating 1.0% Bi_2S_3 attained the ZT of n-type PbS of 1.1 at 923 K (the thermoelectric performance is seen in Figure 4.33[100,101]). Wang et al.[102] enhanced the figure of merit of n-type

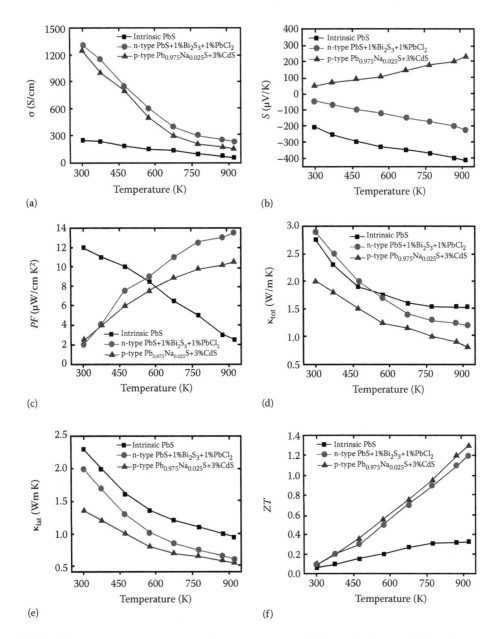

FIGURE 4.33 High thermoelectric performance of n-type PbS + 1%Bi$_2$S$_3$ + 1%PbCl$_2$ and p-type Pb$_{0.975}$Na$_{0.025}$S + 3%CdS: (a) electrical conductivity; (b) Seebeck coefficient; (c) power factor; (d) total thermal conductivity; (e) lattice thermal conductivity; (f) ZT figures as a function of temperature. (Reprinted with permission from Zhao, L. D. et al., *Journal of the American Chemical Society*, 133, 20476–20487, 2011. Copyright 2011. American Chemical Society; reprinted with permission from Zhao, L. D. et al., *Journal of the American Chemical Society*, 134(39), 16327–16336, 2012. Copyright 2012. American Chemical Society.)

PbS to 0.7 at 850 K in the doped n-type $PbS_{1-x}Cl_x(x \leq 0.008)$ approximately two times as much as the binary PbS. The ZT was predicted to be about 1 at 1000 K according to the relationship of ZT and Hall carrier density at temperatures between 500 and 1000 K.

4.4.4.2 p-Type PbS

Zhao et al.[98] also reported high TE performance in Na-doped p-type PbS with SrS and CaS nanostructuring. The lattice thermal conductivity can be reduced to as low as 0.57 W/m K at 923 K by introducing nanoscale precipitates and point defects (Na doping). As a result, the maximum ZT values reached 1.2 and 1.1 at 923 K for $Pb_{0.975}Na_{0.025}S$ with 3.0% SrS and CaS, respectively. Nanostructures indeed work for the lattice thermal conductivity reduction; however, nanostructures will also deteriorate the carrier mobility, thus in electrical transport properties. To improve the carrier mobility for nanostructuring-enhanced thermoelectric PbS, Zhao et al.[101] used the idea of band alignment between nanoprecipitates and matrix; a record high ZT of about 1.3 at 923 K was achieved for 2.5% Na-doped p-type PbS with nanostructured 3.0% CdS. The high ZT, as seen from Figure 4.33, benefited from two main factors, the combination of broad-based phonon scattering on multiple length scales to reduce the lattice thermal conductivity and favorable charge transport through coherent interfaces between the PbS matrix and CdS. The valence band offsets between PbS and CdS are at 0.13 eV. The small offsets allow facile hole transport through the material by presenting small energy barriers for carrier crossing from matrix to nanostructure and vice versa. This work confirmed that bulk p-type PbS with endotaxial placement of judiciously chosen second phases to engineer proper electronic band alignment was a good method to obtain better thermoelectric performance. Parker et al.[103] made estimates for the ZT at 1000 and 1200 K as a function of doping level, for both p-type and n-type. They found that the potential maximum ZT values could be 1.5 for both p-type and n-type PbS. Therefore, there is a promotion potential for PbTe as high performance thermoelectrics.

4.4.5 Summary and Outlook

The intrinsic PbS shares the same crystal structure and rock-salt structure with other lead chalcogenides (PbSe and PbTe), and similar two valence band structure. PbTe has been studied for years, and its ZT figure has been enhanced a lot. This makes lead chalcogenides promising high-performance thermoelectric materials. Due to the high electrical resistivity and thermal conductivity, the ZT of intrinsic PbS only achieve 0.3 in 750 K. Solid solutions, doping, and nanostructuring could introduce disorders to decrease the thermal conductivity. Tuning the carrier concentration and band structure engineering changes the electrical transport properties and optimize the electrical performance. As a result, some exciting research results about PbS thermoelectric performance already appeared. Zhao et al.[11] reported that a high level of Na-doped p-type PbS had ZT values as high as 1.2 and 1.1 at 923 K for nominal $Pb_{0.975}Na_{0.025}S$ with 3.0% SrS and CaS, respectively.

Therefore, with proper means of enhancing the ZT values, PbS has the potential of ZT enhancement and could become comparable to PbTe and PbSe. Another advantage of PbS over the two other lead chalcogenides is its highest melting point and lowest vapor pressure and thus a possible higher working temperature.[99] Combined with economic cost considerations, the low-cost sulfur, the possible higher operation temperature, and the enhanced ZT figures, these advantages can make PbS one of the excellent thermoelectric materials and applications.

In n-type and p-type PbS, both nanoscale precipitates and point defects play an important role in reducing the lattice thermal conductivity. The contribution from nanoscale precipitates of metal sulfides could significantly reduce lattice thermal conductivities. The lowest lattice thermal conductivity for the best performing PbS samples reported is ~0.57 W/m K at 923 K, which is the lowest value ever reported for PbS. This is still higher, however, than the "minimal lattice thermal conductivity" value of ~0.36 W/m K for bulk PbTe calculated by Koh et al.[104] Therefore, even better thermoelectric performance from PbS can be expected by (1) using more effective methods for reducing the lattice thermal conductivity and (2) increasing the Seebeck coefficient and the power factor through the incorporation of elements creating electronic band structure modifications.[105] With both n- and p-type PbS-based thermoelectrics reaching ZT_{max} of at least ~1.1 and 1.2, respectively, at ~925 K, the path is opened for contemplating thermoelectric generators made of these truly inexpensive materials.

4.5 Thermoelectric Properties of Lead-Free SnTe-Based Materials

Qian Zhang

4.5.1 Introduction

Tin telluride (SnTe), a lead-free rock salt IV-VI narrow bandgap semiconductor, was always used to alloy with other tellurides for better TE properties.[106–117] The pristine SnTe has not been favorably considered as a good thermoelectric material because of its low ZT due to the relatively low Seebeck coefficient and high electronic thermal conductivity caused by the very large number of intrinsic Sn vacancies.[118–121] However, the similarity of the electronic band structure of SnTe (the two valence bands contribute to the hole DOSs) with that of PbTe and PbSe suggests that it has the potential to be a good thermoelectric material.[122–126] Recently, with the contribution from the resonant states created close to the Fermi level in SnTe,[127–129] and by effective band convergence,[130–136] great improvement has been achieved in SnTe-based materials. Considering the environmental issue, lead-free SnTe would be a possible substitute for lead chalcogenides for solid-state waste heat recovery if the ZT could be optimized close to the lead-containing chalcogenides. In this section, we will talk about the nonstoichiometry and

Sn vacancies, discuss the resonant states, review the successful band convergence, and present the nanostructure studies in this system hopefully to instruct the further optimization.

4.5.2 Nonstoichiometry and Sn Vacancies

SnTe has ionic–covalent–metallic chemical bonds due to the difference between the electronegativity of Sn and Te atoms, atomic p-orbitals, and the existence of a metallic component caused by nonstoichiometry.[120] The formation energy of the anion vacancy is significantly higher than that for the cation vacancy, so the predominant defects in SnTe are cation Sn vacancies. Figure 4.34[118,120,137–142] shows the phase diagram of the Sn–Te system. The homogeneity region is at 50.1–50.9 atm.% Te ($Sn_{0.996}Te$–$Sn_{0.965}Te$) and the maximum in the melting curves corresponds to 50.4 atm.% Te ($Sn_{0.984}Te$). Then, the possibility of the formation of vacancy superstructure $Sn_{0.984}Te$ is suggested.[120] These intrinsic Sn vacancies always cause

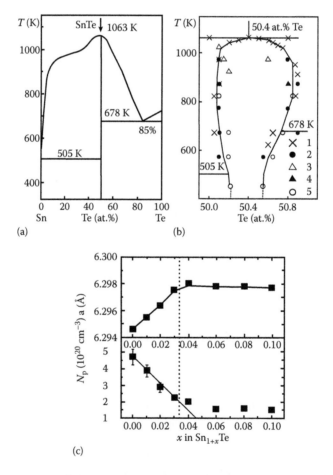

FIGURE 4.34 (a) Sn–Te system phase diagram,[120] and (b) homogeneity region of SnTe compound.[120] Data 1,[118,137,138] data 2,[139] data 3,[140] data 4,[141] data 5[142] (c) lattice parameters and carrier concentration as the function of *x* at room temperature. (Reprinted with permission from Tan, G. J. et al., *Journal of the American Chemical Society*, 136, 7006, 2014. Copyright 2014. American Chemical Society.)

p-type conductivity with high hole concentration increases with increasing Te content from $N_p = 2 \times 10^{20}$ cm^{-3} up to $N_p = 1.8 \times 10^{21}$ cm^{-3} at 300 K, which leads to the low Seebeck coefficient and high electronic thermal conductivity, making SnTe a poor thermoelectric material.[120,122] We must notice that the presence of vacancy point defects will also cause geometrical distortion of the crystal lattice, change the character of lattice vibrations and the phonon spectrum of crystal with an increase in the anharmonicity of atom vibrations, and affect the mechanical properties of crystals. So more studies are needed to clearly make and balance the affection of the Sn vacancies. Figure 4.34c presents a Sn self-compensation study with x changing from 0 to 0.1 in Sn$_{1+x}$Te. The hole carrier concentration decreased with increasing Sn content to a Sn self-compensation limit at about $x = 0.3$, resulting in increased thermoelectric properties (not shown here).[130] However, compared with other system, more enhancements are expected.

4.5.3 Resonant States in SnTe-Based Materials

One way to increase the Seebeck coefficient is to create the resonant states in SnTe. In 1984, Bushmarina et al.[143] found that the transport behaviors of In-doped SnTe were related to resonant scattering, similar to Tl-doped PbTe.[6] Recently, Zhang et al.[127] prepared In-doped SnTe by BM and HP. The room-temperature Pisarenko relations [Seebeck coefficient versus carrier concentration (Figure 4.35a)] showed a deviation of In-doped SnTe from the valence band model (VBM). This model is a two band $\mathbf{k} \cdot \mathbf{p}$ model describing the light nonparabolic hole band ($m_{\mathrm{lh}} = 0.168$) around the L-point combined with a parabolic heavy hole band ($m_{\mathrm{hh}} = 1.92$) lying ΔE ($\Delta E = 0.35$ eV) below the valence band edge.[122,124,144] The L-point energy gap E_g is 0.18 eV.[145,146] The Seebeck coefficient for the nonparabolic light hole band at the L-point S_{lh} is

$$S_{\mathrm{lh}} = \frac{k_{\mathrm{B}}}{e} \left[\frac{{}^1F^1_{-2}(\eta,\alpha)}{{}^0F^1_{-2}(\eta,\alpha)} - \eta \right], \tag{4.4}$$

where k_{B} is the Boltzmann constant, e is the electron charge, ${}^nF^m_k$ is the generalized Fermi function, η is the reduced chemical potential, and $\alpha = k_{\mathrm{B}}T/E_g$ is the nonparabolicity parameter, and we assumed that the deformation potential scattering of holes by acoustic phonons is the main carrier scattering mechanism.[122,124,144] The light hole carrier concentration p_{lh} is

$$P_{\mathrm{lh}} = \frac{1}{3\pi^2} \left(\frac{2m^*_{\mathrm{lh}}k_{\mathrm{B}}T}{\hbar^2} \right)^{3/2} {}^0F^{3/2}_0(\eta,\alpha), \tag{4.5}$$

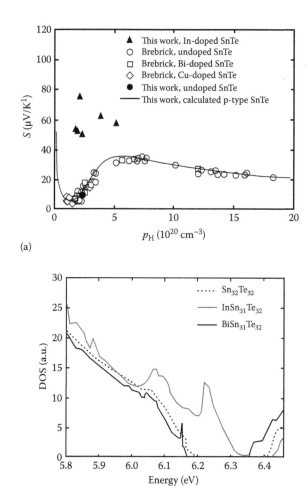

(a)

(b)

FIGURE 4.35 (a) Room-temperature Pisarenko plot for ball milled and hot pressed In$_x$Sn$_{1-x}$Te ($x =$ 0, 0.001, 0.0015, 0.0025, 0.005, 0.0075, and 0.01) in comparison with reported data on undoped SnTe, Bi-doped SnTe, and Cu-doped SnTe. Solid curve is based on VBM. (b) Comparison of DOS for undoped SnTe, Bi-doped SnTe, and In-doped SnTe. The Fermi level in the simulation resides at 6.207 eV, slightly below the DOS hump. (Zhang, Q. et al., *Proceedings of the National Academy of Sciences USA*, 110, 13261, 2013. Copyright 2013 National Academy of Sciences, USA.)

where ℏ is Planck's constant, m_{lh}^* is the light hole DOS effective mass. The Seebeck coefficient and carrier concentration of the parabolic heavy hole band are[144]

$$S_{hh} = \frac{k_B}{e}\left[\frac{{}^1F_{-2}^1(\eta - \Delta_v, 0)}{{}^0F_{-2}^1(\eta - \Delta_v, 0)} - (\eta - \Delta_v)\right], \qquad (4.6)$$

$$P_{hh} = \frac{1}{3\pi^2}\left(\frac{2m_{hh}^* k_B T}{\hbar^2}\right)^{3/2} {}^0F_0^{3/2}(\eta - \Delta_v, 0). \qquad (4.7)$$

Here m_{hh}^{*} is the heavy hole DOS effective mass and $\Delta_V = \Delta E/(k_B T)$. The total Seebeck coefficient from both hole bands S_{total} is taken to be[144]

$$S_{total} = (\sigma_{lh} S_{lh} + \sigma_{hh} S_{hh})/(\sigma_{lh} + \sigma_{hh}). \tag{4.8}$$

The electrical conductivity of the lh band is σ_{lh}

$$\sigma_{lh} = \frac{C}{m_{lh} D_{lh}^2} {}^0F_{-2}^1(\eta, \alpha), \tag{4.9}$$

where D_{lh} is the lh acoustic deformation potential and C is a constant. The heavy hole conductivity σ_{hh} is obtained from Equation 4.7 by replacing D_{lh} with D_{hh}, the hh acoustic deformation potential, and by setting $\alpha = 0$ and replacing η with $\eta - \Delta_v$. Using these equations gives[144]

$$S_{total} = \frac{k_B}{e} \frac{\xi\left({}^1F_{-2}^1(\eta, \alpha) - \eta {}^0F_{-2}^1(\eta, \alpha)\right) + \left({}^1F_{-2}^1(\eta - \eta_v, 0) - (\eta - \eta_v){}^0F_{-2}^1(\eta - \eta_v, 0)\right)}{\xi {}^0F_{-2}^1(\eta, \alpha) + {}^0F_{-2}^1(\eta - \eta_v, 0)} \tag{4.10}$$

where $\xi = m_{hh} D_{hh}^2 / m_{lh} D_{lh}^2$. We take the ratio $D_{hh}/D_{lh} = 0.5$. For a single light hole band, the Hall concentration p_H is related to the actual $p_{lh,H} = p_{lh}/A_{lh}$, where A_{lh} is the Hall factor for the lh band. The total Hall carrier concentration for a two-band system p_H is related to the carrier concentrations in each band p_{lh} and p_{hh}, as[144]

$$p_H = [b p_{lh} + p_{hh}]^2 / [A_{lh} b^2 p_{lh} + A_{hh} p_{hh}], \tag{4.11}$$

where A_{hh} is the Hall factor for the hh band. The parameter b is the mobility ratio of the lh *and* hh bands. Here, we take $b = 4$, which has been previously used for PbTe. The results are relatively insensitive to the choice of b.[127]

The VBM provides a quantitative fit to all the obtained undoped SnTe with different hole concentrations ($2 \times 10^{20} - 1.8 \times 10^{21}$ cm^{-3}, caused by different Sn vacancies[120,122]) and other doped samples, such as Bi- and Cu-doped SnTe.[122] The deviation of the In-doped SnTe from the VBM implies a possible mechanism, which was further studied by first-principles DOS calculation[127] (Figure 4.35b). By comparing the DOS curves of undoped SnTe, Bi-doped SnTe and In-doped SnTe near the top of the valance band (calculated with the same supercell size), a well-defined peak is observed in the DOS of In-doped SnTe that can contribute to the large deviation of the Seebeck coefficient from the VBM. Due to the limitation of computing resources, the current simulated supercell configuration corresponds to 3 atm.% In concentration, higher than that achieved in this experiment. Although an alternative simulation method is required for more dilute doping concentrations, the rich features introduced by In atoms are speculated to help increase the thermoelectric properties.

By using melting and the SPS method, Tan et al.[128] prepared $(SnTe)_{3-3x}(In_2Te_3)_x$ (x = 0, 0.02, 0.04, 0.06, 0.08, and 0.10, in mole ratio) and $(SnTe_{1-y}I_y)_{2.88}(In_2Te_{3-3y}I_{3y})_{0.04}$ (y = 0.0.2%, 0.4%, 0.6%, 0.8%, and 1%, in mole ratio) and found a solubility limit of <6 mol% for In_2Te_3 in SnTe. With a higher In concentration, similar deviation could be obtained (see Figure 4.36a), which again well proved the DOS calculation in Figure 4.35b. To optimize the carrier concentration of best performing sample of $(SnTe)_{2.88}(In_2Te_3)_{0.04}$, an n-type dopant iodine substituted Te and reduced the hole carrier densities. As shown in Figure 4.36b, all the samples still have much higher Seebeck coefficients than the predicted data by VBM, suggesting the still presence of resonant levels.

Figure 4.37 shows room-temperature Pisarenko plot (Figure 4.37a) and calculated DOS near the top valence band for SnTe codoped with In and Cd (Figure 4.37b).[129] The Seebeck coefficients of codoped samples are much higher than undoped and Cd-doped SnTe, comparable to those of In-doped SnTe. These results are consistent with the calculated DOS plots,[129] which presents well-defined peaks near the Fermi level inside the valence bands of both In-doped SnTe and In and Cd-codoped SnTe. However, these peaks are absent in undoped SnTe and Cd-doped SnTe. So we can conclude that the resonant levels are only associated with In doping, not Cd doping, but are still present when incorporating Cd atoms together with In in the lattice. Actually, the Cd-doped SnTe samples also have higher Seebeck coefficients than predicted by VBM (see triangles in Figure 4.37a). This enhancement is associated with the band convergence and increases with increasing Cd concentration. The introduction of Cd in SnTe decreased the energy separation between the two valence bands, leading to band convergence. We will

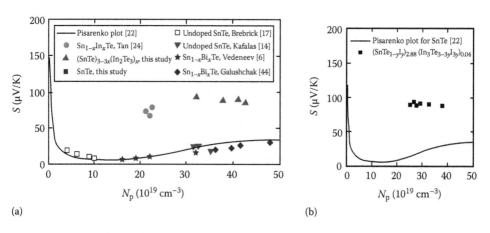

(a)   (b)

FIGURE 4.36 (a) Room-temperature Pisarenko plot for $(SnTe)_{3-3x}(In_2Te_3)_x$ in this study. The data reported on In- and Bi-doped and nondoped SnTe are also included for comparison. (b) Room-temperature Pisarenko plot for $(SnTe_{1-y}I_y)_{2.88}(In_2Te_{3-3y}I_{3y})_{0.04}$. The solid line is a theoretical Pisarenko plot for SnTe based on VBM. (Reprinted with permission from Tan, G. J. et al., *Chemistry of Materials*, 27, 7801, 2015. Copyright 2015. American Chemical Society.)

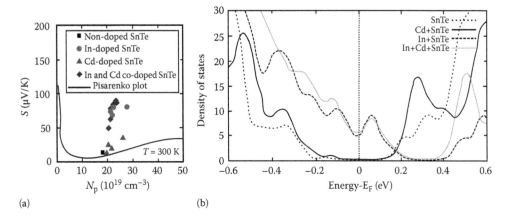

FIGURE 4.37 (a) Room-temperature Pisarenko plot for SnTe with different dopants. The solid line is a Pisarenko plot calculated based on VBM. (b) DOS near the top valence band. (Reprinted with permission from Tan, G. J. et al., *Journal of the American Chemical Society*, 137, 5100–5112, 2015. Copyright 2015. American Chemical Society.)

discuss the details about the two intrinsic valence bands in SnTe and effective band convergence in the next section.

Creating resonant states is an effective way to enhance the thermoelectric properties of materials. By increasing the Seebeck coefficient, the ZT values increased. However, there are still drawbacks that we should consider and resolve: (1) The resonant doping sharply increases the hole effective mass, which can strongly scatter the carriers and significantly decrease the carrier mobility. (2) Normally, resonant states are more beneficial to the thermoelectric properties at room temperature. The high Seebeck coefficient at a higher temperature is more attributed to the two valence bands of SnTe. From the previous study, it is really good to see the coexistence of the resonant states with other doping, which shows that it is possible to combine resonant doping with other methods for the enhancement of the thermoelectric properties. Figure 4.38 presents the temperature dependence of ZTs for undoped SnTe,[127] In-doped SnTe,[127] SnTe-In$_2$Te$_3$,[128] SnTe–In$_2$Te$_3$–I,[128] and Cd-doped SnTe[129] mentioned earlier. Both the peak ZT and the average ZT were increased by resonant doping. The highest ZT value increased to ~1.1 at about 873 K with only In doping and ~1.4 at about 900 K with In and Cd codoping.

4.5.4 Band Convergence in SnTe-Based Materials

Similar to PbTe and PbSe,[144] there are also two valence bands in SnTe [one light hole band at L-point and one heavy hole band at Σ-point (see Figure 4.39)]. So the other way to increase the Seebeck coefficient is to realize the band convergence in SnTe materials system. Figure 4.40 shows the schematic diagram of the electronic structure of SnTe, PbSe, and PbTe near the Fermi level to show the relative positions of the CB, light hole valence band (VB$_L$), and heavy hole valence band (VB$_Σ$). It is obvious to see two major differences among three systems, resulting in the different TE properties in the

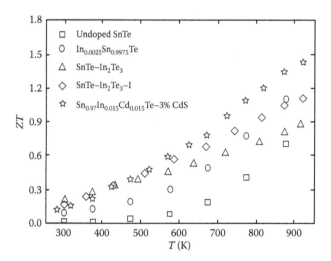

FIGURE 4.38 Temperature dependence of ZTs for undoped SnTe,[127] $In_{0.0025}Sn_{0.9975}Te$,[127] SnTe–In_2Te_3,[128] SnTe–In_2Te_3–I,[128] and $Sn_{0.97}In_{0.015}Cd_{0.015}Te$–3% CdS. (Reprinted with permission from Tan, G. J. et al., *Journal of the American Chemical Society*, 137, 5100–5112, 2015. Copyright 2015. American Chemical Society.)

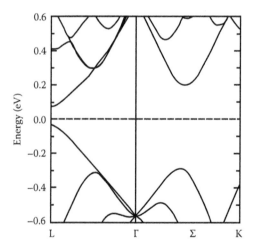

FIGURE 4.39 Calculated band structure for SnTe. (Reprinted from *Journal of Materiomics*, 1, Li, W. et al., Band and scattering tuning for high performance thermoelectric $Sn_{1-x}Mn_xTe$ alloys, 307–315, Copyright 2015, with permission from Elsevier.)

three systems, which are demonstrated in Figure 4.41 predicted by the same model.[127] (1) The smaller L-point energy gap of SnTe makes the nonparabolicity larger and the Seebeck drop faster with increasing concentration. (2) The separation between the light hole and heavy hole band edges in SnTe system is estimated to be ~0.35 eV, larger than those of PbTe (~0.12 eV) or PbSe (~0.26 eV), making the benefit of the heavier band for the Seebeck coefficient less significant. By comparing the predictions from VBM and the TBK model (which ignores the heavy hole band contribution), the

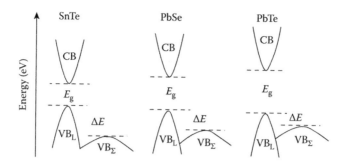

FIGURE 4.40 Schematic diagram of the electronic structure of SnTe, PbSe, and PbTe near the Fermi level to show the relative positions of the CB, light hole valence band (VB$_L$), and heavy hole valence band (VB$_\Sigma$). E_g is the bandgap. ΔE is the light hole–heavy hole band edge energy difference.

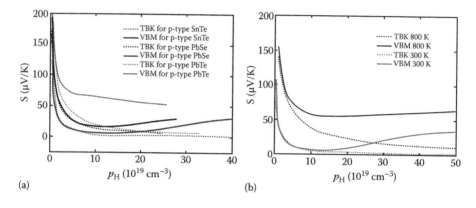

FIGURE 4.41 (a) Room-temperature Pisarenko plots for SnTe (*gray*), PbSe (*black*), and PbTe (*light gray*) modeled by TBK (*dashed line*) and VBM (*solid line*). (b) Room-temperature (*gray*) and 800 K (*black*) Pisarenko plots for SnTe modeled by TBK (*dashed line*) and VBM (*solid line*). In both cases, we take m_{lh} = 0.168 m_e for the light holes and m_{hh} = 1.92 m_e for heavy holes; ΔE = 0.35 eV. (Zhang, Q. et al., *Proceedings of the National Academy of Sciences USA*, 110, 13261, 2013. Copyright 2013 National Academy of Sciences, USA.)

contribution from the heavy hole band happens at a higher carrier concentration and higher temperature in SnTe system.[127]

To better utilize the heavy hole band, several kinds of Sn$_{1-x}$A$_x$Te (A = Hg,[132] Cd,[129,130] Mg,[131] and Mn[133–136]) alloys with enlarged E_g and decreased ΔE (as shown in Figure 4.42) were prepared, making the band convergence easier. Room-temperature Pisarenko relations for optimized Sn$_{1-x}$A$_x$Te samples are presented in Figure 4.43a. All the data deviated from the VBM of SnTe, dropping onto the upper positions with engineered band structures. The Seebeck coefficient dramatically increased at the same carrier concentration, similar with the effect from resonant doping. However, compared with the resonant doping, this method is superior in some aspects. (1) The enlarged bandgap suppressed the bipolar effect. (2) The optimized band structure is beneficial to increase the

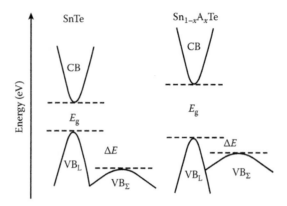

FIGURE 4.42 Schematic diagram of the electronic structure of SnTe and $Sn_{1-x}A_xTe$ near the Fermi level to show the relative positions of the CB, light hole valence band (VB_L), and heavy hole valence band (VB_Σ). E_g is the bandgap. ΔE is the light hole–heavy hole band edge energy difference.

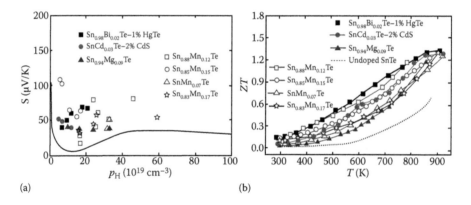

FIGURE 4.43 (a) Room-temperature Pisarenko plot for SnTe samples with band convergence.[130–136] (b) Temperature dependence of ZT for SnTe samples with band convergence.[130–136] The undoped SnTe is presented for comparison. (Zhang, Q. et al., *Proceedings of the National Academy of Sciences USA*, 110, 13261, 2013. Copyright 2013 National Academy of Sciences, USA.)

Seebeck coefficient in the whole temperature range. (3) There is no Hall mobility loss due to the strong resonant scattering. (4) The point defects generated by alloying can help decrease the lattice thermal conductivity. In Figure 4.43b, temperature dependences of ZTs for SnTe samples with band convergence are shown. Both the peak ZT and the average ZT were increased by band convergence. The highest ZT value is ~1.35 at about 910 K for sample alloyed with 1 atm.% Hg.[132] More studies focused on $Sn_{1-x}Mn_xTe$ because of the large solid solution limit of MnTe in SnTe.[133–136] Actually, each optimization has included more strategies, such as resonant doping, band convergence, normal doping for tuning the carrier concentration, defect engineering, nanoengineering, and others. We will present the relative studies on thermal conductivity reduction in the next section.

4.5.5 Thermal Conductivity Reduction of SnTe-Based Materials

Both the electronic thermal conductivity and the lattice thermal conductivity are high in SnTe system. The high electronic part is induced by high intrinsic Sn vacancies, which has been reduced a lot based on the resonant doping and band convergence. However, the lattice thermal conductivity of SnTe is much higher compared with the theoretical lowest lattice thermal conductivity k_{min} (~0.5 W/m K calculated using the model proposed by Cahill et al.[147]). When alloying with HgTe, CdTe, MgTe, or MnTe to engineer the band structure, the alloy defect scattering could effectively decrease the lattice part. MnTe has a large solid solution limit in SnTe (Figure 4.44a), so we show the results of $Sn_{1-x}Mn_xTe$ in this section.[136] Figure 4.44b shows the temperature-dependent total thermal conductivity (κ) and lattice thermal conductivity (κ_L) of $Sn_{1-x}Mn_xTe$. The lowest κ_L in this work is ~0.8 W/m K for $Sn_{0.85}Mn_{0.15}Te$, which can be well predicted by the Debyee–Callaway model (gray solid line) considering the pure alloying effect,[136] indicating the possible room for further reduction by other method. The atomic radium and mass difference between alloying atom and Sn could decide the strength of the scattering.

Then how is the nanoengineering in this system? By BM and HP, single-phased bulk SnTe sample containing nanograins has been prepared.[127] With increasing BM time, the average grain size decreased and the lattice thermal conductivity also decreased. Figure 4.45a shows that the optimized $In_{0.0025}Sn_{0.9975}Te$ samples consist of big grains with diameters of several tens of microns, small grains with diameters of ~1 μm, and nanograins with sizes of around 100 nm.[127] The increased density of grain boundaries effectively lowered the lattice thermal conductivity (as presented in Figure 4.45b). All the grains, whether in microns or nanometers, are single crystals with clean boundaries and good crystallinity, being beneficial to the transport of charge

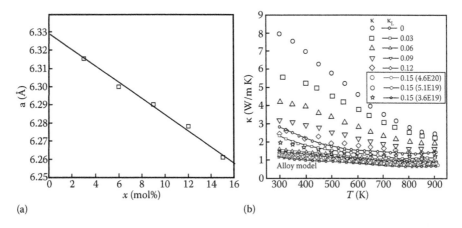

(a)   (b)

FIGURE 4.44 (a) Lattice parameters for $Sn_{1-x}Mn_xTe$ (0 < x < 0.18). (b) Temperature-dependent total thermal conductivity (κ) and lattice thermal conductivity (κ_L) of $Sn_{1-x}Mn_xTe$. The lattice thermal conductivity reduction due to alloy defects scattering can be well predicted by the Debyee–Callaway model (gray solid line). (Reprinted from *Journal of Materiomics*, 1, Li, W. et al., Band and scattering tuning for high performance thermoelectric $Sn_{1-x}Mn_xTe$ alloys, 307–315, Copyright 2015, with permission from Elsevier.)

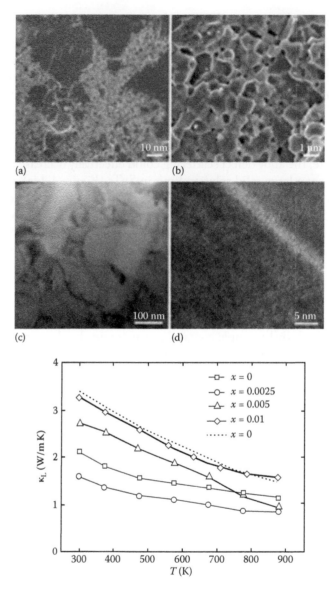

FIGURE 4.45 *Top*: representative (a and b) SEM, (c) TEM, and (d) HRTEM images for as-prepared $In_{0.0025}Sn_{0.9975}Te$ samples by BM and HP. *Bottom*: temperature dependence of lattice thermal conductivity for $In_xSn_{1-x}Te$ ($x = 0$, 0.0025, 0.005, and 0.01). The undoped SnTe prepared by melting and HP is shown by the broken line. (Zhang, Q. et al., *Proceedings of the National Academy of Sciences USA*, 110, 13261, 2013. Copyright 2013 National Academy of Sciences, USA.)

carriers, without degrading the electronic properties. This nanostructure has been combined with resonant doping for the optimization of the thermoelectric properties. Even lower lattice thermal conductivity has been obtained in bulk SnTe-based alloys with nanoprecipitate. Due to the limited solubility and a well-behaved nucleation and growth mechanism, CdS was selected as the second phase in Cd-doped SnTe and In and Cd-codoped SnTe.[129,130] Most nanoscale precipitates range in size from 3 to 4 nm, as shown in Figure 4.46 (top panel).

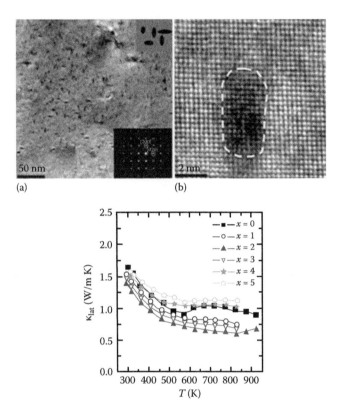

FIGURE 4.46 *Top*: (a) low-magnification TEM image of $SnCd_{0.03}Te$–2%CdS. The inset of the upper panel schematically shows the precipitates (*dark color*) embedded in the matrix (*gray*). The inset of the lower panel is the SAD pattern taken along the [001] direction. (b) HRTEM image of one nanoscale precipitate highlighted by the dashed white line. *Bottom*: lattice thermal conductivity as a function of temperature for $SnCd_{0.03}Te$–x%CdS. (Reprinted with permission from Tan, G. J. et al., *Chemistry of Materials*, 27, 7801, 2015. Copyright 2015. American Chemical Society.)

The lowest lattice thermal conductivity is ~0.58 W/m K for $SnCd_{0.03}Te$–2 atm.%CdS (Figure 4.46 [bottom panel]). This structure was combined with band convergence, resonant doping, and point defect scattering for the high thermoelectric properties. Another multiple phonon scattering in this system has been realized in the composition $Sn_{0.81}Mn_{0.19}Te$ with band convergence.[135] Different kinds of the nanostructures were observed in the bulk samples featured in nanoprecipitate (Figure 4.47a, b), MnTe second phases in the SnTe matrix as laminates (Figure 4.47c, d), and stacking faults inside MnTe laminates (Figure 4.47e, f). All these microstructures along with the mesoscale grains in these samples construct a multiscale hierarchical architecture that can scatter phonons with a broad range of different wavelengths and mean free paths, leading to the very low lattice thermal conductivity close to the theoretical limit.

Considering the high thermal conductivity, SnTe nanowires were grown by thermal vapor deposition via a vapor–liquid–solid process, and the thermoelectric properties (25–300 K) of nanowires with different diameters from ~218 to ~913 nm were studied.[148] The measurements were carried out

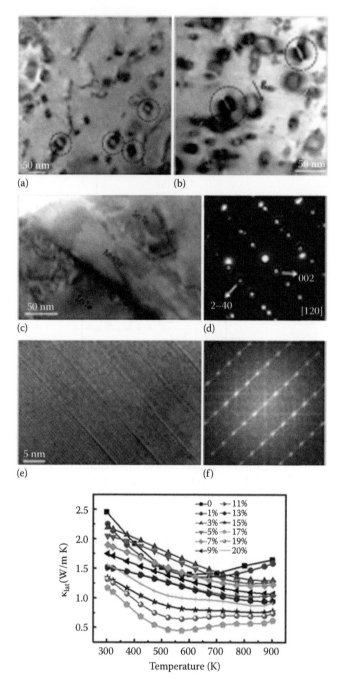

FIGURE 4.47 *Top*: microstructure of $Sn_{0.81}Mn_{0.19}Te$. (a) TEM image, (b) bright-field scanning transmission electron microscope (STEM) image show a high density of nanoscale precipitates, (c) TEM image showing one MnTe laminate inserted into the SnTe matrix and (d) electron diffraction pattern; the SnTe matrix is along the 120 zone axis. (e) HRTEM image of MnTe laminates showing a hierarchical layered structure and (f) FFT image showing diffuse spots along [001] between the main reflection spots. *Bottom*: lattice thermal conductivity as a function of temperature for $Sn_{1-x}Mn_xTe$. (Wu, H. J. et al., *Energy and Environmental Science*, 8, 2015, 3298. Reproduced by permission of The Royal Society of Chemistry.)

FIGURE 4.48 *Left*: (a) high-resolution TEM image of a SnTe nanowire. The inset is an FFT image. (b) Scanning TEM image of a SnTe nanowire. EDX mapping of (c) Sn and (d) Te of the nanowire segment is shown in b. *Right*: (e) SEM image of the platform used for thermoelectric measurement. (f) Magnified SEM image shows the nanowire device in the center of the platform. The Ti/Au electrodes wrap over the SnTe nanowire for both good electrical and thermal contacts. (Xu, E. Z. et al., *Nanoscale*, 7, 2869, 2015. Reproduced by permission of The Royal Society of Chemistry.)

on the single nanowire, which has been transferred onto the thermoelectric platform with good thermal and electrical contacts (see Figure 4.48). The thermal conductivities of the nanowires are found to be lower than that of bulk SnTe with a similar carrier density. The Seebeck coefficient increases with the decrease of nanowire diameter and reaches ~41 μV/K at 300 K for the thinnest nanowire. This study provides another method to enhance the thermoelectric properties of SnTe-based materials.

4.5.6 Conclusion

Different strategies have been synergistically tried to optimize the thermoelectric properties of lead-free SnTe-based materials, which have changed from poor thermoelectric materials to high-performance materials with figure of merits close to those of lead chalcogenides. With the development of preparation and measurement techniques, further enhancements are believed in all the thermoelectric systems, including SnTe. We must also consider the stability and the TE property repeatability of the materials for the real application, which has been experimentally proven in SnTe system with no significant hysteresis between the heating and cooling curves of high-temperature (300–663 K) Young's modulus and shear modulus measurements, indicating the absence of any microcracking or bloating.[134] So in conclusion, SnTe-based materials are promising to be used in cheap, environmentally friendly, and effective thermoelectric generators.

4.6 SnSe

Cheng Chang and Li-Dong Zhao

The well-known interdependence of Seebeck coefficient S, electrical conductivity σ, and thermal conductivity κ complicates the efforts in developing strategies for improving a material's average ZT well above 2.5, especially using less expensive, more earth-abundant materials,[149–151] a feat that could revolutionize the field of thermal energy conversion. Several approaches to enhance ZT have emerged in the last decade including modifying the band structure,[152] heavy valence (conduction) band convergence,[153,154] quantum confinement effects,[155] and electron energy barrier filtering[156] to enhance Seebeck coefficients, nanostructuring,[47] all-scale hierarchical architecturing,[157] and band energy alignment between nanoprecipitate/matrix to maintain hole mobility.[158–160] Most of these approaches aim to maintain a high power factor and/or reduce the lattice thermal conductivity. Alternatively, one can seek high performance in pristine thermoelectric materials with intrinsically low thermal conductivity, which may arise from a large molecular weight,[161] a complex crystal structure,[162] and charge density wave distortions,[163] among other strategies. In this section, one typical thermoelectric candidate, SnSe, will be briefly introduced.

4.6.1 Crystal Structure

Tin selenide crystallizes in a layered orthorhombic crystal structure with lattice parameters a = 11.49 Å, b = 4.44 Å, c = 4.135 Å,[164] which can be considered as layered distorted rock salt structure. Figure 4.49 shows the

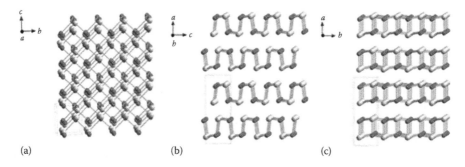

(a)   (b)   (c)

FIGURE 4.49 (a) Crystal structure along the *a* axis: *large*, Sn atoms; *small*, Se atoms. (b) Structure along the *b* axis. (c) Structure along the *c* axis.

perspective views of the room-temperature SnSe crystal structure along the *a*, *b*, and *c* axial directions. The unit cell with eight atoms consists of double two atom-thick layers, which form the zigzag structure. Each layer can be viewed as consisting of six-membered rings in chair conformation.[165] The Sn and Se atoms are bonded with strong covalent bonds within the layers which are perpendicular to the *c* axis, while atoms in different layers are linked with weak force along the *c* axis.[166] Due to the highly layered structure, tin selenide cleave easily along the (001) plane. The structure contains highly distorted $SnSe_7$ coordination polyhedrons, which consists of three short Sn–Se bonds and four long Sn–Se bonds and a lone pair of the Sn^{2+} sterically accommodated between the four long Sn–Se bonds.[164,167] The previous investigations indicate that there is a displacive reversible phase transition in SnSe from the orthorhombic structure (*Pnma*) to the orthorhombic structure (*Cmcm*) at ~750–800 K.[164,168,169] This phase transition can also be observed in SnSe bulks in the constant high pressure environment (~7 GPa)[170] and SnSe nanocrystals at room temperature.[171]

4.6.2 Synthesis and Fabrication

Most conventional synthesis methods in thermoelectrics such as melting method, high-energy milling, and solvothermal method are used to prepare polycrystalline materials. However, recent researches reveal that single-crystalline SnSe outperforms polycrystalline SnSe and show high thermoelectric performance along the *b* axis. Here, we will introduce some synthesis methods of single-crystalline SnSe.

4.6.2.1 Bridgman–Stockbarger Technique

The high-purity elements Sn and Se were mixed in stoichiometric proportion. The mixed powder was then placed in a quartz ampoule, evacuated, and sealed. This charged quartz tube was placed into another bigger, quartz tube, evacuated and flame-sealed. The outer tube is used to prevent the crystal from oxidation by air because the inner tube breaks, owing to the considerable difference in thermal expansion between the crystal and quartz. To get the single crystal, a single crystal furnace with appropriate temperature profile over the length of the ampoule is needed. The mixed powder was slowly heated to 1223 K for 10 h (the melting temperature of SnSe is 1134 K) and soaked at this temperature for 6 h in order to obtain complete reaction between the two components. Then, a mechanical system was used to pull the ampoule through the temperature gradient zone at a very slow speed (~2 mm/h); near fully dense samples are obtained after carefully breaking the ampoule.[164,172,173]

4.6.2.2 Vapor Transport Technique

For the direct vapor transport growth, a special quartz ampoule with a charge end and a growth end, which has reduced diameter, was prepared. The prepared powder was transferred into the special quartz ampoule and was sealed under vacuum. The ampoule was then set in a horizontal furnace to be heated. There was a temperature gradient between the charge end and growth end. The temperature at the charge end and growth end was determined by thermal gravimetric analysis (TGA). SnSe grew in the form of single crystal at the growth end. The ampoule was maintained at this temperature for a period of 7 days. The ampoule was then slowly cooled and brought to room temperature. SnSe single crystals are achieved after breaking the ampoule.[174–176]

4.6.3 SnSe Single Crystals

4.6.3.1 Electrical Transport Properties

The electrical conductivities for pristine SnSe crystals along different crystallographic directions show the same temperature-dependent trend (Figure 4.50a). The first upturn of the electrical conductivity above 525 K is attributed to the thermal excitation of carriers, while the second is related to the phase transition from *Pnma* to the *Cmcm* space group.[171,177] Furthermore, the electrical conductivity of SnSe was increased from ~12 to >1500 S/cm with carrier density of ~4 × 10^{19} cm^{-3} at 300 K by hole doping the material, and the temperature dependence of the electrical conductivity was changed from semiconductor-like to metal-like. With rising temperature, the electrical conductivity of hole-doped SnSe (*b* axis) decreases from 1486 S/cm at 300 K to 148 S/cm at 773 K. It can be readily seen that the electrical conductivities along the *b* and *c* directions are higher than along that along the *a* direction. This anisotropy is due to the higher carrier mobility within the plane of the SnSe slabs than that perpendicular to them.

The Seebeck coefficients show almost isotropic behavior and are independent of crystallographic directions (Figure 4.50b). The gradual decrease of the Seebeck coefficients for pristine single-crystal SnSe above 525 K is consistent with the increasing trend in the electrical conductivity. This behavior is due to the fact that the bandgap (E_g) considerably decreases from *Pnma* (0.61 eV) to *Cmcm* (0.39 eV), and thus, a bipolar conduction process is expected with rising temperature. For Na-doped single-crystal SnSe, the Seebeck coefficient is ~+160 mV/K at 300 K and increases to ~+300 μV/K at 773 K. (Figure 4.50b). The Seebeck coefficient at room temperature is much higher than that for PbTe (70 μV/K), PbSe (60 μV/K), PbS (50 μV/K), and SnTe (25 μV/K) with similar carrier density. The large Seebeck coefficient is the contribution of more than one valence band, which is well supported by the Pisarenko plot (Figure 4.51a) and DFT calculations of the band structure.[31] Figure 4.51b shows the electronic band structure of hole-doped SnSe. There is a very small energy gap (~0.06 eV) between the first two valence bands, and the energy gap between the first and third band is only 0.13 eV. The electronic structure valence bands of SnSe are much more complex than those of lead chalcogenides, and the Fermi level of SnSe even approaches the fourth, fifth, and sixth valence bands for doping levels as high as 5 × 10^{20} cm^{-3}.

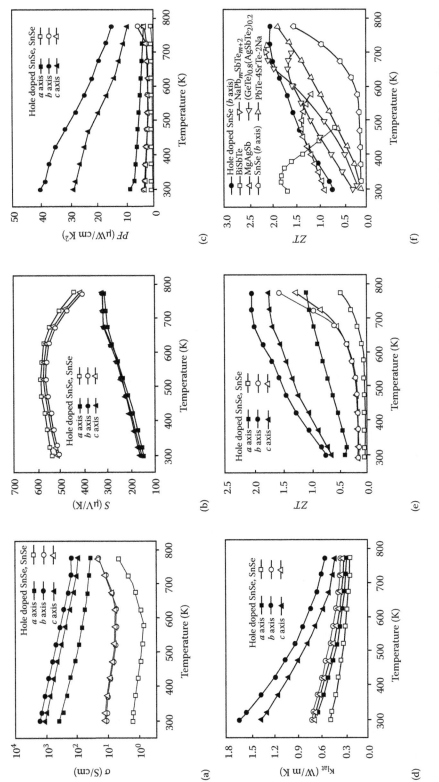

FIGURE 4.50 Thermoelectric properties as a function of temperature for SnSe crystals. (a) Electrical conductivity σ; (b) Seebeck coefficient *S*; (c) power factor *PF*; (d) total thermal conductivity κ_{tot}; (e) *ZT* values; (f) *ZT* value comparisons of hole-doped SnSe (*b* axis) and the current state-of-the-art p-type thermoelectrics.

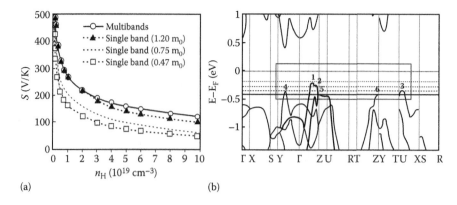

FIGURE 4.51 (a) Calculated Seebeck coefficients as a function of carrier density at 300 K; (b) electronic band structure of hole-doped SnSe.

Compared to other state-of-the-art thermoelectrics, the power factors obtained in SnSe crystals are moderate, but are much higher than in polycrystalline SnSe and those found in other thermoelectrics with intrinsically low thermal conductivity (for example, $Yb_{14}MnSb_{11}$, Ag_6TlTe_5, $AgSbTe_2$).[178–180] Furthermore, the power factors are much enhanced by hole doping. The combination of increased electrical conductivity and high Seebeck coefficient results in a power factor of ~40 μW/cm K^2 for hole-doped SnSe along the *b* axis at 300 K. The power factors remain at a high value of ~14 μW/cm K^2 around 773 K (*b* axis), which is twice as high as that for the undoped SnSe (*b* axis).

4.6.3.2 Thermal Transport Properties

The temperature dependence of the total thermal conductivity (κ_{tot}) is shown in Figure 4.50d. At room temperature, the values of κ_{tot} for pristine single-crystal SnSe are 0.46, 0.70 and 0.68 W/m K along the *a*, *b*, and *c* axis directions, respectively. These low values continue to decrease with rising temperature, the total thermal conductivity is dominated by phonon transport, and κ_{lat} falls as low as 0.30 W/m K at 773 K along the *a* direction. This is a remarkably low value, which is lower than those obtained even by nanostructuring and all-scale hierarchical architecturing of PbTe-based thermoelectric materials. The total thermal conductivity (κ_{tot}) of hole-doped SnSe is also low and shows a decreasing trend with rising temperature. The κ_{tot} of hole-doped SnSe (*b* axis) decreases from ~1.65 W/m K at 300 K to ~0.55 W/m K at 773 K. The lattice thermal conductivity (κ_{lat}) of hole-doped SnSe is as low as that of undoped SnSe.

The low thermal conductivity in SnSe derives from the anharmonicity. In a harmonic solid with three dimensions, if an atom was displaced from its equilibrium position, the force that the atom is subjected to is proportional to its displacement. And the proportionality constant, which is called spring constant, maintains constant. However, in an anharmonic solid, the spring constant does not maintain constant with the atom displacement.[181]

As a result, when atoms in an anharmonic solid are displaced by a phonon, the second phonon will run into a medium with modified elastic properties, resulting in strong interactions.[182] Thus, anharmonicity in bonding dissipates the phonon transport and reduces the thermal conductivity without affecting the electrical transport property in the anharmonic solid.[182–184] Recent research reveals that the exceptionally low lattice thermal conductivity in SnSe derives from the anharmonicity.[164] It has been recently experimentally confirmed that this anharmonicity stems from a bonding instability arising from the long-range resonant p-bond network of Se atoms, which are coupled to stereochemically active Sn 5s orbitals by inelastic neutron scattering measurement.[185]

The strength of the lattice anharmonicity can be estimated from the Grüneisen parameters. The Grüneisen parameters of SnSe along different axes are all very large. The average Grüneisen parameters along the *a*, *b*, and *c* axes are 4.1, 2.1 and 2.3, respectively. Along the *a* axis, the maximum longitudinal acoustic Grüneisen parameter around the Γ-point is extraordinarily high, ~7.2. In contrast, the Grüneisen parameters are 2.05 for $AgSbTe_2$,[183] 3.5 for $AgSbSe_2$,[182] and 1.45 for $PbTe$,[186] corresponding to measured lattice thermal conductivities at room temperature (in watts per meter Kelvin) of 0.68, 0.48, and 2.4, respectively. This indicates that a higher degree of anharmonicity leads to a lower lattice thermal conductivity. The anomalously high Grüneisen parameter of SnSe is a reflection of its crystal structure, which contains very distorted SnSe7 polyhedrons (due to the lone pair of Sn^{2+}) and a zigzag accordion-like geometry of slabs in the *b–c* plane. This implies a soft lattice—and if this lattice were mechanically stressed along the *b* and *c* directions, the Sn–Se bond length would not directly change, but instead the zigzag geometry would be deformed like a retractable spring or an accordion. In addition, along the *a* direction, the weaker bonding between SnSe slabs provides a good stress buffer or cushion, thus dissipating phonon transport laterally. The anomalously high Grüneisen parameter is therefore a consequence of the soft bonding in SnSe, which leads to the very low lattice thermal conductivity.

4.6.3.3 ZT

The combination of moderate power factor and ultralow thermal conductivity results in high *ZT* values of 2.6 and 2.3 at 923 K along the *b* and *c* directions, respectively. However, the *ZT* value at 923 K along the *a* direction is lower because of the lower electrical conductivity along the direction perpendicular to the slabs. Furthermore, the *ZT* of SnSe single crystals are significantly enhanced by using sodium as an effective acceptor. As shown in Figure 4.50e, the *ZT* value measured along the *b* axis is ~0.7 at 300 K and increases to ~2.0 at 773 K, which is much higher compared to the pristine single crystal SnSe and the current state-of-the-art thermoelectric materials (Figure 4.50f). The high power factor and *ZT* result in the highest device *ZT* (ZT_{dev}) from 300 to 773 K known in the field of thermoelectric materials, ~1.34.

4.6.4 SnSe Polycrystals

SnSe was reported to possess the outstanding thermoelectric performance in its single crystal form, benefiting mainly from the complex crystal structure with strong anharmonic and anisotropic bonding.[185,187–189] Since then, the thermoelectric community has been trying different methods to improve the *ZT* values of polycrystalline SnSe. Carrier concentration optimization is a conventional strategy to enhance *ZT*. Polycrystalline SnSe has a very low electrical conductivity compared to single-crystal SnSe along the *b* direction. Individuals in the thermoelectric community tried different dopants to enhance the electrical transport properties and improve the *ZT* values. Recent works suggest that the Ag and Na elements are effective dopants of SnSe, resulting in higher carrier mobilities and *ZT* values than pristine polycrystalline SnSe.[170,190] Band structuring engineering was also tried to enhance the electrical transport properties of polycrystalline SnSe. SnSe has larger bandgap (~0.9 eV) compared to lead chalcogenides, resulting in low electrical transport properties. In recent researches, Te and S element were doped in SnSe to modify the valence band structure and were suggested to successfully enhance the thermoelectric properties of polycrystalline SnSe.[191,192] Figure 4.52 shows the *ZT* values of polycrystalline SnSe. Compared to the pristine polycrystalline SnSe, the *ZT* values are increased to some extent, suggesting that carrier concentration and band structuring engineering are effective strategies to enhance the *ZT* values. However, the values are still lower than those of single-crystal SnSe, and more efforts should be made to improve the thermoelectric properties of polycrystalline SnSe.

The excellent thermoelectric performance of SnSe single crystals leaves a deep impression on the thermoelectric community. However, the reported *ZT* values of polycrystalline SnSe are much lower than that of single SnSe. To solve the problem, it is essential to have a deep understanding of the underlying difference in the electrical and thermal properties between them. The power factors obtained in pristine SnSe single crystals are moderate, reaching

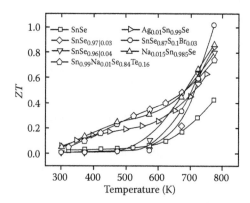

FIGURE 4.52 *ZT* value comparisons of pristine and doped polycrystalline SnSe.

10.1 µW/cm K^2 at 850 K along the *b* direction. Furthermore, the power factor is much enhanced to the value of ~40 µW/cm K^2 at room temperature and remain at a high value of ~14 µW/cm K^2 around 773 K by hole doping. However, the highest power factor values achieved in doped polycrystalline SnSe is only ~5 µW/cm K^2.[193] The origins of the difference come from various contributions. First, unlike the facile doping behavior of Pb-based rock salt chalcogenides, doping SnSe is challenging because of the layered anisotropic structure, where each SnSe layer is two atoms thin, and the locally distorted bonding around the Sn and Se atoms, thus resulting in low carrier concentration. Second, the zigzag structure in SnSe plays an important role in the carrier mobility. It is verified that carrier mobilities are higher within the plane of the SnSe slabs than those perpendicular to them. But the irregular orientation of crystalline grains offsets its function. Third, grain boundaries and voids in polycrystalline SnSe impede the carrier's transportation to some extent.

The intrinsic low lattice thermal conductivity make great contributions to the remarkable high *ZT* of single-crystal SnSe. The polycrystalline SnSe was expected to have lower thermal conductivity because of the phonon scattering introduced by grain boundaries before. However, the thermal conductivity of polycrystalline SnSe was proved to be higher than that of single-crystal SnSe, which exceeds individual's expectation. Recent researches suggest that the intrinsically low thermal conductivity of SnSe is sensitive to the stoichiometric ratio and the sample processing conditions. A recent neutron powder diffraction analysis demonstrated a nearly ideal stoichiometry and an exceptionally high anharmonicity of the chemical bonds of SnSe and reported an ultralow thermal conductivity value close to 0.1 W/m K at room temperature for polycrystalline SnSe.[194] By strictly eliminating exposure to the atmosphere during sample preparation, Zhang et al.[192] observed lattice thermal conductivities of polycrystalline SnSe as low as those of the SnSe single crystals.

What is more is that we found that the thermal conductivity of SnSe could be artificially increased by not carefully handling starting powders. This is an artifact of handling and exposing/grinding the SnSe material (especially powders) in air. We have found that an oxidation layer of SnO_2 can form on the surface of the micron-sized powders, and this artificially raises the thermal conductivity. SnO_2 has over 80 times higher thermal conductivity (~98 W/m K) than that of polycrystalline SnSe. Actually, the thermal diffusivity of polycrystalline SnSe is very low when the SnSe powders are processed in an inert atmosphere glove box, and it approaches the very low 0.26 mm^2/s value at 773 K. This is comparable to 0.2 mm^2/s (*c* axis) and 0.23 mm^2/s (*b* axis) of our SnSe single crystals. We can reproduce the higher thermal conductivity when we add a very small amount of SnO_2 (e.g., 0.01–0.06% by weight) to SnSe and process the powders in the glove box. These thermal diffusivities can be twice as large. Thus, even small surface oxidation can lead to higher thermal conductivity. Crystals handled in air for some time could also develop SnO_2 layers.

4.6.5 Conclusions and Outlook

In this section, we mainly introduced a new promising thermoelectric material SnSe with intrinsically low thermal conductivity. We showed the crystal structure, synthesis methods, and thermoelectric properties of single-crystal and polycrystalline SnSe and made comparisons between single-crystal and polycrystalline SnSe. All the features reveal that SnSe is a state-of-the-art thermoelectric material. However, a multitude of researches still need to be accomplished, including developing high-performance n-type SnSe materials, enhancing the ZT values of polycrystalline SnSe, and improving the mechanical properties to commercial application.

4.7 High-Efficiency $Ag_xSb_xGe_{50-2x}Te_{50}$ Thermoelectric Materials: Effect of Ag and Sb

E. M. Levin

4.7.1 Introduction

One of the most well-known series of thermoelectric materials for midtemperature range are GeTe-based materials where Ge is replaced by Ag and Sb forming $Ag_xSb_xGe_{50-2x}Te_{50}$ (TAGS) series. These materials were developed in the 1970s and reported by Skrabek and Trimmer.[195,196] GeTe is a p-type self-doping semiconductor where the high carrier concentration of free charge carriers is generated by Ge vacancies.[197,198] GeTe as well as TAGS materials experience transformation from the low-temperature rhombohedral ($R3m$) to high-temperature cubic ($Fm\overline{3}m$) structure, which was found to be of second order.[199–201] Due to the peculiarities of the crystal structure and electronic transport, GeTe serves as the base not only for high efficiency thermoelectric materials,[195,196,202,203] but also for phase change materials used in electronic and optical memory devices.[204,205]

TAGS materials are typically shown as $(GeTe)_m(AgSbTe_2)_{100-m}$ and/or by acronym TAGS-m.[196] One of the best materials, typically shown as $(GeTe)_{85}(AgSbTe_2)_{15}$ or $Ag_{6.5}Sb_{6.5}Ge_{37}Te_{50}$ (TAGS-85), has demonstrated high thermoelectric efficiency and was used in various applications.[201,206] The chemical formulas shown earlier are ambiguous because neither GeTe nor $AgSbTe_2$ compounds are used for the synthesis of TAGS materials, but just Ge, Te, Ag, and Sb constituent elements. Hence, the $Ag_xSb_xGe_{50-x}Te_{50}$ chemical equation, where each component is shown in atomic percentage, better represents the composition of TAGS materials and makes it much easier to demonstrate the effects due to alloying. The composition of TAGS-85 can be shown (with approximate coefficients) as $Ag_{6.5}Sb_{6.5}Ge_{37}Te_{50}$, i.e., ~13 at.% of Ge are replaced by [Ag + Sb]; however, the acronym TAGS-m is also used here to show the connection between different representations of their composition.

Although TAGS materials were developed by the National Aeronautics and Space Administration in the 1970s, the nature of their high efficiency was never explained. Here our and literature data for GeTe and TAGS-85 obtained by common (XRD, Seebeck coefficient, electrical resistivity, thermal conductivity) and advanced [^{125}Te nuclear magnetic resonance (NMR)] methods are compared. The main goal is to discuss why the replacement of Ge in GeTe by Ag+Sb enhances the Seebeck coefficient and results in the formation of high efficiency thermoelectric materials.

4.7.2 Materials Synthesis

TAGS materials are typically prepared by the direct reaction of the constituent elements in fused silica ampoules. The impurities can hide true properties of these materials; hence, components of high purity (99.9999%) should be used for synthesis.[207] Ampoules with the constituent elements are heated up to ~1320 K to melt the constituents and periodically shaken to form a homogeneous melt. The melt can be cooled down at a rate of 100°C/h or quenched. The melt did not react with the ampoule, and the ingot can be easily removed from the ampoule after cooling.

Rectangular or disk samples can be prepared by cutting the ingot with a diamond saw and then used for Seebeck coefficient S, electrical resistivity ρ, and thermal conductivity κ_{tot} measurements; typical uncertainties of measurements are ≤5%, 3%, and ~15%, respectively.[208] Note that for the determination of thermal conductivity, the values of thermal diffusivity λ, specific heat capacity c_p, and mass density d are needed because $\kappa_{tot} = \lambda c_p d$. The separation of thermal conductivity due to carriers κ_{car} and lattice κ_{lat}, resulting in $\kappa_{tot} = \kappa_{car} + \kappa_{lat}$, requires to theoretically estimate the temperature dependence of the Lorenz number[200,209] which can be ambiguous.

Measurements of thermoelectric properties can be conducted using samples prepared by SPS[210]; this method allows the preparation of samples with high density and low porosity, but does not necessarily result in better thermoelectric properties. The absolute Seebeck coefficient must be determined and used in analysis[208] while some publications do not show what type of the coefficient, relative or absolute, is reported. For p-type TAGS materials, the relative coefficient is higher than that of the absolute one, particularly if common Pt – (Pt + Rh) thermocouples are used.[208] This results in incorrect (higher) values of the Seebeck coefficient, power factor, and figure of merit.

Phase analysis is typically conducted by XRD using powder samples and Cu-K$_\alpha$ (λ = 1.54 Å) or Co-K$_\alpha$ radiation (λ = 1.79 Å)[199,202,203,208,211]; Rietveld refinement allows to obtain lattice parameters and rhombohedral distortion for GeTe in the temperature range of 300–800 K using neutron[199] and XRD[200] data; similar refinement was used for TAGS materials.

NMR is a very efficient tool for studying complex materials, and it was recently utilized for complex tellurides including GeTe and TAGS materials.[200,208,212,213] ^{125}Te NMR experiments were performed at 126 MHz using a Bruker 400 spectrometer in a magnetic field of 9.4 T at 300 K without sample spinning (static regime).[200,208,212,213] Both ^{125}Te NMR spectra and spin–lattice relaxation measurements are very useful to better understand

GeTe-based materials. Note that the Hall effect is typically used to obtain the carrier concentration, but in some tellurides, it can be misleading,[214] while ^{125}Te NMR is potentially more reliable method to obtain the free (mobile) carrier concentration.[212]

4.7.3 Crystal Structure and Its Evolution versus Composition and Temperature

At 300 K, the IV–VI semiconductor GeTe crystallizes in a rhombohedral structure (R_{3m}).[199,200] This structure can be considered as a distorted NaCl-type cubic structure. The rhombohedral angle shows the degree of the distortion from the cubic structure. The replacement of Ge in GeTe by ~13 at.% Ag + Sb slightly changes the distortion. Figure 4.53 shows XRD patterns for GeTe and $Ag_{6.5}Sb_{6.5}Ge_{37}Te_{50}$ alloys. At 300 K, both materials have rhombohedral lattice; two peaks, (024) and (220), are a signature of the distortion. The smaller distance between the (024) and (220) peaks for $Ag_{6.5}Sb_{6.5}Ge_{37}Te_{50}$ compared to that of GeTe indicates smaller rhombohedral distortion, which is clearly induced by Ag and Sb addition. Table 4.1 shows lattice parameters and rhombohedral angles for GeTe and $Ag_{6.5}Sb_{6.5}Ge_{37}Te_{50}$; the replacement of Ge in GeTe by [Ag + Sb] reduces the rhombohedral distortion.

The rhombohedral distortion in GeTe decreases with temperature, and at T_{cubic}, the crystal lattice becomes mostly cubic ($Fm\bar{3}m$) (Figure 4.54a). T_{cubic} obtained by neutron diffraction[199] is ~705 K, while that from XRD varies in the range between ~660 and ~730 K.[199] T_{cubic} for our GeTe sample is ~660 K (slightly higher that than, ~640 K, initially reported by Levin et al.[200]), while that of $Ag_{6.5}Sb_{6.5}Ge_{37}Te_{50}$ is much lower, ~530 K (Figure 4.54b). Earlier, the phase transition for TAGS-85 was observed by Cook et al.[201] at 510 K (using high-energy synchrotron radiation with λ = 0.123 Å) and by Thompson et al.[215] at 525 K (using neutron diffraction). Hence, it is well established that

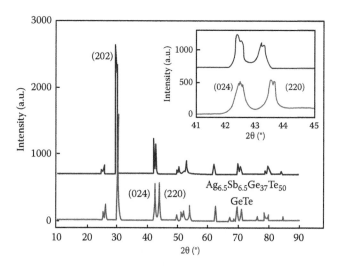

FIGURE 4.53 XRD (Co-K_α radiation) patterns for GeTe and $Ag_{6.5}Sb_{6.5}Ge_{37}Te_{50}$ (TAGS-85) at 300 K. The inset presents an expanded view for 41° ≤ 2θ ≤ 45° showing (024) and (220) Bragg peaks for the same samples and demonstrating rhombohedral lattice distortion.

TABLE 4.1 Lattice Parameters, Rhombohedral Distortion, and T_{cubic} for GeTe and $Ag_{6.5}Sb_{6.5}Ge_{37}Te_{50}$ (TAGS-85)

Material	Lattice Parameters $a = b = c$ 300 K (Å)	Rhombohedral Angle $\alpha = \beta = \gamma$ 300 K (°)	T_{cubic} (K)	Mass Density 300 K (g/cm³)	Specific Heat Capacity ~300 K (J/g K)
GeTe	5.980(7)[a]	88.37(1)[a]	640[a]	6.18[a]	0.256[a]
	5.985(2)[b]	88.17(3)[b]	705[b]		
TAGS-85	5.995(4)[a]	88.99(8)[a]	520[a]	6.18[a]	0.260[a]
		88.15[c]	510[e]	6.13[g]	
		88.12[d]	525[f]	6.45[d]	

[a] Our data (Levin et al.,[200,208,212,213,223,227] Cook et al.[201]).
[b] Chattopadhyay et al.[199]
[c] Schneider et al.[204]
[d] Dong et al.[226]
[e] Cook et al.[201]
[f] Flanders et al.[218]
[g] Salvador et al.[203]

the replacement of ~13 at.% Ge in GeTe by [Ag + Sb] reduces the rhombohedral distortion and thermal stability of rhombohedral structure by ~130–150 K compared to that of GeTe.

4.7.4 Seebeck Coefficient and Electrical Resistivity

Theoretical calculations and optical methods show that GeTe is a semiconductor; the reported energy gap varies between 0.3 and 1 eV.[216] However, GeTe shows p-type conductivity and metallic temperature dependence of the electrical resistivity,[197,200] demonstrating degenerate state. The hole concentration in GeTe at 300 K derived from the Hall effect measurements is high,[197,200,209] 8×10^{20} cm^{-3} (Table 4.2) and is generated by Ge vacancies: two holes per each Ge vacancy.[198]

Figure 4.55a shows one of the most important fundamental questions for TAGS materials: why is the Seebeck coefficient of TAGS-85 much larger compared to that of GeTe? At 300 K, the value of the Seebeck coefficient of GeTe is +34 μV/K (our data [Levin et al.[200]]) which is close to the values from the studies by Alfa Aesar,[207] Gelbstein et al.,[209,210] and Sun,[217] while that of $Ag_{6.5}Sb_{6.5}Ge_{37}Te_{50}$ is +80 μV/K (our data [Levin et al.[200]]) or even larger, +100 μV/K[202] or +110 μV/K.[203] Similarly, a large difference is observed at high temperatures. Reliable data for TAGS materials, including TAGS-85 named as an excellent p-type thermoelectric material, were reported by Flanders et al.[218]

Within a common model for parabolic band and energy-independent charge carrier scattering, $S = \sim m^*/n^{2/3}$,[219] where n and m^* are the carrier concentration and effective mass, respectively. If this model can be applied to TAGS materials, the large enhancement for the Seebeck coefficient can be

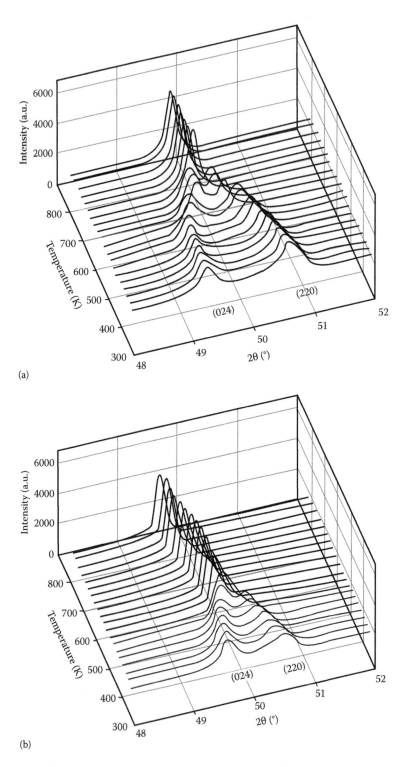

FIGURE 4.54 XRD (Co-K_α radiation) patterns for (a) GeTe and (b) TAGS-85 showing the positions of (024) and (220) Bragg peaks versus temperature and demonstrating disappearance of rhombohedral lattice distortion at T_{cubic}. Note that T_{cubic} for TAGS-85, ~530 K, is much lower than that for GeTe, ~660 K.

TABLE 4.2 Seebeck Coefficient, Electrical Resistivity, Carrier Concentration and Mobility for GeTe and $Ag_{6.5}Sb_{6.5}Ge_{37}Te_{50}$ (TAGS-85)

Material	Seebeck Coefficient (μV/K)		Electrical Resistivity ($\mu\Omega$ m)		Carrier Concentration 300 K $\times 10^{20}$ (cm^{-3})	Carrier Mobility 300 K (cm^2/V s)
	300 K	~720 K	300 K	~720 K		
GeTe	+34[a]	+148[a]	1.4[a]	5.2[a]	8[a]	56[a]
	+28[b]	+135[b]	1.1[b]	–	7[b]	67[b]
	–	+150[c]	–	5.4[c]	7[c]	70[c]
TAGS-85	+82[a]	+178[a]	7.6[a]	13.2[a]	13[g]	5.8[a]
	+110[d]	+160[d]	7.7[d]	9[d]	12[d]	–
	+100[e]	+200[e]	5[e]	12[e]	15[e]	5.7[e]
	+60[f]	+135[f]	3.5[f]	4.3[f]	–	–

[a] Our data (Levin et al.,[200,208,212,213,223,227] Cook et al.[201]).
[b] Gelbstein et al.[210]
[c] Sun et al.[217]
[d] Yang et al.[202]
[e] Salvador et al.[203]
[f] Dong et al.[226]
[g] Our [125]Te NMR data (Levin et al.[223]).

explained by lower carrier concentration or larger effective mass. However, the carrier concentration in TAGS-85 obtained from the Hall effect is even higher, 12×10^{20} or 15×10^{20} cm^{-3},[202,203] than that of GeTe (Table 4.2).

Hence, the Seebeck coefficient of TAGS-85 is larger than that of GeTe, while the carrier concentration is higher. This demonstrates the presence of an additional contribution, which overcomes the expected reduction due to higher carrier concentration and should be attributed to the increase of the effective mass and/or to the energy-dependent charge carrier scattering producing effective energy filtering.[208,220,221]

Figure 4.55a also demonstrates different slopes in S versus T dependencies and the temperature of these slopes: ~660 and ~540 K for GeTe and TAGS-85, respectively. These temperatures are very close to T_{cubic} (see Figure 4.54 and Table 4.1) and can be attributed to the change in crystal lattice symmetry: a larger slope is observed for the low-temperature rhombohedral phase, while a smaller slope is observed for the high-temperature cubic phase.

Based on theoretical calculations, Hoang et al.[222] have suggested that Ag and Sb in GeTe form atomic pairs. In order to experimentally confirm such suggestion, it is important to compare the effects of the replacement of Ge in GeTe by only Ag or Sb with that by [Ag + Sb]. The replacement of Ge in GeTe by Ag ($Ag_xGe_{50-x}Te_{50}$ alloy) decreases the Seebeck coefficient, while that by Sb ($Sb_xGe_{50-x}Te_{50}$) increases it; these effects can be explained by higher or lower carrier concentration, respectively, due to local electron imbalance and/ or higher or lower concentration of Ge vacancies.[223] A comparison of the data for GeTe, $Ag_xGe_{50-x}Te_{50}$, $Sb_xGe_{50-x}Te_{50}$ and TAGS materials supports the

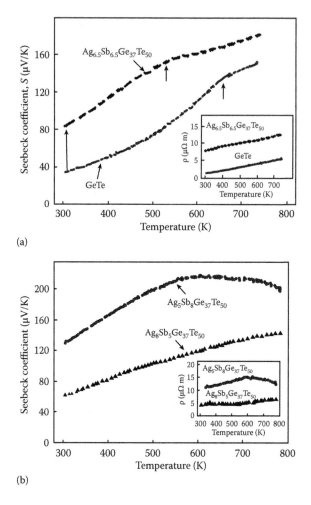

FIGURE 4.55 Temperature dependencies of the absolute Seebeck coefficient of (a) GeTe and TAGS-85 and (b) $Ag_8Sb_5Ge_{37}Te_{50}$ and $Ag_5Sb_8Ge_{37}Te_{50}$ with different Ag/Sb ratio. The insets in (a) and (b) show temperature dependencies of the electrical resistivity of the same samples.

formation of [Ag + Sb] pairs.[223] Hence, it is possible to explain the unusual enhancement of the Seebeck coefficient in TAGS materials by the effect from [Ag + Sb] pairs.

The effects from Ag or Sb can be used for the additional enhancement of the Seebeck coefficient by tuning the carrier concentration. Figure 4.55b shows the Seebeck coefficient versus temperature for different Ag/Sb ratios in $(Ag_xSb_y)Ge_{37}Te_{50}$, where the sum of Ag and Sb is the same as that of TAGS-85, ~13 at.%. A larger amount of Ag (8 at.%) and a smaller amount of Sb (5 at.%) in $Ag_8Sb_5Ge_{37}Te_{50}$ reduce S compared to that in TAGS-85. In contrast, a smaller amount of Ag (5 at.%) and a larger amount of Sb (8 at.%) in $Ag_5Sb_8Ge_{37}Te_{50}$ significantly increases the Seebeck coefficient. Hence, it is likely that in both cases, ~5 at.% of Ag and Sb form pairs, while ~3 at.% of Ag or Sb excess results in higher or lower carrier concentration, respectively. Similar effects of the Ag/Sb ratio on the Seebeck coefficient were reported by Yang et al.[224] and Schröder et al.[225]

The temperature dependencies of the electrical resistivity ρ of both GeTe and TAGS-85 show metallic character. The value of ρ at 300 K of TAGS-85) is larger than that of GeTe by approximately five times. Because the carrier concentration in TAGS-85 is higher than that in GeTe, larger ρ can be explained by a reduction in carrier mobility (Table 4.2). The electrical resistivity of $Ag_8Sb_5Ge_{37}Te_{50}$ is smaller while that of $Ag_5Sb_8Ge_{37}Te_{50}$ is larger than that of TAGS-85 (Figure 4.55b inset).

4.7.5 Thermal Conductivity, Power Factor, and Thermoelectric Figure of Merit

The thermal conductivity in electrically conductive materials is a sum of two contributions due to lattice and mobile charge carriers. Thermal conductivity is dramatically reduced when Ge in GeTe is replaced with [Ag + Sb] (Figure 4.56). At 300 K, the thermal conductivity of GeTe is high, 80 mW/cm K, which can mostly be attributed to the κ_{car} due to high carrier concentration and relatively large carrier mobility. Thermal conductivity of TAGS-85 is much lower, ~19 mW/cm K, which is due to a reduction in both κ_{lat} and κ_{car}. The smaller κ_{car} of TAGS-85 agrees well with a decrease in the carrier mobility (Table 4.2). Table 4.3 shows the values of the thermal conductivity for GeTe and TAGS-85 from various sources.

Figure 4.57 shows the power factors and thermoelectric figures of merit for GeTe and TAGS-85. Note that GeTe has one the highest power factor at 720 K, 42 µW/cm K²,[200,217] while its *ZT* is not large, 0.8, due to high thermal conductivity[201] (Table 4.3). In contrast, TAGS-85 has a smaller power factor, between 25 and 33 µW/cm K², but larger *ZT* of 1.2–1.4.[202,203,208] A very large power factor, ~43 µW/cm K², was reported by Dong et al.[226] for TAGS-85 (sample D) prepared at a pressure of 4 GPa and temperature of 1073 K (Tables 4.2 and 4.3). The Seebeck coefficient of this sample at ~720 K is relatively small, about +135 µV/K, and the enhancement of power factor is

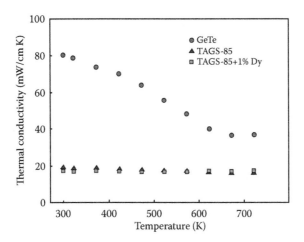

FIGURE 4.56 Temperature dependencies of thermal conductivity of GeTe and TAGS-85, and $Ag_{6.5}Sb_{6.5}Ge_{36}Te_{50}Dy_1$ (TAGS-85+1 at.% Dy). (Levin, E. M. et al.: *Advanced Functional Materials.* 2010. 39. 2049. Copyright Wiley-VCH Verlag GmbH & Co. KGaA. Reproduced with permission.)

TABLE 4.3 Power Factor, Thermal Conductivity, and Figure of Merit of GeTe and $Ag_{6.5}Sb_{6.5}Ge_{37}Te_{50}$ (TAGS-85)

Material	Power Factor (μW/cm K^2)		Thermal Conductivity (mW/cm K)		ZT
	300 K	~750 K	300 K	~750 K	~750 K
GeTe	8[a]	42[a]	80[a]	37[a]	0.8[a]
	–	42[b]	70[b]	32[b]	0.9[b]
TAGS-85	9[a]	25[a]	19[a]	18[a]	1.3[a]
	16[c]	28[c]	17[c]	15[c]	1.4[c]
	20[d]	33[d]	16[d]	16[d]	1.4[d]
	10[e]	43[e]	16[e]	–	–

[a] Our data (Levin et al.,[200,208,212,213,223,227] Cook et al.[201]).
[b] Sun et al.[217]
[c] Yang et al.[202]
[d] Salvador et al.[203]
[e] Dong et al.[226]

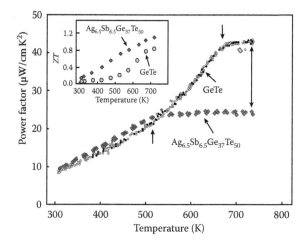

FIGURE 4.57 Temperature dependencies of power factor (*PF*) of GeTe and TAGS-85. The inset shows thermoelectric figure of merit *ZT* of the same materials.

due to low electrical resistivity (4.3 $\mu\Omega$ m) much smaller than that reported for TAGS-85 (Table 4.2); this value was not confirmed yet.

Hence, despite the enhancement of the Seebeck coefficient via the replacement of Ge in GeTe by [Ag + Sb], an increase of electrical resistivity reduces the power factor, while significant reduction in thermal conductivity increases thermoelectric figure of merit.

4.7.6 ^{125}Te NMR

In order to better understand the Seebeck coefficient in complex tellurides, advanced methods are needed. ^{125}Te NMR is a very efficient probe to study local properties of tellurides including the carrier concentration.[200,208,212,213,227,228] The Hall effect shows the integral signal from the free carriers present in a material including from different phases in multiphase materials and can be misleading. NMR is a local probe and enables to obtain a distribution of carrier concentrations. The carrier concentration in tellurides can be derived from ^{125}Te NMR spin–lattice relaxation measurements.[212,213]

Figure 4.58a demonstrates ^{125}Te NMR spectra of GeTe and TAGS-85. The resonance for GeTe is observed at +160 ppm and its width is 460 ppm. For comparison, the ^{125}Te NMR signal for p-type PbTe with hole concentration of 4.0×10^{18} cm^{-3} is observed at –1900 ppm and its width is much smaller, 50 ppm.[212] The resonance position (total shift δ_{total}) relative to the reference material depends on two contributions, $\delta_{total} = \delta_{chem} + K$, where δ_{chem} is the chemical shift due to the chemical environment of Te atoms and K is the Knight shift due to hyperfine interaction between Te nuclei and free charge carriers.[228,229] The resonance for TAGS-85 is observed at +150 ppm and its width is 740 ppm, much larger than that of GeTe. The large width

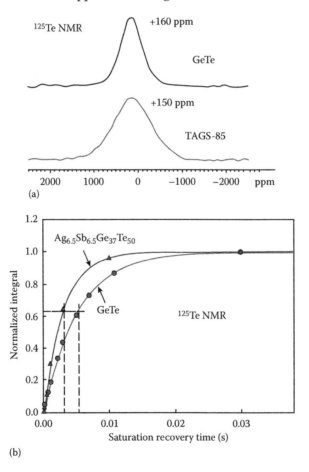

FIGURE 4.58 (a) ^{125}Te NMR spectra and (b) spin–lattice relaxation for GeTe and TAGS-85. Dashed vertical lines in (b) show the values of the spin–lattice relaxation time T_1 for GeTe ($T_1 = 5.3$ ms) and TAGS-85 ($T_1 = 3.1$ ms).

can be attributed to the distribution of the chemical shifts (show the effects from Te environment) or Knight shifts (show the effects from the carrier concentration). Our data demonstrate that the large width of ^{125}Te NMR signal is not related to the distribution of Knight shifts, i.e., it is the effect of Te environment and is due to interactions between Te and [Ag + Sb] atoms.

The ^{125}Te NMR spin–lattice relaxation in GeTe and TAGS-85 shown in Figure 4.58b demonstrates that the relaxation Te nuclei is faster when Ge is partially replaced by [Ag + Sb]. The spin–lattice relaxation time T_1 can be obtained from the relaxation curve; it is 5.3 ms and 3.1 ms for GeTe and TAGS-85, respectively. Based on the theoretical model and relation $1/T_1 \sim n$ (within Maxwell–Boltzmann statistics in an assumption that m^* in GeTe is similar to that in TAGS-85), T_1 can be used to derive the carrier concentration n.[223] Using GeTe as a reference material with the carrier concentration of 8×10^{20} cm^{-3} and $T_1 = 5.3$ ms, the carrier concentration in TAGS-85 can be shown as 13×10^{20} cm^{-3}, close to the values derived from the Hall effect (Table 4.2), i.e., ^{125}Te NMR confirms that the carrier concentration in TAGS-85 is higher than that in GeTe.

^{125}Te NMR signals of $Ag_8Sb_5Ge_{37}Te_{50}$ and $Ag_5Sb_8Ge_{37}Te_{50}$ are observed at +360 ppm and at –350 ppm, respectively. Large positive and negative total shifts compared to that in $Ag_{6.5}Sb_{6.5}Ge_{37}Te_{50}$, as well as shorter and longer T_1, can be mostly attributed to the change in the Knight shift, which reflects higher and lower carrier concentrations due to Ag or Sb excess, respectively. ^{125}Te NMR data allow us to explain the changes observed for the Seebeck coefficient in Figure 4.55b and that reported by Yang et al.,[202] Salvador et al.,[203] and Chen et al.[230] Hence, Ag or Sb excess is an excellent tool to tune the carrier concentration and obtain desired the Seebeck coefficient in TAGS materials.

4.7.7 Effect of Doping TAGS-85 with Rare Earth

Skrabek and Trimmer showed[196] that in order to enhance thermoelectric efficiency of TAGS materials, various elements, e.g., Cu, Sn, In, Ga, Mn, Pb, Se, I, Si, Cd, Fe, Cr, Ni, Ca, Mg, Bi, and Li, were used, but the results are mostly negligible. In contrast, the replacement of Ge or Te in TAGS-85 by rare-earth Dy with large atomic size and localized magnetic moments slightly reduces thermal conductivity and enhances the Seebeck coefficient.

Figure 4.59 shows temperature dependencies of the Seebeck coefficient and thermoelectric figure of merit of TAGS-85, TAGS-85 + 1% Dy, and TAGS-85 + 2% Dy. The largest enhancement of ~20% is observed when Dy replaces Ge in TAGS-85 by 2% Dy.[208] However, because the electrical resistivity of the materials containing 2 at.% Dy is larger than that of the material containing 1% Dy, the values of their figure of merit are similar. ^{125}Te NMR signal position and spin–lattice relaxation time do not significantly change enough to explain the enhancement of the Seebeck coefficient by increased carrier concentration. This allows us to suggest that observed enhancement due to Dy can be attributed to additional mechanism associated with stronger carrier scattering and can be attributed to energy filtering,[208] or to an

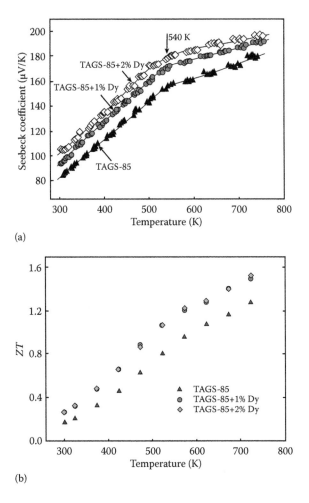

FIGURE 4.59 (a) Seebeck coefficient and (b) thermoelectric figure of merit for TAGS-85 ($Ag_{6.5}Sb_{6.5}Ge_{37}Te_{50}$), TAGS-85 + 1% Dy ($Ag_{6.5}Sb_{6.5}Ge_{36}Te_{50}Dy_1$), and TAGS-85 + 2%Dy ($Ag_{6.5}Sb_{6.5}Ge_{35}Te_{50}Dy_2$). (Levin, E. M. et al.: *Advanced Functional Materials*. 2010. 39. 2049. Copyright Wiley-VCH Verlag GmbH & Co. KGaA. Reproduced with permission.)

increase of the carrier effective mass. Similar enhancement was also observed when Ge in TAGS-85 is replaced with rare-earth Ce or Yb.[227]

4.7.8 Unique Nature of the Seebeck Effect in TAGS Materials

The large thermoelectric figure of merit of TAGS-85 material can be explained by their low thermal conductivity and high power factor. A decrease in thermal conductivity compared to that of GeTe is achieved by a reduction of both components: κ_{lat} due to the presence of [Ag + Sb] atoms in the lattice and κ_{car} due to lowering of the charge carrier mobility. It likely that [Ag + Sb] atoms form atomic pairs.[222,223] and significantly increase both the phonon and free charge carrier scattering. The Seebeck coefficient of TAGS-85 is much larger than that of GeTe, while the electrical resitivity increases too and reduces the power factor. However, the interplay between lower power factor and lower thermal conductivity results in significantly larger thermoelectric figure of merit of TAGS-85 compared to that of GeTe.

Ge vacancy is a major source of free charge carriers in GeTe; impurities present in constituent elements may also affect their concentration. Similarly, free holes in TAGS-85 are generated mostly by Ge vacancies, while some additional contributions, which may increase or decrease the carrier concentration, are expected if Ag and/or Sb are separately present in the material. TAGS materials are not vacancy free as suggested by Schröder et al.,[225] because in this case, it is not possible to explain the high carrier concentration, ~10^{21} cm^{-3}, at 300 K obtained by the Hall effect[202,203] and confirmed by our ^{125}Te NMR measurements (Table 4.2). Schröder et al.[225] used the chemical equations, e.g., $Ge_{0.53}Ag_{0.13}Sb_{0.27}\square_{0.07}Te_1$, where the symbol $\square$ demonstrates the cation vacancies. However, the symbol $\square$ in this equation shows [Ag + Sb + Ge] deficiency of 1.5 at.%, not the vacancy content. Such a deficiency can, in principle, promote Ge vacancies, but some amount of each component may form a separate phase, and there is no direct proportionality between the cation deficiency and cation vacancy concentration in GeTe-based materials.

The enhancement of the Seebeck coefficient in TAGS-85 at 300 K by ~220% compared to that in GeTe, while the carrier concentration is slightly higher, shows the existence of additional strong contribution due to [Ag + Sb] atoms. It can be produced due to an increase of the effective mass,[223,230] but, in principle, it can also be explained by energy-dependent charge carrier scattering. Energy filtering model, where high-energy carriers are scattered at a much smaller extent than low-energy carriers,[220,221] can be used for the phenomenological description of high efficiency of TAGS materials. In this scenario, [Ag + Sb] pairs serve as potential barriers and selectively scatter free charge carriers moving in the opposite direction. The doping of TAGS-85 by rare-earth atoms enhances the Seebeck coefficient and can be used to support energy-filtering mechanism.

Despite a long history of TAGS materials, they still can be considered as a puzzle and used as a model system to better understand the Seebeck effect in complex materials with unstable crystal symmetry. Further improvement of TAGS materials is possible via a search for the optimal [Ag + Sb] content along with rational Ag/Sb ratio and additional doping by various elements.

Acknowledgments

The authors would like to thank K. Schmidt-Rohr (Brandeis University) and M. F. Besser, S. L. Bud'ko, A. Howard, Z. Swanson, and M. Kluckman (Iowa State University and US Department of Energy [DOE]– Ames Laboratory) for help in experiments. This work was supported by the US DOE, Office of Basic Energy Sciences, Division of Materials Sciences and Engineering. The research was performed at Ames Laboratory, which is operated for the US. DOE by the Iowa State University under Contract No. DE-AC02-07CH11358.

References

1. Y. Pei, A. LaLonde, S. Iwanaga, and G. J. Snyder, High thermoelectric figure of merit in heavy hole dominated PbTe. *Energy Environ. Sci.* (2011), 4, 2085.
2. H. Wang, Y. Pei, A. D. LaLonde, and G. J. Snyder, Heavily doped p-type PbSe with high thermoelectric performance: An alternative for PbTe. *Adv. Mater.* (2011), 23, 1366.
3. A. J. Crocker and L. M. Rogers, Valence band structure of PbTe. *J. Phys. Colloq.* (1968), 29, 129.
4. I. A. Chernik, V. I. Kaidanov, M. I. Vinogradova, and N. V. Kolomoets, Investigation of the valence band of lead telluride using transport phenomena. *Sov. Phys. Semicond.* (1968), 2, 645.
5. H. A. Lyden, Temperature dependence of the effective masses in PbTe. *Phys. Rev. A* (1964), 135, A514.
6. J. P. Heremans, V. Jovovic, E. S. Toberer, A. Saramat, K. Kurosaki, A. Charoenphakdee, S. Yamanaka, and G. J. Snyder, Enhancement of thermoelectric efficiency in PbTe by distortion of the electronic density of states. *Science* (2008), 321, 554–557.
7. Q. Y. Zhang et al., Enhancement of thermoelectric figure-of-merit by resonant states of aluminium doping in lead selenide. *Energy Environ. Sci.* (2012), 5, 5246–5251.
8. L. D. Zhao et al., High performance thermoelectrics from earth-abundant materials: Enhanced figure of merit in PbS by second phase Nanostructures. *J. Am. Chem. Soc.* (2011), 133, 20476–20487.
9. A. D. LaLonde, Y. Z. Pei, and G. J. Snyder, Reevaluation of PbTe$_{1-x}$I$_x$ as high performance n-type thermoelectric material. *Energy Environ. Sci.* (2011), 4, 2090–2096.
10. A. D. Lalonde, Y. Pei, H. Wang, and G. J. Snyder, Lead telluride alloy thermoelectrics. *Mater. Today* (2011), 14, 526.
11. L. D. Zhao, J. Q. He, C. Wu, T. P. Hogan, X. Y. Zhou, C. Uher, V. P. Dravid, and M. G. Kanatzidis, Thermoelectrics with earth abundant elements: High performance p-type PbS nanostructured with SrS and CaS. *J. Am. Chem. Soc.* (2012), 134, 7902–7912.
12. Q. Zhang et al., Study of the thermoelectric properties of lead selenide doped with boron, gallium, indium, or thallium. *J. Am. Chem. Soc.* (2012), 134, 17731–17738.
13. H. Wang, Y. Z. Pei, A. D. LaLonde, and G. J. Snyder, Weak electron-phonon coupling contributing to high thermoelectric performance in n-type PbSe. *Proc. Natl. Acad. Sci. USA.* (2012), 109, 9705–5709.
14. H. Wang, E. Schechtel, Y. Z. Pei, and G. J. Snyder, High thermoelectric efficiency of n-type PbS. *Adv. Energy Mater.* (2013), 3, 488–495.
15. Q. Zhang, E. K. Chere, K. McEnaney, M. L. Yao, F. Cao, Y. Z. Ni, S. Chen, C. Opeil, G. Chen, and Z. F. Ren, Enhancement of thermoelectric performance of n-type PbSe by Cr doping with optimized carrier concentration. *Adv. Energy Mater.* (2015), 5, 1401977.
16. R. J. Korkosz et al., High *ZT* in p-type (PbTe)$_{1-2x}$(PbSe)$_x$(PbS)$_x$ thermoelectric materials. *J. Am. Chem. Soc.* (2014), 136, 3225.
17. H. J. Wu, L. D. Zhao, F. S. Zheng, D. Wu, Y. L. Pei, X. Tong, M. G. Kanatizidis, and J. Q. He, Broad temperature plateau for thermoelectric figure of merit *ZT* > 2 in phase-separated PbTe$_{0.7}$S$_{0.3}$. *Nat. Commun.* (2014), 5, 4515.

18. D. Wu et al., Superior thermoelectric performance in PbTe-PbS pseudo-binary: Extremely low thermal conductivity and modulated carrier concentration. *Energy Environ. Sci.* (2015), 8, 2056–2068.

19. Z. T. Tian, J. Garg, K. Esfarjani, T. Shiga, J. Shiomi, and G. Chen, Phonon conduction in PbSe, PbTe, and PbTe$_{1-x}$Se$_x$ from first-principles calculations. *Phys. Rev. B* (2012), 85, 184303.

20. C. M. Jaworski, B. Wiendlocha, V. Jovovic, and J. P. Heremans, Combining alloy scattering of phonons and resonant electronic levels to reach a high thermoelectric figure of merit in PbTeSe and PbTeS alloys. *Energy Environ. Sci.* (2011), 4, 4155–4162.

21. S. Ching-Hua, Thermoelectric properties of Tl-doped PbTeSe crystals grown by directional solidification. *J. Cryst. Growth* (2016), 439, 80.

22. Y. I. Ravich, B. A. Efimova, and I. A. Smirnov, *Semiconducting Lead Chalcogenides*. L. S. Stil'bans (ed.). Plenum, New York, 1970, pp. 165–258.

23. Y. Z. Pei, X. Y. Shi, A. LaLonde, H. Wang, L. D. Chen, and G. J. Snyder, Convergence of electronic bands for high performance bulk thermoelectrics. *Nature* (2011), 473, 66–69.

24. Q. Zhang, F. Cao, W. S. Liu, K. Lukas, B. Yu, S. Chen, C. Opeil, G. Chen, and Z. F. Ren, Heavy doping and band engineering by potassium to improve the thermoelectric figure of merit in p-type PbTe, PbSe, and PbTe$_{1-y}$Se$_y$. *J. Am. Chem. Soc.* (2012), 134, 10031–10038.

25. Y. Noda, M. Orihashi, and I. A. Nishida, Thermoelectric properties of p-type lead telluride doped with silver or potassium. *Mater. Trans. JIM* (1998), 39, 602–605.

26. J. Androulakis, I. Todorov, D. Y. Chung, S. Ballikaya, G. Wang, C. Uher, and M. Kanatzidis, Thermoelectric enhancement in PbTe with K or Na codoping from tuning the interaction of the light- and heavy-hole valence bands. *Phys. Rev. B* (2010), 82, 115209.

27. P. K. Rawat, B. Paul, and P. Banerji, Thermoelectric properties of PbSe$_{0.5}$Te$_{0.5}$: x (PbI$_2$) with endotaxial nanostructures: A promising n-type thermoelectric material. *Nanotechnology* (2013), 24, 215401.

28. E. Grodzicka, W. Dobrowolski, T. Story, A. Wilamowski, and B. Witkowska, The study of a resonant Cr donor in PbTe, PbSe, PbTe$_{1-y}$Se$_y$ and Pb$_{1-x}$Sn$_x$Te. *Ins. Phys. Conf. Ser.* No 144: Section 3 (1995).

29. E. K. Chere, Q. Zhang, K. McEnaney, M. L. Yao, F. Cao, C. Opeil, G. Chen, and Z. F. Ren, Enhancement of thermoelectric performance in n-type PbTe$_{1-y}$Se$_y$ by doping Cr and tuning Te:Se ratio. *Nano Energy* (2015), 13, 355–367.

30. A. Bali, H. Wang, G. J. Snyder, and R. C. Mallik, Thermoelectric properties of indium doped PbTe$_{1-y}$Se$_y$ alloys. *J. Appl. Phys.* (2014), 116, 033707.

31. B. Basu, S. Bhattacharya, R. Bhatt, A. Singh, D. K. Aswal, and S. K. Gupta, Improved thermoelectric properties of Se-doped n-type PbTe$_{1-x}$Se$_x$ ($0 \leq x \leq 1$). *J. Electron. Mater.* (2013), 42, 2292.

32. H. Wang, A. D. LaLonde, Y. Z. Pei, and G. J. Snyder, The criteria for beneficial disorder in thermoelectric solid solutions. *Adv. Funct. Mater.* (2013), 23, 1586–1696.

33. J. Yang, L. L. Xi, W. J. Qiu, L. H. Wu, X. Shi, L. D. Chen, J. H. Yang, W. Q. Zhang, C. Uher, and D. J. Singh, On the tuning of electrical and thermal transport in thermoelectrics: An integrated theory-experiment perspective. *NPG: Computational Mater.* (2016), 2, 15015.

34. J. L. Wang, H. Wang, G. J. Snyder, X. Zhang, Z. H. Ni, and Y. F. Chen, Characteristics of lattice thermal conductivity and carrier mobility of endoped PbSe-PbS solid solutions. *J. Phys. D: Appl. Phys.* (2013), 46, 405301.

35. H. Wang, J. L. Wang, X. L. Cao, and G. J. Snyder, Thermoelectric alloys between PbSe and PbS with effective thermal conductivity reduction and high figure of merit. *J. Mater. Chem. A.* (2014), 2, 3169.

36. J. Androulakis, I. Todorov, J. Q. He, D. Y. Chung, V. Dravid, and M. Kanatzidis, Thermoelectrics from abundant chemical elements: High-performance nano-structured PbSe-PbS. *J. Am. Chem. Soc.* (2011), 133, 10920–10927.

37. A. Volykhov, L. Yashina, and V. Shtanov, Phase equilibria in pseudoternary systems of IV-VI compounds. *Inorg. Mater.* (2010), 26, 464.

38. R. Blachnik and R. Igel, Thermodynamic properties of IV-VI compounds: Lead chalcogenides. *Z. Naturforsch. B.* (1974), 29, 625.

39. S. Aminorroaya Yamini, H. Wang, Z. Gibbs, Y. Pei, S. X. Dou, and G. J. Snyder, Chemical composition tuning in quaternary p-type Pb-chalcogenides—A promising strategy for enhanced thermoelectric performance. *Phys. Chem. Chem. Phys.* (2014), 16(5), 1835–1840.

40. H. Lin, E. S. Bozin, S. J. L. Billinge, J. Androulakis, C. D. Malliakas, C. H. Lin, and M. G. Kanatzidis, Phase separation and nanostructuring in the ther-moelectric material $PbTe_{1-x}S_x$ studied using the atomic pair distribution func-tion technique. *Phys. Rev. B* (2009), 80, 045204.

41. S. N. Girard, K. Schmidt-Rohr, T. C. Chasapis, E. Hatzikraniotis, B. Njegic, E. M. Levin, A. Rawal, K. M. Paraskevopoulos, and M. G. Kanatzidis, Analysis of phase separation in high performance PbTe-PbS thermoelectric materials. *Adv. Funct. Mater.* (2013), 23, 747–757.

42. Q. Zhang, E. K. Chere, Y. M. Wang, H. S. Kim, R. He, F. Cao, K. Dahal, D. Broido, G. Chen, and Z. F. Ren, High hermoelectric performance of n-type $PbTe_{1-y}S_y$ due to deep lying states induced by indium doping and spinodal decomposition. *Nano Energy* (2016), 22, 572–582.

43. S. Johnsen, J. Q. He, J. Androulakis, V. P. Dravid, I. Todorov, D. Y. Chung, and M. G. Kanatzidis, Nanostructures boost the thermoelectric performance of PbS. *J. Am. Chem. Soc.* (2011), 133, 3460–3470.

44. S. Aminorroaya Yamini, H. Wang, Z. Gibbs, Y. Pei, D. Mitchel, S. X. Dou, and G. J. Snyder, Thermoelectric performance of tellurium-reduced quater-nary p-type lead-chalcogenide composites. *Acta Mater.* (2014), 80, 365–372.

45. S. A. Yamini, H. Wang, D. Ginting, D. R. G. Mitchell, S. X. Dou, and G. J. Snyder, Thermoelectric performance of n-type $(PbTe)_{0.75}(PbS)_{0.15}(PbSe)_{0.1}$ composites. *ACS Appl. Mater. Interfaces* (2014), 6, 11476–11483.

46. H. Wang et al., High-performance $Ag_{0.8}Pb_{18+x}SbTe_{20}$ thermoelectric bulk materials fabricated by mechanical alloying and spark plasma sintering. *Appl. Phys. Lett.* (2006), 88(9), 092104.

47. K. F. Hsu et al., Cubic $AgPb_mSbTe_{2+m}$: Bulk thermoelectric materials with high figure of merit. *Science* (2004), 303, 818–821.

48. E. Quarez et al., Nanostructuring, compositional fluctuations, and atomic ordering in the thermoelectric materials $AgPb_mSbTe_{2+m}$: The myth of solid solutions. *J. Am. Chem. Soc.* (2005), 127, 9177–9190.

49. P. F. Poudeu et al., High thermoelectric figure of merit and nanostructuring in bulk p-type $Na_{1-x}Pb_mSb_yTe_{m+2}$. *Angew. Chem.* (2006), 45, 3835–3839.

50. J. Androulakis et al., Nanostructuring and high thermoelectric efficiency in p-type $Ag(Pb_{1-y}Sn_y)(m)SbTe_{2+m}$. *Adv. Mater.* (2006), 18, 1170–1173.

51. M. K. Han et al., Substitution of Bi for Sb and its role in the thermoelectric properties and nanostructuring in $Ag_{1-x}Pb_{18}MTe_{20}$ (M = Bi, Sb) (x = 0, 0.14, 0.3). *Chem. Mater.* (2008), 20, 3512–3520.

52. L. Ahn, C. P. Li, C. Uher, and M. G. Kanatzidis, Thermoelectric properties of the compounds APb_mLaTe_{m+2}. *Chem. Mater.* (2010), 22, 876–882.

53. P. F. P. Poudeu, A. L. Guéguen, C.-I. Wu, T. Hogan, and M. G. Kanatzidis, High figure of merit in nanostructured n-type KPb_mSbTe_{m+2} thermoelectric materials. *Chem. Mater.* (2010), 22, 1046–1053.

54. B. Poudel et al., High-thermoelectric performance of nanostructured bismuth antimony telluride bulk alloys. *Science.* (2008), 320(5876), 634–638.

55. K. F. Cai et al., Preparation and thermoelectric properties of $AgPb_mSbTe_{2+m}$ alloys. *J. Alloys Cmpd.* (2009), 469, 499–503.

56. F. Yan, T. J. Zhu, S. H. Yang, and X. B. Zhao, Microstructure and thermoelectric properties of cubic $AgPb_{18}Sb_{1-x}Te_{20}$ (x = 0.1, 0.3, 0.5) compounds. *Phys. Scr.* (2007), T129, 116–119.

57. D. Bilc et al., Resonant states in the electronic structure of the high performance thermoelectrics $AgPb_mSbTe_{2+m}$: The role of Ag-Sb microstructures. *Phys. Rev. Lett.* (2004), 93, 146403.

58. D. M. Rowe, *CRC Handbook of Thermoelectrics, Macro to Nano*. CRC Press, Taylor & Francis Group, Boca Raton, FL, 2006.

59. B. Poudel et al., High-thermoelectric performance of nanostructured bismuth antimony telluride bulk alloys. *Science*, 2008, 320(5876): 634-638.

60. D. Kraemer et al., *Nat. Mater.* (2011), 10, 532.

61. L. P. Hu, T. J. Zhu, Y. G. Wang, H. H. Xie, Z. J. Xu, and X. B. Zhao, *NPG Asia Mater.* (2014), 6, e88.

62. F. Cao, K. McEnaney, G. Chen, and Z. F. Ren, *Energy Environ. Sci.* (2014), 7, 1615.

63. G. H. Zeng, J. H. Bahk, J. E. Bowers, H. Lu, A. C. Gossard, S. L. Singer, A. Majumdar, Z. X. Bian, M. Zebarjadi, and A. Shakouri, A. *Appl. Phys. Lett.* (2009), 95, 083503.

64. W. S. Liu, K. Lukas, K. McEnaney, S. Lee, Q. Zhang, C. Opeil, G. Chen, and Z. F. Ren, *Energy Environ. Sci.* (2013), 6, 552.

65. C. Hadjistassou, E. Kyriakides, and J. Georgiou, *Energ. Convers. Manage.* (2013), 66, 165.

66. S. Yoon, J. Y. Cho, H. Koo, S. H. Bae, S. Ahn, G. R. Kim, J. S. Kim, and C. Park, *J. Elec. Mater.* (2014), 43, 414.

67. P. H. Ngan, D. V. Christensen, G. J. Snyer, L. T. Hung, S. Linderoth, N. V. Nong, and N. Pryds, *Phys. Status Solidi A*, (2014), 211, 9.

68. Q. Zhang, F. Cao, W. Liu, K. Lukas, B. Yu, S. Chen, C. Opeil, D. Broido, G. Chen, and Z. F. Ren, *J. Am. Chem. Soc.* (2012), 134, 10031–10038.

69. Y. Pei, X. Shi, A. LaLonde, H. Wang, L. Chen, and G. J. Snyder, *Nature* (2011), 473, 66.

70. E. K. Chere, Q. Zhang, K. McEnaney, M. L. Yao, F. Cao, C. Opeil, G. Chen, and Z. F. Ren, *Nano Energy* (2015), 13, 355.

71. Q. Zhang, E. K. Chere, K. McEnaney, M. L. Yao, F. Cao, Y. Ni, S. Chen, C. Opeil, G. Chen, and Z. F. Ren, *Adv. Energy Mater.* (2015), 1401977.

72. X. Yan, B. Poudel, Y. Ma, W. S. Liu, G. Joshi, H. Wang, Y. C. Lan, D. Z. Wang, D. Chen, and Z. F. Ren, *Nano Lett.* (2010), 10, 3373.

73. H. Wang, P. Pei, A. D. LaLonde, and G. J. Snyder, *Adv. Mater.* (2011), 23, 1366.

74. Q. Zhang et al., *J. Am. Chem. Soc.* (2012), 134, 17731.

75. G. T. Alekseeva, E. A. Gurieva, P. P. Konstantinov, L. V. Prokof'eva, M. I. Fedorov, *Semiconductors* (1996), 30, 1125.

76. J. Androulakis, D. Y. Chung, X. L. Su, L. Zhang, C. Uher, and M. G. Kanatzidis, *Phys. Rev. B* (2011), 84, 155207.

77. H. Wang, Y. Z. Pei, A. D. LaLonde, and G. J. Snyder, *Proc. Natl. Acad. Sci. USA.* (2012), 109, 9705.

78. Q. Y. Zhang et al., *Energy Environ. Sci.* 2012, 5, 5246.

79. Y. Pei, J. Lensch-Falk, E. S. Toberer, D. L. Medlin, and G. J. Snyder, *Adv. Funct. Mater.* (2011), 21, 241.

80. Y. Pei, A. LaLonde, S. Iwanaga, and G. J. Snyder, *Energy Environ. Sci.* (2011), 4, 2085.

81. K. Biswas, J. He, Q. Zhang, G. Wang, C. Uher, V. P. Dravid, and M. G. Kanatzidis, *Nat. Chem.* (2011), 3, 160.

82. B. Paul and P. Banerji, The effect of chromium impurity on the thermoelectric properties of PbTe in the temperature range 100–600 K. *J. Appl. Phys.* (2011), 109, 103710.

83. W. S. Liu, K. C. Lukas, K. M. Enaney, S. Lee, Q. Zhang, C. Opeil, G. Chen, and Z. F. Ren, Studies on the Bi_2Te_3–Bi_2Se_3–Bi_2S_3 system for mid-temperature thermoelectric energy conversion. *Energy Environ. Sci.* (2013), 6, 552.

84. G. Joshi, X. Yan, H. Z. Wang, W. S. Liu, G. Chen, and Z. F. Ren, Enhancement in thermoelectric figure-of-merit of an n-type half-heusler compound by the nanocomposite approach. *Adv. Energy Mater.* (2011), 1, 643.

85. X. Shi, J. Yang, J. R. Salvador, M. F. Chi, J. Y. Cho, H. Wang, S. Q. Bai, J. H. Yang, W. Q. Zhang, and L. D. Chen, Multiple-filled skutterudites: High thermoelectric figure of merit through separately optimizing electrical and thermal transports. *J. Am. Chem. Soc.* (2011), 133, 7837.

86. J. Androulakis, I. Todorov, J. Q. He, D. Y. Chung, V. Dravid, and M. J. Kanatzidis, *Am. Chem. Soc.* (2011), 133, 10920.

87. S. A. Yamini, H. Wang, D. Ginting, D. R. G. Mitchell, S. X. Dou, and G. J. Snyder, *ACS Appl. Mater. Interfaces* (2014), 6, 11476.

88. S. Johnsen, J. He, Q. J. Androulakis, V. P. Dravid, I. Todorov, D. Y. Chung, and M. G. Kanatzidis, *J. Am. Chem. Soc.* (2011), 133, 3460.

89. A. Bali, H. Wang, G. J. Snyder, and R. C. Mallik, Thermoelectric properties of indium doped $PbTe_{1-y}Se_y$ alloys. *J. Appl. Phys.* (2014), 116, 033707.

90. Z. Hu and S. Gao, Upper crustal abundances of trace elements: A revision and update. *Chem. Geol.* (2008), 253(3), 205–221.

91. O. Madelung, I-VII compounds. In *Semiconductors: Data Handbook*. Springer, Berlin, 2004, pp. 245–274.

92. Y. Yoğurtçu, A. Miller, and G. Saunders, Pressure dependence of elastic behaviour and force constants of GaP. *J. Phys. Chem. Sol.* (1981), 42(1), 49–56.

93. R. Dalven, A review of the semiconductor properties of PbTe, PbSe, PbS and PbO. *Infrared Phys.* (1969), 9(4), 141–184.

94. M. Beblo et al., Landolt-Börnstein: Numerical data and functional relationships in science and technology—New series. In *Landolt-Bornstein: Group 6: Astronomy*, G. Angenheister, (ed.). Springer-Verlag, Berlin, 1982, 239–253.

95. K. Ahn et al., Improvement in the thermoelectric figure of merit by La/Ag cosubstitution in PbTe. *Chem. Mater.* (2009), 21(7), 1361–1367.

96. L.-D. Zhao, V. P. Dravid, and M. G. Kanatzidis, The panoscopic approach to high performance thermoelectrics. *Energy Environ. Sci.* (2014), 7(1), 251–268.

97. Y. L. Pei and Y. Liu, Electrical and thermal transport properties of Pb-based chalcogenides: PbTe, PbSe, and PbS. *J. Alloys Cmpd.* (2012), 514, 40–44.

98. L. D. Zhao et al., Thermoelectrics with earth abundant elements: High performance p-type PbS nanostructured with SrS and CaS. *J. Am. Chem. Soc.* (2012), 134(18), 7902–7912.

99. S. Johnsen et al., Nanostructures boost the thermoelectric performance of PbS. *J. Am. Chem. Soc.* (2011), 133(10), 3460–3470.

100. L. D. Zhao et al., High performance thermoelectrics from earth-abundant materials: Enhanced figure of merit in PbS by second phase nanostructures. *J. Am. Chem. Soc.* (2011), 133(50), 20476–20487.

101. L. D. Zhao et al., Raising the thermoelectric performance of p-type PbS with endotaxial nanostructuring and valence-band offset engineering using CdS and ZnS. *J. Am. Chem. Soc.* (2012), 134(39), 16327–16336.

102. H. Wang et al., High thermoelectric efficiency of n-type PbS. *Adv. Energy. Mater.* (2013), 3(4), 488–495.

103. D. Parker and D. J. Singh, High temperature thermoelectric properties of rock-salt structure PbS. *Solid State Commun.* (2014), 182, 34–37.

104. Y. K. Koh et al., Lattice thermal conductivity of nanostructured thermoelectric materials based on PbTe. *Appl. Phys. Lett.* (2009), 94(15), 153101-153101-3.

105. J. P. Heremans et al., Enhancement of thermoelectric efficiency in PbTe by distortion of the electronic density of states. *Science* (2008), 321(5888), 554–557.

106. Y. N. Nasirov and Y. S. Feiziev, Effect of small substitutions of tin by neodymium on thermoelectric properties of SnTe. *Phys. Status Solidi* (1967), 24, K157–K159.

107. Y. N. Nasirov, N. P. Sultanova, and T. G. Osmanov, Thermoelectric properties of solid solutions based on SnTe-A^{II}Te-type tin telluride. *Phys. Status Solidi* (1969), 35, K39–K42.

108. N. R. Sultanova, Y. N. Nasirov, M. I. Zargarova, and M. M. Pirzade, Thermoelectric properties of a solid solution of the system SnTe-ZnTe. *Inorg. Mater.* (1974), 10, 1219–1921.

109. P. G. Rustamov, M. A. Alidzhanov, and Y. N. Babaev, The system SnTe-Tl$_2$Te$_3$. *Inorg. Mater.* (1976), 12, 715–717.

110. B. F. Gruzinov, I. A. Drabkin, and E. A. Zakomornaya, Electrical-properties of (PbSe)$_{1-x}$ (SnTe)$_x$ solid-solutions doped with In. *Sov. Phys. Semicond.* (1981), 15, 190–193.

111. V. P. Vedeneev, S. P. Krivoruchko, and E. P. Sabo, Tin telluride based thermoelectrical alloys. *Semiconductors* (1998), 32, 241–244.

112. M. M. Asadov, M. A. Alidzhanov, F. M. Mamedov, and G. I. Kelbaliev, Electrical conductivity and thermoelectric power of SnTe-based alloys doped with Fe. *Inorg. Mater.* (1998), 34, 442–444.

113. J. Androulakis, C. H. Lin, H. J. Kong, C. Uher, C. I. Wu, T. Hogan, B. A. Cook, T. Caillat, K. M. Paraskevopoulos, and M. G. Kanatzidis, Spinodal decomposition and nucleation and growth as a means to bulk nanostructured thermoelectrics: Enhanced performance in Pb$_{1-x}$Sn$_x$Te-PbS. *J. Am. Chem. Soc.* (2007), 129, 9780–9788.

114. I. U. Arachchige and M. G. Kanatzidis, Anomalous band gap evolution from band inversion in Pb$_{1-x}$Sn$_x$Te nanocrystals. *Nano Lett.* (2009), 9, 1583–1587.

115. X. Shi, J. R. Salvador, J. Yang, and H. Wang, Prospective thermoelectric materials: (AgSbTe$_2$)$_{100-x}$(SnTe)$_x$ quaternary system (x = 80, 85, 90, and 95). *Sci. Adv. Mater.* (2011), 3, 667–671.

116. M. K. Han, J. Androulakis, S. J. Kim, and M. G. Kanatzidis, Lead-free thermoelectrics: High figure of merit in p-type $AgSn_mSbTe_{m+2}$. *Adv Energy Mater.* (2012), 2, 157–161.

117. Y. Chen, M. D. Nielsen, Y. B. Gao, T. J. Zhu, X. B. Zhao, and J. P. Heremans, SnTe-AgSbTe$_2$ thermoelectric alloys, *Adv. Energy Mater.* (2012), 2, 58–62.

118. R. F. Brebrick, Deviations from stoichiometry and electrical properties in SnTe. *J. Phys. Chem. Solids* (1963), 24, 27–36.

119. J. A. Kafalas, R. F. Brebrick, and A. J. Strauss, Evidence that SnTe is semiconductor. *Appl. Phys. Lett.* (1964), 4, 93–94.

120. E. Rogacheva, Nonstoichiometry and properties of SnTe semiconductor phase of variable composition. In *Stoichiometry and Materials Science–When Numbers Matter*, A. Innocenti and N. Kamarulzaman (eds). 2011, pp 105–144.

121. M. Zhou, Z. M. Gibbs, H. Wang, Y. M. Han, C. N. Xin, L. F. Li, and G. J. Snyder, Optimization of thermoelectric efficiency in SnTe: The case for the light band. *Phys. Chem. Chem. Phys.* (2014), 16, 20741.

122. R. F. Brebrick and A. J. Strauss, Anomalous thermoelectric power as evidence for two-valence bands in SnTe. *Phys. Rev.* (1963), 131, 104–110.

123. B. A. Efimova, V. I. Kaidanov, B. Y. Moizhes, and I. A. Chernik, Band model of SnTe. *Sov. Phys. Solid State* (1966), 7, 2032–2034.

124. L. M. Roger, Valence band structure of SnTe. *J. Phys. D. Appl. Phys.* (1968), 1, 845–848.

125. S. Santhanam and A. K. Chaudhuri, Transport-properties of SnTe interpreted by means of a 2 valence band model. *Mater. Res. Bull.* (1981), 16, 911–917.

126. D. J. Singh, Thermopower of SnTe from Boltzmann transport calculations. *Funct. Mater. Lett.* (2010), 3, 223–226.

127. Q. Zhang, B. L. Liao, Y. C. Lan, K. Lukas, W. S. Liu, K. Esfarjani, C. Opeil, D. Broido, G. Chen, and Z. F. Ren, High thermoelectric performance by resonant dopant indium in nanostructured SnTe. *Proc. Natl. Acad. Sci. USA.* (2013), 110, 13261.

128. G. J. Tan, W. G. Zeier, F. Y. Shi, P. L. Wang, G. J. Snyder, V. P. Dravid, and M. G. Kanatzidis, High thermoelectric performance SnTe-In$_2$Te$_3$ solid solutions enabled by resonant levels and strong vacancy phonon scattering. *Chem. Mater.* (2015), 27, 7801.

129. G. J. Tan, F. Y. Shi, S. Q. Hao, H. Chi, L. D. Zhao, C. Uher, C. Wolverton, V. P. Dravid, and M. G. Kanatzidis, Codoping in SnTe: Enhancement of thermoelectric performance through synergy of resonance levels and band convergence. *J. Am. Chem. Soc.* (2015), 137, 5100–5112.

130. G. J. Tan, L. D. Zhao, F. Y. Shi, J. W. Doak, S. H. Lo, H. Sun, C. Wolverton, V. P. Dravid, C. Uher, and G. Kanatzidis, High thermoelectric performance of p-type SnTe via a synergistic band engineering and nanostructuring approach. *J. Am. Chem. Soc.* (2014), 136, 7006.

131. A. Banik, U. S. Shenoy, S. Anand, U. V. Waghmare, and K. Biswas, Mg alloying in SnTe facilitates valence band convergence and optimizes thermoelectric properties. *Chem. Mater.* (2015), 27, 581–587.

132. G. J. Tan, F. Y. Shi, J. W. Doak, H. Sun, L. D. Zhao, P. L. Wang, C. Uher, C. Wolveron, V. P. Dravid, and M. G. Kanatzidis, Extraordinary role of Hg in enhancing the thermoelectric performance of p-type SnTe. *Energy Environ. Sci.* (2015), 8, 267–277.

133. J. He et al., Valence band engineering and thermoelectric performance optimization in SnTe by Mn-alloying via a zone-melting method. *J. Mater. Chem. A* (2015), 3, 19974.

134. G. J. Tan, F. Y. Shi, S. Q. Hao, H. Chi, T. P. Bailey, L. D. Zhao, C. Uher, C. Wolverton, V. P. Dravid, and M. G. Kanatzidis, Valence band modification and high thermoelectric performance in SnTe heavily alloyed with MnTe. *J. Am. Chem. Soc.* (2015), 137, 11507–11516.

135. H. J. Wu et al., Synergistically optimized electrical and thermal transport properties of SnTe via alloying high-solubility MnTe. *Energy Environ. Sci.* (2015), 8, 3298.

136. W. Li, Z. W. Chen, S. Q. Lin, Y. J. Chang, B. H. Ge, Y. Chen, and Y. Z. Pei, Band and scattering tuning for high performance thermoelectric $Sn_{1-x}Mn_xTe$ alloys. *J. Materiomics* (2015), 1, 307–315.

137. R. F. Brebrick and A. I. Strauss, Partial pressures in equilibrium with group IV tellurides: II. Tin telluride. *J. Chem. Phys.* (1964), 41, 197–205.

138. R. F. Brebrick, Composition stability limits for the rocksalt-structure phase $(Pb_{1-y}Sn_y)_{1-x}Te_x$ from lattice parameter measurements. *J. Phys. Chem. Solids.* (1971), 32, 551–562.

139. L. E. Shelimova and N. K. Abrikosov, Sn-Te system in the region of SnTe compound. *Zhurnal neorgan. khimii* (1964), 9, 1879–188.

140. S. M. Tairov, B. F. Ormont, and N. O. Sostak, Study of the Pb-Sn-Te system near PbTe-SnTe quasibinary section. *Izv. Akad. Nauk SSSR. Ser. Neorgan. Mater.* (1970), 6, 1584–1588.

141. R. Maselsky and M. S. Lubell, Nonstoichiometry in some group IV tellurides. *Adv. Chem. Ser.* (1963), 39, 210–217.

142. E. I. Rogacheva, G. V. Gorne, N. K. Zhigareva, and A. B. Ivanova, Homogeneity region of tin monotelluride. *Inorg. Mater.* (1991), 27, 194–197.

143. G. S. Bushmarina, B. F. Gruzinov, I. A. Drabkin, E. Y. Lev, and V. M. Yuneev, Characteristics of the effects of In as a dopant in SnTe. *Sov. Phys. Semicond.* (1984), 18, 1374–1377.

144. Y. I. Ravich, B. A. Efimova, I. A. Smirnov, In *Semiconducting Lead Chalcogenides*. L. S. Stil'bans (ed.). Plenum, New York, 1970, pp. 165–258.

145. J. O. Dimmock, I. Melngailis, and A. J. Strauss, Band structure and laser action in $Pb_xSn_{1-x}Te$, *Phys. Rev. Lett.* (1966), 16, 1193–1196.

146. L. Esaki and P. J. Stiles, New type of negative resistance in barrier tunneling. *Phys. Rev. Lett.* (1966), 16, 1108–1111.

147. D. G. Cahill, S. K. Watson and R. O. Pohl, Lower limit to the thermal conductivity of disordered crystals, *Phys. Rev. B* (1992), 46, 6131.

148. E. Z. Xu, Z. Li, J. A. Martinez, N. Sinitsyn, H. Htoon, N. Li, B. Swartzentruber, J. A. Hollingsworth, J. Wang, and S. X. Zhang, Diameter dependent thermoelectric properties of individual SnTe nanowires. *Nanoscale* (2015), 7, 2869.

149. J. R. Sootsman, D. Y. Chung, and M. G. Kanatzidis, New and old concepts in thermoelectric materials. *Angew. Chem. Int. Ed.* (2009), 48, 8616–8639.

150. G. J. Snyder and E. S. Toberer, Complex thermoelectric materials. *Nature Mater.* (2008), 7, 105–114.

151. M. S. Dresselhaus et al., New directions for low-dimensional thermoelectric materials. *Adv. Mater.* (2007), 19, 1043–1053.

152. J. P. Hereman et al., Enhancement of thermoelectric efficiency in PbTe by distortion of the electronic density of states. *Science* (2008), 321, 554–557.

153. Y. Pei et al., Convergence of electronic bands for high performance bulk thermoelectrics. *Nature* (2011), 473, 66–669.

154. W. Liu et al., Convergence of conduction bands as a means of enhancing thermoelectric performance of n-Type $Mg_2Si_{1-x}Sn_x$ solid solutions. *Phys. Rev. Lett.* (2012), 108.

155. T. C. Harman, P. J. Taylor, M. P. Walsh, and B. E. LaForge, Quantum dot superlattice thermoelectric materials and devices. *Science* (2002), 297, 2229–2232.

156. J. P. Heremans, C. M. Thrush, and D. T. Morelli, Thermopower enhancement in lead telluride nanostructures. *Phys. Rev. B* (2004), 70.

157. K. Biswas et al., High-performance bulk thermoelectrics with all-scale hierarchical architectures. *Nature* (2012), 489, 414–418.

158. K. Biswas et al., Strained endotaxial nanostructures with high thermoelectric figure of merit. *Nature Chem.* (2011), 3, 160–166.

159. L.-D. Zhao et al., High thermoelectric performance via hierarchical compositionally alloyed nanostructures. *J. Am. Chem. Soc.* (2013), 135, 7364–7370.

160. L.-D. Zhao et al., Raising the thermoelectric performance of p-type PbS with endotaxial nanostructuring and valence-band offset engineering using CdS and ZnS. *J. Am. Chem. Soc.* (2012), 134, 16327–16336.

161. S. R. Brown, S. M. Kauzlarich, F. Gascoin, and G. J. Snyder, $Yb_{14}MnSb_{11}$: New high efficiency thermoelectric material for power generation. *Chem. Mater.* (2006), 18, 1873–1877.

162. K. Kurosaki, A. Kosuga, H. Muta, M. Uno, and S. Yamanaka, Ag_9TlTe_5: A high-performance thermoelectric bulk material with extremely low thermal conductivity. *Appl. Phys. Lett.* (2005), 87, 061919.

163. J.-S. Rhyee et al., Peierls distortion as a route to high thermoelectric performance in $In_4Se_{3-\delta}$ crystals. *Nature* (2009), 459, 965–968.

164. L. D. Zhao et al., Ultralow thermal conductivity and high thermoelectric figure of merit in SnSe crystals. *Nature* (2014), 508, 373–377.

165. D. J. Xue et al., Anisotropic photoresponse properties of single micrometer-sized GeSe nanosheet. *Adv. Mater.* (2012), 24, 4528–4533.

166. M. J. Peters and L. E. McNeil, High-pressure Mossbauer study of SnSe. *Phys. Rev.* (1990), B41, 5893–5897.

167. K. Adouby et al., Structure and temperature transformation of SnSe: Stabilization of a new cubic phase $Sn_4Bi_2Se_7$. *Zeitschrift für Kristallographie - Crystalline Materials.* (2010), 213.6, 343–349.

168. T. Chattopadhyay, J. Pannetier, and H. G. V. Schnering, Neutron diffraction study of the structural phase transition in SnS and SnSe. *J. Phys. Chem. Solids.* (1986), 47, 879.

169. V. V. Zhdavona, A phase transition of the second kind in SnSe. *Sov. Phys. Solid State.* (1961), 3, 1174–1175.

170. S. Alptekin, Structural phase transition of SnSe under uniaxial stress and hydrostatic pressure: An ab initio study. *J. Mol. Model.* (2011), 17, 2989–2994.

171. W. J. Baumgardner, J. J. Choi, Y. F. Lim, and T. Hanrath, SnSe nanocrystals: Synthesis, structure, optical properties, and surface chemistry. *J. Am. Chem. Soc.* (2010), 132, 9519–9521.

172. V. P. Bhatt, K. Gireesan, and G. R. Pandya, Growth and characterization of SnSe and $SnSe_2$ single crystals. *J. Cryst. Growth*, (1989), 96, 649–651.

173. O. Agnihotri, A. Jain, and B. K. Gupta, Single crystal growth of stannous selenide. *J. Cryst. Growth*, (1989), 96, 649–651.

174. A. B. S. H. C. Agarwal, D. Lakshminarayana, and M. Achimovicova, Growth and thermal studies of SnSe single crystals. *Mater. Lett.* (2007), 61(30), 5188–5190.

175. T. H. Patel, R. Vaidya, and S. G. P. Department, Growth and transport properties of tin mono sulpho selenide single crystals. *J. Cryst. Growth.* (2003), 253(1), 52–58.

176. A. Agarwal, P. D. Patel, and D. Lakshminarayana, Single crystal growth of layered tin monoselenide semiconductor using a direct vapour transport technique. *J. Cryst. Growth.* (1994), 142(3–4), 344–348.

177. T. Chattopadhyay, J. Pannetier, and H. G. Vonschnering, Neutron-diffractionstudy of the structural phase-transition in SnS and SnSe. *J. Phys. Chem. Solids* (1986), 47, 879–885.

178. L. D. Zhao et al., High thermoelectric performance via hierarchical compositionally alloyed nanostructures. *J. Am. Chem. Soc.* (2013), 135, 7364–7370.

179. X. Shi et al., Multiple-filled skutterudites: High thermoelectric figure of merit through separately optimizing electrical and thermal transports. *J. Am. Chem. Soc.* (2012), 134, 2842–2842.

180. H. J. Wu et al., Broad temperature plateau for thermoelectric figure of merit $ZT>2$ in phase-separated $PbTe_{0.7}S_{0.3}$. *Nat. Commun.* (2014), 5, 4515.

181. J. P. Heremans, thermoelectricity: The ugly duckling. *Nature* (2014), 508(7496), 327–328.

182. M. D. Nielsen, V. Ozolins, and J. P. Heremans, Lone pair electrons minimize lattice thermal conductivity. *Energy Environ. Sci.* (2013), 6, 570–578.

183. D. T. Morelli, V. Jovovic, and J. P. Heremans, Intrinsically minimal thermal conductivity in cubic I-V-VI2 semiconductors. *Phys. Rev. Lett.* (2008), 101, 035901.

184. Y. S. Zhang et al., First-principles description of anomalously low lattice thermal conductivity in thermoelectric Cu-Sb-Se ternary semiconductors. *Phys. Rev. B* (2012), 85(5), 054306.

185. C. W. Li et al., Orbitally driven giant phonon anharmonicity in SnSe. *Nature Phys.* (2015), 11, 1063–1069.

186. G. A. Slack, The thermal conductivity of nonmetallic crystals. In *Solid State Physics*. F. Seitz et al. (ed.). Elsevier, Amsterdam, Holland, 1979, pp. 1–71.

187. J. Yang, G. Zhang, G. Yang, C. Wang, and Y. X. Wang, Outstanding thermoelectric performances for both p- and n-type SnSe from first-principles study. *J. Alloys Compd.* (2015), 644, 615–620.

188. Y. Suzuki and H. Nakamura, A supercell approach to the doping effect on the thermoelectric properties of SnSe. *Phys. Chem. Chem. Phys.* (2015), 17, 29647–29654.

189. Z. H. Ge, K. Y. Wei, H. Lewis, J. Martin, and G. S. Nolas, Bottom-up processing and low temperature transport properties of polycrystalline SnSe. *J. Solid State Chem.* (2015), 225, 354–358.

190. C.-L. Chen, H. Wang, Y.-Y. Chen, T. Day, and G. J. Snyder, Thermoelectric properties of p-type polycrystalline SnSe doped with Ag. *J. Mater. Chem. A* (2014), 2, 11171.

191. T. Wei, C. Wu, X. Zhang, Q. Tan, L. Sun, Y. Pan, and J. Li, Thermoelectric transport properties of pristine and Na-doped $SnSe_{1-x}Te_x$ polycrystals. *Phys. Chem. Chem. Phys.* (2015), 17(44), 30102–30109.

192. Q. Zhang et al., Studies on thermoelectric properties of n-type polycrystalline $SnSe_{1-x}S_x$ iodine doping. *Adv. Energy Mater.* (2015), 5(12), 1500360.

193. E. K. Chere et al., Studies on thermoelectric figure of merit of Na-doped p-type polycrystalline SnSe. *J. Mater. Chem. A* (2015), 4, 1848–1854.

194. F. Serrano-Sánchez et al., Record Seebeck coefficient and extremely low thermal conductivity in nanostructured SnSe. *Appl. Phys. Lett.* (2015), 106, 083902.

195. E. A. Skrabek and D. Trimmer, US Patent Specification 3 945 855 (1976).
196. E. A. Skrabek and D. S. Trimmer. In *CRC Handbook of Thermoelectrics*. D. M. Rowe (ed.). CRC Press LLC, Boca Raton, FL, 1995.
197. M. S. Lubell and R. Mazelsky. *J. Electrochem. Soc.* (1963), 110, 520.
198. D. H. Damon, M. S. Lubel, and R. M. Mazelsky. *J. Phys. Chem. Solids* (1967), 28, 520.
199. T. Chattopadhyay, J. X. Boucherle, and H. G. von Schnering. *J. Phys. C: Solid State Phys.* (1987), 20, 1431.
200. E. M. Levin, M. F. Besser, and R. Hanus. *J. Appl. Phys.* (2013), 114, 083713.
201. B. A. Cook, M. J. Kramer, X. Wei, J. L. Harringa, and E. M. Levin. *J. Appl. Phys.* (2007), 101, 053715.
202. S. H. Yang, T. J. Zhu, T. Sun, J. He, S. N. Zhang, and X. B. Zhao. *Nanotechnology* (2008), 19, 245707.
203. J. R. Salvador, J. Yang, X. Shi, H. Wang, and A. A. Wereszczak. *J. Solid State Chem.* (2009), 182, 2088.
204. M. N. Schneider, T. Rosental, C. Stiewe, and O. Oeckler. *Z. Kristallogr.* (2010), 225, 463.
205. H. F. Hammann, M. O'Boyle, Y. C. Martin, M. Rooks, and H. K. Wiskramasinghe. *Nat. Mater.* (2006), 5, 383.
206. C. Woods. *Rep. Prog. Phys.* (1988), 51, 459.
207. Alfa Aesar. https://www.alfa.com/en/.
208. E. M. Levin, S. L. Bud'ko, and K. Schmidt-Rohr. *Adv. Funct. Mater.* (2012), 22, 2766.
209. Y. Gelbstein, B. Dado, O. Ben-Yehuda, Y. Sadia, Z. Dashevsky, and M. P. Dariel. *J. Electron. Mater.* (2010), 39, 2049.
210. Y. Gelbstein, O. Ben-Yehuda, E. Pinhas, T. Edrei, Y. Sadia, Z. Dashevsky, and M. P. Dariel. *J. Electron. Mater.* (2009), 38, 1478.
211. J. Davidow and Y. Gelbstein. *J. Electron. Mater.* (2012), 42, 1542.
212. E. M. Levin, J. P. Heremans, M. G. Kanatzidis, and K. Schmidt-Rohr. *Phys. Rev. B* (2013), 88, 115211.
213. E. M. Levin, R. Hanus, M. Hanson, W. E. Straszheim, and K. Schmidt-Rohr. *Phys. Status Solidi A* (2013), 210, 2628.
214. R. Wolfe, J. H. Wernick, and E. Haszko. *J. Appl. Phys.* (1960), 31, 1959.
215. A. J. Tompson, J. W. Sharp, and C. J. Raw. *J. Electron. Mater.* (2009), 38, 1407.
216. P. M. Nikolic. *Brit. J. Appl. Phys. (J. Phys. D)* (1969), 2, 383.
217. H. Sun, X. Lu, H. Chi, D. T. Morelli, and C. Uher. *Phys. Chem. Chem. Phys.* (2014), 16, 15570.
218. L. Flanders, K. R. Cummer, J. Feinsinger, and B. Heshmatpour. In *Space Technology and Applications International Forum –STAIF*. M. S. El-Genk (ed.). 2006, p. 560.
219. G. J. Snyder and E. S. Toberer. *Nat. Mater.* (2008), 7, 105.
220. Y. I. Ravich. In *CRC Handbook of Thermoelectrics*. D. M. Rowe (ed.). CRC Press LLC, Boca Raton, FL, 1995.
221. G. Benenti and G. Casati. *Philos. T. R. Soc. A* (2011), 369, 466.
222. K. Hoang, S. D. Mahanti, and M. G. Kanatzidis. *Phys. Rev. B* (2010), 81, 115106.
223. E. M. Levin, *Phys. Rev. B* (2016), 93, 045209.
224. S. H. Yang, T. J. Zhu, S. N. Zhang, J. J., Shen, and X. B. Zhao. *J. Electron. Mater.* (2010), 39, 2127.
225. T. Schröder, T. Rosenthal, N. Giesbrecht, M. Nentwig, S. Maier, H. Wang, G. J. Snyder, and O. Oeckler. *Inorg. Chem.* (2014), 53, 7722.

226. Y. Dong, A.-S. Malik, and F. J. Disalvo. *J. Electron Mater.* (2010), 40, 17.
227. E. M. Levin, B. A. Cook, J. L. Harringa, S. L. Budko, R. Venkatasubramanian, and K. Schmidt-Rohr. *Adv. Funct. Mater.* (2011), 21, 441.
228. J. P. Yesinowski. *Top Curr. Chem.* (2012), 306, 229.
229. H. Selbach, O. Kanert, and D. Wolf. *Phys. Rev. B* (1979), 19, 4435.
230. Y. Chen, C. M. Jaworski, Y. B. Gao, H. Wang, T. J. Zhu, G. J. Snyder, J. P. Heremans, and X. B. Zhao. *New J. Phys.* (2014), 16, 013067.

CHAPTER **5**

Thermoelectric Sb-Based Skutterudites for Medium Temperatures

Gerda Rogl, Andriy Grytsiv, Ernst Bauer, and Peter Rogl

Contents

5.1 Introduction

State-of-the-art Sb-based skutterudites are excellent thermoelectric (TE) materials for various reasons: (1) the starting material is cheap (for information, see Figure 5.1) and abundant, (2) they can be used in a wide temperature range (Figure 5.2), (3) they are of high TE quality, and (4) they show reasonably good mechanical performance.

The TE quality of a TE material is characterized by (1) the dimensionless figure of merit ZT and (2) the thermal–electrical conversion efficiency η. The figure of merit $ZT = S^2 T/[\rho(\lambda_e + \lambda_{ph})]$, where S, ρ, and λ are the Seebeck coefficient, the electrical resistivity, and the thermal conductivity, respectively, the last one consisting of an electronic part λ_e and a phonon part λ_{ph}.

$$\eta = \frac{T_h - T_c}{T_h} \frac{\sqrt{1+(ZT)_a} - 1}{\sqrt{1+(ZT)_a} + (T_c/T_h)}, \tag{5.1}$$

195

Price ($/kg)
1000
900
800
700
600
500
400
300
200
100
0
Bi_2Te_3
PbTe
La_3Te_4
$Si_{0.94}Ge_{0.06}$
$Si_{0.8}Ge_{0.2}$
$MnSi_{1.73}$
$Mg_2Si_{0.4}Sn_{0.6}$
Co_4Sb_{12}
$DD_{0.74}Fe_3CoSb_{12}$
TiNiSn
$Ti_{0.5}Zr_{0.25}Hf_{0.25}NiSn$
$Ba_8Ga_{16}Ge_{30}$
$Ba_8Ga_{16}Sn_{30}$
$NaCo_2O_4$
$Yb_{14}MnSb_{11}$
Tellurides
Zn_4Sb_3
Silicides
Skutterudites
Half heusler
Clathrates
Oxide
Zintl compound

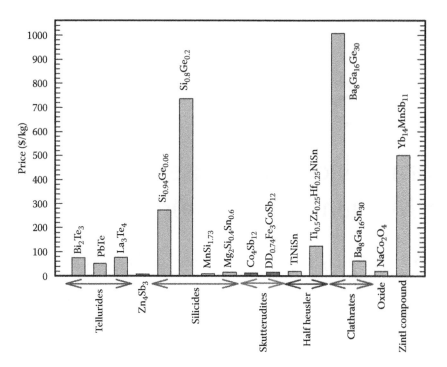

FIGURE 5.1 Prices in US dollars per kilogram for various TE compounds.

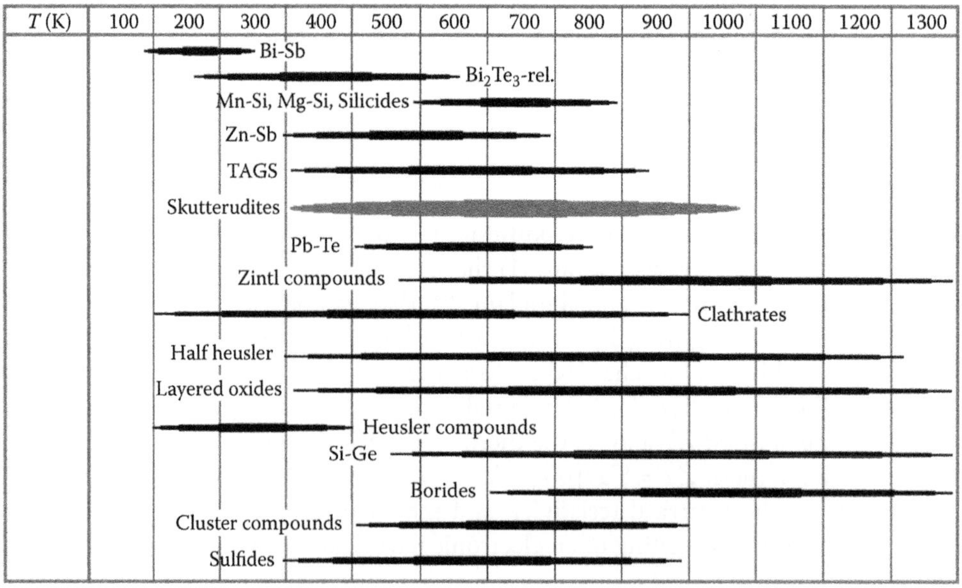

T (K)
100
200
300
400
500
600
700
800
900
1000
1100
1200
1300
Bi-Sb
Bi_2Te_3-rel.
Mn-Si, Mg-Si, Silicides
Zn-Sb
TAGS
Skutterudites
Pb-Te
Zintl compounds
Clathrates
Half heusler
Layered oxides
Heusler compounds
Si-Ge
Borides
Cluster compounds
Sulfides

FIGURE 5.2 Temperature ranges of various TE compounds. (Adapted from Kajikawa, T., Overview of Japanese activities in Thermoelectrics, Thermoelectrics Applications Workshop, San Diego, California, September 29–October 1, 2009.)

where T_h and T_c are the temperatures on the hot and the cold sides, respectively, and $(ZT)_a$ is the average ZT value between T_c and T_h. The variable $(ZT)_a$ has to be carefully evaluated because for $ZT = 0.5$ at 300 K and $ZT = 1$ at 800 K, the average ZT is only 0.75, assuming that all ZT data can be linearly connected (see Figure 5.3b). Therefore, all ZT values have to be taken into account (see Figure 5.3a and c). Generally, that means the higher ZT is over a wide temperature range, the higher the thermal–electric conversion efficiency η is (see Figure 5.3c). A detailed calculation, using ZT_{eng}, is shown in the study by Liu et al.[1]

To our knowledge, the highest ZT values for skutterudites are $ZT > 1.3$ for p-type[2] $DD_{0.59}Fe_{2.7}Co_{1.3}Sb_{11.8}Sn_{0.2}$ (DD stands for didymium, a natural double filler consisting of 4.76% Pr and 95.24% Nd) and $ZT = 1.8$ for n-type[3] $(Sr,Ba,Yb)Co_4Sb_{12} + 9.1$ wt.% $In_{0.4}Co_4Sb_{12}$ (see Figures 5.4[2,5–9] and 5.5[3,4,10–14]) with η(300 – 800 K) = 14.3% and η(400 – 800 K) = 17.6%, respectively. By introducing nanostructuring via severe plastic deformation (for instance, HPT = high-pressure torsion as a tool to further reduce the crystallite size and to increase lattice defects, dislocation density, etc., and thereby to increase the scattering of the heat-carrying phonons), the values of $ZT = 1.4$ (p-type $DD_{0.65}Fe_3Co_1Sb_{12}$[2,90]) or $ZT = 1.9$ (n-type $Sr_{0.09}Ba_{0.11}Yb_{0.05}Co_4Sb_{12} + Yb_2O_3$[4]) have been reached.

The past years have seen intense research devoted to skutterudites, most of them with the goal to enhance ZT, $(ZT)_a$, and the efficiency of this TE

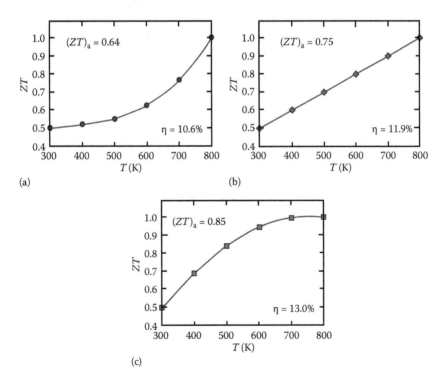

FIGURE 5.3 Variation in ZT_a for different functional dependencies of ZT: (a) positive curvature, (b) linear dependency, (c) negative curvature.

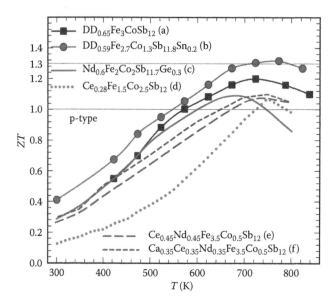

FIGURE 5.4 Temperature-dependent ZT of p-type skutterudites reaching $ZT > 1$: curve (a) (Reprinted from *Intermetallics*, 19, Rogl, G. et al., A new generation of p-type didymium skutterudites with high ZT, 546–555, Copyright (2011), with permission of Elsevier.); curve (b) (Reprinted from *Acta Materialia*, 91, Rogl, G. et al., New bulk p-type skutterudites $DD_{0.7}Fe_{2.7}Co_{1.3}Sb_{12-x}X_x$ (X = Ge, Sn) reaching $ZT > 1.3$, 227–238, Copyright (2015), with permission from Elsevier.); curve (c) (Data reprinted with permission of Zhang, L. et al., *Journal of Applied Physics* 114, 083715, 1–6, 2013, Copyright 2013, American Institute of Physics.); curve (d) (Data reprinted with permission from Tang, X. et al., *Journal of Applied Physics*, 97, 093712, 1–10, 2005. Copyright 2005, American Institute of Physics.); curve (e) (Data from Jie, Q. et al., *Physical Chemistry Chemical Physics*, 15, 6809–6818, 2013. Reproduced by permission of The Royal Society of Chemistry.); and curve (f) (Dahal, T., et al. Effect of triple fillers in thermoelectric performance of p-type skutterudites. Copyright 2015, *J. Alloys Compd.* 623, 104–108.).

material for power generation. The main progress achieved was published in many articles, and particularly in the review articles of Schierning et al.[15] and Rull-Bravo et al.,[16] a thorough update is given, and therefore, in this chapter, the focus is set on the newest research on skutterudites after 2014. For a convenient access, this chapter is organized into the following sections: "Novel Routes of Synthesis" (Section 5.2), "Filler Elements and Optimization of ZT" (Section 5.3), "Mechanical Properties" (Section 5.4), "High-Efficiency Modules" (Section 5.5), "Oxidation Resistance" (Section 5.6), "Long-Term Stability Tests" (Section 5.7), and "Theory" (Section 5.8.).

5.2 Novel Routes of Synthesis

Several groups were dealing with new synthesis methods such as high pressure–high temperature synthesis,[17–20] high-pressure synthesis (HPS)[21,22] or involving high-pressure torsion in the production process,[23] induction melting in the glove box in combination with SPS[24] or pulse plasma sintering,[25] spinodal decomposition,[26] gas atomization,[27,28] or employing a fast chemical synthesis route.[29]

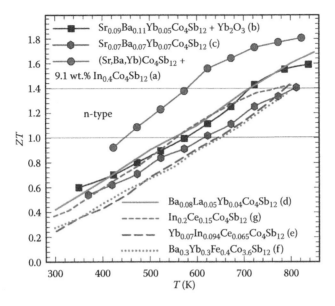

FIGURE 5.5 Temperature-dependent *ZT* of n-type skutterudites reaching *ZT* > 1.4: curve (a) (Reprinted from *Acta Materialia*, 95, Rogl, G. et al., In-doped multifilled n-type skutterudites with *ZT* = 1.8, 201–211, Copyright (2015), with permission from Elsevier.); curve (b) (Reprinted from *Acta Materialia*, 63, Rogl, G. et al., N-type skutterudites (R,Ba,Yb)$_y$Co$_4$Sb$_{12}$ (R = Sr, La, Mm, DD, SrMm, SrDD) approaching ZT~2.0, 30–43, Copyright (2014), with permission from Elsevier.); curve (c) (Data from Rogl, G. et al., *Journal of Physics: Condensed Matter*, 23, 275601, 1–11, 2011, Institute of Physics.); curve (d) (Reprinted with permission from Shi, X. et al., *Journal of the American Chemical Society*, 133, 7837–7846, 2011. Copyright 2011 American Chemical Society.); curve (e) (Reprinted from *Journal of Solid State Chemistry*, 193, Ballikaya, S. et al., High thermoelectric performance of In, Yb, Ce multiplefilled CoSb$_3$ based skutterudite compounds, 31–35, Copyright (2012), with permission from Elsevier.); curve (f) (Reprinted from *Journal of Alloys and Compounds*, 585, Ballikaya, S., and Uher, C., Enhanced thermoelectric performance of optimised Ba, Yb filled and Fe substituted skutterudite compounds, 168–172, Copyright (2014), with permission from Elsevier.); and curve (g) (Reprinted with permission from Li, H. et al., *Applied Physics Letters*, 94, 102114, 1–3, 2009. Copyright 2009, American Institute of Physics.).

Powder metallurgical synthesis under a pressure of 1.86 GPa yielded for In$_{0.5}$Co$_4$Sb$_{11.5}$Te$_{0.5}$ a *ZT* = 0.64 at 673 K,[17] which for n-type skutterudites is not an outstanding high value.

Sun et al.[18,19] synthesized In$_{0.1}$Co$_4$Sb$_{11}$Te$_{0.8}$Ge$_{0.2}$ alloys from the elements by the HTHP (at high temperature and under high pressure) method in an anvil press at a pressure of 1.5–2 GPa and a temperature of 900 K and studied the effect of In filling and Te and Ge codoping on the TE properties. A fairly good *ZT* value of 1.12 at 773 K (n-type) was obtained due to the remarkably enhanced PF and a low thermal conductivity. The same HTHP was applied to synthesize Te,Sn codoped Co$_4$Sb$_{12}$ alloys. With a pressure of 3 GPa for Co$_4$Sb$_{11.2}$Te$_{0.5}$Sn$_{0.3}$, the highest *ZT* was gained (*Z* = 3.05 × 10^{-4} K^{-1} at room temperature; *ZT* = 0.09 at 300 K, n-type).

HPS was also used for the sample preparation of p-type iodine-filled skutterudites by Li et al.[21] and Zhang et al.[22] For preparation, they used two steps (1073 K and 5 GPa, for 0.5 h; and after grinding under Ar, 753 K and

5 GPa, 3 for h) followed by sintering at 5 GPa and 573 K, for 0.5 h. After the first step, the thermal conductivity of $I_yCo_4Sb_{12}$ was very low[21]; in a second step, the highest ZT of 0.4 was observed for $IFe_{0.7}Co_{3.3}Sb_{12}$, which is at least twice that of iron-free $I_{0.79}Co_4Sb_{12}$ or I-free $Fe_{0.65}Co_{3.35}Sb_{12}$.

A superrapid method (3 h) for sample preparation was claimed by Guo et al.[24] by using induction melting in a glove box followed by SPS: $ZT = 1$ for p-type $Nd_{0.9}Fe_3CoSb_{12}$ at 760 K.

Schmitz et al.[27] used gas atomization plus consecutive short-term sintering to produce p- and n-type skutterudites. The ZT achieved with material sintered from atomized powder is comparable to that with the material sintered from preannealed powder of the same composition. This method surely qualifies for industrial production, but TE properties still depend on powder handling and sintering parameters.

Tafti et al.[30] fabricated nanostructured unfilled skutterudites (M_xCo_{1-x} $Sb_{3-y}X_y$, where M is a transition metal such as Fe and Ni and X is Te, Sn, etc.) with high purity using a high yield and green synthesis route. They took low-melting temperature salts of the constituents, melted them, and mixed them homogeneously in a hydrophobic liquid; they dried and ground the filtered product, heated it first under air at 350°C (to convert the material to oxides) and, in a second step, heated the powder under hydrogen at 450°C in order to produce skutterudite which was finally densified via SPS. No ZT was recorded, but thermal conductivity was said to be 3 W/m K at 800 K.

A comparison of TE properties from different synthesis routes was done by Sesselmann et al.,[28] who synthesized $Ce_{0.6}Fe_2Co_2Sb_{12}$ by (1) gas atomization; (2) conventional melting; and (3) from antimonide master alloys, $(Fe_xCo_{1-x}Sb_2)$ and $(CeSb_2)$, by BM and HP. Although starting from the same nominal composition, an electron probe microanalyzer (EPMA) analysis revealed that the final stoichiometry of the skutterudite samples was different. The sample produced via route 3 exhibited the highest ZT value of 0.7 at 700 K.

5.3 Filler Elements and Optimization of *ZT*

The majority of ZT improvements have been achieved through a reduction of thermal conductivity, especially of the lattice part. In a comprehensive analysis on the phonon mode contribution, relaxation time, and mean free path distributions. Guo et al.[31] confirmed that via nanoengineering, the lattice part of the thermal conductivity can be reduced; in addition, they found that a 10% substitution of Sb by As results in a significant reduction of the lattice thermal conductivity by 57%. In general, they demonstrated that elemental substitution combined with boundary scattering is an effective approach for reducing the lattice part of the thermal conductivity of $CoSb_3$.

In the review article by Dehkordi et al.,[32] the authors were looking for methods and strategies (favorable electronic band structure and transport parameters, etc.) to increase ZT by enhancing the PF ($PF = S^2/\rho$).

For further enhancement of ZT, it is of common knowledge that double or multiple filling p-type skutterudites with the formula XFe_4Sb_{12} or n-type skutterudites XCo_4Sb_{12} (where X is a rare-earth or alkaline-earth element) can, due to the different resonant frequencies and masses of the fillers, enhance the phonon scattering and, in this way, reduce the thermal conductivity, which increases ZT. Furthermore, charge compensation by substituting Co or Ni for Fe or Fe or Ni for Co can—hand in hand with a substitution-dependent filler level—control the carrier concentration and influence the TE properties; for instance, Shin and Kim[33] found the highest ZT value of 0.91 at 723 K for n-type $Nd_zFe_{4-x}Co_xSb_{12}$, $x = 0.5$, $z = 0.9$.

For p-type skutterudites, various combinations of In, Ba, La, Ce, Pr, Nd, and Yb as double fillers as well as Mm (Mm = rare-earth mischmetal) in combination with Fe/Co or Fe/Ni substitution were tested.[34–46] The ZT values in the range of 725–800 K of all these compounds are $0.70 \leq ZT \leq 0.87$. Using three fillers, Dahal et al.[9] (Figure 5.4) got a ZT value of 1.1 at 748 K for $Ca_{0.35}Ce_{0.35}Nd_{0.35}Fe_{3.5}Co_{0.5}Sb_{12}$.

Matsubara and Asahi[47] dealt with n-type skutterudites $R_{0.4}X_{0.3}Co_{4-y}Fe_ySb_{12}$ with $R_{0.4} = Ba_{0.1}Yb_{0.2}Al_{0.1}$ and with Al, Ga, In, La, Eu, and Gd as filler elements for $X_{0.3}$. Although Al and Gd reduced the thermal conductivity and enhanced the PF, these elements were not considered useful because of precipitation due to their low solubility limit. After testing various fillers, they finally reached the highest ZT value of 1.5 with $X_{0.3} = La_{0.05}Eu_{0.05}Ga_{0.1}In_{0.1}$, but to compensate for the excess number of carriers introduced by filler addition, Co sites were partially substituted with Fe. Choi et al.[48] studied the influence of In in $Ga_{0.2}In_xCo_4Sb_{12}$ finally reaching $ZT = 0.95$ at 725 K, $x = 0.3$, for the sample with the nominal composition of $Ga_{0.2}In_{0.3}Co_4Sb_{12}$.

An outstanding high $ZT = 1.8$ at 823 K with $\eta(400 - 800 K) = 17.6\%$ was achieved by Rogl et al.,[3] adding 9.1 wt.% of $In_{0.4}Co_4Sb_{12}$ to $(Sr,Ba,Yb)Co_4Sb_{12}$.

Shi et al.[49] investigated the role of Ga as a dopant in Co_4Sb_{12}, where Ga substitutes in Sb sites while also occupying void sites. They showed that by combining quantitative scanning TEM and first-principles calculations, the Ga dual site occupancy breaks the symmetry of the Sb–Sb network, splits the deep triply degenerate conduction bands, drives them downward to the band edge, and increases the overall filling fraction limit. By doping with Yb, $ZT = 1.25$ at 780 K for $Yb_{0.35}Ga_{0.15}Co_4Sb_{12}$ was gained.

Besides all these studies on multifilled skutterudites, there are some new works dealing with p- and n-type single-filled skutterudites as well, e.g., $CaFe_3CoSb_{12}$,[50] $Tl_xFe_{1.5}Co_{2.5}Sb_{12}$,[51] $Sm_yFe_{4-x}Ni_xSb_{12}$,[52] $Sm_xCo_4Sb_{12}$,[53] or $In_xCo_4Sb_{12}$[54]; however, no outstanding high ZT values (all <1.0) have been gained. Hammerschmidt and Paulus[55] from first-principles calculations present a systematic comparison of $Y_xCo_4Sb_{12}$ with Y = Ga, In, and Tl and found a threshold of $x = 0.04$. Below this, threshold agreement with experimental data was found, which is not the case for a larger filling fraction when the filling elements start to interact with the Sb_4 rings. Still, for $Yb_xCo_4Sb_{12}$, Li et al.[56] found via melt-spinning and SPS a way to enhance the filling level of Yb and for x of ~0.21–0.26; they claimed $ZT = 1.2$ at 800 K.

Tang et al.[57,58] studied the equilibrium isothermal sections of the phase diagrams of Yb–Co–Sb and of Ce–Co–Sb and found that a skutterudite material with optimized TE composition can be produced from a range of nominal compositions with appropriate annealing. The highest ZT of 1.3 at 850 K was reached for n-$Yb_{0.30}Co_4Sb_{12}$, whereas for Ce filler (an abundant and cheap filler element), they could enhance the filling level from 0.09 to 0.20, and thus, the corresponding ZT increased to ZT = 1.3 at 850 K for n-$Ce_{0.14}Co_4Sb_{12}$, so far the highest ZT for single-filled skutterudites.[59]

Short carbon fiber-reinforced p-type $CeFe_4Sb_{12}$ composites, with an optimal content of 1 vol.% carbon fibers, led to a ZT value of 0.9 at 800 K.[60]

Several groups tried to enhance ZT by substituting on the antimony 24g site of filled and unfilled skutterudites with Sn,[2,61] Te,[62,63] Te and Sn,[19,64] Se,[65] Te and Se,[65] or Te and Ge[18,19] because these dopant atoms bring perturbation on the vibration of the Sb_4 ring and thus scatter high-frequency phonons. As a result, thermal conductivity can be reduced, but as electrical resistivity and Seebeck coefficient are hardly affected, ZT can be increased. All the substitutions mentioned earlier result in higher ZT values, up to 30% or even 61%.[61,62] For $CoSb_{2.8}Te_{0.15}Se_{0.05}$, ZT = 1.29 at 780 K was the best value for an n-type unfilled skutterudite.

The substitution of Cr for Co increases the mechanical strength and, via ionized impurity scattering, significantly improves the Seebeck coefficient and, as a consequence, the PF.[66] The highest ZT of ~1.3 was achieved for $Ce_{0.25}Co_{3.9}Cr_{0.1}Sb_{12}$, which is an improvement over 30% in comparison to the Cr-free sample and comparable to the best values of single-filled $CoSb_3$ skutterudites. In addition, both the flexural strength and the fracture toughness of $CeFe_4Sb_{12}$ were enhanced with the addition of carbon fibers.

A different route to improve the TE properties of skutterudites was chosen by Battabyal et al.[67] They report about the TE properties of $Ba_{0.4}Co_4Sb_{12}$ and $Ba_{0.4}Co_4Sb_{12}Sn_{0.4}$ (nominal compositions) skutterudites dispersed with Cu_2O NPs. At 573 K, ZT values of 0.31 and 0.12 for $Ba_{0.4}Co_4Sb_{12}$ and $Ba_{0.4}Co_4Sb_{11.6}Sn_{0.4}$, respectively, were obtained, while an enhanced ZT of 0.92 and 0.26 at 573 K was achieved for Cu_2O-dispersed $Ba_{0.4}Co_4Sb_{12}$ and $Ba_{0.4}Co_4Sb_{11.6}Sn_{0.4}$ (there was no clear proof regarding the appropriate lattice site occupied by Sn).

Tan et al.[63] enhanced the TE performance of p-type $FeSb_{2.2}Te_{0.8}$ by introducing in situ formed InSb nanoinclusions, which evenly disperse at the grain boundaries of the skutterudite matrix. This way, the heat-carrying phonons are effectively impeded through enhanced interfacial scattering in conjunction with point defect scattering and mixed valence scattering. The highest ZT of 0.76 at 800 K is achieved for the $FeSb_{2.2}Te_{0.8}$ nanocomposite with 3 mol% InSb, which is an enhancement in comparison to ZT = 0.65 for $FeSb_{2.2}Te_{0.8}$ without any InSb.

Zhao et al.[68] dispersed nanosized WO_3 powder into the $Co_4Sb_{11.7}Te_{0.3}$ matrix by BM and hot press sintering. Phonon scattering by WO_3 inclusions located on the grain boundaries decreased thermal conductivity and yielded ZT = 0.71 at 750 K for the composite with 1.5% WO_3.

Fu et al.[69] chose a similar approach by inducing a "core-shell" microstructure into Yb single-filled skutterudites. This core–shell structure is formed by the thermal diffusion of well-dispersed Ni NPs in the $Yb_{0.2}Co_4Sb_{12}$ powder during HP and is composed of the normal core grains surrounded by Ni-rich nanograin shells. This way, the electrical resistivity is reduced, the PF is enhanced, and ZT is increased from ~0.5 to 1.07 at 723 K for the sample with 0.2 wt.% Ni.

Alinejad et al.[70] showed that oscillating cooling can hinder spontaneous nucleation, improve homogenous nucleation, and impose planar growth, and this way prevents porosity. Therefore, dense single-phase $CoSb_3$ samples can be produced with a ZT value of = 1.33 at 823 K.

Meng et al.[26] demonstrated that spinodal decomposition, a La-poor and a La-rich skutterudite phase within each grain of $La_{0.8}Ti_{0.1}Ga_{0.1}Fe_3CoSb_{12}$, appeared, produced by a combination of water-quenching, long-term annealing, and HP approaches followed by rapid solidification of melting–spinning plus by HP. Lattice thermal conductivity is reduced because of phonon scattering through the coherency strain fields and yields $ZT \sim 1.2$ at 700 K.

It was shown in many publications that skutterudites exhibit a higher ZT when the material is ball milled to very fine (nano)powders, although the postconsolidation of these powders via HP or SPS induces some grain/particle growth. This ZT enhancement is typically attributed to the reduced particle size, which, in turn, decreases the mean free path of phonons and consequently decreases the thermal conductivity. Short et al.[71] investigated whether there is any damage to the crystal structure in the particles formed by BM, which could also affect its thermal conductivity. Using a temperature-dependent extended X-ray absorption fine structure (EXAFS) analysis of a hand-ground and ball milled sample of the skutterudite $Nd_yFe_4Sb_{12}$, the authors could determine that BM causes no significant damage to the local structure around any atom site and that therefore, further improvements in ZT may be possible with even smaller particles.

A guide to optimizing ZT was proposed by Guo et al.,[72] who applied ultrafast spectroscopy to elucidate the carrier dynamics in filled skutterudites. Following the model that the motions of the square rings of the skutterudite structure and of the rattler atoms are coupled, Keiber and Bridges[73] showed via EXAFS that a mixture of NPs could reduce thermal conductivity leading to an improved performance of multiple filled skutterudites by nanodomains with avoided crossings of low-energy optical rattling modes and the acoustic mode at different q vectors. Sergueev et al.[74] carried out a high-pressure study of the lattice dynamics of $Eu_{0.84}Fe_4Sb_{12}$ by means of XPD and nuclear inelastic scattering. They could show that the large anharmonicity of the rattling optical mode that is hybridized with the acoustical phonons at ambient pressure is reduced at a high pressure as the phonon modes decouple and conclude that anharmonic coupling between acoustic and optical phonon modes play a central role in reduced thermal conductivity.

As far as radiation damage is concerned, Chen et al.[75] found that MeV β-rays have no influence on the TE performance of n-type skutterudites for energies lower than the MeV range.

5.4 Mechanical Properties

For TE module applications, not only excellent TE properties are necessary but also certain mechanical standards must be fulfilled; therefore, it is important to provide data on mechanical properties and thermal expansion. In this section, only such properties not yet published in previous papers and review articles[76–79] are discussed.

Li et al.[80,81] studied the intrinsic mechanical behavior of $CoSb_3$ under various shear and tensile loading conditions with the focus on the nature of the brittleness. They found that the calculated data (C_{11}, C_{12}, C_{44} from the stress–strain relationship as a function of various cell distortions from the optimized structure and B [bulk modulus], G [shear modulus], ν [Poisson number] with the Voigt–Reuss–Hill method) of all elastic constants are consistent with previously calculated (ab initio) and measured ones. They found that among all shear and tensile systems, $CoSb_3$ has the lowest ideal strength (7.17 GPa), attributed to the Co–Sb bond deformation and that the Sb rings are always easier to soften than the Co–Sb frameworks finally resulting in structural failure. Filler atoms have little effect on the Sb rings and therefore have no effect on the failure mechanism of $CoSb_3$.

Molecular dynamics simulations by Yan et al.[82] were employed to shed some light on the effect of Ba filling on the uniaxial tensile and compressive mechanical properties of crystalline $CoSb_3$. Under uniaxial tension, a slightly higher stiffness was found but a lower strength and lower ultimate strain for the fully Ba-filled $CoSb_3$ in comparison to binary $CoSb_3$. Under uniaxial compression, a higher stiffness and strength of the fully Ba-filled $CoSb_3$ (hypothetical) exists but a lower ultimate strain in comparison to binary $CoSb_3$. These results show that the skutterudites prefer to be compressive; thus, tension should be avoided in real applications.

He et al.[83] used nanoindentation to compare mechanical properties of TE materials: for p-type $Ce_{0.45}Nd_{0.45}Fe_{3.5}Co_{0.5}Sb_{12}$ and n-type $Yb_{0.35}Co_4Sb_{12}$, a hardness of 5.6 GPa and 5.8 GPa and a Young's modulus of 129.7 and 136.9 GPa, respectively, are slightly lower than those of didymium-filled $CoSb_3$ and other multifilled n-type skutterudites.[79] Dahal et al.[44] also measured with a nanoindenter the hardness (H = 4.2 GPa) and Young's modulus (E = 116 GPa) of $Mm_{0.9}Fe_{3.1}Co_{0.9}Sb_{12}$. Zhang et al.[77] used a microhardness tester for H (H = 4.95 GPa) and the resonant ultrasound spectroscopy method for E (E = 138 GPa) as well as the pulse echo method (E = 136 GPa) to measure $Mm_{0.7}Fe_3CoSb_{12}$. The hardness values measured by Dahal et al.[44] and Zhang et al.[77] can be considered as equal within the error bar, the Young's moduli of Zhang are slightly higher; however, one has to consider that with nanoindentation, only localized areas are measured, whereas resonant ultrasound spectroscopy and the pulse echo method use the elastic response of the entire bulk specimen.

5.5 High-Efficiency Modules

A recent review by Aswal et al.[84] presents an overview on the numerous aspects of the development of highly efficient TE modules based on different classes of the TE compounds. The review summarizes data on skutterudite-based nonsegmented TE modules with efficiencies $\eta_{500\div40°C} = 7\%$,[85] $\eta_{550\div70°C} = 9\%$,[86] and $\eta_{550\div90°C} = 10\%$.[87] The segmentation of the leg via a combination of skutterudites and other TE materials allows to increase the TE performance to $\eta \approx 15\%$ for temperature gradients of $1000 \div 152°C$ and $700 \div 27°C$.[88]

A more recent publication[89] provides data on the durability test of a module prepared from a p-type $(La,Ba,Ga,Ti)_1(Fe,Co)_4Sb_{12}$ ($ZT = 0.75$ at 500°C) and an n-type $(Yb,Ca,Al,Ga,In)_{0.7}(Co,Fe)_4Sb_{12}$ skutterudite ($ZT = 1.0$ at 500°C). The fabricated module with the efficiency $\eta_{600\div30°C} \approx 8\%$ was subjected to the durability test under thermal cycling from 200°C to 600°C for 8000 h. During this test, the power output decreases by 9.5%, while the module's electrical resistance increases by 6.6%.

All these outstanding results, obtained with skutterudites produced on laboratory scale, are already supported by a commercial production of p- and n-type skutterudite powders by Treibacher Industries with $ZT_{800\,K}$ values of 1.1 ± 0.1 (p-type) and 1.3 ± 0.1 (n-type). These ZT values may be increased to 1.3 (p-type) and 1.6 (n-type) by subsequent BM and/or densification conditions as shown by Rogl et al.[90] and confirmed by Nong and Hung[91] and Pryds.[92] The latter built TE modules with an efficiency of 11.2% for a temperature gradient of 533 K. Kober[93] from the German Aerospace Center reported an efficiency of their TE generator, built with the same starting material, of almost 8% for a temperature gradient of 480 K with a mass below 8 kg and a size smaller than $3\ dm^3$. Long-term stability tests on the legs made of Treibacher Industrie AG (TIAG) powder were performed in the Christian Doppler Laboratory in Vienna and are summarized in Section 5.7.1.

5.6 Oxidation Resistance

Although attractive TE efficiencies were obtained for antimony-based p- and n-type skutterudites, important aspects concerning the thermal stability of these materials at high temperatures have to be considered in order to use these materials for power generation. Antimony-based skutterudites decompose at temperatures above 700°C; however, the degradation of the TE properties due to their oxidation on air or sublimation of the antimony may appear as a crucial point for their application. Independent groups extensively investigated the oxidation behavior of various skutterudites on air, and the oxidation mechanism is rather clear.[29,94–107] In all cases, skutterudites exposed to air at temperatures above 400°C show a significant degradation of the TE efficiency.[98,101,108,109]

Table 5.1 summarizes all available data of oxidation temperatures and activation energy on air for various Sb-based skutterudites. One may see that the oxidation on air already occurs at 300–400°C. It was shown that the thermal stability of skutterudites could be enhanced by selective substitutions. Thus, the oxidation of $EuRu_4Sb_{12}$ occurs at a temperature of 460°C that is higher than 360°C reported for $EuFe_4Sb_{12}$.[29] The same tendency of the increase of the oxidation resistance with the concentration of noble elements is reported for the series $CeFe_{4-x}Ru_xSb_{12}$ (x = 1, 2, 3, 4)[100] with a clear increase of the oxidation temperature from 300°C for $CeFe_4Sb_{12}$[97] to 470°C for $CeRu_4Sb_{12}$.[100] The improvement of the oxidation resistance for

TABLE 5.1 Temperature of Onset of Oxidation (T_{oxid}) and Activation Energy (E_{ao}; Temperature Range) for Oxidation of Skutterudites on Air

Compound	T_{oxid} (K)	T_{oxid} (°C)	E_{ao} (kJ/mol); Temperature Range (K)	Reference; comments
$CeFe_4Sb_{12}$	573	300		Sklad et al.[97]
	600	327	40.5; 600–700	Xia et al.[106]
			4.0; 700–800	Xia et al.[106]
			16.0; 800–900	Xia et al.[106]
$EuFe_4Sb_{12}$	633	360		Peddle et al.[29]
$EuRu_4Sb_{12}$	733	460		Peddle et al.[29]
$CeRu_4Sb_{12}$	743	470		Sigrist et al.[100]
$CeFe_3RuSb_{12}$	673	400		Sigrist et al.[100]
$La_{0.3}Fe_3CoSb_{12}$	673	400	14; 673–873	Shin et al.[107]
$Ce_{0.9}Fe_3CoSb_{12}$	650	377		Park et al.[105]
CoP_3	743	470		Leszczynski et al.[95]
$CoSb_3$	653	380		Savchuk et al.[109]
	573	300		Savchuk et al.[109]; Fe-doped
	653	380		Wojciechowski et al.[108]
	653	380		Wojciechowski et al.[108]; 1.5 at% Te
	848	575		Broz et al.[110]
	653	380	160(21); 683–823	stage I[95]
			106(9); 683–823	stage II[95]
	644	371	161(15); 703–823	stage I[99]
			168(21); 703–823	Stage I
			106(9); 703–823	Stage II
			37.4; 773–873	Zhao et al.[98]
	673	400	170; 700–900	Xia et al.[103]
$In_{0.25}Co_3FeSb_{12}$			194; 623–723	Park et al.[102]
$In_{0.25}Co_3MnSb_{12}$			270; 625–725	Park et al.[105]
$Yb_{0.3}Co_4Sb_{12}$	650	377	140–150; <800	Xia et al.[103]
			105–115; >800	Xia et al.[103]

skutterudites with Co content in $CeFe_{4-x}Co_xSb_{12}$ becomes evident considering the data of the oxidation temperature of 337°C (650 K).[104] The same tendency for Fe/Co substituted alloys is reported for Fe-doped $CoSb_3$ thin films for which oxidation was determined to start at about 300°C, while for binary $CoSb_3$, at 380°C.[109]

Values for the activation energies for the oxidation on air (E_{ao}) for Co-based skutterudites are much higher than those for rare-earth-filled Fe-based skutterudites (Table 5.1). This tendency is explained with the high oxygen affinity of rare-earth atoms resulting in the formation of metastable "Fe_4Sb_{12}" that decomposes into $FeSb_2$ and (Sb), which in turn accelerates the oxidation rate.[106] The reduction of the oxidation resistance with increase of the rare-earth contents is also valid for $CoSb_3$-based n-type skutterudites. Xia et al.[103] investigated the influence of the Yb content on the oxidation behavior for the $Yb_yCo_4Sb_{12}$ ($0 \leq y \leq 0.4$), and they report an obvious decrease of E_{ao} with the filling fraction (y). The activation energy decreases in this series from ~170 to ~140 kJ/mol and ~80 kJ/mol, for the temperature ranges from 700 to 800 K and from 800 to 900 K, respectively. An opposite influence of the filler on the oxidation resistance is reported for In-filled skutterudites $In_{0.25}Co_3MnSb_{12}$[105] and $In_{0.25}Co_3FeSb_{12}$.[102] Remarkably high values for the activation energy for the oxidation $E_{ao,623-723\ K}$ = 270 kJ/mol for $In_{0.25}Co_3MnSb_{12}$ and $E_{ao,623-723\ K}$ = 194 kJ/mol for $In_{0.25}Co_3FeSb_{12}$ are reported. Furthermore, the thickness of the oxide layer of these indium-filled skutterudites after aging in air at 823 K was much thinner than that observed after experiments performed at 723 and 623 K. The formation of InSb is believed[102] to retard the oxidation of Sb and act as an obstacle for the growth of the oxide layer. Increased oxidation resistance at a temperature above 800 K is also reported for $CeFe_4Sb_{12}$ by Xia et al.[106] With the increase of the temperature, the oxidation rate decreases from $E_{ao,600-700\ K}$ = 40.5 kJ/mol to $E_{ao,700-800\ K}$ 4.0 kJ/mol, but above 800 K, the activation energy increases to 16.0 kJ/mol. Such behavior is explained by the appearance of the additional inner oxide layer at temperatures above 800 K and the formation of a double-layer oxide structure. A change in the morphology of the oxidized layer is also reported to be responsible for the decrease of the oxidation rate for the $Ce_{0.9}Fe_3CoSb_{12}$ at temperatures above 800 K. Qiu et al.[104] reported on the formation of oxide layers at temperatures below 800 K, but this morphology gradually disappears with the increase of temperature and oxygen needs to diffuse through the thick and more dense oxide scale, and therefore, the oxidation rate is greatly decreased.

Despite a big progress in understanding the oxidation mechanisms for the antimony-based skutterudites, the temperatures for the oxidation on air (300–400 K) are much lower than those where the maximal performance for these materials (700–900 K) is realized. Many authors agree on the fact that a protective coating or encapsulation against oxidation is requested for the practical application of the skutterudite-based modules.

The high vapor pressure of the antimony at elevated temperatures may also limit the practical application of skutterudites at elevated temperatures as shown by Broz et al.[110] To our best knowledge, the thermal decomposition of the skutterudites in protective atmosphere was investigated only for $CoSb_3$.[99,101,111] First data on the thermal stability of $CoSb_3$ in protective atmosphere (dynamic vacuum of 10^{-4} Pa; temperature range from 600°C to 700°C) were reported in 2004 by Caillat et al Snyder and caillat.[112] They were the first to realize the importance of this information; therefore, they investigated the thermal stability of $CoSb_3$ at a temperature range from 600°C to 700°C in a dynamic vacuum of 10^{-4} Pa. The temperature of the decomposition of $CoSb_3$ into $CoSb_2$ and CoSb with the sublimation of antimony was reported to occur at 575°C. The time-dependent weight losses obey a parabolic-like behavior, and the decomposition kinetics were said to be controlled by the antimony diffusion. This parabolic-like time dependence of the weight losses related to the sublimation of the antimony is also reported by Zhao et al.[101] based on the analysis of mass losses for $CoSb_3$ in the temperature range of 600–750°C in vacuum. The effective activation energy for the sublimation of antimony in $CoSb_3$ was estimated to be 44.5 kJ/mol. After a thermal duration test of 16 days at 750°C, the TE performance of $CoSb_3$ declined and the *ZT* value decreased from 0.24 to 0.16 at 327°C. Leszczynski et al.[99] employed nonisothermal differential thermal analysis and thermography in order to investigate the decomposition kinetic of $CoSb_3$. They report that the decomposition starts at 420°C and proceeds in several stages with additional temperature onsets at approximately 580°C, 691°C, 871°C, and 929°C. This temperature reported for the beginning of the decomposition is about 150° lower than the one provided by Snyder and Caillat.[112] The authors suggest that this discrepancy is originated by the use of powder samples instead of bulk samples.

These studies show that the decomposition of the skutterudites due to the sublimation of antimony is a very important issue for the practical application of this material and investigations on thermal stability for the highly efficient filled skutterudites have to be performed. One also has to take into account that TE modules operate under a temperature gradient, and they consist of two types of materials with different activity (partial pressure) of antimony. Therefore, at least two processes of gas transport have to be taken into account: (1) the sublimation of antimony at the cold parts of the module and (2) the mass exchange between p- and n-type TE legs.

5.7 Long-Term Stability Tests

Thermal stability is a very important issue in respect to application. Shin et al.[113] examined the thermal stability of $La_{0.9}Fe_3CoSb_{12}$ by varying the aging conditions of temperature, time, and atmosphere (air and vacuum). They found that $La_{0.9}Fe_3CoSb_{12}$ was significantly oxidized in air above 673 K, the thicker the oxide layers, the longer the aging temperature and time. In vacuum, no oxides formed. With these results, they recommend a protective coating for high-temperature applications. Shi et al.[114] prepared Ni/Ti/SKD/Ni (SKD = skutterudite) thermoelements and studied their interface stability with the result that all interfaces maintained good bonding strength after long-term aging and multiround shock tests as well as satisfactory stability.

5.7.1 Long-Term Stability Tests on Commercially Produced TE Powders

Thermal stability tests (TSTs) were performed at the Christian Doppler Laboratory for Thermoelectrics on TE materials, hot pressed from the powders produced by TIAG (powder production: 2012). Two disks with a diameter of ~100 mm and a height of ~10 mm were hot pressed by a TIAG partner in a commercial hot press, and samples shaped from these disks were labeled as TST1. The reference samples (TST2) were hot pressed from the same powders under laboratory conditions. For the densification under laboratory conditions, the powders were loaded in a glove box (≤5 ppm of H_2O and O_2) into individual graphite dies and hot pressed in a uniaxial hot press system (5N Argon, FCT-HPW 200/250-2200-200KS) at 650°C (p-type) and 700°C (n-type) for 1 h. The samples for TST1 were shaped as shown in Figure 5.6.

Parallelepipeds with three different geometries were cut: ~7 × 8 × 10 mm³ (labeled as geometry A), ~3.2 × 3.2 × 10 mm³ (geometry B), and ~1.8 × 1.8 × 10 mm³ (geometry C). In order to check the isotropy of the hot pressed samples, the specimens with geometry C were cut along and perpendicular to the pressing direction. All specimens were labeled; their dimensions (with accuracy ±0.01 mm) and weights (with accuracy ±0.1 mg) were recorded. The room-temperature electrical resistivity values (averaged), measured for all samples with geometry C, were 601 ± 45 μΩ cm (n-type) and 826 ± 25 μΩ cm (p-type) and show a high isotropy of the hot pressed disks. The couples of p- and n-type materials were sealed in evacuated Ar-filled (0.4 bar at room temperature) silica tubes with a length of 40 ± 1 cm and an inner diameter of 1.2 cm. The specimens were subjected to a temperature gradient T_{hot} = 600°C and T_{cold} = 80°C (Figure 5.6). The samples were located in the hot zone of the tubes, while the opposite ends of

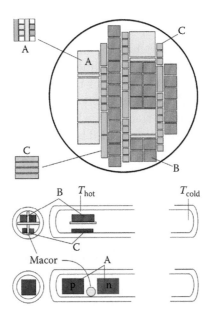

FIGURE 5.6 Schemes for shaping the samples (*top*) and for the experimental setup (*bottom*).

the tubes poked out of the resistance furnace. In order to simulate the conditions for the gas transport of antimony that could occur in TE modules, the experiments included couples of p- and n-type skutterudites, placed in the same silica tube. For the samples with geometry A, the whole set of the TE parameters and resulting *ZT* values could be derived, while for the smaller samples (geometries B and C), only electrical resistivity and thermopower were measured. Additionally, one couple with geometry A was sealed under vacuum in a tube with a length of ~7 cm and heat-treated without any temperature gradient. The specimens of type B and C were characterized after every 4–30 days by controlling their weight and electrical resistivity at room temperature. The sublimation of antimony was calculated as the weight loss for all samples sealed in the tube. Additionally, the samples after every 400–600 h of TST1 were characterized by XRD, SEM, and electron probe microanalysis. The whole set of TE properties (thermopower, electrical, and thermal conductivity) was measured within the temperature range from 300 to 850 K for the samples after TST for 504, 768, 2475, 4363, and 8800 h (argon) and 5323 h (vacuum) and compared with the respective data of the reference sample before TST (called 0 h). Additional experiments (called TST1a and TST1b) were performed with uncoupled p- and n-samples: the specimens with geometries B and C were sealed under vacuum in individual tubes and subjected to heat treatments for 612 h (TST1a: without temperature gradient) and 623 h (TST1b with a temperature gradient). Disks with a diameter and a height of 10 mm, hot pressed under laboratory conditions, were shaped to four parallelepipeds of $3.2 \times 3.2 \times 10$ mm (geometry B) and used for TST2. One couple of p- and n-materials was tested under conditions that are identical with those used for the main series TST1 (0.4 bar Ar, 504 h with temperature gradient) and called TST2a, while a second couple was tested under vacuum for 498 h and a temperature gradient (TST2b).

SEM, EPMA, and XRD were used for the characterization of the starting compacts used for TST1 and TST2. The chemical quality of the starting n-type materials was found to be almost identical for samples hot pressed under commercial and laboratory conditions. No secondary phases were detected by XRD, and only about 3 vol% of rare-earth bi-antimonides was detected by SEM and EPMA; however, the quality of p-type material, prepared with the commercial equipment, appears to be richer in secondary phases. A significant amount of $(Fe,Co)Sb_2$ is recorded by XRD. This observation can be explained because of partial oxidation during processing of the p-type materials via the schematic reaction: $R_y(FeCo)_4Sb_{12} + O_2 \rightarrow (R,Fe,Co,Sb)O_x + 4(Fe,Co)Sb_2$.

Besides this degradation, one may also expect a further degradation of the TE performance during the long-term exploitation of such materials in TE modules. During a big-scale production, it is surely difficult to provide as clean HP conditions as in a laboratory; therefore, the data of thermal stability of partially oxidized materials are also important. Besides a variety of complex oxides, known for the multicomponent system R–Fe–Co–Sb–O, the easily detectable $(Fe,Co)Sb_2$ also forms during this oxidation process. The chemical quality of the samples is consistent with their measured TE properties (Figure 5.7). It is obvious that those TE properties for n-type materials

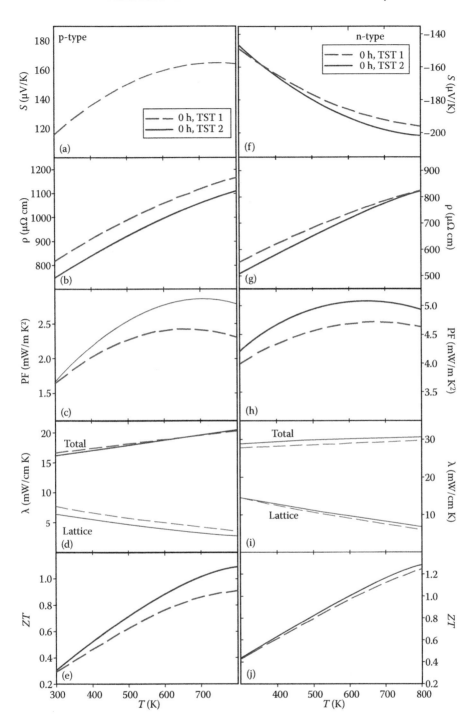

FIGURE 5.7 Comparison of temperature-dependent TE properties for the starting materials, used for TST1 and TST2: (a) and (f)—Seebeck coeficient, S; (b) and (g)—electrical resistivity, ρ; (c) and (h)—power factor, PF; (d) and (i)—thermal conductivity, λ; (e) and (j)—figure of merit, ZT.

are very similar for both TST1 and TST2 samples and that ZT values are above 1.2 at 800 K. The p-type material TST1 shows a significant degradation of the TE performance: the figure of merit at 800 K for this sample is reduced to 0.9 in comparison to the reference sample ($ZT_{800 K} = 1.1$).

5.7.1.1 Weight Changes and Sublimation of Antimony during TST

Change in the weight for the samples during 8880 h of the TST1 is shown in Figure 5.8. First, we noted an unexpected weight exchange between p- and n-type materials: p-type materials lost weight while the weight of n-type materials increased. The weight losses for the p-type materials on a level of ~1.2 wt% stabilize after ~1000 h. The same tendency is observed for the sublimation of antimony (Figure 5.8c). The second unexpected behavior can be clearly seen from the time-dependent weight change (normalized to the surface of the samples, expressed as milligrams per square centimeter) (Figure 5.8a), as there is a definite correlation to the size of the samples: the bigger the samples are, the higher is the change of the weight. This correlation is particularly observed for the n-type materials, while for p-type materials, the scattering of the data slightly hides this tendency.

The increase of the weight for n-type materials is in contradiction to the change of the density of the samples, measured by the Archimedes method (Figure 5.9). The density for both materials decreases during the first 1500–2000 h, and afterward, it stabilizes to values of 99.0(4)% and 98.5(6)% for n- and p-type materials, respectively. The observed decrease of the density for p-type materials by ~1.5% correlates well with the respective decrease of the weight (~1.2 wt%) (Figure 5.8b).

The experiments performed with uncoupled p- and n-samples: TST1a—612 h under vacuum, without temperature gradient (open squares and up triangles in Figure 5.8d through f) and TST1b—623 h with temperature gradient (filled squares and up triangles in Figure 5.8d through f) show different tendencies. One can see that almost no weight losses were observed for both materials for TST1a, while for TST1b, we observe weight losses for both types of skutterudites: n-type materials lost only ~0.1 wt%, but the weight of the p-type samples decreases to ~1.5 wt%. Different behaviors are observed for TST2 samples that were hot pressed under laboratory conditions. For this experiment, one couple of p- and n-type materials (labeled as TST2a) was tested under conditions identical with those used for the main series TST1 (0.4 bar Ar, with temperature gradient), and a second couple (TST2b) was tested under vacuum and temperature gradient. The weight losses during TST2a (filled rhombus and down triangle in Figure 5.8d through f) show that the weight of both p- and n-type samples after 504 h decreases by 0.2 wt% (~2 mg/cm²), whereas the sublimation of the antimony remains close to the level as for the samples hot pressed under commercial conditions (TST1) (Figure 5.8f). The couple, tested under vacuum for 495 h and temperature gradient (TST2b: open rhombus and down triangle in Figure 5.8d through f), shows about six times higher sublimation (Figure 5.8f), and weight losses increase to 4.2 mg/cm² for n-type and

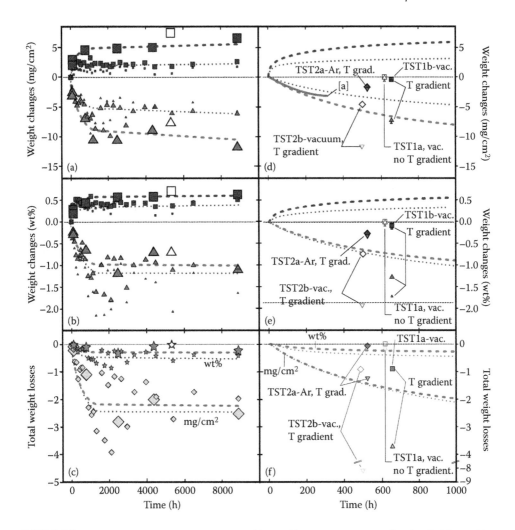

FIGURE 5.8 Time-dependent weight changes. (a) and (d)—weight changes [mg/cm²]; (b) and (e)—weight changes [wt%]; (c) and (f)—total weight losses [mg/cm² and wt%]. Dashed lines are guides for the eyes. Left panel represents data for the whole time of the TST1 (*big symbols*, samples with geometry A; *medium symbols*, samples with geometry B; *small symbols*, samples with geometry C; *open symbols*, samples annealed in vacuum without temperature gradient). Right panel zooms the guidelines for the time interval below 1000 h and represents the data for the TST1a and TST2. TST1a: commercial p- and n-type materials (with the geometries B and C) were sealed under vacuum in separate tubes. TST2: data for the reference samples prepared under laboratory conditions. [a], - weight losses for $CoSb_3$ at 600°C. (Reprinted from *Journal of Alloys and Compounds*, 509, Zhao, D. et al., High temperature sublimation behavior of antimony in $CoSb_3$ thermoelectric material during thermal duration test, 3166–3171, Copyright (2011), with permission from Elsevier.)

to 11.6 mg/cm² for p-type, respectively. The weight losses obtained during TST1 and TST2 are compared in Figure 5.8d with literature data reported for $CoSb_3$ in vacuum without temperature gradient[101] at 600°C. We note that at the same conditions (TST1a), we observe almost no losses for n- and p-type materials (Figure 5.8d, right panel open squares and up triangle). On the other hand, the losses for unfilled $CoSb_3$ tested under vacuum and without a temperature gradient[101] are even higher than for the filled n-type

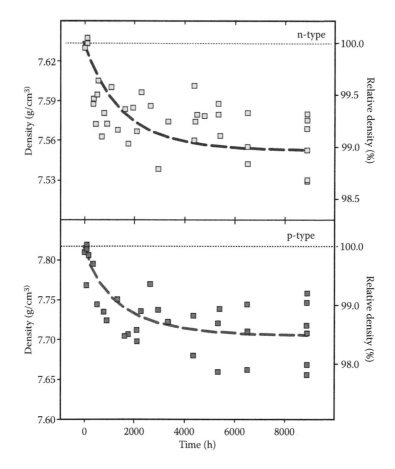

FIGURE 5.9 Time-dependent density of n- and p-type skutterudites (TST1).

skutterudite tested in this work under a temperature gradient in Ar-filled tubes, but they are less than for samples sealed in evacuated tubes. The difference in these observations indicates higher thermal stability of the filled skutterudites against sublimation of antimony.

Considering the absence of the weight exchange for the uncoupled samples and for the samples that are hot pressed under laboratory conditions, we suggest that the mass transport between p- and n-type materials for TST1 samples is related to the partial oxidation of commercial p-type samples. Thus, a significant amount of oxygen was measured by energy dispersive x-ray spectroscopy (EDX) on the surface on beginning after TST1. Despite the fact that EDX is not able to reliably measure oxygen, we clearly see that the surface of n-type materials already after 887 h of the TST1 contains about two times more oxygen (about 8 wt.%) than the p-type material, and these values remain almost unchanged during the whole TST1. On the other hand, the oxygen content, measured from the surface zone to the centre of the samples on a freshly polished surface, shows almost constant and much smaller amounts of oxygen (~1 wt.% O for both p- and n-types) typical for the instrumental noise of our EDX.

The increased amount of oxygen on the surface of the n-type material appears to be strange because of the thickness of this zone (<3 μm), which is much smaller than the 25–50 μm established for the p-type material (Figure 5.10a and b). One may see a formation of the nicely shaped crystals of $(Fe,Co)Sb_2$ on the rough surface of the p-type material (Figure 5.10d through f), while the surface of n-type materials remains much less affected by TST, and the decomposition of n-type materials with the formation of $CoSb_2$ becomes evident only in the last steps of the experiments (Figure 5.10h).

Although we observe during TST a strong change on the surface of the p-type specimens, the morphology of the samples in the inner part appears to be very similar to that observed in the sample before TST1. Only a slight increase of the porosity was detected on a distance of 100–300 μm from the surface. EPMA measurements of the composition of p- and n-skutterudites do not develop clear changes neither with time of TST nor with the distance from the surface zone to the centre of the samples.

5.7.1.2 Time-Dependent Changes of the TE Properties

The decomposition of the p-type skutterudite and the increased porosity in the surface zone have a clear effect on the electrical resistance of the samples. Figure 5.11 presents the time-dependent electrical resistivity at room temperature for the small samples from TST1 with geometry C. The resistivity of these samples was measured in two states: (1) as they come from the TST (with surface zone) (small open symbols in Figure 5.11) and after the surface with thickness of 25–50 μm was removed by grinding (filled small symbols). The resistivity of the samples with the virgin surface zone is always higher than for the samples with a cleaned surface, and this difference increases with increasing test time. After about 1000 h, the surface area of p-type materials becomes insulating.

In order to perform the measurements, we had to break the surface zone by the employment of an additional force on the voltage sensors, but even then, the measurements appear to be nonreproducible, and therefore, the values obtained cannot be considered as relevant. In contrast, the n-type samples with a thin surface zone show only a small difference (below 3%) for the electrical resistivity measured before and after grinding. The surface zone for the n-type material remains conductive at least during 5000 h of the experiment; however, the resistivity of n-type materials increases with time much stronger than for the p-type (see Figure 5.11) counterpart, suggesting a stronger degradation of the TE performance during TST.

All TE properties (ρ, S, and λ) for samples with geometry A were measured in the temperature range of 300–850 K (Figure 5.12). These samples after 768 and 8880 h of TST1 were cut (Figure 5.6) so that we could perform two sets of measurements representing the properties for materials close to the surface (called as outside) (Figure 5.13) and at the center of the samples (called as inside). The comparison of these TE properties for the inner and outer parts of the samples after 8880 TST1 (Figure 5.13) shows that the

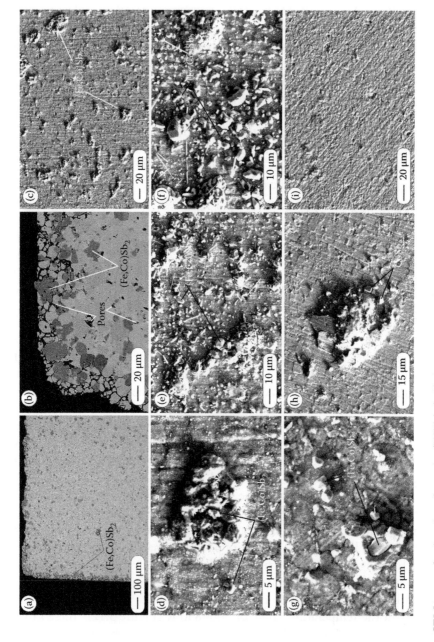

FIGURE 5.10 SEM images for the selected samples (TST1): (a, b) the cut is perpendicular to the surface for p-type sample (type C, 1391 h) that shows the biggest weight losses (above 2 wt%). The surfaces for p-type samples after (c, d) 887, (e) 2952, (f, g) 5392 h and for n-type sample after (h) 5392 and (i) 8880 h.

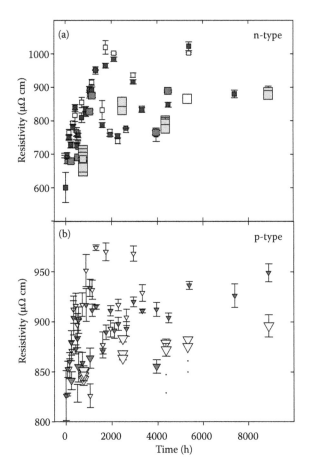

FIGURE 5.11 Time-dependent electrical resistivity at 300 K (TST1) for (a) n- and (b) p-type materials. *Small open symbols,* resistivity for samples without cleaning of the surface (type C); *big open symbols,* samples annealed in vacuum without temperature gradient (type A).

difference is only within the accuracy of the methods. Due to the small size of sample C, the TE temperature dependencies were recorded only for the electrical resistivity and Seebeck coefficient. Comparing these parameters for sample C with those measured for the sample with geometry A (Figure 5.13), one can see that they have very close thermopower values but that sample C shows a higher electrical resistivity, most likely originated by a higher porosity of the small sample. In order to estimate ZT values for the samples with geometry C, we used the lattice thermal conductivity calculated for samples with geometry A. The electronic part of thermal conductivity was calculated from the W-F law $\lambda_e = L_0 T/\rho$ considering a temperature-independent Lorenz number $L_0 = 2.44 \times 10^{-8}$ W Ω/K^2. The figure of merit (ZT), calculated by this approach for the sample with geometry C, is very close to that established for the sample with geometry A (Figure 5.13e and j). The same procedure was carried out for the samples with geometry B that were used for TST2 for 504 h under Ar and vacuum (Figure 5.14). Summarizing these comparisons, we see that the increase of the porosity for p- and n-samples

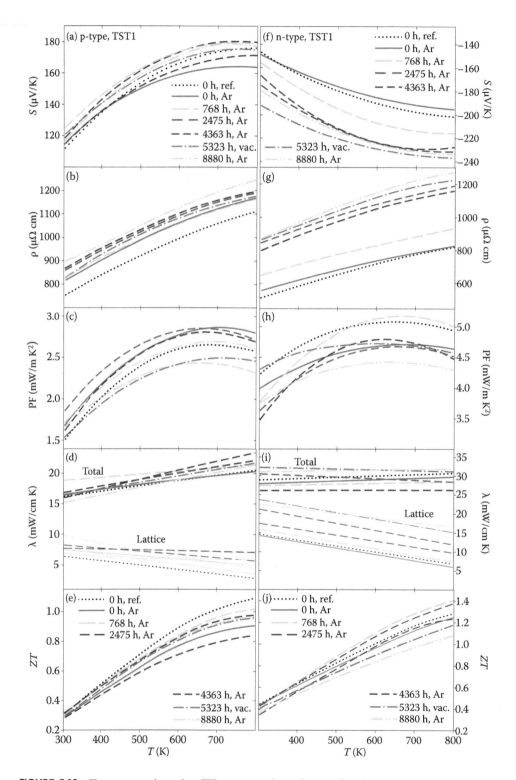

FIGURE 5.12 Temperature-dependent TE properties for p- (*left panel*) and n-type (*right panel*) samples after different time steps of the TST1: (a, f) Seebeck coeficient, S; (b, g) electrical resistivity, ρ; (c, h) power factor, PF; (d, i) thermal conductivity, λ; (e, j) figure of merit, ZT.

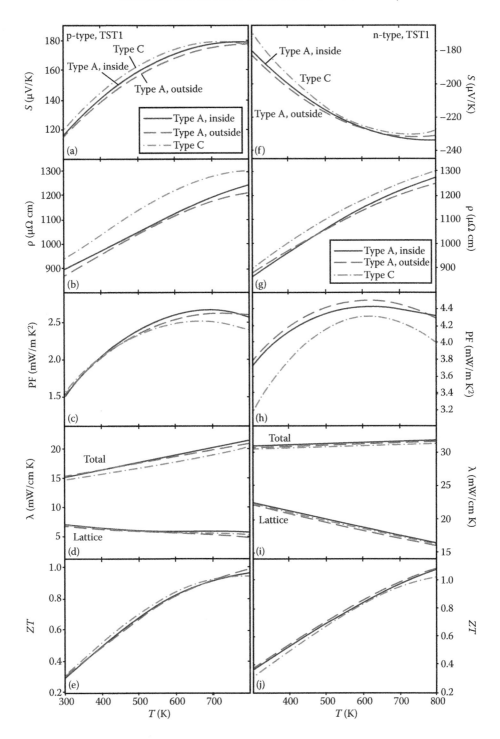

FIGURE 5.13 Comparison of temperature-dependent TE properties for samples of type A and C: (a) and (f)—Seebeck coeficient, S; (b) and (g)—electrical resistivity, ρ; (c) and (h)—power factor, PF; (d) and (i)—thermal conductivity, λ; (e) and (j)—figure of merit, ZT. The thermal conductivity for the samples of type C was calculated from the lattice thermal conductivity and calculated for the samples with geometry A.

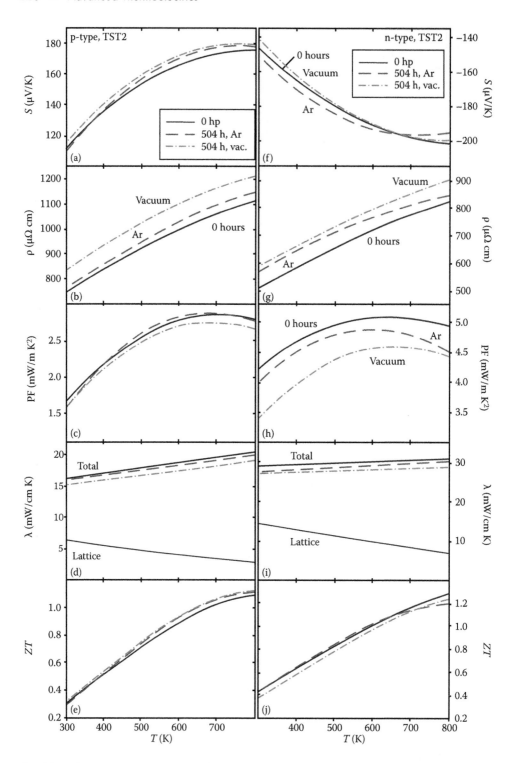

FIGURE 5.14 Temperature-dependent TE properties for p- (*left panel*) and n-type (*right panel*) samples after different time steps of the TST2: (a) and (f)—Seebeck coeficient, S; (b) and (g)—electrical resistivity, ρ; (c) and (h)—power factor, PF; (d) and (i)—thermal conductivity, λ; (e) and (j)—figure of merit, ZT.

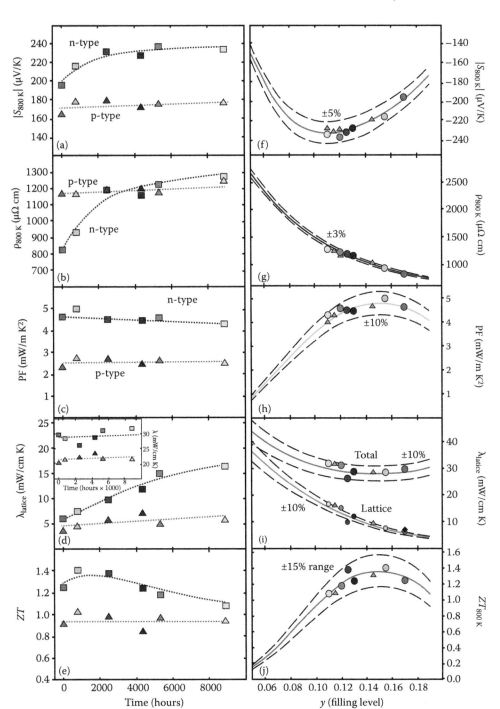

FIGURE 5.15 Time-dependent TE properties of the TST1 (*right panel*) and on the filling level in the n-type skutterudites (*left panel*) (a) and (f)—Seebeck coeficient, *S*; (b) and (g)—electrical resistivity, ρ; (c) and (h)—power factor, PF; (d) and (i)—thermal conductivity, λ; (e) and (j)—figure of merit, *ZT*.

with smaller geometry has an effect on the electrical and thermal conductivities, while *ZT* remains unaffected.

In order to evaluate the influence of time on high-temperature TE properties, the values at 800 K for the TST1 samples with geometry A were plotted versus time (Figure 5.15). The dependencies show that TE parameters for p-type materials remain almost constant during the whole TST1, while for n-type skutterudites, they show a clear dependence on time. Seebeck coefficient (Figure 5.15a) and electrical resistivity (Figure 5.15b) both increase, whereas the resulting PF (Figure 5.15c) remains almost constant. The total thermal conductivity (inset in Figure 5.15d) does not show any significant changes with time, but the lattice contribution to the thermal conductivity increases (Figure 5.15d).

The EPMA of compositions for the n-type skutterudite after different steps of TST1 do not elucidate any clear dependence from time. This is originated in the relatively low accuracy of EPMA for the determination of five rare earths (La,Ce,Pr,Nd, and Sm) with a total amount below 1 at%. On the other hand, the lattice parameters, measured with internal Si-standard, provide a much higher accuracy, and we found that they clearly decrease during TST1 indicating a decrease of the filling level in n-type skutterudite. Considering a compositional dependence of the lattice parameter on the filling level in $(MM_{0.5}Sm_{0.5})_yCo_4Sb_{12}$, this could serve as a more accurate method for the determination of the filling level. We successfully used this methodology for the determination of the filling level in $In_yCo_4Sb_{12}$ skutterudites.[115]

The filling level (y) of the $(MM_{0.5}Sm_{0.5})_yCo_4Sb_{12}$ skutterudites was determined by Rietveld refinement with constraint on an equiatomic occupation of Ce:Sm for the 2a site of the structure. The developed compositional dependence for lattice parameters in Figure 5.16 includes the samples discussed in this work, as well as other samples, investigated during the development of this n-type material. It was found that the filling level, obtained

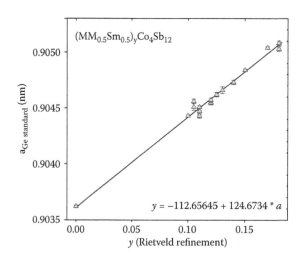

FIGURE 5.16 Compositional dependence of lattice parameters of $(MM_{0.5}Sm_{0.5})_yCo_4Sb_{12}$.

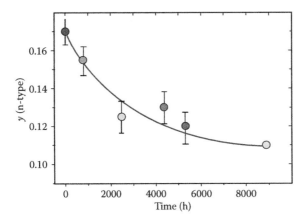

FIGURE 5.17 Time-dependent filling level (y) for n-type skutterudites.

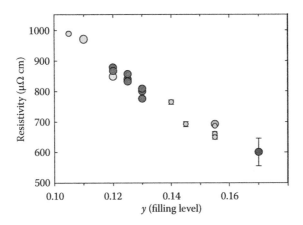

FIGURE 5.18 Dependence of electrical resistivity (300 K) on the filling level (y) in n-type skutterudites (*big symbols*, sample A; *small symbols*, sample C).

from the linear fits, increases with time (Figure 5.17) and correlates very well with electrical resistivity of the samples (Figure 5.18). Following the same strategy, the compositional dependencies of the TE properties at 800 K were plotted (Figure 5.15f through j), developing tendencies similar to those obtained for n-type skutterudites $Ba_yCo_4Sb_{12}$ and $(Sr,Ba,Yb)_yCo_4Sb_{12}$.[10,116] The only difference is that the maximal figure of merit for the skutterudites with bivalent fillers appears at $y \approx 0.25$, while in the discussed skutterudite system formed by three-valent rare earths, the region with the maximal TE performance shifts to a lower filling level of $y \approx 0.17$. Considering this observation, the maximal TE performance in all these n-type skutterudites is realized close to 0.5 extra electrons per formula considering the Zintl–Klemm concept.[117]

In conclusion, it follows that long-term (8880 h) TSTs for p- and n-type commercial skutterudites were performed at a temperature of 600°C. The experimental setup was designed in a way that allows (1) a gas transport between the n- and p-type legs located at 600°C and (2) the sublimation of the antimony from this couple on the cold side of the silica tubes, located at a temperature of 80°C. The influence of argon pressure was investigated, and it was found to have a beneficial effect on the suppression of the Sb evaporation compared with experiments performed in vacuum. The results obtained show an enhanced stability of n-type against thermal decomposition. P-type materials show stronger Sb evaporation that results in a weight loss of ~0.0001 wt%/h (or 0.001 mg/cm^2); however, the decomposition occurs only in the surface zone of the specimens with a thickness of 25–50 μm, and this insulating layer does not affect the TE performance of the legs. Unusual weight exchange between p- and n-legs, observed for the specimens hot pressed in commercial equipment, is attributed to the partial oxidation of p-type materials. The samples hot pressed under reference conditions do not show the weight exchange and exhibit a higher TE performance.

The figure of merit $ZT_{800\,K}$ during the whole TST period remains in the ranges of 0.9 ± 0.1 for p-type, but it slightly decreases from 1.2 to 1.0 for n-type materials. The decrease of the TE performance for n-type materials is explained with a gradual decrease of the filling level from $y = 0.17$ (the beginning of the TST) to $y = 0.11$ (after 8880 h of the TST). Considering that in TE modules, only a very small part of the legs will be subjected to high temperatures (600°C), the observed decrease of ZT for the n-type materials will have only a very small effect on the TE performance of the TE modules built on the base of these materials.

5.8 Theory

There are several recent theoretical papers, e.g., dealing with band structures of binary skutterudites[118] or low-energy phonon dispersion[119] or with ultralow lattice thermal conductivity in the fully filled YbFe$_4$Sb$_{12}$ (ab initio calculations).[120] Another group[121,122] studied the effect of X (X = P,As,Sb) on the transport properties of LaFe$_4$X$_{12}$ compounds. Tan et al.[123] dedicated their work to spin polarization showing that a secondary conduction band with 12 conduction carrier pockets is responsible for the extraordinary TE performance of n-type CoSb$_3$ skutterudites backed up by experiments. Zhao and et al.[124] demonstrated that in bulk TE materials, TE properties of filled skutterudites can simultaneously be optimized through three types of coexisting multilocalization transport behaviors in an independent way. First-principles calculations of structural, elastic, thermodynamic, electronic, and magnetic behavior of UFe$_4$Sb$_{12}$ was carried out by Dah et al.,[125] and a multiband analysis of TE properties of n-type Co$_{1-x}$Ni$_x$Sb$_3$ (0 < x <0.01) in a temperature range of 10–773 K by Kajikawa.[126]

Luo et al.[127] represented the importance of te Zintl concept and, by applying it, discovered a large family of filled skutterudites based on group 9 transition metals Co, Rh, and Ir; the alkali, alkaline-earth, rare-earth elements; and Sb$_4$ polyanions. They report 43 new filled skutterudites with 63 compositional variations, which can be extended to the synthesis of hundreds of additional new compounds.

References

1. Liu, W., Kim, H. S., Jie, Q. et al. 2016. Importance of high power factor in thermoelectric materials for power generation application: A perspective. *Scripta Mater.* 111:3–9.
2. Rogl, G., Grytsiv, A., Heinrich, P. et al. 2015. New bulk p-type skutterudites $DD_{0.7}Fe_{2.7}Co_{1.3}Sb_{12-x}X_x$ (X = Ge, Sn) reaching *ZT* >1.3. *Acta Mater.* 91:227–238.
3. Rogl, G., Grytsiv, A., Yubuta, K. et al. 2015. In-doped multifilled n-type skutterudites with *ZT*=1.8. *Acta Mater.* 95:201–211.
4. Rogl, G., Grytsiv, A., Rogl, P. et al. 2014. N-type skutterudites $(R,BaYb)_yCo_4Sb_{12}$ (R = Sr, La, Mm, DD, SrMm, SrDD) approaching *ZT*~2.0. *Acta Mater.* 63:30–43.
5. Rogl, G., Grytsiv, A., Rogl, P. et al. 2011. A new generation of p-type didymium skutterudites with high *ZT*. *Intermetallics* 19:546–555.
6. Zhang, L., Duan, F., Li, X. et al. 2013. Intensive suppression of thermal conductivity in $Nd_{0.6}Fe_2Co_2Sb_{12-x}Ge_x$ through spontaneous precipitates. *J. Appl. Phys.* 114:083715, 1–6.
7. Tang, X., Zhang, Q., Chen, L. et al. 2005. Synthesis and thermoelectric properties of p-type- and n-type-filled skutterudite $R_yM_xCo_{4-x}Sb_{12}$; R: Ce,Ba,Y; M:Fe,Ni. *J. Appl. Phys.* 97:093712, 1–10.
8. Jie, Q., Wang, H., Liu, W. et al. 2013. Fast phase formation of double-filled p-type skutterudites by ball-milling and hot-pressing. *Phys. Chem. Chem. Phys.* 15:6809–6816.
9. Dahal, T., Jie, Q., Liu, W. et al. 2015. Effect of triple fillers in thermoelectric performance of p-type skutterudites. *J. Alloys Compd.* 623:104–108.
10. Rogl, G., Grytsiv, A., Melnychenko-Koblyuk, N. et al. 2011. Compositional dependence of the thermoelectric properties of $(Sr_xBa_xYb_{1-2x})_yCo_4Sb_{12}$ skutterudites. *J. Phys. Condens. Mat.* 23:275601, 1–11.
11. Shi, X., Yang, J., Salvador, J. R. et al. 2011. Multiple-filled skutterudites: High thermoelectric figure of merit through separately optimizing electrical and thermal transports. *J. Am. Chem. Soc.* 133:7837–7846.
12. Ballikaya, S., Uzar, N., Yildirim, S. et al. 2012. High thermoelectric performance of In, Yb, Ce multiplefilled $CoSb_3$ based skutterudite compounds. *J. Solid State Chem.* 193:31–35.
13. Ballikaya, S., and Uher, C. 2014. Enhanced thermoelectric performance of optimised Ba, Yb filled and Fe substituted skutterudite compounds. *J. Alloys Compd.* 585:168–172.
14. Li, H., Tang, X., Zhang, Q. et al. 2009. High performance $In_xCe_yCo_4Sb_{12}$ thermoelectric materials with *in situ* forming nanostructured InSb phase. *Appl. Phys. Lett.* 94:102114, 1–3.
15. Schierning, G., Chavez, R., Schmechel, R. et al. 2015. Concepts for medium-high to high temperature thermoelectric heat-to-electricity conversion: A review of selected materials and basic considerations of module design. *Translat. Mater. Res.* 2:025001, 1–27.
16. Rull-Bravo, M., Moure, A., Fernandez, J. F. et al. 2015. Skutterudites as thermoelectric material. *RSC Adv.* 5:41653–41667.

17. Deng, L., Wang, L. B., Qin, J. M. et al. 2015. Thermoelectric properties of In-filled and Te-doped $CoSb_3$ prepared by high-pressure and high-temperature. *Mod. Phys. Lett. B* 29: 1550095, 1–7.

18. Sun, H., Jia, X., Lv, P. et al. 2015. Rapid synthesis and thermoelectric properties of $In_{0.1}Co_4Sb_{11}Te_{0.8}Ge_{0.2}$ alloys via high temperature and high pressure. *Scripta Mater.* 105:38–41.

19. Sun, H., Jia, X., Lv, P. et al. 2015. Effect of HPHT processing on the structure, and thermoelectric properties of Co_4Sb_{12} co-doped with Te and Sn. *J. Mater. Chem. A* 3(8):4637–4641.

20. Serrano, F., Gharsallah, M., Cherif, W. et al. 2015. Facile preparation of state-of-the art thermoelectric materials by high-pressure synthesis. *Mater. Today: Proc.* 2:661–668.

21. Li, X., Xu, B. Zhang, L. et al. 2014. Synthesis of iodine filled $CoSb_3$ with extremely low thermal conductivity. *J. Alloys Compd.* 615:177–180.

22. Zhang, L., Xu, B., Li, X. et al. 2015. Iodine-filled $Fe_xCo_{4-x}Sb_{12}$ polycrystals: Synthesis, structure, and thermoelectric properties. *Mater. Lett.* 139:249–251.

23. Eilertsen, J., Surace, Y., Balog, S. et al. 2015. Synthesis of skutterudite nanocomposites in situ. *Z. Anorg. Allg. Chem.* 641(8–9):1495–1502.

24. Guo, J. Q., Geng, H. Y., Ochi, T. et al. 2012. Development of skutterudite thermoelectric materials and modules. *J. Electron Mater.* 41:1036–1042.

25. Kruszewski, M. J., Zyba, R., Ciupinski, L. et al. 2016. Microstructure and thermoelectric properties of bulk cobalt antimonide ($CoSb_3$) skutterudites obtained by pulse plasma sintering. *J. Electron Mater.* 45(3):1369–1376.

26. Meng, X., Cai, W., Liu, Z. et al. 2015. Enhanced thermoelectric performance of p-type filled skutterudites via the coherency strain fields from spinodal decomposition. *Acta Mater.* 98:405–415.

27. Schmitz, A., Schmid, C., Stiewe, C. et al. 2015. Annealing and sintering effects in thermoelectric skutterudites synthesized by gas atomisation. *Phys. Status Solidi A* 213(3):758–765.

28. Sesselmann, A., Skomedal, G., Middleton, H. et al. 2016. The influence of synthesis procedure on the microstructure and thermoelectric properties of p-type skutterudite $Ce_{0.6}Fe_2Co_2Sb_{12}$. *J. Electron Mater.* 45(3):1397–1407.

29. Peddle, J. M., Gaultois, W., Michael, P. et al. 2011. On the oxidation of $EuFe_4Sb_{12}$ and $EuRu_4Sb_{12}$. *Inorg. Chem.* 50(13):6263.

30. Tafti, M. Y., Saleemi, M., Han, L. et al. 2016. On the chemical synthesis route to bulk-scale skutterudite materials. *Ceram. Int.* 42(4):5312–5318.

31. Guo, L., Zhang, Y., Zheng, Y. et al. 2016. Super-rapid preparation of nanostructured $Nd_xFe_3CoSb_{12}$, compounds and their improved thermoelectric performance. *J. Electron. Mater.* 45(3):1271–1277.

32. Dehkordi, A. M., Zebarjadi, M., He, J. et al. 2015. Thermoelectric power factor: Enhancement mechanisms and strategies for higher performance thermoelectric materials. *Mat. Sci. Eng. R.* 97:1–22.

33. Shin, D. K., and Kim, I. H. 2016. Electronic transport and thermoelectric properties of p-type $Nd_zFe_{4-x}Co_xSb_{12}$ skutterudites. *J. Electron Mater.* 45(3):1234–1239.

34. Jeon, B. J., Shin, D. K., and Kim, I. H. 2016. Synthesis and thermoelectric properties of $La_{1-z}Yb_zFe_{4-x}Ni_xSb_{12}$ skutterudites. *J. Electron. Mater.* 45(3):1907–1913.

35. Jeon, B. J., Shin, D. K., and Kim, I. H. 2015. Transport and thermoelectric properties of $La_{1-z}Ce_zFe_{4-x}Ni_xSb_{12}$ skutterudites. *J. Korean Phys. Soc.* 66(12):1862–1867.

36. Jeon, B. J., Shin, D. K., and Kim I. H. 2015. Synthesis and thermoelectric properties of $Ce_{1-z}Yb_zFe_{4-x}Ni_xSb_{12}$ skutterudites. *J. Electron. Mater.* 44(6):1388–1393.

37. Joo, G. S., Shin, D. K., and Kim, I. H. 2016. Synthesis and thermoelectric properties of type double-filled $Ce_{1-z}Yb_zFe_{4-x}Co_xSb_{12}$ skutterudites. *J. Electron. Mater.* 45(3):1251–1256.

38. Joo, G. S., Shin, D. K., and Kim, I. H. 2015. Synthesis and thermoelectric properties of p-type double-filled $La_{1-z}Ce_zFe_{4-x}Co_xSb_{12}$ skutterudites. *J. Korean Phys. Soc.* 67(2):360–365.

39. Joo, G. S., Shin, D. K., and Kim, I. H. 2015. Thermoelectric properties of double-filled p-type $La_{1-z}Yb_zFe_{4-x}Co_xSb_{12}$ skutterudites. *J. Electron. Mater.* 44(6):1383–1387.

40. Sesselmann, A., Klobes, B., Dasgupta, T. et al. 2016. Neutron diffraction and thermoelectric properties of indium filled $In_xCo_4Sb_{12}$ (x=0.05, 0.2) and indium cerium filled $Ce_{0.05}In_{0.1}Co_4Sb_{12}$ skutterudites. *Phys. Status Solidi A* 213(3):766–773.

41. Shin, D. K., and Kim, I. H. 2015. Electronic transport and thermoelectric properties of double-filled $Pr_{1-z}Yb_zFe_{4-x}Co_xSb_{12}$ skutterudites. *J. Korean Phys. Soc.* 67(7):1208–1213.

42. Song, K.-M., Shin, D.-K., and Kim, I.-H. 2015. Synthesis and thermoelectric properties of double-filled $La_{1-z}Nd_zFe_{4-x}Co_xSb_{12}$ skutterudites. *J. Korean Phys. Soc.* 67(9):1597–1602.

43. Song, K.-M., Shin, D.-K., and Kim, I.-H. 2016. Thermoelectric properties of p-type $La_{1-z}Pr_zFe_{4-x}Co_xSb_{12}$ skutterudites. *J. Electron Mater.* 45(3):127–133.

44. Dahal, T., Gahlawat, S., Jie, Q. et al. 2015. Thermoelectric and mechanical properties on mischmetal filled p-type skutterudites $Mm_{0.9}Fe_{4-x}Co_xSb_{12}$. *J. Appl. Phys.* 117:055101, 1–8.

45. Dong, Y., Nolas, G. S., Zeng, X. et al. 2015. High temperature thermoelectric properties of $Ba_xYb_yFe_3CoSb_{12}$ p-type skutterudites. *J. Mat. Res.* 30(17):2558–2563.

46. Lee, W. M., Shin, D. K., and Kim, I. H. 2016. Thermoelectric and transport properties of $Ce_zFe_{4-x}Ni_xSb_{12}$. *J. Electron Mater.* 45(3):1245–1250.

47. Matsubara, M., and Asahi, R. 2016. Optimization of filler elements in $CoSb_3$-based skutterudites for high-performance n-type thermoelectric materials. *J. Electron Mater.* 45(3):1669–1678.

48. Choi, S., Kurosaki K., Li, G. et al. 2016. Enchanted thermoelectric properties of Ga and In Co-added $CoSb_3$-based skutterudites with optimized chemical and microstructure. *AIP Advance J2 - AIP Adv.* 6(12): 125015 (9 pp.)

49. Shi, X., Yang, J., Wu, L. et al. 2015. Band structure engineering and thermoelectric properties of charge-compensated filled skutterudites, *Sci. Rep.* 5:14641, 1–10.

50. Thompson, D. R., Liu, C., Yang, J. et al. 2015. Rare-earth free p-type filled skutterudites: Mechanisms for low thermal conductivity and effects of Fe/Co ratio on the band structure and charge transport. *Acta Mater.* 92:152–162.

51. Choi, S., Kurosaki, K., Yusufu, A. et al. 2015. Thermoelectric properties of p-type Tl-filled skutterudites. *J. Electron. Mater.* 44(6):1743–1749.

52. Carlini, R., Khan, A. U., Ricciardi, R. et al. 2016. Synthesis, characterization and thermoelectric properties of Sm filled $Fe_{4-x}Ni_xSb_{12}$ skutterudites. *J. Alloys Compd.* 655:321–326.

53. Zhang, Q., Chen, C., Kang, Y. et al. 2015. Structural and thermoelectric characterizations of samarium filled $CoSb_3$ skutterudites. *Mater. Lett.* 143:41–43.

54. Visnow, E. Heinrich, C. P., Schmitz, A. et al. 2015. On the true indium content of In-filled skutterudites. *Inorg. Chem.* 54:7818–7827.

55. Hammerschmidt, L., and Paulus, B. 2015. Electronic structure and transport properties of filled $CoSb_3$, skutterudites by first principles. *Phys. Status Solidi A* 213(3):750–757.

56. Li, Y., Qiu, P., Xiong, Z. et al. 2015. Electrical and thermal transport properties of $Yb_xCo_4Sb_{12}$ filled skutterudites with ultrahigh carrier concentrations. *AIP Advances* 5:117239, 1–9.

57. Tang, Y., Chen, S. W., Snyder, G. J. 2015. Temperature dependent solubility of Yb in $YbCoSb_3$ skutterudite and its effect on preparation, optimization and lifetime of thermoelectrics. *J. Materiomics* 1:75–84.

58. Tang, Y., Hanus, R., Chen, S.-W. et al. 2015. Solubility design leading to high figure of merit in low-cost $Ce-CoSb_3$ skutterudites. *Nature Commun.* 6:7584, 1–7.

59. Morelli, D. T., Meisner, G. P., Chen, B. et al. 1997. Cerium filling and doping of cobalt triantimonide. *Phys. Rev. B* 56:7376–7383.

60. Wan, S., Huang, X., Qiu, P. et al. 2015. The effect of short carbon fibers on the thermoelectric and mechanical properties of p-type $CeFe_4Sb_{12}$ skutterudite composites. *Mater. Design* 67:379–384.

61. Duan, F., Zhang, L., Dong, J. et al. 2015. Thermoelectric properties of Sn substituted p-type Nd filled skutterudites. *J. Alloys Compd.* 639:68–73.

62. Fu, L., Yang, J., Xiao, Y. et al. 2016. Thermoelectric performance enhancement of $CeFe_4Sb_{12}$ p-Type skutterudite by disorder on the Sb_4-rings induced by Te doping and nanopores, *J. Electron Mater.* 45(3):1240–1244.

63. Tan, G., Chi, H., Liu, W. et al. 2015. Toward high thermoelectric performance p-type $FeSb_{2.2}Te_{0.8}$ via in situ formation of InSb nanoinclusions. *J. Mater. Chem. C* 3:8372–8380.

64. Navratil, J., Plecha, T., Drasar, C. et al. 2016. The influence of Sn additions on the thermoelectric and transport properties of $FeSb_2$ Te-based ternary skutterudites. *J. Electron Mater* 45(6):2904–2913.

65. Dong, J., Yang, K., Xu, B. et al. 2015. Structure and thermoelectric properties of Se- and Se/Te-doped $CoSb_3$ skutterudites synthesized by high-pressure technique. *J. Alloys Compd.* 647:295–302.

66. Wang, S., Yang, J., Wu, L. et al. 2015. On intensifying carrier impurity scattering to enhance thermoelectric performance in Cr-doped $Ce_yCo_4Sb_{12}$. *Adv. Funct. Mater.* 25(42):6660–6670.

67. Battabyal, M., Priyadarshini, B., Sivaprahasam, D. et al. 2015. The effect of Cu_2O nanoparticle dispersion on the thermoelectric properties of n-type skutterudites. *J. Phys. D: Appl. Phys.* 48:455309, 1–8.

68. Zhao, D., Zuo, M., Wang, Z. et al. 2015. Effects of WO_3 micro/nano-inclusions on the thermoelectric properties of $Co_4Sb_{11.7}Te_{0.3}$ skutterudite. *J. Nanosci. Nanotech.* 15(4):3076–3080.

69. Fu, L., Yang, J., Peng, J. et al. 2015. Enhancement of thermoelectric properties of Yb-filled skutterudites by an Ni-induced "core-shell" structure. *J. Mater. Chem. A* 3(3):1010–1016.

70. Alinejad, B., Castellero, A., and Baricco, M. 2016. Full dense $CoSb_3$ single phase with high thermoelectric performance prepared by oscillated cooling method. *Scripta Mater.* 113:110–113.

71. Short, M., Bridges, F., Keiber, T. et al. 2015. A comparison of the local structure in ball-milled and hand ground skutterudite samples using EXAFS. *Intermetallics* 63:80–85.

72. Guo, R., Wang, X., and Booling, H. 2015. Thermal conductivity of skutterudite $CoSb_3$ from first principles: Substitution and nanoengineering effects. *Sci. Rep.* 5:7806, 1–9.
73. Keiber, T., and Bridges, F. 2015. Modeling correlated motion in filled skutterudites. *Phys. Rev. B* 92(13):134111, 1–10.
74. Sergueev, I., Glazyrin, K., Kantor, I. et al. 2015. Quenching rattling modes in skutterudites with pressure. *Phys. Rev. B* 91: 224304, 1–7.
75. Chen, J., Zha, H., Xia, X. et al. 2015. Influence of high energy β-radiation on thermoelectric performance of filled skutterudites compounds. *J. Alloys Compd.* 640:388–392.
76. Rogl, G., Zhang, L., Rogl, P. et al. 2010. Thermal expansion of skutterudites. *J. Appl. Phys.* 107:043507, 1–10.
77. Zhang, L., Rogl, G., Grytsiv, A. et al. 2010. Mechanical properties of filled antimonide skutterudites. *Mater. Sci. Eng. B* 170:26–31.
78. Rogl, G., Puchegger, S., Zehetbauer, M. et al. 2011. Dependence of the elastic moduli of skutterudites on density and temperature. *MRS Symp. Proc.* 1325:13–19.
79. Rogl G., and Rogl, P. 2011. Mechanical properties of skutterudites. *Sci. Adv. Mater.* 3:517–538.
80. Li, G., An, Q., Li, W. et al. 2015. Brittle failure mechanism in thermoelectric skutterudite $CoSb_3$. *Chem. Mater.* 27:6329–6336.
81. Li, G., An, Q., Goddard III, W. A. et al. 2016. Atomistic explanation of brittle failure of thermoelectric skutterudite $CoSb_3$. *Acta Mater.* 103:775–780.
82. Yang, X. Q., Li, W. J., Chen, G. et al. 2015. Ba-filling effect on the uniaxial tensile and compressive mechanical behavior of crystalline $CoSb_3$: A molecular dynamics study. *J. Electron Mater.* 44(6):1438–1443.
83. He, R., Gahlawat, S., Guo, C. et al. 2015. Studies on mechanical properties of thermoelectric materials by nanoindentation. *Phys. Status Solidi A* 212(10):2191–2195.
84. Aswal, D. K., Basu, R., and Singh, A. 2016. Key issues in development of thermoelectric power generators: High figure-of-merit materials and their highly conducting interfaces with metallic interconnects. *Energy Convers. Manage.* 114:50–67.
85. Salvador, J. R., Cho, J. Y., Ye, Z. et al. 2014. Conversion efficiency of skutterudite-based thermoelectric modules. *Phys. Chem. Chem. Phys.* 16(24): 12510–12520.
86. Muto, A., Yang, J., Poudel, B. et al. 2013. Skutterudite unicouple characterization for energy harvesting applications. *Adv. Energy Mater.* 3:245–251.
87. Guo, L., Xu, X., and Salvador, J. R. 2015. Ultrafast carriers dynamics in filled-skutterudites. *Appl. Phys. Lett.* 106:231902, 1–4.
88. Caillat, T., Borshchevsky, A., Snyder J. G. et al. 2001. High efficiency segmented thermoelectric unicouples. *Proc. AIP Conf.* 552:1107–1112.
89. Ochi, T., Nie, G., Suzuki, S. et al. 2014. Power-generation performance and durability of a skutterudite thermoelectric generator. *J. Electron Mater.* 43(6):2344–2347.
90. Rogl, G., Grytsiv, A., Rogl, P. et al. 2014. Nanostructuring of p- and n-type skutterudites reaching figures of merit of approximately 1.3 and 1.6, respectively. *Acta Mater.* 76:434–448.
91. Nong, N. V., and Hung, L. T. 2016. Development of high performance thermoelectric modules for waste heat harvesting. Oral presentation at EMN, Thermoelectric Materials Meeting, Orlando, FL, February 21–24, 2016.

92. Pryds, N. 2015. High temperature thermoelectric generators for efficient power generation: Modelling and performance. Oral presentation at the Autumn School Thermoelectrics, Duisburg, October 5–8, 2015.

93. Kober, M. 2016. Thermoelectric generators with high power density for application in hybrid cars. Oral Presentation at EMN, Thermoelectric Materials Meeting, Orlando, FL, February 21–24, 2016.

94. Hara, R., Inoue, S., Kaibe, H. T. et al. 2003. Aging effects of large-size n-type $CoSb_3$ prepared by spark plasma sintering. *J. Alloys Compd.* 349:297–301.

95. Leszczynski, J., Malecki, A. L., and Wojciechowski, K. T. 2007. Comparison of thermal oxidation behavior of $CoSb_3$ and CoP_3. Proceedings of the 5th European Conference on Thermoelectrics, Odessa, September 10–12, (2007), 202–206.

96. Godlewska, E., Zawadzka, K., Adamczyk, A. et al. 2010. Degradation of $CoSb_3$ in air at elevated temperatures. *Oxid. Met.* 74(3):113–124.

97. Sklad, A. C., Gaultois, M. W., and Grosvenor A. P. 2010. Examination of $CeFe_4Sb_{12}$ upon exposure to air: Is this material appropriate for use in terrestrial, high-temperature thermoelectric devices? *J. Alloys Compd.* 505(1):L6–L9.

98. Zhao, D., Tian, C., Tang, S. et al. 2010. High temperature oxidation behavior of cobalt triantimonide thermoelectric material. *J. Alloys Compd.* 504(2):552–558.

99. Leszczynski, J., Wojciechowski, K. T., and Malecki, A. L. 2011. Studies on thermal decomposition and oxidation of $CoSb_3$. *J. Therm. Anal. Calorim.* 105(1):211–222.

100. Sigrist, J. A., Walker, J. D. S., Hayes, J. R. et al. 2011. Determining the effect of Ru substitution on the thermal stability of $CeFe_{4-x}Ru_xSb_{12}$. *Solid State Sci.* 13(11):2041–2048.

101. Zhao, D., Tian, C., Liu, Y. et al. 2011. High temperature sublimation behavior of antimony in $CoSb_3$ thermoelectric material during thermal duration test. *J. Alloys Compd.* 509:3166–3171.

102. Park, K. H., You, S. W., Ur, S. C. et al. 2012. High-temperature stability of thermoelectric skutterudite $In_{0.25}Co_3FeSb_{12}$. *J. Electron. Mater.* 41(6):1051–1056.

103. Xia, X., Qiu, P., Shi, X. et al. 2012. High-temperature oxidation behavior of filled skutterudites $Yb_yCo_4Sb_{12}$. *J. Electron. Mater.* 41(8):2225–2231.

104. Qiu, P., Xia, X., Huang, X. et al. 2014. "Pesting"-like oxidation phenomenon of p-type filled skutterudite $Ce_{0.9}Fe_3CoSb_{12}$. *J. Alloys Compd.* 612:365–371.

105. Park, K. H., Seo, W. S., Choi, S.-M. et al. 2014. Thermal stability of the thermoelectric skutterudite $In_{0.25}Co_3MnSb_{12}$. *J. Korea Phys. Soc.* 64(1):79–83.

106. Xia, X., Qiu, P., Huang, X. et al. 2014. Oxidation behavior of filled skutterudite $CeFe_4Sb_{12}$ in air. *J. Electron Mater.* 43(6) 1639–1644.

107. Shin, D. K., Kim, I. H., Park, K. H. et al. 2015. Thermal stability of $La_{0.9}Fe_3CoSb_{12}$ skutterudite. *J. Electron Mater.* 44(6):1858–1863.

108. Wojciechowski, K. T., Leszczynski, J., and Gajerski, R. 2002. Thermal durability properties of pure and Te-doped $CoSb_3$, Proceedings of the 7th European Workshop on Thermoelectrics, Pamplona, 2002.

109. Savchuk, V., Schumann, J., Schupp, B. et al. 2003. Formation and thermal stability of the skutterudite phase in films sputtered from $Co_{20}Sb_{80}$ targets. *J. Alloys Compd.* 351(1–2):248–254.

110. Broz, P., and Zelenka, F. 2015 Specially-adapted type of Netzsch STA instrument as a tool for Knudsen effusion mass spectrometry. *Int. J. Mass Spectrom.* 383–384:13–22.

111. Caillat, T., Sakamoto, J., Lara, L. et al. Progress status of skutterudite-based segmented thermoelectric technology developments. Oral presentation at the 23rd International Conference on Thermoelectrics. Adelaide, July 25–30, 2004.

112. Snyder, G. J., and Caillat, T. 2004. High efficiency thermoelectrics. Paper presented at the Thermoelectric Workshop, San Diego, February 17-20, 2004.

113. Shin, D. K., Kim, I. H., Park, K.-H. et al. 2015. Thermal stability of $La_{0.9}Fe_3CoSb_{12}$ skutterudite. *J. Electron.Mater.* 44(6):1858–1863.

114. Shi, L., Huang, X., Gu, M. et al. 2016. Interfacial structure and stability in Ni/SKD/Ti/Ni skutterudite thermoelements. *Surf. Coat. Tech.* 285:312–317.

115. Grytsiv, A., Rogl, P., Michor, H. et al. 2013. $In_yCo_4Sb_{12}$ skutterudite: Phase equilibria and crystal structure. *J. Electron. Mater.* 42(10):2940–2952.

116. Zhang, L., Grytsiv, A., and Rogl, P. 2009. High thermoelectric performance of triple-filled n-type skutterudites $(Sr,Ba,Yb)_yCo_4Sb_{12}$. *J. Phys. D Appl. Phys.* 42(22) 225405, 1–9.

117. Schäfer, H. 1985. On the problem of polar intermetallic compounds: The stimulation of E. Zintl's work for the modern chemistry of intermetallics. *Ann Rev. Mater. Sci.* 15:1–12.

118. Khan, B., Rahnamaye Aliabad, H. A., Saifullah. et al. 2015. Electronic band structures of binary skutterudites. *J. Alloys Compd.* 647:364–368.

119. Koza, M. M., Boehm, M., Sischka, E. et al. 2015. Low-energy phonon dispersion in $LaFe_4Sb_{12}$. *Phys. Rev. B* 91(1):014305, 1–8.

120. Li, W., and Mingo, N. 2015. Ultralow lattice thermal conductivity of the fully filled skutterudite $YbFe_4Sb_{12}$ due to the flat avoided-crossing filler mode. *Phys. Rev. B* 91(14) 144304, 1–6.

121. Reshak, A. H. 2015. Effect of X on the transport properties of skutterudites $LaFe_4X_{12}$ (X = P, As and Sb) compounds. *J. Alloys Compd.* 651:176–183.

122. Reshak, A. H. 2016. Spin-polarization in filled-skutterudites $LaFe_4Pn_{12}$ (Pn = P, As and Sb). *J. Magn. Magn. Mater.* 40:684–694.

123. Tang, Y., Gibbs, Z. M., Agapito, L. A. et al. 2015. Convergence of multi-valley bands as the electronic origin of high thermoelectric performance in $CoSb_3$ skutterudites. *Nature Mater.* 14:1223–1228.

124. Zhao, W., Wei, P., Zhang, Q. et al. 2015. Multi-localization transport behaviour in bulk thermoelectric materials. *Nature Commun.* 6:6197, 1–7.

125. Dah, S., Ameri, M., Al Douri, Y. et al. 2016. First-principles calculations of structural, elastic, thermodynamic, electronic and magnetic investigations of the filled skutterudite alloy UFe_4Sb_{12}. *Mat. Sci. Semicond. Proc.* 41:102–108.

126. Kajikawa, Y. 2016. Multi-band analysis of thermoelectric properties of n-type $Co_{1-x}Ni_xSb_3$ (0≤x≤0.01) over a wide temperature range of 10–773 K. *J. Alloys Compd.* 664:338–350.

127. Luo, H., Krizan, J. W., Muechler, L. et al. 2015. A large family of filled skutterudites stabilized by electron count. *Nature Commun.* 6:6489, 1–10.

CHAPTER 6

Mg$_2$B^{IV} for Medium Temperatures

Tiejun Zhu, Guanting Yu, Xinbing Zhao, Weishu Liu,
Zhifeng Ren, Johannes de Boor, and Udara Saparamadu

Contents

6.1 Thermoelectric Properties of $Mg_2Si_{1-x}Sn_x$ Solid Solutions

Tiejun Zhu, Guanting Yu, and Xinbing Zhao

6.1.1 Introduction

For the past 60 years, massive investigations have been carried out regarding electrical, optical, and thermal properties of Mg_2B^{IV} (B^{IV} = Si, Ge, Sn) compounds.[1–3] In the recent decades, more attention has been paid to the $Mg_2Si_{1-x}Sn_x$ solid solutions as potential TE materials, because of their high TE performance and environmentally friendly feature with cheap, abundant, and nontoxic constituent elements.[4,5]

The family of Mg_2B^{IV} compounds, including their alloys, crystallize in the antifluorite structure (space group $Fm\bar{3}m$) with B^{IV} in face-centered cubic positions and Mg in tetrahedral sites. The crystal parameters of Mg_2B^{IV} compounds are presented in Table 6.1,[4,6] as well as some basic physical properties.

Generally, three strategies have been used to improve the figure of merit ZT of Mg_2B^{IV} materials. The first one is to optimize carrier concentration through doping. Sb, Bi, Al, La, and Ca have been studied as n-type dopants,[7–11] and Ag, Na, Ga, and Li as p-type dopants.[12–15] The second is to form solid solutions. The ZT can be enhanced due to the increased PF by band convergence and the reduced thermal conductivity by the strong alloy scattering.[5,16,17] The third is a point defect chemistry approach.[18] By changing the Sb doping ratio and excess Mg contents, intrinsic defects, i.e., Mg vacancies and Mg interstitials, can be adjusted and low thermal conductivity as well as high electrical properties can be obtained. Some typical work is presented in Figure 6.1. For n-type $Mg_2Si_{1-x}Sn_x$, $ZT \approx 1.3$ has been attained in Sb-doped $Mg_2Si_{0.3}Sn_{0.7}$,[16] while the maximum ZT value of p-type $Mg_2Si_{1-x}Sn_x$ materials is only 0.5, obtained in Li-doped $Mg_2Si_{0.3}Sn_{0.7}$.[15] Unlike the n-type solid solutions with band convergence effects, the valence band structure has low band degeneracy, which results in the poor p-type electrical properties. Moreover, the hole mobility is much lower than the electron one in all Mg_2B^{IV} compounds.

TABLE 6.1   Crystal Parameters and Physical Properties of Mg_2B^{IV} Compounds

Compound	Melting Point (K)	Spacing (Å)	Density (g/cm³)	E_g (eV)	u_n (300 K)	u_p (cm²/V s)	m_n/m_0	m_p/m_0
Mg_2Si	1375	6.338	1.88	0.77	405	65	0.50	0.90
Mg_2Ge	1388	6.384	3.08	0.74	530	110	0.18	0.31
Mg_2Sn	1051	6.765	3.59	0.35	320	260	1.20	1.30

Source: Zaitsev, V. et al., *Thermoelectrics Handbook*, CRC Press, Taylor & Francis, Boca Raton, Florida, 2006. With permission; Fedorov, M. I., and Isachenko, G. N., *Japanese Journal of Applied Physics*, 54, 07JA05, 2015, Institute of Physics.

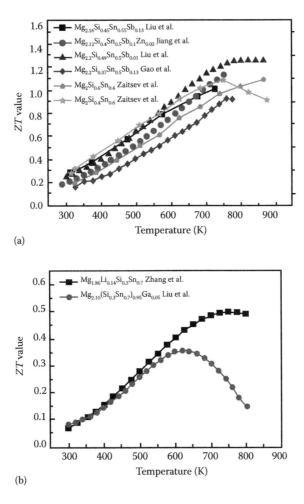

FIGURE 6.1 Temperature dependencies of *ZT* of (a) n-type Mg$_2$(Si, Sn)[5,17,19–21] and (b) p-type Mg$_2$(Si, Sn).[13,15]

6.1.2 Phase Diagram of Mg$_2$Si$_{1-x}$Sn$_x$ Solid Solutions

Phase diagrams of Mg$_2$B^{IV} (B^{IV} = Si, Ge, Sn) compounds are well known.[22] Each binary phase diagram contains only one chemical compound of the Mg$_2$B^{IV}-type and two eutectic points. Mg$_2$Si and Mg$_2$Ge can form continuous solid solutions, but Mg$_2$Si–Mg$_2$Sn and Mg$_2$Ge–Mg$_2$Sn are characterized by a peritectic reaction with a miscibility gap. The widely acknowledged pseudobinary phase diagram for Mg$_2$Si$_{1-x}$Sn$_x$ system with miscibility gap of *x* = 0.4–0.6 was experimentally derived by Nikitin et al.[23] based on the differential thermal analysis techniques and X-ray lattice parameter measurement. The same experiments were performed by Muntyanu et al.,[24] and larger miscibility gap of *x* = 0.3–0.9 was obtained. Vives et al.[25] conducted Mg$_2$Si/Mg$_2$Sn diffusion couple experiment to determine the solubility of the solid solution and confirmed the experimental assessment by Muntyanu et al.[24] Combining experimental work and the computer coupling of phase diagrams and thermochemistry technique, Jung et al.[26] and Kozlov et al.,[27] respectively, generated the phase diagram of Mg_2Si–Mg_2Sn (in Figure 6.2). However, the miscibility gap from the two works is contradictory. With the new synthesis techniques developed, it seems that the miscibility gap depends on the preparation methods shown in Table 6.2. Although some researchers tried to make this issue clear,[28,31] more detailed work on the stability of Mg_2Si–Mg_2Sn solid solution is needed. Two factors should be carefully handled: the slow kinetics for the phase separation of Mg_2Si–Mg_2Sn solution and the actual content of magnesium in the ternary system.[27,28]

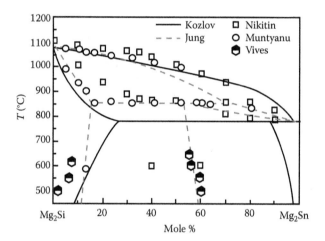

FIGURE 6.2 Pseudobinary diagram of Mg_2Si–Mg_2Sn. (Based on Nikitin, E. et al., *Journal of Inorganic Materials*, 4, 1656, 1970; based on Muntyanu, S. et al., *Izv Akad Nauk Ussr Neorgan Materialy*, 2, 870–875, 1966; reprinted with permission from Vivès, S. et al., *Chemistry of Materials*, 26, 4334–4337. Copyright 2014. American Chemical Society; reprinted from *Journal of Alloys and Compounds*, 509, Kozlov, A. et al., Phase formation in Mg–Sn–Si and Mg–Sn–Si–Ca alloys, 3326–3337, Copyright (2011), with permission from Elsevier; reprinted from *Calphad*, 31, Jung, I.-H. et al., Thermodynamic modeling of the Mg–Si–Sn system, 192–200, Copyright (2007), with permission from Elsevier.)

TABLE 6.2 Reported Different Miscibility Gaps for $Mg_2Si_{1-x}Sn_x$

Synthesis Method	Miscibility Gap x
Flux method[28]	0.2–0.45
Melting[23]	0.4–0.6
Liquid–solid reaction[29]	No gap
Mechanical alloying[30]	0.09–0.72
Sealed corundum crucible[24]	0.08–0.65
Calculated[27]	0.08–0.66

6.1.3 Preparation of Mg$_2$Si$_{1-x}$Sn$_x$ Solid Solutions

Direct melting has long been utilized to produce Mg$_2$B^{IV} compounds and their solid solutions.[32] In the early years, Mg$_2$B^{IV} compounds were usually prepared by the Bridgman method.[1,33] Zaitsev and Nikitin[32] adopted direct melting combined long time annealing to obtain Mg$_2$B^{IV} samples, including solid solutions.

In order to improve the TE performance of Mg$_2$B^{IV} compounds and their solid solutions, more requirements arise toward the synthesis process, e.g., good homogeneity, precisely controlled stoichiometry, reproducibility, and scale-up ability. However, it is quite difficult to obtain the required high-quality samples due to several tricky problems: high vapor pressure and high reactivity of Mg cause the loss of Mg and difficulty in sealing; large mass differences between constituent elements easily lead to inhomogeneous and off-stoichiometric Mg$_2$Si$_{1-x}$Sn$_x$ solid solutions.

In recent years, several different synthesis methods combined with various compaction techniques have been used to fabricate high-quality Mg$_2$B^{IV}-based samples. The focus is to protect the samples from oxygen and to control the Mg loss by proper sealing or low-temperature synthesis condition. The new synthesis methods already proved effective include induction melting,[11] B$_2$O$_3$ flux method,[20] low-temperature solid-state reaction,[34] mechanical alloying,[31,35] and self-propagating high-temperature synthesis.[36] During induction melting, raw materials are heated by the eddy current induced from alternative electromagnetic field. Temperature could be reached above the melting points of the constitute materials within a very short time, but the loss of Mg is quite severe for lack of sealing. B$_2$O$_3$ flux method is featured by the B$_2$O$_3$ liquid sealing to cover the raw materials during the reaction without vacuum or gas protection. A low-temperature solid-state reaction heats the cold pressed raw material mixture at a relatively low temperature to avoid the melting process. By repeating the steps (crushing and milling the resulting pellets to fine powders and cold press again), good homogeneity could be realized, although the whole process is quite complex. Mechanical alloying could also be used for the fabrication of Mg$_2$B^{IV}, and its leading advantages are the greatest reduction of Mg loss and its potential for large-scale production. But particular attention should be paid to the milled active powders. Self-propagating high-temperature synthesis is an ultrafast sintering method, in which the sintering process could be completed within a few seconds after initial ignition, accompanied by a combustion wave. However, the synthesis of solid solutions by this method is unsatisfactory.

As-synthesized compounds are generally milled to powder and then compacted at a high pressure and high temperature for densification and homogeneity. HP and SPS are the two main compaction techniques for Mg$_2$B^{IV}. SPS is a rapid sintering technique and is favorable for reducing the evaporation of Mg. As the high-pulsed current goes through the graphite electrode and the sample powder, electrical discharges occur among adjoined particles, making the boundary highly active for combination and, hence, fast sintering.[37]

6.1.4 Band Features of $Mg_2Si_{1-x}Sn_x$ Solid Solutions

Both Mg_2Si and Mg_2Sn are indirect gap semiconductors with similar band structures.[38,39] Their valence band maximum locates at the Γ point in the Brillouin zone, while the conduction band minimum at the X point.[40] Above the bottom of the conduction band, there is another band separated from the minimum by the gap ΔE, as shown in Figure 6.3. For Mg_2Si, the heavy band (C_H) is above the light one (C_L) with the splitting energy of 0.4 eV,[38] while for Mg_2Sn, the two bands reverse with a 0.16 eV gap.[39] The energy gap dependence on the composition of $Mg_2Si_{1-x}Sn_x$ solid solutions has been studied both experimentally[16,41] and theoretically.[5,42] When $Mg_2Si_{1-x}Sn_x$ solid solution is formed, the relative motion of the two bands can be easily understood by the schematic diagram shown in Figure 6.3. The C_H band monotonically moves down in energy with increasing Sn content, and the C_L band moves first down and then (after $x > 0.4$) up in energy. Thus, at some x, band inversion occurs, and hence, the two conduction bands are at the same distance from the maximum of the valence band. The exact composition x is calculated to be 0.625 by Tan et al.[42] and 0.7 by Zaitsev et al.[41] The two bands may be regarded as effectively converged when their energy separation is small enough, compared with k_BT, and in this splitting range, both of the two low-lying bands practically contribute to the electrical transport. The distance ΔE between the bands of light and heavy electrons can also be found in Table 6.3.

Band convergence is favorable for TE properties because in this case, the DOS is increased due to the increase of band degeneracy. Based on Hall and Seebeck coefficient measurements, the DOS effective mass m^* can be calculated,[45] as listed in Table 6.3. The Seebeck coefficient α versus carrier concentration n_H for the doped $Mg_2Si_{1-x}Sn_x$, in Figure 6.4,[17,21,34,35,46] enables one to directly observe an increase in m^* from $1.1m_e$ to $2.7m_e$ (m_e is the mass

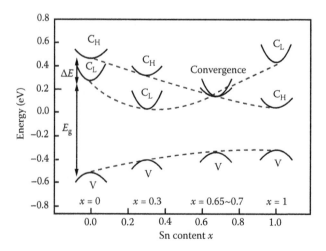

FIGURE 6.3 Schematic diagram for the change of band structure with the Sn content (based on first-principles calculation of Liu et al.[16]).

TABLE 6.3 Parameters for Mg₂Si₁₋ₓSnₓ Band Structure

Nominal composition	ΔE (eV)		m*/mₑ		n (× 10²⁰ cm⁻³)	α (μV/K)
	Theoretical Calculation[42]	Infrared Measurement[41]	Fitting from C_p[16]	Calculated from Transport Properties		
Mg₂Si	0.21	0.4		1.1	1.7[35]	−73[35]
Mg₂Si₀.₈Sn₀.₂	0.31	0.29	0.93	1.3 (1.5)	1.7	−85
Mg₂Si₀.₇Sn₀.₃	0.30	0.24		(1.3)	2.3[43]	−138[43]
Mg₂Si₀.₆Sn₀.₄	0.24	0.18	1.07	1.8 (1.6)	1.8	−108
Mg₂Si₀.₅Sn₀.₅	0.16		0.9	2.1	1.9	−121
Mg₂Si₀.₄Sn₀.₆	0.05	0.07	1.51	2.2 (2.7)	1.7	−132
Mg₂Si₀.₃Sn₀.₇	0.06		1.41	2.8	1.7	−158
Mg₂Si₀.₂Sn₀.₈	0.18		1.26	2.4	1.8	−136
Mg₂Sn	0.39	0.16		2	1.8[44]	−115[44]

Note: Data without reference citations are from Liu et al.[16] The data in parentheses is based on Zaitsev et al.[5]

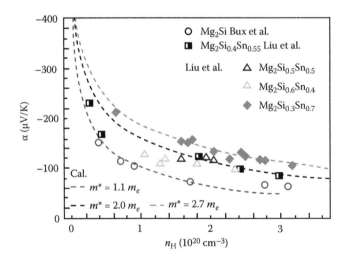

FIGURE 6.4 Pisarenko plot for Mg₂Si₁₋ₓSnₓ. (Liu, X. et al.: Low electron scattering potentials in high performance Mg₂Si₀.₄₅Sn₀.₅₅ based thermoelectric solid solutions with band convergence. *Advanced Energy Materials*. 2013. 3. 1238–1244. Copyright Wiley-VCH Verlag GmbH & Co. KGaA. Reproduced with permission; reprinted with permission from Liu, W. et al., *Chemistry of Materials*, 23, 5256–5263. Copyright 2011 American Chemical Society; Liu, W. et al., *Journal of Physics D: Applied Physics*, 43, 085406, 2010, Institute of Physics; Bux, S. K. et al., *Journal of Materials Chemistry*, 21, 12259, 2011. Reproduced by permission of The Royal Society of Chemistry; Liu, W. et al., *Physical Chemistry Chemical Physics*, 16, 6893–6897, 2014. Reproduced by permission of The Royal Society of Chemistry.)

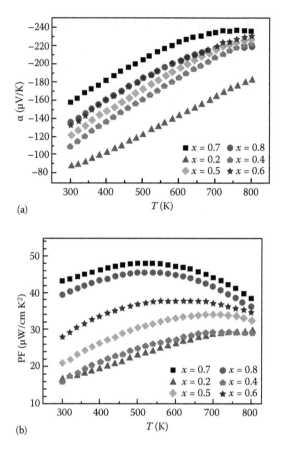

(a)

(b)

FIGURE 6.5 TE properties of Sb-doped $Mg_2Si_{1-x}Sn_x$ at the same carrier concentration (1.8 × 10^{20} cm^{-3}): (a) temperature dependence of Seebeck coefficient and (b) temperature dependence of PF. (Reprinted with permission from Liu, W. et al. *Physical Review Letters*, 108, 166601, 2012. Copyright 2012 by the American Physical Society.)

of an electron) due to the band convergence at $x = 0.6$–0.7. Figure 6.5 presents the temperature dependence of Seebeck coefficient and PF ($\alpha^2\sigma$) for Sb-doped $Mg_2Si_{1-x}Sn_x$ (with carrier concentration at around 1.8 × 10^{20} cm^{-3}). Clearly, Sn content has a strong effect on the magnitude of the Seebeck coefficient and the PF. At $x = 0.7$, the enhancement in the Seebeck coefficient is particularly large and is maintained in the entire range of temperatures from 300 to 800 K.[16]

6.1.5 Electrical and Thermal Transports in $Mg_2Si_{1-x}Sn_x$ Solid Solutions

Similar to most of the known TE materials, acoustic phonon (AP) scattering is the dominant carrier scattering mechanism for Mg_2B^{IV} above room temperature. In this case, carrier mobility interacts with effective mass and deformation potential E_{def} via $\mu \propto m_I^{*-1} m_b^{*-3/2} E_{def}^{-2}$ (m_b^*, the average single-valley DOS effective mass; m_I^*, the inertial mass of carriers along the conduction direction).[47] The deformation potential E_{def} characterizes the change in energy of the electronic band with elastic deformation and thus describes the coupling between phonons and electrons. When a solid solution is formed, alloy disorders cause local potential energy fluctuation that induces additional carrier scattering, the strength of which is described by alloy scattering potential E_{al}. From Figure 6.6a, the temperature dependence of Mg_2Si mobility show $\mu_H \propto T^{-1}$ tendency, indicating the dominant AP scattering. For $Mg_2Si_{0.45}Sn_{0.55}$ and $Mg_2Si_{0.3}Sn_{0.7}$, a relationship $\mu_H \propto T^{-0.5}$ was observed near room temperature, implying that the alloy scattering plays an important role. With the increase of temperature, AP scattering mechanism becomes dominant in the charge carrier scattering process. As presented in Figure 6.6b, by modeling the mobility and fitting experimental data, it is found that $Mg_2Si_{1-x}Sn_x$ solid solutions have both low deformation potential (13 eV[17] and

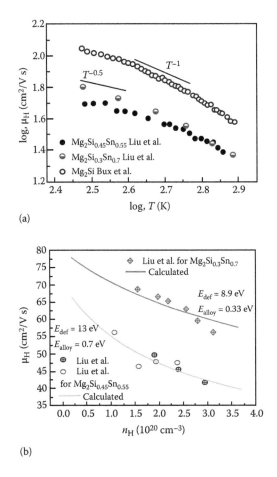

FIGURE 6.6 Electrical transport properties of $Mg_2Si_{1-x}Sn_x$: (a) log T dependence of log μ_H and (b) Hall mobility versus carrier concentration at 300 K. (Liu, X. et al.: Low electron scattering potentials in high performance $Mg_2Si_{0.45}Sn_{0.55}$ based thermoelectric solid solutions with band convergence. *Advanced Energy Materials*. 2013. 3. 1238–1244; Bux, S. K. et al., *Journal of Materials Chemistry*, 21, 12259, 2011. Reproduced by permission of The Royal Society of Chemistry; Liu, W. et al., *Physical Chemistry Chemical Physics*, 16, 6893–6897, 2014. Reproduced by permission of The Royal Society of Chemistry.)

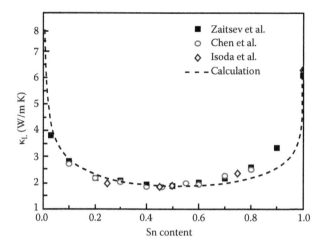

FIGURE 6.7 Lattice thermal conductivity of $Mg_2Si_{1-x}Sn_x$ at room temperature.[28,48,49] The dashed line was calculated based on the Callaway–Klemens model.

8.9 eV[46]) and low alloy scattering potential (0.7 eV[17] and 0.33 eV[46]), which are beneficial for the system to keep relatively high mobility.

Forming $Mg_2Si_{1-x}Sn_x$ solid solution is effective in decreasing thermal conductivity. Mass and strain fluctuation caused by alloying can effectively scatter phonons. Due to the large difference in mass and size between Si and Sn, the lattice thermal conductivity of Mg_2Si decreases from 7.9 to 1.9 W/m K at $x = 0.5$, as shown in Figure 6.7. The dashed line in Figure 6.7 is calculated based on the Callaway–Klemens model,[50,51] considering Umklapp and alloy scattering. The reduced lattice thermal conductivity of $Mg_2Si_{1-x}Sn_x$ is dominantly ascribed to the large disorder due to the alloying. Based on the κ_L reduction in pseudobinary solid solutions, the pseudoternary solid solutions, including Mg_2Si, Mg_2Ge, and Mg_2Sn, are also prepared to further reduce the lattice thermal conductivity.[52,53] For example, Du et al.[52] reported a κ_L of ~1.8 W/m K at room temperature for $Mg_2Si_{0.2}Ge_{0.1}Sn_{0.7}$.

6.1.6 Point Defects in $Mg_2Si_{1-x}Sn_x$ Solid Solutions

The intrinsic conduction type of Mg_2B^{IV} related to the preparation method and chemical composition is summarized in Table 6.4. It is clear that the intrinsic conduction of Mg_2Si is always n-type and the density of electron carrier ranges from 10^{16} to 10^{18} cm^{-3}.[1,54] In contrast, the intrinsic conduction type of Mg_2Ge and Mg_2Sn depends on the Mg contents and the preparation methods used. When Mg content is rich, both Mg_2Ge and Mg_2Sn tend to behave as n-type conduction.[56,59]

First-principles calculations show that the chemical potential of Mg has a decisive impact on the intrinsic conduction type of Mg_2B^{IV} compounds.[60–63] Mg interstitial and Mg vacancy are the two dominant defects in Mg_2B^{IV} and behave as donor and acceptor, respectively. The formation energies of two defects strongly depend on the Mg chemical potential, and thus, their concentrations are affected by the Mg content. The persistent n-type conduction

TABLE 6.4   Intrinsic Conduction Type of Mg₂B^{IV}

System	Preparation Method	Nominal Composition			n/p	Ref.
Mg₂Si	Melting	/	Stoich.	/	$n = 8.0 \times 10^{16}$ cm⁻³	Morris et al.[1]
	Alloying + sintering	/	Stoich.	/	n-Type	Bose et al.[2]
	Vertical Bridgman	Mg-poor	Stoich.	Mg-rich	n-Type	Yoshinaga et al.54
Mg₂Ge	Melting	/	Stoich.	/	$n = 3.0 \times 10^{15}$ cm⁻³	Redin et al.[55]
	Bridgman method	/	Stoich.	/	$p = 1.0 \times 10^{16}$ cm⁻³	Shanks[56]
	Bridgman method	/	/	Mg-rich	$n = 3.0 \times 10^{17}$ cm⁻³	Shanks[56]
Mg₂Sn	Vertical Bridgman	/	Stoich.	/	p-Type to n-type	Chen and Savvides[57]
	Vertical Bridgman	/	/	Mg-rich	n-Type	Chen and Savvides[57]
	RF induction melting	/	Stoich.	/	p-Type to n-type	Chen and Savvides[58]

Note: "/" means that the references haven't mentioned this situation, below the table.

of Mg₂Si originates from the dominant donor-like interstitial Mg. The conduction type of Mg₂Ge and Mg₂Sn can change from p-type to n-type, as is shown in Table 6.4, when the dominant defect varies from Mg vacancy to interstitial Mg.

The intrinsic defects have significant effects on the TE properties of Mg₂B^{IV} and their solid solutions, and their kinds and concentration can be tuned by controlling the composition and synthesis technology.[18,19,21,64,65] Liu et al.[21] and Du et al.[19] carefully studied the dependence of carrier concentration on the Mg excess in Mg₂Si₁₋ₓSnₓ system. Figure 6.8a shows that the measured carrier concentration is much higher than the predicting line, which assumes that Sb/Bi is the sole dopant and supplies one electron per atom. Figure 6.8b shows that electron carrier concentration is indeed enhanced with nominal content of excess Mg. Therefore, the interstitial Mg plays a very important role, acting as a donor, in increasing the carrier concentration, especially when Mg is overstoichiometric.[17,19–21] By Sb doping and tuning Mg excess, carrier concentration can be optimized and a *ZT* value of >1.0 has been obtained.[17,21]

Nolas et al.[65] and Dasgupta et al.[64] investigated the role of Sb dopant, from low to relatively high contents, in binary Mg₂Si. They find that antimony acts as a donor at a low concentration, but facilitates the formation of Mg vacancies at a higher concentration (>10%), shown in Figure 6.9a. Jiang et al.[18] shows that the concentration of Mg vacancies is decided by the content of Sb, following the rule of $\left[V''_{Mg} \right] = 1/2 \left[Sb^{\bullet}_{Sn} \right]$, as is shown in Figure 6.9b. With increasing Sb content, the real Mg content in Mg₂Si₁₋ₓSbₓ

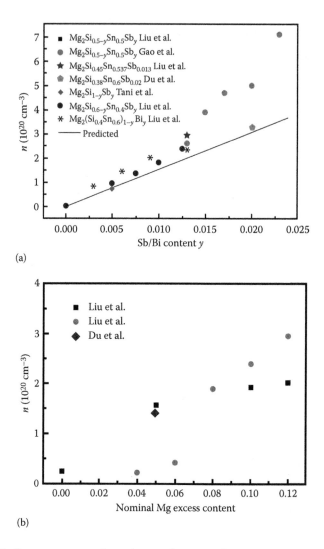

FIGURE 6.8 (a) Dopant content dependence of measured carrier concentration of Mg_2Si_{1-x} Sn_x,[7,17,19–21,34,66] along with the predicted line; (b) the nominal excess Mg dependence of measured carrier concentration.[17,19,21]

and $Mg_2Si_{0.4}Sn_{0.6-x}Sb_x$ monotonically decays, and the room-temperature lattice thermal conductivity decreases. In addition to the mass fluctuation and strain field contribution from Sb substitution for Si/Sn, vacancies also result in a large localized lattice strain and thus have a larger effect on κ_L.[65] Similar effects can also be found in $Mg_2Si_{1-x}Sn_x$ solid solutions.[18] As shown in Figure 6.9, κ_L decreases from ~8 W/m K for Mg_2Si to 1.85 W/m K for $Mg_{1.81}Si_{0.63}Sb_{0.37}$ with high density of Mg vacancies. Later, Jiang et al.[18] find that three types of point defects, i.e., Sb substitution, Mg vacancies, and Mg interstitials, can coexist in Sb-doped $Mg_2Si_{1-x}Sn_x$ solid solutions and can be synergistically implemented to significantly reduce the lattice thermal

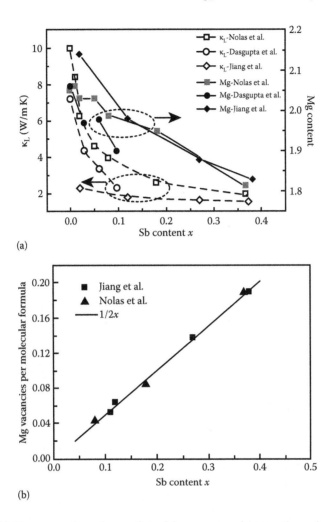

FIGURE 6.9 (a) Sb content dependence of the Mg content and lattice thermal conductivity κ_L of $Mg_2Si_{1-x}Sb_x$[64,65] and $Mg_2Si_{0.4}Sn_{0.6-x}Sb_x$[18] (the blue lines are for the solid solution); (b) calculated Mg vacancies per molecular formula as a function of Sb content in $Mg_2Si_{0.4}Sn_{0.5}Sb_{0.1}$.[18,65]

conductivity to approach the minimum limit (Figure 6.10). At the same time, high electrical properties remained after proper doping. Finally, $ZT >$ 1.1 has been attained at 750 K in $Mg_2Si_{0.4}Sn_{0.5}Sb_{0.1}$ specimen.[18] These results demonstrate the promise of point defect chemistry approach in optimizing TE properties in Mg_2Si-based system.

6.1.7 Acknowledgments

This work was supported by the National Basic Research Program of China (2013CB632503), the National Nature Science Foundation of China (11574267, 51571177), and the Program for New Century Excellent Talents in University (NCET-12-0495).

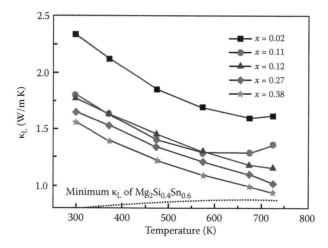

FIGURE 6.10 Temperature dependence of the lattice thermal conductivity of $Mg_2Si_{0.4}Sn_{0.6-x}Sb_x$. (Jiang, G. et al.: High performance $Mg_2(Si,Sn)$ solid solutions: A point defect chemistry approach to enhancing thermoelectric properties. *Advanced Functional Materials*. 2014. 24. 3776–3781. Copyright Wiley-VCH Verlag GmbH & Co. KGaA. Reproduced with permission.)

6.2 Thermoelectric Properties of n-type $Mg_2Sn_{1-x}Ge_x$ Solid Solution

Weishu Liu and Zhifeng Ren

6.2.1 Synthesis of $Mg_2Sn_{1-x}Ge_x$ by BM-HP

In contrast to Mg_2Si and its alloys, $Mg_2Sn_{1-x}Ge_x$ for the medium-temperature range TE applications did not get the attention until the recent works.[44,67] High vapor pressure and chemical activity of the element components could be the challenge in preparing or handling these materials. Conventionally, direct comelting with subsequent annealing, or solid-state reaction with subsequent annealing, or Bridgman growth method was reported to synthesize Mg_2Si-based alloys as mentioned in the Section 6.1. Although the powder metallurgy route, e.g., BM plus HP, has also been used to synthesize the Mg_2Si and its alloys $Mg_2Si_{1-x}Sn_x$,[68–71] it achieves only a low peak ZT (<0.7) compared with that made from conventional ingot metallurgy routes. Figure 6.11 shows the high phase-pure Mg_2Sn synthesized by the BM and HP joint method.[67] No notable oxide impurity was identified from XRD pattern, as shown in Figure 6.11a. It is worthy to point out that the particle size of the starting materials is important to obtain the phase-pure materials. Since both major starting materials Mg and Sn are very soft, they easily stick on the wall of the stainless steel jar. Luckily, the reacted product Mg_2Sn is brittle, which can be ball milled into nanopowders. Urretavizcaya and Meyer[70] noted an over BM could transfer the stable cubic Mg_2Sn into metastable hexagonal Mg_2Sn. The hexagonal Mg_2Sn was also observed by the authors (WL and ZR), which appeared as little impurities and seriously scattered the transport of electrons. Phase-pure $Mg_2Sn_{1-x}Ge_x$ alloys were also successfully synthesized by using the same BM plus HP joint route. The lattice parameter

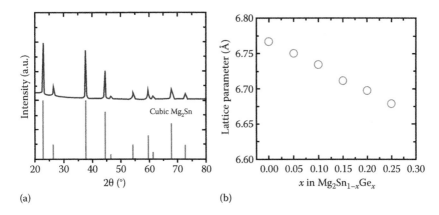

(a)   (b)

FIGURE 6.11 (a) XRD patterns of Mg_2Sn made from the BM and HP joint process. (b) Calculated lattice parameters of cubic $Mg_2Sn_{1-x}Ge_x$ alloys. The standard cubic Mg_2Sn pattern is adapted from PDF No. 03-065-2977. (Liu, W. S., *Proceedings of the National Academy of Sciences USA*, 112, 3269–3274, 2015 and Copyright 2015 National Academy of Sciences, USA.)

TABLE 6.5 EPMA Analysis of Nominal Composition of $Mg_{2+\delta}Sn_{0.73}Ge_{0.25}Bi_{0.02}$ Made by BM and HP

Nominal Composition	Mg (at.%)	Sn (at.%)	Bi (at.%)	Ge (at.%)	Density (g/cm³)
$Mg_{2.02}Sn_{0.73}Ge_{0.25}Bi_{0.02}$	63.545	25.001	1.5737	10.051	3.43
$Mg_{2.04}Sn_{0.73}Ge_{0.25}Bi_{0.02}$	65.645	23.235	1.2667	9.853	3.44
$Mg_{2.06}Sn_{0.73}Ge_{0.25}Bi_{0.02}$	66.908	22.565	1.124	9.403	3.45
$Mg_{2.08}Sn_{0.73}Ge_{0.25}Bi_{0.02}$	69.035	21.604	1.0713	8.289	3.43
$Mg_{2.10}Sn_{0.73}Ge_{0.25}Bi_{0.02}$	68.048	21.669	1.0559	9.227	3.43

Source: (Courtesy of Saparamadu, U.; W. S. Liu, and Ren, Z. F., 2016.)

calculated from the Rietveld refinement of XRD was shown in Figure 6.11b. The near-linear varying lattice parameter with Ge content clearly suggested a well-formed solid solution of $Mg_2Sn_{1-x}Ge_x$ alloys within the current investigated Ge content range ($0 < x < 0.25$). In order to compensate the loss of Mg during the BM and HP process, extra Mg was purposely added. Table 6.5 shows the EPMA analysis of nominal composition of $Mg_{2+\delta}Sn_{0.73}Ge_{0.25}Bi_{0.02}$ ($\delta = 0.02, 0.04, 0.06, 0.08, 0.10$).[72] It was found that the one with extra Mg of $\delta = 0.06$ was closest to the stoichiometric ratio and had the lowest electrical resistivity, highest PF, and hence the highest ZT values. Similar conclusion was also driven from the system of $Mg_{2+\delta}Sn_{0.73}Ge_{0.25}Sb_{0.02}$. However, too much extra Mg (e.g., $\delta = 0.08$ and 0.10 in $Mg_{2+\delta}Sn_{0.73}Ge_{0.25}Bi_{0.02}$) inversely lead to some porous structures, introducing additional scattering to the transport of electrons.

6.2.2 Electrical and Thermal Transports in $Mg_2Sn_{1-x}Ge_x$

Next, we will come to discuss the effect of the alloying effect of the Ge in the $Mg_2Sn_{1-x}Ge_x$. Figure 6.12 shows the temperature-dependent TE properties of $Mg_2Sn_{1-x-y}Ge_xSb_y$, in which 2–3% extra Mg were used to

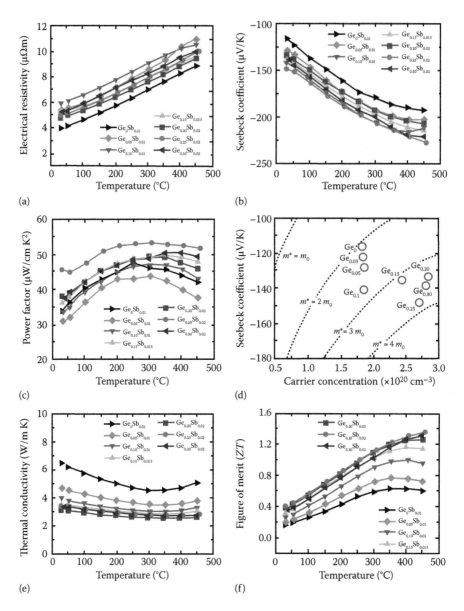

FIGURE 6.12 TE properties of $Mg_2Sn_{1-x-y}Ge_xSb_y$. (a) Electrical resistivity, (b) Seebeck coefficient, (c) PF, (d) Seebeck coefficient versus carrier concentration, (e) thermal conductivity, and (f) *ZT*. (Liu, W. S. et al., *Energy & Environmental Science*, 9, 530–539, 2016. Reproduced by permission of The Royal Society of Chemistry.)

compensate the loss during the BM and HP process.[44] All the samples showed almost linearly increased electrical resistivity and Seebeck coefficient below 300°C, demonstrating a behavior of degenerate semiconductor. One of the notable features of the $Mg_2Sn_{1-x-y}Ge_xSb_y$ materials is the high PF. Among the samples, $Mg_2Sn_{0.73}Ge_{0.25}Sb_{0.02}$ showed the highest PF of 43 μW/cm K² near room temperature and over 50 μW/cm K² in a wide temperature range of 150–450°C. Due to the increased PF and decreased thermal conductivity, the ZT value was significantly enhanced from 0.6 for $x = 0$ to around 1.4 for $x = 0.25$. The sample with higher Ge content ($x = 0.3$) showed a slightly lower ZT value due to a lower PF. Figure 6.12d plots the Seebeck coefficient versus carrier concentration near room temperature for all the $Mg_2Sn_{1-x-y}Ge_xSb_y$, which clearly suggested that the sample with Ge of $x = 0.25$ had the highest effective mass. The Hall measurement confirmed that the effective mass increased from $m^* = 2m_0$ for Mg_2Sn to $m^* = 3.5m_0$ for $Mg_2Sn_{0.75}Ge_{0.25}$ and slightly decreased to $m^* = 3.4m_0$ for $Mg_2Sn_{0.7}Ge_{0.3}$. The evaluation of the carrier effective mass with Ge content resulted from the alloying effect related band convergence. The first-principles calculation confirmed that the conduction band of Mg_2X (X = Sn, Ge) was composed of two subbands: X_H-band and X_L-band, in which the X_H-band was formed by the hybridized Mg 3s and Sn 5d-t_{2g}/Ge 4d-t_{2g} orbitals, while the X_L-band resulted from the hybridized Mg 3p, Sn 6s/Ge 5s, and Sn 5d-e_g/Ge 4d-e_g orbitals.[44] The change of the ionic size r⁺/r⁻ ratio could be the most likely reason for the composition band crossing in the $Mg_2Sn_{1-x}Ge_x$ systems. A direct result of the band convergence effect is the increased S^2n from 2.45×10^{14} V²/K² m³ for x = 0 to 5.40×10^{14} V²/K² m³ for $x = 0.25$ in $Mg_2Sn_{1-x}Ge_x$. However, only an increased S^2n cannot guarantee the improvement of the PF if the carrier mobility is decreased too much due to the alloying scattering. The alloying scattering to charge carriers could cancel the contribution of S^2n to the PF, i.e., $S^2\sigma$. The Hall measurement showed the carrier mobility of $Mg_2Sn_{1-x}Ge_x$ changed from 86.0 cm²/V s, to 56.9 cm²/V s, 46.5 cm²/V s, and 42.8 cm²/V s for Ge content of x = 0, 0.1, 0.2, and 0.3, respectively. The gentle decrease in the carrier mobility with Ge also suggested a smaller deformed potential energy fluctuation U of 0.2 eV. This was much smaller than 0.7 eV for U obtained in $Mg_2Sn_{0.55}Si_{0.45}$. The weak alloying scattering to electrons was another important reason for the high PF. Furthermore, the alloying effect of Ge also resulted in a significant decrease in both the lattice thermal conductivity and the bipolar thermal conductivity, as shown in the Figure 6.13. The κ_{lat} at room temperature of $Mg_2Sn_{1-x-y}Ge_xSb_y$ was 4.96, 3.59, 3.10, 2.70, 2.35, 2.33, and 2.27 W/m K for x = 0, 0.05, 0.10, 0.15, 0.20, 0.25, and 0.30, respectively. The continuous decrease in κ_{lat} with increasing Ge content demonstrated a strong alloying scattering to phonon transport because of the mass difference between Ge and Sn. The plot of μ/κ_{lat} suggested that there was an optimized Ge content. On the other hand, the significantly reduced bipolar thermal conductivity of $Mg_2Sn_{1-x}Ge_x$ system with increasing Ge content was resulted from the widening bandgap. It was noted that, at a given Ge content, the measured bandgap was slightly smaller than the value obtained in the early works for $Mg_2Sn_{1-x}Ge_x$,[5,73] which could be caused by the heavy doping level of Sb dopant and other defects.

Furthermore, the benefits of the Ge in $Mg_2Sn_{1-x}Ge_x$ were summarized as increasing weighted mobility (U), decreasing lattice thermal conductivity (κ_{lat}), and the widening bandgap (E_g). For a given material (e.g., Mg_2Sn or $Mg_2Sn_{0.75}Ge_{0.25}$), the peak ZT usually corresponds to the maximum value of

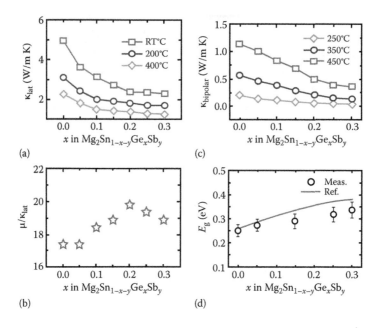

(a) (b) (c) (d)

FIGURE 6.13 (a) Lattice thermal conductivity of $Mg_2Sn_{1-x-y}Ge_xSb_y$, (b) the μ/κ_{lat} ratio of $Mg_2Sn_{1-x-y}Ge_xSb_y$ near room temperature, (c) bipolar thermal conductivity of $Mg_2Sn_{1-x-y}Ge_xSb_y$, and (d) bandgap of $Mg_2Sn_{1-x-y}Ge_xSb_y$. In b, the units of μ and κ_{lat} are $(cm^2/V\ s)$ and $(W/m\ K)$, respectively. (Liu, W. S. et al., *Energy & Environmental Science*, 9, 530–539, 2016. Reproduced by permission of The Royal Society of Chemistry.)

this material in a dual-dimensional space of doping concentration and temperature. Under a two-band theoretical basis, U, κ_{lat}, and E_g could further combined into a new materials parameter B^* as following:

$$ZT = \frac{\left(\delta_e - \xi_f - \dfrac{\delta_e + \delta_h + \xi_g}{1+1/\gamma}\right)^2 (1+\gamma)}{\left((B^*)\dfrac{F_{1/2}(\xi_f)/\Gamma(3/2)}{\xi_g}\right)^{-1} + \dfrac{(\delta_e + \delta_h + \xi_g)^2}{1+1/\gamma} + \left(\dfrac{e}{k_B}\right)^2 (L_e + \gamma L_h)}$$

(6.1)

$$B^* = 6.668 \times 10^{-2}\frac{U^*}{\kappa_{lat}}E_{g\_eV},$$

(6.2)

$$U^* = \mu\left(m^*/m_o\right)^{3/2} T^{3/2},$$

(6.3)

where the ξ_f and ξ_g are the reduced Fermi energy and reduced bandgap which are the equivalent variable to the doping concentration and temperature. It was noted that a new parameter U^*, rather than conventional U, was used because the term of $\mu T^{3/2}$ was theoretically suggested to be temperature independent by Wang et al.[74] The definitions of the Fermi integral-related symbols were given as the following:

$$\delta_i = \frac{(s+5/2)F_{s+3/2}(\xi_{f\_i})}{(s+3/2)F_{s+1/2}(\xi_{f\_i})}, \quad i = e, h, \tag{6.4}$$

$$L_i = \left(\frac{k_B}{e}\right)^2 \left[\frac{(s+7/2)F_{s+3/2}(\xi_{f\_i})}{(s+3/2)F_{s+1/2}(\xi_{f\_i})} - \left(\frac{(s+5/2)F_{s+3/2}(\xi_{f\_i})}{(s+3/2)F_{s+1/2}(\xi_{f\_i})}\right)^2\right] \tag{6.5}$$

$$\gamma = \frac{\mu_h (m_h^*)^{3/2}}{\mu_e (m_e^*)^{3/2}} \frac{F_{1/2}(\xi_{f\_e})}{F_{1/2}(\xi_{f\_h})} = \frac{U_h}{U_e} \frac{F_{1/2}(\xi_{f\_e})}{F_{1/2}(\xi_{f\_h})}, \tag{6.6}$$

where $F_n(\xi_f)$ is the nth-order Fermi integral defined as

$$F_n(\xi_f) = \int_0^\infty \frac{\chi^n}{1 + e^{\chi - \xi_f}} d\chi. \tag{6.7}$$

The subscript $i = e$ or h represents the electrons and holes, respectively, with the reduced Fermi level ξ_{f_e} (ξ_{f_h}) measured from conduction band edge E_c (valence band edge E_v). Figure 6.14a plots the peak $(ZT)_{max}$ of $Mg_2Sn_{1-x-y}Ge_xSb_y$ as a function of Ge content. It showed a peak value was reached as $x = 0.25$ for the Ge content. In order to explore the connection between ZT and the fundamental parameter B^*, defined in Equation 6.2, B^* was calculated using U^* and E_g at room temperature and κ_{lat} at temperature corresponding to peak $(ZT)_{max}$, as shown in Figure 6.14b. The calculated B^* showed a continuous increase from 0.85 to 2.73 with increasing Ge content from $x = 0–0.25$. Such an enhancement in B^* was contributed by ~25% enhancement in weighted mobility U^*, ~27% increase in E_g, and ~50% decrease in κ_{lat}. Compared with $x = 0.25$, the sample with Ge of $x = 0.3$ had smaller B^* of 2.56 due to a smaller U^*. The Ge content-dependent behavior of B^* is quite similar to that of $(ZT)_{max}$. As a result, a near-linear relationship between B^* and $(ZT)_{max}$ was displayed in the Figure 6.14b.

6.2.3 Band convergence line in Mg₂Sn-Mg₂Ge-Mg₂Si

Compared with $Mg_2Sn_{1-x}Si_x$, the advantage of $Mg_2Sn_{1-x}Ge_x$ is higher PF and hence larger output power density. Mao et al.[75] conducted an

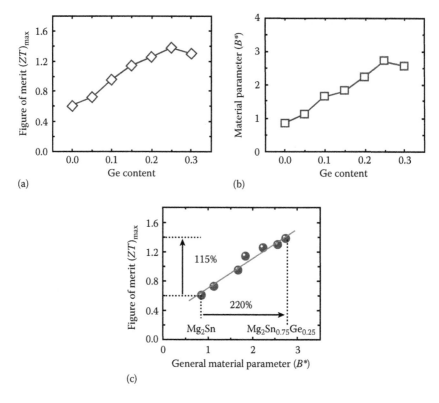

FIGURE 6.14 (a) Ge content-dependent peak $(ZT)_{max}$ for $Mg_2Sn_{1-x-y}Ge_xSb_y$; (b) Ge content-dependent material parameter B^* for $Mg_2Sn_{1-x-y}Ge_xSb_y$; (c) $(ZT)_{max}$ versus B^* for $Mg_2Sn_{1-x-y}Ge_xSb_y$. (Liu, W. S. et al., *Energy & Environmental Science*, 9, 530–539, 2016. Reproduced by permission of The Royal Society of Chemistry.)

investigation along the possible band convergence line with in the ternary $Mg_2Sn–Mg_2Ge–Mg_2Si$ system, i.e., from the $Mg_2Sn_{0.78}Ge_{0.22}$ (MSG) to $Mg_2Sn_{0.7}Si_{0.3}$ (MSS), the band convergence composition in binary $Mg_2Sn–Mg_2Ge$ system and $Mg_2Sn–Mg_2Si$ system, respectively. The new materials $(MSG)_{1-x}(MSS)_x$ showed a consistent carrier effective mass m^* of $2.77–2.91m_0$ without notable trend with the change in the content of MSS, which suggested that the band structures of the materials along the band crossing line would be very close. However, the carrier mobility of $(MSG)_{1-x}(MSS)_x$ showed a continuous decrease with increased content of MSS, which was also consistent with the change of the PF, as shown in the Table 6.6. This was consistent with the previous conclusion that the Si atom may create more serious scattering to the transport of electrons in the Mg_2Sn matrix than the Ge atom.[67] Additionally, Saparamadu et al.[72] suggested that the carrier donor also plays an important role for the high power of $Mg_2Sn_{0.75}Ge_{0.25}$ alloys. It was found that the dopant at the Sn site, such as Sb and Bi, has much less side effect to the transport of electrons than the one on the Mg site, such as Y and La.

TABLE 6.6   TE Transport Properties of (MSG)$_{1-x}$(MSS)$_x$ Made by BM and HP

Composition	n_H (× 10^{20} cm^{-3})	μ_H (cm^2/V s)	m* (m_0)	S (μV/K)	ρ (μΩ m)	PF (μW/m K^2)
x = 0	1.81	72.6	2.86	−148	4.74	46.2
x = 0.2	1.94	68.2	2.91	−145	4.71	44.6
x = 0.4	1.95	67.0	2.81	−141	4.78	41.6
x = 0.6	1.97	66.7	2.82	−141	4.75	41.8
x = 0.8	2.11	60.6	2.93	−140	4.87	40.2
x = 1	2.20	59.3	2.77	−132	4.78	36.5

Source: *Acta Materialia*, 103, Mao, J. et al., Thermoelectric properties of materials near the band crossing line in Mg$_2$Sn-Mg$_2$Ge-Mg$_2$Si system, 633–642, Copyright (2016), with permission from Elsevier.

Note: MSG = Mg$_2$Sn$_{0.765}$Ge$_{0.22}$Sb$_{0.015}$; MSS = Mg$_2$Sn$_{0.685}$Si$_{0.3}$Sb$_{0.015}$.

6.3 Thermoelectric Properties of p-type Mg$_2$(Si, Ge, Sn) Materials

Johannes de Boor, Udara Saparamadu, and Zhifeng Ren

6.3.1 Introduction

Besides nanostructured PbTe,[76] CoSb$_3$-based skutterudites,[77] half-Heusler,[78] SnSe,[79] and Cu$_2$Se-based matertials,[80] Mg$_2$Si-based solid solutions have been widely investigated,[5,16,17,34,81–83] as discussed Mg$_2$(Si,Sn) in Section 6.1 and Mg$_2$(Sn,Ge) in Section 6.2. Here, p-type Mg$_2$(Si,Ge,Sn) will be discussed, which has potential large-scale application[84] because of economical attractiveness and low mass density.[35,85,86]

The contacts of magnesium silicides have been systematically studied.[16,17,34,35,81–91] TE modules of the advanced materials have been fabricated as prototypes and analyzed in laboratories.[92–97] Up to now, Mg$_2$Si-based TE generators were commercialized by some companies, such as Alphabet Energy and Romny Scientific.

The excellent TE properties of Mg$_2$Si-based solid solutions are usually contributed to a reasonably low thermal conductivity due to alloying with Sn and Ge, to a high carrier mobility, and to an excellent PF due to a convergence of conduction bands of Mg$_2$X with compositions around Mg$_2$Si$_{0.4}$Sn$_{0.6}$ and Mg$_2$Ge$_{0.25}$Sn$_{0.75}$.[5,16,67] However, excellent TE properties with 1.1 < ZT_{max} < 1.5 have been reported only for the n-type material, with the p-type being clearly inferior.[98,99] TE generators require both n-type and p-type materials to work efficiently, ideally with similar chemical and thermomechanical properties. As good p-type Mg$_2$X (X = Si, Ge, Sn) has not been available for a long time, n-type Mg$_2$X has often been combined with higher manganese silicide (HMS) to build TE generators. However, the thermomechanical match between n-type Mg$_2$X and p-type HMS is relatively poor, leading to long-term instability issues of these generators.[93,96] Consequently, an efficient p-type Mg$_2$X is highly sought. Fortunately, there has been some progress in the recent years making p-type Mg₂X that is comparable to HMS and competitive Mg₂X-based thermoelectric power generator (TEG) realistic. In this section, we will first give a quick overview of the general optimization criteria for TE materials with a focus on p-type Mg₂X TE materials and compare with the n-tpye Mg₂X materials in Section 6.3.2. Identifying good dopants is much more challenging for the p-type than for the n-type material; we will therefore address this issue specifically in Section 6.3.3. In Section 6.3.4., we will analyze the existing transport data to shed light on the TE transport properties, the band structure, and the applicability of the single parabolic band (SPB) model. We will finally address future challenges in Section 6.3.5. This section focuses on p-type-specific questions. Points that are related to Mg₂X in general or the n-type material can be found in available reviews; see, e.g., Zaitsev et al.[98] and de Boor et al.[100]

6.3.2 Optimization of Thermoelectric Materials and Application to p-Type Mg₂X

In this section, we will firstly discuss the general prerequisites for having a good TE material and what has been realized or predicted for p-type Mg₂X. As the n-type counterpart is an excellent TE material, some findings can be deduced from a direct comparison.

Basically, a good TE material needs a large figure of merit $ZT = \dfrac{\sigma S^2}{\kappa} T$, where σ is the electrical conductivity, κ is the thermal conductivity, S is the Seebeck coefficient, and T is the absolute temperature. Strictly speaking, ZT evaluates the TE properties at a certain temperature, while for practical device efficiencies, integral quantities such as the engineering figure of merit,

$$(ZT)_{eng} = \frac{\left(\displaystyle\int_{T_c}^{T_h} S(T)\,dT \right)^2}{\displaystyle\int_{T_c}^{T_h} \rho(T)\,dT * \int_{T_c}^{T_h} \kappa(T)\,dT} \Delta T \,,$$

are a better measure (with $\rho = \sigma^{-1}$, T_c is the cold-side temperature, and T_h is the hot-side temperature and $\Delta T_s = T_h - T_c$).[101] However, the optimization conditions are essentially the same, so we will restrict the discussion to ZT.

Electrical conductivity, Seebeck coefficient, and the electronic part of the thermal conductivity strongly depend on the carrier concentration n. One fundamental requirement for a good TE material is therefore the practical accessibility of the optimum carrier concentration, e.g., by means of doping. Assuming that this requirement is fulfilled, the potential of a TE material is often assessed using the material parameter $\beta = \dfrac{\left(m_D^*\right)^{1.5} \mu}{\kappa_{lat}}$, where $\left(m_D^*\right)^{1.5}\mu$ is also known as weighted mobility and m_D^* is the DOS effective mass and μ is the carrier mobility.[15,75,102] The DOS effective mass is related to the single-valley effective mass by $m_D^* = N_V^{2/3} m_{SV}^*$, where N_V is the number of energy valleys. Obviously, a low lattice thermal conductivity favors a good TE material. The dependence on the effective mass is more complex since $\mu = \mu(m^*)$. Indeed, it can be shown that for AP scattering as dominant scattering mechanism, $\mu_{AP} \propto (m_{SV}^*)^{-2.5}$, and thus, $\beta \propto \dfrac{N_v}{m_{SV}^*}$, where the difference between m_{SV}^* and the inertial mass m_I^* has been neglected. This indicates that a low effective mass is beneficial as well as a high number of degenerate valleys. The mobilities are furthermore controlled by the interaction potentials, i.e., the (phonon) deformation potential Ξ for scattering by APs and the alloy scattering potential U caused by atomic level disorder. From the considerations mentioned earlier, the following requirements for getting a p-type Mg_2X can be deduced:

1. *Optimum carrier concentration* n_{opt}: While doping of the n-type material has been repeatedly shown, attempted p-doping usually leads to carrier concentrations of orders of magnitude lower than expected and required for optimal TE properties; see, e.g., Tani and Kido,[60,103] Sakamoto et al.,[104] Ihou-Mouko et al.,[105] Akasaka et al.,[106] and Isoda et al.[107] This is particularly severe for the p-type material as ZT maximizes at higher n due to the higher effective mass. This is one of the reasons why p-type material with properties inferior to the n-type was often found. However, it has been recently shown that optimum hole concentrations can be obtained when some dopants are employed, see the discussion on suitable dopants for p-type Mg_2X in Section 6.3.3. It has also been noted that the achievable carrier concentration level depends not only on the employed dopant type and amount, but also on the Si:Ge:Sn ratio in Mg_2X.[108]

2. *Low lattice thermal conductivity* κ_{lat}: The reduction of the lattice thermal conductivity by solid solution formation is possible in p-type Mg_2X in the same manner as in the n-type material; see, e.g., Ihou-Mouko et al.,[105] Mars et al.,[109] and Isachenko et al.[110] It has been shown that the scattering of the carriers on ionized impurities is relatively small, and the influence of the individual dopant on κ_{lat} is thus negligible. Further means to reduce κ_{lat} are by the introduction of further scattering, e.g., due to grain boundaries. However, the reduction of κ_{lat} by the introduction of additional phonon scattering mechanisms often leads to additional charge carrier scattering, corresponding to (undesired) increased *scattering potentials*. This trade-off makes this approach presumably not effective for binary Mg_2Si[111]; however, a slight improvement due to nanostructuring has been predicted for the n-type solid solutions.[112] For the p-type with its intrinsically lower carrier mobility, a larger net gain is plausible but has not been systematically studied experimentally or theoretically.

3. *Valley degeneracy and effective mass:* These are directly related to the band structure. A recently calculated band structure of Mg_2Si is shown in Figure 6.15, together with a zoom in around the valence band maximum for Mg_2Si, Mg_2Ge, and Mg_2Sn. It was obtained by Kutorasinski et al.[113] by using the Korringa–Kohn–Rostoker (KKR) method and is in qualitative agreement, e.g., with Bourgeois et al.[114]

It can be seen that there are three bands possibly contributing to hole transport: the heavy hole band (HH), the light hole band (LH), and a

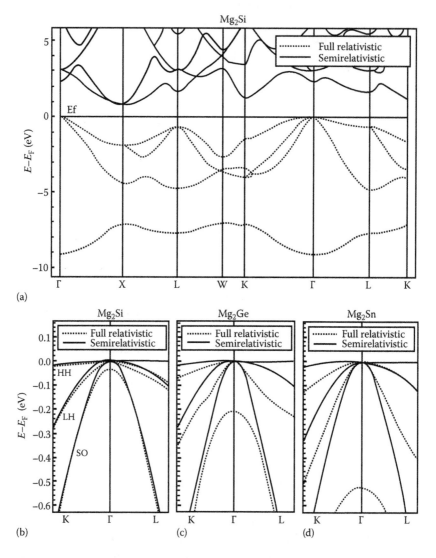

(a)

(b)   (c)   (d)

FIGURE 6.15 (a) Band structure of Mg_2Si from Kutorasinski et al.[113] (b–d) Zoom in at the valence band maxima for Mg_2Si, Mg_2Ge, and Mg_2Sn. The consideration of spin orbit coupling leads to a split-off of the lower band with the energy difference increasing from Si to Sn. (Reprinted with permission from Kutorasinski, K. et al., *Physical Review B*, 89, 115205-1-8, 2014. Copyright 2014 by the American Physical Society.)

split-off band (SO). The latter one is energetically separated, and the split-off energy increases drastically with increasing atomic number of the group 4 elements. The degeneracy factor is thus only $N_{V,p} = 2$ for the p-type material but $N_{V,n} = 3$ for n-type material. It should also be mentioned that Zhang et al.[15] found a removed degeneracy of the two bands at Γ point, with a small but finite energy difference. For the n-type, it has been shown that the two lowest lying bands can converge, increasing $N_{V,n}$ to 6, giving the n-type a clear advantage over the p-type material.

Calculations consistently indicate flatter bands and larger (single-valley) effective masses for the valence band compared to the conduction band. By fitting experimental data for $Mg_2(Si,Sn)$ to multiband solutions of the Boltzmann transport equations, Bahk et al.[115] obtained $m_{LH}^* = 1.0\,m_0$ and $m_{HH}^* = 1.5\,m_0$, but $m_{X1}^* \approx 0.4 - 0.5\,m_0$ and $m_{X3}^* \approx 0.38\,m_0$, where m_0 is the free electron rest mass and X1 and X3 denominate the two lowest conduction bands in Mg_2X. Similarly, Satyala and Vashaee[116] used $m_{LH}^* = 1.0\,m_0$ and $m_{HH}^* = 2.0\,m_0$ and $m_{X1}^* = m_{X3}^* \approx 0.26\,m_0$ (directionally averaged) for their calculations of binary Mg_2Si. The higher effective mass leads to a significantly lowered mobility: Employing ab initio calculations for the band structure Han and Shao[117] calculated a ratio of roughly 1:8 between p and n in binary Mg_2Si, in reasonable agreement with experimental data.

The higher effective mass also causes a peak of ZT at higher carrier concentration values n_{opt}, see, e.g., discussion by Flage-Larsen and Lovvik.[118] This is confirmed by calculations by Satyala and Vashaee,[116] who obtained $ZT_{max} \approx 0.8$ for both n and p-type binary Mg_2Si, however, with $n_{opt} \approx 5 \times 10^{19}$ cm^{-3} and $p_{opt} \approx 4 \times 10^{20}$ cm^{-3}. It has furthermore been reported that some of the relevant p-dopants significantly change the band structure, which means that the often assumed rigid band picture is not applicable anymore. The most prominent examples are Ga and Ag,[13,105,109,114] while, e.g., Li doping leaves the band structure almost untouched.[15]

With respect to the influence of the group 4 element, Kutorasinski et al.[113] found a decrease in m^* going to the heavier compositions. This is in agreement with findings for the n-type where a change of effective mass with composition has been predicted and experimentally observed.[16,42] It is thus plausible that this is also a possible lever for the p-type material; however, experimental evidence for this is still missing.

6.3.3 Dopants for p-Type Mg₂X

Confirmed p-type conduction of Mg_2X has been achieved with several elements, among which are In, Al, K, Ag, Ga, Na, and Li.[15,60,103,119] Al appears to be of amphoteric nature and has been used as n-type dopant,[9,120] and for In and K, we are not aware of reports with good TE properties; we will therefore focus the discussion on Ga, Ag, Li, and Na.

Ga has been successfully employed as p-type dopant in several papers[13,105,121]; it substitutes at the Si position. Liu et al.[13] systematically varied the Ga content in $Mg_{2.1}(Si_{0.3}\,Sn_{0.7})_{1-y}\,Ga_y$ and were able to achieve high hole concentrations of $p_{max} \approx 3 \times 10^{20}$ cm^{-3} and carrier mobilities of

20 cm^2/V s. With these high carrier concentrations, a peak figure of merit of 0.35 was obtained at 650 K. Ga-containing impurity phases were observed for $y \geq 0.05$, indicating finite solubility for Ga in the material.

Theoretical and experimental studies indicate that Ag substitutes at the Mg site[14,117] and the dopant is thus able to provide holes. In particular, Kim et al.[14] reported the electronic structure and the TE properties of Ag-doped Mg_2Sn and $Mg_2Sn_{1-x}Si_x$ ($x = 0.05, 0.1$). The electronic structure study for the Ag-doped $Mg_2Sn_{1-x}Si_x$ ($x = 0, 0.1$) was conducted using the KKR method with the coherent potential approximation applied to account for the chemical disorder. According to the calculations, if Ag substitutes to the Sn site, it is likely to form a resonant-like state. If Ag goes to the Mg site, a rigid band-like behavior is predicated by the calculations. The experimental results reveal no indications for a resonant state behavior which suggests that Ag is most likely in the Mg site. For Ag doping to Mg_2Sn, the PF reaches more than 20 µW/cm K^2 at 300 K with an optimum level doping of 6×10^{19} cm^{-3}.

Tang et al.[122] used melt spinning to synthesize Ag-doped $Mg_2Sn_{0.6}Si_{0.4}$. Here Ag concentrations of 1%, 2%, 5%, and 7% has been substituted to the Mg site of $Mg_2Sn_{0.6}Si_{0.4}$. Ag doping effectively reduces the resistivity of all the samples and provides sufficient acceptors to change the n-type behavior of $Mg_2Sn_{0.6}Si_{0.4}$ (MSS) to p-type. In conclusion, a peak figure of merit of 0.45 at 690 K has been achieved for 5% Ag doping at a carrier concentration of 6×10^{19} cm^{-3}.

However, there is also some debate on the stability of Ag in Mg_2X. Prytuliak et al.[123] showed that Ag can precipitate out of the Mg_2Si lattice at intermediate temperatures. These effects lead to the observed small values for *ZT*. Several papers reported relatively low carrier concentration levels for Ag-doped samples.[14,104,107] The observed low carrier concentration can partially be attributed to insufficient dopant activation. A further reason is given by a theoretical study from Han and Shao[117] who showed that the p-type conductivity of Ag-doped Mg_2Si is achieved through linear clustering of Ag with Mg interstitials ($Ag-Mg_{int}-Ag$). The positive charge of the Ag substitutions is compensated to a large extend by the negative charge of the Mg_{int}, resulting in relatively low carrier concentrations.

As a group 1 element, Na can be a p-dopant if it substitutes at the Mg site.[15,119] Tada et al.[124] used metallic Na and CH_3COONa as dopants to $Mg_2Sn_{0.75}Si_{0.25}$. The highest carrier concentration was found for the CH_3COONa doping, and the CH_3COONa doped materials showed better TE performances than the (metallic) Na doped samples. At 597 K, CH_3COONa decomposes into Na, CH_4, and CO_2. The decomposed Na acts as dopant for $Mg_2Sn_{0.75}Si_{0.25}$. Here, the sample doped with pure (metallic) Na contains detectable MgO and Mg_2SiSn impurity phases. MgO impurity phases are common in n-type Mg_2Sn system due to high vapor pressure of Mg.[66,125,126] For CH_3COONa doping, there are no significant impurity phases detectable. Due to the purity of the synthesized sample, the resistivity of the CH_3COONa doped sample is one order less compared to the pure

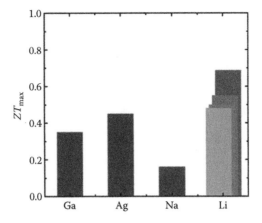

FIGURE 6.16 Best reported figure of merit values for the p-type $Mg_2Sn_{1-x}Si_x$ (with $0.20 < x < 0.4$). Data for Ga taken from Ref. 13, Ag from Ref. 122, Na from Ref. 125, and Li from Refs. 15,108,127,128.

Na doping. The Hall mobility for all the CH_3COONa doping samples is higher compared to the metallic Na-doped samples. Tada et al.[124] concludes that due to the difficulty of handling (metallic) Na, CH_3COONa may be a superior dopant for MSS compared to metallic Na.

Several of the recent papers focused on Li as dopant.[15,108,110,127,128] Zhang et al.[15] demonstrated a better figure of merit of 0.5 and a PF of ~15 µW/cm K² at 750 K by Li doping to the Mg site of MSS. Later reports by Gao et al.,[128] Tang et al.,[127] and Isachenko et al.[110] confirmed the reported good TE properties for the same or similar compositions. Gao et al.[128] used Li_2CO_3 instead of using Li metal as a dopant for $Mg_2Sn_{0.6}Si_{0.4}$ due to the highly reactive nature of Li. Here, the reaction with Mg with Li_2CO_3 will give some impurity phases which can be determined by X-ray diffraction. Nevertheless, the reported carrier concentration and the hole mobility are comparable with the best reported data of other doped materials. de Boor and Saparamadu et al.[108] showed that Li is also an effective dopant for $Mg_2Ge_{0.4}Sn_{0.6}$ and that this material has a comparable figure of merit.

A comparison of the available experimental data indicates that substitution on the Mg site yields better TE properties than substitution on the Si/Ge/Sn site. The best properties have been obtained for Li doping due to higher carrier concentrations and higher carrier mobilities. We will therefore focus on Li-doped samples in the data analysis in Section 6.3.4. In a direct comparison, Zhang et al.[15] found much better dopant activation for Li than for Na. A comparison of the p-dopants is summarized in Figure 6.16, where the best reported ZT_{max} is plotted for each kind of dopant.

6.3.4 Thermoelectric Transport

Although not as intensively studied as the n-type material, there are several reports on the TE properties of Mg_2X. The reported transport data sets differ significantly, which can be due to subtle differences in sample composition, processing parameters, or other potentially not fully understood experimental conditions. One aim of this section is therefore to give an overview of existing literature results. The second aim is to analyze the reported transport data in the framework of a SPB model, compare different studies, and finally try to extract fundamental material parameters. These can be a guideline for further material optimization.

6.3.4.1 Short Introduction to the SPB Model

The SPB model has been extensively employed for the n- and p-type materials[15,17,46,86] and has been presented in detail, e.g., by May and Snyder[129] as well as Flage-Larsen and Lovvik.[118] Using the experimental data for the Seebeck coefficient, the reduced chemical potential η can be extracted from

$$S = -\frac{k_B}{e}\left(\frac{2F_1(\eta)}{F_0(\eta)} - \eta\right),\tag{6.8}$$

where $F_j(\eta)$ is the Fermi integral of order i, k_B is the Boltzmann constant, and e is the elementary charge. The reduced chemical potential, in turn, can be employed to calculate the DOS effective mass m_D^* by using

$$n = 4\pi\left(\frac{2m_D^* k_B T}{h^2}\right)^{3/2} F_{1/2}(\eta).\tag{6.9}$$

The true carrier concentration n differs from the experimentally accessible Hall carrier density $n_H = n/r_H$ by the Hall factor $r_H = \frac{1.5F_{0.5}F_{-0.5}}{2F_0^2}$.

From η, the Lorenz number $L = \frac{k_B^2}{e^2}\frac{3F_0F_2 - 4F_1^2}{F_0^2}$ can be calculated, which allows for the determination of the lattice conductivity $\kappa_{lat} + \kappa_{bi} = \kappa - \kappa_e + \kappa_{bi} = \kappa - L\sigma T + \kappa_{bi}$, provided that the bipolar contribution κ_{bi} is negligible.

The TE figure of merit can be written in terms of more fundamental quantities

$$ZT = \frac{S^2}{L + (\psi\beta)^{-1}}\tag{6.10}$$

with $\Psi = \frac{8\pi e}{3}\left(\frac{2m_0 k_b}{h^2}\right)^{1.5} F_0$, $\beta = \mu_0 \frac{\left(\frac{m_D^*}{m}\right)^{1.5} T^{2.5}}{\kappa_{lat}}$. β and μ_0 are material parameters independent of carrier concentration, and μ_0 is related to the measured Hall mobility by $\mu_0 = \mu_H * 2F_0/F_{-0.5}$.

The equations given earlier are valid in the limit that the scattering of the charge carriers by APs and the potential fluctuations due to the solid solution formation (alloy scattering) are the dominant scattering mechanisms. This is a reasonable assumption for Mg_2X at room temperature and above. The incorporation of dopants will initiate some ionized impurity scattering. However, several calculations show that ionized impurity scattering is very small compared to AP scattering for Mg_2Si at the usual doping levels.[115,116]

For comparison of the data sets from several sources, we have graphically extracted the experimental results (σ, S, κ, n_H) and employed the equations given above to calculate ZT, κ_{lat}, m_D^*, μ_H, and μ_0. For the calculation of μ, a temperature-independent carrier concentration was assumed. Sometimes, it was necessary to back-calculate the experimental results, e.g., if κ_{lat} was given or S and the PF.

The SPB has several conditions that limits in its applicability. First of all, one implicit assumption is a rigid band structure, i.e., that the band structure itself is not changed by the incorporation of dopants, only the Fermi level shifted with respect to the band structure. Theoretical calculations show this assumption to be valid for Li as dopant, but not, e.g,. for Ag and Ga. Bourgeois et al.[114] found a sharp DOS peak near the Fermi level if Ag substitutes Si, confirming earlier work by Mars et al.[109] For Ga, several authors found a prominent band structure modification if Ga substitutes Si, again violating the assumed rigid band structure.[15,105,114] On the other hand, several calculations show that for Mg substitution by Li, the band structure remains relatively unaffected.[15,114] As Li-doped samples show the best reported TE properties, we will focus on Li-doped samples in the following.

Secondly, as it is a single band model, the SPB cannot account for bipolar transport after the onset of the intrinsic conduction. Therefore, for high temperature or low carrier concentration the predictive power of the model decreases.

Thirdly, the band structure shows two relevant valence bands. If these two bands are degenerated as shown in Figure 6.15, they form one "effective" band and the SPB model is applicable. Zhang et al.[15] showed the results of a density functional theory (DFT) calculation where the degeneration was removed. In this case, the contribution of the two bands varies with the chemical potential (hence with n and T) and the SPB picture breaks down. However, this band splitting has not been confirmed yet.

We will therefore employ the SPB model, keeping its limitations in mind and discussing deviations where applicable.

6.3.4.2 Analysis of Thermal Transport

We will start with a discussion of the thermal transport. Solid solution formation can be used to suppress the (lattice) thermal conductivity as for the n-type material. This is visualized in Figure 6.17, where lattice thermal conductivity data from binary p-type Mg_2Si and various solid solutions are presented.

Some variation between the data sets is due to different synthesis techniques (compare, e.g., de Boor et al.[108] and Tang et al.,[127] who investigated the same composition), but the overall trend is as expected. The data from the binary

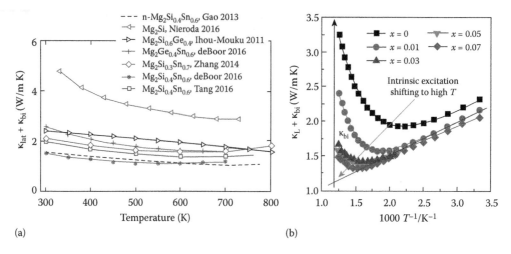

FIGURE 6.17 (a) Lattice thermal conductivity (plus bipolar contribution) for binary Mg_2Si and various solid solutions. The data were extracted from several sources (Mg_2Si,[130] $Mg_2Si_{0.6}Ge_{0.4}$,[105] $Mg_2Ge_{0.4}Sn_{0.6}$,[108] $Mg_2Si_{0.4}Sn_{0.6}$,[108,128] $Mg_2Si_{0.3}Sn_{0.7}$,[15] n-type $Mg_2Si_{0.4}Sn_{0.6}$[82]), and for the solid solutions, the data from the sample with the best TE properties are shown. Solid solution formation drastically reduces the (lattice) thermal conductivity. (b) The influence of the bipolar contribution is more relevant in p-type, and high carrier concentrations are required to suppress it. The shift of the onset with increasing doping/carrier concentration is shown in b. (Zhang, Q. et al., *Physical Chemistry Chemical Physics*, 16, 23576–23583, 2014. Reproduced by permission of The Royal Society of Chemistry.)

material have the highest κ_{lat}, followed by $Mg_2Si_{0.6}Ge_{0.4}$, $Mg_2Ge_{0.4}Sn_{0.6}$, and $Mg_2Si_{0.4}Sn_{0.6}$, i.e., it follows the increasing mass contrast of the employed elements. For compositions with the same elements κ_{lat} is lowest for Si:Sn closest to 50:50, compare $Mg_2Si_{0.4}Sn_{0.6}$ vs. $Mg_2Si_{0.3}Sn_{0.7}$.[15,127] Thermal conductivity reduction due to ionized impurity scattering has been observed for p-Mg_2Si,[130] but the overall influence is expected to be small.[100] However, there is also a significant difference between n-type and p-types. Figure 6.17 also shows κ_{lat} for n-type $Mg_2Si_{0.4}Sn_{0.6}$ with optimized TE properties,[82] and it can be seen that the onset of the bipolar contribution to the thermal conductivity is at a higher temperature and somewhat less pronounced for the n-type. This is due to the higher mobility of the electrons compared to the holes and the higher (partial) conductivity of the electrons.[131] In fact, bipolar thermal transport is one of the current limitations of the p-type TE performance and requires high doping levels to suppress it (see Figure 6.17b). This is more severe as the higher m^* of the holes requires large n for optimization as well, and achieving high doping levels is a practical challenge. Addressing bipolar transport is therefore a significant challenge that might be addressed by compositional variation; see, e.g., the work of Zhang et al.[125] on the n-type material.

6.3.4.3 Analysis of Electrical Transport

One key quantity for electrical transport is the effective mass of the charge carriers m^*. It can be obtained from Equation 6.9 in which DOS effective mass m_D^* is related to the single-valley effective mass m_i by $m_D^* = N_V^{2/3} m_i$, where N_V is the valley degeneracy and $N_V = 2$ if the two valence bands in Figure 6.15 are converged.

The results from the authors as well as from the literature for Li-doped samples are presented in Figure 6.18. The Pisarenko plot in Figure 6.18a shows that most of the data approximately follow the SPB prediction with $m_D^* = 1.4 m_0$. The data from Isoda et al.[132] ($\triangleleft$) for $Mg_2Si_{0.25}Sn_{0.75}$ show lower m_D^* values, but the samples are relatively low doped and close to the onset of intrinsic excitation even at room temperature which might influence the results. The data of Zhang et al.[15] for $Mg_2Si_{0.3}Sn_{0.7}$ also clearly deviate from the SPB prediction, showing a very low effective mass at a low carrier concentration that triples to high carrier concentrations. As the data show some scatter, it is challenging to draw conclusions with respect to composition. However, the average m_D^* values for the $Mg_2(Ge,Sn)$ samples is 1.3 m_0, while for the $Mg_2(Si,Sn)$ samples, it is 1.5 m_0 for 6×10^{19} cm^{-3} < n < 2×10^{20} cm^{-3}. This is in line with theoretical predictions from Kutorasinski et al.,[113] who predicted a decreasing m_D^* from Mg_2Si to Mg_2Ge to Mg_2Sn. No clear trend is discernible with respect to higher Sn content in the samples. The effective mass is plotted in Figure 6.18b as a function of carrier concentration. Disregarding the data from Zhang et al.[15] and Isoda et al.,[107] a small increase

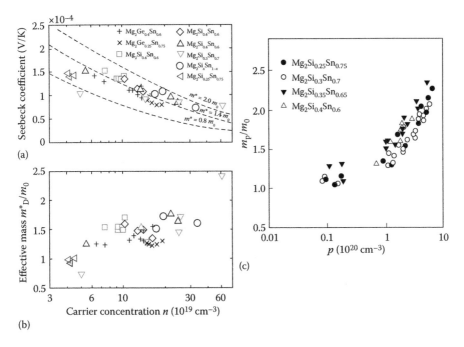

FIGURE 6.18 (a) Pisarenko plot for samples with a composition close to $Mg_2X_ySn_{1-y}$ (X = Si, Ge; $0.20 < y < 0.4$) from the authors and literature reports ("+," "×," square,[108] diamond,[129] up triangle,[128] down triangle,[15] circle,[110] "◁"[134]). The dashed lines correspond to the theoretical prediction within the SPB model for fixed values of m^*. Most of the data groups around $m^* = 1.4 m_0$. (b) Effective mass versus carrier concentration shows a slight increase for Li-doped samples. In contrast to this, a very strong increase for the employed group 3 dopants from the work of Isachenko et al.[121] is shown in (c). (With kind permission from Springer Science+Business Media: *Physics of the Solid State*, Kinetic properties of p-$Mg_2Si_xSn_{1-x}$ solid solutions for x < 0.4, 21, 2009, 1796–1799, Isachenko, G. N. et al.)

of the effective mass with increasing carrier concentration can be discerned. This first indicates that the SPB can be employed with reasonable accuracy for Li-doped $Mg_2(Si,Ge,Sn)$ and, secondly, that the (effective) band is not strictly parabolic or the gap between the LH and the HH band is small but not vanishing.

Figure 6.18c shows $m_D^*(n)$ from Isachenko et al.[121] In this case, the employed dopant was not Li but is specified as group 3 elements such as, e.g., Ga. It shows a much stronger increase of $m_D^*(n)$, in agreement with the theoretically predicted violation of the rigid band picture for Ga doping[15,105,114] and experimental results from Liu et al.[13]

A further fundamental parameter in the SPB is the mobility parameter μ_0. It is defined by $\mu_0 = e\tau_0/m^*$, with the lifetime parameter τ_0 given by τ_0 $(E/k_bT)^{\lambda-0.5}$; here λ is the scattering parameter with $\lambda = 0$ for the considered AP scattering. The related Hall mobility $\mu_0 = \mu_H*2F_0/F_{-0.5}$ decreases with increasing n due to increasing carrier–carrier scattering. The mobility parameter μ_0 accounts for that and is supposed to be independent of n. The Hall mobility of the sample series discussed in Figure 6.18 is shown in Figure 6.19a. For each report/composition, only data from the sample with the best TE properties are shown. It can be seen that all the data follow $\mu_0 \propto T^{-p}$, with $p \leq 1$,

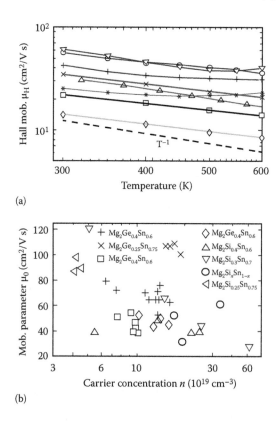

FIGURE 6.19 (a) Hall mobility versus temperature and (b) mobility parameter μ_0 at room temperature as a function of carrier concentration. The legend in (a) is as shown in Figure 6.18; for each composition, only the data of the sample with the best TE properties are shown.

which is consistent with assumed AP phonon scattering as the dominant mechanism plus some influence of alloy scattering. At higher temperatures, a deviation from the linear regime is visible for several samples. This is due to the onset of intrinsic excitation and thus the breaking down of the assumption of a single carrier type that was used to obtain the carrier concentration (and mobility) from the Hall measurement data. Figure 6.19b shows the values of the mobility parameter μ_0 at room temperature. The presented data comprise $Mg_2(Ge,Sn)$ as well as $Mg_2(Si,Sn)$ samples with some variation in the Sn content. However, the data show a large variation among the different reports, emphasizing the influence of sample preparation and synthesis procedure. A decrease of the mobility parameter with increasing carrier concentration is visible for the majority of the data sets, which is not expected in the framework of the SPB model. This indicates that mobility is affected not only by increasing carrier–carrier scattering but also by other mechanisms. The observed slightly increasing $m_D^*(n)$ might be one reason for the observation.

With respect to composition, it can be discerned that for a given Sn content, the Ge-containing compositions have a higher mobility. This is in agreement with the observed lower effective mass for the Ge-containing samples as $\mu_H \propto (m^*)^{-2.5}$.

The TE figure of merit is presented in Figure 6.20a as a function of temperature. It can be seen that the highest figures of merit have been obtained for the compositions around $Mg_2Si_{0.4}Sn_{0.6}$. Also, it is clear that with Li doping, $ZT \geq 0.5$ has been repeatedly obtained in recent years which is a significant improvement to earlier reports with other dopants, where $ZT < 0.35$. The Ge-containing samples seem to have slightly inferior properties, which are due to their higher (lattice) thermal conductivity; however, the difference is relatively small. It should also be pointed out that a higher PF is more important than ZT if power output rather than efficiency is the main concern. This can be the case in waste heat recovery scenarios, where heat may be "free."

Compositions without Sn are not shown, but the reports show clearly inferior properties.[13,105,109,130] Figure 6.20b shows $ZT(n)$ at $T = 700$ K as this is the maximum temperature for some reports and the temperature close to the peak ZT. Within the SPB, one optimum carrier concentration is expected, see, e.g., Bux et al.[35] and Isachenko et al.[121] From Figure 6.20b, a broad plateau is observed between 1×10^{20} cm⁻³ $< n_{opt} < 6 \times 10^{20}$ cm⁻³, with the optimal values of the individual reports significantly differing. de Boor et al.[108] obtained the best TE properties at $n_{opt} \approx 10^{20}$ cm⁻³ for $Mg_2Si_{0.4}Sn_{0.6}$, while Tang et al.[127] obtained these at $n_{opt} \approx 2 \times 10^{20}$ cm⁻³ for the same composition, and Zhang et al.[15] found their maximum at $n_{opt} \approx 6 \times 10^{20}$ cm⁻³ for $Mg_2Si_{0.3}Sn_{0.7}$. Some differences between different reports can be expected as, e.g., synthesis conditions influence μ_0 and κ_{lat}; however, the spread for the p-type is much larger than for the n-type binary or ternary, see, e.g., discussion by de Boor et al.[100] It should also be noted that at 700 K, the samples are usually close to or already in the intrinsic regime, outside the applicability of the SPB model.

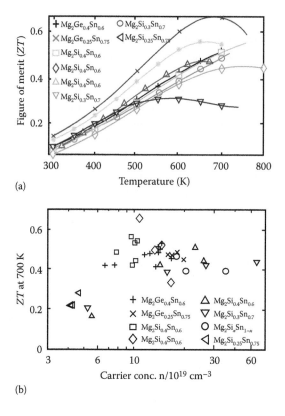

(a)

(b)

FIGURE 6.20 (a) TE figure of merit versus temperature and (b) carrier concentration for Li-doped samples with a composition close to $Mg_2X_ySn_{1-y}$ (X = Si, Ge; 0.20 < y < 0.4). The legend is as shown in Figure 6.18; in (a), only the best samples of each report are shown. In (a), the decrease of ZT due to bipolar transport is clearly visible for all samples at higher temperatures. In (b), a broad carrier concentration range with ZT > 0.5 is visible, but no clear optimum.

6.3.4.4 Transport Analysis Summary

In summary, the preceding analysis shows that the SPB model is not sufficient to completely understand and predict the properties of p-type Mg_2X. For Li-doped samples, the effective mass shows only a slight dependence on carrier concentration for compositions around $Mg_2(Si,Ge)_xSn_{1-x}$ with 0.25 < x < 0.4 and is thus in reasonable agreement with the SPB model. On the other hand, the effective mass for other dopants shows a strong dependence on carrier concentration, and hence, the rigid band picture and the SPB model break down. Lattice thermal conductivity and carrier mobility significantly vary between the different reports. Both are influenced by microstructure and hence synthesis details, but the strong variations in mobility and the strong dependence on composition indicates that other mechanisms strongly contribute to carrier mobility. Due to that, the predictive power of the SPB model with respect to optimum carrier concentration and figure of merit is limited. Experimentally, $ZT_{max} \geq 0.5$ at 700 K is observed for a broad optimum carrier concentration range of 1×10^{20} cm⁻³ $< n_{opt} < 6 \times 10^{20}$ cm⁻³ for compositions around $Mg_2(Si,Ge)_xSn_{1-x}$ with $0.25 < x < 0.4$.

With respect to composition, some general trends and trade-offs can be deduced:

■ Heavier compositions have the advantage of lower lattice thermal conductivity and a much better dopant activation leading to higher achievable carrier concentration; see Figure 6.21 or the discussion by de Boor et al.[108] As the TE properties are strongly controlled by carrier concentration, compositions with high Sn content have shown much better properties. This is in distinct difference from the n-type material where, e.g., binary doped Mg_2Si is also inferior to $Mg_2(Si,Sn)$, but the difference is not so large as optimum carrier concentration can be achieved in both systems.

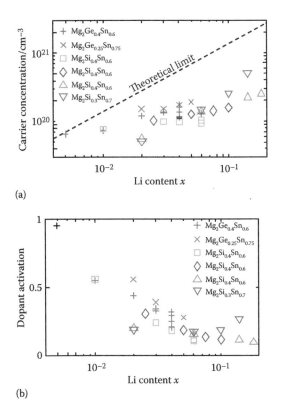

(a)

(b)

FIGURE 6.21 (a) Li content x versus experimentally determined carrier concentration for samples with a composition close to $Mg_2X_ySn_{1-y}$ (X = Si, Ge; $0.25 < y < 0.4$). The dashed line is obtained under the assumption that each Li atom provides one free hole and a lattice constant of $a = 6.6 \times 10^{-10}$ m. Although Li doping allows for high carrier concentrations, the experimentally observed values remain far below the limiting value, especially for higher x values. (b) The ratio of experimental and limiting value indicates that the dopant activation increases with increasing Sn content of the samples and that replacing Si by Ge leads to a higher dopant activation. (Reprinted from *Acta Materialia*, 120, 273–280 de Boor, J. et al., Thermoelectric performance of Li doped, p-type $Mg_2(Ge,Sn)$ and comparison with $Mg_2(Si,Sn)$, 273–280, Copyright (2016), with permission from Elsevier.)

- On the other hand, with increasing Sn content, the lattice thermal conductivity increases (if above $Sn_{0.6}$) due to decreasing alloy scattering of the phonons. Furthermore, the bandgap decreases and the bipolar transport contributes to both thermal conductivity and PF becomes dominating, deteriorating the TE performance.
- Few data for the effective mass as function of composition have been obtained so far, so no trends with respect to, e.g., Sn content can be drawn. By comparing $Mg_2Si_{0.4}Sn_{0.6}$ and $Mg_2Ge_{0.4}Sn_{0.6}$ from the same synthesis route, a slightly reduced effective mass for the Ge-containing samples is found; however, further experimental work is required to confirm this as a general trend. Lower effective masses are favorable for optimized PFs, and replacing of Si by Ge is thus a potential lever for further optimization.
- Charge carrier mobility is heavily affected by the synthesis procedure and not an inherent material parameter. Nevertheless, from the existing data, it can be concluded that increasing the Sn content above $Sn_{0.6}$ and replacing Si by Ge enhance the hole mobility. This is in qualitative agreement with the expectations due to reduced strain and mass fluctuations and in line with the observed lower effective mass of the Ge-containing samples. However, further mechanisms are likely necessary for a complete understanding as exemplified, e.g., by the very low mobilities for doped binary Mg_2Si.

In summary, it appears that the material optimization is more complex for the p-type material than n-type Mg_2X. Besides the usual optimization with optimum carrier concentration and solid solution composition, the dopant type, composition on dopant activation, optimum carrier concentration, and mobility and bandgap establish not only additional challenges, but also additional means for the optimization of the TE properties. More and more systematic investigations of the composition are therefore highly desired and a prerequisite for a truly competitive p-type Mg_2X. This should be accompanied by theoretical calculations especially as the predictive power of the SPB model appears to be limited.

6.3.5 Future Challenges

In this section, we briefly discuss challenges that need to be addressed to make p-type Mg_2X a technological material with realistic application scenarios, in addition to the already discussed somewhat complex optimization of the material with respect to the TE properties.

The first issue is the thermal stability of the material. For the n-type, some investigations for binary and ternary compounds are available,[83,114,133–136] and while the absolute numbers differ, it is clear that Sn-rich compositions are less stable than Sn-poor compositions or binary Mg_2Si. As the best p-type properties are found for Sn-rich compositions, this is also relevant for the p-type. Furthermore, the best properties have been found using Li as a dopant,[13,15] but several authors mention a blackening of the samples after a measurement cycle.[15,128] Even if no immediate degradation of the TE properties was observed, an analysis of the black layer and the formation kinetics remain to be done. It should also be noted that the maximum measurement temperature in the reports (for similar compositions) somewhat differs[15,108,127,128] Nieroda et al.[130] compared the oxidation of Mg_2Si and $Mg_{1.9}Li_{0.1}Si$ and found a much stronger oxidation for the Li-containing sample and an onset at significantly lower temperatures. This indicates that Li doping might affect the thermal stability of the sample and requires thorough and systematic studies, as long-term stability is indispensable for a technological application of the material. If thermal stability is indeed a problem, sample coating might be part of the solution.[130,137]

A second unclear point is the position of Li in the crystal or the compound. Calculations show that Li substitutes Mg rather than Si,[130] and experimental results on both binary Mg_2Si and ternary Mg_2X indicate a quite high solubility of Li in Mg_2X[15,130] Comparison between nominal composition and elemental analysis by inductively coupled plasma spectrometry (ICP) further indicates that most of the Li is indeed in the compound and not lost, e.g., during synthesis.[15,128] However, all reports also show consistently quite low dopant activation, with the measured carrier concentration being systematically lower than the one expected if each Li atom substitutes one Mg atom and provides one free hole; see Figure 6.21.

So the question is where the Li goes in the compound and what the doping mechanism is. Gao et al.[128] suggested that Li partially might go into interstitial positions; however, this is in disagreement with DFT calculations and refinement of neutron diffraction data.[130] It is in principle also possible that finely dispersed Li-containing secondary phases are formed, but these have not been observed so far.[127] Poor dopant activation (and low mobility) has also been experimentally found for Ag-doped Mg_2X. Han and Shao[117] investigated defect formation in Ag-doped Mg_2Si and found that Mg substitution by Ag favors the formation of Mg interstitial Mg_i, which leads to a formation of $Ag–Mg_i–Ag$ chains and a charge compensating effect, resulting in the experimentally observed reduced carrier concentration.[117] Further studies need to show if such a mechanism is also plausible for Li-doped Mg_2X.

References

1. Morris, R. G., R. D. Redin, and G. C. Danielson, Semiconducting properties of Mg_2Si single crystals. *Physical Review*, 1958. **109**(6): pp. 1909–1915.
2. Bose, S., H. Acharya, and H. Banerjee, Electrical, thermal, thermoelectric and related properties of magnesium silicide semiconductor prepared from rice husk. *Journal of Materials Science*, 1993. **28**(20): pp. 5461–5468.
3. Tani, J.-I., and H. Kido, Lattice dynamics of Mg_2Si and Mg_2Ge compounds from first-principles calculations. *Computational Materials Science*, 2008. **42**(3): pp. 531–536.
4. Zaitsev, V. et al., *Thermoelectrics Handbook*. CRC Press, Taylor & Francis, Boca Raton, FL, 2006: pp. 29–31.

5. Zaitsev, V. K. et al., Highly effective $Mg_2Si_{1-x}Sn_x$ thermoelectrics. *Physical Review B*, 2006. **74**(4): pp. 045207/1–045207/5.

6. Fedorov, M. I., and G. N. Isachenko, Silicides: Materials for thermoelectric energy conversion. *Japanese Journal of Applied Physics*, 2015. **54**(7S2): p. 07JA05.

7. Tani, J.-I., and H. Kido, Thermoelectric properties of Sb-doped Mg_2Si semiconductors. *Intermetallics*, 2007. **15**(9): pp. 1202–1207.

8. Tani, J.-I., and H. Kido, Thermoelectric properties of Bi-doped Mg_2Si semiconductors. *Physica B: Condensed Matter*, 2005. **364**(1–4): pp. 218–224.

9. Battiston, S. et al., Synthesis and characterization of Al-doped Mg_2Si thermoelectric materials. *Journal of Electronic Materials*, 2013. **42**(7): pp. 1956–1959.

10. Zhang, Q. et al., Thermoelectric performance of $Mg_{2-x}Ca_xSi$ compounds. *Journal of Alloys and Compounds*, 2008. **464**(1–2): pp. 9–12.

11. Zhang, Q. et al., In situ synthesis and thermoelectric properties of La-doped $Mg_2(Si, Sn)$ composites. *Journal of Physics D: Applied Physics*, 2008. **41**(18): p. 185103.

12. Tada, S. et al., Thermoelectric properties of p-type $Mg_2Si_{0.25}Sn_{0.75}$ doped with sodium acetate and metallic sodium. *Journal of Electronic Materials*, 2013. **43**(6): pp. 1580–1584.

13. Liu, W. et al., Enhanced hole concentration through Ga doping and excess of Mg and thermoelectric properties of p-type $Mg_{2(1+z)}(Si_{0.3}Sn_{0.7})_{1-y}Ga_y$. *Intermetallics*, 2013. **32**: pp. 352–361.

14. Kim, S. et al., Electronic structure and thermoelectric properties of p-type Ag-doped Mg_2Sn and $Mg_2Sn_{1-x}Si_x$ (x = 0.05, 0.1). *Journal of Applied Physics*, 2014. **116**(15): p. 153706.

15. Zhang, Q. et al., Low effective mass and carrier concentration optimization for high performance p-type $Mg_{2(1-x)}Li_{2x}Si_{0.3}Sn_{0.7}$ solid solutions. *Physical Chemistry Chemical Physics*, 2014. **16**(43): pp. 23576–23583.

16. Liu, W. et al., Convergence of conduction bands as a means of enhancing thermoelectric performance of n-type $Mg_2Si_{1-x}Sn_x$ solid solutions. *Physical Review Letters*, 2012. **108**(16): p. 166601.

17. Liu, X. et al., Low electron scattering potentials in high performance $Mg_2Si_{0.45}Sn_{0.55}$ based thermoelectric solid solutions with band convergence. *Advanced Energy Materials*, 2013. **3**(9): pp. 1238–1244.

18. Jiang, G. et al., High performance $Mg_2(Si,Sn)$ solid solutions: A point defect chemistry approach to enhancing thermoelectric properties. *Advanced Functional Materials*, 2014. **24**(24): pp. 3776–3781.

19. Du, Z. et al., Roles of interstitial Mg in improving thermoelectric properties of Sb-doped $Mg_2Si_{0.4}Sn_{0.6}$ solid solutions. *Journal of Materials Chemistry*, 2012. **22**(14): pp. 6838–6844.

20. Gao, H. et al., Flux synthesis and thermoelectric properties of eco-friendly Sb doped $Mg_2Si_{0.5}Sn_{0.5}$ solid solutions for energy harvesting. *Journal of Materials Chemistry*, 2011. **21**(16): pp. 5933–5937.

21. Liu, W., et al., Optimized thermoelectric properties of Sb-doped $Mg_{2(1+z)}Si_{0.5-y}Sn_{0.5}Sb_y$ through adjustment of the Mg content. *Chemistry of Materials*, 2011. **23**(23): pp. 5256–5263.

22. Hansen, M., and K. Anderko, Structure of binary alloys. *Metallurgizdat*, 1962. **2**: pp. 1111–1115.

23. Nikitin, E. et al., A study of the phase diagram for the Mg_2Si-Mg_2Sn system and the properties of certain of its solid solutions. *Journal of Inorganic Materials*, 1970. **4**: p. 1656.

24. Muntyanu, S., E. Sokolov, and E. Makarov, Study of the Mg_2Sn-Mg_2Si system. *Izv Akad Nauk Ussr Neorgan Materialy*, 1966. **2**(5): pp. 870–875.

25. Vivès, S. et al., Combinatorial approach based on interdiffusion experiments for the design of thermoelectrics: Application to the $Mg_2(Si,Sn)$ alloys. *Chemistry of Materials*, 2014. **26**(15): pp. 4334–4337.

26. Jung, I.-H. et al., Thermodynamic modeling of the Mg–Si–Sn system. *Calphad*, 2007. **31**(2): pp. 192–200.

27. Kozlov, A., J. Gröbner, and R. Schmid-Fetzer, Phase formation in Mg–Sn–Si and Mg–Sn–Si–Ca alloys. *Journal of Alloys and Compounds*, 2011. **509**(7): pp. 3326–3337.

28. Chen, L. et al., Miscibility gap and thermoelectric properties of ecofriendly $Mg_2Si_{1-x}Sn_x$ ($0.1 \le x \le 0.8$) solid solutions by flux method. *Journal of Materials Research*, 2011. **26**(24): pp. 3038–3043.

29. Isoda, Y. et al., Thermoelectric properties of Sb-doped $Mg_2Si_{0.5}Sn_{0.5}$. In *ICT '06. 25th International Conference on Thermoelectrics*. Institute of Electrical and Electronics Engineers, Piscataway, NJ, 2006.

30. Riffel, M., and J. Schilz, Mechanically alloyed $Mg_2Si_{1-x}Sn_x$ solid solutions as thermoelectric materials. *Fifteenth International Conference on Thermoelectrics*, 1996. pp. 133–136.

31. Viennois, R. et al., Phase stability of ternary antifluorite type compounds in the quasi-binary systems $Mg_2X–Mg_2Y$ (X, Y = Si, Ge, Sn) via ab-initio calculations. *Intermetallics*, 2012. **31**: pp. 145–151.

32. Zaitsev, V., and E. Nikitin, Electrical properties thermal conductivity and forbidden band width of Mg_2Sn at high temperatures. *Fizika Tverdogo Tela*, 1970. **12**(2): pp. 357–361.

33. Lichter, B. D., Electrical properties of Mg_2Sn crystals grown from non-stoichiometric melts. *Journal of the Electrochemical Society*, 1962. **109**(9): pp. 819–824.

34. Liu, W., X. Tang, and J. Sharp, Low-temperature solid state reaction synthesis and thermoelectric properties of high-performance and low-cost Sb-doped $Mg_2Si_{0.6}Sn_{0.4}$. *Journal of Physics D: Applied Physics*, 2010. **43**(8): p. 085406.

35. Bux, S. K. et al., Mechanochemical synthesis and thermoelectric properties of high quality magnesium silicide. *Journal of Materials Chemistry*, 2011. **21**(33): p. 12259–12266.

36. Zhang, Q. et al., Phase segregation and superior thermoelectric properties of $Mg_2Si_{1-x}Sb_x$ ($0 \le x \le 0.025$) prepared by ultrafast self-propagating high-temperature synthesis. *ACS Applied Materials & Interfaces*, 2016. **8**(5): pp. 3268–3276.

37. Guillon, O. et al., Field-assisted sintering technology/spark plasma sintering: mechanisms, materials, and technology developments. *Advanced Engineering Materials*, 2014. **16**(7): pp. 830–849.

38. Koenig, P., D. W. Lynch, and G. C. Danielson, Infrared absorption in magnesium silicide and magnesium germanide. *Journal of Physics and Chemistry of Solids*, 1961. **20**(1): pp. 122–126.

39. Lipson, H. G., and A. Kahan, Infrared absorption of magnesium stannide. *Physical Review*, 1964. **133**(3A): pp. A800–A810.

40. Au-Yang, M. Y., and M. L. Cohen, Electronic structure and optical properties of Mg_2Si, Mg_2Ge, and Mg_2Sn. *Physical Review*, 1969. **178**(3): pp. 1358–1364.

41. Zaitsev, V., E. Nikitin, and E. Tkalenko, Width of the forbidden band in magnesium silicide-magnesium stannide(Mg_2Si-Mg_2Sn) solid solutions. *Fizika Tverdogo Tela*, 1969. **11**(12): pp. 3584–3587.

42. Tan, X. J. et al., Multiscale calculations of thermoelectric properties of n-type $Mg_2Si_{1-x}Sn_x$ solid solutions. *Physical Review B*, 2012. **85**(20): p. 205212.

43. Zaitsev, V. K. et al., Some features of the conduction band structure, transport and optical properties of n-type Mg_2Si-Mg_2Sn alloys. *Proceedings ICT '02: Twenty-First International Conference on Thermoelectrics*, 2002.

44. Liu, W. S. et al., New insight into the material parameter B to understand the enhanced thermoelectric performance of $Mg_2Sn_{1-x-y}Ge_xSb_y$. *Energy & Environmental Science*, 2016. **9**(2): pp. 530–539.

45. Goldsmid, H., Thermoelectric refrigeration. 1964. Springer, New York. ISBN: 978-1-4899-5725-2.

46. Liu, W. et al., Advanced thermoelectrics governed by a single parabolic band: $Mg_2Si_{0.3}Sn_{0.7}$, a canonical example. *Physical Chemistry Chemical Physics*, 2014. **16**(15): pp. 6893–6897.

47. Bardeen, J., and W. Shockley, Deformation potentials and mobilities in non-polar crystals. *Physical Review*, 1950. **80**(1): pp. 72–80.

48. Zaitsev, V., E. Tkalenko, and E. Nikitin, Lattice thermal conductivity of Mg_2Si-Mg_2Sn, Mg_2Ge-Mg_2Sn, and Mg_2Si-Mg_2Ge solid solutions. *Soviet Physics-Solid State*, 1969. **11**(2): pp. 221–224.

49. Isoda, Y. et al., Thermoelectric properties of sintered $Mg_2Si_{1-x}Sn_x$. *Proceedings of the 23rd International conference on Thermoelectrics, Adelaide, Australia*, 2004. p. 124.

50. Klemens, P. G., Thermal resistance due to point defects at high temperatures. *Physical Review*, 1960. **119**(2): pp. 507–509.

51. Callaway, J., and H. C. von Baeyer, Effect of point imperfections on lattice thermal conductivity. *Physical Review*, 1960. **120**(4): pp. 1149–1154.

52. Du, Z. L., H. L. Gao, and J. L. Cui, Thermoelectric performance of quaternary $Mg_{2(1+x)}Si_{0.2}Ge_{0.1}Sn_{0.7}$ ($0.06 \leq x \leq 0.12$) solid solutions with band convergence. *Current Applied Physics*, 2015. **15**(7): pp. 784–788.

53. Tada, S. et al., Preparation and thermoelectric properties of $Mg_2Si_{0.9-x}Sn_xGe_{0.1}$. *Physica Status Solidi (C)*, 2013. **10**(12): pp. 1704–1707.

54. Yoshinaga, M. et al., Bulk crystal growth of Mg_2Si by the vertical Bridgman method. *Thin Solid Films*, 2004. **461**(1): pp. 86–89.

55. Redin, R. D., R. G. Morris, and G. C. Danielson, Semiconducting properties of Mg_2Ge single crystals. *Physical Review*, 1958. **109**(6): pp. 1916–1920.

56. Shanks, H. R., The growth of magnesium germanide crystals. *Journal of Crystal Growth*, 1974. **23**(3): pp. 190–194.

57. Chen, H. Y., and N. Savvides, Microstructure and thermoelectric properties of n- and p-type doped Mg_2Sn compounds prepared by the modified Bridgman method. *Journal of Electronic Materials*, 2009. **38**(7): pp. 1056–1060.

58. Chen, H. Y., and N. Savvides, High quality Mg_2Sn crystals prepared by RF induction melting. *Journal of Crystal Growth*, 2010. **312**(16–17): pp. 2328–2334.

59. Chen, H., and N. Savvides. Thermoelectric properties and microstructure of large-grain Mg_2Sn doped with Ag. In *Materials Research Society Symposium Proceedings*. Cambridge University Press, Cambridge, UK, 2009, 1166.

60. Tani, J.-I., and H. Kido, First-principles and experimental studies of impurity doping into Mg_2Si. *Intermetallics*, 2008. **16**(3): pp. 418–423.

61. Kato, A., T. Yagi, and N. Fukusako, First-principles studies of intrinsic point defects in magnesium silicide. *Journal of Physics: Condensed Matter*, 2009. **21**(20): p. 205801.

62. Jund, P. et al., Lattice stability and formation energies of intrinsic defects in Mg_2Si and Mg_2Ge via first principles simulations. *Journal of Physics: Condensed Matter*, 2013. **25**(3): p. 035403.

63. Liu, X. et al., Significant roles of intrinsic point defects in Mg_2X (X = Si, Ge, Sn) thermoelectric materials. *Advanced Electronic Materials*, 2015. **2**(2): p. 1500284.

64. Dasgupta, T. et al., Effect of vacancies on the thermoelectric properties of $Mg_2Si_{1-x}Sb_x$($0 \leq x < 0.1$). *Physical Review B*, 2011. **83**(23): p. 235207.

65. Nolas, G., D. Wang, and M. Beekman, Transport properties of polycrystalline $Mg_2Si_{1-y}Sb_y$ ($0 \leq y < 0.4$). *Physical Review B*, 2007. **76**(23): p. 235204.

66. Liu, W. et al., High figure of merit and thermoelectric properties of Bi-doped $Mg_2Si_{0.4}Sn_{0.6}$ solid solutions. *Journal of Solid State Chemistry*, 2013. **203**: pp. 333–339.

67. Liu, W. S. et al., N-type thermoelectric material $Mg_2Sn_{0.75}Ge_{0.25}$ for high power generation. *Proceedings of the National Academy of Sciences USA*, 2015. **112**: pp. 3269–3274.

68. Riffel, M., and J. Schilz, Mechanical alloying of Mg_2Si. *Scripta Metallurgica et Materialia*, 1995. **32**: pp. 1951–1956.

69. Jung, J. Y., and I. H. Kim, Synthesis of thermoelectric Mg_2Si by mechanical alloying. *Journal of the Korean Physical Society*, 2010. **57**: pp. 1005–1009.

70. Urretavizcaya, G., and G. O. Meyer, Metastable hexagonal Mg_2Sn obtained by mechanical alloying. *Journal of Alloys and Compounds*, 2002. **339**: pp. 211–215.

71. You, S. W. et al., Solid-state synthesis and thermoelectric properties of $Mg_{2+x}Si_{0.7}Sn_{0.3}Sb_m$. *Journal of Nanomaterials*, 2013. **2013**: p. 815925.

72. Saparamadu, U. et al., The effect of charge carrier and doping site on thermoelectric properties of $Mg_2Sn_{0.75}Ge_{0.25}$. *Acta Materialia*, 2017. **124**: pp. 528–535

73. Zaitsev, V. K. et al., Thermoelectrics on the base of solid solutions of Mg_2B^{IV} compounds (B^{IV} = Si, Ge, Sn). In *Thermoelectric Handbook: Macro to Nano Structure Materials*. CRC Press, Boca Raton, FL, 2005.

74. Wang, H. et al., Material design considerations based on thermoelectric quality factor. In *Thermoelectric Nanomaterials*, K. Koumoto and T. Mori, editors. Springer Series in Materials Science, Springer-Verlag, Berlin Heidelberg, 182, 2013.

75. Mao, J. et al., Thermoelectric properties of materials near the band crossing line in Mg_2Sn-Mg_2Ge-Mg_2Si system. *Acta Materialia*, 2016. **103**: pp. 633–642.

76. Biswas, K. et al., High-performance bulk thermoelectrics with all-scale hierarchical architectures. *Nature*, 2012. **489**(7416): pp. 414–418.

77. Rogl, G. et al., n-Type skutterudites $(R,Ba,Yb)_yCo_4Sb_{12}$ (R = Sr, La, Mm, DD, SrMm, SrDD) approaching ZT approximate to 2.0. *Acta Materialia*, 2014. **63**: pp. 30–43.

78. Sakurada, S., and N. Shutoh, Effect of Ti substitution on the thermoelectric properties of (Zr,Hf)NiSn half-Heusler compounds. *Applied Physics Letters*, 2005. **86**(8): p. 082105.

79. Zhao, L. D. et al., Ultralow thermal conductivity and high thermoelectric figure of merit in SnSe crystals. *Nature*, 2014. **508**(7496): pp. 373–377.

80. Brown, D. R. et al., Chemical stability of $(Ag,Cu)_2Se$: A historical overview. *Journal of Electronic Materials*, 2013. **42**(7): pp. 2014–2019.

81. Khan, A. U. et al., Thermoelectric properties of highly efficient Bi-doped $Mg_2Si_{1-x-y}Sn_xGe_y$ materials. *Acta Materialia*, 2014. **77**: pp. 43–53.

82. Gao, P. et al., Transport and mechanical properties of high-ZT $Mg_{2.08}Si_{0.4-x}$ $Sn_{0.6}Sb_x$ thermoelectric materials. *Journal of Electronic Materials*, 2013. **43**(6): pp. 1790–1803.

83. Dasgupta, T. et al., Influence of power factor enhancement on the thermoelectric figure of merit in $Mg_2Si_{0.4}Sn_{0.6}$ based materials. *Physica Status Solidi (A)*, 2014. **211**: pp. 1250–1254.

84. LeBlanc, S. et al., Material and manufacturing cost considerations for thermoelectrics. *Renewable and Sustainable Energy Reviews*, 2014. **32**: pp. 313–327.

85. Zhao, J. B. et al., Thermoelectric and electrical transport properties of Mg_2Si multi-doped with Sb, Al and Zn. *Journal of Materials Chemistry A*, 2015. **3**(39): pp. 19774–19782.

86. de Boor, J. et al., Thermoelectric transport and microstructure of optimized $Mg_2Si_{0.8}Sn_{0.2}$. *Journal of Materials Chemistry C*, 2015. **3**(40): pp. 10467–10475.

87. de Boor, J. et al., Fabrication and characterization of nickel contacts for magnesium silicide based thermoelectric generators. *Journal of Alloys and Compounds*, 2015. **632**(0): pp. 348–353.

88. Ferrario, A. et al., Mechanical and electrical characterization of low-resistivity contact materials for Mg_2Si. *Materials Today: Proceedings*, 2015. **2**(2): pp. 573–582.

89. Sakamoto, T. et al., Investigation of barrier-layer materials for Mg_2Si/Ni Interfaces. *Journal of Electronic Materials*, 2015. **45**(3): pp. 1321–1327.

90. Sakamoto, T. et al., The use of transition-metal silicides to reduce the contact resistance between the electrode and sintered-type Mg_2Si. *Journal of Electronic Materials*, 2012. **41**(6): pp. 1805–1810.

91. de Boor, J. et al., Thermal stability of magnesium silicide/nickel contacts. *Journal of Electronic Materials*, 2016. **45**: p. 5313.

92. Kim, H. S. et al., Design of segmented thermoelectric generator based on cost-effective and light-weight thermoelectric alloys. *Materials Science and Engineering B—Advanced Functional Solid-State Materials*, 2014. **185**: pp. 45–52.

93. Nakamura, T. et al., Power-generation performance of a π-structured thermoelectric module containing Mg_2Si and $MnSi_{1.73}$. *Journal of Electronic Materials*, 2015. **44**(10): pp. 3592–3597.

94. Nemoto, T. et al., Improvement in the durability and heat conduction of uni-leg thermoelectric modules using n-type Mg_2Si legs. *Journal of Electronic Materials*, 2014. **43**(6): pp. 1890–1895.

95. Sakamoto, T. et al., Stress analysis and output power measurement of an n-Mg_2Si thermoelectric power generator with an unconventional structure. *Journal of Electronic Materials*, 2014. **43**(6): pp. 1620–1629.

96. Skomedal, G. et al., Design, assembly and characterization of silicide-based thermoelectric modules. *Energy Conversion and Management*, 2016. **110**: pp. 13–21.

97. Tarantik, K. R. et al., Thermoelectric modules based on silicides—Development and characterization. *Materials Today: Proceedings*, 2015. **2**(2): pp. 588–595.

98. Zaitsev, V. K. et al., Thermoelectrics on the base of solid solutions of Mg_2B^{IV} Compounds. In *Thermoelectrics Handbook: Macro to Nano*, D. M. Rowe, editor. CRC Press, Boca Raton, FL, 2005.

99. Fedorov, M. I., and V. K. Zaitsev, Silicide thermoelectrics: State of the art and prospects. In *Thermoelectrics and its Energy Harvesting: Modules, Systems, and Applications in Thermoelectrics*, D. M. Rowe, editor. CRC Press, Taylor & Francis, Boca Raton, FL, 2012.

100. de Boor, J., T. Dasgupta, and E. Mueller, Thermoelectric properties of magnesium silicide based solid solutions and higher manganese silicides. In *Materials Aspect of Thermoelectricty*, C. Uher, editor, Taylor & Francis, Boca Raton, FL, 2016.

101. Kim, H. S. et al., Relationship between thermoelectric figure of merit and energy conversion efficiency. *Proceedings of the National Academy of Sciences*, 2015. **112**(27): pp. 8205–8210.

102. Wood, C., Materials for thermoelectric energy-conversion. *Reports on Progress in Physics*, 1988. **51**(4): pp. 459–539.

103. Tani, J.-I., and H. Kido, Fabrication and thermoelectric properties of Mg_2Si-based composites using reduction reaction with additives. *Intermetallics*, 2013. **32**: pp. 72–80.

104. Sakamoto, T. et al., Thermoelectric characteristics of a commercialized Mg_2Si source doped with Al, Bi, Ag, and Cu. *Journal of Electronic Materials*, 2010. **39**(9): pp. 1708–1713.

105. Ihou-Mouko, H. et al., Thermoelectric properties and electronic structure of p-type Mg_2Si and $Mg_2Si_{0.6}Ge_{0.4}$ compounds doped with Ga. *Journal of Alloys and Compounds*, 2011. **509**(23): pp. 6503–6508.

106. Akasaka, M. et al., The thermoelectric properties of bulk crystalline n- and p-type Mg_2Si prepared by the vertical Bridgman method. *Journal of Applied Physics*, 2008. **104**(1).

107. Isoda, Y. et al., Thermoelectric properties of p-type $Mg_{2.00}Si_{0.25}Sn_{0.75}$ with Li and Ag double doping. *Journal of Electronic Materials*, 2010. **39**(9): pp. 1531–1535.

108. de Boor, J. et al., Thermoelectric performance of Li doped, p-type $Mg_2(Ge,Sn)$ and comparison with $Mg_2(Si,Sn)$. *Acta Materialia*, 2016. **120**: pp. 273–280.

109. Mars, K. et al., Thermoelectric properties and electronic structure of Bi- and Ag-doped $Mg_2Si_{1-x}Ge_x$ compounds. *Journal of Electronic Materials*, 2009. **38**(7): pp. 1360–1364.

110. Isachenko, G. N. et al., Thermoelectric properties of nanostructured p-$Mg_2Si_xSn_{1-x}$ (x = 0.2 to 0.4) solid solutions. *Journal of Electronic Materials*, 2016. **45**(3): pp. 1982–1986.

111. Satyala, N., and D. Vashaee, Detrimental influence of nanostructuring on the thermoelectric properties of magnesium silicide. *Journal of Applied Physics*, 2012. **112**(9).

112. Pshenai-Severin, D. A., M. I. Fedorov, and A. Y. Samunin, The influence of grain boundary scattering on thermoelectric properties of Mg_2Si and $Mg_2Si_{0.8}Sn_{0.2}$. *Journal of Electronic Materials*, 2013. **42**: pp. 1707–1710.

113. Kutorasinski, K. et al., Importance of relativistic effects in electronic structure and thermopower calculations for Mg_2Si, Mg_2Ge, and Mg_2Sn. *Physical Review B*, 2014. **89**(11): pp. 115205/1–115205/8.

114. Bourgeois, J. et al., Study of electron, phonon and crystal stability versus thermoelectric properties in Mg_2X (X = Si, Sn) compounds and their alloys. *Functional Materials Letters*, 2013. **06**(05): p. 1340005.

115. Bahk, J. H., Z. X. Bian, and A. Shakouri, Electron transport modeling and energy filtering for efficient thermoelectric $Mg_2Si_{1-x}Sn_x$ solid solutions. *Physical Review B*, 2014. **89**(7): p. 075204.

116. Satyala, N., and D. Vashaee, Modeling of thermoelectric properties of magnesium silicide (Mg_2Si). *Journal of Electronic Materials*, 2012. **41**: pp. 1785–1791.

117. Han, X. P., and G. S. Shao, Interplay between Ag and interstitial Mg on the p-type characteristics of Ag-doped Mg_2Si: Challenges for high hole conductivity. *Journal of Materials Chemistry C*, 2015. **3**(3): pp. 530–537.

118. Flage-Larsen, E., and O. M. Lovvik, Band structure guidelines for higher figure-of-merit: Analytic band generation and energy filtering. In *Thermoelectrics and Its Energy Harvesting: Materials, Preparation, and Characterization in Thermoelectrics*, D. M. Rowe, editor. CRC Press, Boca Raton, FL, 2012.

119. Fedorov, M. I. et al., Transport properties of $Mg_2X_{0.4}Sn_{0.6}$ solid solutions (X = Si, Ge) with p-type conductivity. *Physics of the Solid State*, 2006. **48**(8): pp. 1486–1490.

120. Tani, J.-I., and H. Kido, Thermoelectric properties of Al-doped $Mg_2Si_{1-x}Sn_x$ ($x \leq 0.1$). *Journal of Alloys and Compounds*, 2008. **466**(1–2): pp. 335–340.

121. Isachenko, G. N. et al., Kinetic properties of p-$Mg_2Si_xSn_{1-x}$ solid solutions for $x < 0.4$. *Physics of the Solid State*, 2009. **51**(9): pp. 1796–1799.

122. Tang, X. et al., Improving thermoelectric performance of p-type Ag-doped $Mg_2Si_{0.4}Sn_{0.6}$ prepared by unique melt spinning method. *Applied Thermal Engineering*, 2017. **111**: pp. 1396–1400.

123. Prytuliak, A. et al., Synchrotron study of Ag-doped Mg_2Si: Correlation between properties and structure. *Journal of Electronic Materials*, 2014. **43**(10): pp. 3746–3752.

124. Tada, S. et al., Thermoelectric properties of p-type $Mg_2Si_{0.25}Sn_{0.75}$ doped with sodium acetate and metallic sodium. *Journal of Electronic Materials*, 2014. **43**(6): pp. 1580–1584.

125. Zhang, L. et al., Suppressing the bipolar contribution to the thermoelectric properties of $Mg_2Si_{0.4}Sn_{0.6}$ by Ge substitution. *Journal of Applied Physics*, 2015. **117**(15): p. 155103.

126. Khan, A. U., N. Vlachos, and T. Kyratsi, High thermoelectric figure of merit of $Mg_2Si_{0.55}Sn_{0.4}Ge_{0.05}$ materials doped with Bi and Sb. *Scripta Materialia*, 2013. **69**(8): pp. 606–609.

127. Tang, X. et al., Ultra rapid fabrication of p-type Li-doped $Mg_2Si_{0.4}Sn_{0.6}$ synthesized by unique melt spinning method. *Scripta Materialia*, 2016. **115**: pp. 52–56.

128. Gao, P. et al., The p-type $Mg_2Li_xSi_{0.4}Sn_{0.6}$ thermoelectric materials synthesized by a B_2O_3 encapsulation method using Li_2CO_3 as the doping agent. *Journal of Materials Chemistry C*, 2016. **4**(5): pp. 929–934.

129. May, A. F., and G. J. Snyder, Introduction to modeling thermoelectric transport at high temperatures. In *Thermoelectrics and its Energy Harvesting: Materials, Preparation, and Characterization in Thermoelectrics*, D. M. Rowe, editor. CRC Press, CRC Press, Boca Raton, FL, 2012.

130. Nieroda, P. et al., Structural and thermoelectric properties of polycrystalline p-type $Mg_{2-x}Li_xSi$. *Journal of Electronic Materials*, 2016. **45**(7): pp. 3418–3426.

131. Wang, S. et al., Conductivity-limiting bipolar thermal conductivity in semiconductors. *Scientific Reports*, 2015. **5**: p. 10136.

132. Isoda, Y. et al., Thermoelectric performance of p-type $Mg_2Si_{0.25}Sn_{0.75}$ with Li and Ag double doping. *Materials Transactions*, 2010. **51**(5): pp. 868–871.

133. Stathokostopoulos, D. et al., Oxidation resistance of magnesium silicide under high-temperature air exposure. *Journal of Thermal Analysis and Calorimetry*, 2015. **121**(1): pp. 169–175.

134. Sondergaard, M. et al., Thermal stability and thermoelectric properties of $Mg_2Si_{0.4}Sn_{0.6}$ and $Mg_2Si_{0.6}Sn_{0.4}$. *Journal of Materials Science*, 2013. **48**(5): pp. 2002–2008.

135. Yin, K. et al., Thermal stability of $Mg_2Si_{0.3}Sn_{0.7}$ under different heat treatment conditions. *Journal of Materials Chemistry C*, 2015. **3**(40): pp. 10381–10387.

136. Borshchev, V. M. et al., Production of silicon from magnesium silicide. *Russian Journal of Applied Chemistry*, 2013. **86**(4): pp. 493–497.
137. Park, S. H., Y. Kim, and C. Y. Yoo, Oxidation suppression characteristics of the YSZ coating on Mg₂Si thermoelectric legs. *Ceramics International*, 2016. **42**(8): pp. 10279–10288.

CHAPTER 7

$Ca_{1-x}Yb_xMg_2Bi_2$ and $Ca_{1-x}Yb_xZn_2Sb_2$-Related 1–2–2 Zintl Phases

Jing Shuai and Zhifeng Ren

Contents

7.1 Introduction

Zintl phases have enjoyed a rich history of scientific study since their initial investigation by Edward Zintl[1–7] and have recently gained interest for use in TE devices for power generation because of their intrinsic "electron–crystal, phonon–glass" nature.[8] Zintl compounds are made up of cations and anions with a significant difference in electronegativity to completely allow electron transfer. The Zintl anions provide the "electron–crystal" electronic structure through the covalently bonded network of complex anion or metalloids, whereas the cations in Zintl phases act as the "phonon glass" characteristic. The low lattice thermal conductivity can be expected in this complex structure, and the transport properties can be tuned by alloying in terms of a wide variety of compositions in the same type structure.

An outstanding example of a Zintl phase, which has aroused significant interest in TE studies, is the layered $CaAl_2Si_2$-type Zintl phases.[9] The highest ZT of ~1.3 achieved for p-type in this 1–2–2 Zintl family[10] is competitive with other Zintl categories (e.g., $Yb_{14}MnSb_{11}$[11]) and even with other good

279

p-type skutterudites[12] and half-Heuslers[13] within 873 K. Even a higher ZT of ~1.5 has been realized in n-type Mg_3Sb_2-based materials. In this chapter, an overview of what is known of the structure–property relationships of this particular layered Zintl phases is presented. The fundamental structural and chemical bonding characteristic, and band structures, which underpin both electronic and phonon transport properties, will be discussed. Also proposed are the approaches for the significant improvement in the figure of merit and the prospect outlined for this promising Zintl phase as efficient TEs for power conversion application in the middle temperature range.

7.2 Chemical Bonding and Electronic Structure

For the layered AB_2X_2 Zintl phases crystallizing in $CaAl_2Si_2$-type structure (space group $P\bar{3}m1$) reported as the TE candidates, the A site contains alkaline-earth or divalent rare-earth element such as Mg, Ca, Ba, Sr, Yb, and Eu; B is a d^0, d^5, d^{10} transition metal (e.g., Mn, Zn, and Cd) or a main group element such as Mg, while X comes from groups 14 and 15 such as Sb and Bi.[14] Following the Zintl–Klemm concept, B and X form anionic sheets with covalent bonding due to the similar electronegativity, while A^{2+} cations donate electrons to the $(B_2X_2)^{2-}$ framework. The structure is often depicted as in Figure 7.1, clearly showing the bonding within the B–X layers between planes of A^{2+}.

The ternary AB_2X_2 Zintl phases as TE material candidates are intrinsically small bandgap p-type semiconductors with complex structure.[15,16] Their rich crystal chemistry renders them amenable to precise tuning of transport properties, which sensitively depends on the degeneracy of the valence band edges around the Brillouin zone centers. Employing the first principle methods based on DFT, the band structures, charge distribution, and chemical bonding were investigated. Two representative band structures ($CaMg_2Bi_2$, $EuMg_2Bi_2$) investigated are shown in Figure 7.2 for comparison and explanation.[10] The band edge states of the valence bands consist of light and heavy hole bands which peak at the zone center as shown in Figure 7.2a and b. Overlaps of B-s and X-p states in the valence region are observed (Figure 7.2c

FIGURE 7.1 View of the structure of AB_2X_2 showing the bonding within X–B layers between the planes of cation A^{2+}.

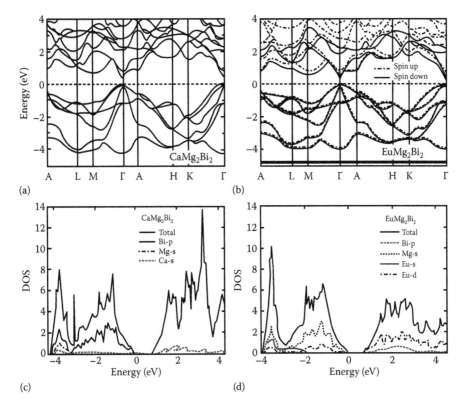

FIGURE 7.2 DFT calculations. Band structures of (a) $CaMg_2Bi_2$ and (b) $EuMg_2Bi_2$. Total and projected DOSs of (c) $CaMg_2Bi_2$ and (d) $EuMg_2Bi_2$.

and d), indicating the hybridization between these states. This hybridization supports the polyanionic $(B_2X_2)^{2-}$ Zintl nature of these materials.

Zhang et al.[16] also described the band structure of this layered structure by drawing a schematic diagram. In Figure 7.3, it indicates that the X-p states dominate the valence bands. Unlike the cubic symmetry resulting in threefold, degenerate p orbitals at the Γ point attributed by the equivalency of the x, y, and z directions in the Brillouin zone, the p_z orbital is separated from the p_x

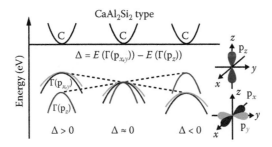

FIGURE 7.3 Electronic bands of $CaAl_2Si_2$-type Zintl compounds. Nondegenerate band Γ (p_z) and doubly degenerate band Γ ($p_{x,y}$) are mainly composed of p_z and $p_{x,y}$ orbitals from anions, respectively. Δ_{cr} is the crystal field splitting energy between $p_{x,y}$ and p_z orbitals at the Γ point.

and p_y orbital in layered structures. As a result, the valence band at the Γ point splits into double degenerate heavy hole band and nondegenerate light hole band as discussed earlier. The energy difference between these two bands is defined as the crystal field splitting energy (Δ_{cr}). The small Δ_{cr} results in the higher band degeneracy near the Fermi level, which is always favorable for the electrical transport properties of TE materials. Taking $EuMg_2Bi_2$ for example, the larger Eu^{2+} ion flattens the $(Mg_2Bi_2)^{2+}$ layer, leading to nearly identical Mg–Bi bond distances. Since the valence band structure is mainly constructed by the Mg–Bi hybridizations, a smaller crystal field splitting energy can be expected, contributing to high electrical transport properties such as high PF.

Besides, Zhang et al.[16] also proposed a selection criterion of $-0.06 < \Delta_{cr} < 0.06$ with bandgap of $E_g < 1.5eV$ for promising Zintl TE candidates in this layered structures. For many years, the energy gap for some promising Zintl phases has been estimated in several papers. However, the estimated bandgap varies from the different calculation method. For example, the bandgaps of $CaMg_2Bi_2$ was reported to be from 0.39 to 0.7 eV.[16,17] Then, Shuai et al. took Fourier transform infrared spectroscopy (FTIR) specular reflectance measurement and used Kramers–Kronig analysis and experimentally show the bandgap to be around 0.42 eV, again demonstrating that this kind of TE candidates is intrinsically small bandgap p-type semiconductors.[10]

Different from the direct band structure reported in ternary AMg_2Bi_2 samples, the band structure of Mg_3Sb_2 has an indirect gap, which contributes to the different transport properties (Figure 7.4). Very recently,

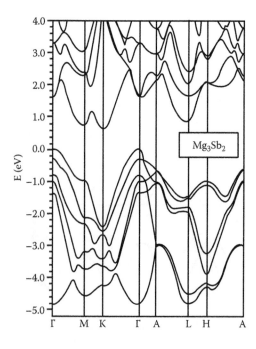

FIGURE 7.4 Calculated band structure of Mg_3Sb_2.

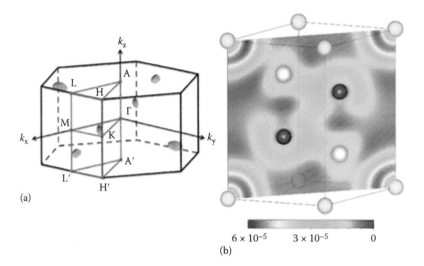

FIGURE 7.5 DFT calculations of n-Mg_3Sb_2. (a) Fermi surface of the conduction band. Six equivalent carrier pockets are located near a midpoint between M and L'. (b) Charge density map of the conduction band electrons obtained by integration from conduction band minimum (CBM) to 0.1 eV above CBM.

researchers at the Panasonic Corporation discovered high TE performance in n-type $Mg_3(Sb,Bi)_2$, which was firstly reported in the *35th International Conference on Thermoelectrics* in Wuhan, China.[18] With extra Mg, the n-type characteristics of Mg_3Sb_2-based compounds have been achieved. They also discovered Zintl compound n-Mg_3Sb_2, the conduction band of which has multivalley Fermi surface ($N_V = 6$) (Figure 7.5a) and light conduction band effective mass ($m^* = 0.19m_e$), is much more beneficial than the p-type to yield extraordinarily high TE performance. As depicted in Figure 7.5b, the charge density map of the conduction band electrons shows that three-dimensional conduction network is formed by the 3s-like orbitals at cationic Mg^{2+} and the weakly hybridized atomic orbitals at anionic $Mg_2Sb_2^{2-}$. Consequently, the inverse average of the effective mass taken over six valleys is nearly isotropic ($m_x = 0.2m_e$, and $m_z = 0.17m_e$). Therefore, the materials design principle of n-Mg_3Sb_2 is quite different from that of p-Mg_3Sb_2, where anisotropic p_z orbital at Sb is dominant because of the crystalline field effect shown in Figure 7.4.

7.3 Thermoelectric Properties

TE candidate ternary Zintl $Ca_{1-x}Yb_xZn_2Sb_2$ was first synthesized by Gascoin et al.[9] in 2005 and investigated to have a promising *ZT* values of ~0.56. Inspired by this work, many other $CaAl_2Si_2$-type Zintl phases have been studied within the last decade. Table 7.1 summarizes most of the explorations on this layered 1–2–2 structures with promising TE properties (compositions with *ZT* above 0.3 are listed). Among those compounds,

TABLE 7.1 *ZT* Values of Selected p-Type TEs Based on $CaAl_2Si_2$-Structured Zintl Phases

Year	AB_2X_2	ZT	T (K)
2005	$Ca_{0.25}Yb_{0.75}Zn_2Sb_2$ (Gascoin[9])	0.56	773
2007	$BaZn_2Sb_2$ (Wang[19])	0.33	673
2008	$YbZn_{1.9}Mn_{0.1}Sb_2$ (Yu[20])	0.65	726
2008	$EuZn_2Sb_2$ (Zhang[21])	0.9	713
2009	$YbCd_{1.6}Zn_{0.4}Sb_2$ (Wang[22])	1.2	650
2010	$Yb_{0.6}Ca_{0.4}Cd_2Sb_2$ (Cao[23])	0.96	700
2010	$Yb_{0.75}Eu_{0.25}Cd_2Sb_2$ (Zhang[24])	0.97	650
2010	$EuZn_{1.8}Cd_{0.2}Sb_2$ (Zhang[25])	1.06	650
2011	$YbCd_{1.85}Mn_{0.15}Sb_2$ (Guo[26])	1.14	650
2012	$YbMg_2Bi_2$ (May[27])	0.44	650
2013	$Mg_3Bi_{0.2}Sb_{1.8}$ (Ponnambalam[28])	0.6	750
2014	$Mg_3Pb_{0.2}Sb_{1.8}$ (Bhardwaj[29])	0.84	773
2014	$Yb_{0.99}Zn_2Sb_2$ (Zevalkink[30])	0.85	800
2015	$Mg_{2.9875}Na_{0.0125}Sb_2$ (Shuai[31])	0.6	773
2016	$YbCd_{1.9}Mg_{0.1}Sb_2$ (Cao[32])	1.08	650
2016	$Ca_{0.995}Na_{0.005}Mg_2Bi_2$ (Shuai[33])	0.9	873
2016	$Eu_{0.2}Yb_{0.2}Ca_{0.6}Mg_2Bi_2$ (Shuai[10])	1.3	873

Zintl antimonides are extensively explored. The ZT above unity are normally observed in the Cd-based antimonides such as $YbCd_{1.6}Zn_{0.4}Sb_2$[22] and $EuZn_{1.8}Cd_{0.2}Sb_2$.[25] Recently, Shuai et al. reported the high TE performance of the rarely studied bismuth-based solid solution Zintl phase $(Eu_{0.5}Yb_{0.5})_{1-x}Ca_xMg_2Bi_2$ with the record ZT as high as 1.3 at 873 K.[10] Moreover, the ZT values of other Bi-based Zintl phases $Ca_{0.5}Yb_{0.5}Mg_2Bi_2$, $Eu_{0.5}Ca_{0.5}Mg_2Bi_2$, and $Eu_{0.5}Yb_{0.5}Mg_2Bi_2$ are also above unity.[10] The TE transport properties of some representative Bi-based alloyed samples are shown in Figure 7.6. Specifically, the optimized $Eu_{0.2}Yb_{0.2}Ca_{0.6}Mg_2Bi_2$ sample exhibits extremely low thermal conductivity (1.5 and 0.9 W/m K at 300 and 873 K, respectively) and high Seebeck coefficient (~200 μV/K) and low electrical resistivity (8–10 mΩ cm). This highest ZT of ~1.3 is competitive with the reported Sb-based p-type Zintl phases and even other typical p-type skutterudites and half-Heusler in this temperature range.

For the A = B = Mg antimonide, as mentioned before, because of the different band structure, the binary Mg_3Sb_2 has the different transport properties from those of the ternary $CaAl_2Si_2$-type Zintls. Although it exhibits low thermal conductivity (1.4 and 0.8 W/mK at 300 and 723 K, respectively) and high Seebeck coefficient (above 400 μV/K), ZT for the parent phase Mg_3Sb_2 is too small (~0.3) due to its high electrical resistivity (~1 Ω m at 300 K).[34] Mg_3Sb_2 has moderate ZT at a high temperature as is also the case for alloyed or doped samples. The highest ZT value of ~0.84 for p-type was achieved in $Mg_3Pb_{0.2}Sb_{1.8}$ by Bhardwaj and Misra.[29]

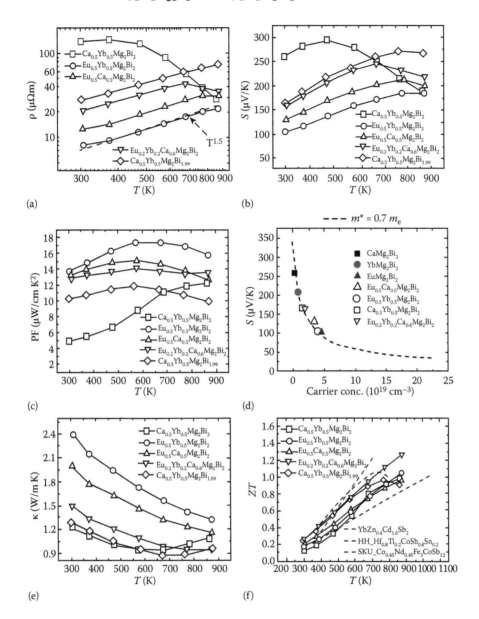

FIGURE 7.6 Temperature-dependent TE properties of $(Ca_{0.5}Yb_{0.5}, Yb_{0.5}Eu_{0.5}, Eu_{0.5}Ca_{0.5})Mg_2Bi_2$, $Ca_{0.5}Yb_{0.5}Mg_2Bi_{1.99}$ and $Eu_{0.2}Yb_{0.2}Ca_{0.6}Mg_2Bi_2$: (a) electrical resistivity; (b) Seebeck coefficient; (c) PF; (e) thermal conductivity; and (f) ZT values in comparison with the Sb-based $YbZn_{0.4}Cd_{1.6}Sb_2$ Zintl phase, p-type half-Heusler, and skutterudite.[22,35,36] (d) Seebeck coefficient dependence of carrier concentration n in accordance with a single band model with $m^* = 0.7\ m_e$ at 300 K.

7.4 Thermal Transport Properties

As observed in Table 7.1, the better performance of the $CaAl_2Si_2$-type Zintl phases with ZT above unity is directly related to strong phonon scattering, arising as a consequence of alloy or solid solution formation. For example, complete solid solutions based on $YbZn_2Sb_2$ and $YbMg_2Bi_2$ have been formed between the isoelectronic pairs of Yb–Ca, Yb–Eu, Zn–Cd, and Zn–Mn. In most cases at room temperature, the lattice thermal conductivity is nearly half of the end compounds due to the mass and size fluctuation.[9,10] Figure 7.7 shows the experimental and the calculated results in the lattice thermal conductivity of $Ca_xYb_{1-x}Mg_2Bi_2$ using a point defect scattering theory.[37] The same case is also observed in the $Yb_{1-x}Ca_xZn_2Sb_2$ series.[9] At a higher temperature, the contribution from alloying effect gets smaller compared to low temperature because the Umklapp scattering becomes more and more efficient. The microstructure of the optimized $Eu_{0.2}Yb_{0.2}Ca_{0.6}Mg_2Bi_2$ sample with the record ZT value has been investigated.[10] The presence of nanoscale grains around 200–500 nm is evident in Figure 7.8. The black

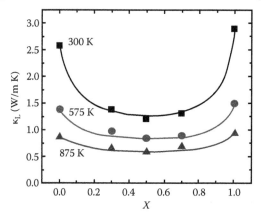

FIGURE 7.7 Significant reduction in lattice thermal conductivity is observed due to point defect scattering in $Yb_{1-x}Ca_xMg_2Bi_2$ at 300, 575, and 875 K. Lines show predicted reduction in κ_L based on mass contrast between Yb and Ca for the respective temperatures.

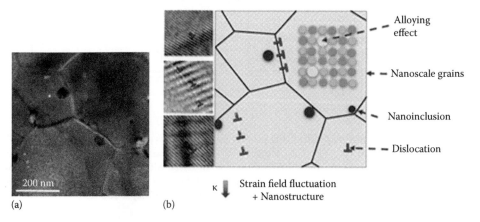

FIGURE 7.8 Typical microstructures for $Eu_{0.2}Yb_{0.2}Ca_{0.6}Mg_2Bi_2$ and schematic describing the various scattering effects, especially strain field fluctuation and nanostructure, leading to decreased thermal conductivity. The microscopy images on the left represent different strains in the three types of interactions. (a) Medium-magnification TEM image showing mesoscale grains with black precipitates and (b) schematic describing the various scattering effects, especially strain field fluctuation and nanostructure, leading to decreased thermal conductivity. The microscopy images on the left in (b) represent different strains in the three types of interactions.

nanoprecipitates around 5–50 nm can also be observed inside grains or near boundaries. Elemental analysis by EDS indicates that Bi concentration is higher in the precipitates. A number of different defect strains observed through TEM might be evidence of point defects in an alloyed structure caused by the mass and size differences as observed in Figure 7.8.

7.5 Electrical Transport Performance

Another significant reason for high TE performance is based on the relative high PF in comparison with other Zintl families. As discussed earlier in the electrical structure part, the good electrical transport performance in this layered structure might rely on the degeneracy of the valence band edge by realizing the ideal Δ_{cr} value around zero. In Figure 7.9, it is shown that the PF can reach as high as 25 μW/cm K^2 when Δ_{cr} approaches zero.[16] For example, solid solutions Eu$_{0.5}$Yb$_{0.5}$Mg$_2$Bi$_2$, YbCd$_{1.6}$Zn$_{0.4}$Sb$_2$, and EuZn$_{1.8}$Cd$_{0.2}$Sb$_2$ with ZT values above unity are marked in solid circle and their Δ_{cr} values are all close to zero. High PFs combined with low thermal conductivity make the CaAl$_2$Si$_2$-type Zintl phases excellent intermediate- to high-temperature TEs.

Zintl chemistry suggested that the materials' properties may be adjusted by defects, e.g., vacancies. The observed high experimental carrier concentrations (10^{19}–10^{20} cm^{-3}) and extrinsic semiconducting behavior in AZn$_2$Sb$_2$ might be consistent with vacancies on the A or B sites. In order to investigate the vacancy effect and optimize the electrical properties, Zevalkink et al.[30] chose Yb$_x$Zn$_2$Sb$_2$ to study because Yb has the largest predicted vacancy concentration among AZn$_2$Sb$_2$, predicated by DFT calculation. The experimental Hall carrier concentration increases for $x < 1$ and remains constant when

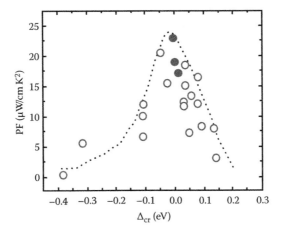

FIGURE 7.9 Experimental PFs as a function of the crystal field splitting energy in CaAl$_2$Si$_2$-type Zintl compounds. Solid solutions Eu$_{0.5}$Yb$_{0.5}$Mg$_2$Bi$_2$, YbCd$_{1.6}$Zn$_{0.4}$Sb$_2$, and EuZn$_{1.8}$Cd$_{0.2}$Sb$_2$ with ZT values above unity are marked in solid circle, and others in Table 7.1 are marked in hollow circle. The curve serves as guide to the eye, showing the best values corresponding to optimum carrier concentrations.

$x > 1$. As a result, the electric transport properties such as electric resistivity and Seebeck coefficient could be adjusted accordingly. The peak ZT value of 0.85 is obtained in the sample with $x = 0.99$. In the other samples with higher Yb content, the average ZT is greatly improved (Figure 7.10).

One should also note that the mobility found in the two-dimensional ternary AB_2X_2 structures is extremely high (up to 200 cm²/V s) in comparison with zero-dimensional Zintls (e.g., ~5 cm²/V s in $Yb_{14}Mn_{0.4}Al_{0.6}Sb_{11}$[11]) and one-dimensional Zintls (e.g., ~17 cm²/V s in $Sr_3Ga_{0.93}Zn_{0.07}Sb_3$[38]) at room temperature. Generally, the carrier mobility would drop in solid solutions due to the chemical disorder scattering, which would badly affect the electrical transport properties. DFT calculation shows that the hole states in these solid solutions are much less influenced than the electron states in the whole band structure, which is beneficial in maintaining the high carrier mobility in the p-type disordered solutions like $(Eu_{0.5}Yb_{0.5})_{1-x}Ca_xMg_2Bi_2$. Based on the a dimensionless material parameter $B = 5.745 \times 10^{-6}[\mu(m^*/m_0)^{3/2}/\kappa_{Lat}]$ $T^{5/2}$, as an alternative to evaluating the maximum ZT,[39] one can conclude that better electronic TE performance could be achieved in the alloyed $CaAl_2Si_2$-type Zintl phases.

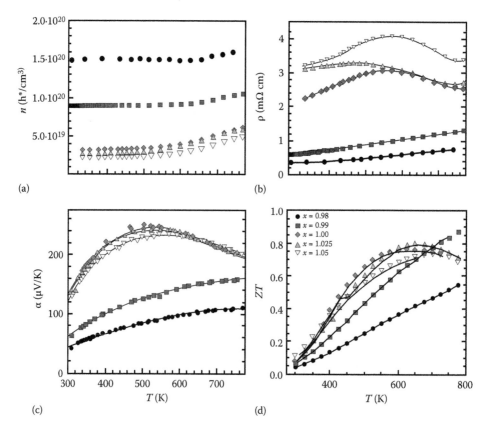

FIGURE 7.10 (a) Hall carrier concentration, (b) resistivity, (c) Seebeck coefficients, and (d) ZT of $Yb_xZn_2Sb_2$.

On the contrary, the A = B = Mg antimonide Mg_3Sb_2 sample exhibits poor electrical transport properties with low carrier concentration and high electrical resistivity. Singh and Parker[40] theoretically investigated the detailed electronic structure and transport properties of Zinlt phase Mg_3Sb_2 and a series of alloy $(AMg_2)X_2$ (A = Ca, Sr, Ba; X = As, Sb, Bi) compounds in relation to their TE performance. They claimed that the several promising compositions in this family are not fully optimized in terms of carrier concentration, especially in Mg_3Sb_2.[40] Several dopants have been investigated in order to optimize the electric transport performance. Earlier investigations include Zn-substituted $Mg_{3-x}Zn_xSb_2$[41] and Cd- and Ag-substituted $Mg_{3-x-y}Cd_xAg_ySb_2$.[42] However, the electrical resistivity is still too high to be a good TE material. Recently, Bi doped into Sb site reported by Ponnambalam and Morelli[28] was approved to be more appealing by effectively decreasing the resistivity. Moreover, Bhardwaj and Misra[29] reported an enhancement in the electrical conductivity while maintaining the Seebeck coefficient by substituting Pb^{4-} on Sb^{3-} site in $Mg_3Sb_{2-x}Pb_x$ alloys. Another effective dopant reported is Na doping into Mg site.[31] The carrier concentration increases from 2.6×10^{15} cm^{-3} for x = 0–1.7×10^{20} cm^{-3} for x = 0.25. As shown in Figure 7.11, the electrical transport properties enhanced by dramatically decreasing the electric resistivity, optimizing Seebeck coefficient, and increasing the PF.

However, when introducing Mn into the Mg site, the Seebeck coefficient becomes negative, but the absolute value remained almost the same as that of the parent phase (Figure 7.12a).[43] Since the measured electrical resistivity of $Mg_{2.6}Mn_{0.4}Sb_2$ is 0.09 Ω cm at room temperature, which is still too high (Figure 7.12b), the highest *ZT* of this Mn-doped n-type material is just 0.1 at 500 K. As we mentioned before, another n-type Mg_3Sb_2-based material with the *ZT* value of ~1.5 at 723 K was reported by the researchers very recently. It was realized by the defect chemistry approach and the introduction of Sb/Bi disorder. The Mg-rich composition stabilizes the n-type carrier transport in the Te-doped samples. The clear transition from high resistivity p-type to low resistivity n-type is observed with excess Mg content from 0.1 to 0.2 along with Te doping (Figure 7.12c). Because of the dramatically reduced electrical resistivity, high Seebeck coefficient, and low thermal conductivity, much enhanced TE performance is realized in this layered Zintl phase n-Mg_3Sb_2.

7.6 Growth Methods for Achieving Pure Phases

It is well known that TE properties are dependent on chemical composition and microstructure that is sensitive to the preparation method. Since the promising $Ca_{1-x}Yb_xZn_2Sb_2$ Zintl compounds were reported, the dominant preparation method for AB_2X_2 Zintl compounds has been melting.[9] In previous reports, the samples are first heated above 1000°C and held at the temperature between 12 and 96 h before being slowly cooled. The total processing time is more than 200 h. Moreover, all the previous reports on AZn_2Sb_2 Zintl phases via melting method have ZnSb, Zn_4Sb_3, or $M_9Zn_{4.5}Sb_9$

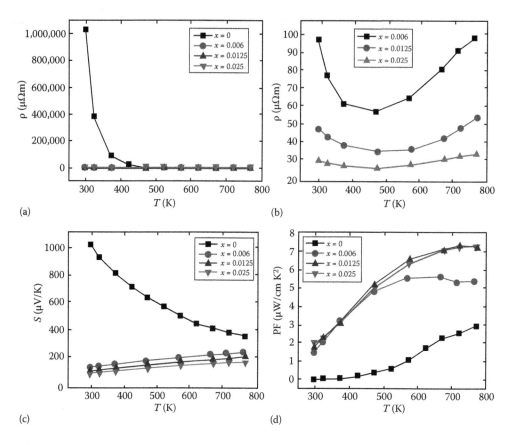

FIGURE 7.11 Temperature-dependent TE properties of $Mg_{3-x}Na_xSb_2$ (x = 0, 0.006, 0.0125, and 0.025): (a) electrical resistivity; (b) electrical resistivity of the doped samples. Temperature-dependent TE properties of $Mg_{3-x}Na_xSb_2$ (x = 0, 0.006, 0.0125, and 0.025); (c) Seebeck coefficient; and (d) PF.

(M = Yb, Eu) as the secondary phases,[9,22,25,21,23,26] and the impurity Bi was found in AMg_2Bi_2 Zintl phases.[28] The reason might be that the initial high-temperature step led A in AZn_2Sb_2 and Mg in AMg_2Bi_2 to vaporize or react with the container.[30]

In AZn_2Sb_2 system, the DFT calculation also predicated that vacancies on the A site are the most energetically favorable point defects and concentration of vacancies strongly depends on the electronegativity of A. The abnormal high carrier concentration found in $YbZn_2Sb_2$ might be an evidence of A vacancies.

Alternatively, mechanical alloying by a high-energy BM process is studied for the synthesis of these Zintl phases. With accurate controlling of the molar ratio, in AZn_2Sb_2 Zintl phases, the pure phases were achieved without impurity phases with the detection limit of XRD.[44] Samples prepared by BM and HP methods showed lower carrier concentrations, with ~8.75 × 10^{18} cm^{-3} in $Ca_{0.5}Yb_{0.5}Zn_2Sb_2$ as the highest, about half of that in $YbZn_2Sb_2$ by melting, as shown in Figure 7.13. The highest ZT value of ~0.9 was achieved

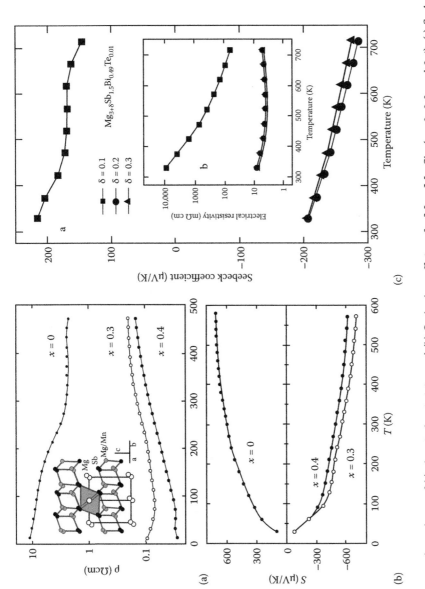

FIGURE 7.12 Temperature dependence of (a) electrical resistivity and (b) Seebeck coefficient for $Mg_{3-x}Mn_xSb_2$ ($x = 0$, 0.3, and 0.4). (c) Seebeck coefficient and electrical resistivity of samples with different excess Mg compositions, $Mg_{3+\delta}Sb_{1.5}Bi_{0.49}Te_{0.01}$ ($\delta = 0.1$, 0.2, 0.3).

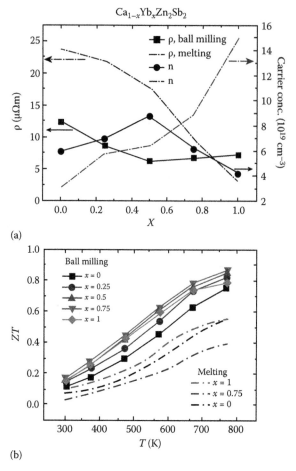

(a)

(b)

FIGURE 7.13 (a) Room-temperature resistivity and carrier concentration comparison between samples of $Ca_{1-x}Yb_xZn_2Sb_2$ (x = 0, 0.25, 0.5, 0.75, and 1) made by BM and melting; (b) temperature-dependent ZT for $Ca_{1-x}Yb_xZn_2Sb_2$ (x = 0, 0.25, 0.5, 0.75, and 1) of BM samples (*solid line*) and melting samples (*dashed line*).

in $Ca_{0.25}Yb_{0.75}Zn_2Sb_2$ prepared by BM, an improvement of ~50% over that of the best reported sample by melting. For Bi-based Zintl phases, Bi deficiency in $CaMg_2Bi_{1.98}$ and $Ca_{0.5}Yb_{0.5}Mg_2Bi_{1.99}$ resulted in pure-phase Zintl phases.[46] The electrical resistivity and Seebeck coefficient of pure-phase $CaMg_2Bi_{1.98}$ and $Ca_{0.5}Yb_{0.5}Mg_2Bi_{1.99}$ decrease because of enhanced carrier concentration and almost unchanged hall mobility. It is also very creative to find 0.5% Bi deficiency in pure-phase $Ca_{0.5}Yb_{0.5}Mg_2Bi_{1.99}$ almost successfully suppress the bipolar effect and the peak for Seebeck coefficient shifts to higher temperature. Similarly, as depicted in Figure 7.14, the pure phases prepared through BM are proven to have much higher TE properties (e.g., average ZT) than those of the samples prepared by melting method.

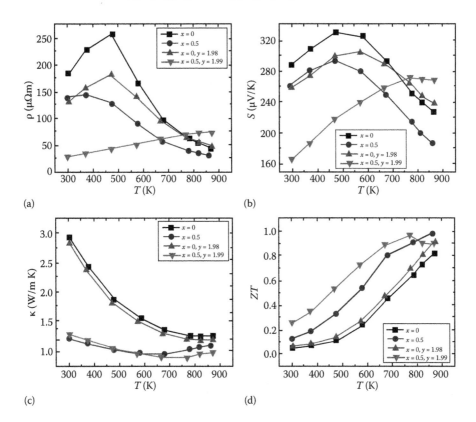

FIGURE 7.14 Temperature-dependent TE properties of $CaMg_2Bi_2$, $Ca_{0.5}Yb_{0.5}Mg_2Bi_2$, $CaMg_2Bi_{1.98}$, and $Ca_{0.5}Yb_{0.5}Mg_2Bi_{1.99}$: (a) electrical resistivity; (b) Seebeck coefficient; (c) thermal conductivity; and (d) ZT values.

7.7 Conclusion

Research efforts by several groups have unequivocally demonstrated that the $CaAl_2Si_2$-type Zintl phases are promising TE materials, and some of them exceeded the important barrier—ZT of the magnitude of unity. The samples manufactured via BM-based technique had enhanced properties compared to the similar materials via solidification-based techniques. Both theory and experiment prove that the cation in Zintl phases provide regions that can be alloyed or precisely doped with weakly affected high carrier mobility and disrupt phonon transport through alloy scattering. Through band engineering and strain fluctuation, the new record ZT of ~1.3 of p-type has been realized in the rarely studied Bi-based Zintl phases. A high ZT of ~1.5 of the n-Mg_3Sb_2 was realized by defect chemistry approach and the introduction of Sb/Bi disorder. There are good reasons to believe that the wide variety of compositions in this same type structure makes it possible for further manipulation of the structures and optimization of the TE properties.

References

1. E. Zintl and W. Dullenkopf, *Z. für Phys. Chem.* **1932**, *B16*, 183.
2. E. Zintl and G. Woltersdorf, *Z. Elektrochem.* **1935**, *41*, 876.
3. E. Zintl, *Angew. Chem.* **1939**, *52*, 1.
4. H. Schäfer, B. Eisenmann, and W. Müller, *Angew. Chem.* **1973**, *12*, 694.
5. H. Schäfer, B. Eisenmann, and W. Müller, *Angew. Chem.* **1973**, *85*, 742.
6. H. Schäfer, *Annu. Rev. Mater. Sci.* **1985**, *15*, 1.
7. F. Laves, *Naturwissenschaften* **1941**, *29*, 244.
8. S. M. Kauzlarich, S. R. Brown, and G. Jeffrey Snyder, *Dalton Trans.* **2007**, *21*, 2099.
9. F. Gascoin, S. Ottensmann, D. Stark, S. M. Haïle, and G. J. Snyder, *Adv. Funct. Mater.* **2005**, *15*, 1860.
10. J. Shuai, H. Geng, Y. Lan, Z. Zhu, C. Wang, Z. Liu, J. Bao, C. W. Chu, J. Sui, and Z. Ren, *Proc. Natl. Acad. Sci. USA* **2016**, *113*, E4125.
11. E. S. Toberer, C. A. Cox, S. R. Brown, T. Ikeda, A. F. May, S. M. Kauzlarich, and G. J. Snyder, *Adv. Funct. Mater.* **2008**, *18*, 2795.
12. G. Rogl, A. Grytsiv, P. Rogl, E. Bauer, and M. Zehetbauer, *Intermetallics* **2011**, *19*, 546.
13. C. Fu, S. Bai, Y. Liu, Y. Tang, L. Chen, X. Zhao, and T. Zhu, *Nat. Commun.* **2015**, *6*, 1229.
14. C. Zheng, R. Hoffmann, R. Nesper, and H. G. Von Schnering, *J. Am. Chem. Soc.* **1986**, *108*, 1876.
15. E. S. Toberer, A. F. May, B. C. Melot, E. Flage-Larsen, and G. J. Snyder, *Dalton Trans.* **2010**, *39*, 1046.
16. J. Zhang, L. Song, K. F. F. Fischer, W. Zhang, X. Shi, G. K. H. Madsen, and B. B. Iversen, *Nat. Commun.* **2016**, *7*, 1.
17. A. F. May, M. A. McGuire, D. J. Singh, R. Custelcean, and G. E. Jellison Jr., *Inorg. Chem.* **2011**, *50*, 11127.
18. H. Tamaki, H. K. Sato, and T. Kanno, *Adv. Mater.*, **2016**, *28*, 10182.
19. X. J. Wang, M. B. Tang, J. T. Zhao, H. H. Chen, and X. X. Yang, *Appl. Phys. Lett.* **2007**, *90*, 232107.
20. C. Yu, T. J. Zhu, S. N. Zhang, X. B. Zhao, J. He, Z. Su, and T. M. Tritt, *J. Appl. Phys.* **2008**, *104*, 013705.
21. H. Zhang, J. T. Zhao, Y. Grin, X. J. Wang, M. B. Tang, Z. Y. Man, H. H. Chen, and X. X. Yang, *J. Chem. Phys.* **2008**, *129*, 164713.
22. X. J. Wang, M. B. Tang, H. H. Chen, X. X. Yang, J. T. Zhao, U. Burkhardt, and Y. Grin, *Appl. Phys. Lett.* **2009**, *94*, 092106.
23. Q. G. Cao, H. Zhang, M. B. Tang, H. H. Chen, X. X. Yang, Y. Grin, and J. T. Zhao, *J. Appl. Phys.* **2010**, *107*, 053714.
24. H. Zhang, L. Fang, M. B. Tang, Z. Y. Man, H. H. Chen, X. X. Yang, M. Baitinger, Y. Grin, and J. T. Zhao, *J. Chem. Phys.* **2010**, *133*, 194701.
25. H. Zhang, M. Baitinger, M. B. Tang, Z. Y. Man, H. H. Chen, X. X. Yang, Y. Liu, L. Chen, Y. Grin, and J. T. Zhao, *Dalton Trans.* **2010**, *39*, 1101.
26. K. Guo, Q. G. Cao, X. J. Feng, M. B. Tang, H. H. Chen, X. Guo, L. Chen, Y. Grin, and J. T. Zhao, *Eur. J. Inorg. Chem.* **2011**, *2011*, 4043.
27. A. F. May, M. A. McGuire, D. J. Singh, J. Ma, O. Delaire, A. Huq, W. Cai, and H. Wang, *Phys. Rev. B* **2012**, *85*, 035202.
28. V. Ponnambalam and D. T. Morelli, *J. Electron. Mater.* **2013**, *42*, 1307.
29. A. Bhardwaj and D. K. Misra, *RSC Advances* **2014**, *4*, 34552.

30. A. Zevalkink, W. G. Zeier, E. Cheng, J. Snyder, J. P. Fleurial, and S. Bux, *Chem. Mater.* **2014**, *26*, 5710.
31. J. Shuai, Y. Wang, H. S. Kim, Z. Liu, J. Sun, S. Chen, J. Sui, and Z. Ren, *Acta Materialia* **2015**, *93*, 187.
32. Q. Cao, J. Zheng, K. Zhang, and G. Ma, *J. Alloys Compd.* **2016**, *680*, 278.
33. J. Shuai, H. S. Kim, Z. Liu, R. He, J. Sui, and Z. Ren, *Appl. Phys. Lett.* **2016**, *108*, 183901.
34. C. L. Condron, S. M. Kauzlarich, F. Gascoin, and G. J. Snyder, *J. Solid State Chem.* **2006**, *179*, 2252.
35. R. Liu, J. Yang, X. Chen, X. Shi, L. Chen, and C. Uher, *Intermetallics* **2011**, *19*, 1747.
36. X. Yan, W. Liu, H. Wang, S. Chen, J. Shiomi, K. Esfarjani, H. Wang, D. Wang, G. Chen, and Z. Ren, *Energy Environ. Sci.* **2012**, *5*, 7543.
37. J. Callaway, *Phys. Rev.* **1959**, *113*, 1046.
38. A. Zevalkink, W. G. Zeier, G. Pomrehn, E. Schechtel, W. Tremel, G. J. Snyder, *Energy Environ. Sci.* **2012**, *5*, 9121.
39. D. Tuomi, *J. Electrochem. Soc.* **1984**, *131*, 2101.
40. D. J. Singh and D. Parker, *J. Appl. Phys.* **2013**, *114*, 143703.
41. H. X. Xin, X. Y. Qin, J. H. Jia, C. J. Song, K. X. Zhang, and J. Zhang, *J. Phys. D: Appl. Phys.* **2009**, *42*, 165403.
42. K. X. Zhang, X. Y. Qin, H. X. Xin, H. J. Li, and J. Zhang, *J. Alloys Compd* **2009**, *484*, 498.
43. S. Kim, C. Kim, Y. K. Hong, T. Onimaru, K. Suekuni, T. Takabatake, and M. H. Jung, *J. Mater. Chem. A* **2014**, *2*, 12311.
44. J. Shuai, Y. Wang, Z. Liu, H. S. Kim, J. Mao, J. Sui, and Z. Ren, *Nano Energy* **2016**, *25*, 136.
45. J. Shuai, Z. Liu, H. S. Kim, Y. Wang, J. Mao, R. He, J. Sui, and Z. Ren, *J. Mater. Chem. A* **2016**, *4*, 4312.

CHAPTER 8

Half-Heuslers for High Temperatures

Hao Zhang, Zhensong Ren, Shuo Chen, Zhifeng Ren,
Ran He, Yucheng Lan, and Lihong Huang

Contents

8.1 n-Type Half-Heusler

Hao Zhang, Zhensong Ren, Shuo Chen, and Zhifeng Ren

8.1.1 Overview

Half-Heusler (HH) compounds are named after Friedrich Heusler, a German chemist and mining engineer. More than 100 HH compounds can be found in Pearson's handbook and the Inorganic Crystal Structural Database. As shown in Figure 8.1, a HH compound consists of XYZ as the main chemical composition, where X is a transitional metal, a noble metal, or a rare-earth element; Y can be a transitional metal or a noble metal; and Z is a main group element.[1] XYZ forms a MgAgAs type of structure (space group $F\bar{4}3m$), where X, Y, and Z atoms form three interpenetrating face-centered cubic sublattices by occupying Wyckoff positions 4b (1/2, 1/2, 1/2), 4c (1/4, 1/4, 1/4), and 4a (0, 0, 0) positions, respectively.[2] The remaining 4d (3/4, 3/4, 3/4) positions are vacant. With such a crystal structure, the strong hybridization of d states of X and Y atoms induces a bandgap in half-Heusler.[3–5] Theoretically, calculations suggest that the gap values vary from about 0.1 to 3.7 eV upon different compositions, depending on the valence electron count (VEC) per unit cell as well as averaged atomic number.[2,6–8] Calculations also show that HHs with 18 VEC per unit cell are stable and have a bandgap in the range of 0–1.1 eV, which is suitable for moderate temperature TE applications.[2,4,6,7,9–11] The criterion of 18 VEC narrows down the selection of semiconducting HH compounds as TE materials to about 30 possible compositions.[7] During the past decade, most researches have been focused on the classical systems MCoSb and MNiSn (M = Ti, Zr, and Hf). The electrical resistivity at room temperature of these intrinsic semiconducting compounds is within the range of 10^{-3}–10^{-5} Ω m.[12] Elements with different valence electrons can be introduced to change the VEC or the major carrier type, so that both n-type and p-type HH materials can be produced for TE devices. The major challenge for HH compounds as TE materials is their high thermal conductivity (6.7–20 W/m K at room temperature).[12–16] In this section, we will focus on the classical n-type HH TE material—MNiSn. The review on classical p-type HH and recently emerged new HH TE compounds are discussed in Sections 8.2 and 8.3, respectively. In addition, the readers can also check out several recent review papers on HH compounds as TE materials.[17–20]

8.1.2 Synthesis of n-Type Half-Heusler Materials

n-Type MNiSn can be obtained through a series of synthesis steps similar to other TE compounds.[17,21] In general, the first step is alloying. The alloying methods include arc melting,[8,22–34] optical floating zone melting,[35–38] induction melting,[18,39–44] and microwave heating.[45,46] Subsequently, the alloys are either measured after annealing (micrograin samples) or undergo pulverization (microngrain samples) or BM (nanograin samples) and then sintering before measurements. The performance of the sample is very sensitive to the sample processing.[21] Some representative experimental work is summarized in Table 8.1, where different synthesis routes lead to quite different microstructural features and thus performance.

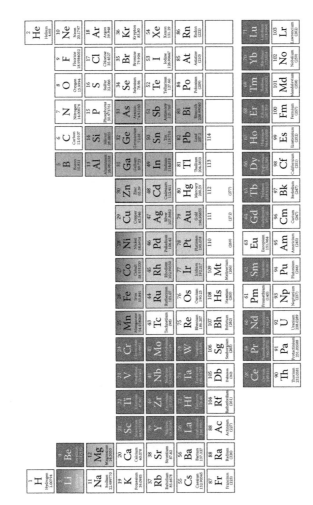

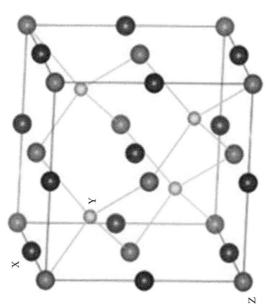

FIGURE 8.1 Crystal structure of HH where X, Y, and Z atoms occupy the Wyckoff positions 4b, 4c, and 4a, respectively. The candidate elements for X, Y, and Z are highlighted in the periodic table of elements with the corresponding grayscale.

TABLE 8.1 Summary of Composition, Synthesis Techniques, Structural Feature, Maximum *ZT*, and Key Conclusion for Representative MNiSn-Based HH Compounds

Composition	Synthesis Techniques	Structural Feature	ZT_{max}	Key Conclusion
$Hf_{0.75}Zr_{0.25}Ni_{1+x}Sn$[26]	Solid-state sintering and annealing	Coherent phase boundaries between the HH matrix and FH nanoinclusions	0.6 at 775 K	Large enhancement in the Seebeck coefficient and electrical conductivity simultaneously at high temperatures.
$ZrNiSn$[47]	Arc melting, pulverization, and SPS	Zr/Sn antisite defects	0.64 at 800 K	Maximum ZT is reached in unannealed ZrNiSn without exterior doping.
$(Hf_{0.25}Zr_{0.75})_{1-x}Nb_xNiSn$[30] (x between 0.018 and 0.022)	Arc melting, BM, and SPS	Nanosized grains	0.9 at 973 K	Nb is an effective electron donor with the advantage of avoiding material loss during arc melting compared to Sb doping.
$Hf_{0.25}Zr_{0.75}NiSn_{0.99}Sb_{0.01}$[28]	Arc melting, BM, and SPS	Nanosized grains	1 at 873 K	Peak ZT value of 1 is maintained in Hf-reduced composition.
$Hf_{0.6}Zr_{0.4}NiSn_{0.98}Sb_{0.02}$[18,41]	Levitation melting, pulverization, and SPS	Uniform and single phase	1.0 at 1000 K	Strong electromagnetic stirring during the levitation melting produces uniform products within several minutes.
$Hf_{0.25}Zr_{0.25}Ti_{0.5}NiSn_{0.98}Sb_{0.002}$[48]	Arc melting, BM, annealing, and SPS	Microscale phase separation	1.2 at 830 K	Microscale phase separation strongly scatters long-wavelength phonons, thus reducing the lattice thermal conductivity.
$Hf_{0.25}Zr_{0.25}Ti_{0.50}NiSn_{0.998}Sb_{0.02}$ and $Zr_{0.5}Ti_{0.5}NiSn_{0.98}Sb_{0.02}$[49]	Arc melting, induction melting, annealing, BM, and HP with densification aid (DA)	Single-phase HH, densified by DA	~1.2 at 820 K	Hot pressing at >1373 K can produce single-phase MNiSn. ZT value of ~1 is obtained in Hf-free $Zr_{0.5}Ti_{0.5}NiSn_{0.98}Sb_{0.02}$.
$Hf_{0.65}Zr_{0.25}Ti_{0.15}NiSn_{0.95}Sb_{0.005}$/nano-$ZrO_2$[50]	Arc melting, pulverization, and SPS	Single-phase HH, micron-sized grains with ZrO_2 nanoinclusions	~1.3 at 875 K	Phonon scattering is enhanced by Ti substitution and ZrO_2 nanoinclusions, which also scatter low-energy carriers and enhance the Seebeck coefficient.

Common to all synthesis techniques mentioned earlier, problems due to the required high energy input and starting element evaporation have to be resolved.[17,21] The very first challenge is alloying high-melting point elements such as Hf (melting point: 2233°C) and Zr (melting point: 1855°C) together with low-melting point and high-vapor pressure elements such as Sb (melting point: 631°C) and Sn (melting point: 232°C) (see Table 8.2[51] for comparison). Although simple ball milling of starting elements can minimize the loss of high-vapor pressure elements, it does not supply high enough energy to ensure the formation of the desired pure HH phase.[52] Therefore, on one hand, a synthesis method with high enough energy input has to be used to obtain a pure phase. On the other hand, attention has to be paid to suppress the evaporation of elements.

Taking arc melting as an example, an alloy ingot is obtained by weighing the constitute elements according to the desired stoichiometric ratio and followed by arc melting in an Ar environment. It should be noted that, depending on the operation of arc melting, the low melting-point and highly volatile elements such as Sb and Sn will be much more significantly evaporated than other elements, leading to an off-stoichiometric alloy with very different electronic properties. To avoid this issue, two approaches have been reported to produce nearly single-phase HHs. One is to simply add an extra 5% of Sb or Sn to the starting materials and strategically place the electric arc only on high-melting point elements to minimize the loss of the starting elements.[53,54] One issue with the arc melting method is nonuniform composition of the as melted sample. For example, the solubility of Ti within the $Hf_{1-x}Zr_xNiSn$ solid solution depends on the Zr to Hf ratio and the temperature (Figure 8.2).[55] At temperatures above 850 K, there is no miscibility gap; thus, a uniform solid solution can be formed. However, after cooling down to room temperature, the final sample may contain varied amounts of Ti-rich and a Ti-poor HH phase due to different processing routes. Even after flipping the reacted chunk three to five times between the two melting processes, impurity phases such as $(HfZr)_5Sn_3$, $TiNi_2Sn$, Ti_6Sn_5, and unalloyed Sn are still observed.[32,33,36,38,56] Annealing at above 1073 K for more than 72 h will effectively eliminate these impurity phases.[32–34,56] In addition, alloying the elements at a lower temperature can also reduce the material loss through evaporation. For example, starting elements in powder form can be sintered at about 1173 K for up to two weeks to obtain the HH phase.[22–26,57–61] It should also be noted that even if pure HH phases are obtained, there still exists disorders in the materials, including Ni occupying 4d vacant sites and Zr/Sn antisite.[18] These defects can modify the band structure and affect the transport properties, which will be discussed in more detail in the following sections.

Nanostructures can significantly scatter phonons without deteriorating much electrical properties, and are widely applied to improve the TE performance. Nanosized powders can be produced via ball milling the arc melted ingot for a few hours.[27–31,62] A rapid hot pressing, which can well

TABLE 8.2 Melting Points and Vapor Pressures of Starting Elements for TE HH Compounds

Element	Hf	Zr	Ti	Ni	Fe	Co	Sb	Sn	Bi
Melting point (K)	2506	2128	1939	1728	1811	1768	904	505	776
Vapor pressure (Pa) at 1800 K	1.63×10^{-6}	8.68×10^{-6}	4.93×10^{-2}	1.37	3.10	1.15	1×10^{5}	56.2	1×10^{5}

Source: Vapor Pressure Calculator. https://www.iap.tuwien.ac.at/www/surface/vapor_pressure.

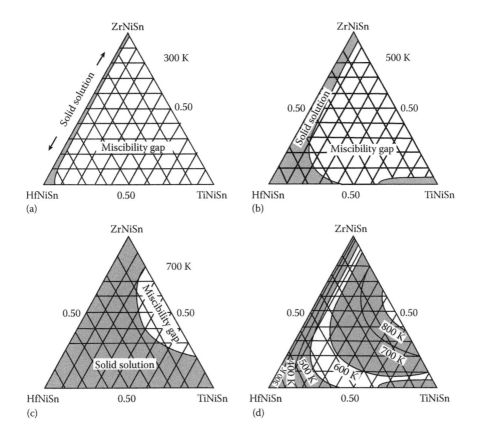

FIGURE 8.2 Pseudoternary $Hf_{1-x-y}Zr_xTi_yNiSn$ phase diagrams were calculated at (a) 300, (b) 500, and (c) 700 K. (d) shows a summary of phase boundaries calculated from 300 to 800 K. (Page, A. et al., *Journal of Materials Chemistry A*, 4, 13949–13956, 2016. Reproduced by permission of The Royal Society of Chemistry.)

preserve the nanostructures, is then used to compact the powder into a bulk sample with density close to their theoretical values. The hot pressed samples can be further machined to the desired shape for measurements and device applications.

Another type of nanosized feature is nanoinclusions or nanoprecipitates, such as InSb,[63] ZrO_2,[50] or full-Heusler (FH) compound.[25,26,44,64] TEM studies indicate that FH nanoinclusions can form coherent lattice with the matrix.[25,26,64] As a result, these nanoinclusions or nanoprecipitates can enhance carrier concentration and retain carrier mobility, thus leading to high electrical conductivity. Meanwhile, the nanoinclusions or nanoprecipitates serve as phonon scattering centers and lower the thermal conductivity.

8.1.3 Individual Thermoelectric Properties

8.1.3.1 Electrical Conductivity

The bandgap of MNiSn (M = Hf, Zr, and Ti) is calculated as 0.42, 0.51, and 0.40 eV for TiNiSn,[10] ZrNiSn,[11] and HfNiSn,[7] respectively. In contrast, the measured values are 0.1–0.3 eV.[65] The lower experimental values may be attributed to atomic disorders, such as M/Sn antisites.[11,18] Ni/interstitial site disorder is also found in MNiSn to create the in-gap electronic states close to the Fermi energy.[66,67] The electrical resistivity at room temperature of intrinsic MNiSn is within the range of 10^{-3}–10^{-5} Ω m.[12] The electrical conductivity (σ) and electrical resistivity (ρ) are related to carrier concentration (n) through the carrier mobility (μ): $1/\rho = \sigma = ne\mu$.[68] To enhance the carrier mobility and carrier concentration, thus leading to high electrical conductivity and PF, two levels of external dopants were introduced into HH compounds.

At the atomic scale, all the three crystallography sites of HHs can be substituted.[69] Table 8.3 summarizes the reported substitution studies for MNiSn-based TE materials.

Isoelectronic alloying is originally considered to have little effect on electrical conductivity.[11,13,70,71] However, it was recently shown that such intermixing can still play a role in the electrical conductivity. $Hf_{0.5}Ti_{0.5}NiSn$ is calculated to have lower carrier concentration than both TiNiSn and HfNiSn.[72] The carrier mobility is likely reduced due to the introduced mass and stain fluctuation and, thus, results in a lower electrical conductivity. Partial substitution of Ni by Pd in ZrNiSn noticeably improves the conductivity, which is probably due to a narrower bandgap.[58]

Elements with different valence electrons can be introduced to change the VEC or even the major carrier type. Specifically for MNiSn based n-type HHs, the M site can be effectively doped by Ce, La, Nb, Ta, and V[13,30,73,74]; at the Ni site, Co, Cr, Cu, and Fe are efficient dopants[13,75]; and at the Sn site, Bi, Ge, Pb, and Sb can all significantly reduce the electrical resistivity.[13,14,24] Although MCoSb is an n-type semimetal at room temperature,[12,15] a p-type conductor can be obtained by partially replacing Co with Fe,[76] or replacing Sb with Ge[77] or Sn.[8,15,16,29,53,78–80]

Beyond doping at the atomic scale, various nanosized features such as metallic nanoparticles can be incorporated into the structure to provide charge carriers. Xie et al.[63] prepared (Ti,Zr,Hf)(Co,Ni)Sb–InSb nanocomposites through a combined high-frequency induction melting method and a subsequent SPS. By introducing 1 at.% InSb NPs at the grain boundaries, all three individual TE properties can be simultaneously improved, and eventually a 160% improvement over the sample containing no nanoinclusions can be achieved.[64] Other types of secondary phases such as FH[25,26,46] and submicron lamellae eutectic phase[81] were also investigated to improve the electrical

TABLE 8.3 Typical Conductivity Tuning Methods for MNiSn-Based Compound

	Isoelectronic Doping	n-Type Doping	p-Type Doping
M site	Intermixing Hf, Zr, and Ti	Ce, La, V, Nb, Ta	Y, Sc
Ni site	Pd, Pt	Cr, Co, Fe, Cu	Co, Ir
Sn site	N/A	Ge, Pb, Sb, B	

transport performances. The coherent grain boundaries in such secondary phases is the key to preserve high carrier mobility.

After effective doping, optimized HH compounds usually have carrier concentration on the order of 10^{20} cm^{-3} and room temperature resistivity value in the range of 10^{-5} to 10^{-6} Ω m,[17,18] which is comparable to other state-of-the-art TE materials, such as Bi_2Te_3 and PbTe.[82,83] For example, one of the optimized compositions $Hf_xZr_{1-x}NiSn_{0.99}Sb_{0.01}$ with peak ZT value of 1 at 873 K has a room-temperature conductivity value of 1.6×10^5 S/m (Figure 8.3a) and carrier concentration of about 2.5×10^{20} cm^{-3} (Figure 8.3b).[28] In general, the electrical conductivity varies with temperature by $\sigma \sim T^{-0.5}$, which indicates alloy scattering (from structural disorder or dopants) as the dominant scattering mechanism.[58,67,84–86] It is also shown that

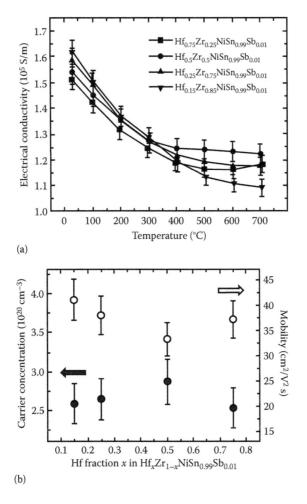

(a)

(b)

FIGURE 8.3 (a) Electrical conductivity, (b) carrier concentration, and mobility of $Hf_xZr_{1-x}NiSn_{0.99}Sb_{0.01}$ nanocomposites (x = 0.15, 0.25, 0.5, and 0.75). (Chen, S. et al.: Effect of Hf concentration on thermoelectric properties of nanostructured n-type half-Heusler materials $Hf_xZr_{1-x}NiSn_{0.99}Sb_{0.01}$. *Advanced Energy Materials*. 2013. 3. 1210–1214. Copyright Wiley-VCH Verlag GmbH & Co. KGaA. Reproduced with permission.)

annealing near the melting point of $Hf_{0.6}Zr_{0.4}NiSn_{0.995}Sb_{0.005}$ can reduce the structure disorder and lattice strain.[87] As a result, the electron Hall mobility can be enhanced and the room temperature electrical conductivity can reach 8×10^5 S/m.[87]

8.1.3.2 Seebeck Coefficient

For degenerate semiconductors, such as heavily doped HHs, under the parabolic band and energy-independent scattering approximation, the Seebeck coefficient is given as[69]

$$S = \frac{8\pi^2 k_B^2}{3eh^2} m^* T \left(\frac{\pi}{3n} \right)^{2/3}, \tag{8.1}$$

where m^* is the DOS effective mass and n is the charge carrier concentration. Based on this equation, large DOS effective mass is favorable for high Seebeck coefficient. For the n-type HH MNiSn, due to its large DOS effective mass of $m^* = 2.8 \pm 0.2$ m_e,[41,67] high Seebeck coefficient above –200 µV/K is obtained at room temperature.[14] On the other hand, the Seebeck coefficient is inversely proportional to the carrier concentration. Therefore, increasing the carrier concentration by dopants to enhance the electrical conductivity will simultaneously decrease the Seebeck coefficient. As a result, the carrier concentration should be tuned to achieve high electrical conductivity without degrading the Seebeck coefficient too much. Again, the optimized carrier concentration at room temperature for MNiSn compounds is found to be on the order of 10^{20} cm^{-3} via different doping methods.[30,88–90]

At a given carrier concentration, the Seebeck coefficient can be enhanced by increasing the derivative of DOS near the Fermi energy. Defects such as Zr/Sn antisites can introduce sharp features in the DOS near the Fermi level.[47] Moreover, resonant states introduced through doping can increase the derivative of DOS as well.[91] As illustrated in Figure 8.4, up to 0.78 at% vanadium doping in $(Hf_{0.75}Zr_{0.25})_{1-x}V_xNiSn$ is reported to be able to create such resonant states, and the DOS near the Fermi level is increased. Therefore, an enhanced Seebeck coefficient is achieved at room temperature.[91] In addition, it is reported that for the nanostructured MNiSn, 1.5% increase of the lattice parameter compared with that of single crystal MNiSn is observed by refinement of X-ray diffraction data. Such a unit cell volume expansion will facilitate band narrowing effects leading to the increase of effective mass. As a result, the Seebeck coefficient is increased by 30% compared to its bulk counterpart over the entire temperature range from 300 to 800 K.[92]

In addition, introducing some selective scattering mechanism to filter the low-energy charge carriers (energy filtering effect) can also increase the Seebeck coefficient. Makongo et al.[25,26] reported that by adding 2 at% extra Ni in HH composition, FH nanoinclusions were formed, which could provide

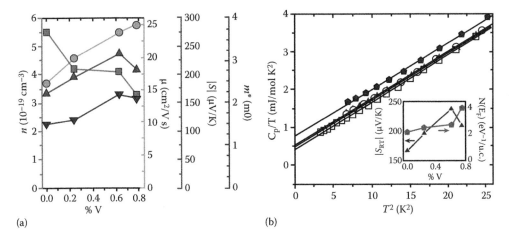

FIGURE 8.4 (a) Carrier concentration (*squares*), carrier mobility (*circles*), absolute value of Seebeck coefficient (*up triangles*), and carrier effective mass (*down triangles*) of V-doped $Hf_{0.75}Zr_{0.25}NiSn$ as a function of V concentration at room temperature. (b) Cp as a function of temperature following Debye regime. The inset shows that absolute value of Seebeck coefficient at room temperature increases with electron DOS up to 0.78 V at% in $(Hf_{0.75}Zr_{0.25})_{1-x}V_xNiSn$. (Reprinted with permission from Simonson, J. et al., *Physical Review B*, 83, 235211, 2011. Copyright 2011 by the American Physical Society.)

additional scattering for the charge carriers at the grain boundaries between the HH matrix and FH. As a result, the low-energy charge carriers would be filtered out, resulting in 67% increase of Seebeck coefficient compared to the HH matrix, as shown in Figure 8.5.[26] Similarly, by incorporating other phases of nanoinclusions such as InSb and ZrO_2, the interface between HH and the nanoinclusions can also provide similar energy-filtering effects, contributing to the improvement of the Seebeck coefficient.[23,50,63] For instance, when the n-type $Ti_{0.5}Zr_{0.25}Hf_{0.25}Co_{0.95}Ni_{0.05}Sb$ sample is incorporated with 1 at% InSb nanoinclusions, the Seebeck coefficient is increased from –146 to –192 μV/K at room temperature.[62]

8.1.3.3 Thermal Conductivity

To achieve high *ZT*, low thermal conductivity is always desired. The total thermal conductivity is the sum of the lattice thermal conductivity and the electronic thermal conductivity from the charge carriers. The n-type HH compounds MNiSn usually have high thermal conductivity due to their high lattice thermal conductivity (above 4 W/m K at room temperature).[14] The lattice thermal conductivity of HH is relatively high compared to other thermoelectic compounds such as Bi_2Te_3[82] and PbTe.[93] Therefore, suppressing the lattice thermal conductivity of HHs is important for the further enhancement of *ZT*.

Alloying effects, which can produce more point defect scattering for phonons, have been proven to be effective in reducing the lattice thermal conductivity. The binary or ternary combination of different ratios of Ti, Zr, and Hf at the M site of MNiSn can reduce approximately 50% of the

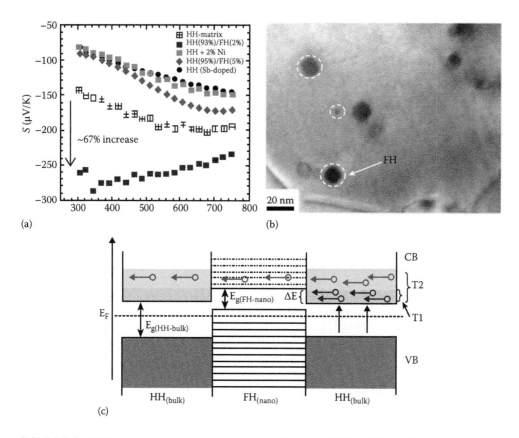

FIGURE 8.5 (a) Temperature-dependent Seebeck coefficient of $HH_{(1-x)}/FH_{(x)}$ bulk nanocomposites (x = 0.02, 0.05) compared to that of $Zr_{0.25}Hf_{0.75}NiSn$ matrix and Sb-doped and nanostructure-free $Zr_{0.25}Hf_{0.75}NiSn_{0.975}Sb_{0.025}$ bulk HH alloy. (b) TEM image of $HH_{(1-x)}/FH_{(x)}$ bulk nanocomposites with x = 0.3. (c) Illustration of the low-energy electron filtering mechanism. ΔE represents the energy barrier between bulk HH matrix and nanometer-scale FH inclusion. T_1 and T_2 represent different temperatures ($T_2 > T_1$), as well as the corresponding energy distribution of electrons. (Reprinted with permission from Makongo, J. P. A. et al., *Journal of the American Chemical Society*, 133, 18843–18852, 2011. Copyright 2011. American Chemical Society.)

lattice thermal conductivity at room temperature compared to the unalloyed MNiSn.[14,27,28,37,41,94] As shown in Figure 8.6a, by alloying Zr at the Hf site in $Hf_{1-x}Zr_xNiSn_{0.98}Sb_{0.02}$, the lattice thermal conductivity is decreased in the whole temperature range.[41] Compared to the sample without Zr, more than 50% reduction of lattice thermal conductivity is achieved when half of the Hf is substituted by Zr. Ternary alloying of Hf, Zr, and Ti can further suppress the lattice thermal conductivity. In Figure 8.6b, $Hf_{0.65}Zr_{0.25}Ti_{0.15}$ $NiSn_{0.995}Sb_{0.005}$ has a room temperature lattice thermal conductivity of 2.6 W/m K, much lower than that of $Hf_{0.6}Zr_{0.4}NiSn_{0.995}Sb_{0.005}$ (3.09 W/m K).[50] Isoelectronic alloy of Pt and Pd at the Ni site can also decrease the lattice thermal conductivity.[39,58] Xie et al.[39] reported that when 15% of Ni is substituted by Pt in the $Hf_{0.65}Zr_{0.35}NiSn_{0.98}Sb_{0.02}$ sample made by levitation melting and SPS, the lattice thermal conductivity decreased from 6 to around

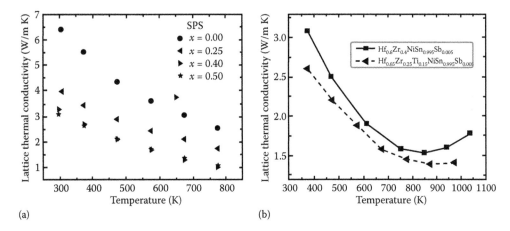

FIGURE 8.6 (a) Temperature-dependent lattice thermal conductivity of $Hf_{1-x}Zr_x NiSn_{0.98}Sb_{0.02}$ with $x = 0$, 0.25, 0.40, and 0.50 prepared by the levitation melt and SPS. (Reprinted from *Acta Materialia*, 57, Yu, C. et al., High-performance half-Heusler thermoelectric materials $Hf_{1-x}Zr_xNiSn_{1-y}Sb_y$ prepared by levitation melting and spark plasma sintering, 2757–2764, Copyright (2009), with permission from Elsevier.) (b) Lattice thermal conductivity of $Hf_{0.6}Zr_{0.4}NiSn_{0.995}Sb_{0.005}$ and $Hf_{0.65}Zr_{0.25}Ti_{0.15}NiSn_{0.995}Sb_{0.005}$.[87] The samples were prepared by arc melting, pulverization, and SPS. All samples in (a) and (b) contain micron-sized grains.

3.2 W/m K. Similarly, the $Zr_{0.5}Hf_{0.5}Ni_{0.5}Pd_{0.5}Sn_{0.99}Sb_{0.01}$ sample made by solid-state reaction and SPS has a much lower room temperature thermal conductivity of 3.1 W/m K compared to the sample without Pd of which the thermal conductivity is about 6 W/m K.[58] However, since Pt and Pd are expensive and rare-earth elements, alloying Pt and Pd is not practical in large-scale applications.

Another approach to suppress the lattice thermal conductivity is enhancing the boundary scattering for phonons by reducing the grain sizes to nanoscale.[27,28,30,31,62,95] Bhattacharya et al.[94] reported that by BM and shock compaction, the grain size of $TiNiSn_{1-x}Sb_x$ decreased from more than 10 μm to less than 1 μm, which leads to a dramatic reduction of lattice thermal conductivity to 3.7 W/m K.[95] Ren's group managed to make $Hf_{0.75}Zr_{0.25}NiSn_{0.99}Sb_{0.01}$ nanopowders with a grain size of around 50 nm by high energy BM (Figure 8.7a).[62] With direct current-assisted rapid HP, the nanostructures can be preserved in the bulk sample with grain size in the range of 200–300 nm (Figure 8.7b). As shown in Figure 8.7c, the room-temperature lattice thermal conductivity of the nanostructured sample is around 3.5 W/m K, which is about 15% reduction compared to the ingot with micron-sized grain. Combining alloying and nanostructures, when Ti substitutes 25 or 50% of Hf in $Hf_{0.75}Zr_{0.25}NiSn_{0.99}Sb_{0.01}$ with nanosized grains, the room-temperature thermal conductivity can be reduced to 3 W/m K.[27] Ideally, if grain growth can be eliminated, the grain size will be comparable with that of as-ball milled nanopowders (~50 nm), and the lattice thermal conductivity will be further reduced due to more grain boundary scattering. In order to do so, some improvement in the densification process

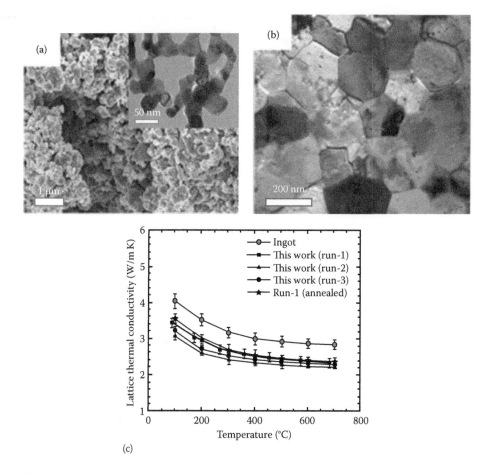

FIGURE 8.7 (a) SEM image of ball-milled $Hf_{0.75}Zr_{0.25}NiSn_{0.99}Sb_{0.01}$ nanopowder with TEM image as inset. (b) Low-magnification TEM image of nanostructured $Hf_{0.75}Zr_{0.25}NiSn_{0.99}Sb_{0.01}$ sample after HP. (c) Lattice thermal conductivity of nanostructured $Hf_{0.75}Zr_{0.25}NiSn_{0.99}Sb_{0.01}$ sample compared to the ingot. (Joshi, G. et al.: Enhancement in thermoelectric figure-of-merit of an n-type half-Heusler compound by the nanocomposite approach. *Advanced Energy Materials*. 2011. 1. 643–647. Copyright Wiley-VCH Verlag GmbH & Co. KGaA. Reproduced with permission.)

is necessary. One possible way is to shorten the hot press duration time and prevent the fast growing of grain size at high temperatures. Another method is to create some secondary phases with high melting temperature to the grain boundaries to stop the grain growth during sintering.

Incorporating secondary phase nanoinclusions into the HH matrix can also create additional boundary scattering for phonons at the interface between the nanoinclusion and HH, leading to the reduction of lattice thermal conductivity. Adding high-melting oxides such as ZrO_2, Al_2O_3, and HfO_2 to HH samples has been reported to be able to suppress the thermal conductivity.[50,59,96–98] For example, it was reported that with 2 and 6 vol% ZrO_2 NPs, the thermal conductivities of ZrNiSn composites were reduced by 8 and 35% over the entire temperature range, respectively.[59] Moreover, by increasing the ratio of

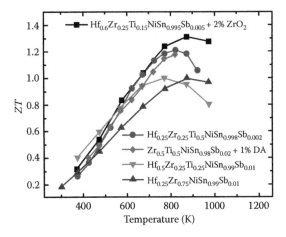

FIGURE 8.8 State-of-the-art ZT values as a function of temperature for MNiSn. The data uncertainty is approximately 10%. (Reproduced from Joshi, G. et al., *Nano Energy*, 2, 82–87, 2013; Chen, S. et al., *Adv Energy Mater*, 3, 1210–1214, 2013; Schwall, M. and Balke, B., *Phys Chem Chem Phys*, 15, 1868–1872, 2013. Reproduced by permission of The Royal Society of Chemistry; Reprinted from *Acta Mater*, 104, Gurth, M., Rogl, G., Romaka, V. V., Grytsiv, A., Bauer, E., Rogl, P., Thermoelectric high ZT half-Heusler alloys $Ti_{1-x-y}Zr_xHf_yNiSn$ ($0 \leq x \leq 1$; $0 \leq y \leq 1$), 210–222, Copyright 2016, with permission from Elsevier; With kind permission from Springer Science+Business Media: *J Electron Mater*, Half-Heusler alloys for efficient thermoelectric power conversion, 45(11), 2016, 5554–5560, Chen, L., Zeng, X. Y., Tritt, T. M., and Poon, S. J.)

Ni in MNiSn, FH nanoinclusions were observed which can also act as the phonon scattering centers to reduce the lattice thermal conductivity.[25,26,46,66,99]

8.1.3.4 Thermoelectric Figure of Merit (ZT)

Combining electrical conductivity (σ), Seebeck coefficient (S), thermal conductivity (κ), and absolution temperature (T), the TE figure of merit (ZT) is calculated as $ZT = (\sigma S^2/\kappa)T$. Figure 8.8 plots recent state-of-the-art values of ZT as a function of temperature for MNiSn.[27,28,48–50] The uncertainty in these ZT values is typically around 10%. The peak ZT appears at 770–870 K and can reach 1.3 on $Hf_{0.65}Zr_{0.25}Ti_{0.15}NiSn_{0.995}Sb_{0.005}$ composited with ZrO_2 NPs.[50] The key factor for high ZT is reduced thermal conductivity by different scattering mechanisms (alloying, grain boundaries, phase separation, and nanoinclusion).

8.1.4 Summary and Perspective

We have reviewed recent progress on the MNiSn-based n-type HH compounds. For electrical conductivity and Seebeck coefficient, the determinant role is played by the band structure, which can be effectively tuned by dopants, structural disorder, and strain. Therefore, synthesis parameters must be carefully adjusted to optimize the electronic structure and properties. For example, a reliable way for rapid alloying of single-phase materials without losing high vapor-pressure elements is desired. A more controllable approach is also needed to engineer grain boundary barriers for effective filtering of low-energy carriers and enhancing the Seebeck coefficient. For thermal conductivity, various features contribute to the phonon scattering, such as grain boundaries, secondary phases, and nanoinclusions. A room-temperature thermal conductivity of ~3 W/m K has been reported in several compounds, which is more than twofold reduction compared to that of single-crystal MNiSn. Innovative techniques are required to further reduce the grain size or induce higher number density of smaller features to effectively scatter phonons with short mean free paths (less than 200 nm). Finally, with the current ZT values of above 1, more efforts should be devoted to device development, such as contact materials and long-term stability under operation temperature (~800 K).

8.2 p-Type Half-Heusler

Ran He, Yucheng Lan, and Zhifeng Ren

8.2.1 Introduction

The TE performance of materials is characterized by the dimensionless figure of merit (ZT)

$$ZT = \frac{S^2 \sigma}{\kappa_L + \kappa_e} T, \tag{8.2}$$

where S, σ, T, κ_L and κ_e are the Seebeck coefficient, electrical conductivity, absolute temperature, lattice thermal conductivity, and electronic thermal conductivity, respectively. Higher ZT corresponds to higher conversion efficiency (η) through the following relation[100]:

$$\eta = \frac{T_H - T_C}{T_H} \cdot \frac{\sqrt{1 + \overline{ZT}} - 1}{\sqrt{1 + \overline{ZT}} + \frac{T_C}{T_H}}. \tag{8.3}$$

Among the various TE materials, HH is a very promising candidate for power generation applications. The HH compounds are crystallized in the space group $F\bar{4}3m$. They possess a usual formula of XYZ, where X, Y, and Z occupies the Wyckoff position 4b, 4c, and 4a, respectively, and leaving the 4d position as voids (Figure 8.9).[17,69] There are hundreds of reported HH compounds, and the ones with VECs of 8 and 18 are semiconductors.[101]

For applications, both the n-type and the p-type TE materials are required. The most widely studied p-type HH is the MCoSb system with M = Hf, Zr, and Ti. The base compound itself is n-type, while the p-type doping can tune the Seebeck coefficient to positive, for example, Sn at Sb, site Fe at Co site, etc.[76,102] The HH compounds attract great interest in the TE field due to their decent ZT,[8,29,78,103] relatively low cost,[17] high PF,[104,105] high mechanical

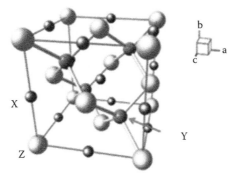

FIGURE 8.9 Crystal structure of HH XYZ, where X, Y, and Z occupy the Wyckoff position 4b (1/2, 1/2, 1/2), 4c (1/4, 1/4, 1/4), and 4a (0, 0, 0), respectively.

reliability,[106] excellent thermal stability, and nontoxicity. The only drawback of HH is the relativity high thermal conductivity, especially the lattice thermal conductivity. Suppressing the phonon transport is crucial for obtaining high-performance TE properties in HH. For this purpose, several effective approaches were developed, including nanoinclusions, nanostructuring, and phase separation. In addition, it is also interesting to search for new compositions since there are many HH that have not been fully investigated yet. Recently, excellent p-type TE properties were reported in the NbFeSb-based HH. A high ZT (~1.5 at 1200 K) and an ultrahigh PF (~106 uW/cm K^2 at room temperature) were reported with M (M = Hf, Zr, Ti) substitution at the Nb site.[101,107] In addition, the ErNiSb-based and the YNiBi-based compositions are also found to possess decent ZT and deserve further attention.

Considering the variety of topics, we arrange this section in the following way: starting with several earlier efforts in enhancing the TE performances in the p-type MCoSb system, followed by several recent concepts that boost the ZT in the p-type MCoSb systems, and finishing with several other p-type HH, especially the NbFeSb-based compounds.

8.2.2 Early Work on p-Type Half-Heusler MCoSb

The substitution of Sn at Sb site yields high electrical conductivity and Seebeck coefficient, as well as a high PF of ~30 uW/cm K^2.[79] However, as mentioned earlier, the thermal conductivity of HH is much higher than other common TE materials. For example, the thermal conductivity at room temperature are ~10 W/m K for $ZrCoSb_{1-x}Sn_x$ and ~4 W/m K for $Hf_{0.5}Zr_{0.5}CoSb_{1-x}Sn_x$,[53,79] while the thermal conductivity of Bi_2Te_3-based TE material are usually on the order of ~1 W/m K.[82] The high thermal conductivity in HH hinders the improvement of ZT, and various efforts were devoted to suppress the thermal conductivity. One way to reduce the thermal conductivity is by using the isoelectronic substitution among Hf, Zr, and Ti.[53,80] This approach enhances the point defect scattering of phonon by increasing the difference of atomic mass and radius. An improved ZT of ~0.5 was reported at 700°C in $Hf_{0.5}Zr_{0.5}CoSb_{0.8}Sn_{0.2}$.[108]

Another way to decrease the thermal conductivity is by introducing secondary precipitates, usually the MO_2,[80,96,97,109] with M being Hf, Zr, and Ti. The oxides could be introduced either *ex situ* or *in situ*. For example, the in situ formation of TiO_2 precipitates in $TiCo_{1-x}Fe_xSb$ are realized by using partially oxidized Fe as the starting material.[76] During arc melting, the following reaction occurs:

$$2Fe_2O_3 + 3Ti \rightarrow 3TiO_2 + 4Fe.$$

The TiO_2 nanoprecipitates are embedded in the HH matrix and serve as extra phonon scatters center. As shown in Figure 8.10a, the thermal conductivity significantly decreased with increased nominal oxidized Fe concentration. Especially, the effects of TiO_2 in phonon scattering was confirmed by comparing two samples with identical nominal composition: $TiFe_{0.3}Co_{0.7}Sb$

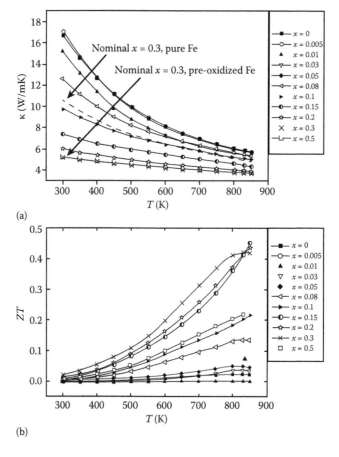

FIGURE 8.10 Temperature-dependent (a) thermal conductivity and (b) ZT of $TiCo_{1-x}Fe_xSb$. The dashed line in a shows the thermal conductivity of $TiCo_{0.7}Fe_{0.3}Sb$ sample without TiO_2 prepared by using pure Fe. (Wu, T. et al., *Journal of Applied Physics*, 102, 103705, 2007, Institute of Physics.)

with one sample used pure Fe as the starting material, the other used the pre-oxidized Fe. As a result, the room temperature thermal conductivity almost doubled in the first case, as indicated by the arrows in Figure 8.10a. The peak ZT of ~0.45 was obtained at ~850 K with 15% nominal Fe substitution at Co site, as shown in Figure 8.10b.

Except for TiO_2, the in situ-formed HfO_2 nanoinclusions and ex situ- introduced ZrO_2 nanopowders were also reported as effective secondary phases in phonon scattering.[96,97] These oxide inclusions suppress the thermal conductivity and improve the ZT. However, the oxides also deteriorate the electron transport and usually yield lower electrical conductivity and PF. Therefore, the improvement in ZT in these cases is not significant.

8.2.3 Nanostructuring

The breakthrough in p-type HH $Hf_{0.5}Zr_{0.5}CoSb_{0.8}Sn_{0.2}$ was achieved by nanostructuring.[8] The thermal conductivity was significantly suppressed due to the enhanced grain boundary phonon scattering in the nanostructured compounds. This was done by applying a high-energy BM technique to the

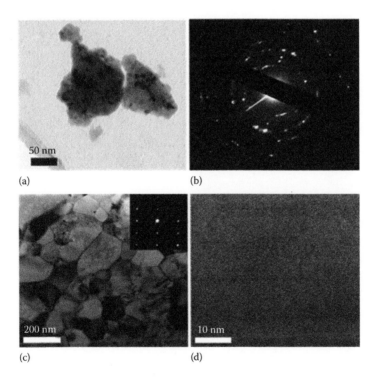

(a)

(b)

(c)

(d)

FIGURE 8.11 TEM characterizations of nanostructured HH. (a) TEM image of the powders of $Hf_{0.5}Zr_{0.5}CoSb_{0.8}Sn_{0.2}$ after BM. (b) Selected area electron diffraction (SAED) pattern showing the polycrystalline nature of an agglomerated cluster in (a). (c) Low-magnification image of the hot pressed $Hf_{0.5}Zr_{0.5}CoSb_{0.8}Sn_{0.2}$, showing the grain size of ~200 nm. *Inset*, SAED of one grain showing single crystallization. (d) HRTEM image showing crystallinity at atomic level. (Reprinted with permission from Yan, X. et al., *Nano Letters*, 11, 556–560, 2011. Copyright 2011. American Chemical Society.)

alloy ingots that were obtained by arc melting the raw elements. As can be seen from Figure 8.11a and b, the grains after BM are usually in the size of ~10 nm. Then, the powders were sintered into a dense sample at elevated temperatures (~1100–1150°C) and external pressure (~80 MPa) by using a direct current HP. The final products are well crystallized with an average grain size of ~200 nm, as shown in Figure 8.11c and d. The grain size is much smaller compared with the samples prepared using other approaches, where the grain sizes are usually in the order of 10–100 μm.

The compared TE properties of bulk and nanostructured $Hf_{0.5}Zr_{0.5}$ $CoSb_{0.8}Sn_{0.2}$ are shown in Figure 8.12. Nanostructured compounds possess higher Seebeck coefficient and lower electrical conductivity, as shown in Figure 8.12a and b. The PF of nanostructured compounds are slightly higher than the bulk counterpart, as shown in Figure 8.12c. On the other hand, a significant drop of the thermal conductivity occurs in the nanostructured sample, as shown in Figure 8.12d. Figure 8.12e shows a similar decrease in the lattice thermal conductivity (κ_L) after subtracting the electronic thermal conductivity (κ_e) from the total thermal conductivity (κ_{tot}) using the single parabolic band model

$$\kappa_e = L\sigma T, \tag{8.4}$$

$$L = \left(\frac{k_B}{e}\right)^2 \left[\frac{3F_2(\eta)}{F_0(\eta)} - \left(\frac{2F_1(\eta)}{F_0(\eta)}\right)^2\right], \tag{8.5}$$

$$S = +\left(\frac{k_B}{e}\right)\left[\frac{2F_1(\eta)}{F_0(\eta)} - \eta\right], \tag{8.6}$$

where L, σ, T, k_B, e, η, and S are the Lorenz number, the electrical conductivity, the absolute temperature, the Boltzmann constant, the carrier charge, the reduced Fermi level, and the Seebeck coefficient, respectively. The functions $F_n(\eta)$ are the Fermi integrals of order n,

$$F_n(\eta) = \int_0^\infty \frac{\chi^n}{1+e^{\chi-\eta}} d\chi. \tag{8.7}$$

As a result, Figure 8.12f shows an enhanced ZT from 0.5 to 0.8 by using nanostructuring, as a result of the decreased thermal conductivity.

Following this approach, the TE properties of the nanostructured binary $Hf_{1-x}Zr_xCoSb_{0.8}Sn_{0.2}$ and $Hf_{1-y}Ti_yCoSb_{0.8}Sn_{0.2}$ and ternary Hf_{1-x-y} $Zr_xTi_yCoSb_{0.8}Sn_{0.2}$ are subsequently investigated.[28,78,113] The TE properties of several nanostructured $MCoSb_{0.8}Sn_{0.2}$ are shown in Figure 8.13. Figure 8.13a

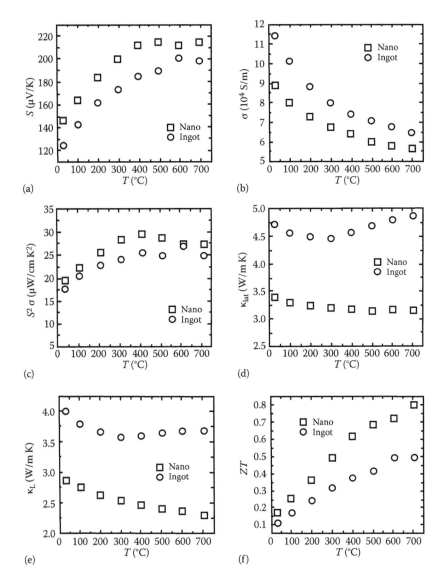

FIGURE 8.12 Temperature-dependent (a) Seebeck coefficient, (b) electrical conductivity, (c) PF, (d) total thermal conductivity, (e) lattice part of thermal conductivity, and (f) ZT of ball-milled and hot pressed sample in comparison with that of the ingot. (Reproduced from Yan, X. A. et al., *Nano Lett*, 11, 556–560, 2011.)

and b shows the Seebeck coefficient and the electrical conductivity, respectively. In a general trend, the higher Hf concentration yields lower Seebeck coefficient in each of the three systems, as shown in Figure 8.13a. This suggests a higher carrier concentration when Hf is higher, as further experimentally examined in the $(Hf,Ti)CoSb_{0.85}Sb_{0.15}$ systems.[109] Thus, the increased Hf concentration elevates the doping efficiency since the nominal dopant concentration is identical among all the compositions. On the other hand, as shown in Figure 8.13b, the lowest electrical conductivity are observed in the

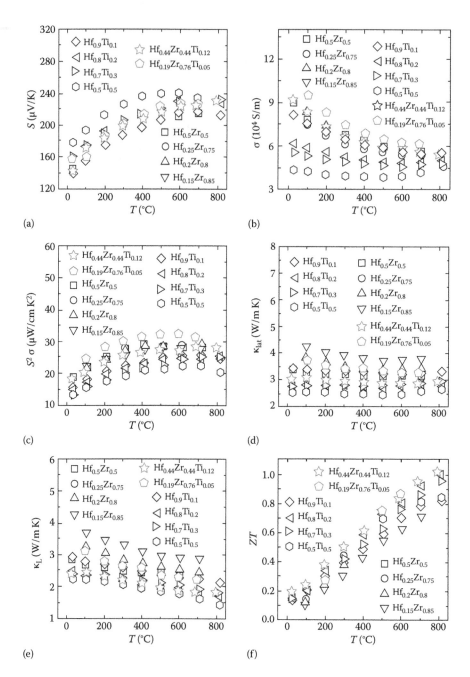

FIGURE 8.13 Temperature-dependent (a) Seebeck coefficient, (b) electrical conductivity, (c) PF, (d) total thermal conductivity, (e) lattice part of thermal conductivity, and (f) ZT of nanostructured $MCoSb_{0.8}Sn_{0.2}$.[111–113] (Yan, X. et al., *Energy and Environmental Science*, 5, 7543, 2012. Reproduced by permission of The Royal Society of Chemistry; Yan, X. et al.: Thermoelectric property study of nanostructured p-type half-heuslers (Hf, Zr, Ti)$CoSb_{0.8}Sn_{0.2}$. *Advanced Energy Materials*. 2013. 3. 1195–1200. Copyright Wiley-VCH Verlag GmbH & Co. KGaA. Reproduced with permission.)

TABLE 8.4 Prices of Relevant Elements in the Year 2010

Element	Hf	Zr	Ti	Ni	Sn	Sb	Nb[a]	Co
Price ($/kg)	563	99.8	10.7	21.8	27.3	8.8	14.3	46

[a] Price of Nb is for the year 2005.

binary $Hf_{1-y}Ti_yCoSb_{0.8}Sn_{0.2}$ system due to the larger mass difference between Hf and Ti. In addition, the ternary $Hf_{1-x-y}Zr_xTi_yCoSb_{0.8}Sn_{0.2}$ system has relatively higher electrical conductivity. The combined electrical conductivity and the Seebeck coefficient gives the PF, as shown in Figure 8.13c. The PF in the binary $Hf_{1-y}Ti_yCoSb_{0.8}Sn_{0.2}$ is the lowest due to their low electrical conductivity. On the other hand, the total and lattice thermal conductivities in the $Hf_{1-y}Ti_yCoSb_{0.8}Sn_{0.2}$ are also much lower due to the large mass difference between Hf and Ti, as shown in Figure 8.13d and e. As a result, Figure 8.13f shows that there are several compositions that possess a peak ZT of ~1 at 700–800°C, such as $Hf_{0.8}Ti_{0.2}CoSb_{0.8}Sn_{0.2}$, $Hf_{0.44}Zr_{0.44}Ti_{0.12}CoSb_{0.8}Sn_{0.2}$, and $Hf_{0.19}Zr_{0.76}Ti_{0.05}CoSb_{0.8}Sn_{0.2}$. The improved ZT highlights HH as a promising candidate for waste heat recovering applications, especially in the 400–800°C range.

Despite the similar ZT value of ~1, the compositions with lower amounts of Hf are preferable because of its lower abundance and higher price compared with other elements, as shown in Table 8.4.[108] A simple calculation indicates that the material cost of $Hf_{0.8}Ti_{0.2}CoSb_{0.8}Sn_{0.2}$ is ~$255 kg^{-1}, and 95% of the total cost is from Hf. Decreasing the Hf concentration significantly suppresses the materials cost. For example, the price of $Hf_{0.44}Zr_{0.44}Ti_{0.12}CoSb_{0.8}Sn_{0.2}$ and $Hf_{0.19}Zr_{0.76}Ti_{0.05}CoSb_{0.8}Sn_{0.2}$ drops to ~$174 kg^{-1} and ~$106 kg^{-1}, respectively. However, even in $Hf_{0.19}Zr_{0.76}Ti_{0.05}CoSb_{0.8}Sn_{0.2}$, ~63% of the total material cost is still from Hf. Therefore, it is necessary to further decrease the Hf concentration or even find Hf-free compositions. As will be shown in Section 8.2.6, the NbFeSb-based p-type HHs are found to possess similar TE properties without any usage of Hf.

8.2.4 Phase Separation

Another effective approach for phonon scattering is by using phase separation.[28,110–112] In a typical experiment, the intrinsic phase separation was obtained by arc melting the raw elements, following the annealing process at 900°C for 7 days, finishing with ice water cooling. Figure 8.14a and b shows the secondary electron image of samples with nominal composition $Zr_{0.5}Hf_{0.5}CoSb_{0.8}Sn_{0.2}$ and $Ti_{0.5}Hf_{0.5}CoSb_{0.8}Sn_{0.2}$. As shown in Table 8.5, the EDS elemental ratio suggests that the samples are not in single phase. The nominal, the average, and the separated compositions are shown in Table 8.5. Moreover, the phase separation in the $Zr_{0.5}Hf_{0.5}CoSb_{0.8}Sn_{0.2}$ is less obvious than the $Ti_{0.5}Hf_{0.5}CoSb_{0.8}Sn_{0.2}$ due to the similar atomic radii between Hf and Zr. Moreover, several other phases such as pure Sn and Hf were also observed.

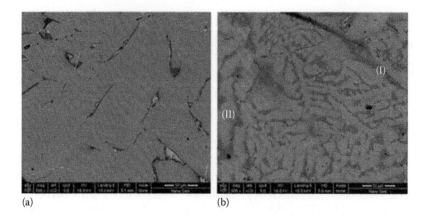

(a)   (b)

FIGURE 8.14 SEM image of (a) $Zr_{0.5}Hf_{0.5}CoSb_{0.8}Sn_{0.2}$ and (b) $Ti_{0.5}Hf_{0.5}CoSb_{0.8}Sn_{0.2}$. (From Rausch, E. et al., *Phys Chem Chem Phys*, 16, 25258–25262, 2014. Reproduced by permission of The Royal Society of Chemistry.)

TABLE 8.5 Nominal Compositions, EDX Measured Average Compositions, and Separated Phases of $MCoSb_{0.8}Sn_{0.2}$

M	EDX Composition	Phases
Ti	$Ti_{0.97}Co_{0.99}Sb_{0.79}Sn_{0.25}$	$Ti_{1.01}Co_{0.97}Sb_{0.91}Sn_{0.1}$
		$Ti_{0.91}Co_{1.19}Sb_{0.23}Sn_{0.67}$
Zr	$Zr_{1.01}Co_{0.96}Sb_{0.77}Sn_{0.25}$	$Zr_{1.02}Co_{0.96}Sb_{0.87}Sn_{0.16}$
		$Zr_{1.01}Co_{0.96}Sb_{0.15}Sn_{0.88}$
Hf	$Hf_{1.07}Co_{0.93}Sb_{0.80}Sn_{0.20}$	$Hf_{1.11}Co_{0.92}Sb_{0.83}Sn_{0.14}$
		$Hf_{0.89}Co_{1.28}Sb_{0.07}Sn_{0.76}$
$Ti_{0.5}Zr_{0.5}$	$Ti_{0.49}Zr_{0.50}Co_{0.94}Sb_{0.89}Sn_{0.18}$	$Ti_{0.45}Zr_{0.56}Co_{0.96}Sb_{0.94}Sn_{0.09}$
		$Ti_{0.64}Zr_{0.38}Co_{0.98}Sb_{0.89}Sn_{0.13}$
$Zr_{0.5}Hf_{0.5}$	$Zr_{0.47}Hf_{0.56}Co_{0.94}Sb_{0.84}Sn_{0.19}$	$Zr_{0.47}Hf_{0.56}Co_{0.93}Sb_{0.88}Sn_{0.15}$
		$Zr_{0.52}Hf_{0.51}Co_{0.97}Sb_{0.72}Sn_{0.28}$
$Ti_{0.5}Hf_{0.5}$	$Ti_{0.52}Hf_{0.56}Co_{0.98}Sb_{0.65}Sn_{0.29}$	$(I)Ti_{0.65}Hf_{0.31}Co_{1.21}Sb_{0.23}Sn_{0.61}$[a]
		$(II)Ti_{0.38}Hf_{0.70}Co_{0.98}Sb_{0.84}Sn_{0.1}$[a]

[a] Roman numbers in parentheses correspond to the areas in Figure 8.14b.

The TE properties of phase-separated $MCoSb_{0.8}Sn_{0.2}$ are shown in Figure 8.15. For comparison, the bulk and nanostructured $Hf_{0.5}Zr_{0.5}CoSb_{0.8}Sn_{0.2}$ are also plotted in Figure 8.15.[8] Both the electrical conductivity and the Seebeck coefficient of phase-separated $Hf_{0.5}Zr_{0.5}CoSb_{0.8}Sn_{0.2}$ are similar to the bulk ingots, as shown in Figure 8.15a and b. On the other hand, Figure 8.15d shows that the thermal conductivity of phase-separated $Hf_{0.5}Zr_{0.5}CoSb_{0.8}Sn_{0.2}$ is much lower than that of the bulk ingot. Consequently, as shown in Figure 8.15e, the phase-separated $Hf_{0.5}Zr_{0.5}CoSb_{0.8}Sn_{0.2}$ obtains a similar ZT compared with the nanostructured compound.

The TE properties of phase-separated p-type HH were further optimized by tuning the dopant concentration $MCoSb_{1-x}Sn_x$. A high ZT value of ~1.15

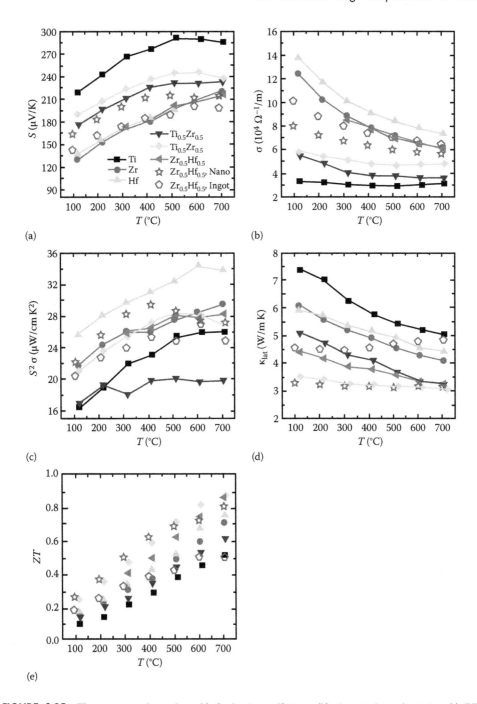

FIGURE 8.15 Temperature-dependent (a) Seebeck coefficient, (b) electrical conductivity, (c) PF, (d) lattice thermal conductivity, and (e) ZT of phase-separated $MCoSb_{0.8}Sn_{0.2}$. For comparison, the TE properties of bulk ingots (*pentagon*) and nanostructured (*star*) $Hf_{0.5}Zr_{0.5}CoSb_{0.8}Sn_{0.2}$ are also plotted. (Reproduced from Yan, X. et al., *Nano Lett*, 11, 556–560, 2011; Rausch, E. et al., *Phys Chem Chem Phys*, 16, 25258–25262, 2014. Reproduced by permission of The Royal Society of Chemistry.)

was reported at 710°C in $Hf_{0.75}Ti_{0.25}CoSb_{0.85}Sn_{0.15}$,[109] which is similar to the optimized nanostructured p-type HH. In comparison with the sample preparation procedure, the advantage of the phase separation approach lies in its simplified procedure since it does not require the high-energy BM and the HP process. However, the phase separation approach is also time-consuming, since it requires a long-time annealing.

8.2.5 Nanoinclusions

Unlike oxide inclusions that deteriorate the electrical conductivity and the power factor, as introduced in Section 8.2.2, another type of nanoinclusion, such as indium–antimony (InSb) and FH, could simultaneously scatter phonons and benefit electron transport. An example is in $TiCo_{0.85}Fe_{0.15}Sb + x\%$ InSb.[113] With up to 1.0 at% in situ-formed InSb, all the TE parameters are simultaneously improved, i.e., the electrical conductivity and Seebeck coefficient become higher, and the lattice thermal conductivity becomes lower. As shown in Figure 8.16a and b, the improved electrical conductivity are due to the enhanced mobility since the carrier mobility in TiCoSb is as low as ~0.5 cm²/V s, while InSb has a very high mobility on the order of ~10^4 cm²/V s for electrons and ~10^3 cm²/V s for holes.[114,115] In addition, Figure 8.16b and c shows the decreased carrier concentration and the improved Seebeck coefficient due to the energy-filtering effects.[116,117] Thus, the PF is significantly enhanced, as shown in Figure 8.16d. On the other hand, Figure 8.16e shows a decreased thermal conductivity. The thermal conductivity decreases not only because InSb is a secondary phase, but also because the precipitated InSb on the grain boundaries impedes the growth of grains, therefore enhancing the grain boundary scattering of phonon.

Figure 8.16f shows the peak ZT value ~0.3 at ~925 K with 1 at% InSb inclusions. Although the ZT is low, the observation of the simultaneous improvement of all the three TE parameters is still quite interesting. Note that with InSb, more than 1.0 at%, the TE properties became worse, showing the existence of an optimum InSb concentration. As seen from Figure 8.16b, with InSb content exceeding 1.5 at%, the mobility dropped, and the carrier concentration increased. This suggests the disappearing of the energy-filtering effect. The mechanisms behind the variation trend were not completely revealed, and further investigations are necessary.

The FH nanoinclusions were recently found to be beneficial for the TE properties in p-type $MCo_{1+x}Sb_{0.9}Sn_{0.1}$, where the extra Co were weighed at the beginning of sample processing.[118,119] The FH precipitations generated in the HH matrix are in the size of ~5–60 nm. Similar to the InSb inclusions, the FH inclusions also boost the Seebeck coefficient due to the energy-filtering effect. As schematically shown in Figure 8.17, the bandgap of FH nanoinclusions enlarges as a result of the quantum refinement when the grain size is small, thus yielding an energy offset of the valance band maxima between the HH matrix and FH inclusions. This energy offset

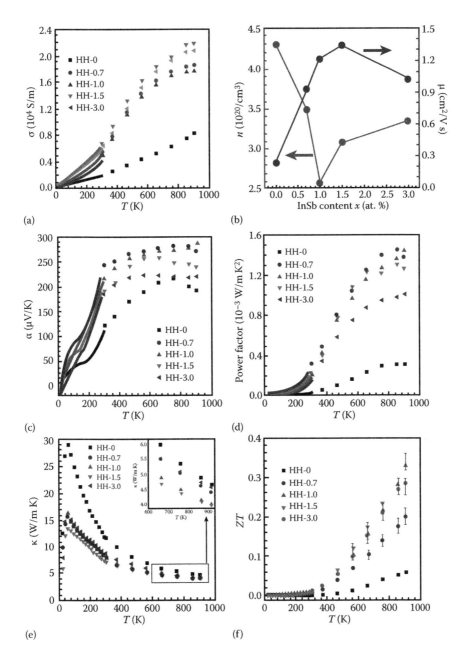

FIGURE 8.16 (a) Temperature-dependent electrical conductivity; (b) InSb concentration dependence of carrier concentration and mobility. Temperature-dependent (c) Seebeck coefficient, (d) PF, (e) thermal conductivity, and (f) ZT of $TiCo_{0.85}Fe_{0.15}Sb$ with different InSb nanoinclusions. (Reprinted from *Acta Mater*, 61, Xie, W. J. et al., Significant ZT enhancement in p-type Ti(Co,Fe)Sb–InSb nanocomposites via a synergistic high mobility electron injection, energy filtering and boundary-scattering approach, 2087–2094, Copyright 2013, with permission from Elsevier.)

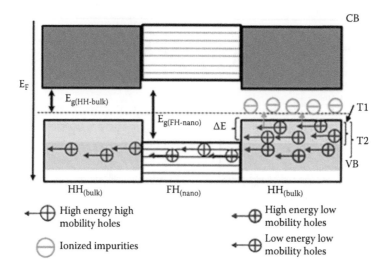

FIGURE 8.17 Schematic illustration of band structure of a p-type HH matrix with the FH nanoinclusion. (From Sahoo, P. et al., *Nanoscale*, 5, 9419–9427, 2013. With permission.)

impedes the transport of low-energy carriers and results in the carrier filtering effect and ultimately yields a lower electrical conductivity and a higher Seebeck coefficient.

Similar to the InSb nanoinclusion, there exists an optimized concentration of the FH phase. The overall TE performances degrade if the extra Co exceeds 5%. One possible explanation is that with higher Co concentration, the FH inclusions become bigger in size and the quantum confinement effect disappears. Subsequently, the bandgap in FH shrinks and the low-energy carrier filtering effects no longer exist. Therefore, the Seebeck coefficient drops since the FH phase is intrinsically a metal. Indeed, the decreasing in the Seebeck coefficient is seen in compositions with excessive extra Co exceeding 5%.[118,119] For example, in the set of compounds $Hf_{0.5}Ti_{0.5}Co_{1+x}Sb_{0.9}Sn_{0.1}$, the Seebeck coefficients at room temperature increases from ~100 to ~240 μV/K with the excess Co being 0 and 5%, respectively; upon further increasing the excess Co to 8%, the Seebeck coefficient dropped to ~140 μV/K. Thus, it might be interesting to investigate size-dependent TE properties of the FH inclusions.

8.2.6 Other p-Type Half-Heuslers

Several other p-type HHs were investigated, such as the ErNiSb- and YNiBi-based compounds. Peak ZT values of ~0.3 and ~0.12 were reported in $ErNi_{0.5}Pd_{0.5}Sb$ and YNiBi.[33,34] The ZT of these compounds are lower than those of MCoSb-based compounds. However, there are still rooms for further improvement if the approaches introduced earlier (nanostructuring, phase separation, etc.) can be properly applied. Recently, the NbFeSb-based HH compounds are reported to possess very good p-type TE properties. Joshi et al.[105] and Fu et al.[104] reported the TE properties of p-type NbFeSb with up to 40% Ti substitution at Nb site. Peak ZT values of ~1 were realized in several compositions, such as $Nb_{0.8}Ti_{0.2}FeSb$ and $Nb_{0.6}Ti_{0.4}FeSb_{0.95}Sn_{0.05}$ at 700°C. He et al. reported that in $Nb_{0.95}Ti_{0.05}FeSb$, by applying the HP method close to the phase transition point, the average grain sizes dramatically increased, yielding an increased carrier mobility and electrical conductivity.[107] Benefiting from the improved electrical conductivity and the almost unchanged Seebeck coefficient, the PFs improve from room temperature up to 700°C. On the other hand, the thermal conductivity is barely affected by the grain sizes in the range from ~0.3 to 4.5 μm. Thus, the ZT is subsequently improved. A peak PF of ~106 μW/cm K^2 was obtained in $Nb_{0.95}Ti_{0.05}FeSb$ at room temperature, while a peak ZT value of ~1.1 was achieved in $Nb_{0.8}Ti_{0.2}FeSb$ at 700°C. The TE properties are shown in Figure 8.18.

In a simplified prediction that omits the contact resistance, the higher ZT yields higher conversion efficiency, while higher PF gives higher power density. This is experimentally proven in the comparison between $Nb_{0.95}Ti_{0.05}FeSb$ and $Nb_{0.8}Ti_{0.2}FeSb$. The output power density, heat to power conversion efficiency, PF, and ZT of $Nb_{0.95}Ti_{0.05}FeSb$ and $Nb_{0.8}Ti_{0.2}FeSb$ are presented in Figure 8.19a through d. Clearly, when comparing Figure 8.19a and c, higher PF leads to higher output power density with the same leg length. On the other hand, as shown in Figures 8.19b and d, higher ZT results in higher efficiency. In addition, as shown in Figure 8.19a, an ultrahigh output power density of ~22 W/cm^2 as obtained in $Nb_{0.95}Ti_{0.05}FeSb$ with cold-side temperature and hot-side temperatures of 293 and 868 K, respectively. This is the highest reported output power density for bulk TE devices to the best of our knowledge.

Although the PFs in the Ti-doped NbFeSb HH are very high, the high thermal conductivity impedes the ZT from being further improved. Thus, it is important to suppress the thermal conductivity while still keeping the PFs high. One approach was investigated by using dopant with heavier atomic mass, i.e., Hf and Zr instead of Ti.[101] The electrical conductivity, the Seebeck coefficient, the thermal conductivity, and the ZT of Hf- and Zr-doped NbFeSb are shown in Figure 8.20a and b. For the electrical conductivity and the Seebeck coefficient, no significant difference was observed using the two dopants. Meanwhile, the thermal conductivity was obviously lower with Hf doping than that with Zr doping due to its heavier mass, as shown in Figure 8.20c. The ZT reached ~1.2 at 973 K and ~1.5 at 1200 K in $Nb_{0.88}Hf_{0.12}FeSb$.

The obtained high ZT in the Hf-doped NbFeSb HH compound suggests its application potentials for heat-to-power conversion. Yet the usage of Hf is not favorable considering its high price and low abundance. Therefore, it deserves active investigation to further decrease the thermal conductivity through other approaches in the NbFeSb-based system.

8.2.7 Summary

The p-type HH usually possesses very high PFs, with the high thermal conductivity being the only drawback. The TE properties of the p-type HH

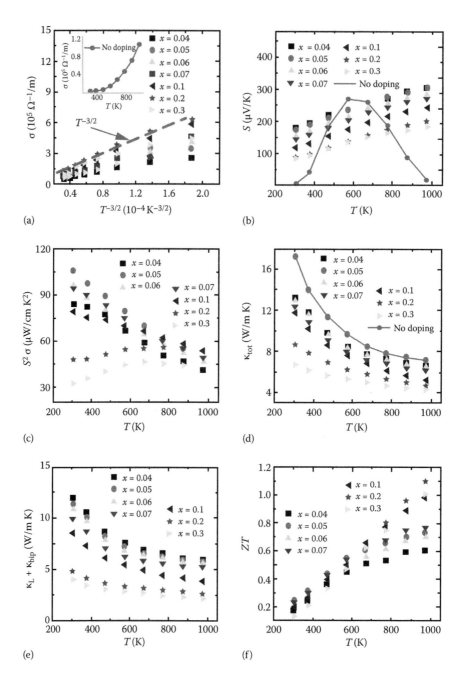

FIGURE 8.18 TE property dependence on temperature for $Nb_{1-x}Ti_xFeSb$ with $x = 0, 0.04, 0.05, 0.06,$ 0.07, 0.1, 0.2, and 0.3. (a) Electrical conductivity, (b) Seebeck coefficient, (c) PF, (d) total thermal conductivity, (e) lattice and bipolar thermal conductivities, and (f) ZT. In (a), the dashed line and the inset show the $\sim T^{-3/2}$ relation and the measured conductivity of undoped NbFeSb, respectively. (From He, R. et al., *Proceedings of the National Academy of Sciences*, 113, 13576–13581, 2016. With permission.)

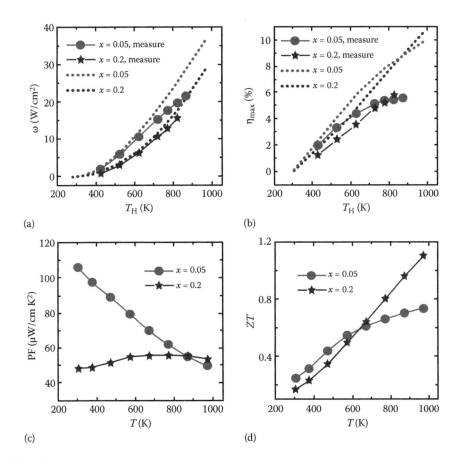

FIGURE 8.19 Calculated (dotted lines) and measured (symbols) (a) output power density and (b) conversion efficiency of samples $Nb_{1-x}Ti_xFeSb$ (x = 0.05 and 0.2) with the cold side temperature at ~20°C and the leg length ~2 mm. Comparison of (c) *PF* and (d) *ZT* of $Nb_{0.95}Ti_{0.05}FeSb$ and $Nb_{0.8}Ti_{0.2}FeSb$. (From He, R. et al., *Proceedings of the National Academy of Sciences*, 113, 13576–13581, 2016. With permission.)

compounds were effectively improved by successfully suppressing the thermal conductivity through various approaches, such as nanostructuring, phase separation, and nanoinclusions. Yet there are still plenty of room for further enhancement since the thermal conductivity is still high in comparison with other common TE materials.

8.3 New Half-Heuslers

Lihong Huang and Zhifeng Ren

Based on the former calculation, the basic requirement to be a TE material is that HH compounds need to satisfy 18 VEC per unit cell. Otherwise, the Fermi level will enter the valence band or conduction band, and the compound will show p-type or n-type metallic properties, respectively.[7,41,45,62,120] Among the HH family with 18 VEC, most research has focused on systems such as MNiSn for n-type[27,28] and MCoSb for p-type,[78,97] where M is an individual

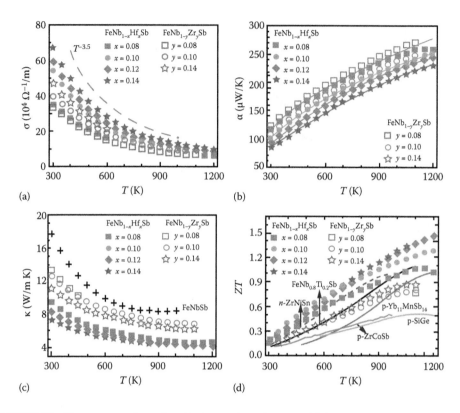

FIGURE 8.20 Temperature-dependent (a) electrical conductivity, (b) Seebeck coefficient, (c) thermal conductivity, and (d) ZT of $Nb_{1-x}Hf_xFeSb$ and $Nb_{1-y}Zr_yFeSb$. (From Fu, C. et al., *Nat Commun*, 6, 8144, 2015. With permission.)

or a combination of Hf, Zr, and Ti, as we mentioned in Sections 8.1 and 8.2. Besides MNiSn and MCoSb (M = Hf, Zr, Ti), other HH TE materials with 18 VEC, such as NbCoSn (ZT = 0.3 at 850 K for $Nb_{0.99}Ti_{0.01}CoSn_{0.9}Sb_{0.1}$),[121,122] VFeSb ($ZT$ = 0.25 at 550 K for pure VFeSb),[123,124] and so on, were also investigated in recent years. Fu et al.[101] reported a high ZT of about 1.5 at 1200 K for the p-type HH alloy $FeNb_{0.88}Hf_{0.12}Sb$ in 2015.

The electronic structure calculations of many HH compounds showed that HHs with VEC = 19 such as VCoSb, NbCoSb, and TiNiSb are metallic, not suitable for semiconductor or TE applications.[125–127] However, to our surprise, we recently discovered that NbCoSb with VEC = 19 could be made into HH alloys with decent TE properties[128] and so is VCoSb.[129] NbCoSb is a n-type TE material with a fairly good ZT value, meaning Nb is a very strong electron donor in comparison with the p-type MCoSb, where M is Hf, Zr, or Ti, which opens a new route to develop new HH TE materials not following the traditional VEC of 18.

In this subsection, we discuss on new HH TE materials, such as NbCoSb and VCoSb, including their preparation, crystal structure and defect, TE performance, and so on.

8.3.1 NbCoSb Microstructure and TE Performance of NbCoSb without Doping

All the samples of NbCoSb-based TE materials were prepared by arc melting, BM, and then HP, please see the details in Huang et al.[130] Figure 8.21 shows the X-ray diffraction patterns of NbCoSb samples hot pressed at different temperatures. All diffraction peaks are well matched with the cubic HH phase of NbCoSb (Joint Committee on Powder Diffraction Standards [JCPDS] 51-1247, a = 0.5897 nm). Very weak diffraction peaks for Nb_3Sb are also identified, which resulted from the peritectic reaction during the arc-melting process. As shown in Table 8.6, lattice constants of all the samples are consistent with that of JCPDS card. The calculated theoretical densities are listed in Table 8.6 together with the measured values and relative percentages.

Figure 8.22 shows the typical SEM images on the freshly fractured surfaces of the NbCoSb samples hot pressed at different temperatures. There are clearly holes inside the samples hot pressed at lower temperature

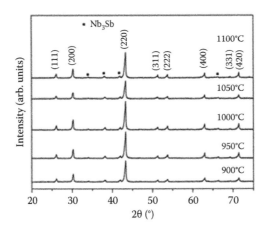

FIGURE 8.21 X-ray diffraction patterns of NbCoSb samples hot pressed at different temperatures. (Reprinted from *Materials Research Bulletin*, 70, Huang, L. H. et al., A new n-type half-Heusler thermoelectric material NbCoSb, 773–778, Copyright [2015], with permission from Elsevier.)

TABLE 8.6 Lattice Constant and Theoretical, Experimental, and Relative Densities of NbCoSb Samples Hot Pressed at Different Temperatures

Sample Name	HPT (°C)	Lattice Constant (nm)	Theoretical Density (g/cm³)	Experimental Density (g/cm³)	Relative Density (%)
HP900	900	0.5901	8.844	7.805	88.25
HP950	950	0.5900	8.848	7.931	89.64
HP1000	1000	0.5899	8.853	8.329	94.08
HP1050	1050	0.5898	8.857	8.332	94.07
HP1100	1100	0.5896	8.866	8.376	94.47

Source: Reprinted from *Materials Research Bulletin*, 70, Huang, L. H. et al., A new n-type half-Heusler thermoelectric material NbCoSb, 773–778, Copyright 2015, with permission from Elsevier.

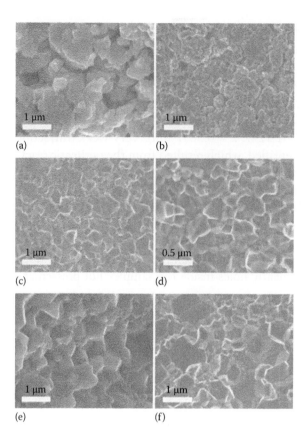

FIGURE 8.22 SEM images of NbCoSb samples hot pressed at different temperatures. (a) 900°C, (b) 950°C, (c, d) 1000°C, (e) 1050°C, and (f) 1100°C. (Reprinted from *Materials Research Bulletin*, 70, Huang, L. H. et al., A new n-type half-Heusler thermoelectric material NbCoSb, 773–778, Copyright [2015], with permission from Elsevier.)

(Figure 8.22a and b); samples with HP temperature (HPT) ≥ 1000°C look densely packed (Figure 8.22c through f). However, the density measurement shows a relative density of only about 94%, which is probably due to the loss of Sb that caused the theoretical density lower than the calculated values shown in Table 8.6. The effects of HPT on the grain size and distribution are clearly shown in the SEM images. The particle size increases with the increasing HPT, while it is not uniform when HPT is higher than 1050°C (Figure 8.22e and f).

The TEM study of the NbCoSb samples hot pressed at 1000°C has been performed. It is clearly shown in Figure 8.23a that the sample is a densely packed polycrystalline and contains grains of around 200–400 nm, which is consistent with the SEM study (Figure 8.22c and d). Figure 8.23b shows an HRTEM image of the lattice fringes and grain boundaries between three adjacent grains. It clearly indicates that individual grains are highly crystalline and the grain boundaries are clean.

The thermal properties were investigated from room temperature to 700°C, as shown in Figure 8.24. The thermal diffusivity (*D*) increases with the

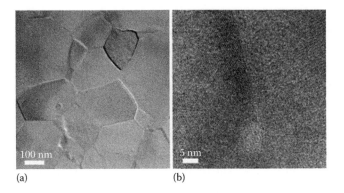

FIGURE 8.23 (a) Medium- and (b) high-resolution TEM images of NbCoSb sample hot pressed at 1000°C. (Reprinted from *Materials Research Bulletin*, 70, Huang, L. H. et al., A new n-type half-Heusler thermoelectric material NbCoSb, 773-778, Copyright [2015], with permission from Elsevier.)

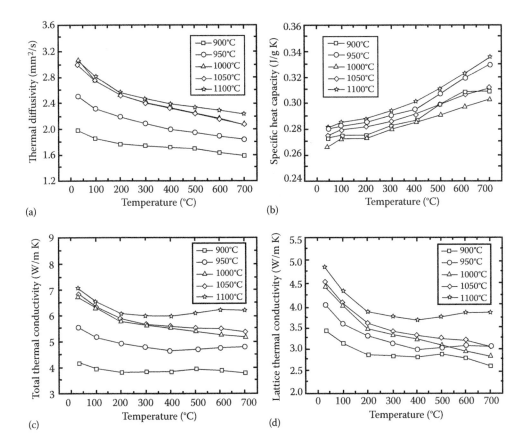

FIGURE 8.24 Temperature-dependent thermal properties of NbCoSb samples hot pressed at different temperatures. (a) Thermal diffusivity, (b) C_p, (c) thermal conductivity, (d) lattice thermal conductivity. (Reprinted from *Materials Research Bulletin*, 70, Huang, L. H. et al., A new n-type half-Heusler thermoelectric material NbCoSb, 773–778, Copyright [2015], with permission from Elsevier.).

inceasing HPT, because of the larger grain size. As plotted in Figure 8.24b, among all the samples, the sample hot pressed at 1000°C shows the lowest specific heat capacity (Cp). Normally, Cp should not change if the composition is fixed. Here, we have a little bit difference in Cps for samples hot pressed at different temperatures, which is the result of different compositions resulted from Sb loss at the high HPTs. When Sb is lost during hot pressing, the Cp should increase since Sb has the lowest Cp among all the three elements. The thermal conductivity is then calculated by multiplying the density, specific heat capacity, and thermal diffusivity of each sample. As shown in Figure 8.24c, the total thermal conductivity increases with increasing HPT. We further tried to estimate the lattice thermal conductivity from the relationship: $\kappa = \kappa_L + \kappa_e + \kappa_{bi}$. Here, the electronic thermal conductivity (κ_e) is calculated according to the W-F relation,[112] $\kappa_e = L\sigma T$, where L is the Lorenz number and T the absolute temperature. To calculate the Lorenz number, we used the electrical conductivity and Seebeck coefficient shown in Figure 8.25a and b. The Lorenz number is calculated from the reduced Fermi energy, which is deduced from the Seebeck coefficient and based on

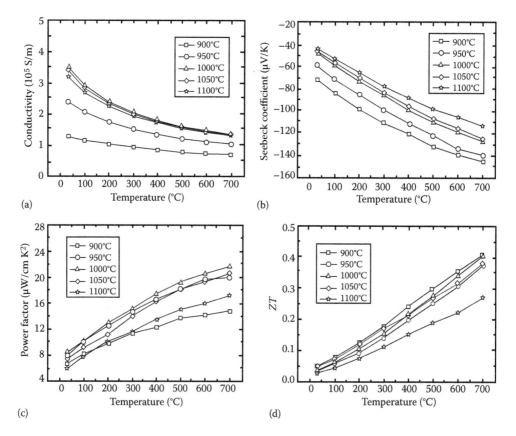

FIGURE 8.25 (a) Electrical conductivity, (b) Seebeck coefficient, (c) PF, and (d) ZT value of NbCoSb samples hot pressed at different temperatures. (Reprinted from *Materials Research Bulletin*, 70, Huang, L. H. et al., A new n-type half-Heusler thermoelectric material NbCoSb, 773–778, Copyright [2015], with permission from Elsevier.).

a single-band model for all samples.[131,132] The Lorenz number is found to be 2.24×10^{-8} W Ω/K^2 at room temperature, which decreases to 1.81×10^{-8} W Ω/K^2 at 700°C for sample HPT1000. Then, the calculated electronic thermal conductivity is subtracted from the measured total thermal conductivity to obtain the lattice thermal conductivity, and the bipolar thermal conductivity is assumed to be negligibly small because the temperature-dependent Seebeck coefficient is monotonically increasing. As shown in Figure 8.24d, the lattice thermal conductivity increases with increasing HPT. The lowest lattice thermal conductivity in sample HPT900 is related to the significant phonon scattering effect at grain boundaries owing to its smallest grain size and pores in the sample.[68,133,134]

The temperature dependence of the electrical conductivity σ, Seebeck coefficient S, PF $S^2\sigma$, and ZT of hot pressed NbCoSb samples are plotted in Figure 8.25. All samples show metallic behavior as the electrical conductivity decreases with increasing temperature, which indicates a stronger electron scattering at higher temperatures (Figure 8.25a). Figure 8.25b shows that the Seebeck coefficient decreases with increasing HPT. Here, all measured Seebeck coefficients are negative, indicating n-type transport behavior, which is a big surprise to us since MCoSb are all strong p-types when M is Ti, Zr, and Hf. It seems Nb is a very strong electron donor. Another feature we noted is that the temperature-dependent Seebeck shows almost a linear increase of up to 700°C, suggesting no bipolar effect in our samples before 700°C. Consequently, the PF is calculated and presented in Figure 8.25c, and the maximum PF is 21.6 μW/cm K^2 at 700°C for sample HPT1000. Finally, a maximal ZT of approximately 0.4 was obtained at 700°C for the sample HPT1000 (Figure 8.25d). Clearly, the best HPT is between 950°C and 1000°C. When it is too low or high, the ZT is lower due to lower density or larger grains, respectively.

The Hall voltage V_H at room temperature is measured by the Physical PropertyMeasurement System (PPMS, Quantum Design) under a 3 T magnetic feld. The carrier concentration n, Hall coefficient R_H, and carrier mobility μ_H are calculated by the formulas $n = IB/V_H qt$, $R_H = -(1/qn)$, and $\mu_H = \sigma R_H$, respectively, where I is the current (8 mA), B is the magnetic field intensity (3 T), $q = -e = 1.6 \times 10^{-19}$ C, t is the thickness of sample, and σ is the electrical conductivity obtained from a commercial system (ULVAC, ZEM-3) before. Table 8.7 summarizes the transport properties of NbCoSb samples at room temperature. Hall coefficients have negative values for all the samples, indicating n-type conduction with electrons as major carriers, in agreement with the Seebeck measurement. It can be observed that the resistivity decreases with increasing HPT of up to 1000°C and then increases, it is mainly because of the variation of mobility.

In short, we have surprisingly found a new n-type HH TE material, NbCoSb with the VEC of 19, different from the traditionally thought of 18, which opens up a new route to develop new HH TE materials not following the VEC of 18. Even though the peak ZT value of ~0.4 is still too low, it is worth pointing out that all the traditional HH materials without any doping

TABLE 8.7 Resistivity, Carrier Concentration, and Mobility at Room Temperature of NbCoSb Samples Hot Pressed at Different Temperatures

Sample Name	HPT (°C)	Hall Coefficient ($\times 10^{-9}$ m^3/C)	Resistivity ($\times 10^{-6}$ Ω m)	Carrier Concentration ($\times 10^{21}$ cm^{-3})	Mobility (cm^2/V s)
HP900	900	−1.14	7.84	5.48	1.45
HP950	950	−0.83	4.16	7.49	2.01
HP1000	1000	−1.03	2.85	6.04	3.62
HP1050	1050	−1.01	2.92	6.21	3.45
HP1100	1100	−1.02	3.11	6.09	3.30

Source: Reprinted from *Materials Research Bulletin*, 70, Huang, L. H. et al., A new n-type half-Heusler thermoelectric material NbCoSb, 773–778, Copyright 2015, with permission from Elsevier.

such as TiCoSb, ZrCoSb, and HfCoSb have peak *ZT*s of only about 0.03,[128] which is much smaller than 0.4. Therefore, it is very interesting to see that the traditionally thought VEC of 18 is not required. And we believe that it is very hopeful to further improve the TE properties of NbCoSb with proper doping. More promising is that the working temperature can be even higher than 700°C for NbCoSb HH compound.

8.3.1.1 Band Structure of NbCoSb

To understand why NbCoSb with VEC = 19 displays such TE properties, we did DOS and band structure calculations using DFT,[135] as shown in Figure 8.26, compared with TiCoSb with 18 valence electrons. TiCoSb shows a semiconducting band structure; a bandgap of 1.0 eV is found at the Fermi edge; and the Fermi level is located at the top of the valence band, which is consistent with Frederick's calculation.[126] NbCoSb shows a bandgap of about 1.0 eV, while the Fermi level is located in the conduction band, meaning the NbCoSb should show more metallic or semimetallic behavior, which is consistent with its high conductivity at room temperature.

The gap width in the XYZ compounds is correlated with the electronegativity difference between the Y and Z elements.[136] TiCoSb and NbCoSb have almost the same gap width because of the same Co and Sb elements in the Y and Z positions, respectively. However, the Fermi levels of the two compounds are very different: for TiCoSb, it is above the valence band edge, but for NbCoSb, it is in the conduction band, meaning the NbCoSb should show more metallic or semimetallic behavior, very different from TiCoSb.

8.3.1.2 TE Performances of NbCoSb after Isoelectronic Doping by V and Ta

To reduce the lattice thermal conductivity κ_L, the effectiveness of isoelectronic alloying on the M sublattice in HH compounds of MNiSn and MCoSb (M = Hf, Zr, Ti) has been successfully demonstrated.[29,77,112,137,138] Point defect disorder caused by strain and mass differences between alloying atoms and host atoms acts as scattering centers for phonons, which reduces κ_L. So we further focus on the effectiveness of isoelectronic alloying on reducing

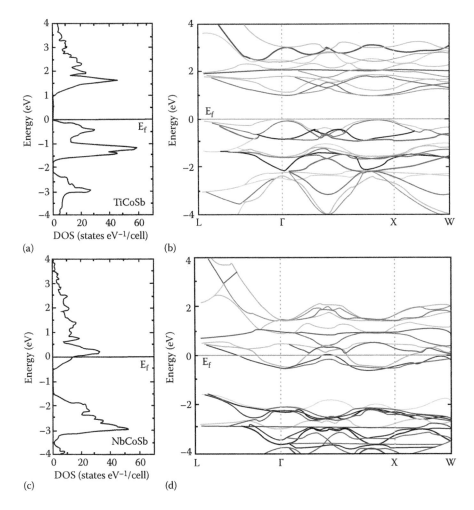

FIGURE 8.26 DOS of (a) TiCoSb and (c) NbCoSb, and the calculated band structure of (b) TiCoSb and (d) NbCoSb. (Huang, L. H. et al., *RSC Advances*, 5, 102469–102476, 2015. Reproduced by permission of The Royal Society of Chemistry.)

the thermal conductivity of the new HH TE material NbCoSb with VEC = 19.[135] All the samples were prepared as the same method as NbCoSb; just the HPT was applied at 1000°C. According to our results, the isoelectronic substitution of Nb by V and Ta in NbCoSb can dramatically decrease the thermal conductivity from 7.0 to 3.3 W/m K for $Nb_{0.44}V_{0.44}Ta_{0.12}$ CoSb at room temperature, but unfortunately, a large PF reduction also occurred. Consequently, the cosubstitution of V and Ta for Nb in NbCoSb is proved effective in suppressing thermal conductivity, ~50% reduction at room temperature, and 40% reduction at 700°C. A maximum *ZT* value of ~0.5 is obtained at 700°C for $Nb_{0.44}V_{0.44}Ta_{0.12}CoSb$ due to the lower thermal conductivity, ~25% increase compared with that of NbCoSb.[130]

Figure 8.27 shows the thermal transport properties, i.e., the thermal diffusivity, specific heat capacity (*Cp*), total thermal conductivity (κ), and lattice

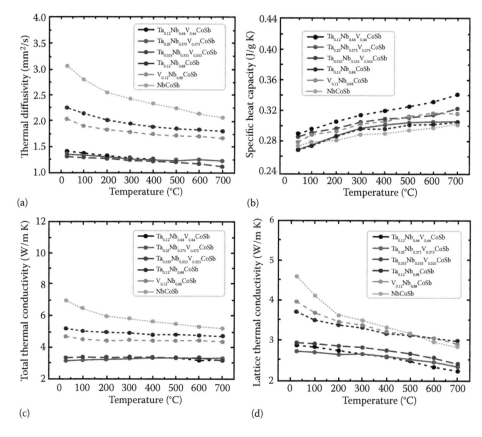

FIGURE 8.27 Temperature dependence of (a) thermal diffusivity, (b) *Cp*, (c) thermal conductivity, and (d) lattice thermal conductivity for HH compounds $Nb_{(1-x)/2}V_{(1-x)/2}Ta_xCoSb$ (x = 0.12, 0.25, and 0.33), $Nb_{0.88}Ta_{0.12}CoSb$, $Nb_{0.88}V_{0.12}CoSb$, and NbCoSb. (Huang, L. H. et al., *RSC Advances*, 5, 102469–102476, 2015. Reproduced by permission of The Royal Society of Chemistry.)

thermal conductivity (κ_L) of the doping samples comparing with NbCoSb. As shown in Figure 8.27c, the samples containing V and Ta show the lowest total thermal conductivity. It is clear that the room temperature κ of sample $Nb_{0.44}V_{0.44}Ta_{0.12}CoSb$ decreased by ~50% compared with that of NbCoSb. As shown in Figure 8.27d, the lattice thermal conductivity decreased a little bit below 500°C for samples partially substituting of Nb by just V or Ta, but the κ_L is largely reduced for samples containing both V and Ta in the whole measured temperature range, almost 40% reduction at room temperature for sample $Nb_{0.44}V_{0.44}Ta_{0.12}CoSb$ in comparison with NbCoSb. The large reduction of lattice thermal conductivity of samples containing both V and Ta is related to the significant phonon scattering effect owing to the additional mass difference and point defects caused by isoelectronic substitution, as shown in HRTEM spectrum (Figure 8.28d and f).

The measured electrical conductivity σ, Seebeck coefficient S, PF $S^2\sigma$, and ZT are shown in Figure 8.29 as a function of temperature. It is generally observed that cosubstitution by V and Ta produces lower electrical conductivity and results in higher Seebeck coefficient than the undoped ones. However,

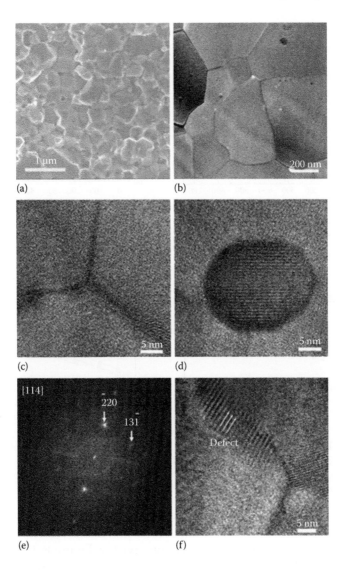

FIGURE 8.28 (a) Representative SEM image; (b) low-magnification TEM image; (c) HRTEM images showing triple grain boundary; (d) HRTEM images showing nanoinclusion inside the individual grain; (e) corresponding fast Fourier transform; (f) HRTEM images showing crystal defect of hot pressed sample $Nb_{0.44}V_{0.44}Ta_{0.12}CoSb$. (Huang, L. H. et al., *RSC Advances*, 5, 102469–102476, 2015. Reproduced by permission of The Royal Society of Chemistry.)

the V- and Ta-cosubstituted samples show lower PF (Figure 8.29c). Among all the cosubstituted samples, $Nb_{0.44}V_{0.44}Ta_{0.12}CoSb$ showed the highest PF of ~16.3 µW/cm K^2 at 700°C. Consequently, a maximum *ZT* value of ~0.5 was obtained at 700°C for samples $Nb_{0.44}V_{0.44}Ta_{0.12}CoSb$, an increase of ~25% compared with NbCoSb.

The SEM and TEM studies reveal that the hot pressed sample $Nb_{0.44}V_{0.44}Ta_{0.12}CoSb$ is a densely packed polycrystalline and consists of grains with sizes of around 200–600 nm (Figure 8.28a and b). Furthermore, some

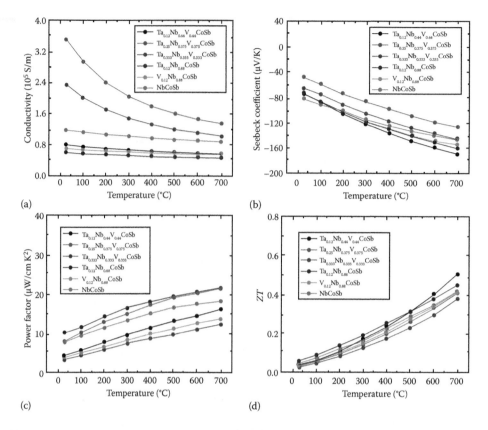

FIGURE 8.29 Temperature dependence of (a) electrical conductivity, (b) Seebeck coefficient, (c) PF, and (d) *ZT* value for HH compounds $Nb_{(1-x)/2}V_{(1-x)/2}Ta_xCoSb$ (x = 0.12, 0.25, and 0.33), $Nb_{0.88}Ta_{0.12}CoSb$, $Nb_{0.88}V_{0.12}CoSb$, and NbCoSb. (Huang, L. H. et al., *RSC Advances*, 5, 102469–102476, 2015. Reproduced by permission of The Royal Society of Chemistry.)

nano inclusions of 10–20 nm were observed in Figure 8.28b and identified in the HRTEM image in Figure 8.28d. Such image contrast are most likely due to the slight composition difference with the single crystalline grain matrix, similar to the Ge-doped Mg_2Sn[139] and Sn/Te-codoped $CoSb_3$.[129] These nanoinclusions are the active phonon-scattering centers. Figure 8.28f shows crystal defects embedded inside the grains along the grain boundary (indicated by white arrows), which are beneficial to lower the thermal conductivity.

Table 8.8 summarizes the transport properties of all the samples at room temperature. The carrier concentration is obviously increased when both V and Ta are used to partially substitute Nb in NbCoSb, implying that even isoelectronic substitution could significantly change the carrier concentration, which is probably due to the differences of the detailed electronic structures of these elements. Meanwhile, the mobility of the cosubstituted samples is largely decreased, which resulted from the enhanced alloy scattering caused by additional disorder.

TABLE 8.8 Resistivity, Carrier Concentration, Hall Mobility, and Effective Mass at Room Temperature of HH Compounds $Nb_{(1-x)/2}V_{(1-x)/2}Ta_xCoSb$ ($x = 0.12$, 0.25, and 0.33), $Nb_{0.88}Ta_{0.12}CoSb$, $Nb_{0.88}V_{0.12}CoSb$, and $NbCoSb$

Nominal Composition	Resistivity ($\times 10^{-5}$ Ω m)	Carrier Concentration ($\times 10^{21}$ cm^{-3})	Hall Mobility (cm^2/V s)
$Nb_{0.44}V_{0.44}Ta_{0.12}CoSb$	1.272	10.416	0.472
$Nb_{0.375}V_{0.375}Ta_{0.25}CoSb$	1.446	14.536	0.297
$Nb_{0.333}V_{0.333}Ta_{0.333}CoSb$	1.673	10.235	0.365
$Nb_{0.88}Ta_{0.12}CoSb$	0.427	4.775	3.065
$Nb_{0.88}V_{0.12}CoSb$	0.846	4.218	1.752
$NbCoSb$	0.285	6.113	3.582

Source: Huang, L. et al., *RSC Advances*, 5, 102469, 2015. Reproduced by permission of The Royal Society of Chemistry.

Compared with NbCoSb, $Nb_{0.44}V_{0.44}Ta_{0.12}CoSb$ has lower thermal conductivity, higher Seebeck coefficient and higher ZT. Therefore, it is clear that the TE properties of NbCoSb could be improved through proper composition tuning. It is necessary to point out that these samples still have a small amount of impurities according to the XRD results. These impurities are very detrimental to the individual TE properties because they affect the thermal and electrical conductivities in a detrimental way. We believe that it is highly possible to double the ZT when all the impurities are completely eliminated, which requires many composition tuning or even a totally different preparing method.

8.3.2 VCoSb Microstructure and Thermoelectric Properties of VCoSb

Another HH compound VCoSb with VEC = 19 was also studied.[31] The VCoSb samples were made by arc-melting and BM, followed by HP at 750, 800, and 900°C, respectively. The microstructures are presented in Figure 8.30. The SEM images show that the grain size has a wide distribution, from less than 100 nm to micrometers, and the higher the HPT, the denser the sample is. The TEM image of the sample hot pressed at 800°C demonstrates the clear grain boundaries and good crystallinity of individual grains. Fast Fourier transformation diffractogram of the selected area is presented in the inset in Figure 8.30e, which is indexed as the HH structure oriented along the [213] zone axis.

The TE properties of VCoSb with different HPTare compared to the arc-melted ingot. As shown in Figure 8.31a, the specific heat capacities (Cp) of all the four samples are almost the same. The Cp steadily increases from 0.32 J/g K at room temperature to 0.35 J/g K at 500°C, but above 600°C, Cp sharply increases from 0.36 J/g K at 600°C to 0.41 J/g K at 700°C. Figure 8.31b is the temperature-dependent thermal diffusivity of the samples. The thermal diffusivity decreases as the temperature

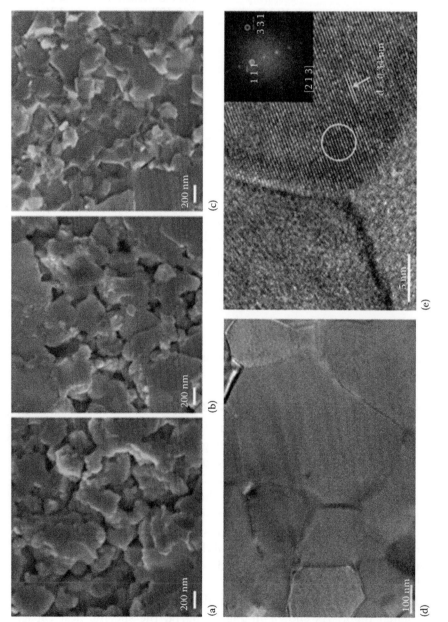

FIGURE 8.30 SEM images of the *VCoSb* samples hot pressed at (a) 750, (b) 800, and (c) 900°C; TEM images at (d) medium- and (e) high-resolution of the sample hot pressed at 800°C. The inset in e is the fast Fourier transformation diffractogram of the selected area. (Reprinted from *Journal of Alloys and Compounds*, 654, Zhang H. et al., 321–326, Copyright 2016, with permission Elsevier.)

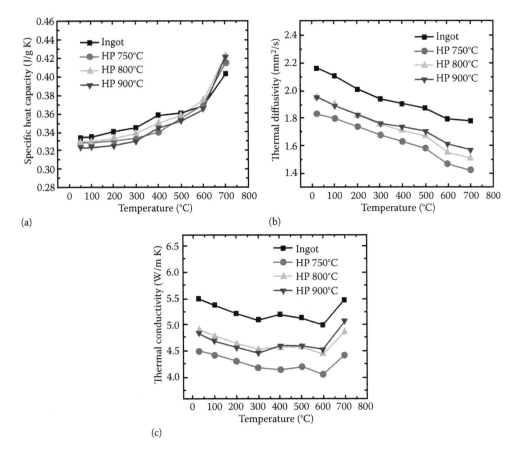

FIGURE 8.31 Temperature-dependent (a) C_p, (b) thermal diffusivity, (c) and thermal conductivity of ingot and three VCoSb samples hot pressed at 750, 800, and 900°C. (Reprinted from *Journal of Alloys and Compounds*, 654, Zhang H. et al., 321–326, Copyright 2016, with permission Elsevier.)

increases. Compared to the ingot sample, the thermal diffusivities of the nanostructured samples are much lower, which demonstrates that by the BM and HP processes, the grain size can be effectively reduced to increase the boundary scattering of phonons. The thermal conductivity is shown in Figure 8.31c. The thermal conductivity drops at 500°C and then rises after 600°C, owing to the decreas of the diffusivity and the sharp increas of C_p.

The temperature-dependent electrical conductivity, Seebeck coefficient, PF, and the calculated ZT are shown in Figure 8.32. The electrical conductivities of nanostructured samples are lower than those of the ingot because of the additional scattering of the charge carriers by the nanograin boundaries. Compared to the ingot, the Seebeck coefficient is also enhanced for the ball-milled and hot pressed samples. As a result, the PFs of the nanostructured samples have improved values than those of the ingot. The samples hot pressed at 800 and 900°C have similar peak values of around 25 μW/cm K² at 700°C. The ZT values are presented in Figure 8.32d. Due to the

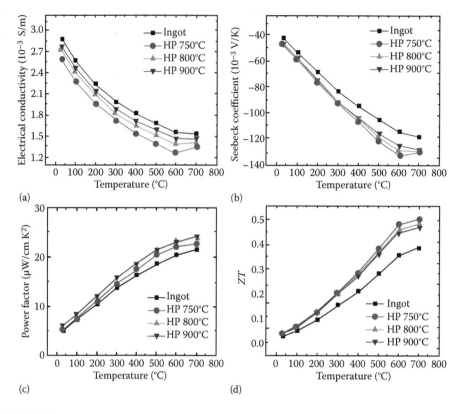

FIGURE 8.32 Temperature-dependent (a) electrical conductivity, (b) Seebeck coefficient, (c) PF, and (d) *ZT* of ingot and three VCoSb samples hot pressed at 750, 800, and 900°C. (Reprinted from *Journal of Alloys and Compounds*, 654, Zhang H. et al., 321–326, Copyright 2016, with permission Elsevier.)

reduced thermal conductivity and the improved PF, nearly 30% enhancement is noticed for the nanostructured samples compared to the ingot. The peak *ZT* about 0.5 is achieved at above 600°C for all the nanostructured VCoSb samples.

References

1. Schierning, G., Chavez, R., Schmechel, R., Balke, B., Rogl, G., and Rogl, P., Concepts for medium-high to high temperature thermoelectric heat-to-electricity conversion: A review of selected materials and basic considerations of module design. *Transl Mater Res* **2015**, *2* (2), 025001.
2. Casper, F., Graf, T., Chadov, S., Balke, B., and Felser, C., Half-Heusler compounds: Novel materials for energy and spintronic applications. *Semicond Sci Tech* **2012**, *27* (6), 063001.
3. Aliev, F. G., Brandt, N. B., Moshchalkov, V. V., Kozyrkov, V. V., Skolozdra, R. V., and Belogorokhov, A. I., Gap at the Fermi level in the ntermetallic vacancy system TiNiSn, ZrNiSn, HfNiSn. *Z Phys B Con Mat* **1989**, *75* (2), 167–171.

4. Aliev, F. G., Kozyrkov, V. V., Moshchalkov, V. V., Scolozdra, R. V., and Durczewski, K., Narrow-band in the intermetallic compounds TiNiSn, ZrNiSn, HfNiSn. *Z Phys B Con Mat* **1990**, *80* (3), 353–357.

5. Galanakis, I., and Mavropoulos, P., Spin-polarization and electronic properties of half-metallic Heusler alloys calculated from first principles. *J Phys-Condens Mat* **2007**, *19* (31), 486002.

6. Galanakis, I., Dederichs, P. H., and Papanikolaou, N., Origin and properties of the gap in the half-ferromagnetic Heusler alloys. *Phys Rev B* **2002**, *66* (13), 134428.

7. Yang, J., Li, H. M., Wu, T., Zhang, W. Q., Chen, L. D., and Yang, J. H., Evaluation of half-Heusler compounds as thermoelectric materials based on the calculated electrical transport properties. *Adv Funct Mater* **2008**, *18* (19), 2880–2888.

8. Yan, X. et al., Enhanced thermoelectric figure of merit of p-type half-Heuslers. *Nano Lett* **2011**, *11* (2), 556–560.

9. Simonson, J. W., and Poon, S. J., Electronic structure of transition metal-doped XNiSn and XCoSb (X = Hf, Zr) phases in the vicinity of the band gap. *J Phys-Condens Mat* **2008**, *20* (25), 255220.

10. Colinet, C., Jund, P., and Tedenac, J. C., NiTiSn a material of technological interest: Ab initio calculations of phase stability and defects. *Intermetallics* **2014**, *46*, 103–110.

11. Ogut, S., and Rabe, K. M., Band-gap and stability in the ternary intermetallic compounds NiSnM (M=Ti,Zr,Hf)—A first-principles study. *Phys Rev B* **1995**, *51* (16), 10443–10453.

12. Poon, S. J., Electronic and thermoelectric properties of half-Heusler alloys. *Semiconduct Semimet* **2001**, *70*, 37–75.

13. Hohl, H., Ramirez, A. P., Goldmann, C., Ernst, G., Wolfing, B., and Bucher, E., Efficient dopants for ZrNiSn-based thermoelectric materials. *J Phys-Condens Mat* **1999**, *11* (7), 1697–1709.

14. Uher, C., Yang, J., Hu, S., Morelli, D. T., and Meisner, G. P., Transport properties of pure and doped MNiSn (M=Zr, Hf). *Phys Rev B* **1999**, *59* (13), 8615–8621.

15. Xia, Y., Bhattacharya, S., Ponnambalam, V., Pope, A. L., Poon, S. J., and Tritt, T. M., Thermoelectric properties of semimetallic (Zr, Hf)CoSb half-Heusler phases. *J Appl Phys* **2000**, *88* (4), 1952–1955.

16. Sekimoto, T., Kurosaki, K., Muta, H., and Yamanaka, S., Thermoelectric and thermophysical properties of TiCoSb-ZrCoSb-HfCoSb pseudo ternary system prepared by spark plasma sintering. *Mater Trans* **2006**, *47* (6), 1445–1448.

17. Chen, S., and Ren, Z., Recent progress of half-Heusler for moderate temperature thermoelectric applications. *Materials Today* **2013**, *16* (10), 387–395.

18. Zhu, T. J., Fu, C. G., Xie, H. H., Liu, Y. T., and Zhao, X. B., High efficiency half-Heusler thermoelectric materials for energy harvesting. *Adv Energy Mater* **2015**, *5* (19), 1500588.

19. Xie, W. J., Weidenkaff, A., Tang, X. F., Zhang, Q. J., Poon, J., and Tritt, T. M., Recent advances in nanostructured thermoelectric half-Heusler compounds. *Nanomaterials (Basel)* **2012**, *2* (4), 379–412.

20. Huang, L. H., Zhang, Q. Y., Yuan, B., Lai, X., Yan, X., and Ren, Z. F., Recent progress in half-Heusler thermoelectric materials. *Mater Res Bull* **2016**, *76*, 107–112.

21. Bos, J.-W. G., and Downie, R. A., Half-Heusler thermoelectrics: A complex class of materials. *J Phys: Condens Matter* **2014**, *26* (43), 433201.

22. Zhou, M., Feng, C. D., Chen, L. D., and Huang, X. Y., Effects of partial substitution of Co by Ni on the high-temperature thermoelectric properties of TiCoSb-based half-Heusler compounds. *J Alloy Compd* **2005**, *391* (1–2), 194–197.

23. Chen, L. D., Huang, X. Y., Zhou, M., Shi, X., and Zhang, W. B., The high temperature thermoelectric performances of $Zr_{0.5}Hf_{0.5}Ni_{0.8}Pd_{0.2}Sn_{0.99}Sb_{0.01}$ alloy with nanophase inclusions. *J Appl Phys* **2006**, *99* (6), 064305.

24. Ponnambalam, V., Zhang, B., Tritt, T. M., and Poon, S. J., Thermoelectric properties of half-Heusler bismuthides $ZrCo_{1-x}Ni_xBi$ (x=0.0 to 0.1). *J Electron Mater* **2007**, *36* (7), 732–735.

25. Makongo, J. P. A. et al., Thermal and electronic charge transport in bulk nanostructured $Zr_{0.25}Hf_{0.75}NiSn$ composites with full-Heusler inclusions. *J Solid State Chem* **2011**, *184* (11), 2948–2960.

26. Makongo, J. P. A. et al., Simultaneous large enhancements in thermopower and electrical conductivity of bulk nanostructured half-Heusler alloys. *J Am Chem Soc* **2011**, *133* (46), 18843–18852.

27. Joshi, G., Dahal, T., Chen, S., Wang, H. Z., Shiomi, J., Chen, G., and Ren, Z. F., Enhancement of thermoelectric figure-of-merit at low temperatures by titanium substitution for hafnium in n-type half-Heuslers $Hf_{0.75-x}Ti_xZr_{0.25}NiSn_{0.99}Sb_{0.01}$. *Nano Energy* **2013**, *2* (1), 82–87.

28. Chen, S., Lukas, K. C., Liu, W. S., Opeil, C. P., Chen, G., and Ren, Z. F., Effect of Hf concentration on thermoelectric properties of nanostructured n-type half-Heusler materials $Hf_xZr_{1-x}NiSn_{0.99}Sb_{0.01}$. *Adv Energy Mater* **2013**, *3* (9), 1210–1214.

29. Yan, X. et al., Stronger phonon scattering by larger differences in atomic mass and size in p-type half-Heuslers $Hf_{1-x}Ti_xCoSb_{0.8}Sn_{0.2}$. *Energ Environ Sci* **2012**, *5* (6), 7543–7548.

30. Zhang, H., Wang, Y. M., Dahal, K., Mao, J., Huang, L. H., Zhang, Q. Y., and Ren, Z. F., Thermoelectric properties of n-type half-Heusler compounds $(Hf_{0.25}Zr_{0.75})_{1-x}Nb_xNiSn$. *Acta Mater* **2016**, *113*, 41–47.

31. Zhang, H., Wang, Y. M., Huang, L. H., Chen, S., Dahal, H., Wang, D. Z., and Ren, Z. F., Synthesis and thermoelectric properties of n-type half-Heusler compound VCoSb with valence electron count of 19. *J Alloy Compd* **2016**, *654*, 321–326.

32. Lee, P. J., and Chao, L. S., High-temperature thermoelectric properties of $Ti_{0.5}(ZrHf)_{0.5-x}Nb_xNi_{0.9}Pd_{0.1}Sn_{0.98}Sb_{0.02}$ half-Heusler alloys. *J Alloy Compd* **2010**, *504* (1), 192–196.

33. Gelbstein, Y. et al., Thermoelectric properties of spark plasma sintered composites based on TiNiSn half-Heusler alloys. *J Mater Res* **2011**, *26* (15), 1919–1924.

34. Appel, O., Schwall, M., Mogilyansky, D., Kohne, M., Balke, B., and Gelbstein, Y., Effects of Microstructural evolution on the thermoelectric properties of spark-plasma-sintered $Ti_{0.3}Zr_{0.35}Hf_{0.35}NiSn$ half-Heusler compound. *J Electron Mater* **2013**, *42* (7), 1340–1345.

35. Ouardi, S. et al., Transport and thermal properties of single- and polycrystalline $NiZr_{0.5}Hf_{0.5}Sn$. *Appl Phys Lett* **2011**, *99* (15), 152112.

36. Kimura, Y., Kuji, T., Zama, A., Shibata, Y., and Mishima, Y., High-performance of half-Heusler MNiSn (M=Hf,Zr) single-phase thermoelectric alloys fabricated using optical floating zone melting. *Mater Res Soc Symp P* **2006**, *886*, 331–336.

37. Kimura, Y., Ueno, H., and Mishima, Y., Thermoelectric properties of directionally solidified half-Heusler $\left(M_{0.5}^{a}, M_{0.5}^{b}\right)$ NiSn (M^a, M^b = Hf, Zr, Ti) alloys. *J Electron Mater* **2009**, *38* (7), 934–939.

38. Kimura, Y., Tanoguchi, T., and Kita, T., Vacancy site occupation by Co and Ir in half-Heusler ZrNiSn and conversion of the thermoelectric properties from n-type to p-type. *Acta Mater* **2010**, *58* (13), 4354–4361.

39. Xie, H. H. et al., Beneficial contribution of alloy disorder to electron and phonon transport in half-Heusler thermoelectric materials. *Adv Funct Mater* **2013**, *23* (41), 5123–5130.

40. Xie, W. J., Jin, Q., and Tang, X. F., The preparation and thermoelectric properties of $Ti_{0.5}Zr_{0.25}Hf_{0.25}Co_{1-x}Ni_xSb$ half-Heusler compounds. *J Appl Phys* **2008**, *103* (4), 043711.

41. Yu, C., Zhu, T.-J., Shi, R.-Z., Zhang, Y., Zhao, X.-B., and He, J., High-performance half-Heusler thermoelectric materials $Hf_{1-x}Zr_xNiSn_{1-y}Sb_y$ prepared by levitation melting and spark plasma sintering. *Acta Mater* **2009**, *57* (9), 2757–2764.

42. Yu, C., Zhu, T. J., Xiao, K., Shen, J. J., and Zhao, X. B., Microstructure and thermoelectric properties of (Zr,Hf)NiSn-based half-Heusler alloys by melt spinning and spark plasma sintering. *Funct Mater Lett* **2010**, *3* (4), 227–231.

43. Zhu, T. J. et al., Effects of yttrium doping on the thermoelectric properties of $Hf_{0.6}Zr_{0.4}NiSn_{0.98}Sb_{0.02}$ half-Heusler alloys. *J Appl Phys* **2010**, *108* (4), 044903.

44. Xie, H.-H., Yu, C., Zhu, T.-J., Fu, C.-G., Snyder, G. J., and Zhao, X.-B., Increased electrical conductivity in fine-grained (Zr,Hf)NiSn based thermoelectric materials with nanoscale precipitates. *Appl Phys Lett* **2012**, *100* (25), 254104.

45. Birkel, C. S. et al., Rapid microwave preparation of thermoelectric TiNiSn and TiCoSb half-Heusler compounds. *Chem Mater* **2012**, *24* (13), 2558–2565.

46. Birkel, C. S. et al., Improving the thermoelectric properties of half-Heusler TiNiSn through inclusion of a second full-Heusler phase: Microwave preparation and spark plasma sintering of $TiNi_{1+x}Sn$. *Phys Chem Chem Phys* **2013**, *15* (18), 6990–6997.

47. Qiu, P., Yang, J., Huang, X., Chen, X., and Chen, L., Effect of antisite defects on band structure and thermoelectric performance of ZrNiSn half-Heusler alloys. *Appl Phys Lett* **2010**, *96* (15), 152105.

48. Schwall, M., and Balke, B., Phase separation as a key to a thermoelectric high efficiency. *Phys Chem Chem Phys* **2013**, *15* (6), 1868–1872.

49. Gurth, M., Rogl, G., Romaka, V. V., Grytsiv, A., Bauer, E., Rogl, P., Thermoelectric high ZT half-Heusler alloys $Ti_{1-x-y}Zr_xHf_yNiSn$ (0 ≤ x ≤ 1; 0 ≤ y ≤ 1). *Acta Mater* **2016**, *104*, 210–222.

50. Chen, L., Zeng, X. Y., Tritt, T. M., and Poon, S. J., Half-Heusler alloys for efficient thermoelectric power conversion. *J Electron Mater* **2016**, *45* (11), 5554–5560.

51. Vapor Pressure Calculator. https://www.iap.tuwien.ac.at/www/surface/vapor_pressure.

52. Zou, M. M., Li, J. F., Du, B., Liu, D. W., and Kita, T., Fabrication and thermoelectric properties of fine-grained TiNiSn compounds. *J Solid State Chem* **2009**, *182* (11), 3138–3142.

53. Culp, S. R., Simonson, J. W., Poon, S. J., Ponnambalam, V., Edwards, J., and Tritt, T. M., (Zr,Hf)Co(Sb,Sn) half-Heusler phases as high-temperature (> 700° C) p-type thermoelectric materials. *Appl Phys Lett* **2008**, *93* (2), 022105.

54. Xia, Y., Ponnambalam, V., Bhattacharya, S., Pope, A. L., Poon, S. J., and Tritt, T. M., Electrical transport properties of TiCoSb half-Heusler phases that exhibit high resistivity. *J Phys-Condens Mat* **2001**, *13* (1), 77–89.

55. Page, A., Van der Ven, A., Poudeu, P. F. P., and Uher, C., Origins of phase separation in thermoelectric (Ti,Zr,Hf) NiSn half-Heusler alloys from first principles. *J Mater Chem A* **2016**, *4* (36), 13949–13956.

56. Katayama, T., Kim, S. W., Kimura, Y., and Mishima, Y., The effects of quaternary additions on thermoelectric properties of TiNiSn-based half-Heusler alloys. *J Electron Mater* **2003**, *32* (11), 1160–1165.

57. Zhou, M., Chen, L., Feng, C., Wang, D., and Li, J.-F., Moderate-temperature thermoelectric properties of TiCoSb-based half-Heusler compounds $Ti_{1-x}Ta_x$ CoSb. *J Appl Phys* **2007**, *101* (11), 113714.

58. Shen, Q., Chen, L., Goto, T., Hirai, T., Yang, J., Meisner, G. P., and Uher, C., Effects of partial substitution of Ni by Pd on the thermoelectric properties of ZrNiSn-based half-Heusler compounds. *Appl Phys Lett* **2001**, *79* (25), 4165–4167.

59. Huang, X. Y., Xu, Z., and Chen, L. D., The thermoelectric performance of $ZrNiSnVZrO_2$ composites. *Solid State Commun* **2004**, *130* (3–4), 181–185.

60. Kawaharada, Y., and Kurosaki, K., High temperature thermoelectric properties of CoTiSb half-Heusler compounds. *J Alloy Compd* **2004**, *384* (1–2), 308–311.

61. Maji, P., Takas, N. J., Misra, D. K., Gabrisch, H., Stokes, K., and Poudeu, P. F. P., Effects of Rh on the thermoelectric performance of the p-type $Zr_{0.5}Hf_{0.5}Co_{1-x}Rh_xSb_{0.99}Sn_{0.01}$ half-Heusler alloys. *J Solid State Chem* **2010**, *183* (5), 1120–1126.

62. Joshi, G., Yan, X., Wang, H., Liu, W., Chen, G., and Ren, Z., Enhancement in thermoelectric figure-of-merit of an n-type half-Heusler compound by the nanocomposite approach. *Adv Energy Mater* **2011**, *1* (4), 643–647.

63. Xie, W. J. et al., Simultaneously optimizing the independent thermoelectric properties in (Ti,Zr,Hf)(Co,Ni)Sb alloy by in situ forming InSb nanoinclusions. *Acta Mater* **2010**, *58* (14), 4705–4713.

64. Liu, Y., Page, A., Sahoo, P., Chi, H., Uher, C., and Poudeu, P. F., Electronic and phonon transport in Sb-doped $Ti_{0.1}Zr_{0.9}Ni_{1+x}Sn_{0.975}Sb_{0.025}$ nanocomposites. *Dalton Trans* **2014**, *43* (21), 8094–8101.

65. Schmitt, J., Gibbs, Z. M., Snyder, G. J., and Felser, C., Resolving the true band gap of ZrNiSn half-Heusler thermoelectric materials. *Mater Horizons* **2015**, *2* (1), 68–75.

66. Miyamoto, K. et al., In-gap electronic states responsible for the excellent thermoelectric properties of Ni-based half-Heusler alloys. *Appl Phys. Express* **2008**, *1* (8), 081901.

67. Xie, H., Wang, H., Fu, C., Liu, Y., Snyder, G. J., Zhao, X., and Zhu, T., The intrinsic disorder related alloy scattering in ZrNiSn half-Heusler thermoelectric materials. *Sci Rep* **2014**, *4*, 6888.

68. Snyder, G. J., and Toberer, E. S., Complex thermoelectric materials. *Nat Mater* **2008**, *7* (2), 105–114.
69. Graf, T., Felser, C., and Parkin, S. S. P., Simple rules for the understanding of Heusler compounds. *Prog Solid State Chem* **2011**, *39* (1), 1–50.
70. Larson, P., Mahanti, S. D., Sportouch, S., and Kanatzidis, M. G., Electronic structure of rare-earth nickel pnictides: Narrow-gap thermoelectric materials. *Phys Rev B* **1999**, *59* (24), 15660–15668.
71. Lee, M.-S., and Mahanti, S. D., Validity of the rigid band approximation in the study of the thermopower of narrow band gap semiconductors. *Phys Rev B* **2012**, *85* (16), 165149.
72. Chaput, L., Tobola, J., Pecheur, P., and Scherrer, H., Electronic structure and thermopower of $Ni(Ti_{0.5}Hf_{0.5})Sn$ and related half-Heusler phases. *Phys Rev B* **2006**, *73* (4), 045121.
73. Culp, S. R., Poon, S. J., Hickman, N., Tritt, T. M., and Blumm, J., Effect of substitutions on the thermoelectric figure of merit of half-Heusler phases at 800°C. *Appl Phys Lett* **2006**, *88* (4), 042106.
74. Li, X.-G., Huo, D.-X., He, C.-J., Zhao, S.-C., and Lue, Y.-F., Effect of rare-earth doping on the thermoelectric properties of the tin-based half-Heusler alloys. *J Inorg Mater* **2010**, *25* (6), 573–576.
75. Katsuyama, S., Matsushima, H., and Ito, M., Effect of substitution for Ni by Co and/or Cu on the thermoelectric properties of half-Heusler ZrNiSn. *J Alloy Compd* **2004**, *385* (1–2), 232–237.
76. Wu, T., Jiang, W., Li, X., Zhou, Y., and Chen, L., Thermoelectric properties of p-type Fe-doped TiCoSb half-Heusler compounds. *J Appl Phys* **2007**, *102* (10), 103705.
77. Wu, T., Jiang, W., Li, X., Bai, S., Liufu, S., and Chen, L., Effects of Ge doping on the thermoelectric properties of TiCoSb-based p-type half-Heusler compounds. *J Alloy Compd* **2009**, *467* (1–2), 590–594.
78. Yan, X., Liu, W., Chen, S., Wang, H., Zhang, Q., Chen, G., and Ren, Z., Thermoelectric property study of nanostructured p-type half-Heuslers (Hf, Zr, Ti)$CoSb_{0.8}Sn_{0.2}$. *Adv Energy Mater* **2013**, *3* (9), 1195–1200.
79. Sekimoto, T., Kurosaki, K., Muta, H., and Yamanaka, S., High-thermoelectric figure of merit realized in p-type half-Heusler compounds: $ZrCoSn_xSb_{1-x}$. *Jpn J Appl Phys Part 2—Lett Express Lett* **2007**, *46* (25–28), L673–L675.
80. Ponnambalam, V., Alboni, P. N., Edwards, J., Tritt, T. M., Culp, S. R., and Poon, S. J., Thermoelectric properties of p-type half-Heusler alloys $Zr_{1-x}Ti_xCoSn_ySb_{1-y}$ (0.0 < x < 0.5, y = 0.15 and 0.3). *J Appl Phys* **2008**, *103* (6), 063716.
81. Bhardwaj, A., and Misra, D. K., Improving the thermoelectric performance of TiNiSn half-Heusler via incorporating submicron lamellae eutectic phase of $Ti_{70.5}Fe_{29.5}$: A new strategy for enhancing the power factor and reducing the thermal conductivity. *J Mater Chem A* **2014**, *2* (48), 20980–20989.
82. Poudel, B. et al., High-thermoelectric performance of nanostructured bismuth antimony telluride bulk alloys. *Science* **2008**, *320* (5876), 634–638.
83. Heremans, J. P. et al., Enhancement of thermoelectric efficiency in PbTe by distortion of the electronic density of states. *Science* **2008**, *321* (5888), 554–557.
84. Kim, S. W., Kimura, Y., and Mishima, Y., High temperature thermoelectric properties of TiNiSn-based half-Heusler compounds. *Intermetallics* **2007**, *15* (3), 349–356.

85. Kawaharada, Y., Uneda, H., Muta, H., Kurosaki, K., and Yamanaka, S., High temperature thermoelectric properties of NiZrSn half-Heusler compounds. *J Alloy Compd* **2004**, *364* (1–2), 59–63.

86. Zhao, D. G., Zuo, M., Wang, Z. Q., Teng, X. Y., and Geng, H. R., Synthesis and thermoelectric properties of tantalum-doped ZrNiSn half-Heusler alloys. *Funct Mater Lett* **2014**, *7* (3), 1450032.

87. Chen, L., Gao, S., Zeng, X., Dehkordi, A. M., Tritt, T. M., and Poon, S. J., Uncovering high thermoelectric figure of merit in (Hf,Zr)NiSn half-Heusler alloys. *Appl Phys Lett* **2015**, *107* (4), 041902.

88. Krez, J., Schmitt, J., Snyder, G. J., Felser, C., Hermes, W., and Schwind, M., Optimization of the carrier concentration in phase-separated half-Heusler compounds. *J Mater Chem A* **2014**, *2* (33), 13513–13518.

89. Gałązka, K., Populoh, S., Xie, W., Yoon, S., Saucke, G., Hulliger, J., and Weidenkaff, A., Improved thermoelectric performance of $(Zr_{0.3}Hf_{0.7})NiSn$ half-Heusler compounds by Ta substitution. *J Appl Phys* **2014**, *115* (18), 183704.

90. Appel, O., and Gelbstein, Y., A Comparison between the effects of Sb and Bi doping on the thermoelectric properties of the $Ti_{0.3}Zr_{0.35}Hf_{0.35}NiSn$ half-Heusler Alloy. *J Electron Mater* **2014**, *43* (6), 1976–1982.

91. Simonson, J., Wu, D., Xie, W., Tritt, T. M., and Poon, S., Introduction of resonant states and enhancement of thermoelectric properties in half-Heusler alloys. *Phys Rev B* **2011**, *83* (23), 235211.

92. Bhardwaj, A., Misra, D., Pulikkotil, J., Auluck, S., Dhar, A., and Budhani, R., Implications of nanostructuring on the thermoelectric properties in half-Heusler alloys. *Appl Phys Lett* **2012**, *101* (13), 133103.

93. Zhang, Q. et al., Heavy doping and band engineering by potassium to improve the thermoelectric figure of merit in p-type PbTe, PbSe, and $PbTe_{1-y}Se_y$. *J Am Chem Soc* **2012**, *134* (24), 10031–10038.

94. Bhattacharya, S. et al., Effect of boundary scattering on the thermal conductivity of TiNiSn-based half-Heusler alloys. *Phys Rev B* **2008**, *77* (18), 184203.

95. Bhattacharya, S., Tritt, T. M., Xia, Y., Ponnambalam, V., Poon, S., and Thadhani, N., Grain structure effects on the lattice thermal conductivity of Ti-based half-Heusler alloys. *Appl Phys Lett* **2002**, *81* (1), 43.

96. Hsu, C.-C., Liu, Y.-N., and Ma, H.-K., Effect of the $Zr_{0.5}Hf_{0.5}CoSb_{1-x}Sn_x$/ HfO_2 half-Heusler nanocomposites on the *ZT* value. *J Alloy Compd* **2014**, *597*, 217–222.

97. Poon, S. J., Wu, D., Zhu, S., Xie, W., Tritt, T. M., Thomas, P., and Venkatasubramanian, R., Half-Heusler phases and nanocomposites as emerging high-*ZT* thermoelectric materials. *J Mater Res* **2011**, *26* (22), 2795–2802.

98. Huang, X. Y., Xu, Z., Chen, L. D., and Tang, X. F. Effect of γ-Al_2O_3 content on the thermoelectric performance of ZrNiSn/γ-Al_2O_3 composites In *Key Engineering Materials*, Trans Tech Publications, Zurich: 2003, pp 79–82.

99. Chai, Y. W., and Kimura, Y., Microstructure evolution of nanoprecipitates in half-Heusler TiNiSn alloys. *Acta Mater* **2013**, *61* (18), 6684–6697.

100. Ioffe, F., *Semiconductor Thermoelements and Thermoelectric Cooling*, Infosearch, Chennai: 1957.

101. Fu, C. G., Bai, S. Q., Liu, Y. T., Tang, Y. S., Chen, L. D., Zhao, X. B. and Zhu, T. J., Realizing high figure of merit in heavy-band p-type half-Heusler thermoelectric materials. *Nat Commun* **2015**, *6*, 1, 8144.

102. Sekimoto, T. et al., Thermoelectric properties of Sn-doped TiCoSb half-Heusler compounds. *J Alloys Compd* **2006**, *407*, 326–329.

103. He, R. et al., Investigating the thermoelectric properties of p-type half-Heusler $Hf_x(ZrTi)_{1-x}CoSb_{0.8}Sn_{0.2}$ by reducing Hf concentration for power generation. *RSC Adv* **2014**, *4*, 64711–64716.

104. Fu, C. et al., Band engineering of high performance p-type FeNbSb based half-Heusler thermoelectric materials for figure of merit $zT > 1$. *Energy Environ Sci* **2015**, *8*, 216–220.

105. Joshi, G. et al., NbFeSb-based p-type half-Heusler for power generation applications. *Energy Environ Sci* **2014**, *7*, 4070–4076.

106. He, R. et al., Studies on mechanical properties of thermoelectric materials by nanoindentation. *Phys Status Solidi* **2015**, *212*, 2191–2195.

107. He, R., Kraemer, D. Mao, J. Zeng, L. Jie, Q. Lan, Y. Li, C. Shuai, J. Kim, H. S. Liu, Y. Broido, D. Chu, C.-W. Chen G. and Ren Z., Achieving high power factor and output power density in p-type half-Heuslers $Nb_{1-x}Ti_xFeSb$. *Proceedings of the National Academy of Sciences*, **2016**, 113, 13576–13581.

108. US Geological Survey. *Metal Prices in the United States through 2010*. US Geological Survey Publications Warehouse, Reston, VA: 2012, pp. 1–204.

109. Rausch, E. et al., Fine tuning of thermoelectric performance in phase-separated half-Heusler compounds. *J Mater Chem C* **2015**, *3*, 10409–10414.

110. Rausch, E. et al., Long-term stability of $(Ti/Zr/Hf)CoSb_{1-x}Sn_x$ thermoelectric p-type half-Heusler compounds upon thermal cycling. *Energy Technol* **2015**, *3*, 1217–1224.

111. Rausch, E. et al. Charge carrier concentration optimization of thermoelectric p-type half-Heusler compounds. *APL Mater* **2015**, *3*, 041516.

112. Rausch, E. et al., Enhanced thermoelectric performance in the p-type half-Heusler $(Ti/Zr/Hf)CoSb_{0.8}Sn_{0.2}$ system via phase separation. *Phys Chem Chem Phys* **2014**, *16*, 25258–25262.

113. Xie, W. J. et al., Significant *ZT* enhancement in p-type Ti(Co,Fe)Sb-InSb nanocomposites via a synergistic high-mobility electron injection, energy-filtering and boundary-scattering approach. *Acta Mater* **2013**, *61*, 2087–2094.

114. Hrostowski, H. J. et al., Hall effect and conductivity of InSb. *Phys Rev* **1955**, *100*, 1672–1676.

115. Breckenridge, R. G. et al., Electrical and optical properties of intermetallic compounds: I. Indium antimonide. *Phys Rev* **1954**, *96*, 571–575.

116. Faleev, S. V., and Léonard, F., Theory of enhancement of thermoelectric properties of materials with nanoinclusions. *Phys Rev B* **2008**, *77*, 214304.

117. Heremans, J. P. et al., Thermopower enhancement in PbTe with Pb precipitates. *J Appl Phys* **2005**, *98*, 063703.

118. Sahoo, P. et al., Enhancing thermopower and hole mobility in bulk p-type half-Heuslers using full-Heusler nanostructures. *Nanoscale* **2013**, *5*, 9419–9427.

119. Chauhan, N. S. et al., A synergistic combination of atomic scale structural engineering and panoscopic approach in p-type ZrCoSb-based half-Heusler thermoelectric materials for achieving high *ZT*. *J Mater Chem C* **2016**, *4*, 5766–5778.

120. Liu, W. S., Yan, X., Chen, G., and Ren, Z. F., Recent advances in thermoelectric nanocomposites. *Nano Energy* **2012**, *1*, 42–56.

121. Ono, Y., Inayama, S., Adachi, H., Kajitani, T., Thermoelectric properties of doped half-Heuslers $NbCoSn_{1-x}Sb_x$ and $Nb_{0.99}Ti_{0.01}CoSn_{1-x}Sb_x$. *Jpn J Appl Phys* **2006**, *45*, 8740–8743.

122. Kouacou, M. A., Koua, A. A., Zoueu, J. T., Konan, K., and J. Pierre, Onset of itinerant ferromagnetism associated with semiconductor-metal transition in $Ti_xNb_{1-x}CoSn$ half-Heusler solid solution compounds. *Pramana—J Phys* **2008**, *71*, 157–166.

123. Zou, M. M., Li, J. F., Guo, P. J., and Kita, T., Thermoelectric properties of fine-grained FeVSb half-Heusler alloys tuned to p-type by substituting vanadium with titanium. *J Solid State Chem* **2013**, *198*, 125–130.

124. Fu, C. G., Xie, H. H., Liu, Y. T., Zhu, T. J., Xie, J., Zhao, X. B., Thermoelectric properties of FeVSb half-Heusler compounds by levitation melting and spark plasma sintering. *Intermetallics* **2013**, *32*, 39–43.

125. Tobola, J., and Pierre, J., Electronic phase diagram of the XTZ (X= Fe, Co, Ni; T= Ti, V, Zr, Nb, Mn; Z= Sn, Sb) semi-Heusler compounds. *J Alloys Compd* **2000**, *296*, 243–252.

126. Evers, C. B. H., Richter, C. G., Hartjes, K., and Jeitschko, W., Ternary transition metal antimonides and bismuthides with MgAgAs-type and filled NiAs-type structure. *J Alloys Compd* **1997**, *252*, 93–97.

127. Melnyk, G., Bauer, E., Rogl, P., Skolozdra, R., and Seidl, E., Thermoelectric properties of ternary transition metal antimonides. *J Alloy Compd* **2000**, *296*, 235–242.

128. Sekimoto, T., Kurosaki, K., Muta, H. and Yamanaka, S., Thermoelectric properties of (Ti, Zr, Hf) CoSb type half-heusler compounds. *Mater Trans* **2005**, *46*, 1481–1484.

129. Ashcroft, N. W., and Mermin, D. N., *Solid State Physics*, Brooks Cole, Stamford, CT: 1976.

130. Huang, L. et al., A new n-type half-Heusler thermoelectric material NbCoSb. *Mater Res Bull* **2015**, *70*, 773.

131. Rowe, D. M., and Bhandari, C. M., *Modern Thermoelectric*. Reston, Reston, VA: 1983.

132. Liu, W. S. et al., Thermoelectric property studies on Cu-doped n-type $Cu_xBi_2Te_{2.7}Se_{0.3}$ Nanocomposites. *Adv Energy Mater* **2011**, *1*, 577–587.

133. Yu, C., Xie, H., Fu, C., Zhu, T., and Zhao, X., High performance half-Heusler thermoelectric materials with refined grains and nanoscale precipitates. *J Mater Res* **2012**, *27*, 2457–2465.

134. He, J. Q. et al., On the origin of increased phonon scattering in nanostructured PbTe based thermoelectric materials. *J Am Chem Soc* **2010**, *132*, 8669–8675.

135. Huang, L., Wang, Y., Shuai, J., Zhang, H., Yang, S., Zhang, Q., and Ren, Z., Thermal conductivity reduction by isoelectronic elements V and Ta for partial substitution of Nb in half-Heusler $Nb_{1-x/2}V_{1-x/2}Ta_xCoSb$. *RSC Adv* **2015**, *5*, 102469.

136. Liu, W. S. et al., New n-type thermoelectric material $Mg_2Sn_{0.75}Ge_{0.25}$ for high power generation. *Proc Nat Acad Sci* **2015**, *112*, 3269.

137. Geng, H. Y., and Zhang, H., Effects of phase separation on the thermoelectric properties of (Ti,Zr,Hf) NiSn half-Heusler alloys. *J Appl Phys* **2014**, *116*, 033708.

138. Liu, Y. F., and Poudeu, P. F. P., Thermoelectric properties of Ge doped n-type $Ti_xZr_{1-x}NiSn_{0.975}Ge_{0.025}$ half-Heusler alloys. *J Mater Chem A* **2015**, *3*, 12507.

139. Liu, W. S., Zhang, B. P., Zhao, L. D., and Li, J. F., Improvement of thermoelectric performance of $CoSb_{3-x}te_x$ skutterudite compounds by additional substitution of IVB-group elements for Sb. *Chem Mater* **2008**, *20*, 7526.

CHAPTER 9

Silicon–Germanium Alloys

Yucheng Lan, Dezhi Wang, and Zhifeng Ren

Contents

9.1 Introduction

Silicon is the second most abundant element in the earth's crust, and it has been widely used in semiconductor devices. The thermal conductivity of single-crystalline silicon is 148 W/m K at 300 K [1]. Heavily doped crystalline silicon possesses most physical properties that identify a good candidate for thermoelectric (TE) power generation applications, such as high melting point, high thermal stability, excellent mechanical properties, high Seebeck coefficient, and high electrical conductivity. However, silicon does not offer a high figure-of-merit (ZT) value due to its high thermal conductivity.

In the late 1950s and early 1960s, it was discovered [2–5] that the high thermal conductivity of silicon could be significantly reduced through Si–Ge *alloying* due to point defects (i.e., mass fluctuation between Si and Ge atoms). The addition of germanium atoms into silicon matrices can effectively reduce the lattice thermal conductivity. Experimental data indicated that [3,5–9] the thermal conductivity of $Si_{1-x}Ge_x$ alloys decreased sharply with increasing Ge concentration when $0 < x < 0.2$ and was almost flat when $0.2 < x < 0.8$. The thermal conductivity of $Si_{1-x}Ge_x$ alloys was typically less than 10 W/m K when x was between 0.2 and 0.8. Thereafter, SiGe alloys have been extensively investigated for TE cooling and power generation applications.

Polycrystalline SiGe TE bulks were first produced by zone leveling in the 1960s [5,10], and later by hot-press (HP) methods since the 1970s [11–16] to shorten manufacture procedures. The pressure-sintered polycrystalline SiGe bulks consisted of microsized particles, and their figure of merits approached that of the zone-leveled grown single-crystalline silicon TE materials [17]. The ZT of SiGe polycrystalline alloys was significantly improved through alloying, with the peak ZT of n-type $Si_{80}Ge_{20}$ reaching about 1 at 900–950°C [3,5,15,16,18–20] and p-type reaching about 0.65 [5,16,21]. The maximum ZT value of 0.74 was reported for boron-doped p-type $Si_{0.85}Ge_{0.15}$ alloy and 1.0 for phosphorus-doped n-type $Si_{0.85}Ge_{0.15}$ alloy at a high temperature of 1200 K [5,22,23].

Besides the direct current (DC) HP methods [24–26], alternating current HP methods [27] and pulse current sintering techniques [28] were also developed. Hot isostatic pressing was also used to consolidate SiGe alloying powders for large-scale production [29]. The ZT values of the produced composites were close to those of DC hot pressed composites.

A significant improvement in the TE properties of SiGe alloys was achieved through nanostructuring in the late 2000s by a ball-milling (BM) and HP method [25,26,30–32]. A peak ZT value of about 0.95 was achieved at 900–950°C in p-type SiGe nanocomposites [26], which was improved by about 90% over RTG samples (ZT = 0.5), and 50% over the highest record of SiGe composites consisting of micrograins (ZT = 0.65 [16]). A peak ZT of about 1.3 was achieved at 900°C in n-type SiGe nanocomposites [25], which was 40% higher than that of RTG samples (ZT = 0.93).

SiGe alloys have been the primary TE materials in power generation devices operating in the temperature range of 600–1000°C [10,33,34]. We will discuss only SiGe nanocomposites in this chapter. The detailed physical properties of SiGe polycrystalline bulks consisting of micrograins can be found in some review literature [34].

9.2 p-Type $Si_{80}Ge_{20}$ Nanocomposites

Boron can easily diffuse into SiGe alloys [35] to form p-type SiGe alloys. p-Type SiGe nanocomposites with higher ZT have been prepared by a high-energy BM/HP technique. The thermal conductivity of p-type SiGe nanocomposites was significantly reduced, and the maximum ZT of p-type nanostructured $Si_{80}Ge_{20}B_5$ was improved from 0.5 to 0.95 [26]. When compared to RTG silicon–germanium ingots, the thermal conductivities of p-type SiGe nanocomposites were reduced by 40% and figure of merits were increased by 90%.

p-Type SiGe nanocomposites were fabricated by two-step methods, such as the BM/HP method. p-Type (boron-doped) single-phase $Si_{80}Ge_{20}$ nanoparticles (NPs) were first prepared by mechanical alloying using high-energy BM technique from elements [26]. The prepared SiGe NPs were single-phase $Si_{80}Ge_{20}B_x$ alloys (Figure 9.1a). The mechanically alloyed $Si_{80}Ge_{20}B_x$ NPs were polycrystalline (Figure 9.1c), consisting of several subnanograins (Figure 9.1d). X-ray diffraction (XRD) and transmission electron microscopy

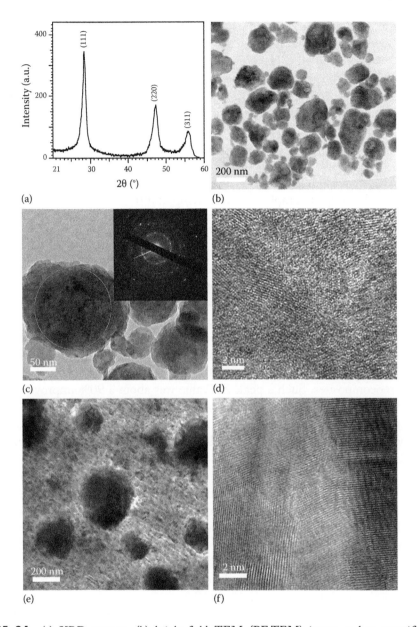

FIGURE 9.1 (a) XRD pattern, (b) bright-field TEM (BF-TEM) image at low magnification, (c) TEM image at medium magnification, and (d) HRTEM image of p-type $Si_{80}Ge_{20}B_x$ NPs prepared by elemental chunks. (e) BF-TEM image and (f) HRTEM image of DC hot pressed p-type $Si_{80}Ge_{20}B_x$ nanocomposites. The inset in (c) is selected-area electron diffraction of an individual grain indicating the polycrystalline nature of the grain. (Reprinted with permission from Joshi, G. et al., *Nano Letters*, 8, 4670–4674, 2008. Copyright 2008. American Chemical Society.)

(TEM) indicated that the crystalline region varied from 20 to 200 nm. The microstructures were explained by the high-energy BM theory [36–39]. The nanograins were formed by a low-temperature mechanical alloying process and not by high-temperature melting and solidification. High strain fields were produced in the ball milled SiGe NPs. Dislocations and atomic level strains were usually created in ball milled NPs, and detailed information can be found in articles on high-energy BM [37,39].

The ball milled SiGe NPs were then hot pressed into nanocomposites. After HP, the grain size in the $Bi_{80}Ge_{20}B_x$ nanocomposites was up to about 20 nm, close to the initial size of the milled nanopowders, indicating no significant grain growth occurring after the DC HP process. These nanograins were highly crystalline, completely random, and closely packed (Figure 9.1f) and had clean boundaries, consistent with a high mass density.

In the SiGe nanocomposites, the mean free path between electrons and phonon is different: about 5 nm for electrons and 2–300 nm for phonons in highly doped samples at room temperature. Thus, nanocomposites can significantly reduce the phonon thermal conductivity (2.5 W/m K in p-type SiGe nanocomposites from 4.6 W/m K in RTG bulks) without significantly reducing the electrical conductivity, resulting in a higher ZT.

Figure 9.2 shows the TE properties of p-type SiGe nanocomposites. The Seebeck coefficient of the $Si_{80}Ge_{20}B_x$ nanocomposites was comparable to that of RTG samples, while the electrical conductivity was higher than that of RTG samples in the whole temperature range. More importantly, the thermal conductivity of the nanostructured bulks was much lower than that of RTG samples over the whole temperature range of up to 1000°C, which led to a peak ZT of about 0.95 in the ball milled and hot pressed $Si_{80}Ge_{20}B_x$ nanocomposites. Such a peak ZT value was about a 90% improvement over that of p-type RTG SiGe alloys ($ZT = 0.5$) currently used in space missions (shown in Figure 9.2d) and 50% above that of the reported highest record value of polycrystalline p-type SiGe composites consisting of micrograins ($ZT = 0.65$ [16]).

The significant reduction in the thermal conductivity in the nanostructured samples was mainly due to the increased phonon scattering at the numerous interfaces of the random nanostructures. The total thermal conductivity (κ) contains two contributions from both the electron carriers (κ_e) and phonon (κ_l), $\kappa = \kappa_e + \kappa_l$. Because the electrical conductivity of the nanocomposite was close to that of RTG samples, the actual phonon thermal conductivity reduction was reduced by at least a factor of 2 based on the experimental data shown in Figure 9.2a.

Modulation-doped p-type $Si_{86}Ge_{14}$ nanocomposites were investigated [31], which enhanced power factors (PFs). The details are discussed in Chapter 11.

Many research groups had fabricated and reported the TE properties of SiGe composites consisting of micrograins before the 2000s [11,23,40]. The grain size of these reported composites was in several micrometers or tens of micrometers. Although the thermal conductivities of these composites were slightly decreased, ZT was not significantly enhanced because of the decreasing electrical conductivity and Seebeck coefficient.

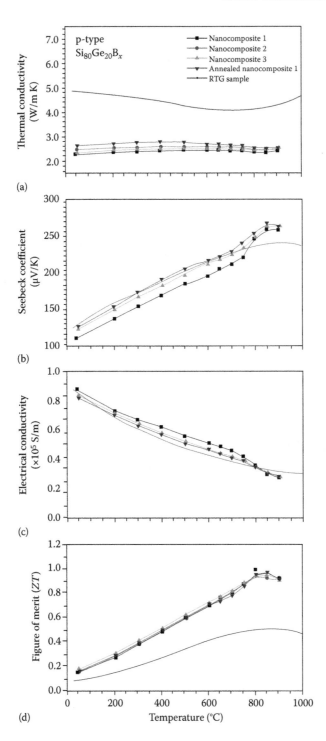

FIGURE 9.2 Temperature dependence of the (a) thermal conductivity, (b) Seebeck coefficient, (c) electrical conductivity, and (d) ZT of three DC hot pressed p-type $Si_{80}Ge_{20}B_x$ nanocomposites (*squares, circles, up triangles*) and a nanocomposite annealed at 1100°C for 7 days in air (*down triangles*) in comparison with the p-type SiGe bulk alloy used in RTGs for space power missions (*dotted line*). (Reprinted with permission from Joshi, G. et al., *Nano Letters*, 8, 4670–4674, 2008. Copyright 2008. American Chemical Society.)

9.3 n-Type Si$_{80}$Ge$_{20}$ Nanocomposites

n-Type SiGe alloys can be produced by phosphorus or arsenic doping [41–43]. Extra valence electrons of the donor atoms become unbounded and are added to SiGe alloys, allowing the alloys to be an electrically conductive n-type semiconductor. Phosphorus is often chosen as an n-type dopant because of its fast diffusion. Phosphorus doping can eliminate strong point defect scattering and inhomogeneity in Si$_{1-x}$Ge$_x$ alloys. n-Type Si$_{63.5}$Ge$_{36.5}$ composites consisting of phosphorus-doped micrograins were reported in the 1980s [15]. Their grain size was in the range of 10–25 μm, and the lattice thermal conductivity was significantly reduced by 20–30%. Arsenic doping usually leads to slightly poorer electron mobility values than phosphorus doping does [44].

Up to now, n-type SiGe nanocomposites have been produced through phosphorus doping and HP fabrication [25,30,45]. These n-type SiGe nanocomposites were usually fabricated by a two-step method. SiGe NPs were first ball milled from SiGe ingots or elements. The microstructure of the mechanically alloyed n-type SiGe NPs was very similar to that of the mechanically alloyed p-type SiGe NPs. Figure 9.3a shows the XRD patterns of Si$_{80}$Ge$_{20}$P$_2$ before (top pattern) and after (bottom pattern) BM. Single-phase NPs were produced by the milling technique. The particle size of the as-prepared ball milled nanopowders was in the range of 30–200 nm, as shown in Figure 9.3b and c. These NPs were composed of many small crystalline subnanograins, similar to p-type mechanically alloyed SiGe NPs. The crystalline size of the subnanograins was in the range of 5–15 nm (Figure 9.3d) with an average size of 12 nm. The SiGe NPs were then hot pressed into nanocomposites. After HP, the average grain size in the hot pressed nanocomposites was 22 nm, indicating that the grain size was almost doubled after hot press, but was still small. These small grains with random crystalline orientations promoted more phonon scattering than the micrometer-sized grains in RTG SiGe bulks. Compared with ball milled NPs, the strain inside the hot pressed nanocomposites was much smaller than that of the as-prepared ball milled nanopowders, with the strain value being 10 times lower in the hot pressed samples. The smaller strain was understandable since the HP was carried out above 1000°C, allowing the buildup of stresses in the nanopowders to be relaxed.

Figure 9.4 shows the TE properties of n-type SiGe nanocomposites. From the comparison of the temperature-dependent thermal conductivity (Figure 9.4a), the n-type SiGe nanocomposites had a much lower thermal conductivity than the RTG reference samples with a grain size of 1–10 μm. From the measured electrical conductivity of the n-type SiGe nanocomposites and RTG samples (Figure 9.4c), κ_e = 0.77 W/m K at room temperature with an electrical conductivity σ = 1.2 × 10^{-5} S/m for RTG samples, whereas κ_e = 0.55 W/m K at room temperature with σ = 0.85 × 10^{-5} S/m for a typical nanocomposite sample. The lattice thermal conductivity (κ_l) of the nanocomposite samples was 1.8 W/m K at room temperature, which was about 47% of the RTG reference samples (κ_l = 3.8 W/m K). The decrease in lattice thermal conductivity κ_l should be mainly due to a stronger boundary phonon scattering in the nanostructured samples.

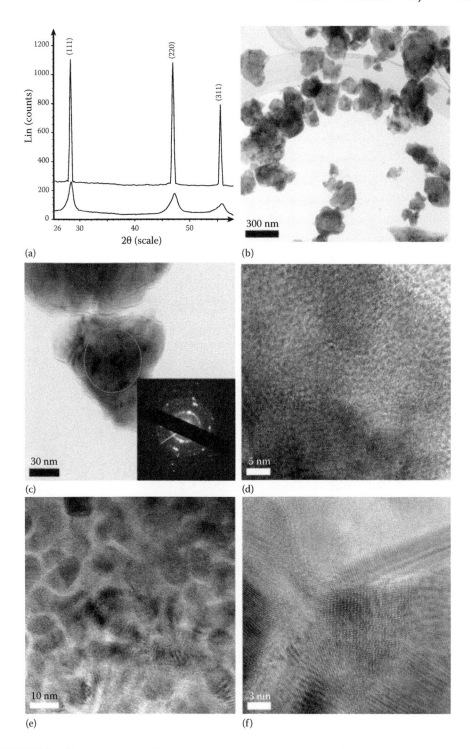

FIGURE 9.3 (a) XRD patterns, (b) BF-TEM image at low magnification, (c) BF-TEM image at medium magnification, and (d) HRTEM image of n-type $Si_{80}Ge_{20}P_2$ NPs. (e) BF-TEM and (f) HRTEM images of hot pressed n-type $Si_{80}Ge_{20}P_2$ nanocomposites made from the NPs shown in (b). The inset of (c) shows a selected-area electron diffraction pattern of an individual NP. (Reprinted with permission from Wang, X. W. et al., *Applied Physics Letters*, 93, 193121, 2008. Copyright 2008, American Institute of Physics.)

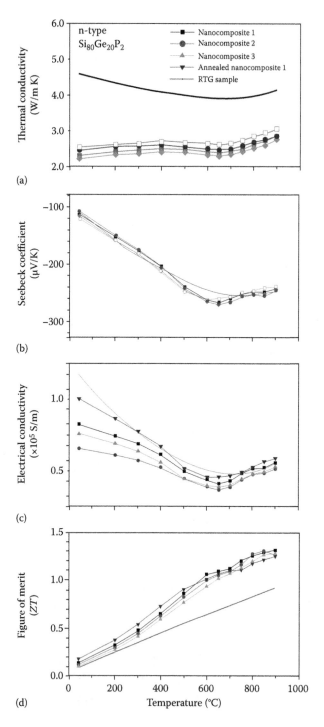

FIGURE 9.4 Temperature dependence of (a) thermal conductivity, (b) Seebeck coefficient, (c) electrical conductivity, and (d) dimensionless ZT on both as-pressed n-type $Si_{80}Ge_{20}P_2$ nanocomposites (*squares, circles, up triangles*) and a nanocomposite annealed at 1050°C for 2 days in air (*down triangles*), in comparison with the RTG reference sample (*dotted curve*). (Reprinted with permission from Wang, X. W. et al., *Applied Physics Letters*, 93, 193121, 2008. Copyright 2008, American Institute of Physics.)

The electrical conductivity of the n-type nanocomposites was normally lower than that of RTG reference samples in the low-temperature region, but was similar above 750°C. The carrier concentrations for both types of samples were almost the same at room temperature (2.2×10^{20} cm^{-3} from the Hall effect measurement). Therefore, the electron mobility should be lower in the nanocomposite samples.

The Seebeck coefficient of the SiGe nanocomposite samples was close to that of RTG reference samples below 400°C and above 700°C, as shown in Figure 9.4b. The Seebeck coefficient of the SiGe nanocomposites was higher than that of RTG reference samples between 400°C and 700°C.

As a result, for the n-type SiGe nanocomposites, the ZT value showed a maximum of about 1.3 at 900°C, as shown in Figure 9.4d. The ZT value was about 40% higher than that of RTG samples ($ZT = 0.93$). The significant enhancement in ZT is mainly attributed to thermal conductivity reduction, which is strongly correlated with the nanostructure features in the n-type $Si_{80}Ge_{20}P_2$ nanocomposites.

Zebarjadi et al. investigated modulation-doped n-type $Si_{84}Ge_{16}$ nanocomposites [31]. Yu et al. studied modulation-doped n-type $Si_{86.25}Ge_{13.75}$ nanocomposites [32]. The PF of the n-type SiGe nanocomposites was significantly enhanced through a modulation doping approach. The details of the modulation-doped SiGe nanocomposites are discussed in Chapter 11.

9.4 SiGe Nanocomposites with Low Ge Content

The TE properties of $Si_{80}Ge_{20}$ alloys are superior over those of pure Si due to the alloying effect and are further enhanced through nanostructuring grain boundary scattering. However, germanium is an expensive and heavy element compared with silicon. Typical material costs for $Si_{1-x}Ge_x$-based TE power-generating devices would be reduced by more than 80% if the Ge fraction was reduced from 0.2 to 0.02, while the elimination of Ge would reduce mass density by 20%, which is essential for mass-constrained deep space missions [44]. Therefore, soon after the success with $Si_{80}Ge_{20}$ nanocomposites [25,26], the nanostructuring concept was applied to $Si_{1-x}Ge_x$ nanocomposites with low Ge content.

9.4.1 n-Type Si₉₅Ge₅ Nanocomposites

A ZT of 0.94 at 900°C was achieved in n-type $Si_{95}Ge_5$ nanocomposites [30]. In this work, chunks of Si and Ge were ball milled down to 5–20 nm crystalline NPs. Figure 9.5a and b shows the XRD pattern and HRTEM images, respectively. The nanopowders were then hot pressed into fully dense nanocomposite samples. Figure 9.5c shows a typical TEM image of the nanocomposites, indicating nanoscaled grain size. HRTEM images (Figure 9.5d) confirmed that the nanocomposites consisted of crystalline nanograins that are 10–30 nm in size.

As shown in Figure 9.6d, the thermal conductivity of the nanostructured SiGe showed about a factor of 10 reductions compared with that of the heavily doped bulk Si, a clear demonstration of the nanosize effect on phonon

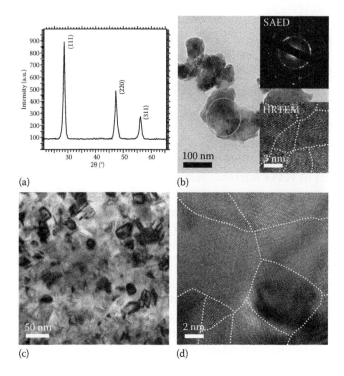

FIGURE 9.5 (a) XRD pattern and (b) TEM image of ball milled n-type $Si_{95}Ge_5$ NPs. The insets in (b) are the electron diffraction pattern and HRTEM image of the circled region. (c) TEM and (d) HRTEM images of the hot pressed $Si_{95}Ge_5$ nanocomposites. (Reprinted with permission from Zhu, G. H. et al., *Physical Review Letters*, 102, 196803, 2009. Copyright 2009 by the American Physical Society.)

scattering. Furthermore, with a 5 atm% replacement of Si by Ge, the thermal conductivity of nanostructured $Si_{95}Ge_5$ was even lower, close to that of $Si_{80}Ge_{20}$ RTG samples, caused by both the nanosize and point defect scattering mechanism. At the same time, the nanostructure approach for $Si_{95}Ge_5$ nanocomposites maintained a high electrical conductivity and PF as shown in Figure 9.6a and c. As a result, the *ZT* was enhanced by a factor of 2 in the nanostructured Si nanocomposites and factor of 4 in nanostructured $Si_{95}Ge_5$ nanocomposites in comparison with large-grained bulk Si.

Zhu et al. further investigated phonon scattering by nanograins and point defects in nanostructured $Si_{95}Ge_5$ [30]. In focusing on the mechanism of reduction in thermal conductivity in nanograined materials by comparison of the phonon scattering process in pure Si and in Si alloys with a low Ge concentration, it was experimentally found that nanograins played an important role in scattering phonons with wavelengths in the nanometer range, while point defect scattering, caused by alloying Ge into Si, was more effective in scattering phonons than just using pure Si nanostructures, especially for scattering phonons with wavelengths of less than 1 nm.

Petermann et al. also investigated $Si_{95}Ge_5$ nanocomposites fabricated by current-assisted sintering [46]. n-Type $Si_{95}Ge_5$ NPs were produced by a gas phase plasma reaction and then sintered to bulks. The *ZT* of 0.75 was achieved at 1000°C in these nanocomposites.

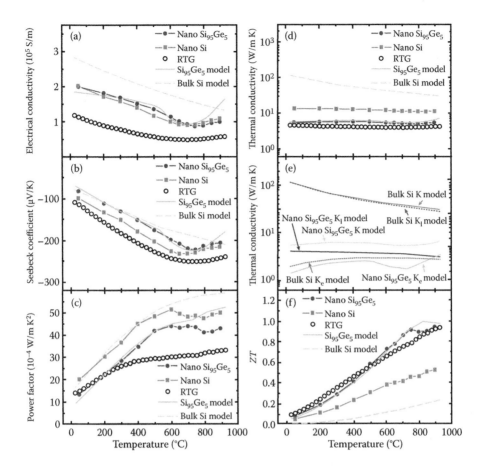

FIGURE 9.6 Temperature-dependent (a) electrical conductivity, (b) Seebeck coefficient, (c) PF, (d) thermal conductivity, (e) electron (κ_e), phonon (κ_l), and total (κ) thermal conductivity by modeling, and (f) ZT of nanostructured Si (*filled squares*), $Si_{95}Ge_5$ nanocomposite (*filled circles*, experiment; *solid line*, model), bulk Si model (*dashed line*), and $Si_{80}Ge_{20}$ RTG samples (*open circles*). (Reprinted with permission from Zhu, G. H. et al., *Physical Review Letters*, 102, 196803, 2009. Copyright 2009 by the American Physical Society.)

9.4.2 n-Type Si Nanocomposites

The Hall coefficients and TE powers of pure silicon were measured as early as the 1910s [47–50]. The thermal conductivity of silicon is around 150 W/m K, too high for it to be a good TE material. This drawback was overcome by reducing its grain size down to nanoscale dimensions. In the 1970s, it was predicted that a grain size of 10 µm would reduce the thermal conductivity of 7% in pure silicon [51]. Some theoretical simulations showed that [52,53] ZT can reach 1.0 at 1173 K if the grain size can be reduced to 10 nm.

Bux et al. reported an experimental approach for phosphorus-doped nanostructured Si bulks [44]. The nanostructured Si nanocomposites were fabricated by high-temperature compaction of high-energy ball milled Si nanopowders with a crystalline size of 5–20 nm. Figure 9.7a shows the TEM image of the ball milled nanopowders. BF TEM images revealed that the crystalline size of the agglomerates was 5–20 nm. The crystalline size of the nanopowders was calculated to be 16 nm based on XRD patterns, in agreement with TEM

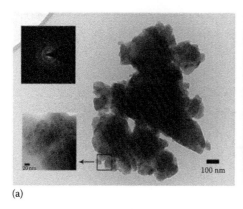
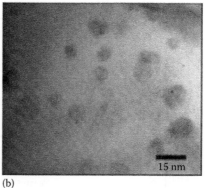

(a)   (b)

FIGURE 9.7 (a) TEM images showing large agglomerates of ball milled Si NPs. (b) TEM image of nanostructured Si nanocomposites. (Reproduced with permission from Bux, S. K. et al.: Nanostructured bulk silicon as an effective thermoelectric material. *Advanced Functional Materials.* 2009. 15. 2445–2452. Copyright 2009, Wiley-VCH Verlag GmbH & Co. KGaA.)

analysis. The grain size of the nanocomposites increased to 50–100 nm after high-temperature compactions, while nanocrystalline domains in 10–15 nm range were still present, as shown in Figure 9.7b.

These nanograins and crystalline domains introduced strong interfacial or boundary phonon scattering, resulting in a dramatic 90% reduction in the lattice thermal conductivity of the n-type Si nanocomposites over large-grained Si. Meanwhile, this very high density of grain boundaries did not affect electron mobility values as much. These combined transport effects produced an unusual increase in the ZT of the nanostructured Si bulks. The temperature-dependent ZT of nanostructured bulk Si is shown in Figure 9.8. The maximum values, ZT_{max}, of the samples ranged from 0.40 to 0.70, which is an unprecedented increase by a factor of 3.5 over that of heavily doped single-crystalline Si.

Therefore, n-type nanostructured bulk Si is an effective high-temperature TE material with about 70% of the performance of the practical $Si_{0.8}Ge_{0.2}$ nanocomposites (with ZT of 1), but without the presence of Ge, an expensive element. It was predicated [44] that maximum ZT values of nearly 1.3 could be achieved at 1275 K if optimization conditions were met.

It is worth noticing that the present experimental record ZT of nanostructured bulk Si was around 0.7 [44], much lower than the theoretical prediction (ZT = 1.0 [53]). The deviation is understandable since the grain size of fabricated Si nanostructured samples is much larger than the theoretical value (10 nm [44]).

Schierning et al. [54] used a different approach in making Si NPs [55], instead of high energy BM. Si NPs were produced in a gas phase process based on the decomposition of silane (SiH_4) in a microwave plasma reactor, as shown in Figure 9.9. n-Type doping was realized by adding phosphine (PH_3) to the precursor gas. The as-prepared Si NPs were further processed by DC sintering by using an SPS machine.

The produced Si nanocomposites preserved the nanocrystalline character with an average crystalline size of 25 nm. A peak figure of merit ZT

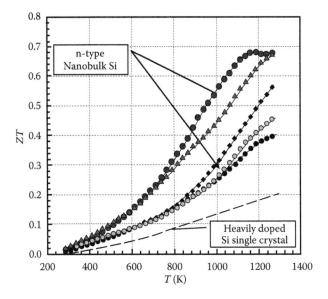

FIGURE 9.8 *ZT* as a function of temperature. Experimental data show a 3.5 increase in *ZT* values by nanostructuring n-type Si, with a peak value of 0.7 at 1275 K. (Reproduced with permission from Bux, S. K. et al.: Nanostructured bulk silicon as an effective thermoelectric material. *Advanced Functional Materials*. 2009. 15. 2445–2452. Copyright 2009, Wiley-VCH Verlag GmbH & Co. KGaA.)

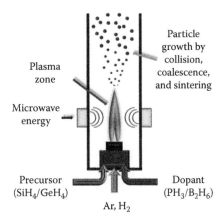

FIGURE 9.9 Sketch of the gas synthesis for doping silicon NPs. The precursors silane (SiH_4) and phosphine (PH_3) decomposed within a microwave-induced Ar/H_2 plasma. The shape of the NPs can be tailored by parameters such as gas pressure, flow rates, microwave energy, and chemical composition of the gas mixture. (Reprinted with permission from Schierning, G. et al., *Journal of Applied Physics*, 110, 113515, 2011. Copyright 2011, American Institute of Physics.)

of 0.5 was achieved at 950°C for a sample sintered at 1060°C with a mean crystalline size of 46 nm [54].

Claudio et al. recently decreased the thermal conductivity of n-type Si nanocomposites through enhanced phonon scattering by nanostructuring technique [56,57]. The Si nanocomposites were prepared by gas phase synthesis, followed by current- and pressure-assisted sintering. Figure 9.10 shows the microstructures of the nanocomposites. TEM images revealed

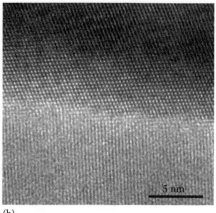

(a)   (b)

FIGURE 9.10 (a) BF-TEM image of Si nanocomposites and (b) HRTEM of a grain boundary between two nanocrystals. (Reproduced by permission from Claudio, T. et al., *Physical Chemistry Chemical Physics*, 16, 25701–25709, 2014. Copyright 2014, The Royal Society of Chemistry.)

grains ranging between 47 and 264 nm with an average of 112–114 nm. XRD analysis indicated an average crystalline size of 34–58 nm. The Si nanocomposites presented a significantly decreased lattice thermal conductivity, 25 W/m K at room temperature and 10 W/m K at 1000°C. The thermal conductivity was high but was compensated for by a very high PF, resulting in a dimensionless figure of merit with a competitive peak value $ZT = 0.57$ at 973°C.

n-Type Si nanocomposites were reviewed recently [58,59]. Readers are referred to the references therein.

9.4.3 p-Type Si Nanocomposites

Kessler et al. produced highly boron-doped silicon NPs by a continuous gas phase process and compacted the NPs into nanocomposites by DC-assisted sintering [60]. The average crystalline size of the yielded p-type Si nanocomposites was between 40 and 80 nm. The maximum ZT value of the p-type Si nanocomposites was 0.30 at 700°C.

9.5 Outlook on SiGe Nanocomposites

The theoretically predicted ZT of SiGe nanocomposites is much higher than the experimental values for the state-of-the-art SiGe nanocomposites. Mingo et al. proposed an NP-in-alloy approach to enhance TE performances of SiGe nanocomposites [61]. With nanosized silicide and germanide fillers in SiGe matrix, a five-fold increase in ZT at room temperature and a 2.5 times increase at 900 K were predicted. A ZT of 1.7 at 900 K can be achieved if optimal NPs could minimize thermal conductivity while not impairing the electrical conductivity. A higher ZT of 2.2 at 800 K was also predicated by Lee et al. [61], who simultaneously examined the thermal conductivity, electrical conductivity, and PF of a single SiGe nanowire. The higher ZT has arisen from the huge reduction in thermal conductivity due to the scatterings of alloying and boundary scattering. All modeled ZT values are higher than the present experimental values ($ZT = 1.3$ [25]).

Hao et al. [52] predicated that the ZT of pure Si nanocomposites [53] (experimental peak $ZT = 0.7$ [44]) is 1.02 at 1173 K if the grain size can be reduced to 10 nm. However, it is challenging to effectively prevent nanograin growth during hot press.

All these theoretical ZT values are much higher than the state-of-the-art experimental values, demonstrating the wide potentials of nanostructured SiGe TE materials. According to their simulation conditions, small crystalline grains with a size of 10 nm or less are required in the nanocomposites. However, it is very challenging to fabricate nanocomposites with such small crystalline grains by using the present fabrication technologies. A novel fabrication method is needed to limit grain growth during fabrication procedures. SiGe nanocomposites with higher ZT might be developed along the way.

References

1. D. R. Lide, ed., *CRC Handbook of Chemistry and Physics*, 85th Edition. CRC Press, Boca Raton, FL, 2004.
2. B. Abeles, D. S. Beers, G. D. Cody, and J. P. Dismukes, Thermal conductivity of Ge-Si alloys at high temperatures, *Physical Review*, vol. 125, pp. 44–46, 1962.
3. B. Abeles, Lattice thermal conductivity of disordered semiconductor alloys at high temperatures, *Physical Review*, vol. 131, pp. 1906–1911, 1963.
4. B. Abeles and R. W. Cohen, Ge-Si thermoelectric power generator, *Journal of Applied Physics*, vol. 35, p. 247, 1964.
5. J. P. Dismukes, L. Ekstrom, E. F. Steigmeier, I. Kudman, and D. S. Beers, Thermal and electrical properties of heavily doped Ge-Si alloys up to 1300 K, *Journal of Applied Physics*, vol. 35, no. 10, pp. 2899–2907, 1964.
6. H. Stöhr and W. Klemm, Über zweistoffsysteme mit germanium: I. Germanium/aluminium, germanium/zinn und germanium/silicium, *Zeitschrift für anorganische und allgemeine Chemie*, vol. 241, no. 4, pp. 305–323, 1939.
7. M. C. Steele and F. D. Rosi, Thermal conductivity and thermoelectric power of germanium-silicon alloys, *Journal of Applied Physics*, vol. 29, no. 11, pp. 1517–1520, 1958.
8. P. Maycock, Thermal conductivity of silicon, germanium, III–V compounds and III–V alloys, *Solid-State Electronics*, vol. 10, no. 3, pp. 161–168, 1967.
9. R. Cheaito, J. C. Duda, T. E. Beechem, K. Hattar, J. F. Ihlefeld, D. L. Medlin, M. A. Rodriguez, M. J. Campion, E. S. Piekos, and P. E. Hopkins, Experimental investigation of size effects on the thermal conductivity of silicon-germanium alloy thin films, *Physical Review Letters*, vol. 109, article 195901, 2012.
10. D. M. Rowe, ed., *CRC Handbook of Thermoelectrics*. CRC Press, Boca Raton, FL, 1995.
11. D. M. Rowe and R. W. Bunce, The thermoelectric properties of heavily doped hot-pressed germanium-silicon alloys, *Journal of Physics D: Applied Physics*, vol. 2, no. 11, p. 1497, 1969.

12. R. A. Lefever, G. L. McVay, and R. J. Baughman, Preparation of hot-pressed silicon-germanium ingots: Part III—Vacuum hot pressing, *Materials Research Bulletin*, vol. 9, no. 7, pp. 863–872, 1974.

13. G. McVay, R. Lefever, and R. Baughman, Preparation of hot-pressed silicon-germanium ingots: Part II—Reduction of chill cast material, *Materials Research Bulletin*, vol. 9, no. 6, pp. 735–744, 1974.

14. D. M. Rowe, Electrical properties of hot-pressed germanium-silicon-boron alloys, *Journal of Physics D: Applied Physics*, vol. 8, no. 9, p. 1092, 1975.

15. D. M. Rowe, V. S. Shukla, and N. Savvides, Phonon scattering at grain boundaries in heavily doped fine-grained silicon-germanium alloys, *Nature*, vol. 290, pp. 765–766, 1981.

16. C. B. Vining, W. Laskow, J. O. Hanson, R. R. Van der Beck, and P. D. Gorsuch, Thermoelectric properties of pressure-sintered $Si_{0.8}Ge_{0.2}$ thermoelectric alloys, *Journal of Applied Physics*, vol. 69, no. 8, pp. 4333–4340, 1991.

17. F. D. Rosi, The research and development of silicon-germanium thermoelements for power generation, *Materials Research Society Symposium Proceedings*, vol. 234, pp. 3/1–3/24, 1991.

18. N. K. Abrikosov, V. S. Zemskov, E. K. Iordanishvili, A. V. Petrov, and V. V. Rozhdestvenskaya, Thermoelectric properties of silicon-germanium-boron alloys, *Fizikai Tekhnika Poluprovodnikov*, vol. 2, pp. 1762–1768, 1968.

19. C. M. Bhandari and D. M. Rowe, Silicon-germanium alloys as high-temperature thermoelectric materials, *Contemporary Physics*, vol. 21, no. 3, pp. 219–242, 1980.

20. G. A. Slack and M. A. Hussain, The maximum possible conversion efficiency of silicon-germanium thermoelectric generators, *Journal of Applied Physics*, vol. 70, no. 5, pp. 2694–2718, 1991.

21. D. M. Rowe, L. W. Fu, and S. G. K. Williams, Comments on the thermoelectric properties of pressure sintered $Si_{0.8}Ge_{0.2}$ thermoelectric alloys, *Journal of Applied Physics*, vol. 73, no. 9, pp. 4683–4685, 1993.

22. J. P. Dismukes and L. Ekstrom, Homogeneous solidification of Ge-Si alloys, *Transactions of the Metallurgical Society of AIME*, vol. 233, p. 672, 1965.

23. N. Savvides and H. J. Goldsmid, Hot-press sintering of Ge-Si alloys, *Journal of Materials Science*, vol. 15, no. 3, pp. 594–600, 1980.

24. B. Poudel, Q. Hao, Y. Ma, Y. Lan, A. Minnich, B. Yu, X. Yan et al., High-thermoelectric performance of nanostructured bismuth antimony telluride bulk alloys, *Science*, vol. 320, no. 5876, pp. 634–638, 2008.

25. X. W. Wang, H. Lee, Y. C. Lan, G. H. Zhu, G. Joshi, D. Z. Wang, J. Yang et al. Enhanced thermoelectric figure of merit in nanostructured n-type silicon germanium bulk alloy, *Applied Physics Letters*, vol. 93, no. 19, p. 193121, 2008.

26. G. Joshi, H. Lee, Y. Lan, X. Wang, G. Zhu, D. Wang, R. W. Gould et al., Enhanced thermoelectric figure-of-merit in nanostructured p-type silicon germanium bulk alloys, *Nano Letters*, vol. 8, no. 12, pp. 4670–4674, 2008.

27. R. W. Bunce and D. M. Rowe, The vacuum hot-pressing of germanium and silicon-germanium alloys, *Journal of Physics D: Applied Physics*, vol. 10, no. 6, p. 941, 1977.

28. M. Otake, K. Sato, O. Sugiyama, and S. Kaneko, Pulse-current sintering and thermoelectric properties of gas-atomized silicon-germanium powders, *Solid State Ionics*, vol. 172, pp. 523–526, 2004.

29. J. L. Harringa and B. A. Cook, Application of hot isostatic pressing for consolidation of n-type silicon-germanium alloys prepared by mechanical alloying, *Materials Science and Engineering: B*, vol. 60, no. 2, pp. 137–142, 1999.

30. G. H. Zhu, H. Lee, Y. C. Lan, X. W. Wang, G. Joshi, D. Z. Wang, J. Yang et al., Increased phonon scattering by nanograins and point defects in nanostructured silicon with a low concentration of germanium, *Physical Review Letters*, vol. 102, no. 19, p. 196803, 2009.

31. M. Zebarjadi, G. Joshi, G. Zhu, B. Yu, A. Minnich, Y. Lan, X. Wang, M. Dresselhaus, Z. Ren, and G. Chen, Power factor enhancement by modulation doping in bulk nanocomposites, *Nano Letters*, vol. 11, no. 6, pp. 2225–2230, 2011.

32. B. Yu, M. Zebarjadi, H. Wang, K. Lukas, H. Wang, D. Wang, C. Opeil, M. Dresselhaus, G. Chen, and Z. Ren, Enhancement of thermoelectric properties by modulation-doping in silicon germanium alloy nanocomposites, *Nano Letters*, vol. 12, pp. 2077–2082, 2012.

33. F. Rosi, Thermoelectricity and thermoelectric power generation, *Solid-State Electronics*, vol. 11, no. 9, pp. 833–868, 1968.

34. C. Wood, Materials for thermoelectric energy conversion, *Reports on Progress in Physics*, vol. 51, no. 4, p. 459, 1988.

35. T. T. Fang, W. T. C. Fang, P. B. Griffin, and J. D. Plummer, Calculation of the fractional interstitial component of boron diffusion and segregation coefficient of boron in $Si_{0.8}Ge_{0.2}$, *Applied Physics Letters*, vol. 68, no. 6, pp. 791–793, 1996.

36. J. S. Benjamin, Mechanical alloying, *Scientific American*, vol. 234, no. 5, pp. 40–49, 1976.

37. C. Suryanarayana, Mechanical alloying and milling, *Progress in Materials Science*, vol. 46, no. 1-2, pp. 1–184, 2001.

38. Y. Lan and Z. Ren, Thermoelectric nanocomposites for thermal energy conversion, *Nanomaterials for Sustainable Energy*, pp. 371–443. Springer International, Cham, Switzerland, 2016.

39. P. S. Gilman and J. S. Benjamin, Mechanical alloying, *Annual Review of Materials Science*, vol. 13, no. 1, pp. 279–300, 1983.

40. D. M. Rowe, Theoretical optimization of the thermoelectric figure of merit of heavily doped hot-pressed germanium-silicon alloys, *Journal of Physics D: Applied Physics*, vol. 7, no. 13, p. 1843, 1974.

41. J. P. Fleurial, C. B. Vining, and A. Borshchevsky, Multiple doping of silicon-germanium alloys for thermoelectric applications, *Proceedings of the 24th Intersociety Energy Conversion Engineering Conference*, vol. 2, pp. 701–705, 1989.

42. B. A. Cook, J. L. Harringa, S. H. Han, and B. J. Beaudry, Parasitic effects of oxygen on the thermoelectric properties of $Si_{80}Ge_{20}$ doped with GaP and P, *Journal of Applied Physics*, vol. 72, no. 4, pp. 1423–1428, 1992.

43. B. A. Cook, J. L. Harringa, S. H. Han, and C. B. Vining, $Si_{80}Ge_{20}$ thermoelectric alloys prepared with GaP additions, *Journal of Applied Physics*, vol. 78, no. 9, pp. 5474–5480, 1995.

44. S. K. Bux, R. G. Blair, P. K. Gogna, H. Lee, G. Chen, M. S. Dresselhaus, R. B. Kaner, and J.-P. Fleurial, Nanostructured bulk silicon as an effective thermoelectric material, *Advanced Functional Materials*, vol. 19, no. 15, pp. 2445–2452, 2009.

45. R. Basu, S. Bhattacharya, R. Bhatt, M. Roy, S. Ahmad, A. Singh, M. Navaneethan, Y. Hayakawa, D. K. Aswal, and S. K. Gupta, Improved thermoelectric performance of hot pressed nanostructured n-type SiGe bulk alloys, *Journal of Materials Chemistry A*, vol. 2, no. 19, pp. 6922–6930, 2014.

46. N. Petermann, T. Schneider, J. Stötzel, N. Stein, C. Weise, I. Wlokas, G. Schierning, and H. Wiggers, Microwave plasma synthesis of Si/Ge and Si/WSi$_2$ nanoparticles for thermoelectric applications, *Journal of Physics D: Applied Physics*, vol. 48, no. 31, p. 314010, 2015.

47. F. G. Wick, Some electrical properties of silicon: I. Thermo-electric behavior of metallic silicon, *Physical Review (Series I)*, vol. 25, pp. 382–390, 1907.

48. F. G. Wick, Some electrical properties of silicon: III. The Hall effect in silicon at ordinary and low temperatures, *Physical Review (Series I)*, vol. 27, pp. 76–86, 1908.

49. A. W. Smith, The variation of the Hall effect in metals with change of temperature, *Physical Review (Series I)*, vol. 30, pp. 1–34, 1910.

50. O. E. Buckley, The Hall effect and allied phenomena in silicon, *Physical Review*, vol. 4, pp. 482–490, 1914.

51. C. M. Bhandari and D. M. Rowe, Boundary scattering of phonons, *Journal of Physical Chemistry*, vol. 11, no. 9, pp. 17–87, 1978.

52. Q. Hao, G. Chen, and M.-S. Jeng, Frequency-dependent Monte Carlo simulations of phonon transport in two-dimensional porous silicon with aligned pores, *Journal of Applied Physics*, vol. 106, no. 11, p. 114321, 2009.

53. Q. Hao, G. Zhu, G. Joshi, X. Wang, A. Minnich, Z. Ren, and G. Chen, Theoretical studies on the thermoelectric figure of merit of nanograined bulk silicon, *Applied Physics Letters*, vol. 97, no. 6, pp. 063–109, 2010.

54. G. Schierning, R. Theissmann, N. Stein, N. Petermann, A. Becker, M. Engenhorst, V. Kessler et al., Role of oxygen on microstructure and thermoelectric properties of silicon nanocomposites, *Journal of Applied Physics*, vol. 110, no. 11, p. 113515, 2011.

55. G. Schierning, J. Stoetzel, R. Chavez, V. Kessler, J. Hall, R. Schmechel, T. Schneider et al., Silicon-based nanocomposites for thermoelectric application, *Physica Status Solidi (a)*, vol. 213, no. 3, pp. 497–514, 2016.

56. T. Claudio, N. Stein, D. G. Stroppa, B. Klobes, M. M. Koza, P. Kudejova, N. Petermann, H. Wiggers, G. Schierning, and R. P. Hermann, Nanocrystalline silicon: lattice dynamics and enhanced thermoelectric properties, *Physical Chemistry Chemical Physics*, vol. 16, pp. 25701–25709, 2014.

57. T. Claudio, G. Schierning, R. Theissmann, H. Wiggers, H. Schober, M. M. Koza, and R. P. Hermann, Effects of impurities on the lattice dynamics of nanocrystalline silicon for thermoelectric application, *Journal of Materials Science*, vol. 48, no. 7, pp. 2836–2845, 2013.

58. D. Narducci, A special issue on silicon and silicon-related materials for thermoelectricity, *The European Physical Journal B*, vol. 88, no. 7, p. 174, 2015.

59. N. Neophytou, Prospects of low-dimensional and nanostructured silicon-based thermoelectric materials: Findings from theory and simulation, *The European Physical Journal B*, vol. 88, no. 4, p. 86, 2015.

60. V. Kessler, D. Gautam, T. Hülser, M. Spree, R. Theissmann, M. Winterer, H. Wiggers, G. Schierning, and R. Schmechel, Thermoelectric properties of nanocrystalline silicon from a scaled-up synthesis plant, *Advanced Engineering Materials*, vol. 15, no. 5, pp. 379–385, 2013.

61. N. Mingo, D. Hauser, N. P. Kobayashi, M. Plissonnier, and A. Shakouri, "Nanoparticle-in-alloy" approach to efficient thermoelectrics: Silicides in SiGe, *Nano Letters*, vol. 9, pp. 711–715, 02 2009.

CHAPTER **10**

Other Thermoelectric Materials

Pengfei Qiu, Xun Shi, Lidong Chen, Jiehe Sui, Jing Li,
Zihang Liu, Zhifeng Ren, Takao Mori, and Jun Mao

Contents

10.1 Liquid-Like Thermoelectric Materials

Pengfei Qiu, Xun Shi, and Lidong Chen

In principle, a reduction in lattice thermal conductivity κ_L can be achieved by reducing C_V, v, and l based on

$$\kappa_L = \frac{1}{3} C_V v l, \tag{10.1}$$

where C_V is the specific heat per unit volume, v is the speed of sound, and l is the phonon mean-free path. The former two parameters are usually constant in a solid material. Thus, the reduction of κ_L can normally be only realized by lowering l. However, by using the very special and unique structural characteristics of ionic conductors, it was demonstrated that the specific heat C_V in solid materials can also be tuned and reduced down to the liquid limit, i.e., down to the two-thirds of the solid limit ($3Nk_B$). Typically, an ionic conductor has two independent sublattices. One sublattice, called solid sublattice, behaves as a typical sublattice in a solid and maintains a rigid framework. The other sublattice, called the liquid sublattice, resembles a liquid where atoms can freely roam throughout the solid sublattice. Solid-state materials such as traditional glasses and crystals propagate heat through longitudinal and transverse waves in the rigid three-dimensional (3D) framework. However, a liquid cannot support sheer vibrations, and the sound waves can only propagate in two-dimensional (2D) planes. As a result, liquid structures have less freedom to conduct heat, and the C_V is limited to two-thirds of the value of the solid (see Figure 10.1). Taking $Cu_{2-\delta}X$ (X = S, Se, or Te) as an example,[1–3] the structure has a rigid face-centered cubic sublattice X, which provides a crystalline pathway for semiconducting electrons. Copper ions on the other hand are highly disordered around the sublattice X and possess a liquid-like mobility. This extraordinary liquid-like (superionic) behavior of copper ions could eliminate some of the transverse vibrational modes in $Cu_{2-\delta}X$, leading to much reduced heat capacities C_V below the limit value of $3Nk_B$ in a solid (see Figure 10.1). As a consequence, exceptionally low values of the lattice thermal conductivity have been recorded (0.2–0.5 W/m K) contributing to high ZT values between 1.1 and 1.7 observed in this otherwise simple semiconductor. In analogy with the phonon–glass electron–crystal concept of Slack, the combination of solid and liquid-like sublattices and their effect on the thermal conductivity have been referred to as the phonon–liquid electron–crystal character.[1–3]

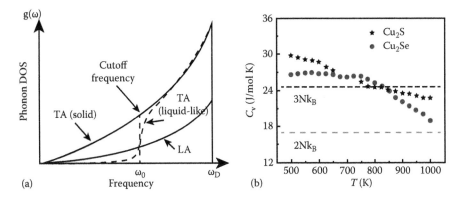

FIGURE 10.1 Schematic diagram of the phonon DOS (a) and the specific heat (b) in liquid-like materials as a function of temperature. The transverse phonons are softened and disappear when the frequency is below the cutoff frequency in liquid-like materials, leading to reduced C_v values. The limiting value of the specific heat of $3Nk_B$ in a solid becomes $2Nk_B$ in a liquid. (Reprinted by permission from Macmillan Publishers Ltd., *Nature Materials*, Liu, H. L. et al., 11, 2012, 422–425, copyright (2012); He, Y. et al.: High thermoelectric performance in non-toxic earth-abundant copper sulfide. *Advanced Materials*. 2014. 26. 3974–3978. Copyright Wiley-VCH Verlag GmbH & Co. KGaA. Reproduced with permission.)

As one family of typical liquid-Like thermoelectric materials, $Cu_{2-\delta}X$ (X = S, Se, and Te) compounds have quite complex crystal structures. At high temperatures, they possess the antifluorite structure (similar to Mg_2X) with X in the face-centered cubic positions and Cu highly disordered in the tetrahedral or other interstitial sites (Figure 10.2a). For different X, $Cu_{2-\delta}X$ compounds form different crystal structures at room temperature. $Cu_{2-\delta}Se$ undergoes a structural phase transition around 400 K,[4] while $Cu_{2-\delta}S$ undergoes two structural transitions between 300 and 900 K,[5] and $Cu_{2-\delta}Te$ has as many as five successive phase transitions between 300 and 900 K.[6] The deficiency of copper also modifies the crystal structure of $Cu_{2-\delta}X$ and makes it more complicated than one would expect based on a simple chemical formula. The highly disordered diffusing copper ions definitely scatter phonons at high temperatures, and their liquid-like nature effectively eliminates transverse phonons from contributing to the heat flow, leading to ultralow thermal conductivity κ_L values as shown in Figure 10.2b.[1] In addition, the study of Cu_2Se revealed the multiformity of Cu ordering in the low-temperature phase, and enormous structural fluctuations during the second-order phase transition also contribute to the extremely low κ_L which approaches the minimal theoretical thermal conductivity κ_{min}.[7] Because the charge state of Cu is +1 and stoichiometric Cu_2X is an intrinsic semiconductor, the electronic transport in $Cu_{2-\delta}X$ can be tuned by altering the deficiency parameter δ. Initial reports indicated that the maximum ZT values[1–3] of 1.5 for Cu_2Se, 1.7 for $Cu_{1.97}S$, and 1.1 for Cu_2Te can be obtained at 1000 K. $Cu_{2-\delta}X$ (X = S, Se and Te) is the only system showing high ZTs in a series of compounds composed of three sequential primary group elements. Various fabrication

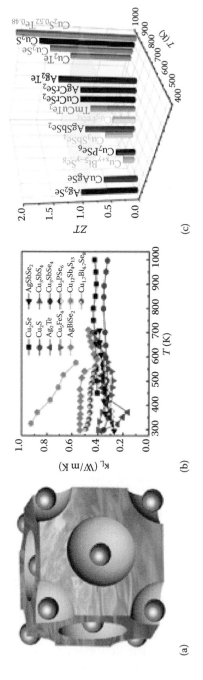

FIGURE 10.2 Liquid-like ionic TE materials. (a) Crystal structure of cubic chalcocite α-phase of $Cu_{2-x}X$. Blue spheres represent X atoms. The liquid-like copper ions freely travel throughout the sulfide sublattice. (b) Extremely low lattice thermal conductivities in liquid-like ionic TE materials. (c) TE figure of merit (*ZT*) in liquid-like ionic TE materials. (Reprinted by permission from Macmillan Publishers Ltd., *Nature Materials*, Liu, H. L. et al., 11, 2012, 422–425, copyright (2012); He, Y. et al.: High thermoelectric performance in non-toxic earth-abundant copper sulfide. *Advanced Materials*. 2014. 26. 3974–3978. Copyright Wiley-VCH Verlag GmbH & Co. KGaA. Reproduced with permission; reprinted by permission from Mi W. et al., *NPG Asia Materials*, He Y. et al. 7, e210, 2015, copyright (2015); Reprinted with permission from Mi W. et al., *Applied Physics Letters*, 104, 133903, 2014. Copyright 2014, American Institute of Physics; Zhu, H. et al. *Journal of Materials Chemistry A*, 3, 10303–10308, 2015. Reproduced by permission of The Royal Society of Chemistry; Wang, X. et al., *Journal of Materials Chemistry A*, 3, 13662–13670, 2015. Reproduced by permission of The Royal Society of Chemistry; Qiu, P. et al., *Energy and Environmental Science*, 7, 4000–4006, 2014. Reproduced by permission of The Royal Society of Chemistry; reprinted from *Materials Chemistry and Physics*, 145, Tewari, G. C. et al., Thermoelectric properties of layered antiferromagnetic CuCrSe$_2$, 156–161, Copyright (2014), with permission from Elsevier; Bhattacharya, S. et al., *Journal of Materials Chemistry A*, 1, 11289, 2013. Reproduced by permission of The Royal Society of Chemistry; reprinted with permission from Gascoin, F., and Maignan, A., *Chemistry of Materials*, 23, 2510–2513, 2011. Copyright 2011. American Chemical Society; reprinted with permission from Weldert, K. S. et al., *Journal of the American Chemical Society*, 136, 12035–12040, 2014. Copyright 2014. American Chemical Society; Lin, H. et al.: Chemical modification and energetically favorable atomic disorder of a layered thermoelectric material TmCuTe$_2$ leading to high performance. *Chemistry*. 2014. 20. 15401–15408. Copyright Wiley-VCH Verlag GmbH & Co. KGaA. Reproduced with permission; Reprinted with permission from Pan, L. et al., *Journal of the American Chemical Society*, 135, 4914–4917, 2013. Copyright 2013. American Chemical Society; Guin, S. N. et al.: High thermoelectric performance in tellurium free p-type AgSbSe$_2$. *Energy and Environmental Science*. 2013. 6. 2603–2608. Copyright Wiley-VCH Verlag GmbH & Co. KGaA. Reproduced with permission.)

methods, including melt-annealing,[1] melt-quenching approach,[8] high-energy BM,[9] self-propagating high-temperature synthesis,[10] and chemical methods such as coprecipitation[11] and solvothermal growth,[12] have been recently applied to synthesize $Cu_{2-\delta}Se$, and the ZT values ranging from 1.5 to 2.1 have been reported. Meanwhile, element alloying such as with Ag at the sites of copper and doping the structure by Al and I have been used to adjust the TE performance.[7,13] Mechanical alloying and melt-solidification techniques have been applied to fabricate $Cu_{2-\delta}S$ compounds with the ZT of up to 1.9 at 970 K.[14] Surprisingly, Cu_2S and Cu_2Te form a complete solid solution, while usually, this is not possible due to the large atomic size difference between S and Te. Moreover, very special mosaic crystal microstructures have been in this unique solid solution, leading to an enhancement in the Seebeck coefficient and a reduction in the lattice thermal conductivity κ_L, simultaneously culminating in a very high ZT value of 2.1 in $Cu_2S_{0.52}Te_{0.48}$ at 1000 K (see Figure 10.2c).[15]

Shortly after $Cu_{2-x}Se$ was reported to show a liquid-like TE transport, a series of Cu- and Ag-containing ionic conductors, including binary compounds $Cu_{2-x}S$,[2,16] $Cu_{2-x}Te$,[3,13] Ag_2Se,[17,18] and Ag_2Te,[14,19] as well as ternary compounds $CuAgSe$,[20,21] Cu_5FeS_4,[22] $CuCrSe_2$,[23,24] $AgCrSe_2$,[25] Cu_7PSe_6,[26] $Cu_{x+y}Bi_{5-y}Se_8$,[27] and $TmCuTe_2$,[28] have stimulated keen interest in the past few years as prospective TE materials with high ZT values. The most important and interesting feature in these ionic conductors is their intrinsically extremely low κ_L (see Figure 10.2b).[1–3,14,17,20,22–30]

Ag_2Se has also been recently studied. It shows the maximum ZT value near unity around room temperature due to its low bandgap (0.2 eV).[17] The material has a high σ, a moderate S, and a very small κ_L. $CuAgSe$ can be regarded as a mixture of Cu_2Se and Ag_2Se. It has two phases: a high-temperature α-phase and a low-temperature β-phase. The former is similar to the high-temperature superionic Cu_2Se with the cubic antifluorite structure, which has mobile Ag/Cu cations randomly distributed among the face-centered cubic Se-sublattice. The low-temperature β-CuAgSe has a layered structure of low symmetry. The layers of Ag and CuSe alternately stack in β-CuAgSe, with Ag atoms at nearly the same plane, Se atoms forming sheets of flattened tetrahedron-sharing corners, and Cu atoms located within the tetrahedrons. The ZT value of CuAgSe is initially low at 300 K but reaches a high value of 0.95 at 623 K.[31] Further studies have shown that the crystal structure and phase purity of CuAgSe critically depend on the chemical composition, and only a small amount of Ag deficiency is allowed in stable CuAgSe. ZT values of up to 0.6 are achieved in chemically stoichiometric CuAgSe.[20]

Bornite (Cu_5FeS_4) is a widespread copper sulfide mineral that occurs in a variety of ore deposits formed under a wide range of geological conditions. Ultralow κ_L has been observed in this natural sulfide because of its inherent and unusually complex crystal structure. By forming a solid solution with Cu_2S, the electronic transport and the TE performance of bornite are easily tuned to achieve a maximum ZT value of 1.2 at 900 K.[22] High-temperature

TE properties of diselenide ($MCrSe_2$, M = Cu, Ag) materials have also been explored. These compounds are naturally layered materials. ZT values around 1–1.4 have been reported for $CuCrSe_2$ and $AgCrSe_2$ on account of their extremely low κ_L value.[24,25] Cu_7PSe_6 has been reported with an extraordinary low thermal conductivity reaching below the glass limit. This may be associated with the soft phonon modes attributed by the molten copper sublattice. Eventually, the maximum ZT of 0.35 was achieved at 575 K.[26] The monoclinic $Cu_{x+y}Bi_5Se_8$ structure contains multiple disorders such as randomly distributed substitutions and interstitial disorders of Cu as well as asymmetrical disorders of Se, which simultaneously leads to an extremely low κ_L.[27]

10.2 BiCuSeO-Based Thermoelectric Materials

Jiehe Sui and Jing Li

10.2.1 Introduction

TE energy conversion technology, which can be used to convert waste heat into electricity, has received much attention in the past decade. The conversion efficiency of TE materials is determined by the figure of merit: $ZT = (S^2\sigma/\kappa)T$, where S is the Seebeck coefficient, σ electrical conductivity, κ thermal conductivity, and T absolute temperature. The larger the ZT value, the higher the conversion efficiency. In the TE community, the ZT value of ~1 is regarded as a performance benchmark for various TE materials.

In recent years, many new TE materials have been developed.[32] Oxide TE materials possess several advantages compared with the other compounds such as chemical and thermal stability, resource abundance, and simple preparation.[33] Oxide TE materials studied include $Na_xCo_2O_4$,[34] $CaMnO_3$,[35] $Ca_3Co_4O_9$,[36] $Bi_2Sr_2Co_2O_9$,[37] $SrTiO_3$,[38] ZnO,[39] $BiCuSeO$,[40] and so on. Among all the oxide TE materials mentioned earlier, BiCuSeO displays a potential as high-performance TE materials.

10.2.2 Crystal and Band Structure

Figure 10.3 shows the crystal structure of BiCuSeO crystallized with space group of *P4/nmm*, a = 3.921 Å, and c = 8.913 Å. It has a layered structure with $(Bi_2O_2)^{2+}$ layer and $(Cu_2Se_2)^{2-}$ layer alternately stacked along the c axis.[41] The $(Bi_2O_2)^{2+}$ layers are built of slightly distorted Bi_4O tetrahedrons, and the $(Cu_2Se_2)^{2-}$ layers are built of slightly distorted $CuSe_4$ tetrahedrons sharing Se–Se edges. The Bi–O bond has more covalent characteristic, and the Cu–Se bond has more ionic character. The $(Bi_2O_2)^{2+}$ layer and $(Cu_2Se_2)^{2-}$ are connected by the weak Bi–Se bond.

Figure 10.4 shows the band structure of BiCuSeO. The bandgap of BiCuSeO is about 0.8 eV. Bi 6s states (–10 to –12 eV relative to the valence band maximum) and O 2p states (–5 to –7 eV) are located deep in the valence

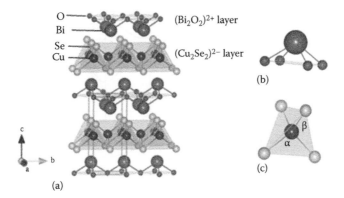

FIGURE 10.3 Crystal structure of BiCuSeO: (a) BiCuSeO, (b) BiO$_4$, and (c) CuSe$_4$.

band.[42] The bottom of the conduction band mainly consists of Bi 6p states. The top of the valence band consists of bonding hybridized Cu 3d–Se 4p states, nonbonding Cu 3d states (–1 to –3 eV) and antibonding hybridized Cu 3d–Se 4p states (close to the valence band maximum). The top of the valence band consists of a hole pocket located on the Γ–M line. The other hole pockets can be observed on the Γ–X line and at the Z point, which are located at about –0.15 eV below the Fermi level, and they participate in the electronic conduction at high hole concentration or at high temperature. A hole pocket is located at the Z point for a 2D characteristics of the crystal structure.

10.2.3 Thermoelectric Properties of BiCuSeO

Table 10.1 shows the crystallographic parameters, TE properties, and elastic properties for BiCuSeO at room temperature.[44] Pristine BiCuSeO is a p-type semiconductor related to the Cu vacancies with the hole concentration of ~10^{18} cm^{-3}. The carrier mobility is low due to the strong carrier scattering of the layered crystal structure. The electrical conductivity of pristine BiCuSeO is rather low, 1.1 S/cm, and the Seebeck coefficient of BiCuSeO is moderate, 349 μV/K. The thermal conductivity of BiCuSeO is rather low, 0.9 W/m K. This value is far lower than that of the other TE materials. The possible origin may be related to the following reasons: heavy atomic mass; layer structure, which scatter phonons; and weak bonds inside the compound.[45–47] Although BiCuSeO possesses low thermal conductivity and moderate Seebeck coefficient, the electrical conductivity is very low, which leads to a low *ZT* value, ~0.45 at 900 K. Therefore, increasing the electrical conductivity will be an efficient method to improve the TE properties of BiCuSeO. According to the formula of the electrical conductivity $\sigma = en\mu$, where *e* is the electric charge, *n* is the carrier concentration, and μ is the carrier mobility, so the *ZT* value can be improved by increasing the *n* and μ.

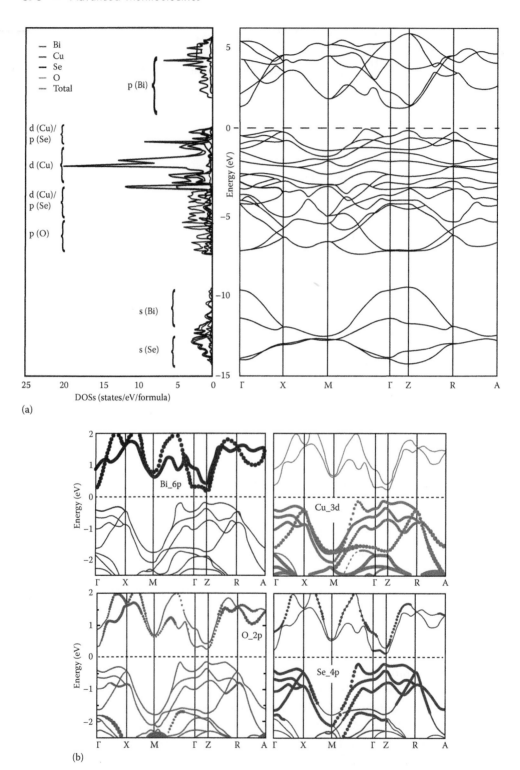

(a)

(b)

FIGURE 10.4 (a) Electronic band structure (*right*) and projected DOS (*left*) for the undoped BiCuSeO.[42] (b) The orbital-decomposed band structure of BiCuOSe compound. (Zou, D. et al., *Journal of Materials Chemistry A*, 1, 8888–8896, 2013. Reproduced by permission of The Royal Society of Chemistry.)

TABLE 10.1 Crystallographic Parameters, TE Properties, and Elastic Properties of BiCuSeO at Room Temperature

Formula weight	367.4858
Crystal system	Tetragonal
Space group	P4/nmm
Unit cell dimensions	$a = 3.921$ Å, $\alpha = 90°$
	$b = 3.921$ Å, $\beta = 90°$
	$c = 8.913$ Å, $\gamma = 90°$
Unit cell volume	137.06 Å³
Z	2
Theoretical density	8.9 g cm⁻³
Carrier concentration	1×10^{18} cm⁻³
Carrier mobility	22 cm²/V s
Band effective mass	Light hole band (0.18m_e)
	Heavy hole band (1.1m_e)
Seebeck coefficient	349 μV/K
Electrical conductivity	1.12 S/cm
Lattice thermal conductivity	0.55 W/m K
Longitudinal sound velocity	3290 m/s
Transverse sound velocity	1900 m/s
Average sound velocity	2107 m/s
Young's modulus	76.5 GPa
Debye temperature	243 K
Poisson ratio	0.25
Grüneisen parameter	1.5

Source: Zhao, L. et al., *Energy and Environmental Science, 7,* 2900–2924, 2014. Reproduced by permission of The Royal Society of Chemistry.

The methods to improve the electrical conductivity include acceptor doping or Cu vacancy to increase the carrier concentration, texturation, or modulation doping (MD) to increase the carrier mobility and band engineering to tune the electrical conductivity.

10.2.3.1 Acceptor Doping to Increase Carrier Concentration

BiCuSeO is a p-type semiconductor, so acceptor doping can increase the carrier concentration. Mg^{2+},[48,49] Ca^{2+},[45,50] Sr^{2+},[44,51] Ba^{2+},[52] Pb^{2+},[53–55] Na^+,[56] K^+,[57] and Ag^+[58] as acceptor doping at Bi^{3+} site can induce holes in BiCuSeO to increase the carrier concentration, thereby leading to improvement in the electrical conductivity and *ZT* value.

Doping with 10% of Ca^{2+}, Sr^{2+}, Ba^{2+}, and Pb^{2+} dramatically increases carrier concentration from 10^{18} to 10^{20} cm⁻³. However, carrier mobility decreases from 23 to 2 cm²/V s due to the increased phonon scattering caused by doping. Electrical conductivity increases from 1.12 to 500 S/cm at room temperature. The Seebeck coefficient decreases from 349 to 100 μV/K due to the increased carrier concentration. Finally, the doping balances the electrical conductivity and Seebeck coefficient, and the *ZT* value is increased. Among all the M^{2+} doping work at the Bi^{3+} site, Ba^{2+} doped BiCuSeO shows the highest *ZT* value of ~1.1 at 923K,[21] which is the first reported *ZT* value of over 1 among the polycrystal oxide TE materials. Mg^{2+} is not so efficient compared with the other doping with valence of 2+. Mg doping (12.5%) leads to an increase of carrier concentration from 1×10^{18} to 4×10^{18} cm^{-3}, electrical conductivity from 1.12 to 32 S/cm at room temperature, and the corresponding *ZT* from 0.45 to 0.67 at 923 K.

For Na^+ doping at the Bi^{3+} site, the carrier concentration and electrical conductivity increase from 10^{18} cm^{-3} and 1.12 S/cm to 10^{20} cm^{-3} and 125 S/cm, respectively. Finally, the *ZT* value is improved to 0.9 at 923 K when the Na doping content is 1.5%. It should be noted that carrier mobility is slightly decreased using Na^+ doping compared with that using M^{2+} doping due to the weak carrier scattering by the small amount of doping. For K^+ doping at the Bi^{3+} site, both the carrier concentration and mobility are increased.[57] For Ag^+ doping at the Bi^{3+} site, carrier concentration, electrical conductivity, and *ZT* are slightly increased compared with those using Na and K doping, which is ascribed to the appearance of impure phase.

10.2.3.2 Inducing Vacancies to Increase Carrier Concentration

Inducing vacancies is another method to improve carrier concentration in BiCuSeO. BiCuSeO with Cu vacancies[59,60] displays higher carrier concentration, electrical conductivity, and *ZT* value than that of pristine BiCuSeO. Cu vacancies in the charge conductive selenide $(Cu_2Se_2)^{2-}$ layers induce holes and increase carrier concentration. This can be explained by the following equation:

$$2BiCu_{1-x}SeO = (Bi_2O_2)^{2+} + (Cu_{2(1-x)}Se_2)^{(2+x)-} + xV_{\overline{Cu}} 2xh^+, \quad (10.2)$$

where $V_{\overline{Cu}}$ is the Cu vacancy and h^+ is the holes. The electrical conductivity is increased from 4.7 to 30 S/cm, and *ZT* increased to 0.81 at 923 K.

Recently, Li et al.[61] induced Bi and Cu dual vacancies as an effective strategy to realize the synergistic optimization of TE properties of BiCuSeO as shown in Figure 10.5. In Bi/Cu dual vacancies $Bi_{0.975}Cu_{0.975}SeO$, induced holes transfer from Bi vacancy center to Cu vacancy center, resulting in much higher electrical conductivity of 4700 S/m at 750 K. What is more, Bi/Cu dual vacancies in the BiCuSeO lead to much intensive phonon scattering in the whole system; therefore, a significant reduction of thermal conductivity is obtained as low as 0.37 W/m K at 750 K. Therefore, a high *ZT* value of 0.84 has been obtained at 750 K in BiCuSeO sample with Bi/Cu dual vacancies.

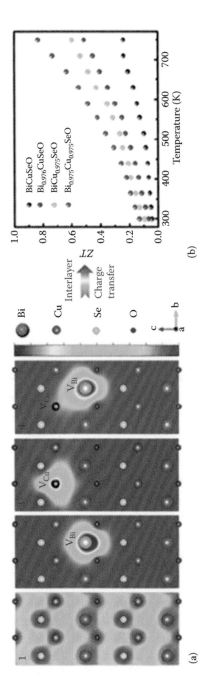

FIGURE 10.5 (a) Schematic representation of trapped positrons for $Bi_{1-x}Cu_{1-y}SeO$ samples in (100) plane; (b) temperature-dependent ZT of $Bi_{1-x}Cu_{1-y}SeO$. (Reprinted with permission from Li, Z. et al., *Journal of the American Chemical Society*, 137, 6587–6593, 2015. Copyright 2015. American Chemical Society.)

10.2.3.3 Texturation to Increase Carrier Mobility

Acceptor doping and inducing vacancies are effective ways to enhance the electrical conductivity via increasing the carrier concentration, but high-level dopants also deteriorate the carrier mobility. Therefore, increasing the carrier mobility is an effective way to improve electrical conductivity and ZT value of pristine BiCuSeO. BiCuSeO is constituted by conductive $(Cu_2Se_2)^{2-}$ layers alternately stacked with insulating $(Bi_2O_2)^{2+}$ layers along the c axis, displaying anisotropic characteristic. Therefore, carrier mobility on the a and b planes (in-plane) is higher than that in the c direction (cross-plane), which is similar to the anisotropy observed in Bi_2Te_3 and $CaCoO_3$ compounds.[62–64] Thus, based on the Ba acceptor doping, the hot forging process is applied to induce texturation to improve the carrier mobility as shown in Figure 10.6a. $Bi_{0.875}Ba_{0.125}CuSeO$ is hot forged for three times to produce textured microstructures with the grains preferentially oriented with the in-plane perpendicular to the pressing direction as shown in Figure 10.6b. The carrier mobility along the direction perpendicular to the pressing direction was increased, increasing the electrical conductivity and maximizing the PF at 923 K from 6.3 µW/cm K² for the sample before hot forging to 8.1 µW/cm K² after the hot forging process. Therefore, the maximum ZT was significantly increased from 1.1 for $Bi_{0.875}Ba_{0.125}CuSeO$ to 1.4 through texturation, which is the highest ZT ever reported among oxygen-containing materials as shown in Figure 10.6c.[65]

10.2.3.4 Modulation Doping to Increase the Carrier Mobility

MD is widely used in 2D electron gas thin film devices to improve the carrier mobility and the electrical conductivity. In the case of MgZnO/ZnO heterostructures, the electrons are generated in the MgZnO-doped layers and transferred to ZnO-undoped layers, which effectively avoids the scattering of the ionized defects in the ZnO layers.[66] Enhanced PFs were achieved in $Si_{1-x}Ge_x$ alloys via constructing heterostructures consisting of one highly doped component and an undoped component. The carrier mobility and PF were found to be significantly enhanced, and the underlying mechanism was attributed to MD approach.[67,68] MD is further confirmed as an efficient method to increase carrier mobility and improve TE properties in BiCuSeO.[69] As shown in Figure 10.7, the heterostructures of the modulation-doped sample preferentially make charge carriers transport in the low carrier concentration area, which increases carrier mobility by a factor of 2 while maintaining the carrier concentration similar to that in the uniformly doped sample. The improved electrical conductivity and retained Seebeck coefficient synergistically lead to a high PF ranging from 5 to 10 µW/cm K². A high ZT value of ~1.4 is achieved in a BiCuSeO system.[69]

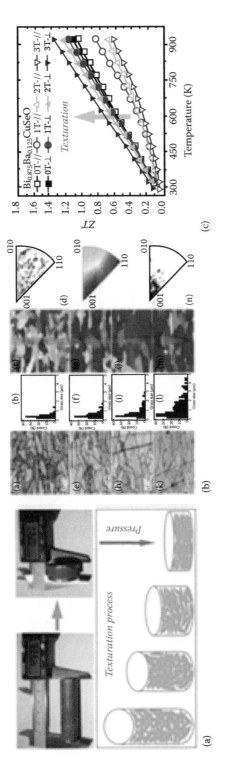

FIGURE 10.6 (a) Hot-forged process, (b) electron backscatter diffraction microstructure, and (c) *ZT* value of textured $Bi_{0.875}Ba_{0.125}CuSeO$. (Sui, J. et al., *Energy and Environmental Science*, 6, 2916, 2013. Reproduced by permission of The Royal Society of Chemistry.)

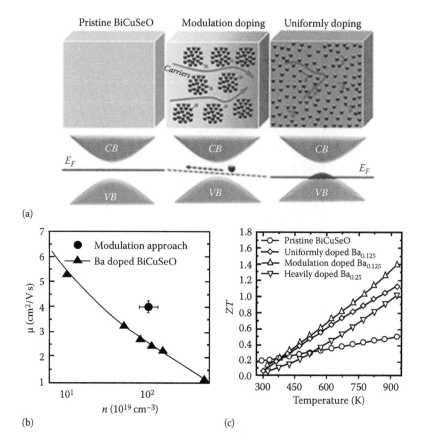

FIGURE 10.7 (a) 3D schematic showing the band structures and Fermi energy levels for the pristine BiCuSeO, modulation-doped $Bi_{0.875}Ba_{0.125}CuSeO$, and uniformly doped $Bi_{0.875}Ba_{0.125}CuSeO$; (b) carrier mobility as a function of carrier concentration; and (c) ZT for the uniformly doped $Bi_{1-x}Ba_xCuSeO$ and modulation-doped $Bi_{0.875}Ba_{0.125}CuSeO$. (Reprinted with permission from Pei, Y. et al., *Journal of the American Chemical Society*, 136, 13902, 2014. Copyright 2014. American Chemical Society.)

10.2.3.5 Band Engineering to Tune the Electrical Conductivity

The bandgap is ~0.8 eV for BiCuSeO, while the analog BiCuTeO is ~0.4 eV. Therefore, the bandgap of BiCuSeO can be decreased by substituting Te at the Se site.[70,71] Figure 10.8 shows the electronic absorption spectra, the schematic figure for bandgap tuning, and TE properties of $BiCuSe_{1-x}Te_xO$. As the bandgap decreased with increasing Te fraction, the electrical conductivity is increased from ~15 to 40 S/cm at 923 K for $BiCuSe_{0.85}Te_{0.15}O$. Coupling with improved electrical conductivity and low thermal conductivity, a high ZT of 0.71 at 923 K is obtained in $BiCuSe_{0.94}Te_{0.06}O$.[70]

La substitution at Bi site also tunes the band structure of BiCuSeO.[72] The experiment result shows that the bandgap is increased with the increase of La content as shown in Figure 10.9a, while the effective mass m^* is decreased. The first-principles calculation shows that energy separation ΔE

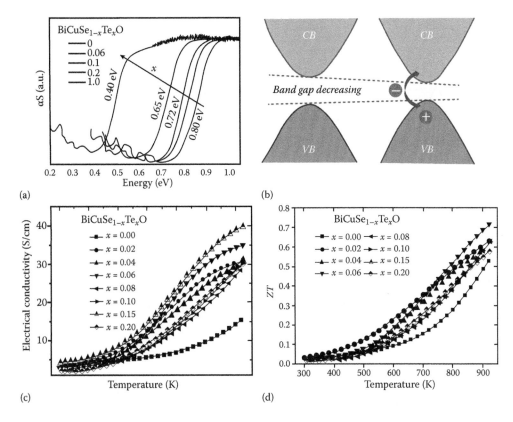

FIGURE 10.8 (a) Electronic absorption spectra for BiCuSe$_{1-x}$Te$_x$O samples, (b) the schematic figure for bandgap tuning, temperature dependence of (c) electric conductivity and (d)ZT for BiCuSe$_{1-x}$Te$_x$O. (Liu, Y. et al., *Chemical Communications*, 49, 8075–8077, 2013. Reproduced by permission of The Royal Society of Chemistry.)

between L bands (heavy band) and Σ bands (light band) is decreased with the increase of the La content. It is known that the participation of the light band in the carrier transport can increase the carrier mobility.[72] Therefore, the carrier mobility is increased from 3 to 40 cm^2/V s with the increase of La substitution content. Finally, the electrical conductivity is increased from 4 to 40 S/cm, and the ZT value is increased from 0.5 to 0.74 at 923 K for La-substituted BiCuSeO, as shown in Figure 10.9c.

10.2.4 Conclusions

BiCuSeO is a promising oxide TE material with the intrinsically low thermal conductivity and moderate Seebeck coefficient. However, low electrical conductivity leads to a low ZT value. Acceptor doping, inducing vacancies, and band engineering are adopted to tune the electrical conductivity. Finally, a record ZT value of ~1.4 is obtained for BiCuSeO-based oxide TE materials.

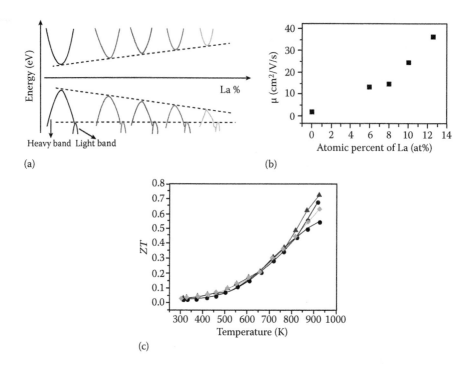

(a)

(b)

(c)

FIGURE 10.9 (a) Band structure, (b) carrier mobility, and (c) ZT of $Bi_{1-x}La_xCuSeO$. (Reprinted with permission from Liu, Y. et al., *Applied Physics Letters*, 106, 233903, 2015. Copyright 2015, American Institute of Physics.)

10.3 High Power Factor of CoSbS

Zihang Liu and Zhifeng Ren

10.3.1 Introduction

Most advanced TE materials primarily consist of expensive elements (Te, etc.) or toxic elements (Pb, etc.), such as Bi_2Te_3 and PbTe.[73–76] Since sulfur (S) is non-toxic and inexpensive as well as highly abundant in the earth's crust,[77] sulfur-based TE materials have attracted some interest, and the highest ZT values larger than unit have been achieved mostly in p-type sulfide compounds.[78–84] Historically, the ZT of corresponding n-type lead-free sulfide materials is still low,[78,82–84] which needs to be enhanced in order to pair up with the p-type to make modules. In addition, most sulfide TE materials including PbS exhibit a low or modest PF,[78,80–84] compared with other advanced TEs. Therefore, more attention should be focused on optimizing the TE performance, including both PF and ZT, of n-type sulfide materials in the future.

Due to the high PF and modest ZT in the mediate temperature range, CoSbS compound has been recently considered as a good n-type sulfide TE material.[84–87] Carlini et al.[88] reported the synthesis and microstructure of CoSbS compound. Thereafter, Parker et al.[85] and Liu et al.[84] achieved modest PFs of ~15 and 20 μW/cm K^2 at 773 K via Ni doping. Very recently, Chmielowski et al.[87] reported a high PF of around 27 μW/cm K^2 at a broad temperature range from 543 to 730 K, which is the record of high PF in sulfide TE materials. The studies mentioned earlier strongly indicate that CoSbS is a promising sulfide TE material for power generation application. In the following, we review the recent progress made in understanding the structure, synthesis, and TE properties of CoSbS.

10.3.2 Crystal Structure and Theoretical Calculation

10.3.2.1 Crystal Structure

CoSbS crystallizes in orthorhombic space group 61 (*Pbca*) with a lower symmetry. The unit cell parameters are a = 584.2(3) pm, b = 595.1(3) pm, and c = 1166.6(4) pm. There are eight octahedrons [CoSb$_3$S$_3$] in one unit cell, where one Co is octahedrally coordinated to three Sb and three S atoms as shown in Figure 10.10, which is isoelectronic and isostructural with respect to the FeS$_2$ structure. Here, Co replaces Fe and Sb replaces one of the S atoms.

10.3.2.2 Theoretical Calculation

Parker et al.[85] first calculated the band structure of CoSbS using the linearized augmented plane wave code WIEN2K. The obtained bandgap is around 0.75 eV, consistent with the reported large Seebeck coefficient. The conduction band of CoSbS is degenerate, with five minima within 0.2 eV of the conduction band minimum, including three at Γ, which is beneficial for the large Seebeck coefficient. However, it does not exhibit the magnetic properties, confirmed by the spin-polarized calculations.

Later, Chmielowski et al.[87] report the band structure and DOS, which are almost identical with the previous results. The calculated DOSs and

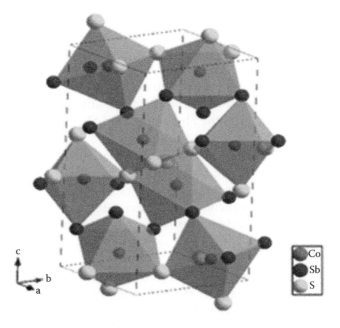

FIGURE 10.10 Crystal structure of CoSbS. (Liu, Z. et al., *Journal of Materials Chemistry C*, 3, 10442–10450, 2015. Reproduced by permission of The Royal Society of Chemistry.)

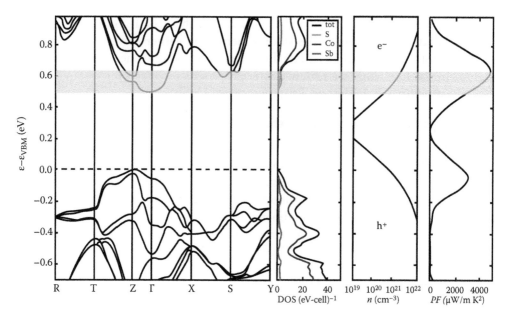

FIGURE 10.11 Band structure, partial DOS, carrier concentration, and PF of CoSbS evaluated from DFT. In the right panel, $\tau = 10^{-14}$ s. The choice of τ is only to obtain an illustration of the dependence of the PF on the electronic structure of CoSbS. (Chmielowski, R. et al., *Journal of Materials Chemistry C*, 4, 3094–3100, 2016. Reproduced by permission of The Royal Society of Chemistry.)

projections for CoSbS are shown in Figure 10.11. The valence bands are of substantial mass with the total DOS increasing rapidly from the band edges, governed by the transition metal element Co. They also note that the width of the Fermi distribution function at 600 K would normally extend more than 150 meV into the conduction band in n-type CoSbS, as represented by the shaded region in the Figure 10.11. The high band degeneracy is known to result in favorable performances. The optimum chemical potential for PF is estimated at approximately 150 meV within the conduction band. Moreover, the energy of formation of intrinsic defects (vacancies and interstitial and antisite defects) is calculated, and the dominating intrinsic defects are the antisite defects S_{Sb} (S on the Sb position) and Sb_S. They also investigate the effect of 49 extrinsic dopants, and the high-throughput DFT defect calculations reveal Ni, Pd, and Te as promising dopant candidates for CoSbS.

Bhattacharya et al.[86] used two materials design strategies for high throughput identification of new ternary sulfides of $M_x M1_y S_z$,[86] where M1 is Sn and Sb, including band structure details and defect thermochemistry. Finally, only CoSbS where the desired carriers can be produced is left (Figure 10.12).

10.3.3 Synthesis Method

Carlini et al.[88] used two steps to prepare CoSbS, first synthesizing CoSb alloy and then adding S. Parker et al.[85] used a method of self–Sb–flux to prepare the single crystals of CoSbS. They also use a very complicated method of tube vacuum melting to prepare polycrystalline Ni-doped CoSbS. Chmielowski et al.[87] used a similar method to prepare Ni-, Te-, and Pd-doped CoSbS

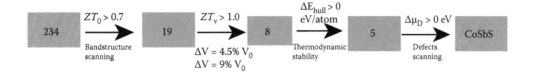

FIGURE 10.12 Schematic flowchart of all the ternary sulfides investigated in this work and the descriptors on the basis of which the candidates were selected. (Bhattacharya, R. et al., *Journal of Materials Chemistry A*, 4, 11086–11093, 2016. Reproduced by permission of The Royal Society of Chemistry.)

polycrystal. Liu et al.[84] directly used a simple mechanical alloying method to successfully prepare Ni-doped CoSbS polycrystal.

10.3.4 Thermoelectric Properties

Since Ni doping and Te doping lead to the highest ZT and PF in the CoSbS system,[84,87] respectively, the following section mainly focuses on their corresponding TE properties.

10.3.4.1 Ni doping

Figure 10.13 shows the electrical transport properties of $Co_{1-x}Ni_xSbS$ samples. The carrier concentration monotonously increases with Ni doping by nearly four orders of magnitude from 3.09×10^{17} cm^{-3} for CoSbS sample to 1.13×10^{21} cm^{-3} for $Co_{0.92}Ni_{0.08}SbS$ sample. This variation is consistent with the fact that Ni has one more electron in its outer shell than Co. The low mobility (from 2 to 4 cm^2/V s) is related to the large effective mass originated from the complex band structure that will be discussed later. As shown the Figure 10.13b, the electrical resistivity decreases by increasing the Ni doping concentration, especially at low temperature. Figure 10.13c shows the large Seebeck coefficient for CoSbS system, namely, –565 μV/K at 300 K and –200 μV/K at 873 K, which is related to the high effective mass at the Fermi level (m^*). The obtained effective mass (m^*) based on the single parabolic band model for CoSbS is 1.1 m_e, which could be ascribed to the substantial band degeneracy near the conduction band edges. More significantly, the m^* for Ni-doped samples $Co_{1-x}Ni_xSbS$ (x = 0.02, 0.04, 0.06, and 0.08) are $6.2m_e$, $5.8m_e$, $5.9m_e$, and $5.9m_e$, respectively. This large increment is fully consistent with the room-temperature Pisarenko plots and Seebeck coefficient as a function of carrier concentration. In Figure 10.13d, the dashed line and solid line represent the $S–n$ relationship with different effective masses (m^* = 1.1m_e and m^* = 6.0m_e). The $S–n$ point of undoped CoSbS locates well on the Pisarenko plots with m^* = 1.1m_e, while the $S–n$ points of Ni-doped CoSbS lay around the Pisarenko plots with m^* = 6.0m_e, contributing to keeping relatively high Seebeck coefficients for Ni-doped samples. This difference indicates that Ni doping in the CoSbS system exhibits a beneficial influence on the band structure, which further needs to be clarified by theoretical calculations. Finally, a maximum PF of 20 μW/cm K^2 at 873 K is obtained for $Co_{0.92}Ni_{0.08}SbS$ sample, which is two times than that for the CoSbS sample.

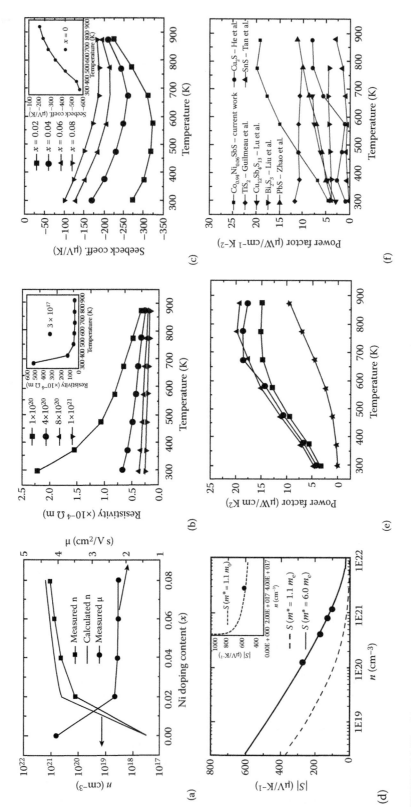

FIGURE 10.13 (a) Room-temperature carrier concentration and mobility as a function of Ni doping content. (b) Electrical resistivity. (c) Seebeck coefficient, (d) room-temperature Pisarenko plots, and (e) PF. (f) PF of $Co_{0.94}Ni_{0.06}SbS$ in comparison with other sulfide TE materials. The insets separately show the electrical transport properties of undoped CoSbS. (Liu, Z. et al., *Journal of Materials Chemistry C*, 3, 10442–10450, 2015. Reproduced by permission of The Royal Society of Chemistry.)

As shown in Figure 10.13f, the PF of $Co_{0.94}Ni_{0.06}SbS$ is much larger than that of other sulfide TE materials, including PbS, Cu_2S, $Cu_{12}Sb_4S_{13}$.

The κ_{total} of nanostructured CoSbS decreases with increasing temperature from 7.9 W/m K at 300 K to 3.9 W/m K at 873 K. The inverse temperature dependence of κ_{total} reveals the predominant phonon contribution to the thermal conductivity. As shown in Figure 10.14a, the κ_{lat} exhibits significant reduction upon Ni doping. For instance, the κ_{lat} at room temperature has decreased almost 20% from 6.3 W/m K for 2% Ni-doped sample to 5.0 W/m K for 8% Ni doping. Because the difference in the atomic mass and radius between Co (58.93 g/mol, 1.25 Å) and Ni (58.69 g/mol, 1.24 Å) is quite small, the phonon scattering caused by point defects from mass and strain fluctuation could not be qualified to explain the reduction of lattice thermal conductivity. According to the Callaway model, the theoretical model fits the experimental data well, using point defect (PD) scattering, electron–phonon (PE) scattering, phonon–phonon (U) scattering, and grain boundary(GB) scattering. It is clear that both the PE scattering and PD scattering drastically increase upon Ni doping. To clarify the contribution of each scattering process to the κ_{lat}, the theoretical and experimental κ_{lat} values of $Co_{0.92}Ni_{0.08}SbS$ are shown in Figure 10.14c. The short dot line, dash line, dot line and solid line represent the simulated κ_{lat} values considering the related scattering process—U, U + PD, U + PD + GB, and U + PD + GB + PE,

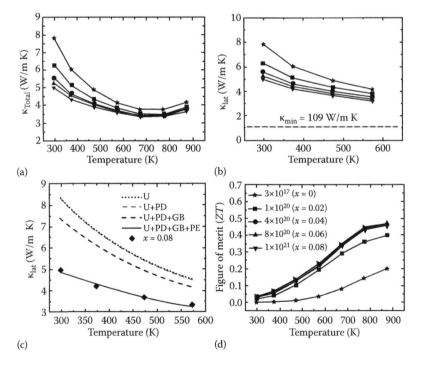

(a) (b) (c) (d)

FIGURE 10.14 (a) Total thermal conductivity κ_{total}. (b) Lattice thermal conductivity κ_{lat}. (c) Contribution of various scattering processes to the lattice thermal conductivity κ_{lat} of $Co_{0.92}Ni_{0.08}SbS$ sample. (d) ZT. (Liu, Z. et al., *Journal of Materials Chemistry C*, 3, 10442–10450, 2015. Reproduced by permission of The Royal Society of Chemistry.)

respectively. As expected, the effect of PD scattering on κ_L is negligible. GB scattering reduces κ_{lat} by 10%. Surprisingly, PE scattering becomes the most important and dominant part in CoSbS with high Ni doping concentration. PE scattering, which is usually assumed as a negligible part for pure semiconductors at high temperature, may become a main contribution to the reduction of lattice thermal conductivity in heavily doped semiconductors with a high m^*, including n-type skutterudite and HH alloys.[89–91] With Ni doping to tune the carrier concentration in the range of 3.5×10^{20}–1.3×10^{21} cm^{-3}, the highest ZT values are about 0.5 at 873 K for three samples ($x =$ 0.04, 0.06, and 0.08), which is two times than that for undoped CoSbS.

10.3.4.2 Te Doping

Table 10.2 shows the room-temperature transport measurements with different dopants.[87] Surprisingly, effective doping leads to increased mobility, which is contradictory with the result of the study by Liu et al.[84] The room-temperature PF of Te-doped CoSbS is about 5 µW/cm K, higher than that of Ni- and Pd-doped CoSbS. However, compared with the effect of Ni doping, Te doping leads to the lower resistivity, higher Seebeck coefficient, and thus higher PF. This phenomenon is as a result of reduced compensating defects Sb$_S$ upon Te doping. For Te doping, the PF increases from 0.7 to 15 µW/cm K^2 and from 7.7 to 27 µW/cm K^2 at 330 and 725 K, respectively. The highest PF of CoSb$_{0.96}$Te$_{0.04}$S is the record in sulfide TE materials, which provide the promising prospect of power generation. Due to the slightly decreased κ_{total}, the highest ZT of CoSb$_{0.96}$Te$_{0.04}$S nearly reaches 0.5 at 725 K, which is five times that of undoped CoSbS (Figure 10.15).

10.3.5 Conclusion

Due to the beneficial valence band structure, CoSbS-based compounds exhibit a high potential of high PF. High-throughput DFT defect calculations reveal Ni, Pd, and Te as promising dopant candidates for CoSbS. Ni doping and Te doping could achieve a high PF around 20 and 27 µW/cm K^2 at a high temperature, respectively, which is the record of high PF in sulfide TE materials. In addition, the strong PE scattering will also keep

TABLE 10.2 Room-Temperature Transport Measurements Doped with Effective Dopants

Dopant	n (cm^{-3})	μ (cm^2/V m)	ρ (mΩ cm)	S (µV/cm)	PF (µW/m K^2)
Undoped	-4.75×10^{18}	5.3	293 ± 2.2	−459 ± 36	72 ± 0.5
+2 at% Bi	-3.78×10^{18}	11.0	160 ± 8.2	−459 ± 16	132 ± 7
+2 at% Se	-1.52×10^{19}	5.9	91 ± 0.4	−354 ± 30	138 ± 0.5
+2 at% Pd	-7.7×10^{19}	9.7	10 ± 0.02	−185 ± 8	342 ± 0.8
+2 at% Ni	-6.94×10^{19}	7.2	13 ± 0.1	−213 ± 19	349 ± 2.4
+2 at% Te	-1.05×10^{20}	15.3	4.4 ± 0.02	−198 ± 7	891 ± 4.2

Source: Chmielowski, R. et al., *Journal of Materials Chemistry C*, 4, 3094–3100, 2016. Reproduced by permission of The Royal Society of Chemistry.

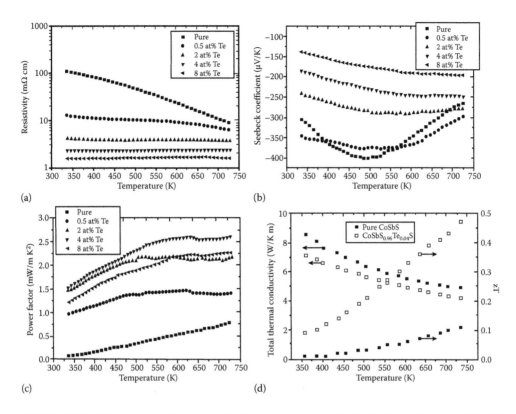

FIGURE 10.15 Temperature-dependent (a) electrical resistivity, (b) Seebeck coefficient, and (c) PF for Te-doped CoSbS. (d) Total thermal conductivity and ZT of $CoSb_{0.96}Te_{0.04}S$. (Chmielowski, R. et al., *Journal of Materials Chemistry C*, 4, 3094–3100, 2016. Reproduced by permission of The Royal Society of Chemistry.)

a lower thermal conductivity. Finally, a ZT of 0.5 can be obtained for both Ni doping and Te doping, which highlights the potential prospect for power generation for CoSbS-based TE materials.

10.4 Borides and Carbides

Takao Mori

10.4.1 Introduction: The Need for High-Temperature Thermoelectric Materials

Several interesting high-temperature TE applications have been recently proposed. Yazawa et al.[92] and Yazawa and Shakouri[93] have made comprehensive simulations and demonstrated high-temperature TE topping cycles which can yield, for example, a 6% increase of output from a 520 MW class Rankine cycle power plant with a relatively modest ZT of 0.7. There are also many examples of factory waste heat which can benefit from high-temperature TE materials, such as wire manufacturing furnaces at ~1120 K[94] etc. In steelworks, a significant 10 kW TE power generation from ~1170 K slag radiative heat has been reported for low-temperature Bi_2Te_3-type modules,[95] which can likely be enhanced by using high-temperature TE modules. The conversion of concentrated high-temperature solar power by a solar TE converter has also been carried out.[96] These applications are all stationary applications with applicative requirements appearing to be less strict than vehicular applications, for example. With the development of viable, good-performance high-temperature TE materials, the implementation of some of these applications has the potential to be relatively quick and of high impact.

In considering high-temperature TE materials to target, the first requirement is for the materials to have high thermal stability. Refractory compounds, such as borides and carbides, typically have double the melting points/synthesis temperatures of the working temperatures of these applications and will therefore be very safe in this regard. I will review the TE properties of some prospective compounds together with recent advances.

10.4.2 Borides

10.4.2.1 Lower Borides

10.4.2.1.1 General Considerations

So-called lower borides with relatively low boron-to-metal ratio of $[B]/[M] \leqq 12$ (M = metal) are typically very good metallic compounds,[97,98] with assumedly not so high TE performance. The reason that they are metallic can be understood when considering the boron network framework. Taking the binary borides as examples, in MB_2, the boron forms graphitic sheets $[B]_\infty$, and as the boron content increases, the basic unit of the boron network becomes clusters, i.e., the B_6 octahedron for MB_4 and MB_6 and the B_{12} cubooctahedron for MB_{12}. All these structural components, the B_2 net, B_6 octahedron, and B_{12} cubooctahedron, have been calculated to be two-electron deficient when bonding in solids, and therefore, combination with trivalent or more electron-rich metal atoms will result in excess electrons in the conduction band and, therefore, good metallic compounds.[97,98]

However, there are at least two systems which are of particular TE interest.

10.4.2.1.2 Alkaline-Earth (Divalent) Hexaborides

First is the alkaline-earth AB_6 (A = Ca, Sr) hexaboride. Despite the relatively high thermal conductivity, the ZT value of ~0.35 have been reported for some bulk samples.[99–102] However, one of the drawbacks of this system is the apparent difficulty of physical property control. From the electron requirements described earlier, alkaline-earth hexaborides can be expected to be nonmetallic. However, despite CaB_6 being extensively researched because of reports of high-temperature ferromagnetism in *Nature*, etc.,[103] which was later elucidated to be due to iron impurities deposited during flux crystal growth,[104,105] there was still controversy whether CaB_6 is a semimetal or small bandgap semiconductor.[106,107] This is because it appears to be very easy to inadvertently dope CaB_6 with impurities, which have significant effects on the physical properties. However, as research on this system continues, control and reliability should increase. The thermal conductivity of CaB_6 is also high, and various efforts have been made to lower it.[108]

While thin films are not expected to be useful for very high-temperature applications, surprisingly high ZT value of ~3 at room temperature has previously been reported for boron carbide nanofilms.[109] Although this result has not been reproduced yet, there is interest in investigating boride nanofilms. Successful thin film growth of SrB_6 with extremely fast growth rates from elemental strontium and decaborane sources using a chemical vapor deposition (CVD)-based method was recently demonstrated.[110]

As might be expected considering the possibility of carbon impurities, the reports on alkaline-earth hexaborides have been predominantly n-type.[99–102] YbB_6 has been reported to be p-type.[111]

10.4.2.1.3 *RTB₄ (AlB₂-Type Analogs)*

RTB_4 (R = rare earth; T = transition metal or Al) is another thermoelectrically interesting group of compounds. A wide combination of rare-earth and metal (mostly transition metal) elements is possible, and there are several structure types. Most common is the $YCrB_4$-type structure (space group *Pbam*), α-type.[112] These types of structure have been called AlB_2-type analogs,[113–115] because the boron-to-metal ratio is 2:1 and boron 2D atomic nets sandwich metal atoms. Since the sizes of rare-earth elements and metals are typically different, instead of homogeneous graphitic [6]-sided rings, the boron nets are composed of [5]- and [7]-sided rings, each sandwiching metal and rare-earth atoms, respectively. There is also the β-type $ThMoB_4$-type (*Cmmm*),[116] with different symmetric arrangement (tiling) of the [5]- and [7]-sided rings (Figure 10.16).

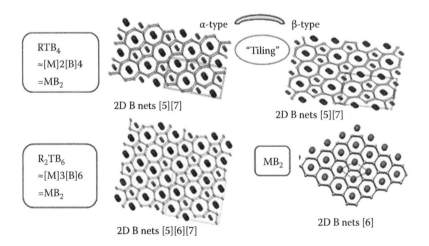

FIGURE 10.16 Schematic view of AlB_2 analog compounds. (Reprinted with permission from Mori, T. et al., *Journal of Applied Physics*, 105, 07E124, 1–3, 2009. Copyright 2009, American Institute of Physics; reprinted with permission from Mori, T. et al., *Physical Review B*, 76, 064404, 1–10, 2007. Copyright 2007 by the American Physical Society; with kind permission from Springer Science+Business Media: *Monatshefte für Chemie*, 105, 1974, 1082, Rogl, P., and Nowotny, H.)

Several different groups have reported theoretical calculations with different methods on the electronic properties of various RMB_4 compounds.[117–120] In terms of electron counting, it is surmised that the boron 2D net of [5]- and [7]-sided rings needs more electrons than the graphitic [6]-ring formation to exist.[118,119] The calculations agree in particular that $YCrB_4$ should be a relatively narrow gap semiconductor with predicted bandgap values varying from 0.05 to 0.17 eV. Sharp features in the DOSs around the Fermi level are predicted which is also attractive for TEs. A simple arc-melt synthesis was done to investigate, and it was reported that $YCrB_4$ has a relatively large PF of 0.6 mW/m K^2 at the midlow temperature of 500 K.[120] Furthermore, it has been proposed that the p and n controls of $YCrB_4$ can be achieved relatively easily through doping.[121]

An interesting phenomena derived from the particular layered structure was also discovered, where in $TmAlB_4$, different tiling patterns (building defects) of α-type were embedded into β-type single crystals and vice versa.[115] The formation of the nanoscale building defects was shown to be controllable,[122] and these were found to have significant effect in lowering the thermal conductivity.[123]

While a good number of RTB_4 compounds are likely to be metallic, over 80 different ones are known to exist, and like $YCrB_4$ with good PF, only a couple so far have been measured regarding TE properties and should be further investigated. Furthermore, ZT has not been determined for the two examples given earlier. Possibly due to measurement constraints of the lab, only the PF was measured for $YCrB_4$,[120] while for $TmAlB_4$ needle crystals, the thermal conductivity was measured with time-domain thermoreflectance, but Seebeck coefficients are unmeasured yet.[123]

10.4.2.2 Higher Borides (Boron Icosahedral Cluster Compounds)

10.4.2.2.1 General Considerations

For the boron-to-metal ratio of $[B]/[M] > 12$ and boron-rich compounds without metal (such as boron carbide), the basic building block of the crystal structure is the B_{12} icosahedral cluster.[6] Similar to the other basic boron components discussed in Section 10.4.2.1.1, the B_{12} icosahedral cluster is also two-electron deficient. Regarding simple electron counting, it was noted[6] that the boron network appears to be electron phobic. Namely, the metal sites in some of the higher borides appear to take partial occupancy, so as not to have excess electrons and not to become metallic.[6] This is an underlying intuitive reason why the higher borides have been found to be predominantly p-type materials.

The higher borides generally follow Mott's variable range hopping (VRH) mechanism for 3D systems[124,125]:

$$\rho \propto \exp[(T_0/T)^{1/4}], \tag{10.3}$$

where T_0 is the so-called characteristic temperature, which is related to the DOSs at the Fermi level $D(E_F)$ and the localization length of the carriers at the Fermi level, ξ, with the following relationship:

$$k_B T_0 = 18.1/[D(E_F)\xi^3]. \tag{10.4}$$

With a fixed $D(E_F)$, a large T_0 is indicative of a strongly disordered system with short localization length.[126]

Assuming a linear DOSs, Zvyagin[127] calculates the Seebeck coefficient in VRH systems:

$$S \propto (T_0 T)^{1/2} d(\ln D(E))/dE \big|_{E_F}. \tag{10.5}$$

Therefore, it can be understood that these VRH systems start to have some advantage as TE materials at high temperatures, as both electrical conductivity and Seebeck coefficient increase with increasing temperatures.

Furthermore, small bipolaron hopping has been proposed as the mechanism for the very good TE properties observed for boron carbide.[128,129] Namely, the Seebeck coefficients can be enhanced through carrier-induced vibrational softening.

10.4.2.2.2 Metal Borides

For the higher borides which form compounds with metals, the metals have almost exclusively been rare-earth elements (disregarding doped materials). An exception is the aluminoborides such as α-AlB_{12} and γ-AlB_{12},[130] where although nominally [B]/[Al] = 12, the actual composition is close to [B]/[Al] = 13.75, and the B_{12} icosahedron is the basic building block of the structure. Vojteer et al.[131] and Ludwig et al.[132] have reported various alkali (e.g., Li) and alkali-earth (e.g., Mg) derivatives of the higher borides such as $Mg_3B_{36}Si_9C$ and $LiB_{12}PC$, although the TE properties of these compounds and α-AlB_{12} and γ-AlB_{12} are still unmeasured. The rare-earth atoms and alkali and alkali-earth atoms have good compatibility with the electron-deficient boron cluster network, because they readily donate valence electrons, and their relatively small shells are convenient for occupying void sites among the boron clusters.[97]

The following compounds are known for the rare-earth higher borides: $RAlB_{14}$, $RB_{15.5}CN$, $RB_{22}C_2N$, $RB_{28.5}C_4$, RB_{25}, $RB_{18}Si_5$ ($R_{3-x}B_{36}Si_8C_2$), $RB_{44}Si_2$, RB_{50}, and RB_{66}, and there are a variety of scandium higher borides which take slightly unique structures because of the particularly small size of scandium.[97]

I will summarize the important known TE properties among these compounds and the very recent developments. $RAlB_{14}$ is striking because it was surprisingly discovered that the Al site occupancy could be controlled at least in the case of $YAlB_{14}$, through synthesis conditions.[133–136] While the metal borides sometimes have metal sites with partial occupancy, the occupancy is usually virtually fixed, with a small homogeneity region. As a result of this control of the Al site occupancy, very large positive and negative Seebeck coefficients ($|\alpha| > 200$ μV/K), i.e., p-type or n-type control, could be achieved in YAl_xB_{14}. This means p and n controls with matching structure (best thermal compatibility) and no foreign doping (i.e., no migration problems).[133–136] It is also valuable because of the general lack of n-type higher borides. However, due to a low electrical conductivity, the overall performance of p-type YAl_xB_{14} was quite low, whereas ZT value just below ~0.1 was very quickly achieved for n-type YAl_xB_{14} around 1000 K.[135] Higher ZT value of ~0.3 can be extrapolated for higher working temperatures of 1500 K, but it is important to begin to investigate the doping and fabrication of composites[137,138] to increase the base of ZT. For example, the lowest measured thermal conductivity of YAl_xB_{14} sample is ~3 W/m K, which is high compared to other metal higher borides, and around three times the estimated value of Cahill's[139] minimum thermal conductivity. This is another indication of a large room to improve ZT.

MgAlB$_{14}$ was also later reported to similarly exhibit p- and n-type behaviors depending on the synthesis starting compositions, although the crystal structure, i.e., occupancy, is not clear.[140] LiAlB$_{14}$ has also been investigated.[141]

$RB_{15.5}CN$, $RB_{22}C_2N$, and $RB_{28.5}C_4$ are another group of higher borides where the elusive n-type behavior was discovered.[142,143] In particular, as will be briefly reviewed in Section 10.4.2.2.3, boron carbide has been a long-time TE champion among the borides but with only p-type, and because they possess a similar structure to boron carbide, $RB_{15.5}CN$, $RB_{22}C_2N$, and $RB_{28.5}C_4$, represent the promising n-type counterparts. However, challenges remain in the difficulty in densification[144] and boron carbide impurities, apparently strongly depressing the n-type characteristics.[145] If boron carbide-free samples could be prepared, it is expected that the performance could be comparable with boron carbide where the ZT value of ~0.6 has been reliably achieved with composites.[146] Indeed, an enhancement of ~100 times was achieved for a $YB_{22}C_2N/VB_2$ composite, although the Seebeck is assumedly depressed and the ZT is still very low, below 10^{-3} due to a density of only 50%.[147]

RB_{25} can be said to be the slightly monoclinically distorted version[148] of empty $RAlB_{14}$ with no Al. The TE properties of YB_{25} show very large positive Seebeck coefficients with very large resistivity.[135] This is a reasonable extrapolation from the low Al-occupied end of YAl_xB_{14}.

$YB_{18}Si_5$ was also measured to be p-type.[149]

$RB_{44}Si_2$ is another p-type group of compounds with a ZT value of ~0.04 around 1000 K and again an extrapolated ZT value of ~0.12 at 1500 K.[150–152] An interesting aspect was recently discovered when the directions of a grown single crystal of $YB_{44}Si_2$ were solved[153] The view of the crystal structure along the crystal growth direction [510] (Figure 10.17a) shows a layered-like structure with dense boron cluster (B$_{12}$ icosahedron and B$_{12}$Si$_3$) layers sandwiching the yttrium atoms. A large anisotropy in the TE properties was observed, with the PF along the crystal growth direction being 3.6 times that of the perpendicular direction at 1000 K, as shown in Figure 10.17b. This difference is driven by the electrical conductivity being ~8 times larger along [510]. It can intuitively be conjectured that the conduction goes well along the metal layers; however, this is interestingly in contrast to what was recently observed for layered boron-graphitic AlB$_2$, which is described in Section 10.4.2.1. In

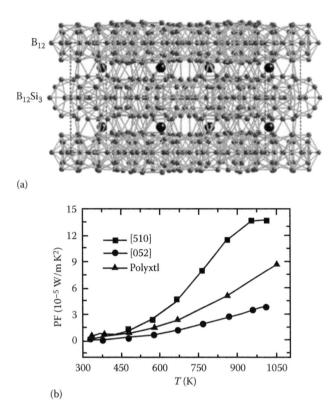

FIGURE 10.17 (a) View of $YB_{44}Si_2$ along the crystal growth direction [510], (b) anisotropic PF of $YB_{44}Si_2$, along [510] (*square*), [052] (*circle*), and polycrystal [150] (*triangle*). (Reprinted from *Journal of Materiomics*, 1, Sussardi, A. et al., 196–204, Copyright [2015], with permission from Elsevier.)

complete contrast to graphite-based materials, the thermal conductivity (and assumedly electrical conductivity because it is a good metal) in AlB_2 took higher values in the cross-plane direction compared to the in-plane direction.[123] These differences can be assumed to be derived from the unique bonding of boron compounds, and further theoretical analysis of these compounds should be carried out to understand this and utilized for TE development.

The RB_{66} compound was for a long time considered to be thermoelectrically unpromising, but recent striking discoveries were made.[138] *ZT* enhancements of 30–40 times were observed for SmB_{62}[154] and YB_{48}; an unexpectedly metal-rich compound of YB_{66}.[155] For these compounds, *ZT* values of ~0.1 to ~0.12 around 1000 K were found, and again an extrapolated *ZT* value of ~0.4 at 1500 K. Considering that these are simple binary compounds not yet doped or made composites of, there appears to be plenty of room for improvement.

$YAlB_{14}$, with p and n controls achieved through Al site occupancy control, and YB_{66}, where a large enhancement was achieved by pushing the boundaries of the homogeneity region, are two recent striking results. They indicate the new possibilities and the direction that efforts should focus on further findings ways to vary the metal site occupancy in the higher borides, and thereby achieving enhanced TE properties.

10.4.2.2.3 Rhombohedral α-Boron-Type Structure Compounds

There are several rhombohedral α-boron-type structure compounds which have gained attention as high-temperature TE materials. All of them are p-type materials. Foremost is boron carbide (so-called B_4C or $B_{12}C_3$). Since very good TE properties were discovered in boron carbide in the United States in the 1980s[156] (early research was also carried out in the former Soviet Union[157]), many investigations have been carried out on these compounds. As mentioned in Section 10.4.2.1.1, it has been proposed that the conduction mechanism in boron carbide is by small bipolaronic hopping,[128,129] and therefore carrier-induced vibrational softening leads to enhanced Seebeck coefficients and TE properties. Very high ZT values indicated to be from greater than 1[158] to even above 3[109] have been reported for different forms of this material. Judging from the results reproduced by different groups measuring at different temperature ranges, ZT value of ~0.6 around 1000 K for boron carbide/TiB_2 composites at the least appears to be very reliable.[137,146,159]

For other related compounds, $B_{12}As_2$,[160,161] $B_{12}P_2$,[162] and $B_{12}O_2$,[163,164] have also been investigated with all exhibiting large Seebeck coefficients. The crystal structure, relatively low temperature synthesis, and TE properties of B_6S_{1-x} ($B_{12}S_2$) were also recently made clear.[165]

Recently, Slack and Morgan[164] have comprehensively reviewed the rhombohedral α-boron-type structure compounds except for boron carbide. From their analysis, it is indicated that $B_{12}O_2$ samples, which have been investigated up to now, are actually oxygen deficient, and very good TE properties can be expected for stoichiometric $B_{12}O_2$.[164] A recipe for synthesis is also given, and while it is not easy, this is another exciting development in the TE borides.

10.4.3 Carbides

Compared to borides, there has not been as much investigation into the TE properties of carbides. Actually, there are only two systems which have been well studied: boron carbide, which is described earlier in Section 10.4.2.2.3, and silicon carbide.

Regarding silicon carbide, a seminal work was carried out by Koumoto et al.[166] In 1987, they reported an investigation into sintering and TE properties of α-SiC and β-SiC. Strikingly, through porosity, they were able to achieve a large reduction in thermal conductivity compared to the electrical conductivity and achieved a ZT value of around 0.06 at ~1300 K for n-type β-SiC sintered in N_2 atmosphere. This is a very interesting result, and although such details are not clear, it may be considered to be a very early example of a kind of nano/microstructuring to enhance TE properties, which has had such a large impact on the field. It is also notable that the p and n characteristics in α-SiC could also be controlled by sintering in Ar and N_2 atmospheres, respectively.

This report sparked various investigations. For example, SiC composites were synthesized to try to further enhance the TE properties. n-typeβ-SiC/

Si_3N_4 composites were fabricated with SPS, and at 7 wt%, an enhancement of ~2 times the PF of a single-phase sample was attained.[167] A maximum PF close to 1 mW/m K² at 1073 K was reported for p-type α-SiC sintered with additives of B_4C, C, and Al_2O_3.[168]

Although nanomaterials and thin films are not likely to be applied at high temperatures, for example, the TE properties of β-SiC thin films fabricated by CVD[169] and laser ablation[170] have been investigated. Large enhancements above the best bulk materials do not appear to be reported yet. Individual β-SiC nanowires were also investigated.[171] A large reduction of the thermal conductivity from the value of ~300 W/m K for typical bulk crystals to 4–12 W/m K was reported. As a result, they report a maximum ZT value of ~3 × 10⁻³ at 370 K, which is actually rather large considering the low temperature. However, there is a relatively large scattering in the properties depending on the individual nanowires. Calculations on exfoliated 2D single layers of Sc_2C terminated by −OH, for example, predict a large Seebeck coefficient above 1000 µV/K at 100 K; although as is noted in the paper, the overall TE properties are unclear.[172] To put this into context, the Seebeck coefficient of conventional YB_{66}, for example, can also reach values near 1000 µV/K at 300 K, but the ZT at this temperature is very low due to poor conductivity.

10.4.4 Outlook

Finally, for the attractive high-temperature TE applications, excellent material stability at these temperatures is an absolute necessary requirement. Refractory materials such as borides and carbides are good candidates, and striking advancements in the TE properties have recently been obtained. While diffusion and electrodes are issues which must of course always be solved in any case, for an application like the topping cycle in power plants, only a high-temperature material is required. This will make the implementation of this application relatively fast, and the race to develop viable refractory TEs is on.

Acknowledgment

The author would like to acknowledge support from Core Research for Evolutional Science and Technology (CREST) under Japan Science and Technology Agency (JST) Grant Number JPMJCR15Q6, Japan and Japan Society for the Promotion of Science (JSPS) KAKENHI Grant Number JP16H06441.

10.5 Constantan Alloy

Jun Mao and Zhifeng Ren

10.5.1 Introduction

The early studies of TEs have mainly focused on metals and their alloys. However, most of the metals have extremely low Seebeck coefficients, with a magnitude of 10 μV/K or less, which prevented them from being efficient TE materials compared to their semiconductors counterpart.[173,174] Nonetheless, it should be pointed out that there are a few exceptions, for example, the constantan alloy, which is the Cu–Ni alloy with ~55 at.% Cu, has anomalously higher Seebeck coefficient compared to other metals, with the magnitude of several tens of microvolts per Kelvin.[175,176] Combining with its low electrical resistivity, a record high TE PF of more than 100 μW/cm K² can be obtained at a high temperature.[177] Although the *ZT* is still very low (~0.2 at 873 K), the high PF enables the constantan alloy to achieve a possibly large output power density, which is quite favorable for certain applications where output power density outweighs the efficiency.

10.5.2 Band Structure of Constantan

The rigid band model was used by Mott[178] to explain the magnetic properties of Cu–Ni alloys. In this model, copper and nickel atoms are assumed to form common *d*-bands, as well as common *s–p* conduction bands. Then when nickel, which has one less electron than copper, is alloyed with copper, this model predicts that the *s–p* bands should progressively empty, decreasing the *d*-band to Fermi level energy separation. At 40–50 at.% nickel (i.e., the constantan alloy), the Fermi level should coincide with the top of the *d*-band.[178,179] However, later optical studies[179–186] and low-temperature specific heat measurements[187–189] have disproved this model. Compared to the rigid-band model, the virtual-bound state (or resonant state) model[190,191] gives a more reasonable description about the band structure of Cu–Ni alloy. In the virtual-bound state model (as shown in Figure 10.18), the nickel *d*-states

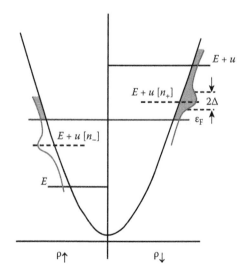

FIGURE 10.18   Schematic view of the virtual-bound state. Virtual *d* levels of widths 2Δ appear at *E* + *u* [*n*_] and *E* + *u* [*n*_], where *E* is the energy of the atomic *d* state. The number of electrons in these levels is computed from the area of the unshaded portion, below the Fermi energy. (Reprinted with permission from Hüfner, S. et al., *Physical Review Letters*, 28, 4881972. Copyright 1972 by the American Physical Society.)

(in a copper-rich alloy) would not form a common band with *d*-states; rather, they would form virtual-bound levels, probably with energy between that of copper *d*-states and of the Fermi level.[179]

Hüfner et al.[192] investigated the DOSs for Cu–Ni alloys by using X-ray photoemission spectroscopy As shown in Figure 10.19, at all concentrations of Ni, except in pure Cu, there is still a considerable DOSs at the Fermi energy. Moreover, the DOSs of Cu–Ni alloys can, to a very good approximation, be made up by superimposing those of Ni and Cu. This means that there is indeed only a very limited sharing of electrons by the two constituents.

The specific heat of a metal at extremely low temperature could be expected to follow the law

$$C_v = \gamma T + b T^3, \tag{10.6}$$

where C_v is the specific heat, γ is the electronic specific heat, b is the lattice contribution of specific heat, and T is the absolute temperature. The coefficient γ provides a direct measure of the DOSs at the Fermi level and hence the DOS effective mass. By plotting C_v/T as a function of T^2, Equation 10.6 is represented by a straight line of slope b and ordinate intercept γ.[187]

Figure 10.20 shows the composition dependence of the coefficient γ for Cu–Ni alloys.[187–189] It is evident that the γ decreases from pure Ni to pure Cu; however, it has a peak around the 55 at.% Cu, which corresponds to the constantan alloy. Considering the virtual-bound state model, it is understandable that with the addition of Ni into pure Cu, the formation of the nickel *d*-states around the Fermi level effectively enhances the DOSs of Cu–Ni alloys, especially around the composition of constantan.

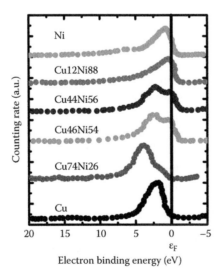

FIGURE 10.19 X-ray photoelectron spectroscopy valence-band spectra for Cu–Ni alloys. (From Heremans, J. P. et al., *Science*, 321, 554–557, 2008. Reprinted with permission of AAAS.)

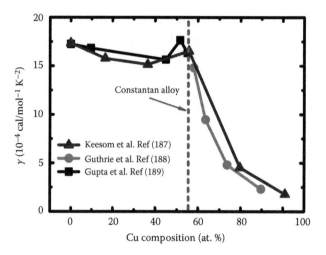

FIGURE 10.20 Composition dependence of electronic specific heat γ of Cu–Ni alloys.

For the semiconductors, Heremans et al.[193] demonstrate that by doping the PbTe with Tl, the resonant level (i.e., virtual-bound state) can be formed near the Fermi level, which effectively enhances the Seebeck coefficient and hence improves the TE properties. Therefore, it is understandable that the anomalously high Seebeck coefficient of constantan should also originate from the virtual-bound state. It should be pointed out that constantan is not the only exception; the metallic YbAl$_3$ also shows a high Seebeck coefficient[194] and hence an ultralarge PF of ~340 μW/cm K^2 around ambient temperature.[195] The TE properties of YbAl$_3$ can also be explained by the virtual-bound state model, which the f-state of Yb atom forms below the Fermi level in the compound.[196,197]

It is important to note that the existence of virtual-bound state in the metallic materials seems have a clear advantage than that in the semiconductors. In the semiconductors, the formation of virtual-bound state will lead to an enhanced band mass and hence a larger Seebeck coefficient. However, the increased band mass will inevitably reduce the carrier mobility, due to the inverse relationship between band mass and carrier mobility when acoustic phonon scattering is the dominant scattering mechanism.[198] In this case, the reduced carrier mobility will generally lead to a decreased PF and hence an inferior TE performance. On the contrary, the metallic materials has very low electrical resistivity; an appreciable increase in the Seebeck coefficient will offset the slightly decreased mobility thus leading to significantly enhancement in PF.

10.5.3 Thermoelectric Properties of Constantan

Several reports of electrical or thermal properties of constantan can be found in the literature,[175,199,200] while detailed TE properties of constantan are rarely reported[201]; this should be due to the reason that the constantan is mainly been considered as a thermocouple rather than an efficient TE material. However, the high PF of constantan alloys makes this material quite interesting for certain TE applications, where large output power density is favorable. In our recent work, the constantan alloys were reinvestigated and prepared by BM and then HP as well as arc-melting method.[177] The Seebeck coefficients of ball milled and hot pressed samples are clearly higher than that of arc-melted ingot, which leads to a record high PF of 102 μW/cm K². The detailed TE properties of constantan alloys are shown in Figure 10.21.

In order to understand the enhancement of Seebeck coefficient in ball milled and hot pressed sample, HRTEM characterizations are carried out. Figure 10.22a shows the TEM image of arc-melted ingot. There are no special microstructures observed in this specimen. However, for the ball milled and hot pressed sample, it is clear that there are many nanoscale twins in the crystalline grains with spacing of about 50–200 nm (Figure 10.20b).

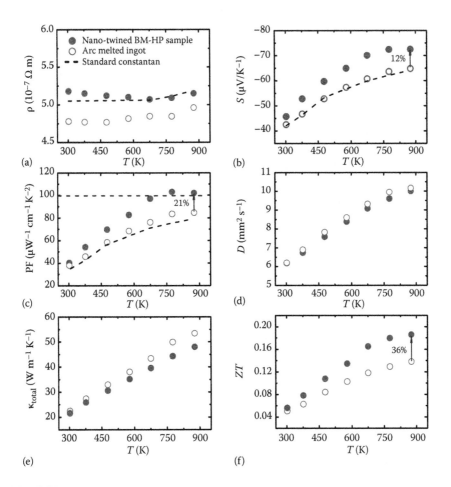

FIGURE 10.21 Comparison of TE properties of arc-melted ingot and nanotwinned sample. (a) Electrical resistivity ρ, (b) Seebeck coefficient S, (c) PF, (d) thermal diffusivity D, (e) total thermal conductivity κ_{total}, and (f) figure of merit ZT.

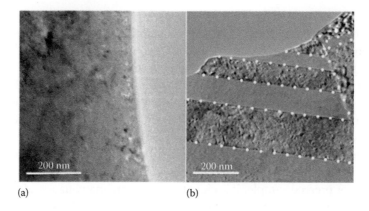

(a)   (b)

FIGURE 10.22 TEM results of arc-melted ingot and BM-HP sample. (a) Arc-melted ingot; (b) nanoscale twins in ball milled and hot pressed sample.

Figure 10.23a shows the TEM image of nanoscale twins in the ball milled and hot pressed sample. Figure 10.23b is a SAED pattern for one of the twins in the ball milled and hot pressed samples, which can be indexed as [101] according to $Fm\overline{3}m$ cubic structure. It is a superposition of two sets of the diffraction spots with [111] direction as twinning axis. The schematic view of Figure 10.23b is shown in Figure 10.23c with the red filled circles and blue unfilled circles representing the diffraction spots from neighboring grains. The rotation angle between these two grains is 120°. HRTEM image corresponding to Figure 10.23b is shown in Figure 10.23d, where the twin boundary is situated in (111) plane marked by the white arrows. The measured angle between the crystal lattice on both sides of the boundary is about 120° as mentioned earlier.

As shown in Figure 10.24, many nanoscale twins exist in the grains of ball milled and hot pressed Cu–Ni alloys. Due to the coherent characteristic of twin boundaries, most of the high-energy carriers will be unaffected by twin boundaries, which enables the nanotwinned samples to maintain a good electrical conductivity. However, for low-energy carriers, with energy lower than the height of potential barrier, they could be effectively scattered. The selective scattering of low-energy electrons will lead to enhanced Seebeck coefficient (i.e., potential barrier scattering). It is noted that the height of the potential barrier plays a decisive role in the scattering mechanism. In order to effectively filter out the low-energy carriers, the height of the potential barrier should not be too high. The noncoherent interfaces (grain boundaries and interfaces between nanoprecipitate and matrix) could severely distort the lattice periodic potential that leads to an unfavorably high potential barrier. On the contrary, due to the coherent characteristic of twin boundaries, the potential barrier by twin boundaries should be much lower and could selectively scatter low-energy carriers while high-energy carriers are unaffected.

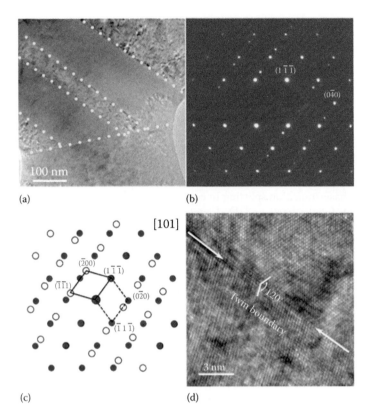

FIGURE 10.23 TEM results of ball milled and hot pressed sample. (a) Nanoscale twins in ball milled and hot pressed sample, (b) [101] zone axis SAED pattern taken from the twin boundary, (c) schematic drawings of the diffraction spots for (b), and (d) HRTEM image corresponding to (b) with the arrows pointing out the twin boundary.

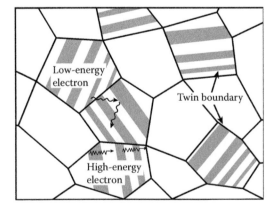

FIGURE 10.24 Schematic view of potential barrier scattering by twin boundaries.

References

1. H. L. Liu, X. Shi, F. F. Xu, L. L. Zhang, W. Q. Zhang, L. D. Chen, Q. Li, C. Uher, T. Day, and G. J. Snyder. 2012. Copper ion liquid-like thermoelectrics. *Nat. Mater.* 11, (5): 422–425.

2. Y. He, T. Day, T. S. Zhang, H. L. Liu, X. Shi, L. D. Chen, and G. J. Snyder. 2014. High thermoelectric performance in non-toxic earth-abundant copper sulfide. *Adv. Mater.* 26, (23): 3974–3978.

3. Y. He, T. Zhang, X. Shi, S.-H. Wei, and L. Chen. 2015. High thermoelectric performance in copper telluride. *NPG Asia Mater.* 7, (8): e210.

4. Z. Vučić, O. Milat, V. Horvatić, and Z. Ogorelec. 1981. Composition-induced phase-transition splitting in cuprous selenide. *Phys. Rev. B* 24, (9): 5398–5401.

5. D. Chakrabarti and D. Laughlin. 1983. The Cu-S (copper-sulfur) system. *Bull. Alloy Phase Diagr.* 4, (3): 254–271.

6. S.-Y. Miyatani, S. Mori, and M. Yanagihara. 1979. Phase diagram and electrical properties of $Cu_{2-\delta}Te$. *J. Phys. Soc. Jpn.* 47, (4): 1152–1158.

7. H. Liu, X. Yuan, P. Lu, X. Shi, F. Xu, Y. He, Y. Tang et al. 2013. Ultrahigh thermoelectric performance by electron and phonon critical scattering in $Cu_2Se_{1-x}I_x$. *Adv. Mater.* 25, (45): 6607–6612.

8. L. L. Zhao, X. L. Wang, J. Y. Wang, Z. X. Cheng, S. X. Dou, J. Wang, and L. Q. Liu. 2015. Superior intrinsic thermoelectric performance with zT of 1.8 in single-crystal and melt-quenched highly dense Cu_{2-x} Se bulks. *Sci. Rep.* 5: 7671.

9. B. Gahtori, S. Bathula, K. Tyagi, M. Jayasimhadri, A. K. Srivastava, S. Singh, R. C. Budhani, and A. Dhar. 2015. Giant enhancement in thermoelectric performance of copper selenide by incorporation of different nanoscale dimensional defect features. *Nano Energy* 13: 36–46.

10. X. Su, F. Fu, Y. Yan, G. Zheng, T. Liang, Q. Zhang, X. Cheng et al. 2014. Self-propagating high-temperature synthesis for compound thermoelectrics and new criterion for combustion processing. *Nat. Commun.* 5: 4908.

11. D. Li, X. Y. Qin, Y. F. Liu, C. J. Song, L. Wang, J. Zhang, H. X. Xin et al. 2014. Chemical synthesis of nanostructured Cu_2Se with high thermoelectric performance. *RSC Adv.* 4, (17): 8638.

12. L. Yang, Z.-G. Chen, G. Han, M. Hong, Y. Zou, and J. Zou. 2015. High-performance thermoelectric Cu_2Se nanoplates through nanostructure engineering. *Nano Energy* 16: 367–374.

13. S. Ballikaya, H. Chi, J. R. Salvador, and C. Uher. 2013. Thermoelectric properties of Ag-doped Cu_2Se and Cu_2Te. *J. Mater. Chem. A* 1, (40): 12478.

14. H. Zhu, J. Luo, H. Zhao, and J. Liang. 2015. Enhanced thermoelectric properties of p-type Ag_2Te by Cu substitution. *J. Mater. Chem. A* 3, (19): 10303–10308.

15. Y. He, P. Lu, X. Shi, F. F. Xu, T. S. Zhang, G. J. Snyder, C. Uher, and L. D. Chen. 2015. Ultrahigh thermoelectric performance in mosaic crystals. *Adv. Mater.* 27, (24): 3639–3644.

16. L. Zhao, X. Wang, F. Y. Fei, J. Wang, Z. Cheng, S. Dou, J. Wang, and G. J. Snyder. 2015. High thermoelectric and mechanical performance in highly dense $Cu_{2-x}S$ bulks prepared by a melt-solidification technique. *J. Mater. Chem. A* 3, (18): 9432–9437.

17. W. Mi, P. Qiu, T. Zhang, Y. Lv, X. Shi, and L. Chen. 2014. Thermoelectric transport of Se-rich Ag_2Se in normal phases and phase transitions. *Appl. Phys. Lett.* 104, (13): 133903.

18. F. F. Aliev, M. B. Jafarov, and V. I. Eminova. 2009. Thermoelectric figure of merit of Ag$_2$Se with Ag and Se excess. *Semiconductors* 43, (8): 977–979.

19. J. Capps, F. Drymiotis, S. Lindsey, and T. M. Tritt. 2010. Significant enhancement of the dimensionless thermoelectric figure of merit of the binary Ag$_2$Te. *Phil. Mag. Lett.* 90, (9):677–681.

20. X. Wang, P. Qiu, T. Zhang, D. Ren, L. Wu, X. Shi, J. Yang, and L. Chen. 2015. Compound defects and thermoelectric properties in ternary CuAgSe-based materials. *J. Mater. Chem. A* 3, (26): 13662–13670.

21. S. Ishiwata, Y. Shiomi, J. S. Lee, M. S. Bahramy, T. Suzuki, M. Uchida, R. Arita, Y. Taguchi, and Y. Tokura. 2013. Extremely high electron mobility in a phonon-glass semimetal. *Nat. Mater.* 12, (6): 512–517.

22. P. Qiu, T. Zhang, Y. Qiu, X. Shi, and L. Chen. 2014. Sulfide bornite thermoelectric material: A natural mineral with ultralow thermal conductivity. *Energy Environ. Sci.* 7, (12): 4000–4006.

23. G. C. Tewari, T. S. Tripathi, H. Yamauchi, and M. Karppinen. 2014. Thermoelectric properties of layered antiferromagnetic CuCrSe$_2$. *Mater. Chem. Phys.* 145, (1–2): 156–161.

24. S. Bhattacharya, R. Basu, R. Bhatt, S. Pitale, A. Singh, D. K. Aswal, S. K. Gupta, M. Navaneethan, and Y. Hayakawa. 2013. CuCrSe$_2$: A high performance phonon glass and electron crystal thermoelectric material. *J. Mater. Chem. A* 1, (37): 11289.

25. F. Gascoin and A. Maignan. 2011. Order–disorder transition in AgCrSe$_2$: A new route to efficient thermoelectrics. *Chem. Mater.* 23, (10): 2510–2513.

26. K. S. Weldert, W. G. Zeier, T. W. Day, M. Panthofer, G. J. Snyder, and W. Tremel. 2014. Thermoelectric transport in Cu$_7$PSe$_6$ with high copper ionic mobility. *J. Am. Chem. Soc.* 136, (34): 12035–12040.

27. J. Y. Hwang, H. A. Mun, S. I. Kim, K. M. Lee, J. Kim, K. H. Lee, and S. W. Kim. 2014. Effects of doping on transport properties in Cu-Bi-Se-based thermoelectric materials. *Inorg. Chem.* 53, (24): 12732–12738.

28. H. Lin, H. Chen, J. N. Shen, L. Chen, and L. M. Wu. 2014. Chemical modification and energetically favorable atomic disorder of a layered thermoelectric material TmCuTe$_2$ leading to high performance. *Chemistry* 20, (47): 15401–15408.

29. L. Pan, D. Bérardan, and N. Dragoe. 2013. High thermoelectric properties of n-Type AgBiSe$_2$. *J. Am. Chem. Soc.* 135: 4914–4917.

30. S. N. Guin, A. Chatterjee, D. S. Negi, R. Datta, and K. Biswas. 2013. High thermoelectric performance in tellurium free p-type AgSbSe$_2$. *Energy Environ. Sci.* 6: 2603–2608.

31. A. J. Hong, L. Li, H. X. Zhu, X. H. Zhou, Q. Y. He, W. S. Liu, Z. B. Yan, J. M. Liu, and Z. F. Ren. 2014. Anomalous transport and thermoelectric performances of CuAgSe compounds. *Solid State Ionics.* 261: 21–25.

32. J. R. Sootsman, D. Y. Chung, and M. G. Kanatzidis. 2009. New and old concepts in thermoelectric materials. *Angew. Chem. Int. Ed.* 48: 8616–8639.

33. M. Ohtaki. 2011. Recent aspects of oxide thermoelectric materials for power generation from mid-to-high temperature heat source. *J. Ceram. Soc. Jpn.* 119: 770–775.

34. Y. Wang, N. S. Rogado, R. J. Cava, and N. P. Ong. 2003. Spin entropy as the likely source of enhanced thermopower in Na$_x$Co$_2$O$_4$. *Nature* 423: 425–428.

35. Y. Zhu, C. Wang, H. Wang, W. Su, J. Liu, and J. Li. 2014. Influence of Dy/Bi dual doping on thermoelectric performance of CaMnO$_3$ ceramics. *Mater. Chem. Phys.* 144: 385–389.

36. D. Kenfaui, B. Lenoir, D. Chateigner, B. Ouladdiaf, M. Gomina, and J. G. Noudem. 2012. Development of multilayer textured $Ca_3Co_4O_9$ materials for thermoelectricgenerators: Influence of the anisotropy on the transport properties. *J. Eur. Ceram. Soc.* 32: 2405–2414.

37. R. Funahashi, I. Matsubara, and S. Sodeoka. 2000. Thermoelectric properties of $Bi_2Sr_2Co_2O_x$ polycrystalline materials. *Appl. Phys. Lett.* 76: 2385–2387.

38. A. Kovalevsky, A. Yaremchenko, S. Populoh, A. Weidenkaff, and J. R. Frade. 2014. Effect of A-site cation deficiency on the thermoelectric performance of donor-substituted strontium titanate. *J. Phys. Chem. C* 118: 4596–4606.

39. M. Ohtaki, T. Tsubota, K. Eguchi, and H. Arai. 1996. High-temperature thermoelectric properties of $(Zn_{1-x}Al_x)O$. *J. Appl. Phys.* 79: 3.

40. L. Zhao, J. He, D. Berardan, Y. Lin, J. Li, C. Nan, and N. Dragoe. 2014. BiCuSeO oxyselenides: New promising thermoelectric materials. *Energy Environ. Sci.* 7, (9): 2900–2924.

41. H. Hiramatsu, H. Yanagi, T. Kamiya, K. Ueda, M. Hirano, and H. Hosono. 2008. Crystal structures, optoelectronic properties, and electronic structures of layered oxychalcogenides *M*CuO*Ch* (*M* = Bi, La; *Ch* = S, Se, Te): Effects of electronic configurations of M^{3+} ions. *Chem. Mater.* 20: 326–334.

42. C. Barreteau, D. Bérardan, E. Amzallag, L. Zhao, and N. Dragoe. 2012. Structural and electronic transport properties in Sr-Doped BiCuSeO. *Chem. Mater.* 24: 3168–3178.

43. D. Zou, S. Xie, Y. Liu, J. Lin, and J. Li. 2013. Electronic structures and thermoelectric properties of layered oxychalcogenides BiCuOCh(Ch = S, Se, and Te): First-principles calculations. *J. Mater. Chem. A* 1: 8888–8896.

44. L. Zhao, J. He, D. Berardan, Y. Lin, J. Li, C. Nan, and N. Dragoe. 2014. BiCuSeO oxyselenides: New promising thermoelectric materials. *Energy Environ. Sci.* 7. (9): 2900–2924.

45. Y. Pei, J. He, J. Li, F. Li, Q. Liu, W. Pan, C. Barreteau, D. Berardan, N. Dragoe, and L. Zhao. 2013. High thermoelectric performance of oxyselenides: intrinsically low thermal conductivity of Ca-doped BiCuSeO. *NPG Asia Mater.* 5: e47.

46. M. Roufosse and P. G. Klemens. 1973. Thermal conductivity of complex dielectric crystals. *Phys. Rev. B* 7: 5379.

47. G. A. Slack. 1979. The Thermal Conductivity of Nonmetallic Crystals. In *Solid State Physics*, edited by H. Ehrenreich, F. Weitz, and D. Turnbull, Academic Press, New York, 34: 1.

48. J. Li, J. Sui, C. Barreteau, D. Berardan, N. Dragoe, W. Cai, Y. Pei, and L. Zhao. 2013. Thermoelectric properties of Mg doped p-type BiCuSeO oxyselenides. *J. Alloys Compd.* 551: 649–653.

49. J. Lan, B. Zhan, Y. Liu, B. Zheng, Y. Liu, Y. Lin, and C. Nan. 2013. Doping for higher thermoelectric properties in p-type BiCuSeO oxyselenide. *Appl. Phys. Lett.* 102: 123905.

50. F. Li, T. Wei, F. Kang, and J. Li. 2013. Enhanced thermoelectric performance of Ca-doped BiCuSeO in a wide temperature range. *J. Mater. Chem. A* 1: 11942–11949.

51. L. Zhao, D. Berardan, Y. Pei, C. Byl, L. P. Gaudart, and N. Dragoe. 2010. $Bi_{1-x}Sr_xCuSeO$ oxyselenides as promising thermoelectric materials. *Appl. Phys. Lett.* 97: 092118.

52. J. Li, J. Sui, Y. Pei, C. Barreteau, D. Bererdan, N. Dragoe, W. Cai, J. He, and L. Zhao. 2012. A high thermoelectric figure of merit ZT > 1 in Ba heavily doped BiCuSeO oxyselenides. *Energy Environ. Sci.* 9, (5): 8543–8547.

53. L. Pan, D. Berardan, L. Zhao, C. Barreteau, and N. Dragoe. 2013. Influence of Pb doping on the electrical transport properties of BiCuSeO. *Appl. Phys. Lett.* 102: 023902.

54. S. D. N. Luu and P. Vaqueiro. 2013. Synthesis, structural characterisation and thermoelectric properties of $Bi_{1-x}Pb_xOCuSe$. *J. Mater. Chem. A* 1: 12270–12275.

55. J. Lan, Y. Liu, B. Zhan, Y. Lin, B. Zhang, X. Yuan, W. Zhang, Wei Xu, and C. Nan. 2013. Enhanced thermoelectric properties of Pb-doped BiCuSeO ceramics. *Adv. Mater.* 44, (51): 5086–5090.

56. J. Li, J. Sui, Y. Pei, X. Meng, D. Berardan, N. Dragoe, W. Cai, and L. Zhao. 2014. The roles of Na doping in BiCuSeO oxyselenides as a thermoelectric material. *J. Mater. Chem. A* 2: 4903–4906.

57. D. S. Lee, T. An, M. Jeong, H. Choi, Y. S. Lim, W. Seo, C. Park, Chan Park, and H. Park. 2013. Density of state effective mass and related charge transport properties in K-doped BiCuOSe. *Appl. Phys. Lett.* 103: 232110.

58. Y. Liu, Y. Zheng, B. Zhan, K. Chen, S. Butt, B. Zhang, and Y. Lin. 2015. Influence of Ag doping on thermoelectric properties of BiCuSeO. *J. Eur. Ceram. Soc.* 35, 845–849.

59. Y. Liu, L. Zhao, Y. Liu, J. Lan, W. Xu, F. Li, B. Zhang et al. 2011. Remarkable enhancement in thermoelectric performance of BiCuSeO by Cu deficiencies. *J. Am. Chem. Soc.* 133: 20112–20115.

60. W. Xu, Y. Liu, L. Zhao, P. An, Y. Lin, A. Marcelli, and Z. Wu. 2013. Evidence of an interlayer charge transfer route in $BiCu_{1-x}SeO$. *J. Mater. Chem. A* 1: 12154.

61. Z. Li, C. Xiao, S. Fan, Y. Deng, W. Zhang, B. Ye, and Y. Xie. 2015. Dual-vacancies: An effective strategy realizing synergistic optimization of thermoelectric property in BiCuSeO. *J. Am. Chem. Soc.* 137: 6587–6593.

62. L. Zhao, B. Zhang, J. Li, H. L. Zhang, and W. Liu. 2008. Enhanced thermoelectric and mechanical properties in textured n-type Bi_2Te_3 prepared by spark plasma sintering. *Solid State Sci.* 10: 651.

63. Y. Xiao, B. Poudel, Y. Ma, W. Liu, G. Joshi, H. Wang, Y. Lan, D. Wang, G. Chen, and Z. Ren. 2010. Experimental studies on anisotropic thermoelectric properties and structures of n-type $Bi_2Te_{2.7}Se_{0.3}$. *Nano Lett.* 10: 3373.

64. J. J. Shen, L. P. Hu, T. Zhu, and X. Zhao. 2011. The texture related anisotropy of thermoelectric properties in bismuth telluride based polycrystalline alloys. *Appl. Phys. Lett.* 99: 124102.

65. J. Sui, J. Li, J. He, Y. Pei, D. Berardan, H. Wu, N. Dragoe, W. Cai, and L. Zhao. 2013. Texturation boosts the thermoelectric performance of BiCuSeO oxyselenides. *Energy Environ. Sci.* 6: 2916.

66. A. Tsukazaki, S. Akasaka, K. Nakahara, Y. Ohno, H. Ohno, D. Maryenko, A. Ohtomo, and M. Kawasaki. 2010. Observation of the fractional quantum Hall effect in an oxide. *Nat. Mater.* 9: 889–893.

67. M. Zebarjadi, G. Joshi, G. Zhu, B. Yu, A. Minnich, Y. Lan, X. Wang, M. Dresselhaus, Z. Ren, and G. Chen. 2011. Power factor enhancement by modulation doping in bulk nanocomposites. *Nano Lett.* 11: 2225.

68. B. Yu, M. Zebarjadi, H. Wang, K. Lukas, H. Wang, D. Wang, C. Opeil, M. Dresselhaus, G. Chen, and Z. Ren. 2012. Enhancement of thermoelectric properties by modulation-doping in silicon germanium alloy nanocomposites. *Nano Lett.* 12: 2077.

69. Y. Pei, H. Wu, D. Wu, F. Zheng, and J. He. 2014. High thermoelectric performance realized in a BiCuSeO system by improving carrier mobility through 3D modulation doping. *J. Am. Chem. Soc.* 136: 13902.

70. Y. Liu, J. Lan, W. Xu, Y. Liu, Y. Pei, B. Cheng, D. Liu, Y. Lin, and L. Zhao. 2013. Enhanced thermoelectric performance of a BiCuSeO system via band gap tuning. *Chem. Commun.* 49: 8075–8077.

71. C. Barreteau, D. Berardan, L. Zhao, and N. Dragoe. 2013. Influence of Te substitution on the structural and electronic properties of thermoelectric BiCuSeO. *J. Mater. Chem. A* 1: 2921–2926.

72. Y. Liu, J. Ding, B. Xu, J. Lan, Y. Zheng, B. Zhan, B. Zhang, Y. Lin, and C. Nan. 2015. Enhanced thermoelectric performance of La-doped BiCuSeO by tuning band structure. *Appl. Phys. Lett.* 106: 233903.

73. B. Poudel, Q. Hao, Y. Ma, Y. Lan, A. Minnich, B. Yu, X. Yan, D. Wang, A. Muto, and D. Vashaee. 2008. High-thermoelectric performance of nanostructured bismuth antimony telluride bulk alloys. *Science* 320, (5876): 634–638.

74. W. Liu, K. C. Lukas, K. McEnaney, S. Lee, Q. Zhang, C. P. Opeil, G. Chen, and Z. Ren. 2013. Studies on the Bi_2Te_3–Bi_2Se_3–Bi_2S_3 system for mid-temperature thermoelectric energy conversion, *Energy Environ. Sci.* 6, (2): 552–560.

75. J. P. Heremans, V. Jovovic, E. S. Toberer, A. Saramat, K. Kurosaki, A. Charoenphakdee, S. Yamanaka, and G. J. Snyder. 2008. Enhancement of thermoelectric efficiency in PbTe by distortion of the electronic density of states. *Science* 321, (5888): 554–557.

76. Y. Pei, X. Shi, A. LaLonde, H. Wang, L. Chen, and G. J. Snyder. 2011. Convergence of electronic bands for high performance bulk thermoelectrics, *Nature* 473, (7345): 66–69.

77. J. Emsley. 2011. *Nature's Building Blocks: An A–Z Guide to the Elements*. Oxford University Press, Oxford.

78. L.-D. Zhao, B.-P. Zhang, W.-S. Liu, H.-L. Zhang, and J.-F. Li. 2008. Enhanced thermoelectric properties of bismuth sulfide polycrystals prepared by mechanical alloying and spark plasma sintering. *J. Solid State Chem.* 181, (12): 3278–3282.

79. C. Wan, Y. Wang, W. Norimatsu, M. Kusunoki, and K. Koumoto. 2012. Nanoscale stacking faults induced low thermal conductivity in thermoelectric layered metal sulfides. *Appl. Phys. Lett.* 100, (10): 101913.

80. L.-D. Zhao, J. He, C.-I. Wu, T. P. Hogan, X. Zhou, C. Uher, V. P. Dravid, and M. G. Kanatzidis. 2012. Thermoelectrics with earth abundant elements: High performance p-type PbS nanostructured with SrS and CaS. *J. Am. Chem. Soc.* 134, (18): 7902–7912.

81. H. Wu, J. Carrete, Z. Zhang, Y. Qu, X. Shen, Z. Wang, L.-D. Zhao, and J. He. 2014. Strong enhancement of phonon scattering through nanoscale grains in lead sulfide thermoelectrics. *NPG Asia Mater.* 6, (6): e108.

82. Q. Tan, L. D. Zhao, J. F. Li, C. F. Wu, T. R. Wei, Z. B. Xing, and M. G. Kanatzidis. 2014. Thermoelectrics with earth abundant elements: Low thermal conductivity and high thermopower in doped SnS. *J. Mater. Chem. A* 2, (41): 17302–17306.

83. Z. Liu, Y. Pei, H. Geng, J. Zhou, X. Meng, W. Cai, W. Liu, and J. Sui. 2015. Enhanced thermoelectric performance of Bi_2S_3 by synergistical action of bromine substitution and copper nanoparticles, *Nano Energy* 13: 554–562.

84. Z. Liu, H. Geng, J. Shuai, Z. Wang, J. Mao, D. Wang, Q. Jie, W. Cai, J. Sui, and Z. Ren. 2015. The effect of nickel doping on electron and phonon transport in the n-type nanostructured thermoelectric material CoSbS. *J. Mater. Chem. C* 3, (40): 10442–10450.

85. D. Parker, A. F. May, H. Wang, M. A. McGuire, B. C. Sales, and D. J. Singh. 2013. Electronic and thermoelectric properties of CoSbS and FeSbS. *Phys. Rev. B* 87, (4): 045205.

86. S. Bhattacharya, R. Chmielowski, G. Dennler, and G. K. Madsen. 2016. Novel ternary sulfide thermoelectric materials from high throughput transport and defect calculations. *J. Mater. Chem. A* 4, (28): 11086–11093.

87. R. Chmielowski, S. Bhattacharya, W. Xie, D. Péré, S. Jacob, R. Stern, K. Moriya, A. Weidenkaff, G. Madsen, and G. Dennler. 2016. High thermoelectric performance of tellurium doped paracostibite. *J. Mater. Chem. C* 4, (15): 3094–3100.

88. R. Carlini, C. Artini, G. Borzone, R. Masini, G. Zanicchi, and G. Costa. 2010. Synthesis and characterisation of the compound CoSbS. *J. Therm. Anal. Calorim.* 103, (1): 23–27.

89. H. Anno, K. Matsubara, Y. Notohara, T. Sakakibara, and H. Tashiro. 1999. Effects of doping on the transport properties of $CoSb_3$. *J. Appl. Phys.* 86, (7): 3780–3786.

90. J. Yang, D. T. Morelli, G. P. Meisner, W. Chen, J. S. Dyck, and C. Uher. 2002. Influence of electron-phonon interaction on the lattice thermal conductivity of $Co_{1-x}Ni_xSb_3$. *Phys. Rev. B Condens. Matter.* 65: 094115.

91. H. H. Xie, H. Wang, Y. Pei, C. Fu, X. Liu, G. J. Snyder, X. Zhao, and T. Zhu. 2013. Beneficial contribution of alloy disorder to electron and phonon transport in half-Heusler thermoelectric materials. *Adv. Func. Mater.* 23: 5123–5130.

92. K. Yazawa, Y. R. Koh, and A. Shakouri. 2013. Optimization of thermoelectric topping combined steam turbine cycles for energy economy. *Appl. Energy* 109: 1.

93. K. Yazawa and A. Shakouri. 2016. Thermoelectric topping cycles with scalable design and temperature dependent material properties. *Scripta Mat.* 111: 58.

94. K. Hatakeyama, T. Nakamura, T. Fujisawa, H. Kobayashi, Y. Hikichi, H. Kurata, N. Saito, and M. Minowa. 2013. The demonstration test on the thermoelectric generator for high temperature industrial furnaces. ICT 2013, Kobe.

95. JFE. http://www.jfe-steel.co.jp/release/2013/07/130710–1.html. Accessed 15 May 2016.

96. A. Weidenkaff, M. Trottmann, P. Tomes, C. Suter, A. Steinfeld, and A. Veziridis. 2013. Solar TE Converter Applications. In *Thermoelectric Nanomaterials*, edited by K. Koumoto and T. Mori, Springer, Heidelberg, 365–382.

97. T. Mori. 2008. Higher borides. In *Handbook on the Physics and Chemistry of Rare Earths*, vol. 38, edited by K. A. Gschneidner Jr., J.-C. Bunzli, and V. Pecharsky, North-Holland, Amsterdam. 105–173.

98. T. Mori. 2012. Rare earth borides, carbides and nitrides. In *The Rare Earth Elements: Fundamentals and Application*, edited by D. Atwood, John Wiley & Sons Ltd, Chichester, 263–279.

99. Yu. B. Paderno. 1969. Electrical properties of hexaborides of alkali- and rare-earth metals. *Poroshk. Metallurgija* 11: 70.

100. K. Yagasaki, S. Notsu, Y. Shimoji, T. Nakama, R. Kaji, T. Yokoo, J. Akimitsu, M. Hedo, and Y. Uwatoko. 2003. Resistivity and thermopower of CaB_6 single crystal. *Physica B* 1259: 329–333.

101. M. Takeda, M. Terui, N. Takahashi, and N. Ueda. 2006. Improvement of thermoelectric properties of alkaline-earth hexaborides. *J. Solid State Chem.* 179: 2823.

102. M. Gürsoy, M. Takeda, and B. Albert. 2015. High-pressure densified solid solutions of alkaline earth hexaborides (Ca/Sr, Ca/Ba, Sr/Ba) and their high-temperature thermoelectric properties. *J. Solid State Chem.* 221: 191.

103. D. P. Young, D. Hall, M. E. Torelli, Z. Fisk, J. L. Sarrao, J. D. Thompson, H. R. Ott, S. B. Oseroff, R. G. Goodrich, and R. Zysler. 1999. High-temperature weak ferromagnetism in a low-density free-electron gas. *Nature* 397: 412.

104. T. Mori and S. Otani. 2002. Ferromagnetism in lanthanum doped CaB_6: Is it Intrinsic? *Solid State Comm.* 123: 287.

105. C. Meegoda, M. Trenary, T. Mori, and S. Otani. 2003. Depth profile of iron in a CaB_6 crystal. *Phys. Rev. B* 67: 172410.

106. D. Hall, D. P. Young, Z. Fisk, T. P. Murphy, E. C. Palm, A. Teklu, and R. G. Goodrich. 2001. Fermi-surface measurements on the low-carrier density ferromagnet $Ca_{1-x}La_xB_6$ and SrB_6. *Phys. Rev. B* 64: 233105.

107. S. Souma, H. Komatsu, T. Takahashi, R. Kaji, T. Sasaki, Y. Yokoo, and J. Akimitsu. 2003. Electronic band structure and Fermi surface of CaB_6 studied by Angle-resolved photoemission spectroscopy. *Phys. Rev. Lett.* 90: 027202.

108. N. Takhashi, M. Terui, N. Ueda, and M. Takeda. 2005. Reduction of thermal conductivity and origin of carrier in alkaline-earth hexaborides. ICT 2005: 24th International Conference on Thermoelectrics 2005, 423–425.

109. S. Ghamaty, J. C. Bass, and N. B. Elsner. 2006. *Thermoelectrics Handbook, Micro to Nano*, edited by D. M. Rowe, Taylor & Francis, London, 57.

110. T. Tynell, T. Aizawa, I. Ohkubo, K. Nakamura, and T. Mori. 2016. Deposition of thermoelectric strontium hexaboride thin films by a low pressure CVD method. *J. Crystal Growth* 449, 10–14.

111. K. Kaymura and M. Takeda. 2011. Thermoelectric and electrical properties of p-type YbB_6. *2011 IOP Conf. Ser.: Mater. Sci. Eng.* 20: 012007.

112. Y. B. Kuzma. 1970. Crystal structure of the compound $YCrB_4$ and its analogs. *Sov. Phys. Crystallogr.* 15: 312.

113. T. Mori, K. Kudou, T. Shishido, and S. Okada. 2011. f-electron dependence of the physical properties of $REAlB_4$; an AlB_2-type analogous "tiling" compound. *J. Appl. Phys.* 109: 07E111.

114. T. Mori, T. Shishido, K. Nakajima, K. Kieffer, and K. Siemensmeyer. 2009. Magnetic properties of the thulium layered compound $Tm_2Al^{11}B_6$, an AlB_2-type analogue. *J. Appl. Phys.* 105: 07E124, 1–3.

115. T. Mori, H. Borrmann, S. Okada, K. Kudou, A. Leithe-Jasper, U. Burkhardt, and Y. Grin. 2007. Crystal structure, chemical bonding, electrical transport, and magnetic behavior of $TmAlB_4$. *Phys. Rev. B* 76: 064404, 1–10.

116. P. Rogl and H. Nowotny. 1974. Ternary complex boride with $ThMoB_4$-type. *Monatsh. Chem.* 105, (5): 1082.

117. N. I. Medvedeva, Yu. E. Medvedeva, and A. L. Ivanovskii Dokl. 2002. Electronic structure of ternary boron-containing phases $YCrB_4$, Y_2ReB_6 and MgC_2B_2. *Phys. Chem.* 383: 75–77.

118. S. Lassoued, R. Gautier, and J.-F. Halet. 2011. The electronic properties of metal borides and borocarbides: Differences and similarities. In *Boron Rich Solids, Sensors, Ultra High Temperature Ceramics, Thermoelectrics, Armor*, edited by N. Orlovskaya and M. Lugovy, Springer, New York, 95–114.

119. S. Lassoued, R. Gautier, A. Boutarfaia, and J.-F. Halet. 2010. Rings and chains in solid-state metal borides and borocarbides. The electron count matters. *J. Organomet. Chem.* 695: 987–993.

120. J. W. Simonson and S. J. Poon. 2010. Applying an electron counting rule to screen prospective thermoelectric alloys: The thermoelectric properties of YCrB$_4$ and Er$_3$CrB$_7$-type phases. *J. Alloys Compd.* 504: 265–272.

121. J.-F. Halet, S. Lassoued, B. Fontaine, R. Gautier, and T. Mori. 2014. *Electronic Structure and Physical Properties of Some Boron-Rich Ternary Compounds.* 18th International Symposium on Boron, Borides, and Other Related Compounds. Invited Talk, Honolulu, HI.

122. T. Mori, I. Kuzmych-Ianchuk, K. Yubuta, T. Shishido, S. Okada, K. Kudou, and Y. Grin. 2012. Direct elucidation of the effect of building defects on the physical properties of α-TmAlB$_4$; an AlB$_2$-type analogous "tiling" compound. *J. Appl. Phys.* 111, (7): 07E127.

123. X. J. Wang, T. Mori, I. Kuzmych-Ianchuk, Y. Michiue, K. Yubuta, T. Shishido, Y. Grin, S. Okada, and D. G. Cahill. 2014. Thermal conductivity of layered borides: The effect of building defects on the thermal conductivity of TmAlB$_4$ and the anisotropic thermal conductivity of AlB$_2$. *APL Mater.* 2: 046113.

124. N. F. Mott. 1968. Conduction in glasses containing transition metal ions. *J. Non-cryst. Solids* 1: 1.

125. A. L. Efros and M. Pollak. 1985. *Electron-Electron Interactions in Disordered Systems*, North-Holland, Amsterdam, 409.

126. T. Mori, J. Martin, and G. Nolas. 2007. Thermal conductivity of YbB$_{44}$Si$_2$. *J. Appl. Phys.* 102: 073510.

127. I. P. Zvyagin. 1973. On the theory of hopping transport in disordered semiconductors. *Phys. Status Solidi B* 58: 443.

128. D. Emin. 1982. Small Polarons. *Phys. Today* 35: 34.

129. I. A. Howard, C. L. Beckel, and D. Emin. 1987. Bipolarons in boron icosahedra. *Phys. Rev. B* 35: 2929–2933.

130. I. Higashi. 2000. Crystal chemistry of α-AlB$_{12}$ and γ-AlB$_{12}$. *J. Solid State Chem.* 154: 168.

131. N. Vojteer, V. Sagawe, J. Stauffer, M. Schroeder, and H. Hillebrecht. 2011. LiB$_{12}$PC, the first boron-rich metal boride with phosphorus - synthesis, crystal structure, hardness, spectroscopic investigations. *Chem. Eur. J.* 17: 3128–3135.

132. T. Ludwig, A. Pediaditakis, V. Sagawe, and H. Hillebrecht. 2013. Synthesis, crystal structure and properties of Mg$_3$B$_{36}$Si$_9$C and related rare earth compounds RE$_{3-x}$B$_{36}$Si$_9$C (RE=Y, Gd–Lu). *J. Solid State Chem.* 204: 113–122.

133. T. Mori, S. Maruyama, Y. Miyazaki, T. Kajitani, and K. Hayashi. 2016. Thermoelectric semiconductor rare earth aluminoboride; synthesis and usage as thermoelectric element, Japanese patent JP6061272.

134. R. Sahara, T. Mori, S. Maruyama, Y. Miyazaki, K. Hayashi, and T. Kajitani. 2014. Theoretical and experimental investigation of the excellent p–n control in yttrium aluminoborides. *Sci. Technol. Adv. Mater.* 15: 035012.

135. S. Maruyama, A. Prytuliak, Y. Miyazaki, K. Hayashi, T. Kajitani, and T. Mori. 2014. Al insertion and additive effects on the thermoelectric properties of yttrium boride. *J. Appl. Phys.* 115: 123702.

136. S. Maruyama, T. Nishimura, Y. Miyazaki, K. Hayashi, T. Kajitani, and T. Mori. 2014. Microstructure and thermoelectric properties of Y$_x$Al$_y$B$_{14}$ samples fabricated through the spark plasma sintering. *Mater. Renew. Sustain. Energy* 3: 1.

137. T. Mori and T. Hara. 2016. Hybrid effect to possibly overcome the tradeoff between Seebeck coefficient and electrical conductivity. *Scripta Mater.* 111: 44.

138. T. Mori. 2016. Perspectives of high temperature thermoelectric applications and p- and n-type aluminoborides. *JOM* 68, 2673–2679.

139. D. G. Cahill, S. K. Watson, and R. O. Pohl. 1992. Thermal properties of boron and boride. *Phys. Rev. B* 46: 6131.

140. S. Miura, H. Sasaki, K. Takagi, and T. Fujima. 2014. Effect of varying mixture ratio of raw material powders on the thermoelectric properties of $AlMgB_{14}$-based materials prepared by spark plasma sintering. *J. Phys. Chem. Solids* 75: 951.

141. L. F. Wan and S. P. Beckman. 2014. Lattice instability in the $AlMgB_{14}$ structure. *Phys. B Condens. Matter* 438: 9.

142. T. Mori and T. Nishimura. 2006. Thermoelectric properties of homologous p- and n-type boron-rich borides. *J. Solid State Chem.* 179: 2908.

143. T. Mori, T. Nishimura, K. Yamaura, and E. Takayama-Muromachi. 2007. High temperature thermoelectric properties of a homologous series of n-type boron icosahedra compounds: A possible counterpart to p-type boron carbide. *J. Appl. Phys.* 101: 093714.

144. D. Berthebaud, T. Nishimura, and T. Mori. 2010. Thermoelectric properties and spark plasma sintering of doped $YB_{22}C_2N$. *J. Mat. Res.* 25: 665.

145. T. Mori, T. Nishimura, W. Schnelle, U. Burkhardt, and Y. Grin. 2014. The origin of the n-type behavior in rare earth borocarbide $Y_{1-x}B_{28.5}C_4$. *Dalton Trans.* 43: 15048.

146. T. Goto. 1998. Boride high temperature thermoelectric material. *Kinzoku* (in Japanese) 68: 1086.

147. A. Prytuliak, S. Maruyama, and T. Mori. 2013. Anomalous effect of vanadium boride seeding on thermoelectric properties of $YB_{22}C_2N$. *Mat. Res. Bull.* 48: 1972.

148. T. Mori, F. Zhang, and T. Tanaka. 2001. Synthesis and magnetic properties of binary boride REB_{25} compounds. *J. Phys. Condens. Matter* 13: L423–L430.

149. T. Mori and H. Nishijima, unpublished. Thermoelectric properties of $YB_{18}Si_5$ sintered with spark plasma sintering.

150. T. Mori. 2005. High temperature thermoelectric properties of B_{12} icosahedral cluster-containing rare earth boride crystals. *J. Appl. Phys.* 97: 093703.

151. T. Mori. 2006. Thermal conductivity of a rare earth B_{12}-icosahedral compound. *Phys. B Condens. Matter* 383: 120.

152. T. Mori, D. Berthebaud, T. Nishimura, A. Nomura, T. Shishido, and K. Nakajima. 2010. Effect of Zn doping on improving crystal quality and thermoelectric properties of borosilicides. *Dalton Trans.* 39: 1027.

153. M. A. Hossain, I. Tanaka, T. Tanaka, A. U. Khan, and T. Mori. 2016. Crystal growth and anisotropy of high temperature thermoelectric properties of yttrium borosilicide single crystals. *J. Solid State Chem.* 233: 1–7.

154. A. Sussardi, T. Tanaka, A. U. Khan, L. Schlapbach, and T. Mori. 2015. Enhanced thermoelectric properties of samarium boride. *J. Materiomics* 1: 196–204.

155. M. A. Hossain, I. Tanaka, T. Tanaka, A. U. Khan, and T. Mori. 2015. YB_{48} the metal rich boundary of YB_{66}; crystal growth and thermoelectric properties. *J. Phys. Chem. Solids* 87: 221.

156. C. Wood and D. Emin. 1984. Conduction mechanism in boron carbide. *Phys. Rev. B* 29: 4582.

157. G. V. Samsonov, G. N. Makarenko, and G. G. Tsebulya. 1960. *Izv. Akad. Nauk SSSR, Otd. Tekhn. Nauk* No. 4.

158. M. Bouchacourt and F. Thevenot. 1985. The correlation between the thermo-electric properties and stoichiometry in the boron carbide phase B4C-B10.5C. *J. Mater. Sci.* 20: 1237.

159. B. Feng, H. P. Martin, and A. Michaelis. 2013. *In situ* preparation and ther-moelectric properties of B_4C_{1-x} – TiB_2 Composites. *J. Electron. Mater.* 42: 2314–2319.

160. Y. Gong, Y. Zhang, M. Dudley, Y Zhang, J. H. Edgar, P. J. Heard, and M. Kuball. 2010. Thermal conductivity and Seebeck coefficients of icosahedral boron arsenide films on silicon carbide. *J. Appl. Phys.* 108: 084906.

161. C. D. Frye, J. H. Edgar, I. Ohkubo, and T. Mori. 2013. Seebeck coefficient and electrical resistivity of single crystal $B_{12}As_2$ at high temperatures. *J. Phys. Soc. Jpn.* 82: 095001.

162. Y. Kumashiro, T. Yokoyama, K. Sato, Y. Ando, S. Nagatani, and K. Kajiyama. 2000. Electrical and thermal properties of $B_{12}P_2$ wafers. *J. Solid State Chem.* 154, 33–38.

163. T. Akashi, T. Itoh, I. Gunjishima, H. Masumoto, and T. Goto. 2002. Thermoelectric properties of hot-pressed boron suboxide (B_6O). *Mater. Trans.* 43: 1719–1723.

164. G. A. Slack and K. E. Morgan. 2014. Some crystallography, chemistry, phys-ics, and thermodynamics of $B_{12}O_2$, $B_{12}P_2$, $B_{12}As_2$, and related alpha-boron type crystals. *J. Phys. Chem. Solids* 75: 1054.

165. O. Sologub, Y. Matsushita, and T. Mori. 2013. An α-rhombohedral boron-related compound with sulfur: Synthesis, structure and thermoelectric prop-erties. *Scripta Mater.* 68: 289.

166. K. Koumoto, M. Shimohigoshi, S. Takeda, and H. Yanagida. 1987. Thermoelectric energy conversion by porous SiC ceramics. *J. Mater. Sci. Lett.* 6: 1453.

167. H. Kitagawa, N. Kado, and Y. Noda. 2002. Preparation of n-type silicon car-bide-based thermoelectric materials by spark plasma sintering. *Mater. Trans.* 43: 3239.

168. Y. Ohba, T. Shimozaki, and H. Era. 2008. Thermoelectric properties of sil-icon carbide sintered with addition of boron carbide, carbon, and alumina. *Mater. Trans.* 49: 1235.

169. J. G. Kim, Y. Y. Choi, D. J. Choi, and S. M. Choi. 2011. Study on Thermoelectric Properties of CVD SiC Deposited with Inert Gases. *J. Electron. Mater.* 40: 840.

170. N. Abu-Ageel, M. Aslam, R. Ager, and L. Rimai. 2000. The Seebeck coef-ficient of monocrystalline alpha-SiC and polycrystalline beta-SiC measured at 300-533 K. *Semicond. Sci. Technol.* 15: 32.

171. L. A. Valentín, J. Betancourt, L. F. Fonseca, M. T. Pettes, L. Shi, M. Soszyński, and A. Huczko. 2013. A comprehensive study of thermoelec-tric and transport properties of β-silicon carbide nanowires. *J. Appl. Phys.* 114: 184301.

172. M. Khazaei, M. Arai, T. Sasaki, C.-H. Chung, N. S. Venkataramanan, M. Estili, Y. Sakka, and Y. Kawazoe. 2013. Novel electronic and magnetic properties of two-dimensional transition metal carbides and nitrides. *Adv. Funct. Mater.* 23: 2185.

173. J. Blatt. 1976. *Thermoelectric Power of Metals*, Springer Science & Business Media, Berlin.

174. D. M. Rowe. 1955. *CRC Handbook of Thermoelectrics*, CRC Press, Boca Raton, FL.

175. F. E. Bash. 1919. *The Manufacture and Electrical Properties of Constantan for Thermocouples*. University of Wisconsin–Madison.

176. B. Coles. Electronic structures and physical properties in the alloy systems nickel-copper and palladium-silver. 1952. *Proc. Phys. Soc. Sect. B* 65: 221.

177. J. Mao, Y. Wang, H. S. Kim, Z. Liu, U. Saparamadu, F. Tian, K. Dahal et al. 2015. High thermoelectric power factor in Cu–Ni alloy originate from potential barrier scattering of twin boundaries. *Nano Energy* 17: 279–289.

178. N. F. Mott. 1936. XXIII. Optical constants of copper-nickel alloys. *London Edinburgh Dublin Philo. Mag. and J. Sci.* 22: 287–290.

179. D. H. Seib and W. E. Spicer. 1968. Experimental evidence for the validity of the virtual bound-state model for Cu-Ni alloys. *Phys. Rev. Lett.* 20: 1441.

180. K. Schröder and D. Önengüt. 1967. Optical absorption of copper and copper-rich copper nickel alloys at room temperature. *Phys. Rev.* 162: 628.

181. J. Feinleib, W. J. Scouler, and J. Hanus. 1969. Optical studies and band structure of Cu–Ni alloys. *J. Appl. Phys.* 40: 1400–1401.

182. L. E. Walldén, D. H. Seib, and W. E. Spicer. 1969. Photoemission and optical studies of Cu–Ni and Ag–Mn Alloys. *J. Appl. Phys.* 40: 1281–1282.

183. G. Stocks, R. Williams, and J. Faulkner. 1971. Densities of states in Cu-rich Ni-Cu alloys by the coherent-potential approximation: comparisons with rigid-band and virtual-crystal approximation. *Phys. Rev. Lett.* 26: 253.

184. H. D. Drew and R. E. Doezema. 1972. Virtual-bound-state induced optical absorptivity in CuNi alloys. *Phys. Rev. Lett.* 28: 1581.

185. D. Beaglehole. 1976. Optical absorption and s and p phase shifts of virtual-bound-state alloys. *Physi. Rev. B* 14: 341.

186. D. Beaglehole and C. Kunz. 1977. Optical studies of dilute CuNi alloys. *J. Phys. F: Metal Phys.* 7: 1923.

187. W. H. Keesom and B. Kurrelmeyer. 1940. Specific heats of alloys of nickel with copper and with iron from 1.2 to 20 K. *Physica* 7: 1003–1024.

188. G. L. Guthrie, S. A. Friedberg, and J. E. Goldman. 1959. Specific heats of some copper-rich copper-nickel alloys at liquid helium temperatures. *Phys. Rev.* 113: 45.

189. K. Gupta, C. Cheng, and P. A. Beck. 1964. Low-temperature specific heat of Ni-base fcc solid solutions with Cu, Zn, Al, Si, and Sb. *Phys. Rev.* 133: A203.

190. J. Friedel. 1956. On some electrical and magnetic properties of metallic solid solutions. *Can. J. Phy.* 34: 1190–1211.

191. P. W. Anderson. 1961. Localized magnetic states in metals. *Phys. Rev.* 124: 41.

192. S. Hüfner, G. K. Wertheim, R. L. Cohen, and J. H. Wernick. 1972. Density of states in CuNi alloys. *Phys. Rev. Lett.* 28: 488.

193. J. P. Heremans, V. Jovovic, E. S. Toberer, A. Saramat, K. Kurosaki, A. Charoenphakdee, S. Yamanaka, and G. J. Snyder. 2008. Enhancement of thermoelectric efficiency in PbTe by distortion of the electronic density of states. *Science* 321: 554–557.

194. H. J. van Daal, P. B. van Aken, and K. H. J. Buschow. 1974. The Seebeck coefficient of YbAl$_2$ and YbAl$_3$. *Phys. Lett. A* 49: 246–248.

195. D. Rowe, V. Kuznetsov, L. Kuznetsova, and G. Min. 2002. Electrical and thermal transport properties of intermediate-valence YbAl$_3$. *J. Phys. D: Appl. Phys.* 35: 2183.

196. E. E. Havinga, K. H. J. Buschow, and H. J. Van Daal. 1973. The ambivalence of Yb in YbAl$_2$ and YbAl$_3$. *Solid State Comm.* 13: 621–627.

197. G. D. Mahan and J. O. Sofo. 1996. The best thermoelectric. *Proc. Natl. Acad. Sci. USA* 93: 7436–7439.

198. Y. Pei, A. D. LaLonde, H. Wang, and G. J. Snyder. 2012. Low effective mass leading to high thermoelectric performance. *Energy Environ. Sci.* 5: 7963–7969.

199. C. Y. Ho, M. W. Ackerman, K. Y. Wu, S. G. Oh, and T. N. Havill. 1978. Thermal conductivity of ten selected binary alloy systems. *J. Phys. Chem.* 7: 959–1178.

200. D. D. Pollock. 1991. *Thermocouples: Theory and Properties*, CRC Press, Boca Raton, FL.

201. H. Muta, K. Kurosaki, M. Uno, and S. Yamanaka. 2003. Thermoelectric properties of constantan/spherical SiO_2 and Al_2O_3 particles composite. *J. Alloys Compd.* 359: 326–329.

CHAPTER 11

Engineering of Materials

Mona Zebarjadi, Gang Chen, Zhifeng Ren, Sunmi Shin,
Renkun Chen, Joseph P. Heremans, Bartlomiej Wiendlocha,
Hyungyu Jin, Bo Wang, and Qinyong Zhang

Contents

11.1 Thermoelectric Materials with Unconventional Doping Schemes

Mona Zebarjadi, Gang Chen, and Zhifeng Ren

11.1.1 Introduction

Heavily-doped semiconductors are perhaps the most promising class of thermoelectric materials. Insulators do not have enough conduction electrons, and metals suffer from very small Seebeck coefficient values. Semiconductors are filling the gap between insulators and metals. The advantage of semiconductors is the ability to tune their chemical potential (μ), i.e., their number of conduction carriers (n). This tuning is achieved via doping. External impurity atoms or, in some cases, vacancies, are introduced in a host matrix to change its carrier concentration.

As the carrier concentration increases, the electrical conductivity (σ) of the semiconductor increases, and the chemical potential initially moves from inside the bandgap closer to the band edge, eventually entering the band. Once the chemical potential is in the band, further increase in the carrier concentrations results in moving away from the band edge and deeper into the band. In most cases, as we move deep in the band, the slope of the density of state (DOS) versus energy decreases. The Seebeck coefficient (S) is proportional to this slope and decreases with further increase in the carrier concentration. As a result of the decreasing Seebeck coefficient and the increasing electrical conductivity, there is an optimum chemical potential and an optimum carrier concentration (see Figure 11.1). The optimum carrier concentration range happens at very large values: 10^{17}–10^{19} cm^{-3}.

At such large carrier concentrations, a significant limiting factor for the carrier mobility is ionized impurity scattering. Let us consider n-type semiconductors. The dopants (external impurity atoms) donate electrons to the conduction band. This process results in positive ionized impurities (ions) left inside the host matrix leading to strong Coulomb interaction between the conduction electrons and the ions, a process that is referred to as ionized impurity scattering. Considering that dopants are essential in a semiconductor in order to have conduction electrons, is it possible to limit the ionized impurity scattering? In this chapter we summarize different proposals to minimize ionized impurity scattering in thermoelectric materials and devices.

11.1.2 Types of Unconventional Doping

11.1.2.1 Intrinsic Carriers

The first idea for limiting ionized impurity scattering is to use intrinsic carriers. These are electrons that are thermally excited from the valance band to the conduction band. There are several problems with this approach. First, the number of intrinsic carriers depends on the temperature and the bandgap of the semiconductor. At low temperatures and at large bandgaps, this number is very low. For example, silicon with a bandgap of 1.1 eV only has an intrinsic carrier concentration of about 10^{10} cm^{-3}, but the optimum carrier

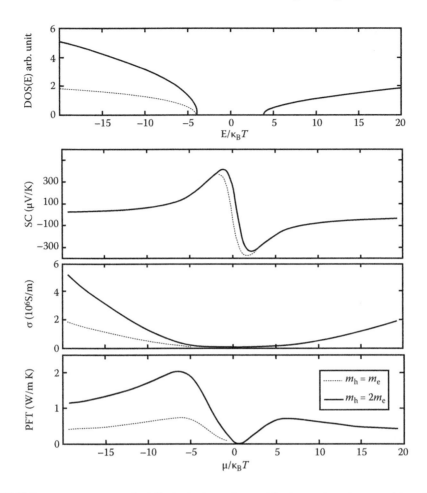

FIGURE 11.1 Density of states (DOS), Seebeck coefficient (*S*), electrical conductivity (σ), and thermoelectric power factor times temperature (*PFT* = σ*S*²*T*) versus chemical potential calculated using a simple two-band model with constant relaxation times approximation to illustrate the dependence of the transport properties to the position of the chemical potential. Two cases are considered; in the first case, the effective mass of electrons and holes are the same (*dotted lines*), and in the second case, the mass of holes is twice the mass of electrons (*solid lines*) and is added to illustrate the importance of asymmetry between electrons and holes.

concentration in silicon is as large as 10^{20} cm^{-3}. Narrow-gap semiconductors have larger intrinsic carrier concentrations. For example, the intrinsic carrier concentration in PbTe is on the order of 10^{17} cm^{-3}, which is quite large. However, if the bands are symmetric (that is, the effective mass of electrons and holes are the same), the chemical potential is in the middle of the bandgap and the resulting Seebeck coefficient is zero (see Figure 11.1). It is possible to benefit from intrinsic carriers using narrow-bandgap semiconductors with extreme asymmetry between conduction and valance bands, a concept that is illustrated in Figure 11.1 and further in Figure 11.2.

Figure 11.1 shows the results of a simple two-band model. In this model, we assume constant relaxation times and Drude-type conduction. We use the same τ value for electrons and holes, and we add their conductivity to

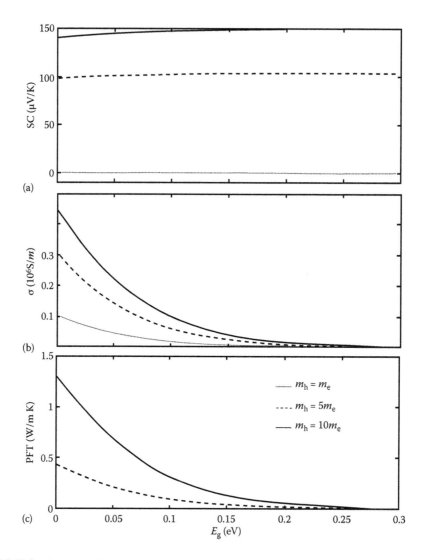

FIGURE 11.2 Potential of intrinsic narrow-gap semiconductors: (a) Seebeck coefficient, (b) electrical conductivity, and (c) power factor times temperature of an intrinsic semiconductor versus bandgap calculated for different effective mass ratios.

find the total conductivity. Under these assumptions, the electrical conductivity and the Seebeck coefficient are calculated using

$$\sigma = \frac{ne^2\tau}{m_e} + \frac{pe^2\tau}{m_h}, \tag{11.1}$$

$$S = \frac{1}{eT} \frac{\displaystyle\int_{-\infty}^{\infty} (\varepsilon - \mu)\,DOS(\varepsilon)\,v_g^2(\varepsilon)\,\frac{\partial f(\varepsilon)}{\partial \varepsilon}\,d\varepsilon}{\displaystyle\int_{-\infty}^{\infty} DOS(\varepsilon)\,v_g^2(\varepsilon)\,\frac{\partial f(\varepsilon)}{\partial \varepsilon}\,d\varepsilon}, \tag{11.2}$$

where n is the electron density, p is the hole density, m_e is the effective mass of electrons, m_h is the effective mass of holes, ε is the energy, τ is the average relaxation time, e is the unit carrier charge, T is the temperature, f is the Fermi–Dirac function, and v_g is the group velocity.

To address intrinsic semiconductors, we can obtain the chemical potential assuming zero doping. Therefore, the electron carrier density is equal to the hole carrier density, and we label them by subindex i, referring to intrinsic ($n_i = p_i$). The results versus bandgap and for different effective mass ratios are shown in Figure 11.2.

However, this model is dependent on using a constant relaxation time $\tau = 10^{-13}$ for all cases to represent the power factor scattering rate at room temperatures. Nevertheless, it shows that even without doping, high power factors are possible when there is large asymmetry between electrons and holes. Of course, further improvement could be achieved when small doping is used to tune the chemical potential optimally.

11.1.2.2 Cluster Doping or Nanoparticle Doping

Next, we discuss the idea of cluster doping or nanoparticle doping. Molecule dopants (e.g., C60),[1] cluster dopants (e.g., clusters of boron atoms in silicon, instead of single boron atoms), or metallic/semimetallic nanoparticle dopants[2] could replace single impurity atoms. The idea is similar in all cases: point charges are replaced by a uniform charge distributed over a finite volume. It is not granted that nanoparticle doping is better than single impurity doping. In fact, in many cases, single impurity doping is a better approach. Assume that the desired carrier density is n. To reach this number of carriers, n single impurity atoms per volume, each donating $Z = 1$ electrons, are needed. If we replace them by N clusters per volume ($N < n$), each cluster should donate $Z = n/N$ electrons. Since the scattering cross-section is proportional to Z^2, the total scattering rate of N clusters is n/N times larger than single impurity atoms. In this argument, we assumed that the clusters are also point charges. If their charge is distributed over a spherical volume with radius r, then when using Born approximation (constant barrier potential), the scattering cross-section scales with r^6 and r^4 for low- and high-energy electrons, respectively. Therefore, small r values are much better. However, we note that r is the effective radius or the radius over which nonzero potential exists, and due to long-range Coulomb interactions, r is usually larger than the nanoparticle or the point-charge radius (see Figure 11.3). In addition, screening effects in some cases can help design nanoparticles with weaker scattering rates compared to conventional impurity doping.[3,4] The result of these considerations is that care should be taken when switching from single impurity doping to nanoparticle doping.

Many groups have used the nanoparticle-doping approach to enhance the thermoelectric properties of thermoelectric materials. In many cases, the main reason for adding nanoparticles was not to improve the carrier mobility. One of the first demonstration of semimetallic nanoparticles in thermoelectric host matrices was the work of Zide et al.[5,6] In this work, ErAs semimetallic nanoparticles were embedded inside $In_{0.53}Ga_{0.47}As$ to introduce

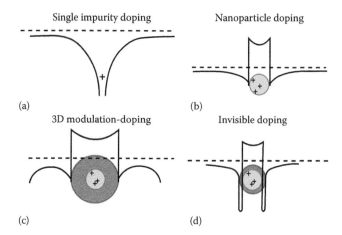

FIGURE 11.3 Schematic of (a) single impurity doping, (b) cluster/nanoparticle doping, (c) modulation doping, and (d) invisible doping potential profile. Inner spheres represent ionized nanoparticles; outer shells represent the coating layers. In the case of modulation doping, a thick coating layer is used as a spacer to increase the distance between the dopants and the conduction electrons. In the case of invisible doping, a thin coating layer is used to cloak the core dopants. Chemical potential is shown using a dashed line in each case.

conduction carriers. However, the main motivation was to improve the Seebeck coefficient via electron filtering. Nanoparticles also are proposed to lower the thermal conductivity.[7,8]

In a more relevant work to our current discussion, Ibáñez et al.[9] demonstrated an interesting strategy to enhance the carrier mobility and thermoelectric power factor using metallic nanoparticles. In their work, silver nanoparticles were added to PbS nanocrystal blocks. In the absence of metallic particles, the carrier mobility of PbS nanocrystals is limited by the scattering potentials at grain boundaries. The addition of silver nanoparticles in the grain boundaries modified the shape of this potential. In the limit of low nanoparticle concentrations, quantum wells formed at the grain boundaries. As the silver nanoparticle concentration increased, the chemical potential increased and moved close to the edge of the well. Eventually, the chemical potential entered the band of the host matrix, resulting in zero effective potential for conduction carriers. Therefore, as the silver nanoparticle density increased, not only the carrier density increased, but also the carrier mobility.

Note that in this discussion, we did not cover quantum dot superlattices.[10,11] This disregard is because the motivation for that work was not to improve the carrier mobility, but to use one-dimensional (1D) band structure to improve the Seebeck coefficient.

11.1.2.3 Modulation Doping

Another approach to limit ionized impurity scattering is to use modulation doping. The idea of modulation doping was first proposed in 2D structures.[12] In modulation doping (also called remote doping and delta-doping), all dopants are concentrated in a thin, doping layer. The doping layer is separated from the main transport channel by a spacer. The band alignment between the doped region, the spacer, and the channel results in carrier transport from the doped region to the channel. The purpose of this transport is to separate carriers from their parent atoms spatially, thereby reducing the ionized impurity scattering. In the 22 years of research between 1978 and 2000, mobility improved by four orders of magnitude. This is a significant enhancement to mobility.[13] The first demonstration of mobility enhancement in 1978 demonstrated a mobility of 10^4 cm²/V s compared to the bulk mobility of 3×10^3 cm²/V s.[12] By 2000, the mobility value of 31 million cm²/V s had been achieved. In the geometry used in 2D modulation doping, charge carriers are trapped at the interface of the channel and the spacer, and move along the interface. Therefore, a smooth and clean interface is crucial to minimize roughness scattering, which is only possible by using molecular beam epitaxy technique.

Thermoelectric materials used in commercial applications are bulk semiconductors, and must be cost effective to be commercially viable. Because of the expense of molecular beam epitaxy, the commercial viability of this strategy is low. There are several possible methods for applying the modulation doping concept to bulk samples. Dopants could be placed in nanoparticles, nanorods, or nanowires, or on cylindrical hole walls. In all cases, ideally, a spacer layer is needed to separate the dopant layer from the charge carriers. It is also reasonable to assume that the active layer feature size (distance between neighboring spacers) should be close to the screening length so that carriers are transporting through the bulk instead of being trapped at the interfaces. A schematic is shown in Figure 11.3c. The extra bending on the sides is a result of limited feature size and is from neighboring nanoparticles.

There has not been any demonstration of the effect of the spacer layer in 3D modulation doping. Of course, the effect of the spacer layer has been investigated before in the 2D devices. In the context of thermoelectric materials, Hou et al.,[14] studied the case of 2D MnSi$_{1.7}$ films. In the studied structure, a heavily doped silicon layer (doped by Al and Cu) was separated from the MnSi$_{1.7}$ film by a thin layer of silicon (<20 nm), which served as the spacer layer. Using the spacer layer, they were able to improve the power factor of MnSi$_{1.7}$ films by a factor of 2.[15] The effect of the spacer layer and modulation doping on thermoelectric transport also was studied experimentally in one dimension, where GaN/AlGaN core–shell nanowires (NWs) showed an improved thermoelectric power factor.[16] Experimentally, it is challenging to add the spacer layer to either of the 3D geometries. Perhaps one way is to use oxide layers. For example, in the case of silicon host matrix (or silicon dopant layers), silicon dioxide can serve as the spacer layer as it is relatively easy to oxidize the surface of silicon. Despite its ease of fabrication, silicon dioxide is not an ideal spacer layer for silicon. Silicon dioxide has a larger bandgap compared to silicon and forms a barrier for carrier transport. Ideally, the spacer layer band should be aligned with the band of the doped layer. In the case of silicon dioxide, the layer must be very thin to allow quantum tunneling of the charges.

The first demonstration of 3D modulation doping in nanostructured SiGe samples did not benefit from the presence of spacer layers. It was shown that by embedding heavily doped silicon nanograins (20–50 nm in size) inside SiGe host matrix, it was possible to enhance the thermoelectric power factor by about 40% compared to conventional SiGe uniform nanostructures doped with boron.[17] In this geometry, charge carriers spill over from the silicon nanoparticles to the host matrix and travel within the host matrix with reduced ionized impurity scattering. However, due to the absence of the spacer layer, the charge carrier can still see the ionized particles and the improvement in mobility was limited.[18] More recently, 3D modulation doping has been used to improve the thermoelectric properties of BiCuSeO,[19] FeSb$_2$,[20] PbTe,[21] and PbSe.[22]

11.1.2.4 Field-Effect Doping

Another way of remote doping is field-effect doping, wherein carriers are induced inside the transport channel by means of gating. This is especially important for device design, as well as for the proof of concept for the effectiveness of remote doping to enhance the thermoelectric power factor. Field-effect doping is not applicable to bulk thermoelectrics, but we can envision the design of fast and efficient 2D and 1D thermoelectric coolers. Liang et al.[23] demonstrated the possibility of tuning the Seebeck coefficient of single PbSe NWs using field-effect-gated devices. Curtin et al.[24] demonstrated the field-effect doping concept in silicon NWs. Gating is done routinely in 2D devices. Graphene, for example, has been the subject of intense study, and its Seebeck coefficient is measured at different temperatures and gating voltages (corresponding to the position of the chemical potential or the carrier density).[25–27] In a recent work,[28] it was shown that by placing graphene on hexagonal boron nitride and using gating, very large thermoelectric power factors could be achieved. The enhancement was the results of several factors including reduced electron hole puddles (compared to graphene on SiO$_2$), low electron–phonon interaction, which is intrinsic to graphene, and remote doping. Measured mobility values were as large as 9600 cm^2/V s at room temperatures, which resulted in a record power factor times temperature value of 10.35 W/m K.

11.1.2.5 Invisible Doping

One main advantage of replacing single-impurity doping with nanoparticle doping is that in the latter case, there are more parameters involved, resulting in more design freedom. Parameters such as nanoparticle size, shape, and material provide opportunities to engineer the electron–nanoparticle scattering rate. One optimistic design for nanoparticle doping is nanoparticles with antiresonance scattering rates. In antiresonance scattering, there is an energy window at which electron–nanoparticle scattering is minimum. Inspired by optical cloaking,[29,30] wherein objects are cloaked in a narrow frequency range of light, we call this energy window the "cloaking window." Surely if the minimum is small enough, electrons with energies in the cloaking window experience only minimal scattering from the nanoparticles. Liao et al.[31] first theoretically demonstrated the possibility of electronic cloaking by using a core–shell nanoparticle. In this design, ionized dopants are incorporated inside the core. The band alignment favors charge transfer from the nanoparticle to the host matrix. A shell layer is then used to coat the ionized core (Figure 11.3d) and to minimize the electron–nanoparticle scattering rate in the cloaking window. When the cloaking window overlaps with the Fermi window, the nanoparticles are invisible to most of the conduction electrons. This approach potentially can result in orders of magnitude improvement in carrier mobility and thermoelectric power factor.[32] The main difficulty is to find proper material combinations of core–shell–host for which cloaking is possible. Since there are three materials involved, the search space is too wide and blind combinatorial search algorithms are impractical.[33] In a recent work, Zebarjadi and Shen[34] used a new combinatorial method called the dimensionless mapping method to find proper core–shell–host combinations. In this method, they eliminated the least sensitive parameters and geometric factors and provided maps for the remaining parameters to accelerate the search. We also have to emphasize that core–shell geometry is not a unique geometry, and it is only one of the many possible geometries for electronic cloaking.[35–37] The applications for electronic cloaking are beyond the thermoelectric field and could be used for design of faster switches, transistors, etc.

11.1.3 Future Directions

Unconventional doping schemes are newly introduced in the thermoelectric field. Their application goes beyond the thermoelectric field as the design of high-mobility semiconductors are important for most semiconductor device applications. There are millions of different material combinations (nanoparticle–host) to explore. For each material combination, there are millions of different geometries and sizes to be considered. There is a need for theoretical guidelines to narrow down the search. Even theoretical, but blind, combinatorial searches are time consuming and impractical. We need to develop smart combinatorial search algorithms and combine them with physical intuitions to find proper combinations. The role of the spacer layer in bulk modulation doping has not been investigated experimentally due to challenges of making core–shell nanoparticles or nanorods. Other possible geometries such as aligned NWs inside a host matrix or porous host matrices filled with dopants are yet to be explored. In the case of invisible doping, there have not been any experimental demonstrations. Since the cloaking window is relatively narrow and should be tuned to overlap the Fermi window, it is extremely sensitive to the materials and geometric parameters involved. Only combinations, for which such parameter sensitivity is weak, have the potential for experimental investigation.

11.1.4 Acknowledgment

The work at UVa is supported by the National Science Foundation (NSF) grant number 1400246.

11.2 Low-Dimensional Thermoelectric Materials

Sunmi Shin and Renkun Chen

11.2.1 Overview

The interdependence of thermoelectric parameters has been a major roadblock for achieving a high thermoelectric figure of merit ($ZT = S^2\sigma T/k$). Tailoring material properties using the classic and quantum-size effect emerged as a promising strategy to decouple the transport parameters for boosting ZT. In fact, in the 2000s, adopting nanostructures has been a major breakthrough to overcome the stagnant ZT values of around 1 since 1950s.[38] In this section, the physical origins of ZT enhancement by engineering thermal and electronic transport in low-dimensional materials will be introduced with focus on the examples of superlattices and NWs for high power factor and low thermal conductivity, respectively.

11.2.2 Electronic Engineering: Improving Power Factor

The idea of using the low-dimensional quantum confinement effect was exploited extensively in the 1990s and early 2000s. In 1993, Hicks and Dresselhaus[39] wrote landmark papers suggesting the quantum confinement effect as a promising strategy to enhance the power factor and thermoelectric figure of merit (ZT) in 2D quantum well structures and 1D conductors.[11] Subsequently, in 1996 Mahan and Sofo[40] introduced the concept of "the best thermoelectrics," which showed that a high power factor can be achieved through a δ-function-like DOS. Low-dimensional materials, such as quantum dots, quantum wires, and quantum wells, generate a highly asymmetric and sharp DOS around the Fermi energy, leading to a high Seebeck coefficient and power factor.

The enhancement of the power factor due to the confinement effect can be understood through the Boltzmann transport equations (BTE) for electrical conductivity σ and Seebeck coefficient S:

$$\sigma = e^2 \int_0^\infty g(E)\upsilon^2(E)\tau(E,T)\left[-\frac{\partial f(E)}{\partial E}\right]dE, \tag{11.3}$$

$$S = \frac{k_B}{e}\left[\frac{\int_0^\infty g(E)\upsilon^2(E)\dfrac{E-E_F}{k_B T}\tau(E,T)\left[-\dfrac{\partial f(E)}{\partial E}\right]dE}{\int_0^\infty g(E)\upsilon^2(E)\tau(E,T)\left[-\dfrac{\partial f(E)}{\partial E}\right]dE}\right], \tag{11.4}$$

where e is the electron charge, $g(E)$ is the total electronic DOS, $f(E)$ is the energy distribution function (i.e., Fermi–Dirac function for charge carriers, $f(E) = 1/\{\exp\left[(E-E_F)/k_B T\right] + 1\})$, E_F is the Fermi energy, k_B is the Boltzmann constant, $\upsilon(E)$ is the electron velocity at energy E, and $\tau(E, T)$ is the relaxation time of electrons at energy E and temperature T. The derivative of the

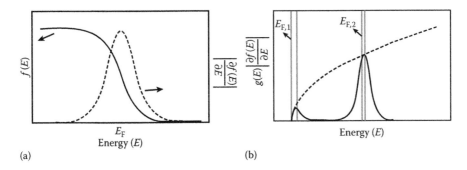

FIGURE 11.4 (a) Fermi–Dirac distribution function $f(E)$ (*solid line*) and its derivative $|\partial f(E)/\partial E|$ (*dotted line*). (b) Differential conductivities $g(E)|\partial f(E)/\partial E|$ (*solid line*) in 3D with different Fermi energies and $g(E)$ for 3D (Equation 11.4) (*dotted line*). Lower Fermi energy ($E_{F,1}$) has a lower differential conductivity, but a more asymmetric $g(E)$ around it, leading to low σ and high S. On the contrary, higher Fermi energy ($E_{F,2}$) has a higher differential conductivity but a less asymmetric $g(E)$ around it, leading to high σ and low S. (Rudolph, R. et al.: Dielectric properties of $Bi_{1-x}Sb_x$ alloys at the semiconductor–semimetal transition. *Physica Status Solidi (B)*. 1980. 102. 295–301. Copyright Wiley-VCH Verlag GmbH & Co. KGaA. Reproduced with permission; Fellmuth, B. et al.: Determination of the cyclotron masses of the conduction band of $Bi_{1-x}Sb_x$ near the semimetal–semiconductor transition. *Physica Status Solidi (B)*. 1980. 160. 695–703. Copyright Wiley-VCH Verlag GmbH & Co. KGaA. Reproduced with permission.)

Fermi–Dirac distribution with respect to energy, $-\partial f(E)/\partial E$, has a peak at the Fermi level and exponentially diminishes when E is $\sim 5k_B T$ away from E_F, as shown in Figure 11.4a, which indicates that electrons with energy close to the Fermi level dominantly contribute to electrical and thermoelectric transport. Equations 11.3 and 11.4 indicate that σ is proportional to the integral of the product $g(E)(-\partial f/\partial E)$, whereas S is proportional to the integral of the product $(E - E_F)g(E)(-\partial f/\partial E)$. In bulk (3D) semiconductors, $g(E)$ is generally proportional to $\sqrt{E}$. Therefore, as E_F increases, e.g., via a higher doping concentration, the amplitude of $g(E)$ increases, leading to a higher σ, but it is less asymmetric around E_F, causing a lower S (Figure 11.4b).

The trade-off between S and σ leads to an optimal doping concentration and a peak power factor value for a given set of material properties. If we explicitly consider the energy dependence of electron relaxation time and assume a power law scattering formula $\tau(E) \sim E^r$ and assume parabolic bands $E = E_c + \hbar^2 k^2/2m^*$ (for electrons) or $E = E_v - \hbar^2 k^2/2m^*$ (for holes), where m^* is the effective mass for electrons or holes, we can reach the following equation for the power factor:

$$S^2 \sigma = \frac{E^r}{m^*}\left[\frac{E_F}{k_B T} + \left(r + \frac{5}{2}\right)^2\right].$$

(11.5)

Equation 11.5 suggests that the maximum power factor is determined by r and m^*. For typical semiconductors, r ranges from –1/2 to 3/2, and m^*/m_0 ranges from 0.01 to 2, where m_0 is the free electron mass. As an example,

Pei et al.[41] showed enhanced S and power factor in Na-doped PbTe due to lighter effective hole mass.

One strategy to break this limit is to use structures or materials with a DOS $g(E)$ deviated from the $\sqrt{E}$ dependence. In particular, if there is a resonance in $g(E)$ at E_F, the Seebeck coefficient and the power factor may be improved over the typical bulk values. In Mahan and Sofo's paper,[40] they showed that a δ-function-like transport distribution $g(E)v_x^2(E)\tau(E)$ would maximize the power factor and ZT. In reality, "a Lorentizan of very narrow width" would most closely resemble the δ-function, as in the case of f-level electronic states in YbAl$_3$. Another notable example is the introduction of resonance in DOS. This resonant-doping idea has been demonstrated in Tl-doped PbTe[42] and In-doped SnTe,[43] and has been reviewed by Heremans et al.[44]

An alternative approach to modify the DOS is based on quantum structures, which offer discrete and sharp $g(E)$ due to the quantum confinement effect. This approach is perhaps more controllable and more desirable for thermoelectrics. In the case of parabolic energy dispersion, $g(E)$ for a single band in 3D, 2D, and 1D systems, including a spin degeneracy factor of 2, are expressed as

$$g_{3D}(E) = \frac{1}{2\pi^2}\left(\frac{2m_e^*}{\hbar^2}\right)^{3/2} E^{1/2}, \tag{11.6}$$

$$g_{2D}(E) = \frac{m_e^*}{L\pi\hbar^2}, \tag{11.7}$$

$$g_{1D}(E) = \frac{1}{L^2\pi}\left(\frac{2m_e^*}{\hbar^2}\right)^{1/2} E^{-1/2}, \tag{11.8}$$

where L is the size of the sample $g(E)$ for the 3 cases are shown in Figure 11.5. In each of these cases, if the E_F is located at an energy level that has

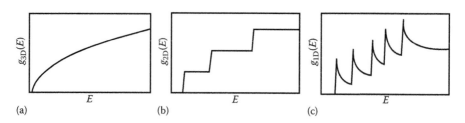

FIGURE 11.5 Plots of DOS as a function of the energy in (a) three dimensions, (b) two dimensions, and (c) one dimension.

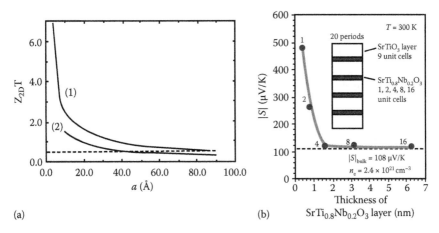

FIGURE 11.6 Plots of (a) ZT as a function of layer thickness in a quantum well structure (*solid lines*) of Bi_2Te_3 for different lattice parameters (i.e., [1] 4.3 Å and [2] 30.5 Å) and in a bulk material (*dashed line*). (Reprinted with permission from Hicks, L. D., and Dresselhaus, M. S., *Physical Review B*, 47, 12727–12731, 1993. Copyright 1993 by the American Physical Society.) (b) Seebeck coefficient as a function of the thickness of $SrTi_{0.8}Nb_{0.2}O_3$ layer in $SrTiO_3/SrTi_{0.8}Nb_{0.2}O_3$ superlattices. (Reprinted by permission from Macmillan Publishers Ltd. *Nature Materials* [Ohta, H. et al., *Nature Materials*, 6, 129–134, 2007], Copyright [2007].)

a drastic change in $g(E)$, namely, below the lowest occupied subband, the power factor could be increased.

Hicks and Dresselhaus[39] showed theoretically an enhanced power factor owing to improved S in quantum-well structures (Figure 11.6a[39,45]). This is due to the narrow carrier distribution near the Fermi energy, leading to the sharp and asymmetric DOS near the Fermi energy, as shown in Figure 11.6. Experimentally, power factor enhancement has been observed in $Si/Si_{1-x}Ge_x$[46] and $PbTe/Pb_{1-x}Eu_xTe$[47] quantum-well superlattices. A large increase in S was observed experimentally with a $SrTi_{0.8}Ni_{0.2}O_3$ layer, where $S = 480\ \mu V/K$ was achieved, which was ~5 times larger than that of the bulk[45] (Figure 11.6b). In addition to large S, quantum-well superlattices also possess high carrier mobility (μ) for high power factor. This is due to the formation of free-carrier accumulation at the interface between large and small bandgap materials. These free carriers are confined to two dimensions (2D electron gas) and are not scattered by the interfaces, leading to high mobility.

Although enhancing power factor in low-dimensional systems has been studied intensively and realized experimentally, this concept is still under investigation for more widespread use. One of the challenges may lie in the fact that the quantum confinement effect does not always lead to an enhanced power factor. To boost the power factor, one has to carefully position the Fermi energy E_F to be within the k_BT below the lowest occupied subband of a quantum structure.[11,39,48] In a series of papers by Cornett and Rabin,[49,50] they showed that, with reducing diameter in NWs, the power factor initially decreases and then increases if multiple subbands are taken into accounts. They also noted the importance of the energy-dependent carrier relaxation time $\tau(E) \sim E^r$, and showed the more dominant beneficial effect of a large r than the quantization of the DOS, because of the preferential scattering of low-energy carriers.[51] Another possible reason is the stringent experimental conditions needed for quantization to occur at room temperature. Moon et al.[52] showed a bulk-like Seebeck coefficient in Ge–Si NWs with the diameter down to ~11 nm, suggesting the absence of the quantization effect for Si and Ge in this size regime at room temperature. Wu et al.[53] observed a large thermoelectric power factor in InAs NWs (with 50–70 nm diameter) below 20 K, but it was attributed to the quantum dot-like states presented in the nonuniform NWs.

On the other hand, the influence of reduced thermal conductivity in nanostructures has been observed more extensively in experiments. Therefore, in the following section, approaches to suppress lattice thermal conductivity using low-dimensional structures, especially NWs, will be discussed.

11.2.3 Lattice Thermal Conductivity in Low-Dimensional Nanostructures

Compared to manipulating charge transport, engineering phonon transport is a more prevailing approach to enhance thermoelectric performance in low-dimensional structures. This is due to the disparity in length scales between charge carriers and phonons in typical thermoelectric materials: mean free path (MFP) for charge carriers usually ranges from 1 to 10 nm, whereas the MFP for phonons can range from 10 to hundreds of nanometers or even over 1 μm. This provides a unique opportunity to use low-dimensional nanostructures to scatter phonons preferentially without significantly affecting charge transport, and has been a major strategy employed for a variety of low-dimensional structures, including quantum dots and NWs. More recently, with the advancement of nanofabrication and synthesis, it is also possible to investigate structures with size comparable to phonon wavelength, which is on the order of a few nanometers at room temperature.[54] This had led to interesting phonon transport phenomena beyond the boundary scattering paradigm. Here, we will review these new developments and also discuss the physical origins behind these results.

Lattice thermal conductivity can be expressed as[55]

$$\kappa = \frac{k_B^4 T^3}{2\pi^2 v_g \hbar^3} \int_0^{\hbar\omega_c/k_B T} \frac{\tau(T,x)e^x}{(e^x - 1)^2} \, dx, \tag{11.9}$$

where $x \equiv \hbar\omega/k_B T$, ω_c is the cutoff frequency (~40 THz for Si[56]), and v_g is the phonon group velocity (assuming Debye dispersion relationships and constant phonon velocity). According to the Matthiessen's rule, the total scattering rate (τ^{-1}) is a combination of different scattering mechanisms due to the *Umklapp* process $\left(\tau_U^{-1}\right)$, impurity $\left(\tau_i^{-1}\right)$, and boundary $\left(\tau_b^{-1}\right)$, namely,

$$\tau^{-1} = \tau_U^{-1} + \tau_i^{-1} + \tau_b^{-1}. \tag{11.10}$$

This indicates that one could utilize phonon boundary or surface scattering in nanostructures to reduce effective phonon MFP and scattering time:

$\tau_b^{-1} = v/D$, where D is the size of the nanostructures such as NW diameter. Thus, the degree of reduction in κ_L due to nanostructuring highly depends on the nanostructure size relative to the bulk phonon MFP, which is typically dictated by *Umklapp* and impurity scattering. Not surprisingly, one would expect substantial thermal conductivity reduction in nanostructures whose bulk phonon MFP is large, such as Si and Ge. In this section, we will review thermal transport studies in NWs of Si, Ge, and their alloys ($Si_{1-x}Ge_x$).

11.2.3.1 Diffusive Phonon Boundary Scattering in Si and Ge Nanowires

One of the most widely studied NW systems for thermoelectrics is Si. In the bulk form, Si has a high lattice thermal conductivity, ~145 W/m K, hence a low *ZT* at room temperature. However, nanostructuring provides an effective route to transform it from a poor thermoelectric material to a good one. The underlying rationale of nanostructuring lies in the long and broad phonon MFP spectra of Si. At room temperature, phonons with MFP greater than

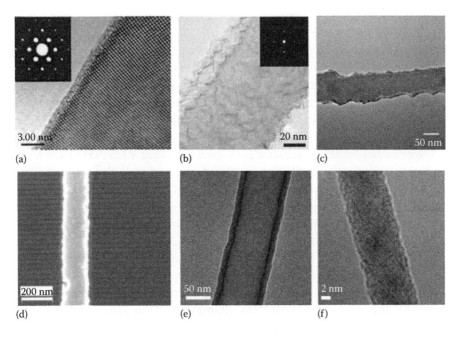

FIGURE 11.7 Electron micrographs (SEM or TEM) of representative Si and Ge nanowires with their thermal conductivity measured. (a) VLS Si NW. (Reprinted with permission from Li, D. et al., *Applied Physics Letters*, 83, 2934–2936, 2003. Copyright 2003, American Institute of Physics.) (b) Electroless-etched Si NW. (Reprinted by permission from Macmillan Publishers Ltd. *Nature* [Hochbaum, A. I. et al., *Nature*, 451, 163–167, 2008], Copyright [2008].) (c) Electron-beam lithography-patterned Si NW. (Reprinted with permission from Hippalgaonkar, K. et al. *Nano Letters*, 10, 4341–4348, 2010. Copyright 2010. American Chemical Society.) (d) Superlattice NW pattern-transferred Si NW. (Reprinted by permission from Macmillan Publishers Ltd. *Nature* [Boukai, A. I. et al. 451, 168–171, 2008], Copyright [2008]. (e) Si NT made from CVD. (Reprinted with permission from Wingert, M. C. et al., *Nano Letters*, 15, 2605–2611, 2015. Copyright 2015. American Chemical Society.) (f) VLS Ge NW. (Reprinted with permission from Wingert, M. C. et al., *Nano Letters*, 11, 5507–5513, 2011. Copyright 2011. American Chemical Society.)

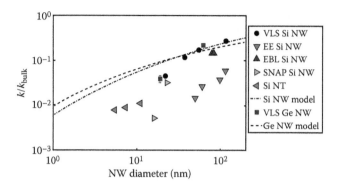

FIGURE 11.8 Thermal conductivity of the Si and Ge NWs shown in Figure 11.7. The Si and Ge NW modeling results are from Kwon et al.[54]

1 μm and 100 nm contribute to ~45 and ~70%, respectively, of the total thermal conductivity.[57–59] Thus, NWs with a diameter less than 100 nm are expected to possess thermal conductivity significantly lower than the bulk value.

Figure 11.7 shows representative electron micrographs of various types of Si NWs, with their measured thermal conductivity shown in Figure 11.8. The thermal conductivity measurements were carried out mainly by using the suspended microdevice technique that was originally developed by Shi et al.[60] and Kim et al.[61] with various subsequent modifications (e.g., usage of Wheatstone bridge to improve the temperature measurement resolution[62]).

The most well-established Si NWs are those synthesized from the vapor liquid solid (VLS) process,[63] which is one of the first and widely used approaches for semiconductor NW growth. These VLS NWs have relatively smooth surfaces, small amount of defects, and well-defined geometry. Li et al.[63] were the first to measure the thermal conductivity of individual VLS Si NWs were (Figure 11.7a). They found that room-temperature thermal conductivity can be reduced significantly, even by one order of magnitude in wires with diameter D less than 40 nm, compared to the bulk value of Si (circles in Figure 11.8). There is also a clear size dependence for room-temperature thermal conductivity. This size dependence can be understood from the phonon boundary-scattering picture, which states that phonon MFP (l_b) is limited by the NW diameter (D). In fact, modeling done by Mingo,[56] Mingo et al.,[64] Martin et al.,[65,66] and McGaughey et al.[67] showed that the thermal conductivity in the large VLS NWs, those with a diameter larger than ~30 nm, can be explained well within the BTE framework with diffusive boundary-scattering condition, namely, $\tau_b^{-1} = v/D$. This is also the regime called the Casimir limit.[56] An interesting consequence of the strong phonon boundary scattering in NWs is that the optical phonon contribution to thermal conductivity, which is usually negligible in bulk semiconductors, becomes relatively more important in NWs. These measurements on VLS NWs, and the associated modeling work thereafter, further confirmed the notion of a long, intrisic phonon MFP, which was first suggested from earlier work on Si thin films.[68] Subsequent advances in ab initio simulations and phonon MFP spectroscopy experiments showed that phonon MFP in Si is not only long, but also exhibits a broad distribution, ranging from ~10 nm to over 1 μm.[57–59]

11.2.3.2 Small-Diameter Si and Ge Nanowires: Below the Diffusive Limit

However, the diffusive boundary scattering (or the Casimir limit) is perhaps not the only picture for phonon transport in semiconductor NWs. In the experiment by Li et al.[63] on VLS Si NWs, they found that the smallest Si NW they measured, with a diameter of 22 nm, showed a thermal conductivity substantially lower than that from the modeling result based on the diffusive transport picture (Figure 11.8). They also observed a quasilinear temperature dependence for thermal conductivity when the temperature was below 100 K. This shows a clear departure from the diffusive boundary-scattering picture where the thermal conductivity should scale as T^3 at low temperature because $\kappa \sim C \sim T^3$ when $\tau_b^{-1} = v/D$. The exact nature of these behaviors has not been understood completely to date, but there has been a large body of subsequent experimental data that showed similar behavior. Measurement by Chen et al.[69] on small diameter VLS Si NWs also showed similar results, namely, lower thermal conductivity than the diffusive boundary-scattering limit and quasilinear temperature dependence for thermal conductance at temperatures below ~100 K. Even smaller NWs, with ~10 nm size, made from a "superlattice nanowire pattern transfer" process pioneered by the Heath group (Figure 11.7d), showed even lower thermal conductivity, about ~100 times lower than that of bulk Si at room temperature[70] (right triangles in Figure 11.8). Similarly, low thermal conductivity in hollow Si nanotubes (NTs) also was measured by Wingert et al.[71] (left triangles in Figure 11.8), where the tubular shell thickness ranges from 5 to 10 nm (Figure 11.7e). The behavior is not unique to Si. In the case of Ge NWs made from the same VLS process (Figure 11.7f), Wingert et al.[72] showed that small diameter Ge NWs ($D \leq 20$ nm) also exhibit thermal conductivity considerably lower than what one would expect from the diffusive scattering model, whereas the thermal conductivity of a larger Ge NW with a diameter of 62 nm agrees well with the model (squares in Figure 11.8).

These results suggest that there are mechanisms beyond the phonon boundary-scattering picture in NWs when the diameter is below approximately 30 nm. A review of experimental data and the possible causes for low thermal conductivity in this small size regime was given by Kwon et al.[54] In a nutshell, two mechanisms are plausible. First, there may exist spatial confinement effects on phonons in small diameter NWs, leading to the modification of phonon spectra and consequently the reduction of phonon group velocity (Figure 11.9a[45,73]). This was studied theoretically by Balandin et al.,[74] Balandin and Wang,[75,76] and Zou and Balandin[73] using the elastic continuum model in a NW geometry, and demonstrated experimentally by Cuffe et al.[77] on thin Si membranes. One should bear in mind that for this effect to work, phonon scattering at the surface of NWs (or thin films) should preserve the coherence. Also, the experimental results shown by Cuffe et al.[77]

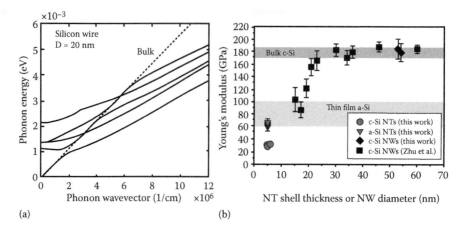

FIGURE 11.9 Two possible reasons responsible for low thermal conductivity of sub-30 nm diameter NWs as shown in Figure 11.8. (a) Modification of phonon dispersion curves due to the spatial confinement effect, leading to reduction of sound speed. (Reprinted with permission from Zou, J., and Balandin, A., *Journal of Applied Physics*, 9, 2932–2938, 2001. Copyright 2001, American Institute of Physics.) (b) Reduced elastic moduli in small-diameter NWs and NTs,[62,80] which also could lead to reduced sound speed. (Reprinted with permission from Wingert, M. C. et al., *Nano Letters*, 15, 2605–2611, 2015. Copyright 2015. American Chemical Society.)

were for fundamental flexural phonon modes. Therefore, it is still an open question regarding whether the confinement effect in thin NWs and membranes is significant for the phonon spectra that are important for thermal transport in thin NWs.

Additionally, modification of elastic properties of the nanostructures may occur through surface effects (Figure 11.9b), which lead to the reduction of phonon velocity. Wingert et al.[71] showed that for the crystalline Si NTs, the tensile moduli were decreased sixfold compared to those of bulk crystalline Si. This elastic-softening effect leads to a ~60% reduction in phonon velocity in the low-frequency limit (or sound speed). Jurgilaitis et al.[78] reported a ~25 and ~35% reduction in sound speed and C_{44} elastic constant, respectively, in 50 nm diameter InSb (111) NWs. Bozyigit et al.[79] observed a softer surface in small-diameter lead chalcogenide nanocrystals using inelastic neutron-scattering measurements and ab initio molecular-dynamics simulations. Generally, it is believed that defects on nanostructure surfaces, such as dangling bonds or oxidation, are responsible for softer surface layers.[80] As the diameters of NWs become smaller, and, hence, the surface-to-volume ratio becomes larger, the surface-softening effect is more prominent, leading to appreciable effects on thermal conductivity. It should be noted that the phonon-confinement and elastic-softening effects are not mutually exclusive. They might coexist, and one could be more important than the other in different parts of the phonon spectra. Also, they both manifest through lower sound speed at low frequency. Further work to delineate these two effects on thermal transport in small diameter NWs and membranes would be desirable.

11.2.3.3 Impact of Surface Morphology

More recent work shows that morphology could also play an important role in phonon transport in NWs, as the wires have a large surface-to-volume ratio. Hochbaum et al.[81] measured the thermal conductivity of Si NWs made from an electroless etching (EE) process, which exhibited a much rougher surface (Figure 11.7b). It was found that these EE Si NWs have much lower thermal conductivity (down triangles in Figure 11.8) compared to smooth VLS Si NWs with comparable diameter, hence lower than the Casimir limit. Subsequent experimental work further showed the important role of surface roughness: Hippalgaonkar et al.[82] on Si NWs fabricated from electron-beam lithography (Figure 11.7c; with data shown as up triangles in Figure 11.8), Lim et al.[83] on roughened VLS Si NWs, Ghossoub et al.[84] on EE Si NWs, Park et al.[85] on VLS-grown rough Si NWs, and Lee et al.[86] on both EE and roughened VLS Si NWs. Theoretically, Martin et al.[65] and Sadhu and Sinha[87] showed the impact of surface roughness on enhanced phonon scattering. Donadio and Galli[88] used molecular simulations and BTE to compute the thermal conductivity of Si NWs with very small diameters (1–3 nm), and showed that the surface coating and disorder lead to a hundredfold reduction in thermal conductivity due to the flattening of phonon dispersion curves and the presence of the nonpropagating modes in the amorphous coating. Neogi et al.[89] combined atomistic modeling and experiments to show that surface roughness and native oxide can tune the thermal conductivity of thin Si membranes (down to 4 nm thickness) over one order of magnitude. Similar effects also have been observed in simulations on core–shell NWs.[90–93] While the exact materials and structures in these studies are different, they all showed a common feature of utilizing surface morphology to tune the thermal conductivity of NWs or membranes. This may become a powerful tool for further thermal conductivity reduction and *ZT* enhancement.

11.2.3.4 SiGe Alloy Nanowires: The Effect of Impurity Scattering

Phonon boundary scattering at the NW surface is effective especially for scattering long MFP (and long wavelength) phonons, but it is not sensitive to phonons that already have short MFP. Therefore, it is possible to combine boundary scattering with impurities to scatter a broad range of phonon spectra, thus creating a synergistic effect to reduce thermal conductivity. The most common impurities are atomistic scale point defects in alloys, such as $Si_{1-x}Ge_x$. Thermal transport in $Si_{1-x}Ge_x$ alloy NWs[94] has been studied both experimentally[95–100] and theoretically.[55,101–103] Figure 11.10 shows representative experimental and theoretical results of the thermal conductivity of $Si_{1-x}Ge_x$ alloy NWs with various Ge concentrations (x) and diameters. Results of the studies by both Lee et al.[97] and Kim et al.[96] showed that the thermal conductivity depends on x, similar to the case of bulk $Si_{1-x}Ge_x$ alloy, namely, there exhibits a minimum thermal conductivity when x possesses an intermediate value. This can be understood from the impurity scattering picture in alloys: the impurity scattering rate is written as,

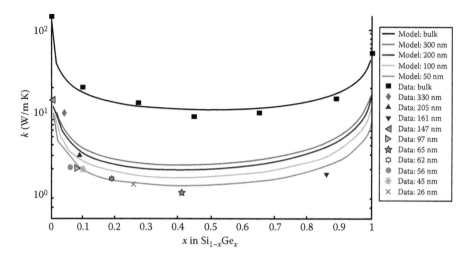

FIGURE 11.10 Thermal conductivity of $Si_{1-x}Ge_x$ alloy NWs as a function of Ge concentration x. The experimental data are from Lee et al.[97] and Kim et al.[96] The modeling results are from Wang et al.[55]

$$\tau_i^{-1} = x(1-x)A\omega^4, \qquad (11.11)$$

where constant A is a parameter that represents the scattering strength of each impurity scattering center,[55] which depends on the contrast in mass and bonding strength between the impurity and host atoms.[104] Equation 11.11 clearly shows that the impurity scattering rate is maximized when x is around 0.5, leading to a minimum thermal conductivity referred to as the alloy limit.[105]

The ω^4 dependence in Equation 11.11 means that the Rayleigh-type impurity scattering is not strong for low-frequency phonons, which results in stronger contribution of long MFP phonons to thermal conductivity in $Si_{1-x}Ge_x$ alloys,[105] compared to either Si or Ge. The introduction of boundary scattering in NWs, which has a nearly frequency-independent scattering rate $\left(\tau_b^{-1} \sim v/D\right)$, leads to stronger scattering of these long MFP (low frequency) phonons and, consequently, further reduction in thermal conductivity. This is shown clearly in Figure 11.10 for $Si_{1-x}Ge_x$ NWs with a diameter less than 100 nm, the thermal conductivity is less than 2.2 W/m K for x ranging from 0.06 to 0.41. This strong reduction in thermal conductivity leads to enhanced ZT in $Si_{1-x}Ge_x$ NWs compared to that in bulk alloys.[97,100] Similar thermal conduction effect has also been shown through computation on alloy NWs made of $Mg_2Si_xSn_{1-x}$.[106]

11.2.4 Summary

In these sections, we have reviewed the basic principles of enhancing the thermoelectric figure of merit ZT in low-dimensional semiconductor nanostructures using two different strategies: enhancement of the power factor and reduction of the thermal conductivity. First, the quantum confinement effect has been suggested to introduce spikes in electron DOS around the

Fermi energy as a powerful approach to yield a high Seebeck coefficient and power factor. Realization of this effect depends on such conditions as the quantization (subband formation), the control of the Fermi energy around the subbands, and the exact nature of band structures and carrier scattering mechanisms. Second, nanostructures such as NWs have been shown to be effective in reducing lattice thermal conductivity, owing to the broad MFP spectra of phonons in semiconductors such as Si, Ge, and $Si_{1-x}Ge_x$ alloys. For large-diameter Si and Ge NWs with smooth surfaces, the thermal conductivity is close to what one would expect from the diffusive phonon boundary scattering picture. For small diameter NWs and thin NTs, thermal conductivity could be lower than this diffusive limit. The causes leading to this lower thermal conductivity could lie in the modification of phonon dispersion curves due to spatial confinement and/or surface-induced elastic softening of NWs and NTs. In addition, thermal conductivity in NWs can be tuned further by the surface morphology (roughness or coating). Finally, alloy NWs such as $Si_{1-x}Ge_x$ show thermal conductivity lower than the alloy limit due to the synergistic effects of impurity scattering (for high-frequency phonons) and boundary scattering (for low-frequency phonons). These studies suggested that low-dimensional semiconductors are promising for further enhancing the thermoelectric figure of merit and will continue to serve as an important platform for probing nanoscale thermal and thermoelectric transport.

11.3 Thermoelectric Materials with Resonant States

Joseph P. Heremans, Bartlomiej Wiendlocha, and Hyungyu Jin

11.3.1 Introduction

In optimizing the thermoelectric figure of merit (ZT), the research has encountered several material properties that are mutually contraindicated. These properties consist of the thermopower (or Seebeck coefficient) S and the electrical (σ) and thermal (κ) conductivities of the material. Contraindication signifies that generally, where mechanisms improve one property, it is in sacrifice of another. Thus, we note that ZT can be written as the product of two sets of these contraindicated properties in the parentheses as follows:

$$ZT \equiv \frac{S^2\sigma}{\kappa} = (S^2 n)\left(\frac{\mu}{\kappa}\right)eT, \qquad (11.12)$$

where $\sigma = ne\mu$ is the electrical conductivity, with charge-carrier concentration n, charge e, and mobility μ. The ratio μ/κ is contraindicated because defects and impurities improve one effect while adversely affecting the other. The product ($S^2 n$) is the other contraindicated property. In fact, generally, per the Pisarenko relation[107] valid for doped semiconductors, the thermopower decreases when the carrier concentration increases. One concerted avenue of research has focused on minimizing lattice thermal conductivity as a means of maximizing μ/κ. In this review, we give our attention to the purely electronic property, $S^2 n$, which is dominated by the electronic band-structure details and those of the scattering mechanisms. The most promising, and concomitantly the most realistic, method for increasing $S^2 n$ is to engineer the electronic density of states (abbreviated as DOS) g near the Fermi energy. The use of semiconductors with engineered band structures is important for world energy issues only if large quantities of thermal and electrical power can be converted in thermoelectric technology. Therefore, approaches that apply to bulk 3D thermoelectric materials are the focus of this review.

This review is organized in four parts. After this introduction, we explain the principles by which resonant impurity levels can increase g locally and thereby increase $S^2 n$, as well as the physical nature of the levels themselves. In the third section, the delocalized nature of the resonant levels is emphasized, including the fact that they have dispersion and contribute to the band structure, which is then shown using calculated Bloch spectral functions for Tl-doped PbTe. In the last section, we mention three experimental examples of resonant levels, Tl in PbTe, In in SnTe, and Sn in Bi_2Te_3, and we present details of a more recently discovered resonant level formed by K dopants in BiSb alloys.

11.3.2 Mahan–Sofo Optimization of the Density of States

The Mahan–Sofo theory[40] predicts the shape of the energy dependence of the transport function $\sigma_M(E)$, which is strongly influenced by the DOS $g(E)$ that maximizes ZT. Before discussing it in detail, note that most good thermoelectric semiconductors exhibit band conduction to conduct electricity, because this mechanism leads to values for mobility that exceed 1 cm²/V s, which are clearly important in Equation 11.12. Band conduction is assured under two conditions. The first is the Anderson–Ioffe–Regel limit[108]: the wave function of the electrons must be longer than their MFP. In a macroscopic sample, it means that the electrical conductivity must exceed the minimum metallic conductivity, of order

$$\sigma_{min} = \frac{\pi^2}{z} \frac{e^2}{h} \frac{1}{a} \sim \frac{61[\Omega^{-1}cm^{-1}]}{a[nm]}, \tag{11.13}$$

where a is the interatomic distance, h is Planck's constant, and z is the coordination number of the atom. The numerical value is valid for octahedrally bonded solids ($z = 6$).

For band conduction in metals and degenerate semiconductors, the Mott criterion specifies under what conditions the electrical conductivity remains finite even at the lowest temperature. The Mott criterion dictates that for the electronic state in question,[109]

$$a_B^* n^{1/3} > \frac{1}{4}\left(\frac{\pi}{3}\right)^{1/3} \simeq 0.25, \tag{11.14}$$

where n is the electron concentration and the effective Bohr radius for the electron is $a_B^* = a_B(m_e/m^*)(\varepsilon/\varepsilon_0)$. Here, a_B is the Bohr radius of the hydrogen atom (0.053 nm), m_e is the free-electron mass, m^* is the effective mass, ε_0 is the permittivity of a vacuum, and ε is the static dielectric constant. In essence, the Mott criterion also measures to what degree electron motion is affected by the presence of random potential wells around impurities and defects in a sample. Since we are going to discuss resonances around impurities in thermoelectric semiconductors, a larger product $a_B^* n^{1/3}$ in Equation 11.14 results in a smaller loss of mobility. This indicates that the magnitude of the loss in mobility that the addition of such an impurity will induce will scale as $a_B^* n^{1/3}$.

Next, we introduce the Mott formula[108] for the thermopower of a degenerate electron gas—a formula that is based solely on Fermi statistics. The electrical conductivity and thermopower of the sample are given by

$$\sigma = \sigma_M(E_F),$$

$$S = \frac{\pi^2}{3}\frac{k_B}{e}k_B T\frac{\partial \ln \sigma_M(E)}{\partial E}, \tag{11.15}$$

where the key part of this equation is the Mott conductivity $\sigma_M(E)$ (also called the transport function). The Mott conductivity should not be confused with electrical conductivity and cannot be accessed experimentally. As explained by Mott himself, "$\sigma_M(E)$ represents the conductivity the system *would have* if the energy at the surface of the Fermi distribution were E; its variation with E is important in the discussion of the thermoelectric phenomena." The Mott conductivity is defined by the fact that the electrical conductivity σ is given by the Mott conductivity's integral over energy $\int_0^\infty \sigma_M(E)(-\partial f_0/\partial E)\,dE$ in nondegenerate statistics (f_0 is the Fermi–Dirac distribution function) or its value at the Fermi energy E_F in degenerate statistics. In essence, $\sigma_M(E)$ is the product of the charge e, the DOS $g(E)$, and the mobility $\mu(E)$ of the carrier at energy E. Therefore Equation 11.15 becomes

$$S = \frac{\pi^2}{3}\frac{k_B}{e}k_B T\left[\frac{1}{g(E_F)}\frac{\partial g(E)}{\partial E}\bigg|_{E_F} + \frac{1}{\mu(E_F)}\frac{\partial \mu(E)}{\partial E}\bigg|_{E_F}\right]. \tag{11.16}$$

It is clear from Equation 11.16 that when $g(E)$ has a sharp dependence on E near E_F, the thermopower will be large. It is not so clear that this will benefit ZT, because it is not clear how the shape of the DOS will affect the electrical conductivity or the mobility and not clear what the role of the "background" DOS at energies further away from E_F will be. The problem of optimizing the shape of $\sigma_M(E)$ in order to maximize ZT has been studied by Mahan and Sofo.[40] They concluded that for a given carrier concentration,

$S(n)$ is higher where σ_M and g are larger and have a stronger dependence on E. The limiting case shown to give the optimal ZT enhancement occurs where $\sigma_M(E)$ is a delta function at $2.4\ k_B T$ above or below E_F and that such a shape of $\sigma_M(E)$ may result from a strongly peaked DOS. Unfortunately, this is not naturally attainable due to local sample composition variations that expand the delta function. The background DOS is a second limitation. Here, it is identified as the presence of available electronic states due to the atoms in the host solid (typically, the background DOS in 3D solids has as energy dependence $g(E) \propto E^{1/2}$), independent of the atoms or nanostructure that give rise to the δ-function-like DOS in the solid. As demonstrated by Mahan and Sofo, the background DOS strongly decreases the optimal ZT. However, in order to control the position of E_F, a nonzero background DOS is required. Thus, distorting and increasing g as much as possible over and above an $E^{1/2}$ function near the Fermi level is the aim of thermoelectric power factor research. Hicks and Dresselhaus[11] predicted that exceptional thermoelectric materials could be made from 1D quantum wires given the shape of $g(E)$ described earlier. Such effects have been the basis of varied experimental work, as has been reviewed previously.[110]

Therefore, the overall goal is to add dilute amounts of chemical impurities to samples that achieve this shape of $g(E)$ without significantly affecting the mobility of the charge carriers. The specific impurities that can do this create resonant impurity levels in the conduction or valence bands of the semiconductors in question, by analogy with the effect of similar impurities in metals.[111] A previous review of the work on the use of resonant levels to maximize $S^2 n$ is given in Heremans et al.[44] In the rigid band model, which is generally accepted in all semiconductor handbooks, conventional dopant atoms added to semiconductors, such as group III elements in Si as acceptors or group V elements as donors, do not change the shape of $g(E)$. Continuing with the example of adding a group V element in a group IV elemental semiconductor, such as P in Si, the energy level that corresponds to the fifth electron on the outer shell of P forms an impurity level a few tens of millielectronvolts (an effective Rydberg E_d) below the conduction band edge. At a temperature such that $k_B T > E_d$, the extra electrons are excited thermally in the conduction band and conduct electricity, but by hypothesis, the presence of a small concentration of P in Si does not affect the shape of the conduction band. Not all impurities do this. Some dilute impurity atoms have $E_d < 0$, so the electrons on that impurity always release in the band. Such impurity creates a distortion in the DOS; a typical example is given by adding transition metal atoms with unfilled d-shells into narrow-gap semiconductors, such as Ti in PbTe.[112] In that case, the impurity level adds to the DOS of the conduction band, but the Ti d-orbitals do not hybridize much with the Pb and Te s- and p-orbitals, the Ti level remains extremely localized, and electrons on Ti d-orbitals do not meet the Mott and Anderson–Ioffe–Regel criteria for band conduction (see also the following discussion).

In many thermoelectric semiconductors, it has been possible to identify impurities that dope the material and distort the DOS by adding excess states that contain parts of the wave function of the impurity, but mostly from the host material itself. Such impurities act almost as catalysts that bring energy levels that would be deep in the valence band of the host semiconductor close to the Fermi energy. These levels are then mostly of a delocalized nature, and it is possible for them to meet the Mott and Anderson–Ioffe–Regel criteria for band conduction: semiconductors doped with such impurities show an increase in the value of S^2n over their undoped counterpart. The simple model that describes how the distortion in DOS enhances S is discussed by Heremans et al.,[44] where a somewhat artificial approach is taken by assuming that the two terms $\partial g(E)/\partial E\big|_{E_F}$ and $\partial \mu(E)/\partial E\big|_{E_F}$ in Equation 11.16 are independent of each other. Due to space constraints, we refer the reader to earlier reviews[44] and mention for completion that the effects of energy-selective scattering on $\partial \mu(E)/\partial E\big|_{E_F}$ can be quite important. They typically occur when the scattering impurity has an additional internal degree of freedom: in the case of the dilute Kondo effect,[113] this is its spin. A similar thing happens when the impurity can move and develops a vibrational degree of freedom: the Nielsen–Taylor effect.[114] Both of these effects also affect the thermopower, but they are beyond the scope of this review.

11.3.3 Spectral Density Functions

The problem of the charge localization or the drop in the carriers' mobility is among the most important issues if one wishes to use the resonant impurities to improve the thermoelectric performance of a semiconductor. The effect that may be detrimental and make the resonant level useless in the thermoelectric context is the formation of an isolated, narrow impurity-induced band. For a very small bandwidth, this is equivalent to the localization of the carriers, since the effective mass becomes very large. In such a case, the thermopower of the system becomes a weighted average of the thermopowers of the host S_h and impurity S_i bands, with the electrical conductivities σ_h and σ_i being the weights. If the band formed by impurity is narrow, $\sigma_i \ll \sigma_h$, then the weighted average $S_{tot} = [(\sigma_i S_i + \sigma_h S_h)/(\sigma_i + \sigma_h)] \approx S_h$ becomes the original thermopower of the host material. This effect is essentially the same as the background DOS effect discussed by Mahan and Sofo,[40] as the two-band combination results in two DOS components: one narrow impurity DOS peak and second, small background DOS of the host semiconductor.

Resonant impurities may be predisposed to form localized states, since they form more or less narrow peaks in the DOS of the doped materials. Recent studies[115–117] showed that this issue may be analyzed by using the spectral density Bloch functions $A_B(\mathbf{k},E)$ (BSF), which generally allows the investigation of how the impurity modifies the electronic dispersion relations in a host material (see Gordon et al.[118] and Ebert et al.[119] for more details about the theory of BSFs). In brief, $A_B(\mathbf{k},E)$ may be considered as a $\mathbf{k}$-resolved DOS function, which, for a perfect crystalline material at selected $\mathbf{k}$, is a delta function of energy for each of the bands ν, $A_B(\mathbf{k},E) = \delta(\mathbf{k}, E - E_{\nu,k})$—it is zero everywhere outside the point $(\mathbf{k}, E_{\nu,k})$, at which in band ν, there is an energy eigenvalue $E_{\nu,k}$, and the energy integral of $A_B(\mathbf{k},E)$ is equal to one for each band ν (one electron occupies the $(\mathbf{k}, E_{\nu,k})$

electronic state) or two, for a spin-degenerated nonmagnetic case. The set of BSFs here shows the position of the energy eigenvalues ($k, E_{v,k}$), and thus allows plotting of the traditional dispersion relations $E(k)$. For the doped material or alloy, due to the presence of impurities and disorder, bands become smeared due to the scattering on impurities; thus, BSFs are broadened, usually taking the form of a Lorentz function, with the width of this function corresponding to the lifetime of a given electronic state. The set of BSFs for the doped (disordered) material also allows the plotting of the analog of dispersion relations, but each band now has a band center and a bandwidth. In other words, BSFs allow the investigation of the k dependence of the electronic state introduced by impurities.

Calculations of BSFs for PbTe:Tl,[115] PbTe:Ti,[116] and $Mg_2Sn:Ag$[120] using the Korringa–Kohn–Rostoker and the coherent potential approximation (KKR-CPA) method[121] were reported recently, and the two first cases are reviewed here. Figure 11.11 shows the comparison of DOS and BSF for Tl- (top panel) and Ti-doped (bottom panel) PbTe. For the thallium case, the 6s-like resonant state strongly distorts the last valence band, which goes between Σ and L k-points, and BSF becomes non-Lorentzian in shape (see Wiendlocha[115] for more details). The number of states near the valence-band edge increases, as seen in the DOS plot, but the additional electronic states do not form an isolated impurity band. These states, being hybridized thallium and PbTe electronic states, gather around the Σ–L band, increasing the number of carriers available for the electronic transport. Thus, the effect of the presence of the Tl resonant impurity is qualitatively similar to the effect of an increase in band degeneracy. Both effects increase the number of carriers available to transfer current and heat, and the thermoelectric performance of the material may benefit from a resonant level as it may from an increase in band degeneracy, as well. The electronic structure of PbTe:Tl cannot be decoupled into host + impurity bands (or, equivalently, resonant + background DOS) due to the strong hybridization of the impurity and host electronic states. On the other hand, such hybridization cannot be too large, since, for such a case, the impurity would lose the resonant character. One can see that the balance between the resonance and full hybridization is a key factor for reaching the goal.

As a counterexample, the spectral functions for the Ti-doped PbTe are presented in Figure 11.11 (bottom panel). Here, the resonant states formed by 3d electronic orbitals of titanium create spin-polarized sharp peaks in the DOS, but the electronic states associated with these peaks are hybridized poorly with the host and do not show any k-dependence, rather than being aligned in a flat impurity band-like localized states. This difference explains why the resonant level associated with Tl in PbTe was found to increase the thermopower and boost the ZT of the material, whereas such a positive effect was absent in Ti-doped PbTe.

The second important issue is the problem of scattering—does such a resonant state deprive the mobility due to the carriers' scattering on the resonant level? This is discussed with a help of the residual resistivity and electronic lifetimes. The basic analysis of the magnitude of resistivity of PbTe:Tl as a function of temperature already shows that the scattering of

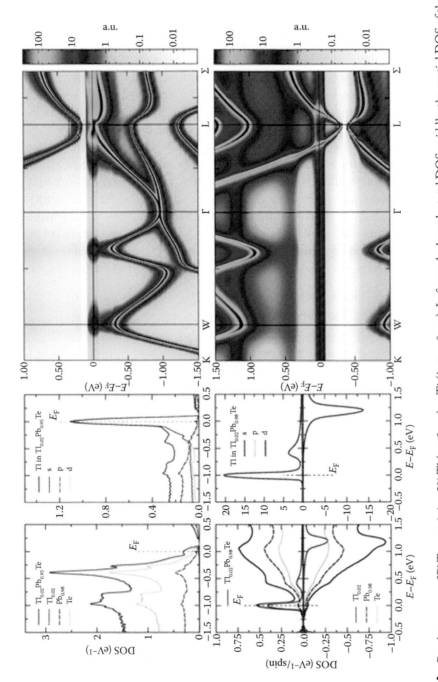

FIGURE 11.11 Band structure of PbTe containing 2% Tl (*top frame*) or Ti (*bottom frame*). Left panels show the total DOS; middle, the partial DOS of the impurity atoms, with angular momentum expansion; and right, the BSF, which, for the spin-polarized Ti case, is summed for both spin directions. The Tl level has a dispersion that encompasses both L and Σ, showing that the Tl level is quite distinct from the heavy hole band. The Ti electronic states are aligned in impurity bands. (Reprinted from *Scripta Materialia*, 111, Wiendlocha, B. et al., Recent progress in calculations of electronic and transport properties of disordered thermoelectric materials, 33, Copyright [2016], with permission from Elsevier.)

mobile carriers due to the presence of a resonant level is not much stronger than the electron–phonon scattering at elevated temperatures ($T > 300$ K). The residual resistivity $\rho_0 = \rho(T = 0)$ for 1.5–2% Tl-doped PbTe is about 1 mΩ cm. When the temperature increases, $\rho(T) = \rho_0 + \rho_{ph}(T)$ is doubled (~2 mΩ cm) at $T = 300$ K and reaches 4–5 mΩ cm at 700 K (see Heremans et al.,[42] Matusiak et al.,[122] Keiber et al.,[123] and Jaworski et al.[124]). In other words, the temperature-induced reduction of mobility becomes stronger than that due to the scattering on the resonant level, which limits the conductivity at the low-temperature region. This simple reasoning is consistent with the analysis of the relaxation times. In PbTe, for the dopings $5 \times 10^{19} - 1 \times 10^{20}$ cm^{-3}, for $T = 700$ K, the electron–phonon relaxation time is on the order of $\tau_{ph} = 1 \times 10^{-14}$ s (see the model developed by Ahmad and Mahanti[125]). Thus, it is on the same order as the lifetime for the resonantly blurred valence band in PbTe:Tl, estimated basing on the BSF width[115] to be $\tau_{res} \sim 10^{-14}$ s. This shows that scattering on the resonant level at elevated temperatures stops being the main scattering mechanism, and phonon scattering may dominate, in agreement with the Nernst-effect measurements of Jaworski et al.[124] The reduction of mobility due to the resonant level does not have to be so strong as to question the idea of the positive influence of the resonant level on thermoelectric properties.

11.3.4 Resonant Levels Identified Experimentally in Thermoelectric Semiconductors

There are several examples of resonant levels identified in the literature on thermoelectric semiconductors: thallium in PbTe,[42] indium in SnTe,[43] tin in Bi$_2$Te$_3$,[126] and potassium in Bi and BiSb alloys. Several new candidates for resonant levels in the tetradymites also were suggested recently.[127] The former examples have been the subject of previous reviews[44]; only the DOS of Sn impurity in Bi$_2$Te$_3$ is shown here (Figure 11.12b). We emphasize here the case of the potassium doping of Bi and BiSb alloys that are reviewed elsewhere in this book.[128] In particular, the bulk Bi$_{1-x}$Sb$_x$ alloys are semiconductors for 6 at.% < x < 22 at.%, while the other alloys are semimetals in which electrons and holes coexist (see Chapter 2 in the present work). Surprisingly, the group III elements are acceptors in Bi and BiSb,[129] even though the Bi 6s-orbitals create bands deep in the valence-band complex, and only the three 6p-orbitals contribute to the valence band near E_F, making the elements in these alloys trivalent, just like the dopants. This effect is due to a formation of a deep, hybridized electronic state, quite similar to what resonant levels do. The substitutional alkali elements also act as acceptors in Bi (interstitial Li is a donor[130]).

We recently reported[131] that potassium is a resonant level in Bi and Bi$_{1-x}$Sb$_x$ alloys, based on earlier theoretical predictions[132,133] that the alkali metals may form resonant levels in Bi and Bi–Sb alloys. The KKR-CPA[134] computed DOS for the K-doped Bi is shown in Figure 11.12a.* The narrow DOS peak just above E_F, which is formed from 4p-states of potassium, suggests that in a semiconducting Bi–Sb alloy, the resonance will be n-type.

---

* Semirelativistic KKR-CPA calculations with technical details the same as by Jin et al.[129] for In-doped Bi.

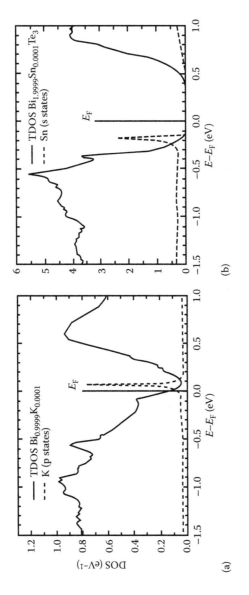

FIGURE 11.12 (a) Calculated electronic DOS of potassium-doped Bi, clearly illustrating the presence of a resonant level near the top of the valence band. A similar calculation was made for $Bi_{88}Sb_{12}$ and shows a similar result. In contrast to PbTe:Tl or Bi_2Te_3:Sn, the resonant level (RL) is on p-like orbitals. (b) DOS of Sn-doped Bi_2Te_3, showing s-like resonance on tin.

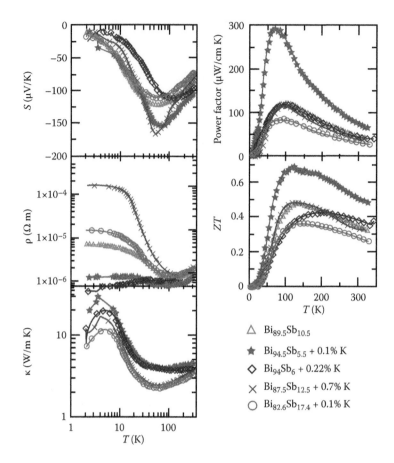

FIGURE 11.13 Experimental temperature of the thermopower (or Seebeck coefficient) (S), resistivity (ρ), and thermal conductivity (κ) of the K-doped BiSb alloys with the nominal compositions indicated. The resulting power factors (S^2/ρ) and ZT calculated from those data are also shown. The sample near the composition of the Dirac point (5.5% Sb) and doped with 0.1 % potassium gives the record ZT.

The result of this particular resonant level is a great increase in ZT, which, in the cryogenic temperature range, established record ZTs. Using the Bridgeman method, a sequence of single crystals of $Bi_{1-x}Sb_x$ alloys with nominal concentrations 5.5% $\leq x \leq$ 17% was grown. These were then doped with varying amounts of K. A reference sample of $Bi_{89.5}Sb_{10.5}$, a composition at the previously reported optimum, also was grown without doping. The thermoelectric properties including the thermopower (Seebeck coefficient) S, resistivity ρ, and thermal conductivity κ of the single crystals were measured along the trigonal direction and are shown in Figure 11.13. The low-temperature resistivity makes clear that K-doped samples with $x > 11\%$ have higher resistivity than the $x = 10.5\%$ sample. The K-doped samples with $x < 10\%$ have a much lower resistivity than the reference sample, but the thermopower does not decrease. The thermoelectric power factor of the K-doped samples is thereby greatly enhanced, as reported in Figure 11.13, reaching nearly 300 μW/cm K². This value is three times higher than the highest

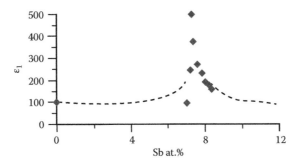

FIGURE 11.14 Static relative dielectric constant ε_1 of $Bi_{1-x}Sb_x$ alloys as a function of x.

previously reported power factor for these alloys. Thus, a "hero" sample of $Bi_{94.5}Sb_{5.5}$ doped with 0.1% K reaches $ZT \approx 0.7$ near 110 K and has a ZT value of >0.5 at all temperatures between 60 and 300 K.

It may appear surprising that the optimal sample has a Sb concentration of 5.5%, at the limit of the semimetal-to-semiconductor transition, whereas the optimal thermoelectric alloys are in the semiconducting regime near $x \approx 11 \pm 1$ at.%. We attribute this fact to the high electrical conductivity (10^6 S/m) of these alloys at low temperature, in spite of the presence of the K impurity. Indeed, the optimal power factor and ZT are observed at the concentration x, where the static dielectric constant of the alloy reaches 500 (it is of the order of 100 in) for $x \approx 7$ at.% as shown in Figure 11.14, compiled from data by Rudolph et al.[135] and Fellmuth et al.[136] This observation is consistent with the large Bohr radius (see Equation 11.14). Therefore, this is the composition at which the electron wave functions mostly probe potentials of Bi and Sb atoms and are particularly insensitive to scattering on potassium atoms, providing a large mobility.

11.3.5 Summary and Acknowledgments

In summary, doping semiconductors with impurities that create resonant levels remains a viable way to increase ZT by increasing the DOS of electrons near the Fermi energy. The resonant levels themselves need to develop delocalized electronic states, with wave functions that widely extend in the space around the impurity. This means that for resonant levels to be effective in increasing ZT, they cannot form narrow impurity bands, and this issue may be analyzed using spectral functions. The positive effect on the thermopower, with moderate decrease in mobility, arises because the resonant level increase the number of electronic states around the original host band, and these states mostly stem from orbital levels of the atoms of the host semiconductor, with little contribution from the electronic orbitals of the impurity. The impurity can be seen as only a catalyst.

JPH acknowledges support of the NSF Emerging Frontiers in Research and Innovation program grant number 1433467. BW was partially supported by the Polish National Science Center (project no. DEC-2011/02/A/ST3/00124).

11.4 Effects of Resonant Scattering on Thermoelectric Properties

Bo Wang and Qinyong Zhang

11.4.1 Introduction

Resonant effects are an alternative to the various traditional ways to suppress lattice thermal conductivity in order to improve the thermoelectric figure of merit.[137] Resonant effects, as a method of enhancing the power factor, were first proposed by Kaidanov[138] and Ravich[139] in the 1980s. Since the successful work of Heremans et al. in PbTe:Tl in 2008,[42] resonant effects have gained significant attention, and work on enhancing the mechanism for resonant effects has been reported for various thermoelectric materials systems.[43,140–144]

As suggested by both Ravich et al.[138,139] and Heremans,[42] the state levels lying in the conduction or valence band caused by the impurities introduced, unlike the shallow impurity levels in the bandgap close to the band edge, are called resonant states.

According to Heremans,[44] impurity-induced resonant states form a hump in the background DOS. In other words, the background DOS is distorted, resulting in a rapid DOS effective mass change, with energy at the corresponding position of the hump.[44] This change is the cause of the reported Seebeck coefficient (S) and ZT improvement in PbTe:Tl.[44] Heremans et al. called this mechanism the DOS effect, which implies that the carriers in the resonant states can participate in the transport process like those in the extended band states.[44] Their calculations demonstrate that the enhancement of S from resonant scattering (RS) is small, especially at high temperatures at which the acoustic phonon scattering (AC) is dominant. Therefore, because of the much higher Fermi level required, the RS mechanism was shown to be impractical in PbTe.

According to Ravich,[139] resonant states, which, unlike extended band states, are not free states, will be broadened into an impurity band with a half maximum width of Γ. This broadening is due mainly to the hybridization of the quasilocalized impurity states and the free extended band states. In other words, the carriers in the extended band states can move to the same energy as the quasilocalized resonant states for a limited residual time. They can then move back to the band states to participate in the transport without reference to their previous momentum. This hybridization process, called resonant scattering, would result in momentum dissipation and an energy-dependent relaxation time.[139,145] Their extensive review of experimental and theoretical works[138,145,146] suggests that RS can be a promising mechanism to improve the thermoelectric properties, although the details of the effects of the RS parameters remain unclear.

In this section, we compare the electrical conductivity (σ), Seebeck coefficient (S), power factor ($S^2\sigma$), Lorenz number (L), and the electrical contribution to the thermal conductivity (κ_e) for a system with and without RS. We use a simple model based on Ravich's work,[139] varying the RS intensity, resonance band location (E_i), and width (Γ). For simplicity, only a single band with a parabolic DOS was considered here.

11.4.2 Calculation Model

As suggested by both Ravich[139] and Heremans,[44] the DOS of the impurity-induced band can be approximated as a bell-shaped Lorentz function:

$$g_i(E) = \frac{N_i \Gamma}{\pi} \left[(E - E_i)^2 + \left(\frac{\Gamma}{2} \right)^2 \right]^{-1}, \tag{11.17}$$

where the N_i is the impurity concentration and E_i is the center of the impurity-induced band relative to the extended band edge concerned.

In this situation, if only AC and RS are considered, the relaxation time τ is the combination of RS relaxation time τ_{rs} and AC relaxation time τ_{ac} through Matthiesen's rule:

$$\frac{1}{\tau} = \frac{1}{\tau_{rs}} + \frac{1}{\tau_{ac}}. \tag{11.18}$$

The energy-dependent relaxation time of RS is[139]

$$\tau_{rs} = \tau_{rs}^{(0)} \left[1 + \left(\frac{E - E_i}{\Gamma/2} \right)^2 \right], \tag{11.19}$$

where $\tau_{rs}^{(0)}$ is the minimum value of τ_{rs} when the Fermi level E_F reaches the resonant band center E_i. If the RS intensity $\tau_{ac}^{(E_F)}/\tau_{rs}^{(0)}$ is defined as A, then the ratio of the whole relaxation time τ to the AC relaxation time τ_{ac} can be written:

$$\psi(E) = \frac{\tau}{\tau_{ac}} = \left(1 + \frac{\tau_{ac}}{\tau_{rs}} \right)^{-1} = \left[1 + \frac{A}{1 + \left[(E - E_i)/(\Gamma/2) \right]^2} \right]^{-1}. \tag{11.20}$$

When $A = 0$, the entire scattering mechanism is actually AC; RS has no effect. For the integral

$$I_{A,n} \propto \int_0^\infty \left(-\frac{df}{dx} \right) \tau_{ac} \psi(x) g(E) x^n \, dx, \tag{11.21}$$

bearing in mind that the AC relaxation time $\tau_{ac} \propto E^{-1/2}$ and the DOS $g(E) \propto E^{1/2}$ for a parabolic band, it can be simplified as

$$I_{A,n} \propto \int_0^\infty \left(-\frac{df}{dx} \right) \psi(x) x^n \, dx. \tag{11.22}$$

We followed the experimental results of Ravich,[139] which indicated that the quasilocalized impurity resonant states do not contribute much to the carrier transport[147] to show distinctly the effects of the RS. This method differs from Heremans' model,[44] which treated the impurity bad states the same as the extended background band states, and would lead to a distorted DOS compared to that without a resonant band. Thus, in the integral $I_{A,n}$, only the DOS of the background band was included; that of the impurity band was neglected given that its contribution to the carrier transport was much smaller. As such, the ratio of the relevant thermoelectric parameters for the system with and without RS can be expressed as

$$\frac{\sigma_{rs+ac}}{\sigma_{ac}} = \frac{I_{A,1}}{I_{0,1}}, \tag{11.23}$$

$$\frac{S_{rs+ac}}{S_{ac}} = \left(\frac{I_{A,2}}{I_{A,1}} - E_F^* \right) \bigg/ \left(\frac{I_{0,2}}{I_{0,1}} - E_F^* \right), \tag{11.24}$$

$$\frac{L_{rs+ac}}{L_{ac}} = \left[\frac{I_{A,3}}{I_{A,1}} - \left(\frac{I_{A,2}}{I_{A,1}} \right)^2 \right] \bigg/ \left[\frac{I_{0,3}}{I_{0,1}} - \left(\frac{I_{0,2}}{I_{0,1}} \right)^2 \right]. \tag{11.25}$$

To evaluate the effects of RS, we can calculate various ratios of thermoelectric parameters and their dependence on the reduced Fermi energy ($E_F^* = E_F/k_B T$, where k_B is the Boltzmann constant) and various RS parameters such as A, E_i, and Γ. The ratios to be calculated include the power factor ($S^2\sigma$) and the electronic contribution to thermal conductivity ($\kappa_e = L\sigma T$).

11.4.3 Results and Discussion

The calculated ratios of the dependence of thermoelectric properties on E_F^* for various scattering intensities A, mentioned above, are shown in Figure 11.15. In that figure, the impurity band center E_i and the half-maximum-width Γ were set to $6k_B T$ and $2k_B T$, respectively, following the experimental results of Tl in PbTe.[148]

The dependence of $\psi(E) = \tau/\tau_{ac}$ on the reduced Fermi energy E_F^* is shown in Figure 11.15a, which demonstrates a lower relaxation time due to RS in an E_F^* range much wider than the impurity bandwidth Γ. The stronger the RS (larger A) involved, the lower the minimum of τ/τ_{ac} that can be reached at E_i. This behavior of the $\psi(E)$ will surely have remarkable effects on the transport parameters from the integral $I_{A,n}$ (Equation 11.22).

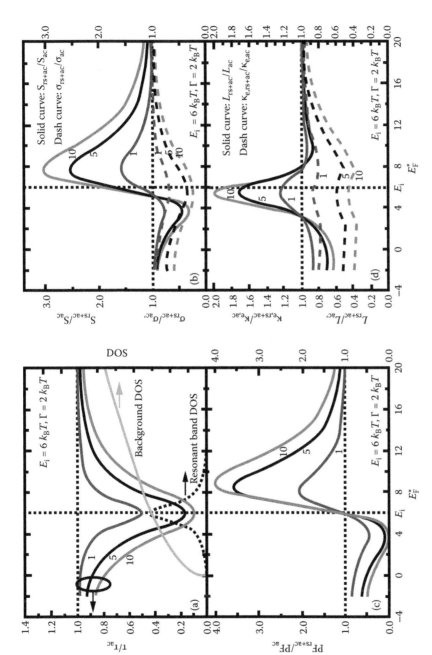

FIGURE 11.15 Dependences of the ratios of relaxation time and thermoelectric properties on E_F^*. The numbers beside the curves represent the scattering intensity A. The vertical dotted line indicates the position of E_i, while the horizontal dashed line indicates the ratio of unity. The dependences of (a) the $\psi(E) = (\tau/\tau_{ac})$, the DOS of the parabolic background extended band (*gray curve*), and the DOS of the impurity-induced resonant band (*dashed curve*); (b) the electrical conductivity ratio $\sigma_{rs+ac}/\sigma_{ac}$ (*dashed curve*) and the Seebeck coefficient ratio S_{rs+ac}/S_{ac} (*solid curve*); (c) the power factor ratio PF_{rs+ac}/PF_{ac}^*; and (d) the Lorenz number ratio L_{rs+ac}/L_{ac} (*solid curve*), and the ratio of the electronic contribution to the thermal conductivity $\kappa_{e,rs+ac}/\kappa_{e,ac}$ (*dashed curve*) on the E_F^* are shown.

The calculated results in Figure 11.15b show the effect of increasing E_F^* from $-2k_BT$ to $20k_BT$: the S_{rs+ac}/S_{ac} first decreased to a minimum of 0.31 and then increased rapidly to a maximum of ~3, followed by a gradual decrease to unity. This complex feature demonstrated that S can be enhanced only when the E_F^* is in the correct position; otherwise, S will deteriorate. Despite Ravich's prediction that RS will enhance S generally,[139] our detailed calculations show that whether the thermoelectric properties are enhanced or deteriorated by the RS mechanism depends on the alignment of E_i and the Fermi level.

In contrast to the complex dependence of S_{rs+ac}/S_{ac} on E_F^*, the electrical conductivity ratio $\sigma_{rs+ac}/\sigma_{ac}$ shown in Figure 11.15b demonstrates ratios no larger than 1, with a minimum at E_F^* near E_i due to the activity of RS. It is clearly understood that a larger A leads to a lower electrical conductivity as shown in Figure 11.15b.

The features of the Seebeck coefficient and electrical conductivity mentioned above result in an interesting dependence of power factor, calculated as $S^2\sigma$, on E_F^*, as shown in Figure 11.15c. The most impressive feature is that when RS is present, the power factor can be improved two to four times when A varies from 1 to 10. For all the scattering intensity factors shown here, the enhancement of the power factor occurs from E_F^* close to E_i, and this enhancement spans a very wide E_F^* range, larger than $10 \ k_BT$. However, when the Fermi energy is located below the impurity band center E_i, the power factor of the system with RS is lower than that with AC, and can even reach zero when $A > 10$. This illustrates that, to enhance the power factor, the effect of RS on the system is critical to make E_F^* higher than E_i at all relevant temperatures.

Another effect of RS on the transport parameter is on the Lorenz number ratio (L_{rs+ac}/L_{ac}), shown in Figure 11.15d. In the E_F^* range of (3–8) k_BT, the Lorenz number will be higher than unity due to RS. However, the Lorenz number is lower than unity outside of that range. This behavior of the Lorenz number is consistent with the conclusions of Prokofieva et al.[148]

Combined with the suppression of the electrical conductivity due to RS (Figure 11.15b), the ratio ($k_{e,rs+ac}/k_{e,ac}$), of the electrical contribution to thermal conductivity ($\kappa_e = L\sigma T$), in all calculated E_F^* is no higher than unity. With higher A, the κ_e would be smaller. With the effects of simultaneous reduction of κ_e and enhancement of the power factor, RS can be a very promising mechanism for improvements to the thermoelectric figure of merit.

To explore the effects of the scattering intensity A further, a wide range of A between 0.1 and 30 was calculated. The maximum power factor ratio (PF_{rs+ac}/PF_{ac}) for each A and the corresponding $k_{e,rs+ac}/k_{e,ac}$ are shown in Figure 11.16a. It can be seen that the power factor can be enhanced even at $A = 0.1$. The maximum power factor ratio first increased rapidly to ~4 at $A = 10$, and after reaching the highest ratio of 4.13 at $A = 18$, the enhancement ratio slowly decreased. Thus, $A = 10$ can be seen as the optimized scattering intensity. For each A, the $k_{e,rs+ac}/k_{e,ac}$, when the maximum PF_{rs+ac}/PF_{ac} is reached, also is shown in Figure 11.16a. It can be seen that the electronic

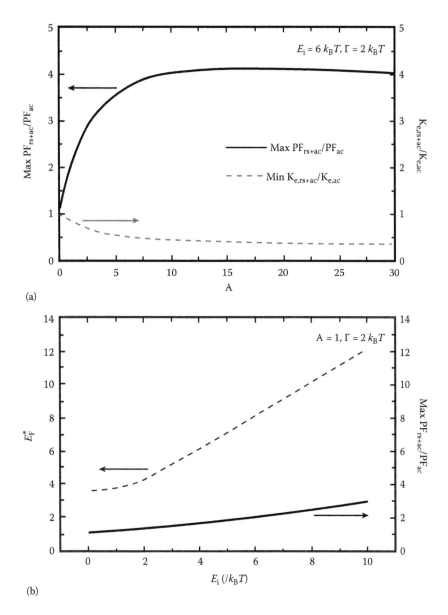

(a)

(b)

FIGURE 11.16 (a) Dependence of the maximum power factor ratio and the corresponding ratio of the electronic contribution to the thermal conductivity on the scattering factor A and (b) the relation of the maximum power factor ratio to the impurity band center E_i and the corresponding E_F^*.

contribution to thermal conductivity monotonically decreases with A, and at $A = 10$, it decreases to 0.45 times that without RS, which indicates the promising potential of this scattering mechanism for ZT improvements.

The effects of the location of the impurity band center E_i on the maximum power factor enhancement were also calculated, and the results are shown in Figure 11.16b. The results show a monotonic increase of the maximum power factor ratio with E_i. Based on these results, we can say that, for the purpose of power factor enhancement, higher E_i is preferred if it is possible. E_i is

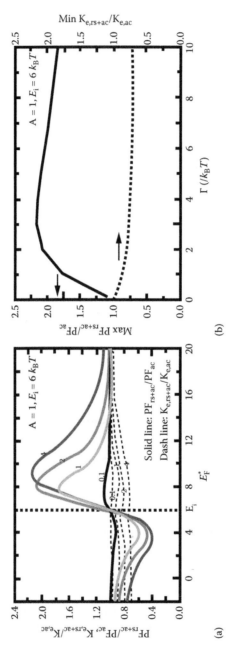

FIGURE 11.17 (a) Dependence of power factor ratios on the E_F^* for various impurity bandwidths Γ. (b) Dependences of maximum power factor ratios and corresponding ratios of electronic contribution to thermal conductivity on Γ. The numbers beside the curves represent the half-maximum-width Γ of the impurity band with the unit of $k_B T$.

element- and material system-dependent, as in the In-created resonant bands located at 0.07, 0.1, and 0.3 eV in PbTe, PbSe, and PbS, respectively, while Al-created resonant states are located ~0.3 eV above the conduction edge in PbTe.[138] This result provides a rule for choosing dopants for specific material systems. Meanwhile, the E_F^* corresponding to the maximum power factor ratio for each E_i is also shown in Figure 11.16b. It can be seen that when E_i is in the range of $(0-1)$ $k_B T$, the E_F^* of the maximum ratio is independent of E_i and stays at ~$3.6 k_B T$, while at higher E_i, the E_F^* of the maximum ratio is at a constant level of $(2.1-2.2)$ $k_B T$ higher than E_i. This indicates that for a given dopant and material system, an appropriate additional dopant is needed to achieve higher power factor enhancement, as Kaidanov et al.[145] discussed.

From Ravich's work, the width (Γ) of the Tl-created impurity band in the valence band of PbTe is $(2-5)$ $k_B T$, depending on the concentration of the Tl in PbTe:Tl,[149] while that of the impurity band induced by indium in PbTe is a magnitude thinner.[42] The different Γ of these two dopants results in quite different behavior in PbTe, i.e., in PbTe:Tl, the RS is pronounced, while in PbTe:In, the RS rarely can be found. Thus, it is necessary to evaluate the effects of Γ to understand the full extent of the RS mechanism. The calculated results for $A = 1$ and $E_i = 6$ $k_B T$ shown in Figure 11.17a demonstrate that when $\Gamma = 0.1$ $k_B T$, there is a very narrow impurity band, similar to that of the PbTe:In system.[145] For this narrow impurity band, there is little change in both the power factor and the electrical conductivity. When the Γ increases to 4 $k_B T$, the power factor can be enhanced ~2.1 times, while the electrical contribution to the thermal conductivity decreases ~0.7 times. Figure 11.17a also shows that, with a wider impurity band, the maximum power factor will be reached at higher E_F^*. Additional calculated results on the effects of Γ shown in Figure 10.17b illustrate that the highest power-factor enhancement can be reached when $\Gamma \sim k_B T$.

The detailed calculation mentioned above shows that, in order to improve thermoelectric properties by the RS mechanism, E_F^* should be kept higher than the impurity band center E_i at all relevant temperatures. Because the E_F^* will continue to approach the band edge in rising temperatures due to the degeneration lift, this $E_F^* > E_i$ requirement is a significant challenge in experimental work. The good news is that Kaidanov and Ravich[138] and Mashkova and Nemov[150] have shown that the impurity band center E_i will also approach the band edge in Tl- or In-doped PbTe. Additionally, double doping with dopants that do not create resonant bands, e.g., Na in PbTe:Tl,[149] can help control the carrier concentration further, and hence E_F^*, although there may be a limitation due to the solubility of the second dopant.

11.4.4 Conclusion

The effects of RS on thermoelectric properties were evaluated in detail by calculating variations of the scattering intensity A, the impurity the bandwidth Γ, and its center E_i. The results show that with the introduction of the RS mechanism, the power factor can be improved up to four times, and the electrical contribution to the thermal conductivity can be suppressed by about 0.45 times that of the system without RS. Simultaneously, these results were compared with those of the ordinary AC at the optimized RS parameters, provided that the Fermi level could be controlled accurately to lie at least above the impurity band center E_i.

Acknowledgments

For this work, the authors would like to acknowledge the financial support of NSF of China, under the project numbers 51372208, 51472207, and 51572226, and the Science and Technology Foundation of Sichuan province under the project numbers 2015GZ0060 and 2015TD0017.

References

1. Popov, M., Buga, S., Vysikaylo, P., Stepanov, P., Skok, V., Medvedev, V., Tatyanin, E. et al. C60-doping of nanostructured Bi-Sb-Te thermoelectrics. *Phys. Status Solidi* **2011**, *208*, 2783–2789.
2. Del Castillo-Castro, T., Larios-Rodriguez, E., Molina-Arenas, Z., Castillo-Ortega, M. M., and Tanori, J. Synthesis and characterization of metallic nanoparticles and their incorporation into electroconductive polymer composites. *Compos. Part A Appl. Sci. Manuf.* **2007**, *38*, 107–113.
3. Faleev, S. V., and Leonard, F. Theory of enhancement of thermoelectric properties of materials with nanoinclusions. *Phys. Rev. B* **2008**, *77*.
4. Zebarjadi, M., Esfarjani, K., Shakouri, A., Bahk, J.-H., Bian, Z., Zeng, G., Bowers, J., Lu, H., Zide, J., and Gossard, A. Effect of nanoparticle scattering on thermoelectric power factor. *Appl. Phys. Lett.* **2009**, *94*, 202105.
5. Zide, J. M., Klenov, D. O., Stemmer, S., Gossard, A. C., Zeng, G., Bowers, J. E., Vashaee, D., and Shakouri, A. Thermoelectric power factor in semiconductors with buried epitaxial semimetallic nanoparticles. *Appl. Phys. Lett.* **2005**, *87*, 112102.
6. Zide, J. M. O., Bahk, J.-H., Singh, R., Zebarjadi, M., Zeng, G., Lu, H., Feser, J. P. et al. High efficiency semimetal/semiconductor nanocomposite thermoelectric materials. *J. Appl. Phys.* **2010**, *108*, 123702.
7. Mingo, N., Hauser, D., Kobayashi, N. P., Plissonnier, M., and Shakouri, A. Nanoparticle-in-alloy approach to efficient thermoelectrics: Silicides in SiGe. *Nano Lett.* **2009**, *9*, 711–715.
8. Alboni, P. N., Ji, X., He, J., Gothard, N., and Tritt, T. M. Thermoelectric properties of $La_{0.9}CoFe_3Sb_{12}$–$CoSb_3$ skutterudite nanocomposites. *J. Appl. Phys.* **2008**, *103*, 113707.
9. Ibáñez, M., Luo, Z., Genç, A., Piveteau, L., Ortega, S., Cadavid, D., Dobrozhan, O. et al. High-performance thermoelectric nanocomposites from nanocrystal building blocks. *Nat. Commun.* **2016**, *7*, 10766.
10. Dresselhaus, M. S., Chen, G., Tang, M. Y., Yang, R. G., Lee, H., Wang, D. Z., Ren, Z. F., Fleurial, J.-P., and Gogna, P. New directions for low-dimensional thermoelectric materials. *Adv. Mater.* **2007**, *19*, 1043–1053.
11. Hicks, L., and Dresselhaus, M. Thermoelectric figure of merit of a one-dimensional conductor. *Phys. Rev. B* **1993**, *47*, 16631–16634.
12. Dingle, R., Störmer, H. L., Gossard, A. C., and Wiegmann, W. Electron mobilities in modulation-doped semiconductor heterojunction superlattices. *Appl. Phys. Lett.* **1978**, *33*, 665.

13. Pfeiffer, L., and West, K. The role of MBE in recent quantum Hall effect physics discoveries. *Phys. E Low Dimens. Syst. Nanostructures* **2003**, *20*, 57–64.

14. Hou, Q. R., Gu, B. F., Chen, Y. B., and He, Y. J. Enhancement of thermoelectric power factor of $MnSi_{1.7}$ films by Si addition and modulation doping. *Phys. Status Solidi* **2012**, *209*, 1307–1312.

15. Hou, Q. R., Gu, B. F., Chen, Y. B., He, Y. J., and Sun, J. L. Enhancement of the thermoelectric power factor of $MnSi_{1.7}$ film by modulation doping of Al and Cu. *Appl. Phys. A* **2013**, *114*, 943–949.

16. Song, E., Li, Q., Swartzentruber, B., Pan, W., Wang, G. T., and Martinez, J. A. Enhanced thermoelectric transport in modulation-doped GaN/AlGaN core/shell nanowires. *Nanotechnology* **2016**, *27*, 15204.

17. Zebarjadi, M., Joshi, G., Zhu, G., Yu, B., Minnich, A., Lan, Y., Wang, X., Dresselhaus, M., Ren, Z., and Chen, G. Power factor enhancement by modulation doping in bulk nanocomposites. *Nano Lett.* **2011**, *11*, 2225–2230.

18. Yu, B., Zebarjadi, M., Wang, H., Lukas, K., Wang, H., Wang, D., Opeil, C., Dresselhaus, M., Chen, G., and Ren, Z. Enhancement of thermoelectric properties by modulation-doping in silicon germanium alloy nanocomposites. *Nano Lett.* **2012**, *12*, 2077–2082.

19. Pei, Y.-L., Wu, H., Wu, D., Zheng, F., and He, J. High thermoelectric performance realized in a BiCuSeO system by improving carrier mobility through 3D modulation doping. *J. Am. Chem. Soc.* **2014**, *136*, 13902–13908.

20. Koirala, M., Zhao, H., Pokharel, M., Chen, S., Dahal, T., Opeil, C., Chen, G., and Ren, Z. Thermoelectric property enhancement by Cu nanoparticles in nanostructured $FeSb_2$. *Appl. Phys. Lett.* **2013**, *102*, 213111.

21. Yamini, S. A., Mitchell, D. R. G., Gibbs, Z. M., Santos, R., Patterson, V., Li, S., Pei, Y. Z., Dou, S. X., and Jeffrey Snyder, G. Heterogeneous distribution of sodium for high thermoelectric performance of p-type multiphase lead-chalcogenides. *Adv. Energy Mater.* **2015**, *5*, 1501047.

22. Wang, H., Cao, X., Takagiwa, Y., and Snyder, G. J. Higher mobility in bulk semiconductors by separating the dopants from the charge-conducting band—A case study of thermoelectric PbSe. *Mater. Horiz.* **2015**, *2*, 323–329.

23. Liang, W., Hochbaum, A. I., Fardy, M., Rabin, O., Zhang, M., and Yang, P. Field-effect modulation of Seebeck coefficient in single PbSe nanowires. *Nano Lett.* **2009**, *9*, 1689–1693.

24. Curtin, B. M., Codecido, E. A., Krämer, S., and Bowers, J. E. Field-effect modulation of thermoelectric properties in multigated silicon nanowires. *Nano Lett.* **2013**, *13*, 5503–5508.

25. Dollfus, P., Hung Nguyen, V., and Saint-Martin, J. Thermoelectric effects in graphene nanostructures. *J. Phys. Condens. Matter* **2015**, *27*, 133204.

26. Zuev, Y. M., Chang, W., and Kim, P. Thermoelectric and magnetothermoelectric transport measurements of graphene. *Phys. Rev. Lett.* **2009**, *102*, 96807.

27. Checkelsky, J. G., and Ong, N. P. Thermopower and Nernst effect in graphene in a magnetic field. *Phys. Rev. B* **2009**, *80*, 81413.

28. Duan, J., Wang, X., Lai, X., Li, G., Watanabe, K., Taniguchi, T., Zebarjadi, M., and Andrei, E. Y. High thermoelectric power factor in graphene/hBN devices. *Proc. Nat. Acad. Sci. USA* **2017**, *113*, 14272-14276.

29. Alu, A., and Engheta, N. Cloaking and transparency for collections of particles with metamaterial and plasmonic covers. *Opt. Express* **2007**, *15*, 7578.

30. Liu, N., Guo, H., Fu, L., Kaiser, S., Schweizer, H., and Giessen, H. Three-dimensional photonic metamaterials at optical frequencies. *Nat. Mater.* **2008**, *7*, 31–37.

31. Liao, B., Zebarjadi, M., Esfarjani, K., and Chen, G. Cloaking core-shell nanoparticles from conducting electrons in solids. *Phys. Rev. Lett.* **2012**, *109*, 126806.

32. Zebarjadi, M., Liao, B., Esfarjani, K., Dresselhaus, M., and Chen, G. Enhancing the thermoelectric power factor by using invisible dopants. *Adv. Mater.* **2013**, *25*, 1577–1582.

33. Shen, W., Tian, T., Liao, B., and Zebarjadi, M. Combinatorial approach to identify electronically cloaked hollow nanoparticles. *Phys. Rev. B* **2014**, *90*, 75301.

34. Zebarjadi, M., and Shen, W. An algorithm to reduce combinatorial search in design of invisible dopants. *Comput. Mater. Sci.* **2016**, *113*, 171–177.

35. Gu, N., Rudner, M., and Levitov, L. Chirality-assisted electronic cloaking of confined states in bilayer graphene. *Phys. Rev. Lett.* **2011**, *107*, 156603.

36. Oliver, D., Garcia, J. H., Rappoport, T. G., Peres, N. M. R., and Pinheiro, F. A. Cloaking resonant scatterers and tuning electron flow in graphene. *Phys. Rev. B* **2015**, *91*, 155416.

37. Fleury, R., and Alù, A. Furtive quantum sensing using matter-wave cloaks. *Phys. Rev. B* **2013**, *87*, 201106.

38. Majumdar, A. Thermoelectricity in semiconductor nanostructures. *Science* **2004**, *303*(5659), 777–778.

39. Hicks, L. D., and Dresselhaus, M. S. Effect of quantum-well structures on the thermoelectric figure of merit. *Phys. Rev. B* **1993**, *47*(19), 12727–12731.

40. Mahan, G. D., and Sofo, J. O. The best thermoelectric. *Proc. Nat. Acad. Sci. USA* **1996**, *93*(15), 7436–7439.

41. Pei, Y. Z., LaLonde, A., Iwanaga, S., and Snyder, G. J. High thermoelectric figure of merit in heavy hole dominated PbTe. *Energy Environ. Sci.* **2011**, *4*(6), 2085–2089.

42. Heremans, J. P., Jovovic, V., Toberer, E. S., Saramat, A., Kurosaki, K., Charoenphakdee, A., Yamanaka, S., and Snyder, G. J. Enhancement of thermoelectric efficiency in PbTe by distortion of the electronic density of states. *Science* **2008**, *321*(5888), 554–558.

43. Zhang, Q., Liao, B. L., Lan, Y. C., Lukas., K., Liu, W. S., Esfarjani, K., Opeil, C., Broido, D., Chen, G., and Ren, Z. F. High thermoelectric performance by resonant dopant indium in nanostructured SnTe. *Proc. Nat. Acad. Sci. USA* **2013**, *110*(33), 13261–13266.

44. Heremans, J. P., Wiendlocha, B., and Chamoire, A. M. Resonant levels in bulk thermoelectric semiconductors. *Energy Environ. Sci.* **2012**, *5*(2), 5510–5530.

45. Ohta, H., Kim., S., Mune, Y., Mizoguchi., T., Nomura. K., Ohta, S., Nomura, T. et al. Giant thermoelectric Seebeck coefficient of two-dimensional electron gas in SrTiO₃. *Nat. Mater.* **2007**, *6*(2), 129–134.

46. Sun, X., Liu., J., Cronin, S. B., Wang, K. L., Chen, G., Koga, T., and Dresselhaus, M. S. Experimental study of the effect of the quantum well structures on the thermoelectric figure of merit in Si/Si₁₋ₓGeₓ system. *Mat. Res. Soc. Symp. Proc.* **1999**, *545*, 369–374.

47. Hicks, L. D., Harman, T. C., Sun, X., and Dresselhaus, M. S. Experimental study of the effect of quantum-well structures on the thermoelectric figure of merit. *Phys. Rev. B* **1996**, *53*(16), 10493–10496.

48. Nakpathomkun, N., Xu, H. Q., and Linke, H. Thermoelectric efficiency at maximum power in low-dimensional systems. *Phys. Rev. B* **2010**, *82*(23), 235428.

49. Cornett, J. E., and Rabin, O. Universal scaling relations for the thermoelectric power factor of semiconducting nanostructures. *Phys. Rev. B* **2011**, *84*(20), 205410.

50. Cornett, J. E., and Rabin, O. Thermoelectric figure of merit calculations for semiconducting nanowires. *Appl. Phys. Lett.* **2011**, *98*(18), 182104.

51. Cornett, J. E., and Rabin, O. Effect of the energy dependence of the carrier scattering time on the thermoelectric power factor of quantum wells and nanowires. *Appl. Phys. Lett.* **2012**, *100*(24), 242106.

52. Moon, J., Kim, J. H., Chen, Z. C. Y., Xiang, J., and Chen, R. K. Gate-modulated thermoelectric power factor of hole gas in Ge-Si core-shell nanowires. *Nano Lett.* **2013**, *13*(3), 1196–1202.

53. Wu, P. M., Gooth, J., Zianni, X., Svensson, S. F., Gluschke, J. G., Dick, K. A., Thelander, C., Nielsch, K., and Linke, H. Large thermoelectric power factor enhancement observed in InAs nanowires. *Nano Lett.* **2013**, *13*(9), 4080–4086.

54. Kwon, S., Wingert, M. C., Zheng, J. L., Xiang. J., and Chen. R. K. Thermal transport in Si and Ge nanostructures in the 'confinement' regime. *Nanoscale* **2016**, *8*(27), 13155–13167.

55. Wang, Z., and Mingo, N. Diameter dependence of SiGe nanowire thermal conductivity. *Appl. Phys. Lett.* **2010**, *97*(10), 101903.

56. Mingo, N. Calculation of Si nanowire thermal conductivity using complete phonon dispersion relations. *Phys. Rev. B* **2003**, *68*(11), 113308.

57. Jain, A., and McGaughey, A. J. H. Effect of exchange-correlation on first-principles-driven lattice thermal conductivity predictions of crystalline silicon. *Comput. Mater. Sci.* **2015**, *110*, 115–120.

58. Regner, K. T., Sellan. D. P., Su, Z. H., Amon, C. H., McGaughey, A. J. H., and Malen, J. A. Broadband phonon mean free path contributions to thermal conductivity measured using frequency domain thermoreflectance. *Nat. Commun.* **2013**, *4*, 1640.

59. Esfarjani, K., Chen, G., and Stokes, H. T. Heat transport in silicon from first-principles calculations. *Phys. Rev. B* **2011**, *84*(8), 085204.

60. Shi, L., Li, D. Y., Yu, C. H., Jang, W. Y., Kim, D. Y., Yao, Z., Kim, P., and Majumdar, A. Measuring thermal and thermoelectric properties of one-dimensional nanostructures using a microfabricated device. *J. Heat Transf.—Trans. ASME* **2003**, *125*(6), 1209–1209.

61. Kim, P., Shi, L., Majumdar, A., and McEuen, P. L. Thermal transport measurements of individual multiwalled nanotubes. *Phys. Rev. Lett.* **2001**, *87*(21), 215502.

62. Wingert, M. C., Chen, Z. C. Y., Kwon, S., Xiang, J., and Chen, R. K. Ultra-sensitive thermal conductance measurement of one-dimensional nanostructures enhanced by differential bridge. *Rev. Sci. Instrum.* **2012**, *83*(2), 024901.

63. Li, D. Y., Wu, Y. Y., Kim, P., Shi, L., Yang, P. D., and Majumdar, A. Thermal conductivity of individual silicon nanowires. *Appl. Phys. Lett.* **2003**, *83*(14), 2934–2936.

64. Mingo, N., Yang, L., Li, D., and Majumdar, A. Predicting the thermal conductivity of Si and Ge nanowires. *Nano Lett.* **2003**, *3*(12), 1713–1716.

65. Martin, P., Aksamija, Z., Pop, E., and Ravaioli, U. Impact of phonon-surface roughness scattering on thermal conductivity of thin Si nanowires. *Phys. Rev. Lett.* **2009**, *102*(12), 125503.

66. Martin, P. N., Aksamija, Z., Pop, E., and Ravaioli, U. Reduced thermal conductivity in nanoengineered rough Ge and GaAs nanowires. *Nano Lett.* **2010**, *10*(4), 1120–1124.

67. McGaughey, A. J. H., Landry, E. S., Sellan, D. P., and Amon, C. H. Size-dependent model for thin film and nanowire thermal conductivity. *Appl. Phys. Lett.* **2011**, *99*(13), 131904.

68. Ju, Y. S., and Goodson, K. E. Phonon scattering in silicon films with thickness of order 100 nm. *Appl. Phys. Lett.* **1999**, *74*(20), 3005–3007.

69. Chen, R., Hochbaum, A. I., Murphy, P., Moore, J., Yang, P. D., and Majumdar, A. Thermal conductance of thin silicon nanowires. *Phys. Rev. Lett.* **2008**, *101*(10), 105501.

70. Boukai, A. I., Bunimovich., Y., Tahir-Kheli, J., Yu, J. K., Goddard, W. A., and Heath, J. R. Silicon nanowires as efficient thermoelectric materials. *Nature* **2008**, *451*(7175), 168–171.

71. Wingert, M. C., Kwon, S., Hu, M., Poulikakos, D., Xiang, J., and Chen, R. K. Sub-amorphous thermal conductivity in ultrathin crystalline silicon nanotubes. *Nano Lett.* **2015**, *15*(4), 2605–2611.

72. Wingert, M. C., Chen, Z. C. Y., Dechaumphai, E., Moon, J., Kim, J. H., Xiang, J., and Chen, R. K. Thermal conductivity of Ge and Ge-Si core-shell nanowires in the phonon confinement regime. *Nano Lett.* **2011**, *11*(12), 5507–5513.

73. Zou, J., and Balandin, A. Phonon heat conduction in a semiconductor nanowire. *J. Appl. Phys.* **2001**, *89*(5): 2932–2938.

74. Balandin, A., Tang. Y. S., and Wang, K. L. Thermal management of ultra-thin SOI devices: Effects of phonon confinement. *Silicon-Based Optoelectron.* **1999**, *3630*, 135–142.

75. Balandin, A., and Wang, K. L. Effect of phonon confinement on the thermoelectric figure of merit of quantum wells. *J. Appl. Phys.* **1998**, *84*(11), 6149–6153.

76. Balandin, A., and Wang, K. L. Significant decrease of the lattice thermal conductivity due to phonon confinement in a free-standing semiconductor quantum well. *Phys. Rev. B* **1998**, *58*(3), 1544–1549.

77. Cuffe, J., Chavez, E., Shchepetov, A., Chapuis, P. O., El Boudouti, E. H., Alzina, F., T. Kehoe, J. Gomis-Bresco, D. Dudek, Y. Pennec et al. Phonons in slow motion: Dispersion relations in ultrathin Si membranes. *Nano Lett.* **2012**, *12*(7), 3569–3573.

78. Jurgilaitis, A., Enquist, H., Andreasson, B. P., Persson, A. I. H., Borg, B. M., Caroff, P., Dick, K. A. et al. Time-resolved X-ray diffraction investigation of the modified phonon dispersion in InSb nanowires. *Nano Lett.* **2014**, *14*(2), 541–546.

79. Bozyigit, D., Yazdani, N., Yarema, M., Yarema, O., Lin, W. M. M., Volk, S., Vuttivorakulchai, K., Luisier. M., Juranyi, F., and Wood, V. Soft surfaces of nanomaterials enable strong phonon interactions. *Nature* **2016**, *531*(7596), 618–622.

80. Zhu, Y., Xu, F., Qin. Q. Q., Fung. W. Y., and Lu, W. Mechanical properties of vapor-liquid-solid synthesized silicon nanowires. *Nano Lett.* **2009**, *9*(11), 3934–3939.

81. Hochbaum, A. I., Chen, R. K., Delgado, R. D., Liang, W. J., Garnett, E. C., Najarian, M., Majumdar, A., and Yang, P. D. Enhanced thermoelectric performance of rough silicon nanowires. *Nature* **2008**, *451*(7175), 163–167.

82. Hippalgaonkar, K., Huang, B. L., Chen, R. K., Sawyer, K., Ercius, P., and Majumdar, A. Fabrication of microdevices with integrated nanowires for investigating low-dimensional phonon transport. *Nano Lett.* **2010**, *10*(11), 4341–4348.

83. Lim, J. W., Hippalgaonkar, K., Andrews, S. C., Majumdar, A., and Yang, P. D. Quantifying surface roughness effects on phonon transport in silicon nanowires. *Nano Lett.* **2012**, *12*(5), 2475–2482.

84. Ghossoub, M. G., Valavala, K. V., Seong, M., Azeredo, B., Hsu, K., Sadhu, J. S., Singh, P. K., and Sinha, S. Spectral phonon scattering from sub-10 nm surface roughness wavelengths in metal-assisted chemically etched Si nanowires. *Nano Lett.* **2013**, *13*(4), 1564–1571.

85. Park, Y. H., Kim, J., Kim, H., Kim. I., Lee, K. Y., Seo, D., Choi. H. J., and Kim. W. Thermal conductivity of VLS-grown rough Si nanowires with various surface roughnesses and diameters. *Appl. Phys. A—Mater. Sci. Process.* **2011**, *104*(1), 7–14.

86. Lee, J., Lee, W., Lim. J., Yu. Y., Kong, Q., Urban, J. J., and Yang. P. D., Thermal transport in silicon nanowires at high temperature up to 700 K. *Nano Lett.* **2016**, *16*(7), 4133–4140.

87. Sadhu, J., and Sinha, S. Room-temperature phonon boundary scattering below the Casimir limit. *Phys. Rev. B* **2011**, *84*(11), 115450.

88. Donadio, D., and Galli, G. Atomistic simulations of heat transport in silicon nanowires. *Phys. Rev. Lett.* **2009**, *102*(19), 195901.

89. Neogi, S., Reparaz, J. S., Pereira, L. F. C., Graczykowski, B., Wagner, M. R., Sledzinska, M., Shchepetov, A. et al. Tuning thermal transport in ultrathin silicon membranes by surface nanoscale engineering. *ACS Nano* **2015**, *9*(4), 3820–3828.

90. Chen, J., Zhang, G., and Li, B. W. Phonon coherent resonance and its effect on thermal transport in core-shell nanowires. *J. Chem. Phys.* **2011**, *135*(10), 104508.

91. Chen, J., Zhang, G., and Li, B. W. Impacts of atomistic coating on thermal conductivity of germanium nanowires. *Nano Lett.* **2012**, *12*(6), 2826–2832.

92. Hu, M., Zhang, X. L., Giapis, K. P., and Poulikakos, D. Thermal conductivity reduction in core-shell nanowires. *Phys. Rev. B* **2011**, *84*(8), 085442.

93. Hu, M., Giapis, K. P., Goicochea. J. V., Zhang, X. L., and Poulikakos. D. Significant reduction of thermal conductivity in Si/Ge core-shell nanowires. *Nano Lett.* **2011**, *11*(2), 618–623.

94. Amato, M., Palummo, M., Rurali, R., and Ossicini, S. Silicon-germanium nanowires: Chemistry and physics in play, from basic principles to advanced applications. *Chem. Rev.* **2014**, *114*(2), 1371–1412.

95. Yin, L., Lee, E. K., Lee, J. W., Whang, D., Choi, B. L., and Yu, C. The influence of phonon scatterings on the thermal conductivity of SiGe nanowires. *Appl. Phys. Lett.* **2012**, *101*(4), 043114.

96. Kim, H., Kim, I., Choi, H. J., and Kim, W. Thermal conductivities of Si1-xGex nanowires with different germanium concentrations and diameters. *Appl. Phys. Lett.* **2010**, *96*(23), 233106.

97. Lee, E. K., Yin, L., Lee. Y., Lee, J. W., Lee, S. J., Lee, J., Cha, S. N. et al. Large thermoelectric figure-of-merits from SiGe nanowires by simultaneously measuring electrical and thermal transport properties. *Nano Lett.* **2012**, *12*(6), 2918–2923.

98. Grauby, S., Puyoo, E., Rampnoux, J. M., Rouviere, E., and Dilhaire, S. Si and SiGe nanowires: Fabrication process and thermal conductivity measurement by 3 omega-scanning thermal microscopy. *J. Phys. Chem. C* **2013**, *117*(17), 9025–9034.

99. Hsiao, T. K., Chang, H. K., Liou. S. C., Chu, M. W., Lee, S. C., and Chang, C. W. Observation of room-temperature ballistic thermal conduction persisting over 8.3 μm SiGe nanowires. *Nat. Nanotechnol.* **2013**, *8*(7), 534–538.101.

100. Martinez, J. A., Provencio, P. P., Picraux, S. T., Sullivan, J. P., and Swartzentruber, B. S. Enhanced thermoelectric figure of merit in SiGe alloy nanowires by boundary and hole-phonon scattering. *J. Appl. Phys.* **2011**, *110*(7), 074317.

101. Chan, M. K. Y., Reed, J., Donadio. D., Mueller, T., Meng, Y. S., Galli, G., and Ceder, G. Cluster expansion and optimization of thermal conductivity in SiGe nanowires. *Phys. Rev. B* **2010**, *81*(17), 174303.

102. Malhotra, A., and Maldovan. M. Impact of phonon surface scattering on thermal energy distribution of Si and SiGe nanowires. *Sci. Rep.* **2016**, *6*, 25818.

103. Xie, G. F., Guo, Y., Wei, X. L., Zhang, K. W., Sun, L. Z., Zhong, J. X., Zhang, G., and Zhang, Y. W. Phonon mean free path spectrum and thermal conductivity for $Si_{1-x}Ge_x$ nanowires. *Appl. Phys. Lett.* **2014**, *104*(23), 233901.

104. Ratsifaritana, C. A., and Klemens, P. G. Scattering of phonons by vacancies. *Int. J. Thermophys.* **1987**, *8*(6), 737–750.

105. Garg, J., Bonini, N., Kozinsky, B., and Marzari, N. Role of disorder and anharmonicity in the thermal conductivity of silicon-germanium alloys: A first-principles study. *Phys. Rev. Lett.* **2011**, *106*(4), 045901.

106. Li, W., Lindsay, L., Broido, D. A., Stewart, D. A., and Mingo, N. Thermal conductivity of bulk and nanowire $Mg_2Si_xSn_{1-x}$ alloys from first principles. *Phys. Rev. B* **2012**, *86*(17): 174307.

107. Ioffe, A. F. *Physics of Semiconductors*. **1960**. New York: Academic Press (translated from Russian. 1957. *Fizika Poluprovodnikov*. Moscow: Russian Academy of Sciences).

108. Mott, N. F. and Davis, E. A. *Electronic Processes in Non-crystalline Materials*. 1979. Oxford: Oxford University Press.

109. Behnia, K. On mobility of electrons in a shallow Fermi sea over a rough seafloor. *J. Phys.: Condens. Matter* **2015**, *27*, 375501.

110. Heremans, J. P. Low-dimensional thermoelectricity. *Acta Phys. Polon. A* **2005**, *108*, 609.

111. Korringa, J., and Gerritsen, A. N. The cooperative electron phenomenon in dilute alloys. *Physica* **1953**, *19*, 457.

112. König, J. D., Nielsen, M. D., Gao, Y. B., Winkler, M., Jacquot, A., Böttner, H., and Heremans, J. P. Titanium forms a resonant level in the conduction band of PbTe. *Phys. Rev. B* **2011**, *84*, 205126.

113. Kondo, J. Resistance minimum in dilute magnetic alloys. *Prog. Theor. Phys.* **1964**, *32*, 37.

114. Nielsen, P. E., and Taylor, P. L. Theory of thermoelectric effects in metals and alloys. *Phys. Rev. Lett.* **1970**, *25*, 371.

115. Wiendlocha, B. Fermi surface and electron dispersion of PbTe doped with resonant Tl impurity from KKR-CPA calculations. *Phys. Rev. B* **2013**, *88*, 205205.

116. Wiendlocha, B. Localization and magnetism of the resonant impurity states in Ti doped PbTe. *Appl. Phys. Lett.* **2014**, *105*, 133901.

117. Wiendlocha, B., Kutorasinski, K., Kaprzyk, S., and Tobola, J. Recent progress in calculations of electronic and transport properties of disordered thermoelectric materials. *Scr. Mater.* **2016**, *111*, 33.

118. Gordon, B. E. A., Temmerman, W. E., and Györffy, B. L. On the Fermi surfaces of paramagnetic Cu_cNi_{1-c} alloys. *J. Phys. F* **1981**, *11*, 821.

119. Ebert, H., Ködderitzsch, D., and Minár, J. Calculating condensed matter properties using the KKR-Green's function method—Recent developments and applications. *Rep. Prog. Phys.* **2011**, *74*, 096501.

120. Kim, S., Wiendlocha, B., Jin, H., Tobola, J., and Heremans, J. P. Electronic structure and thermoelectric properties of p-type Ag-doped Mg_2Sn and $Mg_2Sn_{1-x}Si_x$ (x = 0.05, 0.1). *J. Appl. Phys.* **2014**, *116*, 153706.

121. Ebert, H. et al. *The Munich SPR-KKR Package*, version 6.3.1. 2012. http://ebert.cup.uni-muenchen.de/SPRKKR.

122. Matusiak, M., Tunnicliffe, E. M., Cooper, J. R., Matsushita, Y., and Fisher, I. R. Evidence for a charge Kondo effect in $Pb_{1-x}Tl_xTe$ from measurements of thermoelectric power. *Phys. Rev. B* **2009**, *80*, 220403(R).

123. Keiber, T., Bridges, F., Sales, B. C., and Wang, H. Complex role for thallium in PbTe:Tl from local probe studies. *Phys. Rev. B* **2013**, *87*, 144104.

124. Jaworski, C. M., Wiendlocha, B., Jovovic V., and Heremans, J. P. Combining alloy scattering of phonons and resonant electronic levels to reach a high thermoelectric figure of merit in PbTeSe and PbTeS alloys. *Energy Environ. Sci.* **2011**, *4*, 4155–4162.

125. Ahmad, S., and Mahanti, S. D. Energy and temperature dependence of relaxation time and Wiedemann-Franz law on PbTe. *Phys. Rev. B* **2010**, *81*, 165203.

126. Jaworski, C. M., Kulbachinskii, V. A., and Heremans, J. P. Tin forms a resonant level in Bi_2Te_3 that enhances the room temperature thermoelectric power. *Phys. Rev. B* **2009**, *80*, 233201.

127. Wiendlocha, B. Resonant levels, vacancies, and doping in Bi_2Te_3, Bi_2Te_2Se, and Bi_2Se_3 tetradymites. *J. Elect. Mater.* **2016**, *45*, 3515–3531.

128. Vandaele, K., and Heremans, J. P. Low temperature thermoelectric materials: Bi-Sb alloys. In *Advance Thermoelectrics: Materials, Contacts, Devices, and Systems*, ed. Z. Ren, section 2.1. Boca Raton, FL: CRC Press/Taylor & Francis.

129. Jin, H., Wiendlocha, B., and Heremans, J. P. P-type doping of elemental bismuth with indium, gallium, and tin: A novel doping mechanism in solids. *Energy Environ. Sci.* **2015**, *8*, 2027–2040.

130. Orovets, C. M., Chamoire, A. M., Jin, H., Wiendlocha, B., and Heremans, J. P. Lithium as an interstitial donor in bismuth and bismuth-antimony alloys. *J. Elect. Mater.* **2012**, *41*, 1648–1652.

131. Heremans J. P., and Jin, H. Thermoelectric and spin-caloritronic coolers: From basics to recent developments. *SPIE Photonics West Conf. Proc.* **2016**, *9765*, 1–17.

132. Wiendlocha, B., Jin, H., Tobola, J., Kaprzyk, S., and Heremans, J. P. Search for resonant impurities in bismuth and bismuth antimony alloys—First principles study. *The 30th International Conference on Thermoelectrics*, 2011. July 17–21, Traverse City, MI.

133. Heremans, J. P., Jin, H., and Wiendlocha, B. Potassium is a resonant level in $Bi_{1-x}Sb_x$ alloys. *APS March Meeting*, 2012. February 27–March 2, Boston, MA.

134. Bansil, A., Kaprzyk, S., Mijnarends, P. E., and Tobola, J. Electronic structure and magnetism of $Fe_{3-x}V_xX$ (X=Si, Ga, and Al) alloys by the KKR-CPA method. *Phys. Rev. B* **1999**, *60*, 13396.

135. Rudolph, R., Krueger, H., Fellmuth, B., and Herrmann, R. Dielectric properties of $Bi_{1-x}Sb_x$ alloys at the semiconductor-semimetal transition. *Phys. Stat. Sol. B* **1980**, *102*, 295.

136. Fellmuth, B., Krueger, H., Rudolph, R., and Herrmann, R. Determination of the cyclotron masses of the conduction band of $Bi_{1-x}Sb_x$ near the semimetal-semiconductor transition. *Phys. Stat. Sol. B* **1980**, *101*, 695.

137. Liu, W., Yan, X., Chen, G., and Ren, Z. Recent advances in thermoelectric nanocomposites. *Nano Energy* **2011**, *1*(10), 42–56.

138. Kaidanov, V. I., and Ravich, Y. I. Deep and resonance states in $A^{IV}B^{VI}$ semiconductors. *Sov. Phys. Uspekhi* **1985**, *28*(1), 31–53.

139. Ravich, Y. I. Selective carrier scattering in thermoeletric materials. In *CRC Handbook of Thermoelectrics*, ed. D. M. Rowe. **1995**. CRC Press LLC: New York.

140. Zhang, Q., Wang, H., Zhang, Q., Liu, W., Yu, B., Wang, H., Wang, D., Ni, G., Chen, G., and Ren, Z. Effect of silicon and sodium on thermoelectric properties of thallium doped lead telluride based materials. *Nano Lett.* **2012**, *12*(5), 2324–2330.

141. Zhang, Q. Y., Wang, H., Liu, W. S., Wang, H. Z., Yu, B., Zhang, Q., Tian, Z. et al. Enhancement of thermoelectric figure-of-merit by resonant states of aluminium doping in lead selenide. *Energy Environ. Sci.* **2012**, *5*(1), 5246–5251.

142. Tan, G., Shi, F., Hao, S., Chi, H., Zhao, L.-D., Uher, C., Wolverton, C., Dravid, V. P., and Kanatzidis, M. G. Codoping in SnTe: Enhancement of thermoelectric performance through synergy of resonance levels and band convergence. *J. Am. Chem. Soc.* **2015**, *137*(15), 5100–5112.

143. Tan, G., Zeier, W. G., Shi, F., Wang, P., Snyder, G. J., Dravid, V. P., and Kanatzidis, M. G. High thermoelectric performance SnTe-In_2Te_3 solid solutions enabled by resonant levels and strong vacancy phonon scattering. *Chem. Mater.* **2015**, *27*(22), 7801–7811.

144. Zhao, W., Wei, P., Zhang, Q., Peng, H., Zhu, W., Tang, D., Yu, J. et al. Multi-localization transport behaviour in bulk thermoelectric materials. *Nat. Commun.* **2015**, *6*.

145. Kaidanov, V. I., Nemov, S. A., and Ravich, Y. I. Resonant scattering of carriers in IV-VI semiconductors. *Sov. Phys. Semicond.-USSR* **1992**, *26*(2), 113–125.

146. Prokofieva, L. V., Ravich, Y. I., Pshenay-Severin, D. A., Konstantinov, P. P., and Shabaldin, A. A. Resonance states, heavy quasiparticles, and the thermoelectric figure of merit of IV-VI materials. *Semiconductors* **2012**, *46*(7), 866–872.

147. Ravich, Y. I., Nemov, S. A., and Proshin, V. I. Hopping conduction via highly localized impurity states of indium in $Pb_{0.78}Sn_{0.22}Te$ solid solutions. *Semiconductors* **1995**, *29*(8), 754–756.

148. Prokofieva, L. V., Shabaldin, A. A., Korchagin, V. A., Nemov, S. A., and Ravich, Y. I. Lorentz number and Hall factor in degenerate semiconductors during resonance scattering of charge carriers. *Semiconductors* **2008**, *42*(10), 1161–1170.

149. Mashkova, T. R., and Nemov, S. A. Dependence of the energy position of a band of resonance states on temperature and impurity concentrations in PbTe:Tl,Na. *Sov. Phys. Semicond.-USSR* **1985,** *19*(10), 1148–1149.

150. Kaidanov, V. I., Nemov, S. A., Ravich, Y. I., and Zaitsev, A. M. Influence of resonance states on the Hall effect and electrical conductivity of PbTe doped with both thallium and sodium simultaneously. *Sov. Phys. Semicond.-USSR* **1983,** *17*(9), 1027–1030.

CHAPTER 12

Simulation of Phonons

David Broido, Matt Heine, Natalio Mingo, and Austin J. Minnich

---

Contents

12.1 Phonon Thermal Transport and Lattice Thermal Conductivity from First Principles

David Broido, Matt Heine, and Natalio Mingo

12.1.1 Introduction

The thermal conductivity is a fundamental thermal transport parameter that describes the ability of a material to conduct heat.[1–3] The focus of this chapter is on ab initio theoretical approaches to calculate the lattice thermal conductivity κ, where heat is carried by phonons. Thermal conductivity κ determines the utility of materials for specific thermal management applications. Materials with low thermal conductivity find applicability in, for example, efficient thermoelectric (TE) cooling and power generation devices,[2-7] while materials with high thermal conductivity are needed for passive cooling of microelectronics.[8] In recent years, significant progress has been made in developing predictive ab initio models to describe phonon thermal transport, and some relevant references are cited at the end of this chapter.[9-41] We apologize in advance for any references omitted inadvertently.

Here, we focus on approaches that involve a solution of the Peierls–Boltzmann equation (PBE) for phonon transport in semiconductors and insulators. The theoretical framework for these approaches was developed by Rudolf Peierls in 1929.[42] In this work, Peierls described the phonon–phonon processes that arise from the anharmonicity of the interatomic potential. However, even when retaining only the lowest-order three-phonon scattering, the computational complexity of the PBE precluded rigorous solutions for decades. In fact, in his classic 1960 book, Ziman[1] stated, "the Boltzmann equation is so exceedingly complex that it seems hopeless to expect to generate a solution from it directly." Omini and Sparavigna[43,44] developed an iterative approach to solve the PBE in the mid-1990s, but adjustable parameters were still needed to describe the harmonic interatomic forces giving the phonon modes and the anharmonic interatomic forces describing the phonon-phonon scattering rates that limit the intrinsic thermal conductivity. Density functional theory (DFT) provided a way to accurately determine these interatomic forces, and their use in the PBE solution for three-phonon scattering processes has now given excellent agreement with measured κ data for many materials.[9-41] In the following, we describe the approach used to numerically solve the PBE. We also describe methods to calculate the interatomic force constants (IFCs) using DFT. In particular, the temperature-dependent effective potential (TDEP) approach, an approach relevant for TE materials, is summarized. Then examples are given for some TE materials.

12.1.2 Peierls–Boltzmann Equation

Here, we consider a nonmetallic crystal subjected to a small-temperature gradient ∇T. The nonequilibrium distribution function n_λ deviates from the Bose distribution n_λ^0 by a small deviation n_λ^1, proportional to ∇T. The PBE linearized in ∇T is[1]

$$\mathbf{v}_\lambda \cdot \nabla T \frac{\partial n_\lambda^0}{\partial T} = \frac{\partial n_\lambda}{\partial T}\bigg|_{\text{scatt}}, \qquad (12.1)$$

where $\mathbf{v}_\lambda$ is the phonon group velocity and $\lambda \equiv (\mathbf{q}, s)$ labels the phonon mode with wave vector $\mathbf{q}$ and polarization s. Here, the left-hand side represents phonon drift due to the applied temperature gradient, and the right-hand side is due to phonon scattering. Expressing the deviation function as $n_\lambda^1 = -(\partial n_\lambda^0 / \partial T) \mathbf{F}_\lambda \cdot \nabla T$, the PBE can be rewritten as[23,26]

$$\mathbf{F}_\lambda = \tau_\lambda^0 (\mathbf{v}_\lambda + \Delta_\lambda), \qquad (12.2)$$

with

$$\frac{1}{\tau_\lambda^0} = \sum_{\lambda's''}^{+} \Gamma_{\lambda\lambda'\lambda''}^{+} + \frac{1}{2} \sum_{\lambda's''}^{-} \Gamma_{\lambda\lambda'\lambda''}^{-} + \sum_{\lambda'} \Gamma_{\lambda\lambda'}, \qquad (12.3)$$

$$\Delta_\lambda = \sum_{\lambda's''}^{+} \Gamma_{\lambda\lambda'\lambda''}^{+} (\xi_{\lambda\lambda''} \mathbf{F}_{\lambda''} - \xi_{\lambda\lambda'} \mathbf{F}_{\lambda'}) + \sum_{\lambda's''}^{-} \Gamma_{\lambda\lambda'\lambda''}^{-} (\xi_{\lambda\lambda''} \mathbf{F}_{\lambda''} + \xi_{\lambda\lambda'} \mathbf{F}_{\lambda'}) + \sum_{\lambda'} \Gamma_{\lambda\lambda'} \xi_{\lambda\lambda'} \mathbf{F}_{\lambda'},$$

$$(12.4)$$

where $\xi_{\lambda\lambda'} = \omega_\lambda / \omega_{\lambda'}$ and $\Gamma_{\lambda\lambda'\lambda''}^{\pm}$ are three-phonon scattering probabilities:

$$\Gamma_{\lambda\lambda'\lambda''}^{\pm} = \frac{\hbar\pi}{4N} \frac{n_{\lambda'}^0 \mp n_{\lambda''}^0 + \frac{1}{2} \mp \frac{1}{2}}{\omega_\lambda \omega_{\lambda'} \omega_{\lambda''}} \left| \Phi_{\lambda\lambda'\lambda''}^{\pm} \right|^2 \delta(\omega_\lambda \pm \omega_{\lambda'}, \omega_{\lambda''}) \delta_{\mathbf{q}\pm\mathbf{q}', \mathbf{q}''+\mathbf{G}}. \qquad (12.5)$$

Here, N is the number of unit cells in the crystal, ω_λ is the phonon frequency, and $\Phi_{\lambda\lambda'\lambda''}^{\pm}$ are the three-phonon scattering matrix elements:

$$\Phi_{\lambda\lambda'\lambda''}^{\pm} = \sum_{i \in U.C.} \sum_{j,k} \sum_{\alpha\beta\gamma} \Phi_{ijk}^{\alpha\beta\gamma} \frac{e_{i\alpha}^\lambda e_{j\beta}^{\pm\lambda'} e_{k\gamma}^{-\lambda''}}{\sqrt{M_i M_j M_k}}, \qquad (12.6)$$

In Equation 12.6, i, j, and k designate atoms; α, β, and γ are Cartesian coordinates; $\Phi_{ijk}^{\alpha\beta\gamma}$ are third-order IFCs; and $e_{i\alpha}^\lambda$ is the αth component of the phonon eigenvector for mode λ, with the understanding that $-\lambda \equiv (-\mathbf{q}, s)$. Translational invariance allows restriction of the i summation to just the atoms in a single unit cell.

The delta functions in Equation 12.5 enforce conservation of energy and quasimomentum in each three-phonon scattering process:

$$\omega_\lambda \pm \omega_{\lambda'} = \omega_{\lambda''}, \qquad (12.7a)$$

$$\mathbf{q} \pm \mathbf{q}' = \mathbf{q}'' + \mathbf{G}. \qquad (12.7b)$$

In Equation 12.7b, $\mathbf{G}$ is a reciprocal lattice vector. It is frequently assumed that processes with $\mathbf{G} = 0$, the so-called normal processes, are individually nonresistive, while umklapp processes, which have $\mathbf{G} \neq 0$, are individually resistive. That this is not the case, which can be seen by recognizing that normal processes can become umklapp processes and vice versa for different choices of the first Brillouin zone. A nice discussion of this point has been recently provided.[45] In fact, it is the restriction to all processes being normal processes that removes thermal resistance and gives rise to a nondissipating heat flux.[1,45]

In addition to the intrinsic three-phonon scattering, the last term in Equation 12.3 describes scattering from point defects. For example, isotopes give a mass disorder for which the scattering rates are

$$\Gamma_{\lambda\lambda'} = \frac{\pi}{2N} \omega_\lambda^2 \sum_{i \in U.C.} g_i \left| \mathbf{e}_i^{\lambda*} \cdot \mathbf{e}_i^{\lambda'} \right|^2 \delta(\omega_\lambda - \omega_{\lambda'}), \tag{12.8}$$

where $g_i = \sum_s f_{is}(1 - M_{is}/\bar{M}_i)^2$ is the mass variance parameter, and f_{is} is the fraction of isotopes found on atom i, whose mass is M_{is}, and $\bar{M}_i = \sum_s f_{is} M_{is}$ is the isotope averaged mass of atom i.

Once $\mathbf{F}_\lambda$ is determined from the solution of Equation 12.2, the lattice thermal conductivity tensor is obtained as

$$\kappa_{\alpha\beta} = \frac{k_B}{NV_0} \sum_\lambda n_\lambda^0 \left(n_\lambda^0 + 1 \right) (\hbar\omega_\lambda / k_B T)^2 v_\lambda^\alpha F_\lambda^\beta, \tag{12.9}$$

where V_0 is the unit cell volume.

12.1.3 Numerical Solution of the Peierls–Boltzmann Equation

Equation 12.2 is solved using an iterative approach first implemented by Omini and Sparavigna.[43,44] To start the process, the zeroth-order approximation is taken to be $\mathbf{F}_\lambda^0 = \mathbf{v}_\lambda \tau_\lambda^0$. Here, the inverse of the total scattering rate τ_λ^0 approximates the phonon lifetime in what is commonly referred to as the relaxation time approximation (RTA). As constructed, the RTA treats the normal phonon–phonon scattering processes as independent resistive processes. As a result, the RTA underestimates $\kappa_{\alpha\beta}$. With the increasing number of iterations, $\kappa_{\alpha\beta}$ increases and finally saturates to the converged value. Once the relative change in the conductivity tensor upon successive iterations is less than a chosen convergence criterion, the iteration is stopped.

For semiconductors with moderately high thermal conductivity ($\kappa \sim$ 50–150 W/m K) such as Si, Ge, and GaAs, umklapp scattering is strong enough that the room temperature-converged κ is only about 5% larger than the RTA value. For TE materials, which have much smaller κ, the iterative procedure produces only minor increase in κ, making it unnecessary.

Large increases in κ are seen for high-κ bulk materials such as diamond, where umklapp scattering processes are weak even at room temperature. For diamond, the converged κ is about 50% larger than the RTA value and becomes even larger with decreasing temperature. The measured diamond κ is accurately reproduced by the iterative approach,[10] which remains stable down to below 50 K. For high-quality single crystals of diamond, boundary scattering begins to play an important role below 100 K.[46–48] Its inclusion for diamond retains the stability of the iterative solution over the full temperature range but requires the introduction of sample-dependent adjustable fit parameters that undermine the usefulness of the ab initio approach. For 2D planar layers such as graphene,[35–38] Umklapp scattering is very weak, giving much larger converged κ compared to the RTA values even at 300 K and giving rise to hydrodynamic phonon behavior.[35,36] In this case, the convergence stability of the iterative solution is aided by the inclusion of boundary scattering from finite-size 2D structures, with dimensions of up to tens of microns.[35,37] Other numerical solutions of the PBE have been developed.[36,38,39] The approaches by Cepellotti et al.[36] and Fugallo et al.[38] gave stable solutions of the PBE for infinite-sized 2D planar layers.

Specific approaches to calculating the three-phonon scattering rate integrals in Equation 12.5 include the phase space approach,[43,44] the adaptive Gaussian method,[49] and the tetrahedron approach.[50] A number of platforms are now widely used that implement solutions of the PBE, such as ShengBTE,[26] Phono3py,[51] and ALAMODE.[52] ShengBTE uses the adaptive Gaussian method, thus avoiding the problems inherent to fixed smearing approaches. Phono3py and ALAMODE use a tetrahedron method. The TDEP code gives the option of using either adaptive Gaussian method or tetrahedron method.

The conservation of quasimomentum and energy imposes strict constraints on the three-phonon scattering collisions. For a fixed phonon mode λ in a bulk crystal and for fixed phonon branches of the other two participating phonons (s', s''), the conservation conditions restrict phonon wave vector $\mathbf{q}'$ to lie on a set of 2D surfaces. We refer to the collection of such surfaces as the phase space for three-phonon scattering. For the low-κ crystalline materials needed for TEs, it is desirable to have a large scattering phase space. This can be achieved by the following: (1) Choosing heavy atoms, as well as giving smaller group velocities for heat-carrying acoustic phonons, which lowers the frequency scale, causing increased scattering between acoustic phonon and optic phonons.[53] (2) Choosing complex crystal structures with many atoms in the unit cell, which gives rise to many optic phonon branches; optic phonons typically have smaller phonon group velocities and lower populations, and they can provide scattering channels for acoustic phonons. (3) Engineering the dispersions to increase this phase space. An example of (2) and (3) is the skutterudite $YbFe_4Sb_{12}$[32]; another example of (3) is PbTe, which is an incipient ferroelectric for which the soft transverse optical (TO) phonon branch contributes to strong scattering between acoustic and optic phonons.[16,17,19–21] These cases are discussed further below.

In the numerical approach described, it is possible to separate the contributions to the scattering rates from normal and umklapp processes for a given choice of Brillouin zone. Typically, a Wigner–Seitz zone is chosen that is centered about the Γ point. Then, one explicitly calculates that normal processes dominate over umklapp processes at sufficiently low frequencies for all materials. This has been illustrated in the frequency dependence of the umklapp scattering rates, which have a higher power than that for N scattering at low frequency[11] such that if only umklapp processes were retained, a low frequency divergence of $\kappa_{\alpha\beta}$ would result. Thus, normal as well as umklapp processes are required for convergent thermal conductivity.

12.1.4 Interatomic Force Constants

The lattice potential energy is typically expressed as an expansion in atomic displacements about an assumed equilibrium configuration:

$$\Phi = \frac{1}{2!}\sum_{ij}\sum_{\alpha\beta}\Phi_{ij}^{\alpha\beta}u_{i\alpha}u_{j\beta} + \frac{1}{3!}\sum_{ijk}\sum_{\alpha\beta\gamma}\Phi_{ijk}^{\alpha\beta\gamma}u_{i\alpha}u_{j\beta}u_{k\gamma} + \cdots. \quad (12.10)$$

In the equation, i, and j, and k designate atoms; α, β, and γ are Cartesian coordinates; and $\Phi_{ij\ldots}^{\alpha\beta\ldots} \equiv \partial^n\Phi/\partial u_{i\alpha}\partial u_{j\beta}\ldots\big|_0$ are the IFCs of order n. The only inputs required to solve the PBE are the second-order (harmonic) and third-order anharmonic IFCs, which allow the calculation of phonon modes and intrinsic (i.e., phonon–phonon) and extrinsic (e.g., phonon-defect, phonon-boundary) scattering rates. These IFCs can be obtained from first-principles using DFT codes such as Quantum Espresso[54–56] or the Vienna ab initio simulation package (VASP).[57–60]

12.1.4.1 Standard Approach

For a given crystal structure, the set of lattice parameters $\{a_\ell\}$ are adjusted so as to minimize the total electronic energy $E_{el}(\{a_\ell\})$, while maintaining crystal symmetry. This "relaxation" procedure defines the equilibrium configuration about which the atoms are displaced to calculate IFCs. Both the second- and third-order IFCs can be calculated by using either density functional perturbation theory, which is a linear response approach using a reciprocal space representation[56,61,62] or a real-space finite difference supercell approach within the DFT.[63] For polar materials, long-range Coulomb forces[64] give rise to long-wavelength macroscopic electric fields that result in a nonanalytic part of the dynamical matrix at the Γ-point which leads to the longitudinal optical–TO splitting. Such interactions can be described using Born effective charges and a dielectric tensor.[56,64]

12.1.4.2 Symmetry Constraints on Interatomic Force Constants

Symmetry relations between IFCs significantly reduce the number of independent elements to be calculated. These stem from invariance of the interatomic potential under point/space group operations and of rigid translations and rotations of the lattice. Translational invariance can be enforced in different ways. We have used both least-squares fitting and Lagrange multiplier approaches.[14,23] The invariance of the crystal potential under rigid rotation connects *n*th-order IFCs to those of order $n + 1$.[65] Additional constraints, known as the Huang invariance conditions, arise by requiring vanishing stresses.[64] In principle, vanishing stress is already achieved through structural relaxation. However, the Huang invariance conditions can correct errors introduced in the relaxation procedure.

For an infinite lattice, there are an infinite number of IFCs of each order. In practice, a range of interaction is chosen, for example, by specifying the number of nearest-neighbor atomic shells interacting with central cell atoms to be considered. Standard DFT packages, such as Quantum Espresso[54–56] and VASP,[57–60] typically enforce only the point/space group symmetries on the determined IFCs with translational invariance imposed afterward. However, some previous calculations have taken care to impose all necessary symmetries.[63,66] Recently, an approach to calculate IFCs using internal lattice coordinates has been developed.[67] In this approach, the IFCs satisfy all symmetry requirements by construction. In the TDEP approach described in the following, all symmetries can also be imposed in advance to construct a smaller irreducible set of IFCs that need to be calculated.

12.1.4.3 Quasi-Harmonic Approximation

Conventional ab initio calculations do not capture the explicit dependence of the IFCs upon temperature. At the simplest level, one calculates IFCs by perturbing an ideal lattice, which is the static equilibrium configuration that minimizes the total energy. The obtained T-independent IFCs are then used to calculate temperature-dependent quantities such as phonon modes, specific heat, and thermal conductivity. Temperature dependence can be approximated by using the quasi-harmonic approximation (QHA). In the QHA, the Helmholtz free energy F is minimized with respect to the set of lattice parameters $\{a_\ell\}$ and any other degrees of freedom of a given crystal:

$$F(\{a_\ell\}, T) = E_{\text{el}}(\{a_\ell\}) + F_{\text{vib}}(\{a_\ell\}, T), \tag{12.11}$$

$$F_{\text{vib}}(\{a_\ell\}, T) = \sum_{\lambda} \left[\left(k_{\text{B}} T \ln \left\{ 1 - \exp \left[\frac{-\hbar \omega_\lambda(\{a_\ell\})}{k_{\text{B}} T} \right] \right\} \right) + \frac{1}{2} \hbar \omega_\lambda(\{a_\ell\}) \right]. \tag{12.12}$$

In the (assumed harmonic) vibrational contribution to F, the phonon frequencies are taken to explicitly depend on the lattice parameters but not on the temperature. Schematically, in the QHA, the IFCs then depend only implicitly on T, e.g., $\Phi_{ij}^{\alpha\beta} = \Phi_{ij}^{\alpha\beta}[\{a_\ell(T)\}]$, where $\{a_\ell(T)\}$ gives the set of lattice parameters determined at T, a deficiency that can be remedied using the TDEP approach.

12.1.4.4 Temperature-Dependent Effective Potential Approach

If one imagines a lattice at finite temperature, this static equilibrium configuration is, in fact, a thermodynamically improbable configuration. Furthermore, even at zero temperature, the quantum mechanical zero-point motion dictates nonzero mean-squared ionic displacements. The TDEP approach[68-71] incorporates these effects by sampling the energy surface at thermally relevant (probable) lattice configurations instead and then considering the free energy of the system.

One way to sample this thermally relevant phase space is through ab initio molecular dynamics. In Born–Oppenheimer molecular dynamics (BOMD), a lattice supercell is constructed and the ions are treated as classical particles moving under the influence of quantum mechanically calculated forces (calculated using DFT). To work in the more convenient canonical ensemble, a thermostat is incorporated into these dynamics, which simulates the interaction of the system with a heat bath at a specified temperature T. As the system evolves over many time steps, according to this prescription, it will explore the phase space for the given $\{a_\ell\}$ and T. Symmetrized second- and third-order IFCs are then obtained by adjusting the second- and third-order terms to minimize the mean square deviation between the forces calculated from the potential energy in Equation 12.10 and the BOMD calculated forces over a set of time steps.

Note that higher-order effects enter the BOMD-calculated forces both via the magnitude of ionic displacements *and* in the number of ions displaced. In other words, an ion not only samples anharmonicity by taking larger excursions at higher T, but also moves in a different environment, an environment in which the surrounding ions are also thermally displaced rather than in their equilibrium positions. The latter effect, for example, allows the possibility for thermal displacements of one ionic species to indirectly affect the IFCs of a different species. Thus, the TDEP IFCs are effective force constants which contain *explicit* temperature dependence as well as higher-order anharmonic effects.

Furthermore, since the thermal ionic displacements in these time steps are typically larger than those used in the conventional approach, the calculated forces are also larger in magnitude, resulting in a smaller relative error in forces.[71] Also since all atoms are displaced, each BOMD force calculation contains information about each IFC. Therefore, determining anharmonic IFCs does not present a significant increase in computational effort beyond the determination of harmonic IFCs. Thus, the characteristics of the calculation scheme that afford sampling of higher-order effects also present computational advantages.

A Helmholtz free energy surface is then calculated, which includes both contributions calculated from the IFCs and any contributions present in the BOMD results but not captured by the IFCs. The free energy is then minimized with respect to the $\{a_\ell\}$ at each T to obtain zero-pressure force constants that may be used to calculate physical quantities.

An alternative phase space sampling scheme is the use of the IFCs to construct an ensemble of probable lattice configurations generated from the canonical ensemble. This is done by constructing the ionic displacements for a given configuration by expanding in terms of phonon modes. For each configuration, a DFT calculation is performed, and this calculation takes the place of a BOMD time step. In this way, zero-point motion effects may be explicitly included in the IFCs. As with BOMD sampling, IFCs are determined in a self-consistent manner; IFCs determine configurations which determine new IFCs until self-consistency is reached.

Similar approaches that include thermal motion of atoms beyond the QHA have been independently developed, for example, by Souvatzis et al.,[72] Errea et al.,[73] van Roekeghem et al.,[74] and Antolin et al.[75]

Figure 12.1 gives an example of a temperature-dependent phonon scattering phase space in PbTe captured by the TDEP approach. PbTe is an excellent TE material largely because of its unusually low lattice thermal conductivity. Recently, it was shown that this stems from the fact that PbTe is an incipient ferroelectric with a soft TO mode at Γ.[16,17,19–21,76–80] The strong scattering between this soft TO phonon branch and the acoustic branches gives lower than expected κ. Neutron scattering data show that the TO mode shifts upward with increasing temperature,[76,77,80] a consequence of the strong anharmonicity near the ferroelectric instability (Figure 12.2).[21,76,78,80] Such behavior is not obtained by calculations using the QHA, which instead shows a softening of the TO mode with increasing T.[20,21] On the other hand, recent TDEP calculations do give a stiffening of the TO mode with increasing T, which causes the scattering phase space to precipitously decrease, as illustrated in Figure 12.1. As a result, the thermal resistivity shows a decrease in slope around room temperature.[21]

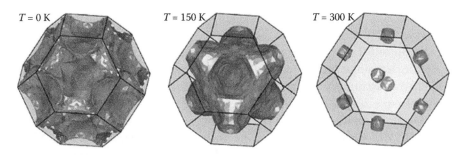

$T = 0$ K $T = 150$ K $T = 300$ K

FIGURE 12.1 Energy and momentum-conserving three-phonon phase space surfaces in PbTe for scattering between the zone center TO mode and LA and TA$_2$ phonon branches, calculated at three increasing temperatures. (Reprinted with permission from Romero, A. H. et al., *Physical Review B*, 91, 214310, 2015. Copyright 2015 by the American Physical Society.)

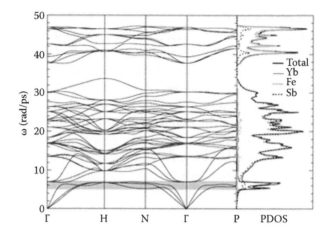

FIGURE 12.2 Phonon dispersion and DOS of $YbFe_4Sb_{12}$ with contributions from each element. (Reprinted with permission from Li, W., and Mingo, N., *Physical Review B*, 91, 035108, 2014. Copyright 2014 by the American Physical Society.)

12.1.4.5 Examples

Early calculations of ab initio phonon thermal transport focused on relatively simple crystals such as silicon, germanium, and diamond,[9–11] as well as zinc blende III–V compounds.[12–15] The good agreement between the calculated thermal conductivities of these materials and measured values provided strong validation of the predictive capability of the first-principles approach. In recent years, more complex structures have been examined, in particular, those of interest for TEs. In the following, we give two examples of these. In the first, a low thermal conductivity is found in a skutterudite structure, $YbFe_4Sb_{12}$.[32] This structure has 17 atoms in the unit cell so 51 phonon branches must be considered in determining the phase space for three-phonon scattering compared to only six for the simple crystal structures that have been predominantly studied in the past. In the second example, thermal transport in the prototypical TE material Bi_2Te_3 is examined using the TDEP approach,[68–71] which is designed to capture the strong anharmonicity that occurs in that system.

12.1.4.5.1 $YbFe_4Sb_{12}$

Skutterudites are bulk crystals with complex unit cells. Their good TE properties are linked to a low thermal conductivity, which was originally qualitatively attributed to the rattling of filler atoms. Ab initio calculations of thermal conductivity in these compounds permit us to go far beyond qualitative models and to obtain a deep insight into the actual physical mechanisms determining phonon conduction. This was illustrated by Li and Mingo.[32] In that paper, it was shown that the low thermal conductivity of $YbFe_4Sb_{12}$ is related to filler modes, but surprisingly, the anharmonicity of the filler atomic bonds plays no role. The analysis leading to this conclusion was made by comparing $YbFe_4Sb_{12}$ with a similar material, $BaFe_4Sb_{12}$, which has a 10 times higher thermal conductivity. The phonon spectrum of $YbFe_4Sb_{12}$ (Figure 12.3) was shown to be very similar to that of $BaFe_4Sb_{12}$, except for

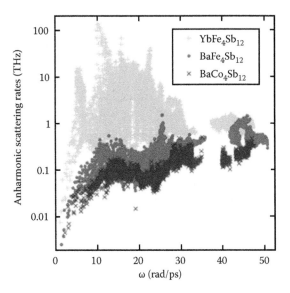

FIGURE 12.3 Anharmonic scattering rates of $YbFe_4Sb_{12}$, $BaFe_4Sb_{12}$, and $BaCo_4Sb_{12}$ at room temperature. (Reprinted with permission from Li, W., and Mingo, N., *Physical Review B*, 91, 035108, 2014. Copyright 2014 by the American Physical Society.)

the presence of a flat dispersion region (marked in gray in the figure) displaying avoided branch crossings, related to vibrations of the filler (Yb) atom.

However, the three-phonon scattering rates of $YbFe_4Sb_{12}$ are much larger than those of $BaFe_4Sb_{12}$. These increased scattering rates could be due either to larger anharmonicity or to having a larger number of allowed processes. The second possibility is a function of the harmonic properties of the crystal only. In order to find out, Ref.[32] computed the scattering rates of a "hybrid" material with the second-order IFCs of $YbFe_4Sb_{12}$, and the third-order IFCs of $BaFe_4Sb_{12}$, which yielded a thermal conductivity very similar to that of the genuine $YbFe_4Sb_{12}$. This showed that increased anharmonicity was not the reason for the smaller κ. A direct calculation of the three-phonon scattering phase space and the anharmonic scattering rates (Figure 12.3) confirmed this conclusion. Furthermore, repeating the calculation while artificially removing the flat branches resulted in a smaller number of allowed processes and smaller scattering rates. A final assessment consisted of completely removing all anharmonicities involving the Yb filler atom, which did not appreciably change the scattering rates. The calculated κ values for $YbFe_4Sb_{12}$ are several times lower than that of the Ba-based compounds lying below 1 W/m K around 300 K.[32] Thus, ab initio calculations were able to determine that filler anharmonicity plays little role in the low thermal conductivity of $YbFe_4Sb_{12}$, whereas the flat modes are at the root of the phenomenon, due to enhanced number of available three-phonon scattering processes. This computational analysis was done by evaluating numerical quantities and building hypothetical mix-and-match materials that are not accessible to experiment. In doing so, a clear picture of the physical phonon transport mechanisms emerged, which would have not been possible to obtain by experiment and qualitative modeling alone.

12.1.4.5.2 Bi₂Te₃

To illustrate the TDEP approach, we have considered Bi_2Te_3 a prototypical TE material.[34] Figure 12.4 shows the calculated phonon dispersions of Bi_2Te_3 along high-symmetry directions (red curves), obtained using the TDEP approach with IFCs determined at 77 K, compared to measured data taken at that same temperature.[81] For both acoustic phonon and optic phonon branches, relatively good agreement is achieved between the calculated results and measurement throughout the Brillouin zone. We note that at 77 K, the IFCs are not much different from those at $T = 0$ K, but at a higher temperature, phonon modes are softened and AP velocities are reduced.

In the TDEP approach, the linear and volume coefficients of thermal expansion (CTEs) can be evaluated in a straightforward way. The Helmholtz free energy minimization procedure described earlier automatically gives temperature-dependent lattice parameters from which the linear CTEs can be immediately determined. Volume CTEs as a function of T can be calculated as $\alpha = (1/V)(\partial V/\partial T) \equiv (1/\Omega)(\partial\Omega/\partial T)$, with Ω being the unit cell volume. For Bi_2Te_3, the calculated α using TDEP is 4.9×10^{-5} K^{-1} at 300 K, closer to the measured value[82] of 5.2×10^{-5} K^{-1} than the value of 4.4×10^{-5} K^{-1} obtained using the QHA.

Figure 12.5 shows the in-plane and cross-plane Bi_2Te_3 lattice thermal conductivities κ_{xx} and κ_{zz}, respectively, as a function of temperature calculated using TDEP. The fields around the curve show the range of thermal conductivity values for ±1% change in volume about the one minimizing the free energy. The calculated κ_{xx} shows good agreement with the measured data from Goldsmid[83] over a wide temperature range. In particular, at room temperature, $\kappa_{xx} = 1.3$ W/m K, which falls nicely between the measured values.

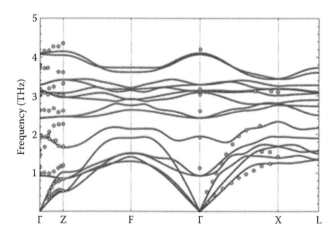

FIGURE 12.4 Bi_2Te_3 phonon dispersions along high-symmetry directions at 77 K. Curves calculated using the TDEP approach[34]; points are measured data from Kullmann et al.[81] (Reprinted with permission from Hellman, O., and Broido, D. A., *Physical Review B*, 90, 134309, 2014. Copyright 2014 by the American Physical Society.)

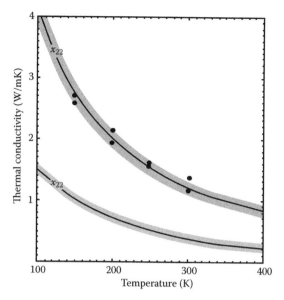

FIGURE 12.5 The in-plane (κ_{xx}) and cross-plane (κ_{zz}) lattice thermal conductivities of Bi_2Te_3 calculated using the TDEP approach.[34] In-plane experimental values from two samples (black circles) are from Goldsmid.[83] (Reprinted with permission from Hellman, O., and Broido, D. A., *Physical Review B*, 90, 134309, 2014. Copyright 2014 by the American Physical Society.)

In the higher-temperature region, κ_{xx} varies as $1/T^{1.3}$. For pure three-phonon scattering with $T = 0$ K IFCs, the thermal conductivity should inversely depend on T. The larger exponent indicates that higher-order anharmonicity is playing a role. We note that calculations done using $T \sim 0$ K IFCs[84] give a somewhat higher room temperature $\kappa_{xx} = 1.7$ W/m K, again reflecting the stronger anharmonicity with increasing T captured by the full TDEP calculations.

Phonon mean free path (MFP) spectroscopy gives a new tool to access the MFPs of phonons contributing to the lattice thermal conductivity.[85–87] The calculated MFP distribution for Bi_2Te_3 shows that about 50% of phonons contributing to κ_{xx} at 300 K have MFPs smaller than 10 nm.[34,88] This is two orders of magnitude smaller than the corresponding point for Si and is reflected in the much smaller Bi_2Te_3 thermal conductivity. Also, low-lying optic modes are found to contribute about 30% to the total κ_{xx} at 300 K.

The Debye temperature of Bi_2Te_3 is only 150 K suggesting strong T-dependent anharmonicity around and above room temperature, which should be reflected in inelastic neutron scattering measurements. Such measurements probe the anharmonic phonon self-energy: $\Sigma_\lambda = \Delta_\lambda + i\Gamma_\lambda$. The neutron scattering cross section is[89]

$$\sigma_\lambda(\Omega) \propto \frac{2\omega_\lambda\Gamma_\lambda(\Omega)}{\left(\Omega^2 - \omega_\lambda^2 - 2\omega_\lambda\Delta_\lambda\right)^2 + 4\omega_\lambda^2\Gamma_\lambda^2(\Omega)}, \tag{12.13}$$

where the imaginary part of the self-energy, which describes the phonon lifetime broadening, is

$$\Gamma_{qs}(\Omega) = \frac{\hbar\pi}{16} \frac{V_0}{(2\pi)^3} \sum_{s's''} \int_{BZ} d\mathbf{q}' \left[\left| \Phi_{ss's''}^{+}(\mathbf{q},\mathbf{q}',\mathbf{q}'') \right|^2 2(n_{\mathbf{q}'s'} - n_{\mathbf{q}''s''}) \right.$$

$$\left. \delta(\Omega + \omega_{\mathbf{q}'s'} - \omega_{\mathbf{q}''s''}), + \left| \Phi_{ss's''}^{-}(\mathbf{q},\mathbf{q}',\mathbf{q}'') \right|^2 (n_{\mathbf{q}'s'} + n_{\mathbf{q}''s''} + 1)\delta(\Omega - \omega_{\mathbf{q}'s'} - \omega_{\mathbf{q}''s''}) \right],$$

$$(12.14)$$

and $\mathbf{q}'' = \mathbf{q} \pm \mathbf{q}' + \mathbf{G}$ is satisfied by the first (+) and second (−) δ-function terms, respectively. The real part of the self-energy, which gives the phonon frequency shifts, is obtained from the Kramers–Kronig transformation:

$$\Delta_{qs}(\Omega) = \frac{1}{\pi} \int \frac{\Gamma_{qs}(\omega)}{\omega - \Omega}. \tag{12.15}$$

For weak anharmonic systems, $\omega_{qs} \gg \Delta_{qs}$, Γ_{qs} and the cross section give well-defined Lorentzians with line widths Γ_{qs} and with peaks centered at $\Omega = \omega_{qs} + \Delta_{qs}$. Such behavior occurs for Bi_2Te_3 at low temperature and is shown in Figure 12.6a for $T = 77$ K. For higher T (Figure 12.6b and c), anharmonicity becomes much stronger, and the phonon spectrum can no longer be described by peaks at single frequencies and single line widths. Instead, $\sigma_{qs}(\Omega)$ shows noticeable frequency shifts and significant broadening with some modes becoming completely diffused.

12.1.5 Concluding Remarks

The field of ab initio thermal transport has matured over the past decade. For many materials, the κ calculated has been shown to be in excellent agreement with measured data,[9–14,16–24,34,40] demonstrating that it is now possible to predict the lattice thermal conductivity, with remarkable accuracy, in the absence of any experimental input. Techniques reviewed here have now been integrated with established ab initio packages[26,51,52] and are now widely used, and new approaches such as the TDEP approach are being developed. This combined with advances in computing power have allowed complex materials to be investigated. Both examples considered earlier present serious computational challenges that would have been impossible to overcome only several years ago.

Recent progress has also been accomplished in including electron–phonon interactions so that that thermal transport in doped semiconductors and metals can now also be studied from first-principles.[40,41,90–92] Computational challenges remain such as incorporating the calculation data for single crystals to nanostructured samples with phonon scattering on multiple length scales.[93–95] It is anticipated that these will be overcome with advances in computing power and algorithm development in the coming years.

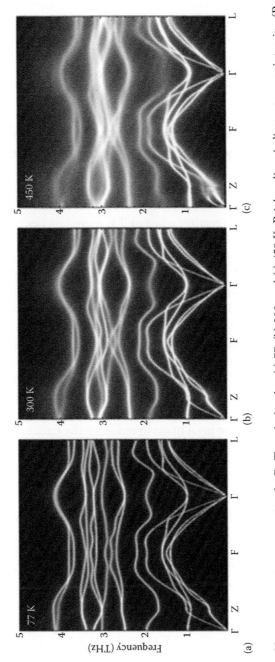

FIGURE 12.6 Neutron scattering cross ection for Bi₂Te₃ calculated at (a) 77, (b) 300, and (c) 450 K. Brighter lines indicate stronger intensity. (Reprinted with permission from Hellman, O., and Broido, D. A., *Physical Review B*, 90, 134309, 2014. Copyright 2014 by the American Physical Society.)

12.1.6 Acknowledgments

We thank Olle Hellman, Lucas Lindsay, Wu Li, Jesús Carrete, and Nebil A. Katcho for their contributions to several of the works cited or summarized in this chapter. DAB acknowledges support from the Office of Basic Energy Sciences under award no. DE-FG02-09ER46577. NM acknowledges support from the European Union Horizon 2020 program (Project ALMA, grant no. 645776), L'Agence Nationale de la Researche (ANR) through Project SIEVE, and the Air Force Office of Scientific Research, US Air Force, under award no. FA9550-15-1-0187 DEF.

12.2 Multiscale Thermal Conductivity Modeling in Complex Thermoelectric Crystals Using the Boltzmann Transport Equation

Austin J. Minnich

12.2.1 Introduction

A key challenge of creating efficient TE materials lies in minimizing parasitic heat conduction by phonons while permitting the flow of electrons. The task of engineering phonon transport is particularly difficult because of the lack of simulation tools to describe heat conduction in the complex microstructure of realistic TE crystals. Nearly all phonon modeling approaches to analyze thermal conductivity measurements are based on works introduced in the 1950s in which simple phonon dispersions and semiempirical expressions for relaxation times were inserted into the kinetic equation for thermal conductivity, and the fitting parameters determined by fitting to experimental data.[96–99] This remains the primary approach by which thermal conductivity measurements are interpreted and from which physical insight is extracted from experimental data. Even with these limited modeling tools, significant gains have been made in reducing the lattice contribution to thermal conductivity in a variety of nanostructured TE material systems.[100–104] These nanostructured materials exploit the much longer MFPs of phonons compared to electrons to scatter phonons while minimally affecting electrical conduction.

The advances achieved in the past decade are impressive, especially when considering the lack of predictive thermal conductivity models available to the TE community. Further improvements require a rational design of material microstructure based on predictive calculations of thermal conductivity. Fortunately, these calculations are now within reach due to advances in thermal conductivity modeling across a wide range of length scales, from the atomistic scale to the mesoscale. Historically, a key computational challenge in multiscale thermal conductivity modeling was describing heat conduction through the complex microstructure of a TE crystal using the Boltzmann transport equation (BTE). In this chapter, we provide an overview of a new stochastic algorithm called variance-reduced MC (VRMC) that overcomes this challenge by solving BTE in a complex geometry many orders of magnitude faster than prior numerical schemes.[105–107] With input from first-principles calculations,[108] this algorithm enables the ab initio calculation of thermal conductivity of a TE crystal with a realistic microstructure without any adjustable parameters.

12.2.2 Background

The general approach for thermal conductivity modeling was introduced in the 1950s by Callaway,[96] Holland,[97] Steigmeier and Abeles[98] and Klemens.[99] In the single-mode RTA, the thermal conductivity tensor $\kappa_{\alpha,\beta}$ of a crystal can be expressed using kinetic theory as[109]

$$\kappa_{\alpha,\beta} = \sum_{\lambda} C_{\lambda} v_{\alpha,\lambda} v_{\beta,\lambda} \tau_{\lambda}, \qquad (12.16)$$

where C_{λ} is the spectral volumetric specific heat, $v_{\alpha,\lambda}$ is the component of the group velocity along crystal axis α, τ_{λ} is the relaxation time, and λ indicates a phonon mode of wave vector **k** and polarization j.

The RTA does not account for normal processes and thus underestimates the thermal conductivity. Callaway's[96] original model contained a term that was intended to account for normal processes, and recent work by Allen[110] has proposed a new form of this term. However, first-principles calculations show that neither of these approximate terms accurately models normal processes.[111] Considering that the RTA is accurate to within 5–10% in many crystals,[112] neglecting normal processes and using Equation 12.16 is usually sufficient.

In principle, calculating thermal conductivity using Equation 12.16 is simple if the phonon dispersion ω_{λ} and relaxation times τ_{λ} are known. The primary challenge with thermal conductivity modeling is that knowledge of the dispersion and lifetimes for all phonon modes in the Brillouin zone is not readily available. In the time period of the early works on thermal conductivity modeling, the dispersion relation of crystals was known only along high-symmetry directions based on neutron scattering measurements. A priori knowledge of phonon lifetimes remained unavailable by either experiment or calculation.

Faced with these challenges, the only tractable path to close the problem was to make approximations. In Callaway's[96] model, the complicated phonon dispersion was replaced with an isotropic Debye model or a constant slope line from the zone center to zone edge, with the slope given by an average sound velocity. Holland[97] later used a bilinear model in an attempt to account for the dispersion of phonons at the zone edge. For relaxation times, general expressions were derived by Pomeranchuk,[113] Klemens,[99] and others for scattering mechanisms such as U scattering, PD scattering, and others. These expressions contained fitting parameters that could be adjusted to match experimental data. These models could then be used to obtain some microscopic properties of the phonons responsible for carrying heat such as their MFPs.

This approach has been used to study thermal phonons for decades. However, it has a number of severe limitations that obscure the fundamental picture of heat conduction that is necessary to engineer the thermal conductivity of crystals. First, the drastic simplification of phonon dispersion to the Debye model is very inaccurate at temperatures around the Debye temperature, where the full spectrum of phonons is occupied, particularly for phonons near the Brillouin zone edge with small group velocities.[114,115] Second, the relaxation time models are derived using approximations that are not valid for a large portion of the phonon spectrum. For example, the long wavelength approximation used for U scattering results in errors of several hundreds percent for phonons above a certain frequency.[116] Finally, the approach can only account for volumetric scattering processes through the scattering time τ_λ; the actual spatial distribution of defects is not incorporated, and thus, this defect scattering rate requires fitting parameters to be determined from the experimental data.

Despite these drawbacks, this basic approach has been used for decades due to the absence of alternatives. In recent years, however, advances in computational methods for thermal conductivity modeling have enabled transformational changes in thermal conductivity modeling. One of the most important of these advances is the introduction of ab initio calculations of phonon dispersions and line widths over the Brillouin zone so that the thermal conductivity of a pure crystal can be calculated without any fitting parameters.[108] This approach has now been applied to dozens of crystals with considerable success, but it is extremely computationally expensive to apply to realistic TE materials with complicated microstructures. Other atomistic simulation methods such as molecular dynamics similarly require exceptionally large computational effort and are impractical to apply to materials with complex microstructures.

The most tractable approach for multiscale thermal conductivity modeling in a mesoscopic domain involves solving the BTE in the actual microstructure of the crystal with the phonon dispersion, intrinsic lifetimes, and defect scattering rates specified. Although less intensive than atomistic calculations, solving the BTE has historically been proven to be prohibitively challenging, especially in a complex 3D geometry that includes the full spectral properties of phonons. However, in the past 5 years, this once formidable calculation has been rendered straightforward by the introduction of VRMC algorithms by Péraud and Hadjiconstantinou,[105,106] Radtke and Hadjiconstantinou,[117] and Hadjiconstantinou et al.[118] These algorithms reduce the computational cost compared to traditional Monte Carlo (MC) by factors on the order of 10^6 and now allow the routine simulation in complex domains with first-principles input rigorously included. As a result, parameter-free, multiscale modeling of thermal phonon transport in the actual microstructure of a material is now within reach.

The required first-principles inputs for predictive thermal conductivity modeling are described in other chapters. Here, we describe the mesoscale modeling approach based on VRMC algorithms.

12.2.3 Variance-Reduced Monte Carlo Algorithms

The BTE is the fundamental governing equation of thermal transport at length scales larger than phonon wavelengths. The equation is given by[109]:

$$\frac{\partial \Psi_\lambda}{\partial t} + \mathbf{v}_\omega \cdot \nabla \Psi_\lambda = -\frac{\Psi_\lambda - \Psi_\lambda^0(T_p(\mathbf{r},t))}{\tau_\omega}, \tag{12.17}$$

where Ψ_λ is the desired distribution function, $\mathbf{v}_\omega$ is the group velocity, τ_ω is the relaxation time, and λ denotes a phonon mode of wavevector $\mathbf{k}$ and polarization j. The equilibrium distribution Ψ_0 is related to Ψ_λ by an integral relation that is used to determine the temperature $T_p(\mathbf{r}, t)$.

Solving this equation is challenging due to the high dimensionality of $\Psi\lambda$ and the integrodifferential nature of the equation. For complex geometries such as those in a TE crystal, numerical solutions are required. The most straightforward strategy is simply to apply finite differences to the derivatives, discretize the integrals, and numerically solve the equation, an approach known as discrete ordinates.[119] This approach for phonons was originally reported by Majumdar[120] and Joshi and Majumdar[121] in a 1D geometry. However, discrete ordinates are challenging to apply to multiple spatial dimensions while including spectral phonon properties due to the substantial memory requirements. Other approaches taken in the thermal transport field include a two-flux method,[122–124] MC,[125–131] finite volume,[132–134] a MFP sampling algorithm,[135] and a lattice Boltzmann solver.[136] However, these algorithms either are computationally costly or make simplifying approximations that limit the validity of the solution.

Recently, VRMC algorithms have been introduced that solve the BTE many orders of magnitude faster than other algorithms in complex 3D geometries while rigorously including the spectral properties of phonons.[105,106,117,118] These deviational variance-reduced algorithms were originally introduced to simulate rarified gas dynamics[137,138] and have been adapted for phonons by Radtke and Hadjiconstantinou[117] and Péraud and Hadjiconstantinou.[105,106] These algorithms enable the BTE to be solved, without approximation, many orders of magnitude faster than prior numerical approaches.

To understand the principle underlying VRMC, we first give an example of an MC calculation using the traditional algorithm. Consider a calculation of the spatial heat flux in a domain with a fixed, localized temperature rise at 301 K at the center while the rest of the domain, which extends to infinity in all directions, is at 300 K. With traditional MC, this problem would be treated by discretizing the domain into cells and adding computational particles representing phonons into each cell with a Bose–Einstein distribution at the local temperature. Particles are present in every cell, and the desired heat flux is calculated by advecting and scattering the particles while sampling the heat flux in each cell after the system reaches steady state.

While this algorithm is the typical one used in many studies, it is clearly inefficient in several respects. First, the difference between Bose–Einstein distributions at 301 and 300 K is extremely small, meaning tremendous numbers of particles are required to adequately resolve the difference in the presence of stochastic noise due to number fluctuations of the particles. Second, in a sufficiently large computational domain that is not subject to artificial boundary effects, most of the domain is at 300 K, meaning that the majority of computational particles are not used to simulate the thermal transport process of interest but are rather needlessly reproducing a 300 K Bose–Einstein distribution that is analytically known. This unnecessary computation dramatically increases the memory requirements, computational effort, and stochastic noise of the simulation.

Deviational MC algorithms are based on this key observation: there is no need to stochastically simulate a quantity, such as the Bose–Einstein distribution, that is analytically known. In deviational MC, this deterministic information is incorporated into the simulation by letting computational particles represent deviations from the given equilibrium distribution, in this example, a Bose–Einstein distribution at 300 K, rather than actual phonons as in traditional MC. Because both positive and negative deviations from the equilibrium distribution can occur, deviational particles also possess a sign, as schematically illustrated in Figure 12.7a. Besides this change, however, the simulation largely proceeds as in traditional MC: particles are advected, scattered, and sampled to obtain the desired thermal quantity such as temperature and heat flux.

This small change in the algorithm drastically improves the computational efficiency in several critical ways. First, while in traditional MC, particles are required in every cell, in deviational MC, no particles are present if the cell is at the equilibrium temperature. Thus, the only particles present in a deviational MC simulation are used to simulate the thermal transport process of interest. Further, deviational MC can resolve arbitrarily small temperature gradients because particles represent deviations from an equilibrium distribution rather than an absolute quantity. This ability is in marked contrast to traditional MC for which computational cost rapidly increases with decreasing temperature gradients.

Note that the deviational algorithm is always applicable regardless of the magnitude of the temperature gradient. However, it is more efficient compared to traditional MC when the maximum temperature variation over the domain ΔT is much smaller than the equilibrium temperature T,[118] with the efficiency enhancement improving as $(T/\Delta T)^2$. For the example given earlier in which $\Delta T = 1$ K and $T = 300$ K, the efficiency improvement is on the order of 10^5 compared to traditional MC algorithms, and thus, a simulation that would take more than a day with traditional MC can be run in 1 s with deviational MC.

In the case when $\Delta T/T \ll 1$, which often occurs in practice, an additional simplification can be made that further reduces the computational cost.[105] In this situation, the distribution to which scattering phonons relax Ψ_0 is nearly

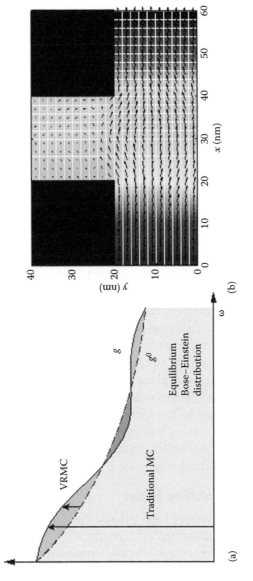

FIGURE 12.7 (a) Schematic of the concept underlying deviational VRMC methods. VRMC achieves reduced variance by only simulating the deviation from the known Bose–Einstein distribution at a given temperature. Green and orange regions represent positive and negative deviations from the equilibrium distribution, respectively. (b) Calculation of the temperature field (*color*) and heat flux field (*arrows*) in the T geometry subjected to hot and cold boundaries on the left and right sides, respectively. VRMC achieves exceptionally low stochastic noise with minimal computational cost.

the same everywhere in the domain due to the small temperature differential, allowing Ψ_0 to be linearized. This linearization decouples the computational particles, allowing them to be completely independently simulated from each other and without any need for temporal and spatial discretizations. This decoupling simplifies parallelization and reduces computational cost and memory requirements by orders of magnitude. Unlike the original deviational algorithm, though, this linearized algorithm is only applicable in the limit $\Delta T/T \ll 1$ due to the linearization of the Bose–Einstein distribution. However, this limit is often satisfied in practice. An example of the exceptionally low stochastic noise results that can be obtained using this algorithm is shown in Figure 12.7b.

The linearized algorithm enables another substantial benefit. A challenge of solving for the heat flux over the broad spectrum of phonon frequencies is that certain phonons may substantially contribute to heat conduction due to their long relaxation times but are unlikely to be sampled because their DOD is small. In the traditional algorithms, the frequencies of the phonons are drawn from a heat flux distribution that is proportional to the DOS. As a result, even if modes with small DOS substantially contribute to heat conduction, as occurs in silicon, these phonons are unlikely to be sampled, leading to large noise for certain phonon frequencies.

The linearized algorithm allows this challenge to be overcome because all the particles can be independently simulated, allowing an approach similar to importance sampling to be implemented. In this approach, phonons are uniformly emitted over the frequency spectrum and subsequently weighted by the original heat flux distribution to provide a variance-reduced estimate of the heat flux. Péraud et al. recently showed that this approach formally corresponds to solving the adjoint BTE.[107] The underlying idea is that when the estimator, for this example, the heat flux of a particular phonon mode, corresponds to a small region of phase space, one can exploit the time reversibility of the BTE to switch the source and estimator: phonons are emitted from the estimator and follow their trajectories backward in time to the source, which may occupy a much larger region of phase space than the estimator. In this way, the probability that a given sample contributes to the desired estimator is considerably increased, reducing the overall noise.

12.2.4 Application to Nanocrystalline Silicon

We now provide an example of how VRMC can be used to provide insight into phonon heat conduction in TE materials.[139] Specifically, we use the algorithm to examine the lattice thermal conductivity of nanocrystalline silicon.[140] Silicon is of interest as a TE material because it is earth abundant, inexpensive, and not toxic. However, the TE performance of Si is considerably lower than that of good TEs, with a peak ZT of around only 0.2. This poor performance is largely due to the high thermal conductivity of Si. A recent experimental work demonstrated much higher ZTs in nanocrystalline Si due to reduced thermal conductivity, but the value of $ZT = 0.7$ is still not competitive with top TE materials.

Therefore, a better understanding of the phonon scattering mechanisms in nanocrystalline silicon, particularly those due to the grain boundaries, is essential to further increase ZT. In fact, a rigorous understanding of grain boundary scattering is lacking as showed by Wang et al.,[140] who reported temperature (T)-dependent measurements of the thermal conductivity of nanocrystalline silicon. The most common model for grain boundary scattering is to simply take the scattering rate to be a constant, implying that the transmission coefficient across the interface is the same for all phonons. With this model, known as the gray model, one should observe a T^3 dependence of thermal conductivity at sufficiently low temperatures, following the trend of the specific heat. However, they instead observed a T^2 dependence, implying that the assumption of constant transmission coefficient is incorrect. A straightforward analysis of the trend instead implies that the transmissivity must decrease with increasing phonon frequency, suggesting that lower frequency modes are less scattered by interfaces compared to higher frequency modes. This nongray boundary scattering would have an important effect on the distribution of heat among the phonon spectrum and hence on strategies to further reduce the thermal conductivity of nanocrystalline Si and SiGe. However, rigorous simulations of phonon transport in nanocrystalline TEs with nongray grain boundary scattering have not been possible due to the computational challenges described in Section 12.2.

In this example, we use VRMC to examine the effects of frequency-dependent grain boundary scattering on the thermal conductivity of nanocrystalline Si. The computational domain is a 3D cube with square grains as illustrated in Figure 12.8a.[140] The domain bisects the grain in each dimension, placing the corner of the grains in the center of the domain. We impose a periodic heat flux boundary conditions along the direction of temperature gradient to model an infinite repeating structure.[130] The other two directions have a specular boundary condition imposed by symmetry. These boundary conditions allow the thermal properties of a polycrystal of infinite extent to be calculated using only one period.

We impose a linearly varying equilibrium temperature $T_{eq}(x)$ that allows the control temperature to follow the physical temperature more closely.[105] This variation in T_{eq} can be implemented as a uniform volumetric source of deviational phonon bundles, each representing a finite amount of deviational energy. The frequencies of the phonon bundles are drawn from a uniform distribution, rather than the heat flux distribution, according to the adjoint BTE. The algorithm then proceeds by stochastically simulating the advection and scattering and sampling of phonon bundles sequentially and completely independently exactly as described by Péraud et al.[105] Finally, the calculated spectral heat flux is weighted by an appropriate distribution to obtain an unbiased calculation of the heat flux.[107]

The critical parameter in our model is the phonon transmissivity across the grain boundaries. Many previous works assumed a gray model, but this model is not consistent with the measurements of Wang et al.[140] Atomistic methods such as atomistic Green's functions could be used to determine the

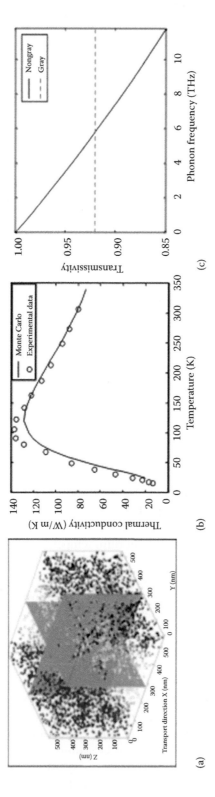

FIGURE 12.8 (a) Schematic of the computational domain. The gray planes indicate grain boundaries at which phonons have a probability to be transmitted or reflected. A periodic heat flux boundary condition is applied along the *x* direction, and periodic boundary conditions are applied in the *y* and *z* directions. The particles represent phonons with color indicating the local temperature. (b) Thermal conductivity data from Wang et al.[140] (*symbols*) and the fit from VRMC (*line*). The model successfully recovers the T^2 dependence of thermal conductivity at low temperatures. (c) The nongray transmissivity required to fit the experimental data (*blue line*) and the original gray transmissivity (*dashed line*). (Hua, C. and Minnich, A. J., *Semiconductor Science and Technology*, 29, 124004, 2014, Institute of Physics.)

transmissivity, but for simplicity, here we use a nongray transmissivity $t(\omega)$ proposed by Wang et al.[140] in the form of

$$t(\omega) = \frac{1}{\gamma\omega/\omega_{max} + 1}, \qquad (12.18)$$

where γ is a fitting parameter and ω_{max} is the maximum phonon frequency. This model is qualitatively consistent with atomistic Green's function calculations.[141] Note that this condition implies that the transmissivity approaches one as frequency goes to zero, which is physically consistent with the expectation that long wavelength phonons are unaffected by atomistic disorder at a grain boundary. In this work, we denote gray and nongray as constant and phonon frequency-dependent transmissivity, respectively. Note that other parameters in the BTE always depend on phonon frequency regardless of the transmissivity model.

With the transmissivity determined, we proceed to the nanocrystalline material by determining the fitting parameter γ in Equation 12.18. The reported material had 550 nm grains and was nominally undoped. By fitting the experimental thermal conductivity with our model, as in Figure 12.8b, we obtain $\gamma = 0.12$, yielding the transmissivity profile shown in Figure 12.8c. The simulated thermal conductivities of nanocrystalline silicon obtained using the nongray transmissivity model are in good agreement with the experiment over the full temperature range.

With the transmissivity determined, we now examine some predictions of the model. The spectral thermal conductivity versus phonon frequency for several different grain boundary scattering models is given in Figure 12.9a. Compared to bulk silicon, the contribution of low-frequency, long-MFP phonons is substantially reduced in both the gray and nongray models due to boundary scattering. However, the gray model overpredicts the reduction in heat flux for phonons with frequencies less than approximately 1 THz. In the nongray model, these low-frequency phonons are still able to contribute to heat conduction due to their high probability for transmitting across the grain boundary. This effect is even more pronounced when other scattering mechanisms, particularly by point defects, are accounted for in Figure 12.9b. Point defects annihilate the contribution of high-frequency phonons to such an extent that a large fraction of the heat is carried by phonons with frequencies less than 4 THz despite the presence of grain boundaries that are typically assumed to scatter these modes.

To determine the key length scales for heat conduction, we calculate the phonon MFPs. We plot this quantity as the thermal conductivity accumulation function, which has been shown to be a useful quantity for understanding thermal transport.[142] We calculate the accumulation function by determining an effective MFP for each mode that incorporates all the scattering mechanisms, including grain boundary scattering, using the spectral

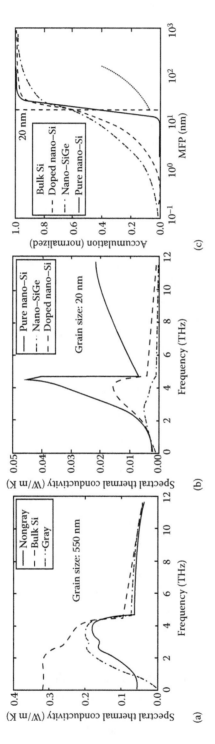

FIGURE 12.9 (a) Room-temperature spectral thermal conductivity versus phonon frequency with 550 nm grain: bulk silicon (*dashed line*), nanocrystalline silicon with gray model (*dash-dotted line*), and nongray model (*solid line*). Low-frequency phonons carry more heat in the nongray model. (b) Room-temperature spectral thermal conductivity versus phonon frequency with 20 nm grain: pure nanocrystalline silicon (*solid line*), doped nanocrystalline silicon (*dashed line*), and nanocrystalline silicon germanium (*dash-dotted line*). High-frequency phonons are scattered by mass defects while low-frequency phonons are scattered by electrons. (c) Thermal conductivity accumulation versus phonon MFP: pure nanocrystalline silicon (*solid line*), doped nanocrystalline silicon (*dashed line*), doped nanocrystalline silicon germanium (*dash-dotted line*), and bulk silicon (*dotted line*). Heat is still carried by long MFP phonons in nanocrystalline Si/SiGe even though many low frequency phonons are scattered by electrons.

thermal conductivity for each phonon frequency and polarization and the kinetic equation: $\Lambda_{eff} = 3k(\omega, j)/C(\omega, j)v(\omega, j)$. The spectral thermal conductivity can then be sorted by MFP, from which the thermal conductivity accumulation is obtained from the cumulative sum of the spectral thermal conductivity.

Our simulations show a surprising result. As shown in Figure 12.9c, at room temperature, about 60% of the total heat in undoped nanocrystalline silicon with a grain size around 20 nm is carried by the phonons with MFPs longer than the grain size, which are denoted as long MFP phonons here. Even in doped nanocrystalline Si and SiGe, for which low-frequency phonons are scattered by electrons, as much as 35% of the total heat is carried by these long MFP phonons.

These observations indicate that nanocrystalline grain boundaries may not be as effective as previously believed at scattering long MFP phonons. If a grain boundary can be designed such that it diffusely scatters all phonons independent of frequency as in the gray model, additional reductions in the phonon thermal conductivity on the order of 30% can be achieved.[139] Therefore, great potential to further increase the TE performance of nanocrystalline Si/SiGe exists if such grain boundaries can be designed. This example shows the important insights into improving TE performance that can be obtained using rigorous thermal conductivity modeling based on the BTE.

12.2.5 Summary

In this section, we have described multiscale thermal conductivity modeling based on VRMC algorithms. With first-principles input, these algorithms enable predictive thermal conductivity modeling of realistic TE crystals with complex microstructures. We presented an example of how we used these calculations to study the lattice thermal conductivity of nanocrystalline silicon. The TE community thus has an exciting time ahead as it begins to use these predictive tools to rationally design microstructures for minimum lattice thermal conductivity and high TE performance.

Acknowledgments

This work was sponsored in part by the NSF under Grant No. CBET CAREER 1254213 and by Boeing under the Boeing–Caltech Strategic Research & Development Relationship Agreement.

References

1. J. M. Ziman, *Electrons and Phonons* (Oxford University Press, London, 1960).
2. D. G. Cahill, W. K. Ford, K. E. Goodson, G. D. Mahan, A. Majumdar, H. J. Maris, R. Merlin, and S. R. Phillpot, Nanoscale thermal transport, *Journal of Applied Physics* **93**, 793 (2003).

3. D. G. Cahill, P. V. Braun, G. Chen, D. R. Clarke, S. Fan, K. E. Goodson, P. Keblinski et al. Nanoscale thermal transport, *Applied Physics Reviews* **1**, 011305 (2014).

4. B. Poudel, Q. Hao, Y. Ma, Y. Lan, A. Minnich, B. Yu, X. Yan et al., High-thermoelectric performance of nanostructured bismuth antimony telluride bulk alloys, *Science* **320**, 634–638 (2008).

5. A. I. Hochbaum, R. Chen, R. D. Delgado, W. Liang, E. C. Garnett, M. Najarian, A. Majumdar, and P. Yang, Enhanced thermoelectric performance of rough silicon nanowires, *Nature* **451**, 163–167 (2008).

6. I. Boukai, Y. Bunimovich, J. Tahir-Kheli, J.-K. Yu, W. A. Goddard III, and J. R. Heath, Silicon nanowires as efficient thermoelectric materials, *Nature* **451**, 168–171 (2008).

7. J.-K. Yu, S. Mitrovic, D. Tham, J. Varghese, and J. R. Heath, Reduction of thermal conductivity in phononic nanomesh structures, *Nature Nanotechnology* **5**, 718–721 (2010).

8. P. Ball, Feeling the heat, *Nature (London)* **492**, 174 (2012).

9. D. A. Broido, M. Malorny, G. Birner, N. Mingo, and D. A. Stewart, Intrinsic lattice thermal conductivity of semiconductors from first principles, *Applied Physics Letters* **91**, 231922 (2007).

10. D. A. Ward, D. A. Broido, Stewart, and G. Deinzer, Ab initio theory of the lattice thermal conductivity in diamond, *Physical Review B* **80**, 125203 (2009).

11. K. Esfarjani, G. Chen, and H. T. Stokes, Heat transport in silicon from first-principles calculations, *Physical Review B* **84**, 085204 (2011).

12. L. Lindsay, D. A. Broido, and T. L. Reinecke, High thermal conductivity and large isotope effect in GaN from first principles, *Physical Review Letters* **109**, 095901 (2012).

13. T. Luo, J. Garg, J. Shiomi, K. Esfarjani, and G. Chen, Gallium arsenide thermal conductivity and optical phonon relaxation times from first-principles calculation, *Europhysics Letters* **101**, 16001 (2013).

14. L. Lindsay, D. A. Broido, and T. L. Reinecke, Ab initio thermal transport in compound semiconductors, *Physical Review B* **87**, 165201 (2013).

15. L. Lindsay, D. A. Broido, and T. L. Reinecke, First-principles determination of ultrahigh thermal conductivity of boron arsenide: A competitor for diamond? *Physical Review Letters* **111**, 025901 (2013).

16. T. Shiga, J. Shiomi, J. Ma, O. Delaire, T. Radzynski, A. Lusakowski, K. Esfarjani, and G. Chen, Microscopic mechanism of low thermal conductivity in lead telluride, *Physical Review B* **85**, 155203 (2012).

17. Z. Tian, J. Garg, K. Esfarjani, T. Shiga, J. Shiomi, and G. Chen, Phonon conduction in PbSe, PbTe, and $PbTe_{1-x}Se_x$ from first-principles calculations, *Physical Review B* **85**, 184303 (2012).

18. J. Shiomi, K. Esfarjani, and G. Chen, Thermal conductivity of half-Heusler compounds from first-principles calculations, *Physical Review B* **84**, 085204 (2011).

19. S. Lee, K. Esfarjani, T. Luo, J. Zhou, Z. Tian, and G. Chen, Resonant bonding leads to low lattice thermal conductivity, *Nature Communications* **5**, 3525 (2014).

20. J. M. Skelton, S. C. Parker, A. Togo, I. Tanaka, and A. Walsh, Thermal physics of the lead chalcogenides PbS, PbSe, and PbTe from first principles, *Physical Review B* **89**, 205203 (2014).

21. A. H. Romero, E. K. U. Gross, M. J. Verstraete, and Olle Hellman, Thermal conductivity in PbTe from first principles, *Physical Review B* **91**, 214310 (2015).

22. X. Tang and J. Dong, Lattice thermal conductivity of MgO at conditions of Earth's interior, *Proceedings of the National Academy of Sciences* **107**, 4539–4543 (2010).

23. W. Li, L. Lindsay, D. A. Broido, D. A. Stewart, and N. Mingo, Thermal conductivity of bulk and nanowire $Mg_2Si_xSn_{1-x}$ alloys from first principles, *Physical Review B* **86**, 174307 (2012).

24. J. Garg, N. Bonini, B. Kozinsky, and N. Marzari, Role of disorder and anharmonicity in the thermal conductivity of silicon-germanium alloys: A first-principles Study, *Physical Review Letters* **106**, 045901 (2011).

25. A. Kundu, N. Mingo, D. A. Broido, and D. A. Stewart, Role of light and heavy embedded nanoparticles on the thermal conductivity of SiGe alloys, *Physical Review B* **84**, 125426 (2011).

26. W. Li, J. Carrete, N. A. Katcho, and N. Mingo, ShengBTE: A solver of the Boltzmann transport equation for phonons, *Computer Physics Communications* **185**, 1747 (2014).

27. L. Lindsay, D. A. Broido, and T. L. Reinecke, Phonon-isotope scattering and thermal conductivity in materials with a large isotope effect: A first-principles study, *Physical Review B* **88**, 144306 (2013).

28. D. A. Broido, L. Lindsay, and A. Ward, Thermal conductivity of diamond under extreme pressure: A first-principles study, *Physical Review B* **86**, 115203 (2012).

29. J. Carrete, W. Li, and N. Mingo, Finding unprecedentedly low-thermal-conductivity half-Heusler semiconductors via high-throughput materials modeling, *Physical Review X* **4**, 011019 (2014).

30. S. Lee, K. Esfarjani, J. Mendoza, M. S. Dresselhaus, and G. Chen, Lattice thermal conductivity of Bi, Sb, and Bi-Sb alloy from first principles, *Physical Review B* **89**, 085206 (1996).

31. B. Liao, S. Lee, K. Esfarjani, and G. Chen, First-principles study of thermal transport in $FeSb_2$, *Physical Review B* **89**, 035108 (2014).

32. W. Li and N. Mingo, Ultralow lattice thermal conductivity of the fully filled skutterudite $YbFe_4Sb_{12}$ due to the flat avoided-crossing filler modes, *Physical Review B* **91**, 035108 (2014).

33. W. Li and N. Mingo, Thermal conductivity of fully filled skutterudites: Role of the filler, *Physical Review B* **89**, 184304 (2014).

34. O. Hellman and D. A. Broido, Phonon thermal transport in Bi_2Te_3 from first principles, *Physical Review B* **90**, 134309 (2014).

35. S. Lee, D. Broido, K. Esfarjani, and G, Chen, Hydrodynamic phonon transport in suspended graphene, *Nature Communications* **6**, 6290 (2015).

36. A. Cepellotti, G. Fugallo, L. Paulatto, M. Lazzeri, F. Mauri, and N. Marzari, Phonon hydrodynamics in two-dimensional materials, *Nature Communications* **6**, 6400 (2015).

37. L. Lindsay, Wu Li, J, Carrete, N. Mingo, D. A. Broido, and T. L. Reinecke, Phonon thermal transport in strained and unstrained graphene from first principles, *Physical Review B* **89**, 15426 (2014).

38. G. Fugallo, A. Cepellotti, L. Paulatto, M. Lazzeri, N. Marzari, and F. Mauri, Thermal conductivity of graphene and graphite: Collective excitations and mean free paths, *Nano Letters* **14**, 6109 (2014).

39. L. Chaput, Direct solution to the linearized phonon Boltzmann equation, *Physical Review Letters* **89**, 035108 (2014).

40. B. Liao, B. Qiu, J. Zhou, S. Huberman, K. Esfarjani, and G. Chen, Significant reduction of lattice thermal conductivity by the electron-phonon interaction in silicon with high carrier concentrations: A first-principles study, *Physical Review Letters* **114**, 115901 (2015).

41. A. Jain and A. J. H. McGaughey, Thermal transport by phonons and electrons in aluminum, silver, and gold from first principles, *Physical Review B* **93**, 081206(R) (2016).

42. R. Peierls, Zur kinetischen Theorie der Wärmeleitung in Kristallen, *Annalen der Physik* **393**, 1055 (1929).

43. M. Omini and A. Sparavigna, Thermal conductivity of dielectric solids with diamond structure: An iterative approach to the phonon Boltzmann equation in the theory of thermal conductivity, *Physica B* **212**, 101 (1995).

44. M. Omini and A. Sparavigna, Beyond the isotropic-model approximation in the theory of thermal conductivity, *Physical Review B* **53**, 9064 (1996).

45. A. A. A. Maznev and O. B. Wright, Demystifying Umklapp vs normal scattering in lattice thermal conductivity, *American Journal of Physics* **82**, 1062 (2014).

46. D. D. G. Onn, A. Witek, Y. Z. Qiu, T. R. Anthony, and W. F. Banholzer, Some aspects of the thermal conductivity of isotopically enriched diamond single crystal, *Physical Review Letters* **68**, 2806 (1992).

47. L. Wei, P. K. Kuo, R. L. Thomas, T. R. Anthony, and W. F. Banholzer, Thermal conductivity of isotopically modified single crystal diamond, *Physical Review Letters* **70**, 3764 (1993).

48. J. R. Olson, R. O. Pohl, J. W. Vandersande, A. Zoltan, T. R. Anthony and W. F. Banholzer, Thermal conductivity of diamond between 170 and 1200 K and the isotope effect, *Physical Review B* **47**, 14850 (1993).

49. J. R. Yates, X. Wang, D. Vanderbilt, and I. Souza, Spectral and Fermi surface properties from Wannier interpolation, *Physical Review B* **75**, 195121 (2007).

50. P. E. Blöchl, O. Jepsen, and O. K. Andersen, Improved tetrahedron method for Brillouin-zone integrations, *Physical Review B* **49**, 16223 (1994).

51. A. Togo, L. Chaput, and I. Tanaka, Distributions of phonon lifetimes in Brillouin zones, *Physical Review B* **91**, 094306 (2015).

52. T. Tadano, Y. Gohda, and S. Tsuneyuki, Anharmonic force constants extracted from first-principles molecular dynamics: Applications to heat transfer simulations, *Journal of Physics: Condensed Matter* **26**, 225402 (2014).

53. L. Lindsay and D. A. Broido, Three-phonon phase space and lattice thermal conductivity in semiconductors, *Journal of Physics: Condensed Matter* **20**, 165209 (2008).

54. Quantum Espresso, http://www.quantum-espresso.org/

55. P. Giannozzi, S. Baroni, N. Bonini, M. Calandra, R. Car, C. Cavazzoni, D. Ceresoli et al., Quantum Espresso: A modular and open-source software project for quantum simulations of materials, *Journal of Physics: Condensed Matter* **21**, 395502 (2009).

56. S. Baroni, S. Gironcoli, A. D. Corso, and P. Giannozzi, Phonons and related crystal properties from density-functional perturbation theory, *Reviews of Modern Physics* **73**, 515–562 (2001).

57. G. Kresse and J. Furthmüller, Efficiency of ab-initio total energy calculations for metals and semiconductors using a plane-wave basis set, *Computational Material Science* **6**, 15 (1996).

58. G. Kresse and D. Joubert, From ultrasoft pseudopotentials to the projector augmented-wave method, *Physical Review B* **59**, 1758 (1999).

59. G. Kresse and J. Furthmüller, Efficient iterative schemes for ab initio total-energy calculations using a plane-wave basis set, *Physical Review B* **54**, 11169 (1996).
60. G. Kresse and J. Hafner, Ab initio molecular dynamics for open-shell transition metals, *Physical Review B* **48**, 13115 (1993).
61. G. Deinzer, G. Birner, G. and D. Strauch, Ab initio calculation of the linewidth of various phonon modes in germanium and silicon, *Physical Review B* **67**, 144304 (2003).
62. G. Deinzer, M. Schmitt, A. P. Mayer, and D. Strauch, Intrinsic lifetimes and anharmonic frequency shifts of long-wavelength optical phonons in polar crystals, *Physical Review B* **69**, 014304 (2004).
63. K. Esfarjani and H. T. Stokes, Method to extract anharmonic force constants from first principles calculations, *Physical Review B* **77**, 144112 (2008).
64. M. Born and K. Huang, *Dynamical Theory of Crystal Lattices* (Clarendon Press, Oxford, 1998).
65. G. Leibfried and W. Ludwig, Theory of anharmonic effects in crystals, *Solid State Physics* **12**, 275–444, 1961.
66. M. H. F. Sluiter, M. Weinert and Y. Kawazoe, Determination of the elastic tensor in low-symmetry structures, *Europhysics Letters* **43**, 183 (1998).
67. J. Carrete, W. Li, L. Lindsay, D. A. Broido, L. J. Gallego and N. Mingo, Physically founded phonon dispersions of few-layer materials and the case of borophene, *Materials Research Letters* **4**, 2166 (2016).
68. O. Hellman, I. A. Abrikosov, and S. I. Simak, Lattice dynamics of anharmonic solids from first principles, *Physical Review B* **84**, 180301(R) (2011).
69. O. Hellman, P. Steneteg, I. A. Abrikosov, and S. I. Simak, Temperature dependent effective potential method for accurate free energy calculations of solids, *Physical Review B* **87**, 104111 (2013).
70. O. Hellman and I. A. Abrikosov, Temperature-dependent effective third-order interatomic force constants from first principles, *Physical Review B* **88**, 144301 (2013).
71. O. Hellman, *Thermal Properties of Materials from First Principles*, PhD thesis, University of Linköping (2012).
72. P. Souvatzis, O. Eriksson, M. I. Katsnelson, and S. P. Rudin, Entropy driven stabilization of energetically unstable crystal structures explained from first principles theory, *Physical Review Letters* **100**, 095901 (2008).
73. I. Errea, M. Calandra, and F. Mauri, First-principles theory of anharmonicity and the inverse isotope effect in superconducting palladium-hydride compounds, *Physical Review Letters* **111**, 177002 (2013).
74. A. van Roekeghem, J. Carrete, and N. Mingo, Anomalous thermal conductivity and suppression of negative thermal expansion in ScF_3, *Physical Review B* **94**, 020303 (2016).
75. N. Antolin, O. D. Restrepo, and W. Windl, Fast free-energy calculations for unstable high-temperature phases, *Physical Review B* **86**, 054119 (2012).
76. O. Delaire, J. Ma, K. Marty, A. F. May, M. A. McGuire, M.-H. Du, D. J. Singh, A. Podlesnyak, G. Ehlers, M. D. Lumsden, and B. C. Sales, Giant anharmonic phonon scattering in PbTe, *Nature Materials* **10**, 614–619 (2011).
77. H. A. Alperin, S. J. Pickart, J. J. Rhyne, and V. J. Minkiewicz, Softening of the transverse-optic mode in PbTe, *Physics Letters A* **40**, 295 (1972).
78. J. An, A. Subedi, and D. J. Singh, Ab initio phonon dispersions for PbTe, *Solid State Communications* **148**, 417 (2008).

79. Y. Zhang, X. Ke, P. R. C. Kent, J. Yang, and C. Chen, Anomalous lattice dynamics near the ferroelectric instability in PbTe, *Physical Review Letters* **107**, 175503 (2011).

80. C. W. Li, O. Hellman, J. Ma, A. F. May, H. B. Cao, X. Chen, A. D. Christianson et al., Phonon self-energy and origin of anomalous neutron scattering spectra in SnTe and PbTe thermoelectrics, *Physical Review Letters* **112**, 175501 (2014).

81. W. Kullmann, G. Eichhorn, H. Rauh, R. Geick, G. Eckold, and U. Steigenberger, Lattice dynamics and phonon dispersion in the narrow gap semiconductor Bi_2Te_3 with sandwich structure, *Physica Status Solidi B* **162**, 125 (1990).

82. D. Bessas, I. Sergueev, H.-C. Wille, J. Perßon, D. Ebling, and R. P. Hermann, Lattice dynamics in Bi_2Te_3 and Sb_2Te_3: Te and Sb density of phonon states, *Physical Review B* **86**, 224301 (2012).

83. H. J. Goldsmid, The thermal conductivity of bismuth telluride, *Proceedings of the Physical Society B* **69**, 203 (1956).

84. The ab initio MD calculations of T~0K IFCs were actually done at about 30K since thermally displaced atoms are required in order to get interatomic forces.

85. A. J. Minnich, J. A. Johnson, A. J. Schmidt, K. Esfarjani, M. S. Dresselhaus, K. A. Nelson, and G. Chen, Thermal conductivity spectroscopy technique to measure phonon mean free paths, *Review Letters* **107**, 095901 (2011).

86. J. A. Johnson, A. A. Maznev, J. Cuffe, J. K. Eliason, A. J. Minnich, T. Kehoe, C. M. Sotomayor Torres, G. Chen, and L. A. Nelson, Direct measurement of room-temperature nondiffusive thermal transport over micron distances in a silicon membrane, *Physical Review Letters* **110**, 025901 (2013).

87. A. J. Minnich, Determining phonon mean free paths from observations of quasiballistic thermal transport, *Physical Review Letters* **110**, 025901 (2012).

88. O. Hellman and D. A. Broido, Erratum: Phonon thermal transport in Bi_2Te_3 from first principles (*Physical Review B* **90**, 134309 [2014]), *Physical Review B* **92**, 219903 (2015).

89. R. A. Cowley, Anharmonic crystals, *Reports on Progress in Physics* **31**, 123 (1968).

90. F. Giustino, M. L. Cohen, and S. G. Louie, Electron-phonon interaction using Wannier functions, *Physical Review B* **76**, 224301 (2007).

91. J. Noffsinger, F. Giustino, B. D. Malone, C.-H. Park, S. G. Louie, and M. L. Cohen, EPW: A program for calculating the electron–phonon coupling using maximally localized Wannier functions, *Computer Physics Communications* **181**, 2140 (2010).

92. S. Poncé, E. R. Margineb, C. Verdia, and F. Giustino, EPW: Electron-phonon coupling, transport and superconducting properties using maximally localized Wannier functions, *Computer Physics Communication* **209**, 116–133 (2016).

93. J.-P. M. Péraud, C. D. Landon, and N. G. Hadjiconstantinou, Monte Carlo methods for solving the Boltzmann transport equation, *Annual Review of Heat Transfer* **17**, 205 (2014).

94. J.-P. M. Péraud and N. G. Hadjiconstantinou, Adjoint-based deviational Monte Carlo methods for phonon transport calculations, *Physical Review B*, **91**, 235321 (2015).

95. G. Romano, K. Esfarjani, D. A. Strubbe, D. Broido, and A. M. Kolpak, Temperature-dependent thermal conductivity in silicon nanostructured materials studied by the Boltzmann transport equation, *Physical Review B* **93**, 035408 (2016).

96. J. Callaway, Model for lattice thermal conductivity at low temperatures. *Physical Review* **113**(4), 1046–1051 (1959).

97. M. G. Holland, Analysis of lattice thermal conductivity, *Physical Review* **132**(6), 2461–2471 (1963).

98. E. F. Steigmeier and B. Abeles, Scattering of phonons by electrons in germanium-silicon alloys.*Physical Review* **136**(4A), A1149–A1155 (1964).

99. P. G. Klemens, Thermal conductivity, in F. Seitz and D. Turnbull, editors, *Solid State Physics* (Academic Press, Cambridge, MA, 1958).

100. M. Zebarjadi, K. Esfarjani, M. S. Dresselhaus, Z. F. Ren, and G. Chen, Perspectives on thermoelectrics: From fundamentals to device applications. *Energy and Environmental Science* **5**, 5147–5162 (2011).

101. A. J. Minnich, M. S. Dresselhaus, Z. F. Ren, and G. Chen, Bulk nano-structured thermoelectric materials: Current research and future prospects, *Energy and Environmental Science* **2**(5), 466–479 (2009).

102. L.-D. Zhao, V. P. Dravid, and M. G. Kanatzidis, The panoscopic approach to high performance thermoelectrics. *Energy and Environmental Science* **7**, 251–268 (2013).

103. J. He, M. G. Kanatzidis, and V. P. Dravid, High performance bulk thermoelectrics via a panoscopic approach, *Materials Today* **16**(5), 166–176 (2013).

104. G. J. Snyder and E. S. Toberer, Complex thermoelectric materials, *Nature Materials* **7**(2), 105–114 (2008).

105. J.-P. M. Péraud and N. G. Hadjiconstantinou, An alternative approach to efficient simulation of micro/nanoscale phonon transport, *Applied Physics Letters* **101**(15), 153114–153114–4 (2012).

106. J.-P. M. Péraud and N. G. Hadjiconstantinou, Efficient simulation of multi-dimensional phonon transport using energy-based variance-reduced monte carlo formulations, *Physical Review B* **84**(20), 205331 (2011).

107. J.-P. M. Péraud and N. G. Hadjiconstantino, Adjoint-based deviational Monte Carlo methods for phonon transport calculations, *Physical Review B* **91**(23), 235321 (2015).

108. D. A. Broido, M. Malorny, G. Birner, N. Mingo, and D. A. Stewart, Intrinsic lattice thermal conductivity of semiconductors from first principles, *Applied Physics Letters* **91**(23), 231922 (2007).

109. G. Chen, *Nanoscale Energy Transport and Conversion* (Oxford University Press, New York, 2005).

110. P. B. Allen. Improved Callaway model for lattice thermal conductivity, *Physical Review B* **88**(14), 144302 (2013).

111. J. Ma, W. Li, and X.Luo, Examining the Callaway model for lattice thermal-conductivity, *Physical Review B* **90**(3), 035203 (2014).

112. A. Ward, D. A. Broido, D. A. Stewart, and G. Deinzer, Ab initio theory of the latticethermal conductivity in diamond, *Physical Review B* **80**(12), 125203 (2009).

113. I. I. Pomeranchuk, On the thermal conductivity of dielectrics at temperatures higher than the debye temperature, *Journal of Physics-USSR* **4**, 259 (1941).

114. A. Jeong, S. Datta, and M. Lundstrom, Full dispersion versus Debye model evaluation of lattice thermal conductivity with a Landauer approach, *Journal of Applied Physics*, **109**(7), 073718–073718–8 (2011).

115. A. M. Zebarjadi, J. Yang, K. Lukas, B. Kozinsky, B. Yu, M. S. Dresselhaus, Opeil, Z. Ren, and G. Chen, Role of phonon dispersion in studying phonon mean free paths in skutterudites, *Journal of Applied Physics* **112**(4), 044305–044305–7 (2012).

116. A. Ward and D. A. Broido, Intrinsic phonon relaxation times from first-principles studies of the thermal conductivities of Si and Ge, *Physical Review B* **81**(8), 085205 (2010).
117. G. A. Radtke and N. G. Hadjiconstantinou, Variance-reduced particle simulation of the Boltzmann transport equation in the relaxation-time approximation, *Physical Review E* **79**(5), 056711 (2009).
118. N. G. Hadjiconstantinou, G. A. Radtke, and L. L. Baker, On variance-reduced simulations of the Boltzmann transport equation for small-scale heat transfer applications, *Journal of Heat Transfer* **132**(11), 112401–8 (2010).
119. S. Chandrasekhar, *Radiative Transfer* (Courier Dover Publications, Mineola, NY, 1950).
120. A. Majumdar, Microscale heat conduction in dielectric thin films, *Journal of Heat Transfer* **115**(1), 7–16 (1993).
121. A. A. Joshi and A. Majumdar, Transient ballistic and diffusive phonon heat transport in thinFilms, *Journal of Applied Physics* **74**(1), 31–39 (1993).
122. J. A. Rowlette and K. E. Goodson, Fully coupled nonequilibrium electron-phonon transport in nanometer-scale silicon FETs, *IEEE Transactions on Electron Devices* **55**(1), 220–232 (2008).
123. S. Sinha, E. Pop, R. W. Dutton, and K. E. Goodson, Non-equilibrium phonon distributions in sub-100 nm silicon transistors, *Journal of Heat Transfer* **128**(7), 638 (2006).
124. S. Sinha and K. E. Goodson, Thermal conduction in sub-100 nm transistors, *Microelectronics Journal* **37**(11), 1148–1157, (2006).
125. T. Klitsner, J. E. VanCleve, H. E. Fischer, and R. O. Pohl, Phonon radiative heattransfer and surface scattering, *Physical Review B* **38**(11), 7576–7594 (1988).
126. R. B. Peterson, Direct simulation of phonon-mediated heat transfer in a Debye crystal, *Journal of Heat Transfer* **116**(4), 815–822 (1994).
127. D. Lacroix, K. Joulain, D. Terris, and D. Lemonnier, Monte Carlo simulation ofphonon confinement in silicon nanostructures: Application to the determination of the thermal conductivity of silicon nanowires, *Applied Physics Letters* **89**(10), 103104–103104–3 (2006).
128. S. Mazumder and A. Majumdar, Monte carlo study of phonon transport in solidthin films including dispersion and polarization, *Journal of Heat Transfer* **123**(4), 749–759 (2001).
129. A. Mittal and S. Mazumder, Monte Carlo study of phonon heat conduction in siliconthin films including contributions of optical phonons, *Journal of Heat Transfer* **132**(5), 052402 (2010).
130. Q. Hao, G. Chen, and M.-S. Jeng, Frequency-dependent monte carlo simulationsof phonon transport in two-dimensional porous silicon with aligned pores, *Journal of Applied Physics* **106**(11), 114321 (2009).
131. M. S. Jeng, R. Yang, D. Song, and G. Chen, Modeling the thermal conductivity and phonon transport in nanoparticle composites using monte carlo simulation, *Journal of Heat Transfer* **130**(4), 042410 (2008).
132. S. V. J. Narumanchi, J. Y. Murthy, and C. H. Amon, Simulation of unsteadysmall heat source effects in sub-micron heat conduction, *Journal of Heat Transfer* **125**(5), 896–903 (2003).
133. S. V. J. Narumanchi, J. Y. Murthy, and C. H. Amon, Comparison of different phonon transport models for predicting heat conduction in silicon-on-insulator transistors, *Journal of Heat Transfer* **127**(7), 713–723 (2005).

134. S. V. J. Narumanchi, J. Y. Murthy, and C. H. Amon, Submicron heat transport model in silicon accounting for phonon dispersion and polarization, *Journal of Heat Transfer* **126**(6), 946–955 (2004).

135. A. J. H. McGaughey and A. Jain, Nanostructure thermal conductivity prediction by Monte Carlo sampling of phonon free paths, *Applied Physics Letters* **100**(6), 061911–061911–3 (2012).

136. P. Heino, Lattice-Boltzmann finite-difference model with optical phonons for nanoscale thermal conduction, *Computers and Mathematics with Applications* (Mesoscopic Methods in Engineering and Science, International Conferences on Mesoscopic Methods in Engineering and Science) **59**(7), 2351–2359 (2010).

137. T. M. M. Homolle and N. G. Hadjiconstantinou, A low-variance deviational simu-lation Monte Carlo for the Boltzmann equation, *Journal of Computational Physics* **226**(2), 2341–2358 (2007).

138. L. L. Baker and N. G. Hadjiconstantinou, Variance reduction for Monte Carlo solutionsof the Boltzmann equation. *Physics of Fluids (1994–present)* **17**(5), 051703 (2005).

139. C. Hua and A. J. Minnich, Importance of frequency-dependent grain boundary scattering in nanocrystalline silicon and silicon–germanium thermoelectric, *Semiconductor Science and Technology* **29**(12), 124004 (2014).

140. Z. Wang, J. E. Alaniz, W. Jang, J. E. Garay, and C. Dames, Thermal conductivity of nanocrystalline silicon: Importance of grain size and frequency-dependent mean free paths, *Nano Letters* **11**(6), 2206–2213 (2011).

141. H. Zhao and J. B. Freund, Phonon scattering at a rough interface between two fcclattices, *Journal of Applied Physics* **105**(1), 013515 (2009).

142. C. Dames and G. Chen, Thermal conductivity of nanostructured thermoelectric materials, in D. Rowe, editor, *CRC Thermoelectrics Handbook: Macro to Nano* (CRC Press, Boca Raton, FL, 2005).

528   Bibliography of Chapter 11: SCC

CHAPTER **13**

Reliable Prediction of Efficiency and Output Power and Balance between Materials and Devices in Thermoelectric Power Generators

Hee Seok Kim, Weishu Liu, and Zhifeng Ren

Contents

13.1 Cumulative Temperature Dependence Model for Homogeneous Materials

In 1909, Altenkirch[1] first derived the conversion efficiency of a TE generator based on the constant property model (CPM) and identified the desired attributes for three physical properties. Good TE materials have a large Seebeck coefficient (S), low electrical resistivity (ρ), and low thermal conductivity (κ). In 1957, Ioffe[2] optimized the TE efficiency as

$$\eta_{CPM} = \eta_c \frac{\sqrt{1 + ZT_{avg}} - 1}{\sqrt{1 + ZT_{avg}} + T_c/T_h},$$ (13.1)

where T_h, T_c, and T_{avg} are the hot-side, cold-side, and average temperatures, respectively. The variable η_c is the Carnot efficiency, $\eta_c = (T_h - T_c)/T_h$. Z is defined as $S^2/(\rho\kappa)$, showing how efficiently a material can convert energy at a specific temperature, and a peak ZT has represented a performance of TE materials since the 1950s.[3-5] This peak ZT shows the performance only at a specific temperature and is not enough to represent the characteristics of a material in operation over broad thermal boundary conditions. The maximum efficiency by Equation 13.1, derived based on the CPM, is inadequate when Z is temperature dependent. Due to the assumption of temperature independence, Equation 13.1 only correctly predicts the maximum efficiency at a small temperature difference between the cold and hot sides, or in limited TE materials[6-8] that have Z almost constant over their operating temperature range. By overlooking the assumption and simply using Equation 13.1, incorrect high efficiency that is not practically achievable[9,10] has been reported. In most TE materials, where S, ρ, and κ are temperature dependent, ZT values do not increase in a linear fashion with temperature.[11-15] In addition, TE materials operate at a large temperature difference, so the prediction by the conventional model cannot be reliable. In order to use Equation 13.1, the average Z is obtained in two ways to predict efficiency: (1) an integration with respect to temperature, $Z_{int} = (1/\Delta T)\int_{T_c}^{T_h} Z(T)dT$, and (2) a Z value corresponding to the average temperature, $Z_{int}T_{avg}$ and $Z_{T_{avg}}T_{avg} = Z(T_{avg})$. The $Z_{int}T_{avg}$ and $Z_{T_{avg}}T_{avg}$ are called the average ZT hereafter. Figure 13.1a shows the temperature-dependent ZT and average ZT values of the p-type Ni-doped MgAgSb,[16] and their corresponding efficiencies by Equation 13.1 as shown in Figure 13.1b are matched with the numerical simulation due to the fact that the Z vs. T is not too much off a constant (Figure 13.1c). The numerical simulations were carried out based on the finite-difference method to account for the temperature dependence of S, ρ, and κ over a large temperature difference between the cold and hot sides.[17,18] In Figure 13.1d, however, n-type In_4Se_{3-x}[10] has larger variations of average ZTs depending on how they are averaged due to the strong temperature dependence (Figure 13.1f),

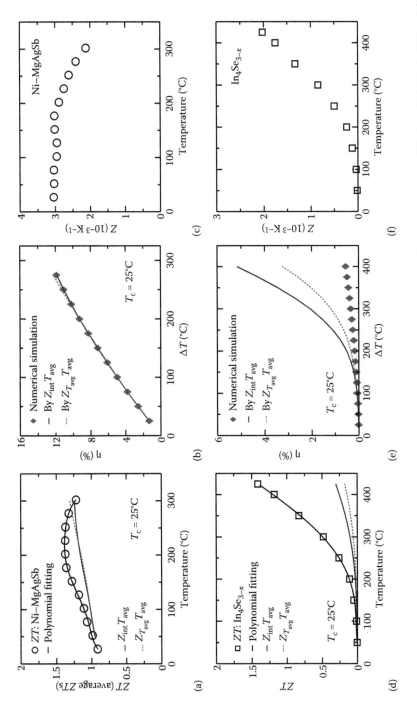

FIGURE 13.1 Inadequacy of the relationship between the conventional ZT and the maximum efficiency. ZT vs. T of Ni-doped (a) MgAgSb[16] and (d) In$_4$Se$_{3-x}$,[10] in which circles are measured data and black line is fitted curves. Solid and dashed lines are average ZT values by $Z_{int}Z_{avg}$ and $Z_{T_{avg}}T_{avg}$, respectively. Efficiencies at $T_c = 25°C$ by numerical simulation (*solid diamonds*) and by CPM using integration (*solid line*) and average temperature (*dashed line*) for Z_{avg} of Ni-doped (b) MgAgSb and (e) In$_4$Se$_{3-x}$. Z vs. T of Ni-doped (c) MgAgSb and (f) In$_4$Se$_{3-x}$.

which results in much different efficiency predictions (Figure 13.1e). Even though the peak ZT of In_4Se_{3-x} is higher than that of Ni-doped MgAgSb, the average ZTs and predicted efficiency are much lower, which means that the ZT peak value is not the right indicator for the efficiency of a TE material. In addition, any average ZT cannot be a correct index due to lack of consistency for the following reasons: (1) average ZTs vary by the averaging techniques (Figure 13.1a and d), (2) $Z_{T_{avg}}$ is estimated to be larger than Z_{int} when the Z curve is in convex upward shape (Figure 13.1c) and vice versa (Figure 13.1f), and (3) analytical prediction of some materials based on the average ZTs is far away from the numerical analysis (Figure 13.1e). For this reason, the conventional efficiency formula (Equation 13.1) often misleads and gives rise to an impractically high efficiency prediction. Thus, it is desirable to establish a new model to predict reliable conversion efficiency based on the temperature-dependent TE properties under a large temperature difference. In this section, an engineering dimensionless figure of merit $(ZT)_{eng}$ and an engineering power factor $(PF)_{eng}$ are defined as realistic indicators associated with large temperature difference including the temperature dependence of TE materials, and a new efficiency formula based on cumulative temperature dependence (CTD) model is derived, which includes the corrected contribution of Joule and Thomson heat.

13.1.1 Efficiency Based on the Engineering Figure of Merit $(ZT)_{eng}$

Missing Thomson effect in the CPM may cause an underestimation or overestimation of the efficiency depending on the degree of Thomson heat at a given temperature gradient. Some studies evaluated the conversion efficiency, analytically accounting for Thomson effect, which is valid in limited conditions by assuming a constant Thomson coefficient[19,20]; linear behavior of S, ρ, and κ[21]; and a temperature-dependent Seebeck coefficient with constant ρ and κ.[22] Kim et al.[23] established the CTD model by taking into account the temperature dependence of S, ρ, κ, and τ at a large temperature difference, where τ is the Thomson coefficient, and no assumption is required to specify the type of temperature dependence of S, ρ, κ, and τ. In addition, practical fractions of Joule and Thomson heat associated with thermal boundaries are evaluated.

The governing equation for energy balance over 1D differential element of TE material is[24]

$$\frac{d}{dx}\left(\kappa(T)\frac{dT}{dx}\right) + J^2\rho(T) - J\tau(T)\frac{dT}{dx} = 0, \tag{13.2}$$

where x and J are the distance from the heat source and current density, respectively, and $\tau(T)$ is the temperature-dependent Thomson coefficient defined as $\tau(T) = TdS(T)/dT$. Integrating Equation 13.2 with respect to x becomes

$$\kappa(T)\frac{dT}{dx} + J^2 \int_0^x \rho(T)dx - J\int_0^x \tau(T)\frac{dT}{dx}dx + C_1 = 0. \tag{13.3}$$

The second time integration with respect to x yields

$$\int_{T_h}^T \kappa(T)dT + J^2 \int_0^x\int_0^x \rho(T)dxdx - J\int_0^x\int_{T_h}^T \tau(T)dT\,dx + C_1x + C_2 = 0. \tag{13.4}$$

By applying the boundary conditions of $T = T_h$ at $x = 0$ and $T = T_c$ at $x = L$, where L is a length in x direction of the TE leg, the integration constants C_1 and C_2 are solved as

$$C_1 = \frac{1}{L}\int_{T_c}^{T_h}\kappa(T)dT - \frac{J^2}{L}\int_0^L\int_0^x \rho(T)dxdx + \frac{J}{L}\int_0^L\int_{T_h}^T \tau(T)dT\,dx, C_2 = 0. \tag{13.5}$$

From Equations 13.3 and 13.5, the conduction heat at $T = T_h$ becomes

$$-\kappa(T)\frac{dT}{dx}\Big|_{T_h} = \frac{1}{L}\int_{T_c}^{T_h}\kappa(T)dT - \frac{J^2}{L}\int_0^L\int_0^x \rho(T)dxdx + \frac{J}{L}\int_0^L\int_{T_h}^T \tau(T)dT\,dx, \tag{13.6}$$

where A is the cross-sectional area. The constitutive relation for heat Q and electric current I is given as

$$Q = -A\kappa\frac{dT}{dx} + ITS. \tag{13.7}$$

By substituting Equation 13.6 into Equation 13.7 and applying the boundary conditions, Q_h at the hot side is obtained as

$$Q_h = \frac{A}{L}\int_{T_c}^{T_h}\kappa(T)dT + IT_hS(T_h) - W_JI^2R - W_TI\int_{T_c}^{T_h}\tau(T)dT. \tag{13.8}$$

The terms on the right-hand side represent conduction heat, Peltier heat, Joule heat, and Thomson heat, respectively, where $S(T_h)$ is the Seebeck coefficient at T_h, and R is the electrical resistance of a TE leg expressed as

$$R = \frac{1}{\Delta T} \frac{L}{A} \int_{T_c}^{T_h} \rho(T) \, dT. \tag{13.9}$$

W_J and W_T are dimensionless weight factors of Joule and Thomson heat, respectively, which are defined as

$$W_J = \frac{\int_0^L \int_0^x \rho(T) \, dx \, dx}{(L^2/\Delta T) \int_{T_c}^{T_h} \rho(T) \, dT}, \tag{13.10}$$

$$W_T = \frac{\int_0^L \int_T^{T_h} \tau(T) \, dT \, dx}{L \int_{T_c}^{T_h} \tau(T) \, dT}. \tag{13.11}$$

The conversion efficiency of TE materials accounting for Thomson effect can be expressed as the ratio of output power to input heat rate:

$$\eta = \frac{P_{out}}{Q_h} = \frac{\left(V_{oc}^2/R\right)\left[m/(1+m)^2\right]}{(A/L)\int_{T_c}^{T_h} \kappa(T) \, dT + IT_h S(T_h) - W_J I^2 R - W_T I \int_{T_c}^{T_h} \tau(T) \, dT}, \tag{13.12}$$

where m is the ratio of external to internal electrical resistance.

Substituting Equation 13.9 into Equation 13.12 and applying $I = V_{oc} R^{-1}(1 + m)^{-1}$ to Equation 13.12 yields

$$\eta = \frac{\dfrac{m}{1+m}}{\dfrac{1+m}{(ZT)_{eng}} + \dfrac{S(T_h)T_h}{\int_{T_c}^{T_h} S(T) \, dT} - \dfrac{W_J}{(1+m)} - \dfrac{W_T \int_{T_c}^{T_h} \tau(T) \, dT}{\int_{T_c}^{T_h} S(T) \, dT}}, \tag{13.13}$$

where $(ZT)_{eng}$ is the engineering figure of merit defined as

$$(ZT)_{\text{eng}} = \frac{\left(\int_{T_c}^{T_h} S(T)\mathrm{d}T\right)^2}{\int_{T_c}^{T_h} \rho(T)\mathrm{d}T \int_{T_c}^{T_h} \kappa(T)\mathrm{d}T} \Delta T = \frac{(PF)_{\text{eng}}}{\int_{T_c}^{T_h} \kappa(T)\mathrm{d}T} \Delta T. \qquad (13.14)$$

$(PF)_{\text{eng}}$ is the engineering PF. By optimizing the ratio m satisfying $\mathrm{d}\eta/\mathrm{d}m = 0$ regarding Equation 13.13, the maximum efficiency by the CTD model and its corresponding optimized m_{CTD} are finalized as

$$\eta_{\text{CTD}} = \eta_c \frac{\sqrt{1+(ZT)_{\text{eng}}\alpha_1\eta_c^{-1}} - 1}{\alpha_0\sqrt{1+(ZT)_{\text{eng}}\alpha_1\eta_c^{-1}} + \alpha_2}, \qquad (13.15)$$

$$m_{\text{CTD}} = \sqrt{1+(ZT)_{\text{eng}}\alpha_1\eta_c^{-1}}, \qquad (13.16)$$

where α_i is defined as

$$\alpha_i = \frac{S(T_h)\Delta T}{\int_{T_c}^{T_h} S(T)\mathrm{d}T} - \frac{\int_{T_c}^{T_h} \tau(T)\mathrm{d}T}{\int_{T_c}^{T_h} S(T)\mathrm{d}T} W_T\eta_c - iW_J\eta_c \quad (i = 0, 1, \text{ and } 2).$$

$$(13.17)$$

Calculating the correct W_J and W_T is essential to obtain an exact prediction of the maximum efficiency by Equation 13.15. Since the cumulative temperature dependence of ρ and τ is considered under a large temperature gradient in this model, the lumped constant 1/2 by the CPM,[25] i.e., half of Joule and Thomson heat to the hot side, is not exact anymore, so practical Joule and Thomson heat influencing on the hot side should be evaluated. As shown in Equations 13.10 and 13.11, a temperature gradient $\mathrm{d}T/\mathrm{d}x$ is required to calculate the exact weight factors. However, since it is a solution of the nonlinear differential equation in Equation 13.2 in case S, ρ, and κ are temperature dependent, the exact W_J and W_T cannot be directly solved. When ignoring radiation/convection heat loss and assuming the temperature gradient dominantly caused by conduction heat, one can express the temperature gradient by applying $Q_{\text{conduction}} = (A/L)\int_{T_c}^{T_h} \kappa(T)\mathrm{d}T$ from the first term in Equation 13.8 and $\kappa_{\text{avg}} = (1/\Delta T)\int_{T_c}^{T_h} \kappa(T)\mathrm{d}T$ into the definition of conduction heat:

$$\frac{dT}{dx} \approx -\frac{Q_{\text{conduction}}}{A\kappa_{\text{avg}}} = -\frac{(A/L)\int_{T_c}^{T_h}\kappa(T)dT}{(A/\Delta T)\int_{T_c}^{T_h}\kappa(T)dT} = -\frac{\Delta T}{L}. \qquad (13.18)$$

Based on the relation $dx \approx -(L/\Delta T)dT$, the weight factors can be rewritten with respect to T as

$$W_J = \frac{\int_{T_c}^{T_h}\int_{T}^{T_h}\rho(T)dT\,dT}{\Delta T\int_{T_c}^{T_h}\rho(T)dT}, \qquad (13.19)$$

$$W_T = \frac{\int_{T_c}^{T_h}\int_{T}^{T_h}\tau(T)dT\,dT}{\Delta T\int_{T_c}^{T_h}\tau(T)dT}. \qquad (13.20)$$

If $\rho(T)$ and $\tau(T)$ are assumed to be temperature independent, the weight factors of Equations 13.19 and 13.20 become 1/2, and then Equation 13.8 is equal to the form based on the CPM.[25] Thus, the W_J and W_T are good general approximations for the corrected contribution of Joule and Thomson heating associated with the hot side of a TE material.

Figure 13.2a shows the maximum efficiency prediction of K-doped $PbTe_{1-y}Se_y$[11] (circles) and $SnSe$[15] (squares) according to a temperature difference by ramping T_h up to maximum operating temperature of each material while $T_c = 50°C$, where the solid line indicates the maximum efficiency computed by the numerical simulation. Figure 13.2b shows how relatively accurate the efficiency calculations by CPM and CTD model are compared with the efficiency η_n by the numerical simulation. The calculation by CPM overestimates the efficiency by 10% of the relative difference for K-doped $PbTe_{1-y}Se_y$ at $T_h = 600°C$ and more than twofold for SnSe at $T_h = 700°C$ compared with the numerical prediction. On the other hand, the maximum efficiency by Equation 13.15 is predicted more accurately by 5 and 9% of the relative difference for K-doped $PbTe_{1-y}Se_y$ and SnSe, respectively, which results from taking not only the cumulative temperature effect but also the Thomson effect of material properties into account. Figure 13.2c and d shows the calculated efficiencies and their accuracy with respect to η_n of HH ($Hf_x(ZrTi)_{1-x}CoSb_{0.8}Sn_{0.2})$[7] and skutterudite (SKU)

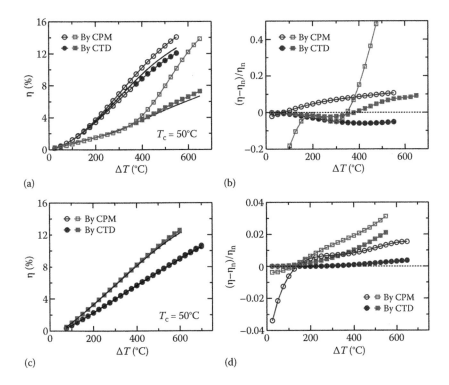

FIGURE 13.2 (a) Predicted efficiencies as a function of ΔT and (b) relative accuracy with respect to numerical results of K-PbTeSe (*circles*) and SnSe (*squares*). (c) Predicted efficiency and (d) relative accuracy of HH (*circles*) and SKU (*squares*).

$(Ce_{0.45}Nd_{0.45}Fe_{3.5}Co_{0.5}Sb_{12})$,[6] in which the calculated efficiencies of both materials by the two models have good agreement within 4% of relative difference compared with those by numerical simulation. The relatively accurate efficiency and small variation of the prediction for HH and SKU compared with K-doped $PbTe_{1-y}Se_y$ and SnSe are caused by the linear-like behavior of their S, ρ, and κ with low rate of change with respect to temperature. Thus, Equation 13.15 leads to the reliable efficiency predictions resulting from the accounting for CTD, Thomson effect, and modified intensity of Joule and Thomson heat.

The accuracy is partially associated with the temperature-dependent $\tau(T)$ representing the intensity of Thomson heat (Figure 13.3a). The positive τ implying $dS/dT > 0$ contributes to the increase of the efficiency compared to the case of $\tau = 0$.[20] In order to examine the ΔT-dependent Thomson heat at practical operating temperature difference, the overall Thomson coefficient at given T_h and T_c, defined as $\tau_{\Delta T} = \displaystyle\int_{T_c}^{T_h} \tau(T)\,dT$, is shown in Figure 13.3b, where positive values of $\tau_{\Delta T}$ for K-doped $PbTe_{1-y}Se_y$, HH, and SKU over the whole temperature range lead to increasing the efficiency as compared when

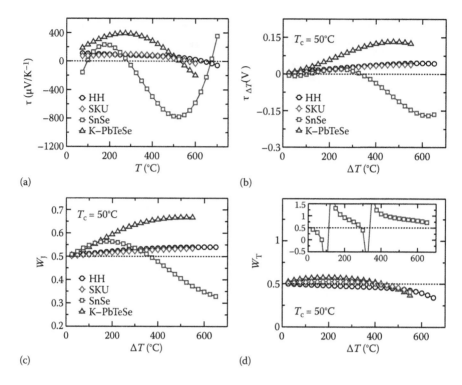

FIGURE 13.3 Contribution for accurate efficiency prediction by overall Thomson coefficient and weight factors. (a) Calculated Thomson coefficient τ at each temperature and (b) the overall Thomson coefficient $\tau_{\Delta T}$ as a function of T_h at $T_c = 50°C$. ΔT-dependent weight factor for (c) Joule heating W_J and (d) Thomson heating W_T, where T_c is fixed at 50°C.

the Thomson effect is ignored, so it is closer to the efficiency predicted by numerical analysis. The overall effect of Thomson heat associated with ΔT for SnSe differs in temperature boundaries where $\tau_{\Delta T}$ fluctuates across zero. However, the efficiency results from the interrelation of conduction and Joule and Thomson heat, as well as the temperature dependence of S, ρ, and κ, so it is difficult to exactly predict the quantified degree of accuracy between the analytical results by the CTD model and the numerical simulations. The difference comes from the linearized expression of dT/dx and the different types of temperature dependence, i.e., cumulative in the CTD model and instantaneous in the numerical model. In Figure 13.3c, W_J of SnSe fluctuates across 1/2 (by the CPM), while the others monotonically increase over 1/2 with ΔT. The increasing W_J at larger ΔT indicates that more fraction of Joule heat is associated with the hot side than half of it due to the increasing trend of $\rho(T)$. The ΔT-dependent W_T is shown in Figure 13.3d, where W_T for K-doped PbTe$_{1-y}$Se$_y$, HH, and SKU have decreasing tendencies and become below 1/2 at certain ΔT, which means the effect by Thomson heat on the heat flux at the hot end gets smaller since its dS/dT decreases. For SnSe, ΔT-dependent W_T (inset in Figure 13.3d) has two diverged points around at 150°C and 400°C at which $\tau_{\Delta T}$ becomes zero (Figure 13.3b), indicating that no overall Thomson heat at that temperature difference is considered, even though Thomson heat exists at each temperature. W_T in Equation 13.20 becomes infinite when $\tau_{\Delta T} = 0$, which seems to make Equation 13.17 invalid, but it does not affect the efficiency analysis since W_T is always paired with $\tau_{\Delta T}$.

13.1.2 Output Power Based on the Engineering Power Factor (PF)$_{eng}$

The output power density (W m^{-2}) at the maximum efficiency based on the $(PF)_{eng}$ is expressed as[23]

$$P_d = \frac{(PF)_{eng} \Delta T}{L} \frac{m_{opt}}{(1 + m_{opt})^2},$$ (13.21)

$$(PF)_{eng} = \frac{\left(\int_{T_c}^{T_h} S(T) dT \right)^2}{\int_{T_c}^{T_h} \rho(T) dT}.$$ (13.22)

The output power density is dependent on the dimensions of the TE leg as well as the material properties, while the efficiency is determined only by the characteristics of the material. Figure 13.4a and b shows the output power densities of HH and SnSe at $T_c = 100$°C with ΔT, where a cubic TE leg is assumed for the output power prediction ($2 \times 2 \times 2$ mm^3), and a simple averaged Seebeck coefficient and electrical resistance at a given temperature difference are used for the conventional *PF*-based prediction by the CPM. The power density by CTD has better agreement with the numerical prediction within 1% and 13% of the relative difference for HH and SnSe, respectively, whereas that by CPM shows 13 and 33% of the relative difference at $\Delta T = 600$°C for HH and SnSe, respectively. In Figure 13.4c, $(PF)_{eng}$ shows a similar tendency to the power density (Figure 13.4a and b), indicating an intrinsic performance for power generation of a TE material operated at practical temperature gradients, which has quite different trends from *PF* (Figure 13.4d). The amount of power generation by HH is more than tenfold as large as that of SnSe in the same TE leg dimensions as shown in Figure 13.4c (inset). By adjusting the leg dimensions for matching similar input heat range to keep the same ΔT, the power density can be comparable,[23] but reducing or increasing a leg length causes thermomechanical structural issues.

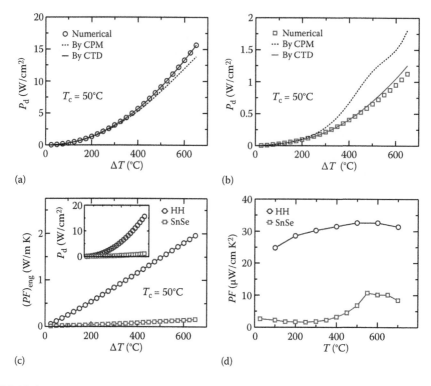

FIGURE 13.4 Output power densities of (a) HH and (b) SnSe along with ΔT in a cubic-shaped TE leg. (c) ΔT-dependent $(PF)_{\text{eng}}$ of HH and SnSe at $T_c = 50°C$ showing the same tendency to the power density P_d curves (inset). (d) Temperature-dependent PF of HH and SnSe.

13.2 Cumulative Temperature Dependence Model for p–n Module Devices

The conversion efficiency of a typical TE module consisting of p-type and n-type materials connected thermally in parallel and electrically in series is a function of the device figure of merit $[Z]^*$ of a p–n pair, defined as[25]

$$[Z]^* = \frac{(S_p - S_n)^2}{\left(\sqrt{\kappa_p \rho_p} + \sqrt{\kappa_n \rho_n}\right)^2},$$ (13.23)

where the subscripts p and n denote p-type and n-type materials, respectively. The square bracket and asterisk indicate p–n module property and the maximum quantity by optimizing the dimension of the material, respectively. The maximum efficiency and output power density of a TE module have been analytically calculated based on an assumption that S, ρ, and κ are temperature independent as[25]

$$\left[\eta\right]^*_{\text{CPM}} = \eta_c \frac{\sqrt{1+[Z]*T_{\text{avg}}}-1}{\sqrt{1+[Z]*T_{\text{avg}}}+T_c/T_h}, \tag{13.24}$$

$$\left[P_d\right]^*_{\text{CPM}} = \frac{(S_p - S_n)^2}{\left(\sqrt{L_p\rho_p}+\sqrt{L_n\rho_n}\right)^2} \frac{\Delta T^2}{4}. \tag{13.25}$$

This is reasonably accurate when its temperature difference ΔT is small or when $[Z]$ is temperature independent over the operating temperature range, but most TE material properties are strongly temperature dependent. In addition, TE modules require larger temperature gradient on TE modules. Hence, Equation 13.24 based on the CPM is not reliable for the prediction of the practical conversion efficiency of TE modules operating at a large temperature difference between hot and cold sides. As TE modules operate with pairing of p-type and n-type materials, a balance of thermal flow and electric current of dissimilar p-type and n-type materials is also required. This section discusses the maximum efficiency and power density of TE modules based on the CTD model leading to more reliable predictions than the CPM and the maximum device engineering figure of merit $[ZT]^*_{\text{eng}}$ and the maximum device engineering PF density $[PF]^*_{\text{eng,d}}$ as direct indicators of a TE module at practical temperature differences. Three different p–n pairings are investigated: (1) SKU: $Ce_{0.45}Nd_{0.45}Fe_{3.5}Co_{0.5}Sb_{12}$ (p-type)[6] and $Ba_{0.08}La_{0.05}Yb_{0.04}Co_4Sb_{12}$ (n-type);[26] (2) chalcogenides: $K_{0.02}Pb_{0.98}Te_{0.75}Se_{0.25}$ (p-type)[11] and $Pb_{0.995}SeCr_{0.005}$ (n-type)[27]; and (3) chalcogenides: SnSe (p-type)[15] and PbSe (n-type).[27]

13.2.1 Cumulative Temperature Dependence Model of p–n Module

The 1D governing equation for a steady-state heat flow in TE module is shown in Equation 13.2. Integrating Equation 13.2 twice with respect to x, with the application of the boundary conditions $T|_{x=0} = T_h$ and $T|_{x=Lp} = T_c$, yields the conduction heat at T_h of p-type leg as

$$-A_p \kappa_p(T) \frac{dT}{dx}\bigg|_{T_h} = \frac{A_p}{L_p} \int_{T_c}^{T_h} \kappa_p(T) dT - \frac{I^2}{A_p L_p} \int_0^{L_p} \int_0^x \rho_p(T) dx\,dx + \frac{I}{L_p} \int_0^{L_p} \int_{T_h}^T \tau_p(T) dT\,dx.$$

(13.26)

The conduction heat of n-type leg is also obtained in the same way. From the constitutive relation for heat Q and electric current I by Equation 13.7, the heat transferred into each TE leg at $T = T_h$ becomes

$$Q_{h,p} = I_p T_h S_p(T_h) + \frac{A_p}{L_p} \int_{T_c}^{T_h} \kappa_p(T) dT - W_{J,p} I_p^2 R_p - W_{T,p} I_p \int_{T_c}^{T_h} \tau_p(T) dT,$$

(13.27)

$$Q_{h,n} = I_n T_h S_n(T_h) + \frac{A_n}{L_n} \int_{T_c}^{T_h} \kappa_n(T) dT - W_{J,n} I_n^2 R_n - W_{T,n} I_n \int_{T_c}^{T_h} \tau_n(T) dT,$$

where R_p and R_n are ΔT-dependent electrical resistance at a given thermal boundary, defined as $R = (LA^{-1}\Delta T^{-1}) \int_{T_c}^{T_h} \rho(T) dT$. The terms on the right-hand side in Equation 13.27 represent Peltier, conduction, Joule, and Thomson heat, respectively. W_J and W_T are ΔT-dependent weight factors for practical fraction to the hot side of the total Joule and Thomson heat associated with the temperature dependence of $\rho(T)$ and $\tau(T)$, defined as[23]

$$W_{J,p\,or\,n} = \frac{\int_{T_c}^{T_h} \int_T^{T_h} \rho_{p\,or\,n}(T) dT\,dT}{\Delta T \int_{T_c}^{T_h} \rho_{p\,or\,n}(T) dT},$$

(13.28)

$$W_{T,p\,or\,n} = \frac{\int_{T_c}^{T_h} \int_T^{T_h} \tau_{p\,or\,n}(T) dT\,dT}{\Delta T \int_{T_c}^{T_h} \tau_{p\,or\,n}(T) dT}.$$

(13.29)

By Equation 13.27, the total input heat at T_h becomes

$$Q_h = Q_{h,p} + Q_{h,n}$$

$$= IT_h \left[S_p(T_h) - S_n(T_h) \right] + \frac{A_p}{L_p} \int_{T_c}^{T_h} \kappa_p(T) dT + \frac{A_n}{L_n} \int_{T_c}^{T_h} \kappa_n(T) dT \tag{13.30}$$

$$- (W_{J,p} R_p + W_{J,n} R_n) I^2 - \left[W_{T,p} \int_{T_c}^{T_h} \tau_p(T) dT - W_{T,n} \int_{T_c}^{T_h} \tau_n(T) dT \right] I,$$

where $I = I_p = -I_n$. The output power accounting for Thomson effect is[28]

$$P_{out} = \frac{V_{oc}^2}{R_{int}} \frac{m}{(1+m)^2}, \tag{13.31}$$

where V_{oc} is the open circuit voltage expressed as

$$V_{oc} = \int_{T_c}^{T_h} S_p(T) dT - \int_{T_c}^{T_h} S_n(T) dT. \tag{13.32}$$

m is the ratio of load (R_L) to internal ($R_{int} = R_p + R_n$) electrical resistance, $m = R_L/R_{int}$. The conversion efficiency of the module is expressed as the ratio of the output power to input heat:

$$\eta = \frac{P_{out}}{Q_h}. \tag{13.33}$$

Equations 13.30 through 13.33 yield

$$\eta = \frac{\dfrac{m}{1+m}}{\dfrac{1+m}{[ZT]_{eng}} + \dfrac{T_h \left(S_p(T_h) - S_n(T_h) \right)}{\displaystyle\int_{T_c}^{T_h} S_p(T) dT - \int_{T_c}^{T_h} S_n(T) dT} - \dfrac{W_{J,p} R_p' + W_{J,n} R_n'}{1+m} - (W_{T,p} \tau_p' + W_{T,n} \tau_n')}, \tag{13.34}$$

where $R'_p = R_p/R_{int}$ and $R'_n = R_n/R_{int}$; τ'_p and τ'_n are expressed as

$$\tau'_{p\,or\,n} = \frac{\int_{T_c}^{T_h} \tau_{p\,or\,n}(T)dT}{\left(\int_{T_c}^{T_h} S_p(T)dT - \int_{T_c}^{T_h} S_n(T)dT\right)}. \tag{13.35}$$

Here, the engineering dimensionless figure of merit of p–n module $[ZT]_{eng}$ is defined as

$$[ZT]_{eng} = \frac{\left(\int_{T_c}^{T_h} S_p(T)dT - \int_{T_c}^{T_h} S_n(T)dT\right)^2}{\left(\dfrac{L_p}{A_p}\int_{T_c}^{T_h} \rho_p(T)dT + \dfrac{L_n}{A_n}\int_{T_c}^{T_h} \rho_n(T)dT\right)\left(\dfrac{A_p}{L_p}\int_{T_c}^{T_h} \kappa_p(T)dT + \dfrac{A_n}{L_n}\int_{T_c}^{T_h} \kappa_n(T)dT\right)}\Delta T. \tag{13.36}$$

The engineering PF of p–n module $[PF]_{eng}$ in watts per Kelvin is defined as

$$[PF]_{eng} = \frac{\left[\int_{T_c}^{T_h} S_p(T)dT - \int_{T_c}^{T_h} S_n(T)dT\right]^2}{(L_p/A_p)\int_{T_c}^{T_h} \rho_p(T)dT + (L_n/A_n)\int_{T_c}^{T_h} \rho_n(T)dT}. \tag{13.37}$$

By optimizing the ratio m satisfying $d\eta/dm = 0$ regarding Equation 13.34, the optimized efficiency and its corresponding $[m]_{CTD}$ are obtained as

$$[\eta]_{CTD} = \eta_c \frac{\sqrt{1+[ZT]_{eng}[\alpha]_1 \eta_c^{-1}} - 1}{[\alpha]_0\sqrt{1+[ZT]_{eng}[\alpha]_1 \eta_c^{-1}} + [\alpha]_2}, \tag{13.38}$$

$$[m]_{CTD} = \sqrt{1+[ZT]_{eng}[\alpha]_1 \eta_c^{-1}}, \tag{13.39}$$

where $[\alpha]_i$ is defined as

$$[\alpha]_i = \frac{\left(S_p(T_h)-S_n(T_h)\right)\Delta T}{\int_{T_c}^{T_h} S_p(T)dT - \int_{T_c}^{T_h} S_n(T)dT} - (W_{T,p}\tau'_p - W_{T,n}\tau'_n)\eta_c - i(W_{J,p}R'_p + W_{J,n}R'_n)\eta_c, \tag{13.40}$$

where $i = 0, 1,$ and 2. If S, ρ, and κ are assumed to be temperature independent by the CPM, Equation 13.40 is reduced to

$$[\alpha]_i = 1 - \frac{\eta_c}{2} i. \tag{13.41}$$

Due to $[ZT]_{eng} = [Z]\Delta T$ from Equation 13.36 for temperature-independent properties, $[m]_{CTD}$ by Equation 13.39 becomes

$$[m]_{CPM} = \sqrt{1 + [Z] \cdot T_{avg}}, \tag{13.42}$$

and its corresponding efficiency is identical to Equation 13.24, which indicates that Equations 13.38 through 13.40 are generic formulas that can be converted to the expression based on conventional as well as the CTD model.

The optimized efficiency by Equation 13.38 can be maximized by matching the leg size of p-type and n-type materials for balancing thermal flow and electric current. Taking the derivative of the denominator of Equation 13.36 and setting it equal to zero yields β_e for the maximum $[ZT]_{eng}$ as

$$\beta_e = \frac{A_n L_p}{A_p L_n} = \sqrt{\frac{\int_{T_c}^{T_h} \rho_n(T) dT \int_{T_c}^{T_h} \kappa_p(T) dT}{\int_{T_c}^{T_h} \rho_p(T) dT \int_{T_c}^{T_h} \kappa_n(T) dT}}, \tag{13.43}$$

and the corresponding maximum engineering figure of merit of the module $[ZT]_{eng}^*$ and the maximum efficiency become

$$[ZT]_{eng}^* = [Z]_{eng}^* \Delta T = \frac{\left(\int_{T_c}^{T_h} S_p(T) dT - \int_{T_c}^{T_h} S_n(T) dT \right)^2}{\left(\sqrt{\int_{T_c}^{T_h} \kappa_p(T) dT \int_{T_c}^{T_h} \rho_p(T) dT} + \sqrt{\int_{T_c}^{T_h} \kappa_n(T) dT \int_{T_c}^{T_h} \rho_n(T) dT} \right)^2} \Delta T, \tag{13.44}$$

$$[\eta]_{CTD}^* = \eta_c \frac{\sqrt{1 + [ZT]_{eng}^* [\alpha]_1 \eta_c^{-1}} - 1}{[\alpha]_0 \sqrt{1 + [ZT]_{eng}^* [\alpha]_1 \eta_c^{-1}} + [\alpha]_2}. \tag{13.45}$$

When S, ρ, and κ are temperature independent, $[Z]^*_{\text{eng}}$ in Equation 13.44 is reduced to

$$[Z]^*_{\text{eng}} = [Z]^* = \frac{(S_p - S_n)^2}{\left(\sqrt{\kappa_p \rho_p} + \sqrt{\kappa_n \rho_n}\right)^2}, \qquad (13.46)$$

where $[Z]^*$ is the optimal figure of merit for the maximum efficiency based on the CPM, and its corresponding maximum efficiency becomes identical to Equation 13.24.

The output power density P_d in watts per square meter by Equations 13.31 and 13.37 is expressed as

$$P_d = \frac{P_{\text{out}}}{A_p + A_n}$$

$$= \frac{V_{oc}^2}{(A_p + A_n) R_{\text{int}}} \frac{m}{(1+m)^2} \Delta T$$

$$= \frac{\left(\int_{T_c}^{T_h} S_p(T)\,dT - \int_{T_c}^{T_h} S_n(T)\,dT\right)^2}{(A_p + A_n)\left[(L_p/A_p)\int_{T_c}^{T_h} \rho_p(T)\,dT + (L_n/A_n)\int_{T_c}^{T_h} \rho_n(T)\,dT\right]} \frac{m}{(1+m)^2} \Delta T$$

$$= \frac{[PF]_{\text{eng}}}{(A_p + A_n)} \frac{m}{(1+m)^2} \Delta T$$

$$= [PF]_{\text{eng,d}} \frac{m}{(1+m)^2} \Delta T. \qquad (13.47)$$

When the optimal value of m is 1 by taking the derivative $m/(1 + m)^2$ with respect to m and setting it to zero, the output power density P_d is proportional to the engineering PF density $[PF]_{\text{eng,d}}$ at a given ΔT, so the maximum P_d and its corresponding condition β_p are obtained by taking the derivative of the denominator in Equation 13.47 with respect to A_n/A_p,

$$[P_d]^*_{\text{CTD}} = [PF]^*_{\text{eng,d}} \frac{\Delta T}{4L}$$

$$= \frac{\left(\int_{T_c}^{T_h} S_p(T)\,dT - \int_{T_c}^{T_h} S_n(T)\,dT\right)^2}{\left(\sqrt{\int_{T_c}^{T_h} \rho_p(T)\,dT} + \sqrt{\int_{T_c}^{T_h} \rho_n(T)\,dT}\right)^2} \frac{T}{4L}, \qquad (13.48)$$

$$\beta_p = \frac{A_n}{A_p} = \sqrt{\frac{\int_{T_c}^{T_h} \rho_n(T)\,dT}{\int_{T_c}^{T_h} \rho_p(T)\,dT}}, \tag{13.49}$$

where $[PF]_{eng,d}^*$ is the maximum engineering PF density of a module, and $L = L_p = L_n$. $[P_d]_{CTD}^*$ (Equation 13.48) is transformed to $[P_d]_{CPM}^*$ (Equation 13.25), assuming that the material properties are constant along with the temperatures. Thus, $[\eta]_{CTD}^*$ and $[P_d]_{CTD}^*$ are generic expressions for the maximum efficiency and output power density of a module device.

13.2.2 Maximum Efficiency and Output Power Density of p–n Module

The ΔT-dependent maximum efficiency for p-SKU/n-SKU is shown in Figure 13.5a, where T_c is fixed at 100°C and T_h is ramped up to 500°C. The efficiency predictions by the CPM and CTD models have good agreement in the whole temperature range, and show 2.6% and 2.5% of relative difference compared with results by the numerical simulation at $\Delta T = 400$°C (Figure 13.5b), respectively. The conventional model can show the reliable efficiency prediction for p-SKU/n-SKU because S, ρ, and κ of p-type and n-type SKUs have weak temperature dependence,[6,26] i.e., linear behavior of ZT. Thus, the conventional model can be simply utilized in such limited material systems. For p-PbTe/n-PbSe, however, the CPM underestimates the maximum efficiency over the whole temperature range (Figure 13.5c) and gives rise to 27% of relative difference compared with the simulation result at $\Delta T = 400$°C (Figure 13.5d), whereas the efficiency prediction by the CTD model becomes accurate over the whole temperature range and leads to only 2.8% of underestimation compared to the simulation result. For p-SnSe/n-PbSe-paired module shown in Figure 13.5e and f, the CTD model leads to the relative difference by 3% while the conventional model overrates it by 17% compared with the simulation. In some cases, however, the conventional model shows a more accurate prediction. For example, in p-SnSe/n-PbSe, if $\Delta T = 500$°C, the CPM only overrates by 2.5% as shown in the inset of Figure 13.5e, while it predicts mostly inaccurate efficiency in the rest of temperature range (Figure 13.5e). This is because the averaged constant values of S, ρ, and κ at $\Delta T = 500$°C are by chance very close to each equivalent value which includes the influence of temperature-dependent properties as well as the Thomson effect taken into account by the numerical simulation. Thus, the prediction by the CPM is not reliable since it sometimes has inadvertent accuracy with lack of analogy.

Figure 13.6 shows the maximum P_d with respect to ΔT for three p–n module configurations in which the temperature gradient is assumed to be 200°C mm^{-1}, where $L_p = L_n$ in a typical π-shape module, and the temperature profiles of each TE leg at $T_h = 500$°C by the numerical computation.

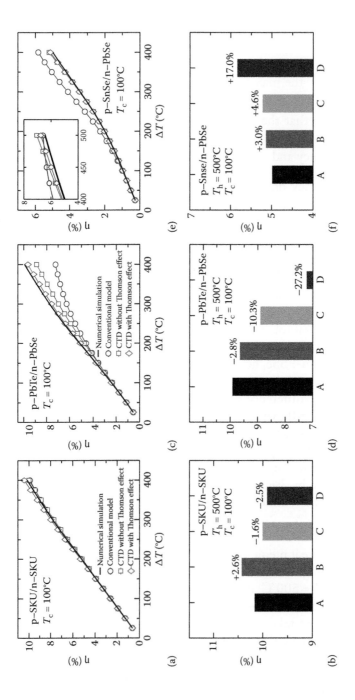

FIGURE 13.5 Maximum efficiency as a function of ΔT and its relative difference compared to the numerical simulation at $\Delta T = 400°C$ for (a, b) p-SKU/n-SKU, (c, d) p-PbTe/n-PbSe, and (e, f) p-SnSe/n-PbSe. The efficiency of p-SnSe/n-PbSe up to $\Delta T = 500°C$ is shown as the inset in e. In (b), (d), and (f), A represents numerical simulation; B, CTD model with Thomson effect; C, CTD model without Thomson effect; and D, conventional model (CPM). The CTD model without Thomson effect takes into account the cumulative effect of material properties but assumes that the accumulated Thomson coefficient is zero to investigate the effect of Thomson heat on the efficiency.

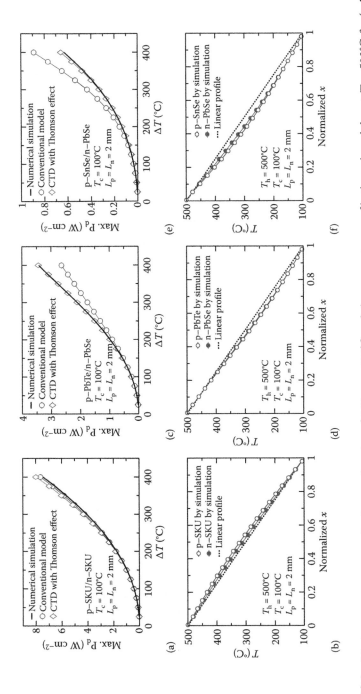

FIGURE 13.6 Maximum output power density as a function of ΔT up to 400°C and the temperature profile through each leg at $T_h = 500$°C for (a, b) p-SKU/n-SKU, (c, d) p-PbTe/n-PbSe, and (e, f) p-SnSe/n-PbSe. The length of legs was determined based on the temperature gradient $\Delta T_{max}/L = 200$°C/mm.

The analytical prediction for p-SKU/n-SKU by the CPM has good agreement with the simulation results by 2.7% at $\Delta T = 400°C$ (Figure 13.6a) due to the high linearity of ZT curves of each SKU material while the maximum P_d predictions for p-PbTe/n-PbSe (Figure 13.6c) and p-SnSe/n-PbSe (Figure 13.6e) lead to –23 and 42% of relative difference at $\Delta T = 400°C$ compared to the simulation results, respectively. As the prediction by the CPM is reliable only when TE properties are not much varied with temperature such as p-type and n-type SKUs, the CTD model computes a more accurate prediction through the whole temperature range (Figure 13.6c and e) and gives rise to 0.5 and 4.9% of relative difference at $\Delta T = 400°C$ for p-PbTe/n-PbSe and p-SnSe/n-PbSe, respectively, which shows that Equation 13.48 is more reliable by taking Thomson effect into consideration. The differences between the CTD model and numerical simulations are mainly caused by the assumption of the linear approximation for temperature gradient in evaluating R_{int} for integrating $\rho(T)$ with respect to x. In p-SKU/n-SKU, the temperature profiles by the simulations is over the linear approximation in terms of x (solid line) in Figure 13.6b, for which lower R_{int} based on the linear approximation is obtained due to $\rho \propto T$ in SKUs.[6,26] This results in the overestimated output power density by the CTD model compared to the numerical simulation as shown in Figure 13.6a. In contrast, since the temperature distributions of p-PbTe/n-PbSe by the simulation are lower than the linear profile (Figure 13.6d), the overestimated R_{int} by this model leads to the lower output power density (Figure 13.6c) as ρ of p-PbTe and n-PbSe is proportional to T.[11,27] For p-SnSe/n-PbSe, the temperature profiles of both p-type and n-type leg are below the linear distribution with respect to x (Figure 13.6f), but ρ of the p-type SnSe is higher than that of n-type PbSe by two orders of magnitude and inversely proportional to T.[15] This causes the overestimated output power density by this model (Figure 13.6e) resulting from the underrated R_{int} that is dominant by R_p of SnSe. Therefore, the CTD model including the Thomson effect more correctly predicts the output power generation than the CPM, and the main difference between the CTD model and the numerical analysis is caused by the linear approximation for the temperature distribution in this model. Another possible reason for the difference is that this model accounts for the cumulative effect of the temperature dependence of each property at thermal boundaries, which is path independent according to the temperature while the numerical model incorporates the instantaneous temperature dependence that is path dependent associated with the temperature.

In Figure 13.7, β_p (solid symbols) for the maximum output power density is obtained by Equation 13.49 and compared with β_e (open symbols). For small variations of S, ρ, and κ between p- and n-type materials such as p-SKU/n-SKU and p-PbTe/n-PbSe, the maximum efficiency is not much different from those at $A_p = A_n$ ($\beta_e = 1$), but for p-SnSe/n-PbSe showing large property variations between p- and n-type materials, the maximum conditions should be examined instead of assuming β_e or $\beta_p = 1$. The relative differences between β_p and β_e for p-SKU/n-SKU, p-PbTe/n-PbSe, and p-SnSe/n-PbSe are 1.5, 14, and 65%, respectively, where the difference

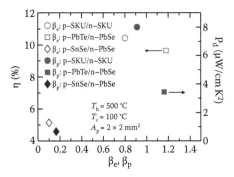

FIGURE 13.7 Comparison of β_e (*open symbols*) and β_p (*solid symbols*) corresponding to the maximum efficiency and maximum output power density, respectively.

between them is caused by the ratio of thermal conductivity of p-type and n-type materials (Equations 13.43 and 13.49). However, the efficiencies led by β_e and β_p are close each other within 1% in p-SKU/n-SKU and p-SnSe/n-PbSe and within 4% in p-SnSe/n-PbSe since they are on the plateau curve near the peak efficiency of each p–n configuration. Thus, finding the range of β_e or β_p is crucial for a thermal and electric balance of p–n configuration, but it is not a critical design parameter of a module to select either β_e or β_p unless a fine tuning of a module is necessary to see what the primary concern of the module is, e.g., its efficiency or output power.

13.2.3 Weight Factors for Practical Contribution of Joule and Thomson Heat in p–n Module

To examine the effect of Thomson heat on the TE module based on the CTD model, Figure 13.8 shows the cumulative effect of practical Joule and Thomson heat on the input heat flux at the hot side, respectively. The CPM assumes half of the Joule heat returning to the hot and cold sides by the temperature-independent ρ. The concept of the CTD provides a practical contribution of Joule heat on each thermal boundary of a single TE leg as

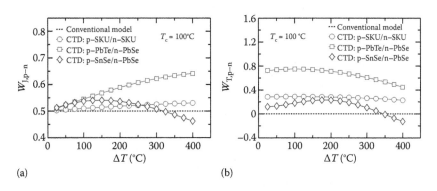

FIGURE 13.8 The cumulative effect of (a) Joule heat and (b) Thomson heat on the input heat flux at the hot side.

W_J, and its combination effect in p–n pair module is defined by referring to Equation 13.34.

$$W_{J,p-n} = W_{J,p}R_p' + W_{J,n}R_n'. \tag{13.50}$$

In Figure 13.8a, $W_{J,p-n}$ of p-SKU/n-SKU and p-PbTe/n-PbSe is over 1/2, indicating more Joule heat on the hot side than that based on the CPM, and it has an increasing trend as ΔT increases due to $d\rho/dT > 0$ of each material. In p-SnSe/n-PbSe, $W_{J,p-n}$ fluctuates and becomes below 1/2 at $\Delta T = 300°C$ due to a large reduction of ρ of SnSe with higher temperature.[15] Figure 13.8b shows the combined effect of Thomson heat for input heat flux, which is also defined as

$$W_{T,p-n} = W_{T,p}\tau_p' - W_{T,n}\tau_n'. \tag{13.51}$$

$W_{T,p-n}$ for p-SKU/n-SKU and p-PbTe/n-PbSe gradually decreases with higher temperature due to the decreasing tendency of $d|S|/dT$, indicating that less Thomson effect is associated. For p-SnSe/n-PbSe, $W_{T,p-n}$ shows a parabolic behavior up to $\Delta T = 350°C$ and decreases below zero. The negative $W_{T,p-n}$ corresponds to the cumulative effect of heat absorbed by Thomson effect and exerts an influence on decreasing the conversion efficiency at the given thermal boundary condition compared with the model without Thomson effect, as shown in Figure 13.5e. In contrast, the positive $W_{T,p-n}$ for p-SKU/n-SKU and p-PbTe/n-PbSe has an effect on increasing the efficiency compared with that in the absence of Thomson heat as shown in Figure 13.5a and c.

13.2.4 Relationship between $(ZT)_{eng}$ and $[ZT]_{eng}^*$

Figure 13.9a shows $[ZT]_{eng}^*$ as a function of ΔT for p-SKU/n-SKU, p-PbTe/n-PbSe, and p-SnSe/n-PbSe, and the maximum efficiency as inset. Note that $[ZT]_{eng}^*$ has a close correspondence to the maximum efficiency (inset), which means that $[ZT]_{eng}^*$ is the practical indicator to judge which p–n configuration is superior in terms of the efficiency. In Figure 13.9b, $[PF]_{eng,d}^*$ is the intrinsic output power density per ΔT as a material property, so it directly represents the power generation at given temperature boundaries and the ratio m by showing similar tendency to P_d at matched load condition (inset). Thus, $[PF]_{eng,d}^*$ is also the practical indicator for the output power generation of each p–n module. Figure 13.9c through e shows $[ZT]_{eng}^*$, $(ZT)_{eng}$ of p- and n-type materials, and their averaged $(ZT)_{eng}$ for each p–n configuration, where the parentheses denotes the engineering figure of merit of a single material. For p-SKU/n-SKU and p-PbTe/n-PbSe, the simply averaged $(ZT)_{eng}$ of each p- and n-type materials has a good agreement with $[ZT]_{eng}^*$ of their corresponding p–n module within 1% as shown in Figure 13.9c and d while the averaged $(ZT)_{eng}$ of p-SnSe/n-PbSe is overrated by 47% compared to $[ZT]_{eng}^*$ of the module at $\Delta T = 400°C$ shown in Figure 13.9e. The

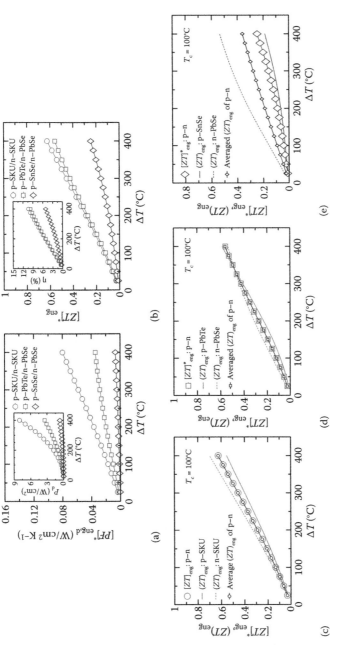

FIGURE 13.9 (a) Maximum engineering figure of merit of p–n module $[ZT]^*_{eng}$ and the efficiency (*inset*) at $T_c = 100°C$ and (b) the maximum engineering PF density $[PF]^*_{eng,d}$ and output power density (*inset*) as a function of ΔT of the p–n configurations, where $T_c = 100°C$ and $L_n = L_p = 2$ mm. Comparison of $[ZT]^*_{eng}$ (*larger symbols*), $(ZT)_{eng}$ of individual p- (*solid lines*) and n-type (*dashed lines*) materials, and their averaged $(ZT)_{eng}$ (*smaller symbols*) in (c) p-SKU/n-SKU, (d) p-PbTe/n-PbSe, and (e) p-SnSe/n-PbSe.

differences is caused by the correspondence of the cumulative temperature-dependent S, ρ, and κ for p- and n-type materials in the p–n module, so one should be careful to predict a module efficiency by simply averaging the ZTs of p- and n-type materials based on the CPM.

13.3 Correlation between Performance Indicators and Materials/Devices Characteristics

13.3.1 *ZT* vs. *(ZT)*$_\text{eng}$

As discussed in the previous sections, due to the necessity of describing the overall performance throughout the operating temperature range of a material, many have reported an average ZT $((ZT)_\text{avg})$ of a material with given thermal boundary conditions.[16,29–31] However, the $(ZT)_\text{avg}$ is insufficient because of its variation depending on how it is averaged. Alternatively, an effective ZT $((ZT)_\text{eff})$ has been used,[32–34] which is more reliable than the $(ZT)_\text{avg}$, but it does not account for the effect of temperature difference, while indicating an equivalent value at an averaged temperature. It has been shown that the engineering figure of merit $(ZT)_\text{eng}$ and engineering PF $(PF)_\text{eng}$ represent a cumulative performance of a material at given boundary temperatures which accounts for the overall temperature dependence of each property. Figure 13.10a shows the temperature-dependent ZT of selected p-type and n-type TE materials, representing instantaneous performance at a temperature point. This conventionally used indicator for TE conversion efficiency does not show which material performs better than the others without additional complicated calculations. Figure 13.10b shows its corresponding $(ZT)_\text{eng}$, which is a ΔT-dependent indicator for efficiency at given temperature boundary conditions. This section shows one-to-one correspondence between $(ZT)_\text{eng}$ and conversion efficiency of a material, and the output power density has a linear relation to $(PF)_\text{eng}$ as well. The advantage is that one can immediately determine the practical efficiency of a material

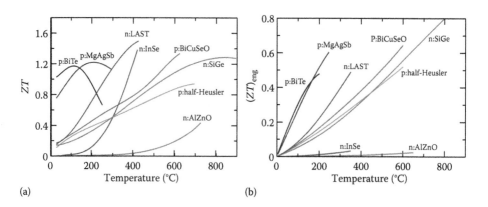

FIGURE 13.10 ZT vs. T and $(ZT)_\text{eng}$ vs. T. (a) ZT dependence of temperature and (b) $(ZT)_\text{eng}$ dependence of ΔT of selected n-type and p-type materials. For $(ZT)_\text{eng}$, a cold-side temperature T_c is assumed at 50°C for all materials.

at a given thermal boundary condition once $(ZT)_{\text{eng}}$ is known, regardless of what the material is and their temperature differences are. However, the average and effective ZT as well as conventional ZT and PF have no such clear indication as to its efficiency and output power. In addition, the use of the efficiency vs. $(ZT)_{\text{eng}}$ and power density vs. $(PF)_{\text{eng}}$ relationships enables one to avoid misdirected TE material exploration.

13.3.2 $(ZT)_{\text{eng}}$ vs. Efficiency

The efficiency of state-of-the-art p-type and n-type materials at its corresponding maximum operable temperature is plotted in Figure 13.11a with respect to its average ZT defined as

$$(ZT)_{\text{avg}} = \frac{1}{\Delta T} \int_{T_{\text{c}}}^{T_{\text{h}}} ZT(T)\,\mathrm{d}T, \tag{13.52}$$

where T_{c} is fixed at 50°C for all materials. The efficiency η_n is numerically computed based on a finite-difference analysis which accounts for the temperature dependence of S, ρ, and κ as well as the Thomson effect. The full list of p-type and n-type materials and their maximum operable temperatures $T_{\text{h,max}}$

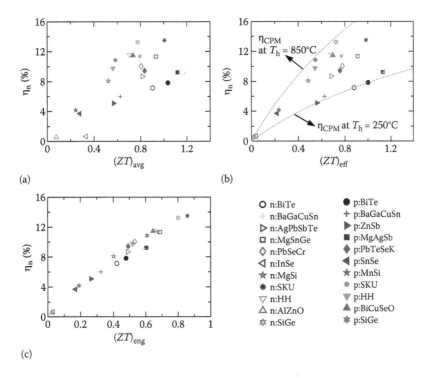

(a)

(b)

(c)

O n:BiTe   ● p:BiTe
+ n:BaGaCuSn   + p:BaGaCuSn
▷ n:AgPbSbTe   ▶ p:ZnSb
□ n:MgSnGe   ■ p:MgAgSb
◇ n:PbSeCr   ◆ p:PbTeSeK
◁ n:InSe   ◀ p:SnSe
★ n:MgSi   ★ p:MnSi
✳ n:SKU   ✳ p:SKU
▽ n:HH   ▼ p:HH
△ n:AlZnO   ▲ p:BiCuSeO
✩ n:SiGe   ✴ p:SiGe

FIGURE 13.11 Efficiency vs. ZT, $(ZT)_{\text{eff}}$, and $(ZT)_{\text{eng}}$. Predicted efficiency of materials as a function of (a) average $(ZT)_{\text{avg}}$, (b) effective $(ZT)_{\text{eff}}$, and (c) engineering $(ZT)_{\text{eng}}$, where all efficiencies are evaluated by numerical simulation accounting for the temperature dependence of each property. T_{c} is fixed at 50°C for all efficiencies.

TABLE 13.1 Maximum Operable Temperature of p-Type and n-Type Materials at Hot Side

p-Type			n-Type		
Material	$T_{h,max}$ (°C)	Ref.	Material	$T_{h,max}$ (°C)	Ref.
$Bi_{0.4}Sb_{1.6}Te_3$	250	a	$Bi_2Te_{2.7}Se_{0.3}S_{0.015}$	250	a
β-Zn_4Sb_3	250	Toberer et al.[35]	$Ba_8Ga_{14}Cu_2Sn_{30}$	275	Saiga et al.[36]
$Ba_8Ga_{15.75}Cu_{0.25}Sn_{30}$	275	Saiga et al.[36]	$AgPb_mSbTe_{m+2}$	400	Zhou et al.[37]
MgAgSb	295	Zhao et al.[38]	In_4Se_{3-x}	400	Rhyee et al.[10]
$Pb_{0.98}Te_{0.75}Se_{0.25}K_{0.02}$	450	Zhang et al.[11]	$Mg_2Sn_{0.75}Ge_{0.25}$	450	Liu et al.[39]
SnSe	450	Zhao et al.[15]	$Pb_{0.995}SeCr_{0.005}$	450	Zhang et al.[27]
$MnSi_{1.78}$	500	Chen et al.[40]	$Ba_{0.08}La_{0.05}Yb_{0.04}$ Co_4Sb_{12}	550	Shi et al.[26]
$Ce_{0.45}Nd_{0.45}Fe_{3.5}$ $Co_{0.5}Sb_{12}$	550	Jie et al.[6]	Mg_2Si-0.5 at% Sb/1.0 at% Zn	550	Oto et al.[41]
$Hf_{0.19}Zr_{0.76}Ti_{0.048}$ $CoSb_{0.8}Sn_{0.2}$	650	He et al.[7]	$Hf_{0.25}Zr_{0.75}NiSn_{0.99}$ $Sb_{0.01}$	650	Chen et al.[42]
$Bi_{0.875}Ba_{0.125}CuSeO$	650	Sui et al.[43]	ZnO-0.25 at% Al	700	Jood et al.[44]
$Si_{80}Ge_{20}B_5$	850	Joshi et al.[45]	$Si_{80}Ge_{20}P_2$	850	Wang et al.[46]

Note: a: unpublished data.

are listed in Table 13.1. From Figure 13.11a, it is very clear from the scattered pattern that $(ZT)_{avg}$ is not an effective indicator of the potential for efficiency of a material. For example, n-type AlZnO and InSe give rise to similar efficiencies, even though the $(ZT)_{avg}$ of InSe is 0.33, about four times larger than that of AlZnO, the peak ZTs of AlZnO and InSe are also quite different: 0.4 at 700°C and 1.4 at 400°C, respectively, as shown in Figure 13.10a. The comparable efficiency yet large difference in $(ZT)_{avg}$ is also observed for p-type HH and MgAgSb. In contrast, n-type SiGe and AgPbSbTe have comparable $(ZT)_{avg}$, 0.78 and 0.82, respectively, but the efficiency of AgPbSbTe is lower than that of SiGe by 40%. Similarly, p-type Bi_2Te_3 has 30% lower efficiency while having 40% higher $(ZT)_{avg}$ compared to BiCuSeO. Thus, $(ZT)_{avg}$ is unable to represent the overall performance of a material, since the direct averaging of ZT leads to incorrect estimates depending on its convexity along with temperature. The $(ZT)_{avg}$ is only reasonable when ZT is perfectly linear to temperature. Instead of the direct averaging of temperature-dependent ZT, $(ZT)_{eff}$ gives rise to proper effective characteristics associated with the averaged temperature $T_{avg} = (T_h + T_c)/2$, defined as[32]

$$(ZT)_{eff} = \frac{\left(\int_{T_c}^{T_h} S(T) dT \right)^2}{\Delta T \int_{T_c}^{T_h} \rho(T)\kappa(T) dT} T_{avg}. \tag{13.53}$$

Figure 13.11b shows the efficiency corresponding to the $(ZT)_{\text{eff}}$ of the materials, in which the relation of η_n and $(ZT)_{\text{eff}}$ still shows the scattered pattern in between the dashed lines. The upper and lower lines indicate the η_{CPM} along with $(ZT)_{\text{eff}}$ when ΔT is 800 °C and 200°C, respectively, at $T_c = 50$°C. $(ZT)_{\text{eff}}$ indicates more practical averaged values than $(ZT)_{\text{avg}}$. For example, n-type InSe shows a reasonable value of $(ZT)_{\text{eff}}$ associated with its conversion efficiency. However, it still lacks the effect of given thermal boundary conditions by evaluating an equivalent ZT at an averaged temperature. $(ZT)_{\text{eff}}$ is also not able to represent the practical performance of a material associated with its operable temperature conditions. Figure 13.11c plots the efficiency of materials according to their $(ZT)_{\text{eng}}$, showing a linear correlation. From this η_n vs. $(ZT)_{\text{eng}}$ relation, one can directly recognize a conversion efficiency of a material by evaluating $(ZT)_{\text{eng}}$ without computing an efficiency. For example, a $(ZT)_{\text{eng}}$ of 0.6 represents about 10% conversion efficiency regardless of the type of material or the operating temperature range, making it is easy to compare the performance of different materials. This almost linear relation is the benefit of taking into account each accumulated property based on ΔT rather than T_{avg}. By the definition of $(ZT)_{\text{eff}}$ and $(ZT)_{\text{eng}}$, $(ZT)_{\text{eff}}$ is an underestimation compared to $(ZT)_{\text{eng}}$ when a given thermal boundary condition is $T_h > 3T_c$ (i.e., $\Delta T > T_{\text{avg}}$) and is overestimated when $T_h < 3T_c$ ($\Delta T < T_{\text{avg}}$).

In Figure 13.12a, the $(ZT)_{\text{eff}}$ of BaGaSn, MgSnGe, SKU, and HH is overestimated, and the relative differences between $(ZT)_{\text{eff}}$ and $(ZT)_{\text{eng}}$ decrease as T_h increases. $(ZT)_{\text{eff}}$ becomes underestimated when T_h increases and reaches over $3T_c$, at which point $(ZT)_{\text{eff}}$ is comparable to $(ZT)_{\text{eng}}$ only when T_h is 969 K at a given $T_c = 323$ K ($T_h = 3T_c$). In SiGe, $(ZT)_{\text{eff}}$ is underestimated compared to $(ZT)_{\text{eng}}$ when $T_h = 1123$ K. The solid line represents the conversion efficiency of materials with similar $(ZT)_{\text{eng}}$ and ΔTs, while $(ZT)_{\text{eff}}$ is not consistent with the efficiency trend.

As shown in Figure 13.11c, variation in the η_n vs. $(ZT)_{\text{eng}}$ correspondence is observed. Such variation from a perfect one-to-one mapping curve comes from the presence of instantaneous temperature dependence in S, ρ, and κ as well as the Thomson effect in the efficiency prediction by the numerical analysis. Figure 13.12b shows the ratio of the efficiency by numerical simulation to analytical prediction (Equation 13.1) of the materials in terms of $(ZT)_{\text{eng}}$, in which a relative difference represents a degree of off linearity in material properties. When ZT goes to infinity, the η_c is only dependent on a thermal boundary condition regardless of material systems (Figure 13.12c). A practical efficiency limit of TE materials is investigated by accounting for the different temperature dependence of each material within the same temperature gradient. Figure 13.12d shows a specific Carnot efficiency η_c^{CTD} that is a ratio of η_c to the degree of temperature dependence of materials. η_c^{CTD} is expressed based on $(ZT)_{\text{eng}}$ in the CTD model as

$$\eta_c^{\text{CTD}} = \lim_{(ZT)_{\text{eng}} \to \infty} \eta_{\text{CTD}} = \frac{\eta_c}{\alpha_0}. \tag{13.54}$$

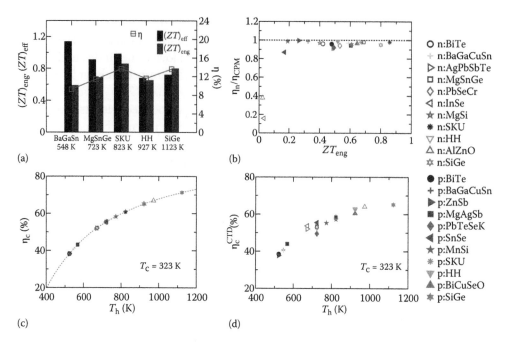

FIGURE 13.12 $(ZT)_{eng}$, $(ZT)_{eff}$, and Carnot efficiency by classic and CTD model. (a) $(ZT)_{eng}$ and $(ZT)_{eff}$ of the selected n-type materials, shown on the x axis in ascending order of the maximum hot-side temperature while the cold side temperature is fixed at 323 K. The solid curve with open square symbols represents the predicted efficiency, which follows the same trend of $(ZT)_{eng}$. (b) Relative efficiency, a ratio of η_n simulation to η_{CPM} shows the temperature dependence effect on efficiency evaluation, showing the efficiency prediction by the CPM is overestimated in most materials. Carnot efficiency according to the hot-side temperature by the (c) CPM and (d) CTD model.

When $ZT \to \infty$, the effect of S becomes significant. Thus, η_c^{CTD} in Equation 13.54 incorporates only the Peltier and Thomson effects associated with temperature-dependent S, represented in α_0. This means that focusing on temperature dependence as well as improving S improve TE efficiency.

13.3.3 $(PF)_{eng}$ vs. Output Power Density

The output power density P_d in terms of the average PF $(PF)_{avg}$ and engineering PF $(PF)_{eng}$ is shown in Figure 13.13, respectively, where

$$(PF)_{avg} = \frac{1}{\Delta T} \int_{T_c}^{T_h} PF(T)\mathrm{d}T. \tag{13.55}$$

The TE leg length was determined to satisfy a constant temperature gradient of 200°C/mm, e.g., 3 mm length of a material for $\Delta T = 600$°C. In Figure 13.13a, the $(PF)_{avg}$ has a scattered relation to P_d, similar to η_n vs. $(ZT)_{avg}$, as shown in Figure 13.11a. This shows that $(PF)_{avg}$ is insufficient in directly indicating the output power performance of materials. For example, p-type Bi_2Te_2 and HH have a comparable $(PF)_{avg}$ of 28 and 30 µW/cm K^2, respectively, but its corresponding P_d is more than four times

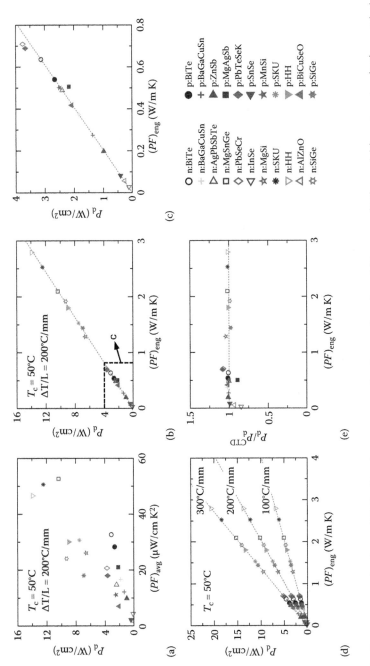

FIGURE 13.13 Output power density *vs. PF*. Predicted output power density as a function of (a) average $(PF)_{avg}$ and (b) engineering $(PF)_{eng}$, with the $(PF)_{eng}$ in 0–0.8 range shown in more detail in c. The dashed lines in b and c represent P_d evaluated by CTD model. $(PF)_{eng}$ *vs.* $(PF)_{eng}$ according to $(PF)_{eng}$ in different temperature gradients based on CTD model. (e) Relative P_d, a ratio of P_d by numerical simulation to $(PF)_{eng}$ by CTD model.

different due to different thermal boundary conditions. In addition, n-type MgSnGe and SiGe generate comparable P_d around 10 W/cm² while the $(PF)_{avg}$ of SiGe is much lower than that of MgSnGe. In contrast, Figure 13.13b and c illustrates the linear correlation between P_d and $(PF)_{eng}$ under a constant temperature gradient. This indicates that the $(PF)_{eng}$ of a material represents its output power density associated with a given temperature difference by accounting for the cumulative effect of the temperature dependence in $S(T)$ and $\rho(T)$. Some degrees of variation off of a perfect linear correspondence are observed in Figure 13.13b and c, in which the dashed lines are the P_d determined by the CTD model along with $(PF)_{eng}$. The output power density according to $(PF)_{eng}$ in different temperature gradients such as 100 and 300 °C/mm based on CTD model is shown in Figure 13.13d. Figure 13.13e shows the relative P_d, a ratio of P_d by numerical computation to P_d by CTD model, which explains that the variation comes from the fact that the prediction by CTD model takes into consideration the cumulative effect of the temperature-dependent properties. The CTD model accounts for path independence according to temperatures while the numerical model accounts for instantaneous temperature dependence.

13.3.4 Direct Indicators for p–n Module Devices

In the same manner, efficiencies and power densities of p–n pair modules can be directly expected by evaluating $[ZT]^*_{eng}$ and $[PF]^*_{eng,d}$, respectively. Figure 13.14a shows module efficiencies based on selected p-type and n-type materials according to $[ZT]^*_{eng}$, which has a linear relationship between $[\eta]^*_{CTD}$ and $[ZT]^*_{eng}$. Such a correlation indicates that $[ZT]^*_{eng}$ effectively represents an efficiency of a module device regardless of the types of p–n configuration as well as thermal boundary conditions, for example, a $[ZT]^*_{eng}$ of 0.6 directly indicates a module efficiency of about 10%. This is the same correlation as homogenous materials as shown in Figure 13.11c. Figure 13.14a shows $[P_d]^*_{CTD}$ according to the $[PF]^*_{eng,d}$ of module devices, in which a

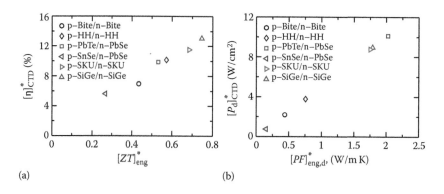

(a)   (b)

FIGURE 13.14 (a) Module efficiency as a function of the device engineering figure of merit $[ZT]^*_{eng}$ and (b) output power density of module devices according to device PF density $[PF]^*_{eng,d}$, where T_c is fixed at 50°C.

fixed temperature gradient is considered as 200°C/mm for all p–n configurations. Specifically, a module device of $[PF]^*_{eng,d} = 1$ is expected to generate the power density of 5 W/cm² at the given temperature gradient. It is shown that $(ZT)_{eng}$ and $(PF)_{eng}$ for homogeneous materials, and $[ZT]^*_{eng}$ and $[PF]^*_{eng,d}$ for p–n module devices have linear correspondence to TE conversion efficiency and output power density of materials and devices, respectively. These represent TE performance as direct indicators rather than conventional ZT and PF, which are characterized at specific temperatures with very small temperature differences. Such direct correlation shows the direct pathway to evaluate the conversion efficiency and output power of a material, leading to correct predictions of the potential of TE materials and devices.

13.4 Engineering Thermal Conductivity κ_{eng}: Dilemma of Thermal Conductivity Reduction in Practical Devices

For the past decade, reducing the thermal conductivity via nanostructuring has been widely adapted to improve ZT.[46–51] Recently, the peak ZT value was continuously raised in several materials due to the extremely low thermal conductivity, especially κ_{lat}, such as $Bi_{0.5}Sb_{1.5}Te_3$ (ZT_{peak} = 1.86; κ_{lat} = 0.3 W/m K),[52] PbTe:SrTe (ZT_{peak} = 2.2; κ_{lat} = 0.5 W/m K),[53] Cu_2Se (ZT_{peak} = 2.3; κ_{total} = 0.15 W/m K),[54] and SnSe (ZT_{peak} = 2.6; κ_{lat} = 0.23 W/m K).[15] Although the low thermal conductivity leads to higher ZT, it brings new challenges in designing a TE generator in device level, which has not been paid attention so far.

The lower thermal conductivity gives rise to higher thermal resistance of TE leg, which can generate larger temperature gradient across a leg with a given length or reduce a leg length corresponding to given boundary temperatures. This may lead to larger power generation and less consumption of a material. However, a reduced leg length due to the lowered thermal conductivity may cause several challenges to fabricate a TE generator: (1) larger shear stress at joining interfaces, (2) higher ratio of an electric contact resistance to an internal resistance of a leg, and (3) larger effect of thermal contact resistance at interfaces. The structural reliability of a TE generator is highly dependent on interfacial structures, in which thermal shear stresses are led under a large temperature gradient between T_h and T_c. Such thermally induced stresses cause some realignments of leg position,[55] cracks on a TE leg,[56] degradation of bonding strength,[57] etc., which eventually result in the deterioration of output power performance. In order to reduce the thermal stresses, variations of structural design such as tapered legs,[58] a linear structure with dovetail-shaped electrode,[59] an angled linear structure,[60] and the optimized dimensions of components[61,62] were investigated. However, none of the research has been focused on how the reduced thermal conductivity influences the device reliability of a TE module as well as how low the thermal conductivity of a TE material can be reduced from the device point of view.

In this section, an engineering thermal conductivity (κ_{eng}) of a TE material is defined, which is an allowable minimum thermal conductivity to keep a module device thermomechanically reliable and thermoelectrically high performable. The κ_{eng} at a steady state highlights the importance of a minimum level of the thermal conductivity for a reliable module design at static operations, and the κ_{eng} at a transient mode guides the sustainability of a TE material to transient thermal impacts. A methodology is presented to evaluate κ_{eng} for a device reliably operating at both steady-state and transient operations, and it is found that the thermal conductivities of the state-of-the-art p-type Bi_2Te_3 have already been close to the κ_{eng} from the device point of view, even though there is still room for further reduction of the thermal conductivity from the material point of view. This tells us that enhancing the *ZT* by reducing thermal conductivity influences the device reliability, so it cannot be without limit, and discuss how to improve the device reliability and TE performance in a balanced way by considering κ_{eng}.

13.4.1 Steady-State Operation Mode

13.4.1.1 Lower Limit of Leg Length for Reliable Thermal Shear Stress

The shearing stresses at a joint interface are evaluated in this section based on the one-pair p–n assembly model developed by Suhir et al.,[63–65] which was shown in Figure 13.15a and validated and compared with 2D and 3D finite-element analysis models.[66] This model assumed that a TE leg provides a mechanical compliance between ceramic plates at hot (top) and cold

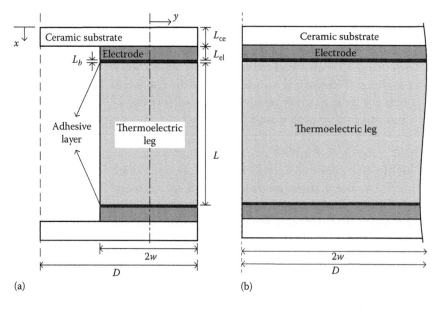

FIGURE 13.15 (a) Schematic illustration of the model for the shear stress evaluation. w is a half width of the leg and D is a half width of ceramic substrate in the one-pair assembly. (b) Infinitely long assembly with a continuous bond with a full fraction factor, the same cross-section area of a leg, electrode, and substrate.

(bottom) sides, and no thermal loading on legs is associated in parallel direction to the joining interface.[65]

The maximum shear stress τ_{max} occurs at the edge of TE legs is expressed as[65]

$$\tau_{max} = \tau_\infty \chi_r,$$

$$\tau_\infty = \frac{\alpha_{ce}\Delta T}{\sqrt{2AB\beta_{TE}^\infty \lambda_{TE}}},$$
(13.56)

$$\chi_r = \tanh kw \left[1 + \left\{2kw(\sqrt{f_{ce}} - 1)\sinh 2kw + \cosh 2kw\right\}^{-1}\right],$$

where τ_∞ is the shearing stress at the edge in an assembly of infinite width with a continuous bonding (Figure 13.15b) and χ_r is a dimensionless reduction factor associated with module design parameters such as a half width of a leg (w) and fraction factor (f_{ce}) showing a ratio of the cross-section areas of a ceramic to a TE leg. kw is the product of parameters of the interfacial shearing stress (k) and a half width of a leg (w):

$$kw = \sqrt{\frac{2B\lambda_{TE}w^2}{\beta_{TE} + (A-1)\beta_{TE}^\infty}}.$$
(13.57)

α_{ce} and λ_{TE} are the CTE of a ceramic plate and compliance of a TE leg, respectively. β_{TE}^∞ and β_{TE} are the interfacial compliance of a TE leg at infinite w (i.e., $f_{ce} = 1$ as shown in Figure 13.15b) and finite w (i.e., $f_{ce} > 1$ as shown in Figure 13.15a), respectively, as a function of leg length (L). A and B are constants associated with the ratio of mechanical properties and length of each component. It is important to figure out the L_{min} of a leg to prevent the deformation of the adhesive layer, because it causes realignment[55] of the leg position, which additionally results in a torsional stress field through a TE leg and deteriorates the output power generation by degrading the quality of the interfacial structures. The L_{min} can also be evaluated based on other yield strength criterion such as that of a diffusion barrier layer or newly formed phase from chemical reactions if it is lower than the yield strength of the solder. In this case, the L_{min} is the minimum length to avoid a failure of bonding between the leg and metallization layer on it before yielding the solder. For a reliable design, the induced shear stress by Equation 13.56 should be less than the yield strength criterion τ_Y, $\tau_\infty\chi_r < \tau_Y$, in which the L_{min} can be determined at $\chi_r = \tau_Y/\tau_\infty$, where dimensionless χ_r is in the range from 0 to 1 and τ_Y/τ_∞ is usually less than 1. In Figure 13.16a, according to increasing L at fixed length of other components, τ_∞ is reduced showing τ_Y/τ_∞ increasing at a certain τ_Y, and χ_r decreases along with L while smaller w influences more reduction of the shear stress. This means that a shorter and/or wider shape of a leg generates larger maximum shear stress. A minimum length of a leg at each width can be evaluated to avoid yielding, where L corresponding to the intersection of

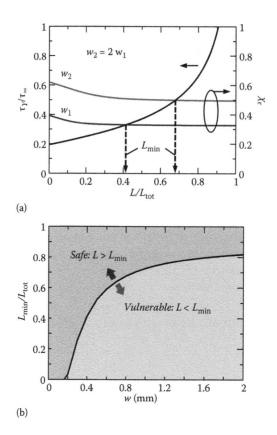

(a)

(b)

FIGURE 13.16 (a) Ratio of a yield strength criterion τ_Y to ultimate maximum shear stress τ_∞ and dimensionless reduction factor χ_r of the shear stress as a function of nondimensional length L/L_{tot} (L: leg length; $L_{tot} = L + L_{sub}$; L_{sub} is the sum of the lengths of ceramics, metal electrodes, and solder layers; here L_{sub} is set as 2.14 mm). The L_{min} is evaluated by intersecting χ_r and τ_Y/τ_∞. (b) The w–L_{min} relation to avoid the yield criteria of the interfacial bonding. The boundary temperatures are set as $T_h = 350°C$ and $T_c = 50°C$.

χ_r and τ_Y/τ_∞ is L_{min}. The w–L_{min} relation is shown in Figure 13.16b, in which the upper region of the curve indicates a safe zone that the interfacial shear stress is lower than the τ_Y, while the lower zone indicates vulnerability to the shearing stress. Thus, L longer than L_{min} is desired for a TE module to sit in the safe zone for its thermomechanically reliable design.

13.4.1.2 Upper Limit of Leg Length for Maximum Power Generation

To figure out the hot-side temperature of the TE leg depending on leg geometries and thermal boundary conditions, a constant property model is considered,[25] in which the effect of ceramic and metal electrodes is ignored. In case of the boundary condition that a heat source is finite and the cold-side temperature is fixed, the input heat Q_{in} can be expressed as

$$Q_{in} = (2w)^2 \bar{\kappa} \frac{\Delta T}{L} + I\bar{S}T_h - \frac{1}{2}I^2\bar{R}, \tag{13.58}$$

where R is the electric resistance, and the overbar denotes an averaged value at a given T_h and T_c. By applying $I = \bar{S}\Delta T/(\bar{R} + R_L)$ into Equation 13.58, where the maximum output power is assumed to be generated at a matched ratio of load to internal resistance, i.e., $\bar{R} = R_L$, Equation 13.58 can be rewritten, and T_h is figured out as

$$T_h = -\left(\frac{4}{3}\bar{\kappa}\bar{P}_F^{-1} + \frac{2}{3}T_c\right) + \left[\left(\frac{4}{3}\bar{\kappa}\bar{P}_F^{-1} + \frac{2}{3}T_c\right)^2 + \frac{8}{3}\bar{P}_F^{-1}q_{in}L\right]^{1/2} + T_c, \quad (13.59)$$

where $q_{in} = Q_{in}/(2w)^2$ and $\bar{P}_F = \bar{S}^2/\bar{\rho}$.

Figure 13.17a shows the L-dependent hot-side temperature according to the thermal conductivity of the leg, where it is assumed that the variations of the thermal conductivities do not affect P_F, i.e., the thermal conductivity is reduced while P_F is kept unchanged to see ZT improvement by reduced

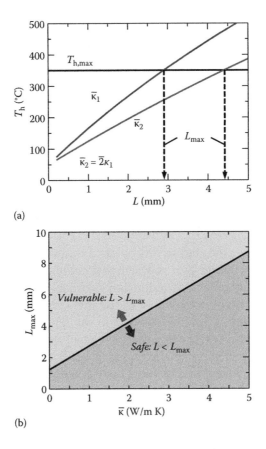

(a)

(b)

FIGURE 13.17 (a) Hot-side temperature with leg length L according to averaged thermal conductivities $\bar{\kappa}$. The horizontal line indicates the allowable maximum hot-side temperature $T_{h,max}$. (b) L_{max} associated with $\bar{\kappa}$ at a given thermal boundary. The q_{in}, T_c, and $\bar{P}_F$ are set to 20 W/cm², 50°C, and 30 μW/cm K², respectively.

thermal conductivity. The lowered thermal conductivity sometimes accompanies a decrease in electric conductivity, but the reduced electric conductivity trades off an increase of Seebeck coefficient, so this assumption is not completely unreasonable. In order to maximize a specific power generation such as watts per kilogram or watts per dollar, the larger temperature difference with minimal leg length is essential where $T_{h,max}$ should be equal to T_h utilized as the hot-side temperature in the shear stress analysis (Figure 13.16) to make thermal boundaries of both criteria identical. Thus, the allowable upper limit of a leg length L_{max} is obtained by setting L-dependent T_h equal to $T_{h,max}$. Once $T_{h,max}$ is selected, the L_{max} is obtained by rearranging Equation 13.59 as

$$L_{max} = \frac{3\bar{P}_F(T_{h,max} - T_c)}{8q_{in}} \left[(T_{h,max} - T_c) + 2\left(\frac{4}{3}\bar{\kappa}\bar{P}_F^{-1} + \frac{2}{3}T_c \right) \right]. \quad (13.60)$$

Figure 13.17b shows L_{max} corresponding to the $\bar{\kappa}$ of a material. In order to achieve the thermoelectrically maximum performance, it is noted that a TE material has its allowable maximum leg length at a given thermal boundary, and a certain length of a leg should be kept associated with a specific $\bar{\kappa}$.

13.4.2 Engineering Thermal Conductivity on Steady-State Mode

According to the discussion of L_{min} and L_{max} in the previous sections, we obtain the length condition satisfying the thermomechanical reliability and thermoelectrical performance of a module device as $L_{min} < L < L_{max}$. Here, L_{min} is determined based on τ_∞, τ_Y, χ_r, and w, i.e., $L_{min} = f_1(\tau_\infty, \tau_Y, \chi_r, w)$, and the L_{max} is figured out by $T_{h,max}$, q_{in}, $\bar{P}_F$, and $\bar{\kappa}$, i.e., $L_{max} = f_2(T_{h,max}, q_{in}, \bar{P}_F, \bar{\kappa})$. Note that L_{min} is $\bar{\kappa}$ independent and L_{max} is w independent. In case of reducing the thermal conductivity for ZT improvement, the L_{max} becomes smaller as shown in Figure 13.17a. By continuing to reduce the thermal conductivity, L_{max} is decreased and equal to the L_{min} at a specific $\bar{\kappa}$ that is defined as the engineering thermal conductivity κ_{eng} of a TE material for steady-state mode: f_2 is reduced to become $f_2 = f_1$ satisfying both thermomechanically reliable criterion and thermoelectrically maximum performance. If the thermal conductivity is further reduced to be smaller than κ_{eng}, its corresponding L becomes lower than the L_{min}, which makes the induced thermal shear stress exceeding the yield strength criterion τ_Y (Figure 13.16). Once the L_{min} is determined by Equation 13.56, the κ_{eng} at steady state is also analytically evaluated by Equation 13.60 where $L_{max} = L_{min}$ as

$$\kappa_{eng} = \frac{3\bar{P}_F}{8} \left[\frac{8q_{in}L_{min}}{3\bar{P}_F\Delta T_{max}} - \frac{4T_c}{3} - T_{max} \right], \quad (13.61)$$

which is plotted in Figure 13.18a, in which the ultimate minimum length $L_{min,u}$ indicates the situation that L_{max} and L_{min} are large and small enough,

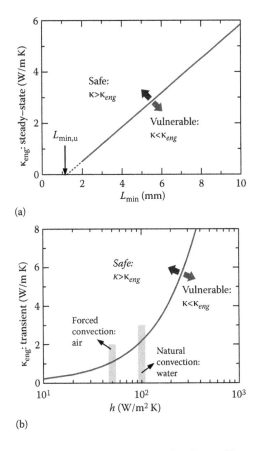

(a)

(b)

FIGURE 13.18 (a) κ_{eng} at a steady-state operation associated with L_{min}. The $L_{min,u}$ is obtained as $L_{min,u} = P_F(T_{h,max} - T_c)(3T_{h,max} + T_c)/(8q_{in})$ from Equation 13.61. (b) The κ_{eng} at a transient operation mode according to heat transfer coefficient h, where ΔT_{max}, α, and σ_u/E are assumed to be 300°C, 1.7×10^{-6} °C^{-1}, and 0.7×10^{-3}, respectively.

respectively, so that L_{max} does not become equal to L_{min} by reducing thermal conductivity, i.e., $L_{max} > L_{min}$ is kept even with $\bar{\kappa}$ approaching zero. In this case, a TE module is robust enough for reducing the thermal conductivity of a TE material in a module not to affect the reliability of a device if $L_{min,u}$ is a reasonable value for module fabrication.

13.4.3 Engineering Thermal Conductivity on Transient Operation Mode

The thermal stresses described in the previous sections are based on the steady-state mode that is a general condition for most TE modules in operation. In some applications, the boundary temperatures change over time and vary from place to place depending on environment temperatures or heat source circumstances. For example, a waste heat recovery system in automobiles or combustion TE generators often operates at irregular thermal fluctuations.[67] Such thermal loadings by the transient temperature change give rise to thermal stresses resulting in an acceleration of crack initiation and/or failure, which exhibits higher impact especially to brittle TE materials.

To avoid a failure of brittle materials without warning, many authors[68–71] were reported on quantifying a parameter for thermal stress resistance of ceramic materials caused by thermal loading/shock, but less attention has been paid to TE materials for this issue. Since TE materials also have a high degree of brittleness that leads to low resistance to thermal impact, it is necessary to examine the thermal stress resistance of a leg experiencing transient thermal loadings in extreme cases. Hasselman[70] shows various types of thermal stress resistance depending on thermal environments and geometries of materials since no single factor can adequately characterize a material for various conditions. Among them, this section is focused on a homogeneous TE leg assuming that it has infinite width and temperature-independent physical properties at a constant heat transfer coefficient h under a mild thermal impact.[68] The maximum allowable temperature change ΔT_{cr} so as to avoid the failure by a transient thermal loading is given by[71]

$$\Delta T_{cr} = \frac{3\sigma_u(1-\nu)\kappa}{\alpha EhL},$$
(13.62)

where σ_u, ν, α, and E are the ultimate strength, Poisson's ratio, CTE, and Young's modulus, respectively. The ratio of σ_u to E of a material usually has a little variation due to their proportional relationship on each other, and α and h are the intrinsic material property and the environmental factor of heat exchangers or thermal boundaries, respectively, so the ratio of κ to L is only susceptible to the ΔT_{cr} against the transient thermal loadings. For a TE material, ΔT_{cr} should exceed the ΔT_{max} ($\Delta T_{cr} \geq \Delta T_{max} = T_{h,max} - T_c$) to avoid failure of a TE material before reaching the maximum performable temperature difference. Here, we define the engineering thermal conductivity κ_{eng} in a transient state at $L = L_{min}$ as

$$\kappa_{eng} = \frac{\alpha EhL_{min}\Delta T_{max}}{3\sigma_u(1-\nu)},$$
(13.63)

which is the minimum thermal conductivity allowing ΔT_{max} without failure of a TE material by a transient thermal loading. Although κ_{eng} for a transient mode is defined under the assumption of homogeneous TE leg, it is the meaningful parameter as a pertinent material property to compare with other materials for its intrinsic reliability against a dynamic thermal loading. Regarding the σ_u in Equation 13.63, tensile and/or shear strength are more dangerous in TE materials than compressive strength, so a smaller value of either tensile or shear strength is taken into account for the κ_{eng} evaluation for a conservative design. Figure 13.18b shows the κ_{eng} at a transient mode by Equation 13.63 according to h, which is beneficial to compare a safety criterion with other materials against thermal loadings because even a single thermal impact over a criterion can cause failure of TE materials. h of 50–100 W/m² K is in the range of exhaust gas or thermal reservoir.[69,72,73] Thus, the κ_{eng} at the transient mode is a type of design factors restricting an operating environment of a TE material against failure by dynamic thermal loadings.

13.4.4 Case Study: Bi₂Te₃-Based Module

13.4.4.1 L_{min} for Thermomechanical Reliability

The ratio of τ_Y to τ_∞ and the shear stress reduction factor χ_r of a Bi_2Te_3-based module is shown in Figure 13.19a, as a function of L/L_{tot}, in which the thickness of ceramic plates, electrodes, and solder layers are 0.7, 0.5, and 0.07 mm, respectively, and f_{ce} = 2.86. The mechanical properties of each component are listed in Table 13.2. The τ_Y in this analysis indicates a yield strength of solder layer ranging from 22.8 to 27.8 MPa at room temperature of various solder alloys applicable to Bi_2Te_3-based module assembly.[74] Another criteria can be applied to τ_Y as a fracture strength of Bi_2Te_3 such as 34 MPa of ultimate tensile strength at room temperature[75] or a bonding strength of Ni metallization–Bi_2Te_3 leg configuration, which was measured up to 30 MPa at room temperature.[76] For a conservative design of the module, the yield strength of solder alloys leads to L_{min} of the TE leg with a cross-sectional area of 1.6×1.6 mm² (w = 0.8 mm),[34] which should be longer than 0.77 mm at a steady-state operation under the maximum ΔT = 150°C at T_c = 50°C. The critical w–L_{min} relation for the thermomechanical reliability is shown in Figure 13.19b, where it is reliable when selected w and L are placed in the upper zone of the border; otherwise, an interfacial structure in a module is vulnerable at the operation temperature.

13.4.4.2 L_{max} for Maximum Power Generation

To figure out the hot-side temperature of the TE leg, the numerical simulation by a finite-difference method was carried out based on a realistic assembly model accounting for the temperature dependence material properties. Figure 13.19c shows the L-dependent hot-side temperature of the $Bi_{0.4}Sb_{1.6}Te_3$ leg according to the average thermal conductivity of the leg, where $T_{h,max}$ = 200°C. In Figure 13.19d, L_{max} is 1.15 mm by the numerical analysis and 1.33 mm by the analytical model of the $Bi_{0.4}Sb_{1.6}Te_3$ ($\bar{\kappa}$ = 1.13 W/m K). It becomes 0.68 and 0.95 mm by the numerical and analytical models, respectively, if the thermal conductivity is further reduced to $\bar{\kappa}$ = 0.6 W/m K, where q_h is assumed to be 20 W/cm², which is in a typical heat flux density

TABLE 13.2 Mechanical Properties of Each Component as Input Data

Component	Young's Modulus (GPa)	Poisson's Ratio	CTE ($\times 10^{-6}$ °C^{-1})	Yield Strength at 25°C (MPa)
AlN plate (Subhash and Ravichandran[77])	320	0.237	4.2	–
Cu electrode (Siewert et al.[74])	115	0.31	17	–
Sn-Sb solder (Suhir and Shakouri[64] and Siewert et al.[74])	44.5	0.33	27	25.7
p-type Bi₂Te₃ (Al-Merbati et al.[58] and Gao et al.[61])	47.9	0.4	16.8	112

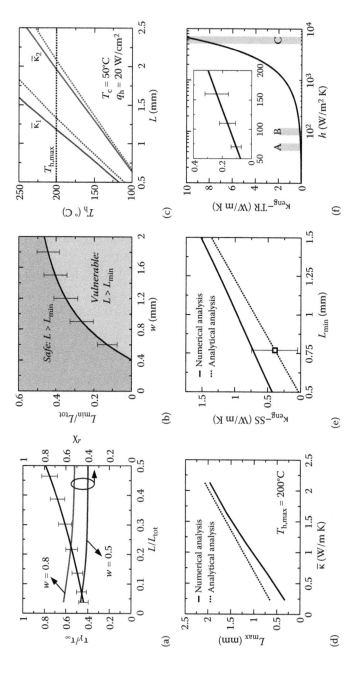

FIGURE 13.19 (a) τ_Y/τ_∞ and reduction factor χ_r of the shear stress as a function of nondimensional length L/L_{tot}. The various yield strengths of solders (23–28 MPa) are shown as error bars. (b) The w vs. L/L_{tot} relation to avoid the yield criteria of the solder. (c) The hot-side temperature with L according to $\bar{\kappa}$. The horizontal line indicates $T_{h,max}$. The solid and dashed curves indicate the numerical and analytical analysis, respectively, where $\bar{\kappa}_1 = 1.13$ and $\bar{\kappa}_2 = 2.13$ W/m K. (d) The L_{max} associated with κ at a given thermal boundary. (e) The κ_{eng} at a steady state with L_{min}. (f) The κ_{eng} at a transient mode as a function of h. A: $h = 50$ W/m² K (forced convection by air), B: $h = 100$ W/m² K (natural convection by water), and C: $h = 6000$ W/m² K (forced convection by water).

range (up to 40 W/cm^2) for most TE applications.[78] The L_{max} according to $\bar{\kappa}$ by the analytical model (Equation 13.60) outweighs the L_{max} by the numerical analysis over the whole $\bar{\kappa}$ range. The difference is mainly because of the absence of Thomson heat and the assumption of temperature independence of S, ρ, and κ in the analytical model. This leads to the underrated κ_{eng} of a steady-state mode by the analytical model as 0.4 W/m K at $L_{min} = 0.77$ mm while the κ_{eng} evaluated by the numerical analysis is more conservative with larger κ_{eng} (0.7 W/m K at $L_{min} = 0.77$ mm) as shown in Figure 13.19e. According to L_{min} of 0.77 mm, its corresponding κ_{eng} at a transient mode is found as 0.08, 0.15, and 9 W/m K by Equation 13.63 at $h = 50$, 100, and 6000 W/m^2 K, respectively (Figure 13.19f).

13.4.4.3 Intrinsic Lower Limit of κ and the Engineering Thermal Conductivity

It has been known that nanostructure approaches suppress the thermal conductivity of most TE materials without the deterioration of P_F by introducing intensive phonon scattering.[15,46,47,49,53,54,79,80] As an example, for $Bi_{0.5}Sb_{1.5}Te_3$,[47] the total thermal conductivity which is the sum of carrier, lattice, and bipolar contribution, was reduced from 1.9 to 1.1 W/m K at 200°C due to the significant reduction of lattice and bipolar parts, where the thermal conductivities of the carrier, lattice, and bipolar effects are 0.6, 0.35, and 0.15 W/m K, respectively. Here, we briefly discuss whether there is any room available for further reduction of thermal conductivity based on each contribution. Firstly, there is not much room to further reduce the carrier thermal conductivity (κ_{carr}) in order to keep high P_F at the optimized carrier concentration, of which most TE materials is in the order of 10^{19} cm^{-3}. For this reason κ_{carr} also has an optimized range not to degrade P_F and ZT due to its interlinked relation with electrical resistivity (ρ) by Lorenz number (L), $\kappa_{carr} = L\rho^{-1}T$, even though intrinsic κ_{carr} can become much lower value than an optimized range. For $Bi_{2-x}Sb_xTe_3$ materials, the κ_{carr} at 200°C was reported as of about 0.6 W/m K for $ZT \sim 1.1$ ($P_F \sim 20$ μW/cm K^2),[47] 0.25 W/m K for $ZT \sim 0.65$ ($P_F \sim 11$ μW/cm K^2),[81] 0.45 W/m K for $ZT \sim 1.0$ ($P_F \sim 17$ μW/cm K^2),[81] and 0.5 W/m K for $ZT \sim 1.0$ ($P_F \sim 25$ μW/cm K^2).[82] Secondly, the lattice thermal conductivity (κ_{latt}) of the nanostructured $Bi_{0.5}Sb_{1.5}Te_3$ at 200°C is 0.35 W/m K, which was significantly reduced compared with that of an ingot sample (0.7 W/m K).[47] However, it still has room for further reduction of the lattice contribution. The boundary for the minimum thermal conductivity of amorphous solid was proposed by Slack[83] and later modified by Cahill et al.,[84] assuming that the phonon scattering length is one-half of the wavelength. The minimum κ_{latt} of some amorphous solids and disordered crystals as "glasslike" was calculated to be as low as 0.23 and 0.35 W/m K, respectively.[85] Chiritescu et al.[86] and Takashiri et al.[87] reported the estimated minimum κ_{latt} for $(Bi,Sb)_2Te_3$ and $Bi_2Te_{3-x}Se_x$, respectively, as ~0.2 W/m K. Thirdly, the bipolar thermal conductivity ($\kappa_{bipolar}$) can also be further reduced if a proper band-widening strategy is applied.[40,88] For an ideal case, $\kappa_{bipolar}$ is suppressed and assumed to be zero. According to the evaluation of κ_{carr}, κ_{latt}, and $\kappa_{bipolar}$, the optimized minimum total thermal conductivity of bulk

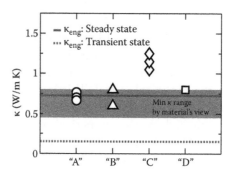

FIGURE 13.20 Averaged thermal conductivities of the state-of-the-art p-type Bi_2Te_3 as compared with κ_{eng} of the p-type $Bi_{0.4}Sb_{1.6}Te_3$ for steady-state (*solid line*), transient mode (*dashed line*), and intrinsic minimum κ range by the point of view (*band*) of the material. The measured property data of the reference materials are from (A) Jimenez et al.,[81] (B) Lee et al.,[90] (C) Hu et al.,[82] and (D) Kim et al.[52]

$Bi_{2-x}Sb_xTe_3$ achievable without sacrificing P_F can be estimated in the range of 0.45–0.8 W/m K (κ_{carr} = 0.25–0.6, κ_{latt} = 0.2, and $\kappa_{bipolar}$ = 0). The $Bi_{2-x}Sb_xTe_3$ material is the well-known laminar structure and has anisotropic transport properties along the crystalline direction of in-plane and out-of-plane, so the carrier thermal conductivity in the out-of-plane direction of the film, such as the Bi_2Te_3/Sb_2Te_3 superlattice film, could be much lower than the values we discussed.[89]

The thermal conductivities of some of the state-of-the-art p-type Bi_2Te_3 materials have been reported as 0.5–1.7 W/m K for ZT value of ~1.4.[47,81,82,90] Figure 13.20 shows the state-of-the-art p-type Bi_2Te_3 has the $\overline{\kappa}$ close to and/or below the κ_{eng} at steady state. This indicates that $\overline{\kappa}$ is saturated to κ_{eng} if the same mechanical properties of each Bi_2Te_3 are assumed, so there is not much room for ZT improvement by reducing the thermal conductivity, and too many efforts have been already put in some cases, even though the material can presumably be tuned for further reduction of thermal conductivity. As discussed, the reduction of thermal conductivity may cause an immediate fracture of a material when it is below the transient κ_{eng}, and the failure of the interfacial structure or degradation of the output power occurs with thermal conductivity lower than the steady state κ_{eng}. Thus, the strategy of reducing thermal conductivity for ZT enhancement should be carefully addressed by considering the mechanical reliability issue.

13.4.5 Engineering Thermal Conductivity in Practical Design

So far, we discussed κ_{eng} based on the TE assembly with the ideal bonding of dissimilar materials for the shear stress analysis with respect to the yield and/or bonding strength measured at room temperature. It should be noted that the yield strength of the solder material is generally temperature dependent, so an actual value associated with the hot-side temperature becomes much lower, e.g., the yield strength of Sn–Pb is reduced to 75% from room temperature to 130°C,[91] and Sn–Ag–Cu has 67% reduction from room temperature to 160°C.[92] This decreased yield strength at higher temperature gives rise to increasing κ_{eng} at steady-state mode since its corresponding L_{min} becomes larger to avoid yielding the interfacial structure.

Another important issue one needs to consider is an actual interfacial structure of the TE assembly. Usually, semiconducting TE materials cannot be directly bonded onto metal electrodes due to the large difference of their work function,[29] so that thin metallization layers are introduced on TE materials. In addition, a diffusion barrier is also required at the interface of the high-temperature side to avoid the formation of unwanted intermetallic compounds that not only cause high contact resistance,[93] but also degrade the mechanical properties of the joint.[94] For Bi_2Te_3 materials, the bonding strength of the interface composed of Ni metallization by sputtering/electroplating and HP varies from 10 to 30 MPa at room temperature,[93] and its bonding strength at higher temperature may become even lower. Such weak bonding strength leads to increasing κ_{eng} at steady-state mode. For example, if the bonding strength is assumed to be 10 MPa by sputtering/electroplating, the L_{min} is 9.5 mm at cross-sectional area of 1.6×1.6 mm^2. Its corresponding κ_{eng} for steady-state and transient mode becomes about 9.3 and 1.9 W/m K at $h = 100$ W/m^2 K, respectively. The weak bonding strength requires high aspect ratio ($L \gg w$) of a leg and larger κ_{eng}, where such a tall leg gives rise to decreasing output power performance, and current Bi_2Te_3 materials are vulnerable due to their much lower thermal conductivity than κ_{eng} by this criterion. A secondary phase formed by chemical reaction between the leg and metal alloy exhibits different characteristics, which may weaken bonding strength. Barako et al.[95] demonstrated the cracking at the hot-side interface of Bi_2Te_3-based commercial module under a generator mode. This might be because κ_{eng} associated with the bonding and/or yielding criteria of the material is higher than TE materials' thermal conductivity. Practically, whichever of the lower value among the yield strength of a solder, the bonding strength at interfaces, and the fracture strength of the TE material at the hot-side temperature should be taken into account for the evaluation of κ_{eng}.

References

1. E. Altenkirck, *Physikalische Zeitschrift*, 1909, **10**, 560–568.
2. A. F. Ioffe, *Semiconductor Thermoelements and Thermoelectric Cooling*, Infosearch, London, 1957.
3. H. J. Goldsmid, A. R. Sheard, and D. A. Wright, *Br. J. Appl. Phys.*, 1958, **9**, 365–370.
4. G. J. Snyder and E. S. Toberer, *Nat. Mater.*, 2008, **7**, 105–114.
5. Y. Lan, A. J. Minnich, G. Chen, and Z. F. Ren, *Adv. Funct. Mater.*, 2010, **20**, 357–376.
6. Q. Jie, H. Wang, W. Liu, H. Wang, G. Chen, and Z. F. Ren, *Phys. Chem. Chem. Phys.*, 2013, **15**, 6809–6816.
7. R. He, H. S. Kim, Y. C. Lan, D. Z. Wang, S. Chen, and Z. F. Ren, *RSC Adv.*, 2014, **4**, 64711–64716.
8. J. Shuai, H. S. Kim, Y. Lan, S. Chen, Y. Liu, H. Zhao, J. Sui, and Z. F. Ren, *Nano Energy*, 2015, **11**, 640–646.

9. K. F. Hsu, S. Loo, F. Guo, W. Chen, J. S. Dyck, C. Uher, T. Hogan, E. K. Polychroniadis, and M. G. Kanatzidis, *Science*, 2004, **303**, 818–821.

10. J.-S. Rhyee, K. H. Lee, S. M. Lee, E. Cho, S. Il Kim, E. Lee, Y. S. Kwon, J. H. Shim, and G. Kotliar, *Nature*, 2009, **459**, 965–968.

11. Q. Zhang, F. Cao, W. Liu, K. Lukas, B. Yu, S. Chen, C. Opeil, D. Broido, G. Chen, and Z. F. Ren, *J. Am. Chem. Soc.*, 2012, **134**, 10031–10038.

12. M. Koirala, H. Zhao, M. Pokharel, S. Chen, T. Dahal, C. Opeil, G. Chen, and Z. F. Ren, *Appl. Phys. Lett.*, 2013, **102**, 213111.

13. S. D. Bhame, D. Pravarthana, W. Prellier, and J. G. Noudem, *Appl. Phys. Lett.*, 2013, **102**, 211901.

14. H. S. Kim, K. Kikuchi, T. Itoh, T. Iida, and M. Taya, *Mater. Sci. Eng. B*, 2014, **185**, 45–52.

15. L.-D. Zhao, S.-H. Lo, Y. Zhang, H. Sun, G. Tan, C. Uher, C. Wolverton, V. P. Dravid, and M. G. Kanatzidis, *Nature*, 2014, **508**, 373–377.

16. D. Kraemer, J. Sui, K. McEnaney, H. Zhao, Q. Jie, Z. F. Ren, and G. Chen, *Energy Environ. Sci.*, 2015, **8**, 1299–1308.

17. T. P. Hogan and T. Shih, *Thermoelectrics Handbook: Macro to Nano*, Taylor & Francis, Abingdon, 2005.

18. G. D. Mahan, *J. Appl. Phys.*, 1991, **70**, 4551–4554.

19. G. Min, D. M. Rowe, and K. Kontostavlakis, *J. Phys. D: Appl. Phys.*, 2004, **37**, 1301–1304.

20. J. E. Sunderland and N. T. Burak, *Solid-State Electron.*, 1964, **7**, 465–471.

21. O. Yamashita, *Energy Convers. Manage.*, 2008, **49**, 3163–3169.

22. J. Chen, Z. Yan, and L. Wu, *J. Appl. Phys.*, 1996, **79**, 8823–8828.

23. H. S. Kim, W. S. Liu, G. Chen, C. W. Chu, and Z. F. Ren, *Proc. Natl. Acad. Sci. U.S.A.*, 2015, **112**, 8205–8210.

24. C. A. Domenicali, *J. Appl. Phys.*, 1954, **25**, 1310–1311.

25. S. W. Angrist, *Direct Energy Conversion*, Allyn and Bacon, Boston, 1965.

26. X. Shi, J. Yang, J. R. Salvador, M. Chi, J. Y. Cho, H. Wang, S. Bai, J. Yang, W. Zhang, and L. Chen, *J. Am. Chem. Soc.*, 2011, **133**, 7837–7846.

27. Q. Zhang, E. K. Chere, K. McEnaney, M. Yao, F. Cao, Y. Ni, S. Chen, C. Opeil, G. Chen, and Z. F. Ren, *Adv. Energy Mater.*, 2015, **5**, 1401977.

28. H. S. Kim, W. Liu, and Z. Ren, *J. Appl. Phys.*, 2015, **118**, 115103.

29. W. Liu, Q. Jie, H. S. Kim, and Z. F. Ren, *Acta Mater.*, 2015, **87**, 357–376.

30. H. J. Wu, L. D. Zhao, F. S. Zheng, D. Wu, Y. L. Pei, X. Tong, M. G. Kanatzidis, and J. Q. He, *Nat. Commun.*, 2014, **5**, 4515.

31. L.-D. Zhao, V. P. Dravid, and M. G. Kanatzidis, *Energy Environ. Sci.*, 2013, **7**, 251–268.

32. A. A. Efremov and A. S. Pushkarsky, *Energy Convers.*, 1971, **11**, 101–104.

33. E. Müller, K. Zabrocki, C. Goupil, G. Snyder, and W. Seifert, Functionally Graded Thermoelectric Generator and Cooler Elements. in *CRC Handbook of Thermoelectrics: Thermoelectrics and Its Energy Harvesting*, ed. D. M. Rowe, Taylor & Francis, Abingdon, 2012, vol. 1, ch. 14.

34. A. Muto, D. Kraemer, Q. Hao, Z. F. Ren, and G. Chen, *Rev. Sci. Instrum.*, 2009, **80**, 093901.

35. E. S. Toberer, P. Rauwel, S. Gariel, J. Tafto, and G. J. Snyder, *J. Mater. Chem.*, 2010, **20**, 9877–9885.

36. Y. Saiga, B. Du, S. K. Deng, K. Kajisa, and T. Takabatake, *J. Alloys Compd.*, 2012, **537**, 303–307.

37. M. Zhou, J.-F. Li, and T. Kita, *J. Am. Chem. Soc.*, 2008, **130**, 4527–4532.

38. H. Zhao, J. Sui, Z. Tang, Y. Lan, Q. Jie, D. Kraemer, K. McEnaney, A. Guloy, G. Chen, and Z. F. Ren, *Nano Energy*, 2014, **7**, 97–103.

39. W. Liu, H. S. Kim, Q. Jie, B. Lv, M. Yao, Z. Ren, C. P. Opeil, S. Wilson, C. W. Chu, and Z. F. Ren, *Proc. Natl. Acad. Sci. U.S.A.*, 2015, **112**, 3269–3274.

40. X. Chen, L. Shi, J. Zhou, and J. B. Goodenough, *J. Alloys Compd.*, 2015, **641**, 30–36.

41. Y. Oto, T. Iida, T. Sakamoto, R. Miyahara, A. Natsui, K. Nishio, Y. Kogo, N. Hirayama, and Y. Takanashi, *Phys. Status Solidi C*, 2013, **10**, 1857–1861.

42. S. Chen, K. C. Lukas, W. Liu, C. P. Opeil, G. Chen, and Z. Ren, *Adv. Energy Mater.*, 2013, **3**, 1210–1214.

43. J. Sui, J. Li, J. He, Y. L. Pei, D. Berardan, and H. Wu, *Energy Environ. Sci.*, 2013, **6**, 2916–2920.

44. P. Jood, R. J. Mehta, Y. Zhang, G. Peleckis, X. Wang, R. W. Siegel, T. Borca-Tasciuc, S. X. Dou, and G. Ramanath, *Nano Lett.*, 2011, **11**, 4337–4342.

45. G. Joshi, H. Lee, Y. Lan, X. Wang, G. Zhu, D. Wang, R. W. Gould et al., *Nano Lett.*, 2008, **8**, 4670–4674.

46. X. W. Wang, H. Lee, Y. C. Lan, G. H. Zhu, G. Joshi, D. Z. Wang et al., *Appl. Phys. Lett.*, 2008, **93**, 193121.

47. B. Poudel, Q. Hao, Y. Ma, Y. Lan, A. Minnich, B. Yu, X. Yan et al., *Science*, 2008, **320**, 634–638.

48. W. Liu, X. Yan, G. Chen, and Z. F. Ren, *Nano Energy*, 2012, **1**, 42–56.

49. X. Yan, G. Joshi, W. Liu, Y. Lan, H. Wang, S. Lee, J. W. Simonson et al., *Nano Lett.*, 2011, **11**, 556–560.

50. Q. Zhang, J. He, T. J. Zhu, S. N. Zhang, X. B. Zhao, and T. M. Tritt, *Appl. Phys. Lett.*, 2008, **93**, 102109.

51. W. Liu, Z. Ren, and G. Chen, Nanostructured Thermoelectric Materials. In *Thermoelectric Nanomaterials*, eds. K. Koumoto and T. Mori, Springer Berlin, Heidelberg, 2013, pp. 255–286.

52. S. I. Kim, K. H. Lee, H. A. Mun, H. S. Kim, S. W. Hwang, J. W. Roh, D. J. Yang et al., *Science*, 2015, **348**, 109–114.

53. K. Biswas, J. He, I. D. Blum, C.-I. Wu, T. P. Hogan, D. N. Seidman, V. P. Dravid, and M. G. Kanatzidis, *Nature*, 2012, **489**, 414–418.

54. H. Liu, X. Yuan, P. Lu, X. Shi, F. Xu, Y. He, Y. Tang et al., *Adv. Mater.*, 2013, **25**, 6607–6612.

55. H.-S. Choi, W.-S. Seo, and D.-K. Choi, *Electron. Mater. Lett.*, 2011, **7**, 271–275.

56. D. Zhao, X. Li, L. He, W. Jiang, and L. Chen, *Intermetallics*, 2009, **17**, 136–141.

57. Y. Hori, D. Kusano, T. Ito, and K. Izumi, presented in part at the 18th Int. Conf. on Thermoelectrics, Baltimore, MD, 1999.

58. A. S. Al-Merbati, B. S. Yilbas and A. Z. Sahin, *Appl. Therm. Eng.*, 2013, **50**, 683–692.

59. H. S. Kim, T. Itoh, T. Iida, M. Taya, and K. Kikuchi, *Mater. Sci. Eng. B*, 2014, **183**, 61–68.

60. T. Sakamoto, T. Iida, Y. Ohno, M. Ishikawa, and Y. Kogo, *J. Electron. Mater.*, 2014, **43**, 1620–1629.

61. J.-L. Gao, Q.-G. Du, X.-D. Zhang, and X.-Q. Jiang, *J. Electron. Mater.*, 2011, **40**, 884–888.

62. T. Clin, S. Turenne, D. Vasilevskiy, and R. A. Masut, *J. Electron. Mater.*, 2009, **38**, 994–1001.

63. E. Suhir, *J. Appl. Mech.*, 1986, **53**, 657–660.

64. E. Suhir and A. Shakouri, *J. Appl. Mech.*, 2012, **79**, 061010.

65. E. Suhir and A. Shakouri, *J. Appl. Mech.*, 2013, **80**, 021012.
66. A. Ziabari, E. Suhir, and A. Shakouri, *Microelectron. J.*, 2014, **45**, 547–553.
67. M. Barth and K. Boriboonsomsin, *ACCESS Magazine*, 2009, **35**, 2–9.
68. S. S. Manson, *NACA Techical Note* 2933, 1953.
69. W. D. Kingery, *J. Am. Ceram. Soc.*, 1955, **38**, 3–15.
70. D. P. H. Hasselman, *Ceramurgia Int.*, 1978, **4**, 147–150.
71. E. Brochen, J. Poetschke, and C. G. Aneziris, *Int. J. Appl. Ceram. Tec.*, 2014, **11**, 371–383.
72. J. H. Lienhard, *A Heat Transfer Textbook*, Courier Dover Publications, Mineola, NY, 2013.
73. P. M. Mayer and R. J. Ram, *Nanosc. Microsc. Therm.*, 2006, **10**, 143–155.
74. T. Siewert, S. Liu, D. R. Smith, and J. C. Madeni, *Properties of Lead-free Solders* NIST and Colorado School of Mines, 2002.
75. D. Vasilevskiy, F. Roy, E. Renaud, R. A. Masut, and S. Turenne, presented in part at the Proc 25th Int. Conf. on Thermoelectrics, Vienna, August 6–10, 2006.
76. W. Liu, K. C. Lukas, K. McEnaney, S. Lee, Q. Zhang, C. P. Opeil, G. Chen, and Z. F. Ren, *Energy Environ. Sci.*, 2013, **6**, 552–560.
77. G. Subhash and G. Ravichandran, *J. Mater. Sci.*, 1998, **33**, 1933–1939.
78. J.-P. Fleurial, *JOM*, 2009, **61**, 79–85.
79. X. Yan, W. Liu, H. Wang, S. Chen, J. Shiomi, K. Esfarjani, H. Wang, D. Wang, G. Chen, and Z. F. Ren, *Energy Environ. Sci.*, 2012, **5**, 7543–7548.
80. G. Zhu, H. Lee, Y. Lan, X. Wang, G. Joshi, D. Wang, J. Yang et al., *Phys. Rev. Lett.*, 2009, **102**, 196803.
81. S. Jimenez, J. G. Perez, T. M. Tritt, S. Zhu, J. L. Sosa-Sanchez, J. Martinez-Juarez, and O. Lopez, *Energy Convers. Manage.*, 2014, **87**, 868–873.
82. L.-P. Hu, T.-J. Zhu, Y.-G. Wang, H.-H. Xie, Z.-J. Xu, and X.-B. Zhao, *NPG Asia Mater.*, 2014, **6**, e88.
83. G. A. Slack, in *Solid State Physics*, eds. F. S. Henry Ehrenreich and T. David, Academic Press, Cambridge, MA, 1979, vol. 34, pp. 1–71.
84. D. G. Cahill and R. O. Pohl, *Ann. Rev. Phys. Chem.*, 1988, **39**, 93–121.
85. D. G. Cahill, S. K. Watson, and R. O. Pohl, *Phys. Rev. B*, 1992, **46**, 6131–6140.
86. C. Chiritescu, C. Mortensen, D. G. Cahill, D. Johnson, and P. Zschack, *J. Appl. Phys.*, 2009, **106**, 073503.
87. M. Takashiri, K. Miyazaki, S. Tanaka, J. Kurosaki, D. Nagai, and H. Tsukamoto, *J. Appl. Phys.*, 2008, **104**, 084302.
88. H. J. Goldsmid, *Introduction to Thermoelectricity*, Springer Berlin, Heidelberg, 2012.
89. R. Venkatasubramanian, E. Siivola, T. Colpitts, and B. O'Quinn, *Nature*, 2001, **413**, 597–602.
90. P.-Y. Lee, J. Hao, T.-Y. Chao, J.-Y. Huang, H.-L. Hsieh, and H.-C. Hsu, *J. Electron. Mater.*, 2014, **43**, 1718–1725.
91. X. Q. Shi, W. Zhou, H. L. J. Pang, and Z. P. Wang, *J. Electron. Packag.*, 1999, **121**, 179–185.
92. P. T. Vianco, J. A. Rejent, and J. J. Martin, *JOM*, 2003, **55**, 50–55.
93. W. S. Liu, H. Wang, L. Wang, X. Wang, G. Joshi, G. Chen, and Z. F. Ren, *J. Mater. Chem. A*, 2013, **1**, 13093–13100.
94. T. Y. Lin, C. N. Liao, and A. T. Wu, *J. Electron. Mater.*, 2011, **41**, 153–158.
95. M. T. Barako, W. Park, A. M. Marconnet, M. Asheghi, and K. E. Goodson, *J. Electron. Mater.*, 2013, **42**, 372–381.

CHAPTER 14

Mechanical Properties of Thermoelectric Materials

Sonika Gahlawat, Kenneth White, Zhifeng Ren,
Yasuo Kogo, and Tsutomu Iida

Contents

14.1 Mechanical Properties of Thermoelectric Materials and Devices

Sonika Gahlawat, Kenneth White, and Zhifeng Ren

14.1.1 Introduction

Over the past few decades, researchers have successfully developed thermo-electric (TE) materials exhibiting exceptionally high peak figure of merit, ZT.[1-6] While such reports suggest a potential for further improved power generation at higher temperatures, some of the underlying problems limiting the commercialization of TE devices still remain unresolved in the literature. This chapter attempts to address such problems and present some viable solutions.

A typical TE module consists of n- and p-type semiconducting TE legs connected electrically in series and thermally in parallel. The module also includes a substrate soldered to an electrical interconnect, which is soldered to the TE leg. Substrate materials must not only electrically insulate and thermally conduct, but also support the mechanical loads imposed by the thermal gradients and thermoelastic mismatch between the substrate, TE legs, and interconnect. Typically, ceramic materials satisfy the first of these requirements. ZT for most TE materials peaks at high temperatures, often leading to mechanical failure. For instance, in radioisotopic TE generators used in space applications, hot-side conditions may induce sublimation. Hot-side sublimation could result in a reduction in the cross-sectional area and a consequent mechanical failure of the device.[7] In automotive waste heat recovery applications, the average exhaust temperature varies between 500°C and 600°C, with operating excursions to 1000°C. The substrate, electrical interconnects, and solder joint material must withstand the hot-side elevated temperatures. When the operational temperatures exceed the reflow temperature of the solder joint material, they imperil the electrical connections in the device. If one chooses materials that are stable for most of the operating temperature range, temperature spikes could have detrimental effects on the device and its components. Thus, while a significantly higher peak ZT at higher temperatures may have been achieved for some TE materials, improvements in power generation expected from higher ZT may not be fully realized as the mechanical integrity of the device itself could become jeopardized at high temperatures.

Secondly, the brittle nature of most TE materials,[8-10] as well as of the ceramic substrates, makes TE module design difficult. This difficulty arises from brittle failure occurring due to a combination of mechanical loading, the flaw population of the ceramic, and the inherent resistance to crack propagation of the ceramic. Additionally, the strength of ceramics exhibits significant scatter compared with that of ductile metals because, in ceramics, fracture initiates from preexisting flaws, which not only exist as a distribution in a specimen, but the distribution also varies from one specimen to another. This variation in flaw population causes a variation in measured strength.[11-14] Section 14.2 addresses the mechanics of brittle failure in detail.

Constrained thermal expansion leads to generation of thermal stresses, which, if large enough, could induce failures from cracking, delamination, fatigue, etc.[15] CTE mismatch between TE materials, interconnects, substrate, and the interface exacerbates the problem of stress concentration.[16–18] The precise estimation of these stresses requires knowledge of elastic properties, such as Poisson's ratio, Young's modulus, and CTE, of the materials. In addition, these properties affect the thermal shock resistance of the materials.[19–21]

Finally, the long-term stability of TE materials must be investigated. Both cyclic and noncyclic temperature gradients may result in microstructural changes such as grain growth, annealing of vacancies and dislocations, and diffusion and may affect the stability of different phases in multiphase TE materials.[22–26] In other words, the microstructure and, hence, the operating thermal and electrical mechanisms may change with time.

Section 14.1.2 reviews studies that involve thermomechanical stresses in TE devices. These studies shed light on how mechanically robust TE materials and other components of TE devices must be. Section 14.1.3 describes some fundamental mechanical properties and summarizes the mechanical properties of commonly used TE materials. Section 14.1.4 highlights the effect of microstructure, and hence, of processing, on mechanical and TE behavior and discusses toughening mechanisms for the improvement of mechanical strength.

14.1.2 Thermomechanical Stresses in Thermoelectric Devices

Numerous studies have investigated thermomechanical stresses in TE devices based on different considerations, such as effects of leg geometry and leg spacing, substrate effects, interfacial effects, cyclic thermal loading, and thermal shock effects.

14.1.2.1 Geometrical Effects

Al-Merbati et al.[27] studied the influence of leg geometry on device performance. They reported high stress levels around the leg corners and that the magnitude of stress generally increased with proximity to the hot side. Additionally, the authors showed that switching to trapezoidal leg geometry reduced thermal stresses. Erturun et al.[28] compared the effects of various leg geometries—rectangular prism, trapezoidal prism, cylindrical, and octagonal prism—on thermal stresses and found maximum stresses for the case of rectangular prismatic and cylindrical legs. Studies also showed that the employment of thinner and longer (slender) legs resulted in significant stress relief.[29–32] For example, the authors of one such study[31] claimed that thinning the legs by a factor of 4 lowered the maximum shear stresses by as much as 80%.

Tachibana and Fang[33] estimated the explicit relationship between thermal stress and temperature difference and CTE and module dimensions. Based on the assumption that the TE device does not bend during testing and the substrate is substantially more rigid than the TE material, the authors offered a first-order estimate of the thermal stress to be proportional to $(L\alpha\Delta T/h)^2$ rather than $(L\alpha\Delta T/h)$, where L, α, ΔT, and h denote module thickness, CTE, temperature difference, and leg height, respectively. Wu et al.[34] and Erturun et al.[28] demonstrated that while higher temperature gradients would increase the TE conversion efficiencies, they would also result in higher thermal stresses.

14.1.2.2 Effect of Boundary Conditions

Clin et al.[35] used finite-element analysis to simulate thermal stresses in a TE module and concluded that the thermal stresses imposed by temperature gradients mainly depended on the mechanical boundary conditions imposed on the module and on CTE mismatch between different components. Maximum stresses developed at the corners of the legs. In another study, Turenne et al.[36] studied stress distribution in large freestanding TE modules and in those rigidly fixed between two surfaces for thermal exchange. While the boundary conditions significantly changed the stress distribution, the authors concluded that the application of external compressive loading on the TE module resulted in the generation of global compressive stresses.

14.1.2.3 Substrate Effects

Jinushi et al.[37] studied thermal stresses in high-temperature TE modules. The authors concluded that implementing a Cu substrate on the cold side and a carbon sheet and mica sheet on the hot side and using silicon grease on the substrate side significantly reduced thermomechanical stresses, as they found no deterioration in the performance or mechanical failure of the module even after 30 heat cycles simulating furnace sintering. Nemoto et al.[38] compared the performance of a conventional unileg TE module with that of a 'half-skeleton'-type unileg module, where a thermally conductive sheet replaced the ceramic substrate soldered to the hot side, and subjected both modules to thermal cycling. The authors reported the following findings: (1) Cracks were observed in the TE legs of the module based on the conventional design, which the authors attributed to CTE mismatch between the substrate, solder, electrode, and TE legs. The interfacial stresses due to CTE mismatch dramatically increased due to large temperature gradients and heat cycles. Cracking, observed after just 26 heat cycles, resulted in a 40% increase in electrical resistance and more than 50% decrease in power output. (2) Low thermal stresses were observed upon replacing the cold-side substrate with a less rigid and pliable thermally conductive sheet. This half-skeleton structure-based module kept continuously working even after 370 heat cycles, and no cracks were observed despite some fluctuations in the output noted up to the first 100 cycles. (3) The use of silicon nitride as a substrate instead of alumina for insulating purposes increased the power output. Additional advantages of using silicon nitride include higher mechanical strength and lower CTE compared with alumina.

In another study conducted by Kambe et al.[39] the authors assessed the effects of the encapsulation of modules on their performance and the use of compliant pads instead of rigid ceramic substrates. The authors stated that while the use of mica sheets for electrical insulation and carbon sheets for accommodating differential thermal expansion led to significant stress relief, it also made the modules extremely fragile. Thus, additional care would have to be taken during module brazing, assembly, and welding. The encapsulated TE modules of SiGe implementing compliant pads could withstand 1400 heat cycles at 550°C and had an efficiency that is 84% higher than that of unencapsulated modules. Encapsulation also provides an answer to humidity, corrosion, and oxidation-related issues. In addition, the bond-free compliant pads effectively accommodated severe CTE mismatch.

14.1.2.4 Effect of Interfaces

According to Morschel and Bastian,[40] large thermal gradients could lead to surface deformation and alter the thermal contact between TE legs and electrodes. The authors proposed a technique based on interferometry to measure thermomechanical deformations in TEGs. Ravi et al.[16] demonstrated the importance of CTE mismatch between different components and choosing appropriate solder materials in reducing thermal stresses. Jia and Gao[32] examined the influence of leg geometry in segmented TEGs and showed that any kind of plastic deformation in the weld strips and electrode could effectively moderate thermal stresses. Salzgeber et al.[41] demonstrated that excessive stresses occurred at interfaces between the hot-side electrical contact and TE legs and suggested that employment of sliding interfaces could lower thermomechanical fatigue loads. Hori et al.[42] also reported that employing appropriate soldering alloy, with a yield stress lower than the strength of the TE material, provides stress relaxation by undergoing plastic deformation. Picard et al.[43] evaluated the effects of TE material properties on the degree of thermal stresses generated by temperature gradients in segmented TEGs and concluded that the use of yielding interfaces in TE legs could help mitigate the thermal stress problem, provided the electrical contact remained intact. High stress concentration was recorded at the corners of TE legs near the hot-side electrodes. According to the authors, the use of the same alloy for soldering the different segments of the TE legs could further reduce thermal stresses. Lin et al.[44] conducted a study to identify the appropriate soldering material for high-temperature TE modules. Severe diffusion of soldering material in the TE material had imposed the temperature limit of 80°C, even though the TE materials could be used at up to 150°C. In order to increase the operational temperature range, the authors experimented with materials including Pd, Ni/Au, Ag, and Ti/Au. All except Ti/Au were reported to diffuse into the TE material and form weaker alloys and other compounds. The authors claimed that Ti/Au bonded well to both the electrode and TE leg; it prevented any interdiffusion between them and did not diffuse into either the leg or the electrode.

Zhao et al.[45] analyzed the effect of aging at 550°C on titanium-based interfaces between SKU and electrodes and reported that upon aging, a multilayer structure composed of intermetallic compound (IMC) phases formed,

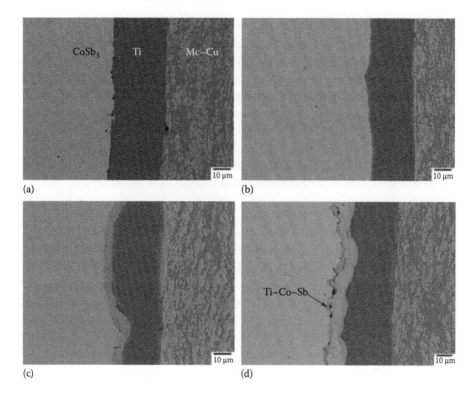

FIGURE 14.1 Backscattered electron imaging of $CoSb_3$/Ti/Mo–Cu interface (a) before thermal aging and after thermal aging for (b) 5, (c) 10, and (d) 20 days. (Zhao, D. et al.: Microstructure contact studies for skutterudite thermoelectric devices. *International Journal of Applied Ceramic Technology.* 2012. 9. 733–741. Copyright Wiley-VCH Verlag GmbH & Co. KGaA. Reproduced with permission.)

as shown in Figure 14.1, which resulted from the diffusion of antimony and titanium atoms. With further aging, the thickness of the brittle IMC layer increased, followed by the development of cracks. After 40 days of aging, the electrode detached from the joint, which was attributed to cracks in IMC layer that led to fracture failure of the joint.

14.1.2.5 Effect of Thermal Fatigue

In 2010, Hatzikranoitis et al.[46] indicated that cyclic thermal loading of TE devices resulted in a decrease in Seebeck coefficient and increase in electrical and thermal resistivity, thereby degrading the performance. In 2012, Park et al.[47] and Barako et al.[48,49] analyzed the performance of a Bi_2Te_3-based TE module subjected to thermal cycling. Barako et al.[48,49] considered a TE system under compressive loading so as to account for the mechanical boundary conditions imposed during the actual operation. This constraint limits the thermal expansion of different components and results in thermomechanical stresses in the module, especially at the interfaces. In order to understand the effects of thermal fatigue, the authors varied the module temperature between 146°C and –20°C and maintained the sink at 23°C. The time period of each cycle, 60s, was considerably smaller than that found in commercial usage. This

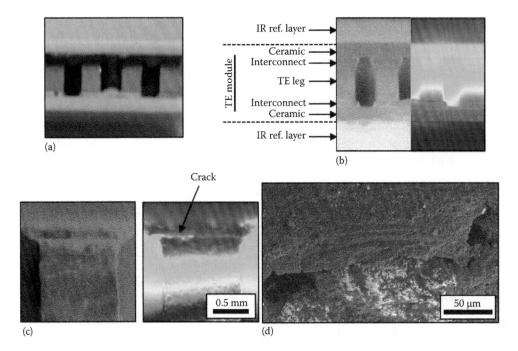

FIGURE 14.2 (a) Optical image of TE module, (b) infrared image of TE module cross section, (c) optical and infrared images of damaged interface between electrical interconnect and TE leg, and (d) SEM micrograph showing crack propagation in the interface. (From Barako, M. T. et al., *Thirteenth InterSociety Conference on Thermal and Thermomechanical Phenomena in Electronic Systems*, 86–92, 2005. © 1995 IEEE. With permission.)

shortened cycle limits the effect that diffusion may have on the failure mechanism compared with real-world use. The authors reported a steady degradation in the module performance during the first 10,000 thermal cycles, followed by a dramatic reduction in ZT. After 45,000 thermal cycles, the ZT dropped to 3.16% of its initial value and the authors reported complete device failure due to interfacial fracture. They detected, using infrared imaging, cracks in the leg–interconnect junction resulting in a substantial increase in electrical resistance and lowered the TE performance. Infrared images of the damaged interface are shown in Figure 14.2. In addition, they discovered that multiple TE legs exhibited thermal signature characteristics of mechanical damage. In short, irrespective of the operating conditions, TE performance degraded with thermal cycling. Dikwa et al.[50] reported a decline in the performance of a unileg structure-based TE module made of $Ca_3Co_4O_9$ TE material on the account of the degradation of TE properties due to thermal cycling.

14.1.2.6 Thermal Shock Studies

These TE materials should expect thermal shock loading not only resulting from service temperature spikes but also due to the metallizing and soldering processes. Metallizing involves coating the TE leg with metal to form the required diffusion barrier. Soldering involves dipping the metallized TE leg in a molten alloy bath for the purpose of joining the TE leg to the interconnect. Pelletier et al.[51] conducted quenching and thermal shock experiments on TE disks and noted that during quenching in a hot medium, the disk surface and periphery developed compressive stresses while the core developed tensile stresses. Thinner disks and anisotropic materials were reported to produce higher maximum stresses. The authors concluded that heating the disks in progressive steps could reduce the final stresses. Further, they reported fracturing of specimens during direct quench in a molten soldering bath from ambient temperature.

14.1.2.7 Existence of Tensile Stresses

Dozens of studies quantified thermal stresses developed in TE modules, but the authors reported only the von Mises stresses.[27–32,35,38,41,43] The von Mises criterion[52–54] defines conditions for plastic yielding and gives no information about the nature of stresses (tensile, compressive, or shear). Only few studies assess the nature of these stresses and report that tensile stresses develop in some areas of TE legs, leading to the critical condition for crack instability.

In 2005, Kaibe et al.[55] reported tensile stress localization at the interface of the hot-side electrode of a cascade-type TE module that used p-Mn–Si and n-Mg–Si as the leg materials for the hot side and Bi–Te-related alloys for the cold side. Li et al.[56] used finite-element method (FEM) to simulate the effect of thermal cyclic loading on a segmented TE unicouple with SKU and bismuth telluride used as leg materials for hot and cold sides, respectively. Their results demonstrated that the maximum stresses in the TE materials occurred at the interface between the hot-side electrode and leg and were tensile in nature. Moreover, the maximum tensile stresses reported were significantly higher than the tensile strength of the SKU, suggesting that the unicouple would have failed due to thermal fatigue. They also showed that inserting a functionally graded material between the leg and the electrode reduced the stresses by as much as 80%. Through the combined use of experimental and simulated heat cycle tests, Hori et al.[42] showed that the output of TE modules declined due to cyclic thermal loading. They attributed the performance degradation to increasing electrical resistance because of the deterioration of the contact condition of the hot-side solder. The FEM analysis showed that when temperatures on the hot and cold sides are 180°C and 30°C, respectively, the modules developed tensile stresses as high as 142.6 MPa.

Sakamoto et al.[57] examined the mechanical stability of an unconventional Mg_2Si-based TEG structure, which employed TE legs inclined at an angle with the electrical interconnects and the substrates. Based on this geometry, the authors calculated the maximum tensile stresses and compared those to the ultimate tensile strength of the different materials used in the device. This approach could be misleading in the sense that ceramics and other brittle materials do not have a traditionally defined tensile strength; a combination of applied stress, geometry, and flaw size determines the crack growth behavior in such materials.

Wereszczak and Case[58] and Chen et al.[59] showed that the corners of the TE legs closer to the hot side developed tensile stresses while the interior region of the legs experienced compressive stresses, as shown in Figure 14.3. The lower side of the substrate near the hot side also developed large tensile stresses. Although the electrical interconnects could, due to their ductile nature, undergo limited plastic deformation and lead to some stress relief, the thermal stresses, if high enough, could nucleate cracks between the substrate and interconnect and lead to device failure.

In summary, optimizing thermal and electrical transport properties of TE materials will not be sufficient. While the development of economically viable TE devices for large-scale applications warrants high ZT, which requires higher operational temperature range, the use of TE devices at higher temperatures makes them more vulnerable to elevated thermal stresses arising from larger temperature gradients. However, operation at lower temperatures can still result in thermal fatigue failure. Having proven that TE legs experience tensile stresses that induce mechanical failure, it becomes necessary to investigate the effects of cyclic loading and temperature spikes on the microstructure of different components of TE devices, as well as on the mechanical and TE behavior of TE devices.

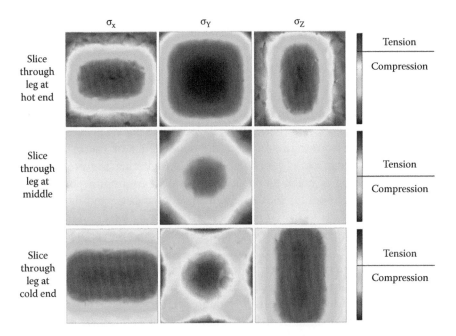

FIGURE 14.3 Stresses in a TE leg subjected to axial temperature gradient. The figure shows existence of tensile stresses in TE legs. (From Wereszczak, A. A., and Case, E. D., *Mechanical Response of Thermoelectric Materials*, Oak Ridge National Laboratory, Oak Ridge, TN, 2015.)

14.1.3 Mechanical Properties of Commonly Used TE Materials

This section summarizes commonly reported mechanical and physical properties, such as hardness, Young's modulus, compressive and flexural strength, and toughness for many TE materials.

An accurate estimation of thermal stresses generated in a TE device requires the knowledge of the elastic properties of all the components employed in the device. Of principal importance, the Young's modulus measures the stiffness or rigidity of bonds in a material. Generally, it varies with crystallographic direction; therefore, single crystals and textured polycrystalline samples exhibit elastic anisotropy. Polycrystalline materials, however, may be considered isotropic due to the randomness in grain orientation. In addition, Young's modulus affects the fracture strength and thermal shock of a material. While higher elastic moduli may imply higher fracture strength, it may deteriorate the thermal shock resistance under certain design conditions. For instance, the fracture strength and, consequently, fracture toughness directly vary with Young's modulus: $\sigma_f = \sqrt{2\gamma E/\pi a}$, where σ_f, γ, E, and a indicate fracture strength, surface energy, Young's modulus, and flaw size, respectively. However, of the five thermal shock resistance parameters, R, R', and R'' signify resistance to fracture initiation and weakly vary with modulus as $(1/E)^{1/2}$, while R''' and R'''' denote thermal shock fracture toughness and have no dependence on E[20,21]:

$$R = \frac{\sigma_f(1-\nu)}{E\alpha} \Rightarrow R \sim E^{-0.5},$$

$$R''' = \frac{E}{\sigma_f^2(1-\nu)} = \frac{\pi a}{2\gamma(1-\nu)} \Rightarrow R''' \sim E^0.$$

Here, R and R''' are thermal shock resistance parameters, α is CTE, and ν denotes Poisson's ratio. In fact, the dependence of thermal shock resistance parameters on E varies depending on whether the problem involves crack initiation or crack propagation. The most commonly employed methods for elastic modulus measurement include acoustic and nanoindentation techniques.

While Young's modulus is a well-defined quantity related to the elastic behavior of the material, hardness lacks a fundamental definition and depends on the type of test method employed for measurement. A wide variety of hardness tests exists, with indentation tests being the most commonly used ones. Indentation tests[60–63] define hardness as a pressure, rebound tests[64–66] give it units of length, damping test or Herbert test[67,68] defines it in units of time, and some other methods express it as a dimensionless quantity. Conceptually, hardness represents the resistance to permanent or plastic deformation. Tabor[69–71] demonstrated that, in case of metals, $\sigma_{YS} = H/3$, where σ_{YS} and H represent yield stress and hardness, respectively.

The explanation lies in the fact that two-thirds of the mean contact pressure in indentation, being hydrostatic in nature, does not generate any plastic flow; only one-third of the mean pressure is deviatoric and contributes in producing plastic flow. The readers must bear in mind that this relation holds only for isotropic solids that have a constant yield stress, i.e., the solid has fully work hardened, and the elastic deformation remains insignificant. Therefore, the equation applies to most metals; the exceptions include very hard steels and other metals for which the ratio of elastic modulus to yield stress drops below 100.

At a microscopic level, plastic deformation and hardness can be explained in terms of dislocation movement. Externally applied loads cause dislocations to move; dislocation movement signifies plastic deformation. The easier the dislocation motion, the lower the yield stress and, hence, the hardness. Indentation causes plastic deformation within a limited volume, termed as plastic zone, underneath the indenter. According to strain gradient plasticity theory,[72–74] the applied loads nucleate two kinds of dislocations within the plastic zone: statistically stored dislocations and geometrically necessary dislocations (GNDs). While the former result from shear and depend on the average strain in indentation, the latter arise from strain gradients due to bending.[75] In other words, the GNDs depend on strain gradients and address the strain compatibility issues. According to the Taylor model of plasticity,[76,77] the nucleated dislocations may interact with one another and form a locked network, known as the Taylor network. A finite externally applied stress, of the order of the Peierls stress,[78] can then force these dislocations to move, thereby initiating plastic flow. Metals, especially those with face-centered cubic crystal structure, have low Peierls stresses due to the nondirectional or delocalized bonding, which results in wide dislocations.[79–81] This allows for the movement of dislocations and, hence, plastic flow, in metals. In ceramics, however, strong directional bonding exists that causes the dislocations to be narrower and the Peierls stress to be higher than in metals. The high Peierls stress limits dislocation mobility in ceramics, thereby causing them to prefer brittle fracture. Restricted dislocation motion in ceramics results in higher hardness compared with that in metals.

Besides hardness, dislocation movement can also be correlated with the ability of a material to resist crack propagation, also known as its fracture toughness. Under external loads, cracks and surface flaws provide ideal nucleation sites for dislocations. In fact, the nucleation of dislocations in the crack tip region controls the fracture behavior of the material. In ductile metals, substantial dislocation nucleation and movement at the tip of the crack may blunt the crack tip. In other words, the dislocation activity occurring at the crack tip partially relieves strain, lowers the localized stress intensity factor, and, thus, shields the tip from the actual applied stress. Fracture toughness corresponds to the critical stress intensity factor (magnitude of stress field associated with the crack tip) given by $K_c = Y\sigma\sqrt{\pi a}$, where σ denotes applied stress, a indicates flaw size, and Y represents a geometric constant. Here,

K_c is referred to as a general critical stress intensity factor, whereas K_{IC}, to be referred to later, represents the mode I critical stress intensity factor. Mode I represents the most common loading configuration of a crack opening in tension. High stresses near the crack tip cause the material to plastically flow or yield, which requires the expenditure of energy and may blunt the crack tip. This process shields the crack tip and reduces the localized stress intensity factor at the crack tip to less than K_c:

$$K_{tip} = K_c - K_{shielding}.$$

On the other hand, ceramics exhibit limited dislocation motion. Ideal brittle materials exhibit no dislocation activity and, consequently, no shielding of the crack tip through plastic deformation. This brittle condition causes K_{tip} to become equal to K_c. With limited dislocation activity, the occurrence of fracture requires K_{tip} to exceed K_c. Fracture toughness is a material property that depends on applied stress as well as flaw size. Even "pristine" materials contain microstructural flaws and those that are generated during manufacturing processes. These include pores, scratches, cracks, grain boundaries, impurity agglomerates, second-phase particles, etc. The presence of these flaws causes the fracture strength of materials to be considerably smaller than their theoretical strength. As a rule of thumb, the strength of polycrystalline ceramics inversely scales with the critical flaw size, which may scale with the grain size.[82,83] Readers must note here that in fine-grained ceramics, usually, the processing or machining flaws initiate fracture. Flaws act as stress concentrators. For a given flaw shape, the largest flaw leads to the highest stress concentration. This necessitates a deeper understanding of the processing routes and their effects on critical flaw size distribution. In addition, thermal expansion or anisotropy can remarkably amplify stress levels near triple points and, thus, induce grain boundary microcracking in ceramics.[84] This results in cracks roughly on the order of size of grains. All these factors illustrate the role the microstructure plays in governing fracture. For more details, readers can refer to any fundamental fracture mechanics book.[85-88]

Standard fracture toughness testing employs single-edge-notched bend specimens, chevron-notched specimens, double cantilever beam specimen, compact specimens, etc.[89,90] A majority of the studies that reported K_{IC} for commonly used TE materials employed the Vickers indentation fracture toughness (VIF) technique.[91,92] An exhaustive description of the VIF technique has been provided by Nose and Fujii[93] and Sakai and Bradt.[94] At this point, the readers should be cautioned about the erratic nature of the VIF technique.[95] More than 30 different equations[96-100] exist for the VIF technique, all differing by an empirical constant, and none can be universally used for all brittle materials. In fact, studies show how remarkably the VIF technique differs from standardized fracture toughness tests.[101-104] While one can use VIF technique for obtaining a ballpark toughness value, it cannot replace the standard toughness test methods.

Structural properties of a material include tensile strength, flexural or bending strength, and compressive strength. As previously mentioned, in ceramics, significant scatter in strength exists due to the variation in flaw population. The highly brittle nature of ceramics makes tensile strength measurements difficult. Flaw population in the entire specimen volume dictates failure in tensile tests; however, in bending tests, only one of the specimen surfaces experiences tensile loading, and the flaws only on that surface, instead of in the entire volume of the specimen, contribute to failure. This causes the bending strength to be higher than the tensile strength. Both three-point and four-point bending tests can be used for measuring bending strength, although three-point bending tests result in higher strength values for a given ceramic material than four-point bending because the four-point bending configuration involves a much larger area fraction being subjected to maximum or near-maximum tensile stress than that in three-point bending configuration.

Compressive strength quantifies the ability of a material to withstand a compressing or crushing force. The compressive strength of ceramics exceeds their bending strength almost by a factor of 5–30. Therefore, compressive forces are much less likely to cause failure. In addition, failure due to compressive stresses does not follow linear elastic fracture mechanics.

Table 14.1 lists the mechanical properties of commonly used TE materials.[8–10,105–144] Hot-pressed nanostructured HHs exhibit the highest hardness, around 12 GPa, as well as Young's modulus, around 220 GPa; second-highest fracture toughness, 1.7 MPa $m^{1/2}$; and flexural strength, 180–275 MPa, in the group. The p-type HH shows marginally better properties than its n-type counterpart. Spark plasma-sintered $Ca_3Co_4O_9$ have maximum toughness of all the TE materials, which ranges between 2.2 and 2.8 MPa $m^{1/2}$, and flexural strength, around 250–320 MPa, but in terms of hardness and Young's modulus, they compete with bismuth telluride and lead telluride systems. Silicon–germanium alloys manifest hardness comparable to that of HH alloys, but their Young's modulus modestly falls behind. Bismuth telluride and lead telluride systems rank the lowest on all counts: hardness, modulus, flexural strength, and toughness, while SKUs rank in the middle. To give the readers an idea, fracture toughness of monolithic structural ceramics such as hot-pressed silicon nitride averages around 4–6 MPa $m^{1/2}$ while its flexural strength ranges from 700 to 1000 MPa.[145] So even the toughest TE materials are still extremely brittle when compared with other structural ceramics. The data also indicate that the hardest materials need not be the toughest; while $Ca_3Co_4O_9$ are the toughest in the table, their hardness compares with that of extremely brittle systems such as bismuth telluride alloys. There is no correlation between hardness and toughness, as shown in Figure 14.4. Therefore, while hardness serves as a quick screening tool, it does not offer enough to characterize the strength of any material.

TABLE 14.1 Mechanical Properties of Commonly Used TE Materials

Composition	Preparation	CTE ($\times 10^{-6}$ K^{-1})	Elastic Constants (GPa)	Poisson's Ratio	Hardness (GPa)	Young's Modulus (GPa)	Flexural Strength (MPa)	Compressive Strength (MPa)	Fracture Toughness (MPa m$^{1/2}$)	Peak ZT	Reference	Additional Information
Half-Heusler												
$Hf_{0.44}Zr_{0.44}Ti_{0.12}CoSb_{0.8}Sn_{0.2}$	BM + HP		c11 = 277, c44 = 83	0.252	12 (nano, HV)	225 (nano)					Gahlawat et al.[105]	
$Hf_{0.44}Zr_{0.44}Ti_{0.12}CoSb_{0.8}Sn_{0.2}$	BM + HP				12.8 ± 0.3 (nano)	221 ± 6.2 (nano)					He et al.[8]	
$Hf_{0.25}Zr_{0.75}NiSn_{0.99}Sb_{0.01}$	BM + HP				9.1 ± 1.2 (nano)	186.5 ± 17.6 (nano)					He et al.[8]	
TiNiSn	HP		c11 = 214.9, c44 = 67.5	0.271		171.6 ± 2.3 (RUS)					Rogl et al.[106]	
TiNiSn	Cast				10.18 ± 0.26 (HV)						Rogl et al.[106]	
ZrNiSn	HP				914.4 ± 0.6 (nano)	241.1 ± 4.0 (nano)					Germond[107]	
ZrNiSn	HP		c11 = 224.8, c44 = 75.1	0.249		187.7 ± 2.7 (RUS)					Rogl et al.[106]	
ZrNiSn	PECS		c11 = 263, c44 = 85	0.262		214 (RUS)					Fan[108]	
HfNiSn	HP		c11 = 184.7, c44 = 65.9	0.223		161.1 ± 0.2 (RUS)					Rogl et al.[106]	
$Ti_{0.75}Zr_{0.25}NiSn$	HP		c11 = 221, c44 = 70.2	0.268		178.2 ± 1.8 (RUS)					Rogl et al.[106]	
$Ti_{0.3}Zr_{0.35}Hf_{0.35}NiSn$	HP				8.18 ± 0.38 (HV)						Krez[109]	
$ZrNiSn_{0.72}Sb_{0.28}$	HP		c44 = 34.5			56.9 (PEM)					Kawaharada et al.[110]	
$Zr_{0.25}Hf_{0.75}NiSn_{0.975}Bi_{0.025}$	HP				5.5 (HV)						O'Connor et al.[111]	
$Zr_{0.5}Hf_{0.5}CoSn_{0.01}Sb_{0.99}$	HP				8.79 ± 0.38 (nano)	246.4 ± 4.5 (nano)					Verges et al.[112,113]	
$Zr_{0.5}Hf_{0.5}Ni_{0.1}Pd_{0.9}Sn_{0.9}Sb_{0.01}$	HP				8.23 ± 0.23 (HV)	204.7 ± 4.2 (nano)					Verges et al.[113]	
$Ti_{0.26}Sc_{0.04}Zr_{0.35}Hf_{0.35}NiSn$	SPS				8.97 ± 0.12 (HV)						Krez[109]	
$TiFe_{1.33}Sb$	HP		c11 = 233.3, c44 = 77.8	0.25	9.5 (nano)	194.4 ± 2.6 (nano)					Rogl et al.[106]	

(Continued)

TABLE 14.1 (CONTINUED) Mechanical Properties of Commonly Used TE Materials

Composition	Preparation	CTE ($\times 10^{-6}$ K^{-1})	Elastic Constants (GPa)	Poisson's Ratio	Hardness (GPa)	Young's Modulus (GPa)	Flexural Strength (MPa)	Compressive Strength (MPa)	Fracture Toughness (MPa m$^{1/2}$)	Peak ZT	Reference	Additional Information
TiFe$_{0.665}$Co$_{0.3}$Sb	HP		c11 = 231.6, c44 = 78.0	0.25	9.2 (nano)	195.1 ± 5.5 (nano)					Rogl et al.[106]	
TiFe$_{0.125}$Co$_{0.875}$Sb$_{0.875}$Sn$_{0.125}$	HP		c11 = 230.2, c44 = 76.7	0.25	8.7 (nano)	191.8 ± 5.9 (nano)					Rogl et al.[106]	
TiCoSb	SPS					196 (PEM)					Sekimoto et al.[114]	
ZrCoSb	SPS					207 (PEM)					Sekimoto et al.[114]	
HfCoSb	SPS					192 (PEM)					Sekimoto et al.[114]	
ErPdSb	SPS					123 (PEM)					Sekimoto et al.[115]	
LaPdBi	SPS					83 (PEM)					Sekimoto et al.[116]	
GdPdBi	SPS					102 (PEM)					Sekimoto et al.[116]	
ErPdBi	SPS					110 (PEM)					Sekimoto et al.[115]	
Bismuth Telluride												
Bi$_{0.4}$Sb$_{1.6}$Te$_3$	BM + HP				1.1 ± 0.1 (nano)	41.5 ± 4.4 (nano)					He et al.[8]	
Bi$_2$Te$_{2.7}$Se$_{0.3}$	BM + HP				1.2 ± 0.1 (nano)	38.8 ± 5.6 (nano)					He et al.[8]	
n-type Bi$_2$Te$_3$	MA + SPS						62 ± 5			0.94	Zhao et al.[117]	
	MA + SPS + forging						120 ± 5			1.18	Zhao et al.[117]	
Bi$_{0.4}$Sb$_{1.6}$Te$_3$	MS + angular extrusion						80.3			0.00333 [Z]	Hayashi et al.[118]	Extrusion temp.: ~773 K
Bi$_{0.4}$Sb$_{1.6}$Te$_3$	MS + angular extrusion						97			0.00306 [Z]	Hayashi et al.[118]	Extrusion temp.: ~643 K
Bi$_{0.5}$Sb$_{1.5}$Te$_3$	Hot ingot extrusion						50.8			0.00270 [Z]	Seo et al.[119]	Extrusion temp.: ~460°C

(Continued)

TABLE 14.1 (CONTINUED) Mechanical Properties of Commonly Used TE Materials

Composition	Preparation	CTE (× 10⁻⁶ K⁻¹)	Elastic Constants (GPa)	Poisson's Ratio	Hardness (GPa)	Young's Modulus (GPa)	Flexural Strength (MPa)	Compressive Strength (MPa)	Fracture Toughness (MPa m$^{1/2}$)	Peak ZT	Reference	Additional Information
$Bi_{0.5}Sb_{1.5}Te_3$	Hot ingot extrusion						31.2			0.00156 [Z]	Seo et al.[119]	Extrusion temp.: ~340°C
0.1 vol.% SiC–Bi_2Te_3	MA + SPS				0.69 ± 0.04 (HV)	37.5 ± 1.3 (sound resonance)			1.35 ± 0.25 (VIF)	1.04 (423 K)	Zhao et al.[120]	
n-type 95% Bi_2Te_3–5% Bi_2Se_3 doped with 0.04% SbI_3	Gas atomization				0.44 (HV)						Bhuiyan et al.[121]	
$Bi_{0.5}Sb_{1.5}Te_3$	BM + HP + hot extrusion						92			0.00294 [Z]	Seo et al.[122]	Extrusion temp.: ~440°C
$Bi_{0.5}Sb_{1.5}Te_3$	MA + SPS				0.81(HV)	40.1 (nano)	65.8		0.82 (VIF)	0.88 (323 K)	Liu et al.[123]	
$Bi_{0.5}Sb_{1.5}Te_3$ ± 0.1% SiC	MA + SPS				0.87 (HV)	44.9 (nano)	73.8		0.91 (VIF)	0.97 (323 K)	Liu et al.[123]	
$Bi_{0.5}$ $Sb_{1.5}$ Te_3	BM + SPS				0.82 (HV)	48 (PEM)	62				Bomshtein et al.[124]	GS: ~0.5 µm
	BM + SPS				0.57 (HV)	45 (PEM)	60				Bomshtein et al.[124]	GS: ~63 µm
$Bi_{0.5}Sb_{1.5}Te_3$	ZM						53.6	37.3		1.05 (360 K)	Zheng et al.[125]	Samples cut parallel to ZM direction
	ZM						48.7 (100°C)	36.8 (100°C)			Zheng et al.[125]	Samples cut parallel to ZM direction
	ZM						50.7 (200°C)	31.3 (200°C)			Zheng et al.[125]	Samples cut parallel to ZM direction
	ZM						9.6	12.7		0.75 (380 K)	Zheng et al.[125]	Samples cut normal to ZM direction
	ZM							10.2 (100°C)			Zheng et al.[125]	Samples cut normal to ZM direction
	ZM							10.7 (200°C)			Zheng et al.[125]	Samples cut normal to ZM direction
$Bi_{0.5}Sb_{1.6}Te_3$ ± 0 wt% MWCNT	BM + HP			0.25 (350 K)		47.1 (RUS)	32 (biaxial)			1.1 (325 K)	Ren et al.[126]	
$Bi_{0.5}Sb_{1.6}Te_3$ ± 0.5 wt% MWCNT	BM + HP			0.25 (350 K)		44.8 (RUS)	90 (biaxial)				Ren et al.[126]	Plastic deformation occurred before fracture

(Continued)

TABLE 14.1 (CONTINUED) Mechanical Properties of Commonly Used TE Materials

Composition	Preparation	CTE ($\times 10^{-6}$ K^{-1})	Elastic Constants (GPa)	Poisson's Ratio	Hardness (GPa)	Young's Modulus (GPa)	Flexural Strength (MPa)	Compressive Strength (MPa)	Fracture Toughness (MPa m$^{1/2}$)	Peak ZT	Reference	Additional Information
$Bi_{0.5}Sb_{1.5}Te_3$	ZM		c11 = 70	0.24	0.26 (HV)	59 (PEM)	10		0.75 (SENB)	0.9 (325 K)	Zheng et al.[127]	Parallel to ZM direction
	ZM		c11 = 74	0.36		44 (PEM)					Zheng et al.[127]	Normal to ZM direction
	MS-PAS		c11 = 66	0.28	0.40 (HV)	51 (PEM)	66	130	1.05 (SENB)	1.22 (340 K)	Zheng et al.[127]	σ_i: >60 at 200°C; σ_c: >108 at 400°C
	HP		c11 = 53	0.28		42 (PEM)				0.6 (320 K)	Zheng et al.[127]	
Silicon–Germanium												
$Si_{0.8}Ge_{0.2}P_2$	HP				10.8 ± 0.7 (nano)	166.3 ± 7.1 (nano)					He et al.[8]	
$Si_{0.8}Ge_{0.2}B_5$	HP				10.7 ± 0.9 (nano)	155.6 ± 4.6 (nano)					He et al.[8]	
n-type $Si_{80}Ge_{20}$	BM + HP			0.235	12.4 14.7 (nano)	135–147			0.98–1.03 (VIF)	1 (800°C)	Kallel et al.[128]	
n-type $Si_{80}Ge_{20}$	BM + SPS				13.7 (nano)	141 (nano)		1180	1.6 (VIF)		Bathula et al.[129]	
Lead Telluride System												
$In_{0.005}PbSe$	HP				0.6 ± 0.1 (nano)	65.9 ± 5.9 (nano)					He et al.[8]	
$Ag_{0.80}Pb_{18}Sb_{1.0}Te_{20}$	Cast						16 ± 3.4 (biaxial)				Ren et al.[9]	
$Ag_{0.86}Pb_{19}Sb_{1.0}Te_{20}$	HP						52.9 (biaxial)				Ren et al.[9]	
$Ag_{0.43}Pb_{18}Sb_{1.2}Te_{20}$	HP			0.28 (up to 600 K)	0.92 ± 0.06 (HV)	57.92 ± 3.12 (nano)	30.23 ± 9.68 (biaxial)		0.34		Ren et al.[9]	
$Ag_{0.43}Pb_{18}Sb_{1.2}Te_{20}$	HP			0.28 (up to 600 K)	1.07 ± 0.13 (nano)						Ren et al.[9]	E linearly decreased to 48 GPa at 523 K (RUS)

(Continued)

TABLE 14.1 (CONTINUED) Mechanical Properties of Commonly Used TE Materials

Composition	Preparation	CTE (×10⁻⁶ K⁻¹)	Elastic Constants (GPa)	Poisson's Ratio	Hardness (GPa)	Young's Modulus (GPa)	Flexural Strength (MPa)	Compressive Strength (MPa)	Fracture Toughness (MPa m$^{1/2}$)	Peak ZT	Reference	Additional Information
$Pb_{0.95}Sn_{0.05}Te$–8%PbS	Cast			0.245 ± 0.003	0.68 ± 0.5 (HV)	47.7 ± 0.2 (RUS)					Ni et al.[130]	
$Pb_{0.95}Sn_{0.05}Te$–8%PbS	Hot-pressed			0.240 ± 0.002	1.18 ± 0.07(HV)	53.1 ± 0.1 (RUS)			0.35 ± 0.03 (VIF)		Ni et al.[130]	
PbS	PECS			0.27	0.72 ± 0.10 (HV)	68.95 ± 0.10 (RUS)					Schmidt et al.[131]	
PbSe	PECS			0.27	0.44 ± 0.02 (HV)	64.3 ± 0.16 (RUS)					Schmidt et al.[131]	
n-type PbTe	Hot extrusion					59 (RUS)					Vasilevskiy et al.[132]	Linearly reduces to 51 GPa at 573 K
SKU												
$Ce_{0.45}Nd_{0.45}Fe_{3.5}Co_{0.5}Sb_{12}$	BM + HP				5.6 ± 0.4 (nano)	129.7 ± 11.0 (nano)					He et al.[8]	
$Yb_{0.35}Co_4Sb_{12}$	BM + HP				5.8 ± 2.2 (nano)	136.9 ± 21.9 (nano)					He et al.[8]	
$CeFe_3RuSb_3$	HP			0.22–0.29		133–139 (IEM)	37.3 ± 6.9 (4-pt)	657 ± 161	1.1–2.8 (SVB, CHEV)		Ravi et al.[133]	E: ~125–132 at 673 K
$CoSb_3$	HP			0.14–0.25		137–141 (IEM)	85.5 ± 13.3 (4-pt)	766 ± 119	1.7		Ravi et al.[133]	E: ~125–132 at 673 K
$CeFe_4Sb_{12}$	Melting–annealing + SPS						62.8 ± 40		1.22 (SENB)	0.85 (750 K)	Wan et al.[134]	
$CeFe_4Sb_{12}$ ± 1 vol% carbon fibers	Melting–annealing + SPS						95.8 ± 25		1.28 ± 0.04 (SENB)	0.9 (800 K)	Wan et al.[134]	
$Yb_{0.13}Ba_{0.10}Co_4Sb_{12}$	MS + SPS			0.2		139 (RUS)	121.6 ± 18.8			1.1 (750 K)	Salvador et al.[135]	
							100.6 ± 15.3 (300°C)				Salvador et al.[135]	

(Continued)

TABLE 14.1 (CONTINUED)   Mechanical Properties of Commonly Used TE Materials

Composition	Preparation	CTE (× 10⁻⁶ K⁻¹)	Elastic Constants (GPa)	Poisson's Ratio	Hardness (GPa)	Young's Modulus (GPa)	Flexural Strength (MPa)	Compressive Strength (MPa)	Fracture Toughness (MPa m^{1/2})	Peak ZT	Reference	Additional Information
Co_4Sb_{12}	HP				3.35 (HV)		61.9 ± 9.7 (500°C)			0.40 (560 K)	Salvador et al.[135]	
									0.82 ± 0.11 (VIF)		Eilertsen et al.[136]	
									0.52 ± 0.04 (COD)		Eilertsen et al.[136]	
									0.51 ± 0.06 (SEVNB)		Eilertsen et al.[136]	
$In_{0.1}Co_4Sb_{12}$	HP				6.10 (HV)				0.46 ± 0.13 (VIF)	0.05 (620K)	Eilertsen et al.[136]	
									0.49 ± 0.03 (COD)		Eilertsen et al.[136]	
									0.57 ± 0.06 (SEVNB)		Eilertsen et al.[136]	
$CoSb_{2.875}Te_{0.125}$ ± 0vol.% nano-TiN	SPS						116.8 ± 5.6		1.06 ± 0.04 (SENB)	0.9 ± 0.1 (800 K)	Duan et al.[137]	
$CoSb_{2.875}Te_{0.125}$ ± 1vol.% nano-TiN	SPS						148.2 ± 5.6		1.47 ± 0.04 (SENB)	1.0 ± 0.1 (800 K)	Duan et al.[137]	
$Ba_{0.09}Sr_{0.02}DD_{0.22}Yb_{0.02}Fe_{2.4}Ni_{1.6}Sb_{12}$	BM + HP + HPT	11.9		0.23 ± 0.02	5.30 ± 0.20 (HV)	104 (RUS)				1.0 (650 K)	Rogl et al.[138]	
$Ba_{0.15}DD_{0.28}Yb_{0.05}Fe_3NiSb_{12}$	BM + HP + HPT	11.8		0.28 ± 0.01	5.49 ± 0.24 (HV)	109.6 (RUS)				1.0 (700 K)	Rogl et al.[138]	
$Ce_{0.9}Fe_{3.5}Co_{0.5}Sb_{12}$	CGSR			0.230 ± 0.008		131.2 ± 0.6 (RUS)					Schmidt et al.[139]	
$Ce_{0.9}Fe_{3.5}Co_{0.5}Sb_{12}$	CGSR + wet milling			0.242 ± 0.011		126.8 ± 0.4 (RUS)					Schmidt et al.[139]	
$Co_{0.95}Pd_{0.05}Te_{0.05}Sb_3$	CGSR + dry, wet milling			0.226 ± 0.001		137.8 ± 1.2 (RUS)					Schmidt et al.[139]	

(Continued)

TABLE 14.1 (CONTINUED) Mechanical Properties of Commonly Used TE Materials

Composition	Preparation	CTE ($\times 10^{-6}$ K^{-1})	Elastic Constants (GPa)	Poisson's Ratio	Hardness (GPa)	Young's Modulus (GPa)	Flexural Strength (MPa)	Compressive Strength (MPa)	Fracture Toughness (MPa m$^{1/2}$)	Peak ZT	Reference	Additional Information
$Co_{0.95}Pd_{0.05}Te_{0.05}Sb_3$ doped with 0.1 at% cerium	CGSR + dry, wet milling	13.3–14.4		0.231 ± 0.006		140.6 ± 0.2 (RUS)					Schmidt et al.[139]	
Others												
$[AgSbTe_2]_{1-x}[GeTe]_x$	SPS			0.24		50 (RUS)				1.36 (700 K)	Salvador et al.[140]	
Cu_2Se	SPS							45	2 ± 0.02 (VIF)		Tyagi et al.[141]	
$La_{3-x}Te_4$	BM ± HP	12.6			3.28 ± 0.06 (HV)				0.70 ± 0.06 (VIF)		Ma et al.[10]	
$Ca_3Co_4O_9$	CS				0.12 ± 0.01 (HV)	10.5 ± 2.5 (nano)	15		0.4 ± 0.02		Kenfaui et al.[142]	
$Ca_3Co_4O_9$	HP	15 (out-of-plane) 7 (in-plane)			1.31 ± 0.12 (HV)	56 ± 7 (nano)	250 ± 12		2.2 ± 0.43		Kenfaui et al.[142]	
$Ca_3Co_4O_9$	SPS	18.8 (out-of-plane) 8 (in-plane)			1.55 ± 0.09 (HV)	84 ± 3 (nano)	320 ± 8		2.8 ± 0.47 (SENB)		Kenfaui et al.[142]	
$Ca_3Co_4O_9$	CS				0.11 ± 0.03 (nano)	10 ± 2 (nano)	18.4 ± 0.5		0.40 ± 0.02 (SENB)		Kenfaui et al.[143]	
$Ca_3Co_4O_9$	PA-SPS				3.2 ± 0.5 (nano)	87 ± 5 (nano)	284 ± 8		2.84 ± 0.40 (SENB)		Kenfaui et al.[143]	
$YbAl_3$	HP			0.182 ± 0.003	4.81 ± 0.73 (HV)	174 ± 2.5 (RUS)			1.69 ± 0.20 (VIF)		Schmidt et al.[144]	
$YbAl_3$	PECS			0.182 ± 0.004	6.81 ± 0.30 (HV)	174 ± 2.5 (RUS)			1.13 ± 0.07		Schmidt et al.[144]	

Note: CHEV = chevron notched flexure specimen; COD = crack opening displacement; GS = grain size; HV = Vickers hardness; IEM = impluse excitation method; nano = nanoindentation; PEM = pulse-echo method; RUS = resonant ultrasound spectroscopy; SENB = single-edge-notched beam; SEVNB = single-edge vee-notched bend; SUB = sharp V-notch beam; VIF = Vickers indentation fracture.

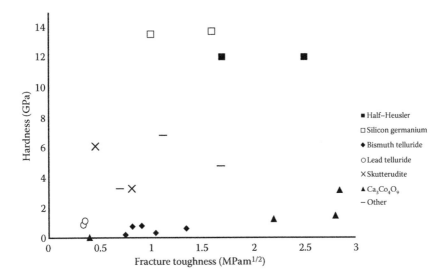

FIGURE 14.4 Hardness and toughness of commonly used TE materials. High hardness need not imply high toughness or high fracture strength.

14.1.4 Microstructure–Property Correlation

This section discusses the case studies that underline the effect of key microstructural features on the mechanical behavior of TE materials. The goal is to understand how the microstructure could be engineered in order to enhance the mechanical properties.

14.1.4.1 Critical Flaw Sizes in Hot-Pressed Half-Heusler

Gahlawat et al.[146] determined the flexural strength of p-type HH at elevated temperatures, up to 600°C, using three-point bend tests. Flexural strength varied between 180 and 275 MPa at all temperatures. The fracture toughness K_{IC} measured at room temperature, using SENB configuration, averaged around 1.7–2.2 MPa m$^{1/2}$. SEM analyses (Figure 14.5) of the fractured surfaces revealed critical flaw size of at least two orders of magnitude larger than the average grain size values (200–400 nm). The toughness and flexural strength data and SEM micrographs suggest that critical flaw size varies between the largest microstructural feature, in this case, the largest grain size (~15 μm), and the processing and machining flaws, which are at least half an order of magnitude higher than the largest grains (~50 μm). Figure 14.6 shows the SEM micrograph of the fractured surface of p-type HH. Had the average grain size in the microstructure initiated fracture, the strength would have been substantially higher than reported. This implies that the larger processing flaws introduced during processing, not the intrinsic microstructural flaws, limit the strength of this TE material.

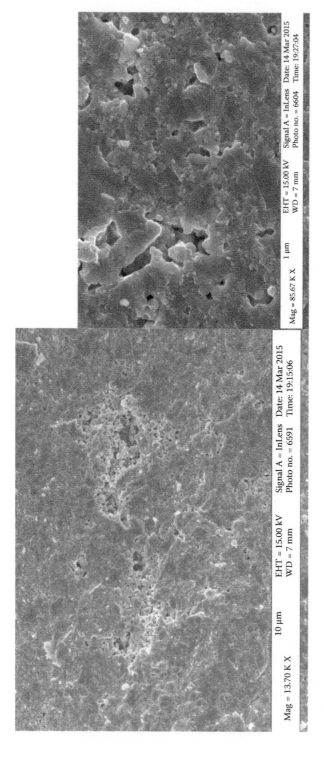

FIGURE 14.5 SEM micrograph showing critical flaw dimensions observed during flexural strength testing of p-type HH at 300°C. This particular specimen had bimodal grain size distribution. (From Gahlawat, S., He, R., Chen, S., Ren, Z. F. and White, K. W., unpublished work.)

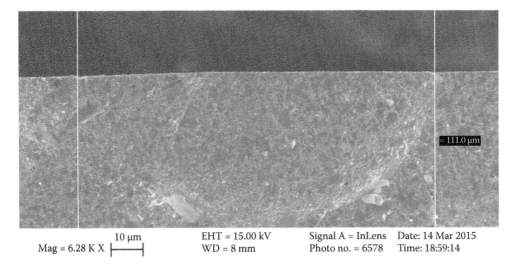

Mag = 6.28 K X	10 μm	EHT = 15.00 kV	Signal A = InLens Date: 14 Mar 2015
		WD = 8 mm	Photo no. = 6578 Time: 18:59:14

FIGURE 14.6 SEM micrographs showing huge unsintered regions in p-type HH. Porosity could be contributing to failure. (From Gahlawat, S., He, R., Chen, S., Ren, Z. F. and White, K. W., unpublished work.)

14.1.4.2 Effects of Structural Hierarchy and Nanoprecipitates in Sintered Bismuth Telluride

Zone melting (ZM) offers a promising alternative for scaling up the production of TE materials such as bismuth telluride alloys. ZM produces polycrystalline-textured ingots that suffer from machining defects such as surface cracks, edge chipping, and delamination. The explanation lies in the van der Waals forces that hold together the Te(1)–Te(1) layers in the bismuth telluride structure, thereby making the layers susceptible to cleavage. Texturing induced by ZM makes K_{IC} remarkably low in the direction parallel to these layers. Even for cracks oriented perpendicular to these planes, K_{IC} hovers around 0.75 MPa

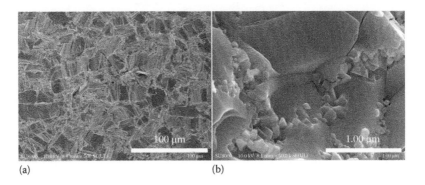

(a)   (b)

FIGURE 14.7 SEM micrograph of $Bi_{0.5}Sb_{1.5}Te_3$ specimen prepared using MS-PAS technique. The micrograph shows the structural hierarchy present in the specimen. (Zheng, Y. et al.: Mechanically robust BiSbTe alloys with superior thermoelectric performance: A case study of stable hierarchical nanostructured thermoelectric materials. *Advanced Energy Materials*. 2015. 5. Copyright Wiley-VCH Verlag GmbH & Co. KGaA. Reproduced with permission.)

m$^{1/2}$. Flexural and compressive strengths of ZM ingots averaged at 10 and 15 MPa, respectively.[125,127]

Research shows that structural hierarchy imparts ultrahigh toughness to natural materials such as bamboo and conch shells.[147–151] Motivated by this concept, Zheng et al.[127] produced Bi$_{0.5}$Sb$_{1.5}$Te$_3$ ingots with hierarchical structures and nanoprecipitates, using melt-spinning and plasma-assisted sintering (MS-PAS) and compared their properties with those of ZM ingots. Figure 14.7 shows the micrograph of the free fracture surface of MS-PAS specimens. In MS-PAS specimens, the authors observed considerable variation in grain size, from submicron level to few tens of microns, and nanoprecipitates embedded in the grain boundaries. Figure 14.8b shows a comparison between the flexural and compressive strengths of ZM ingots

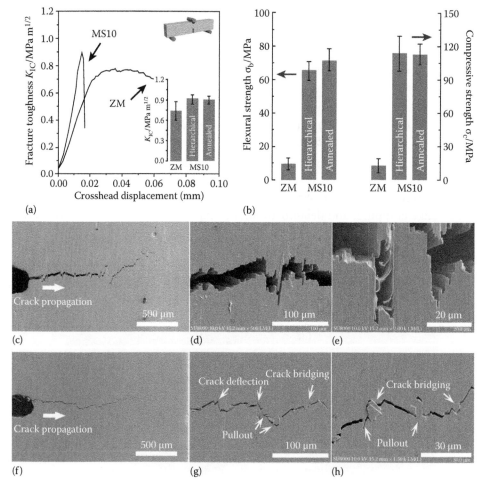

FIGURE 14.8 (a) Fracture toughness of ZM and MS-PAS Bi$_{0.5}$Sb$_{1.5}$Te$_3$, (b) room-temperature flexural strength and compressive strength of ZM and (c–e) MS-PAS Bi$_{0.5}$Sb$_{1.5}$Te$_3$, crack propagation and fracture surfaces of ZM and (f–h) MS-PAS Bi$_{0.5}$Sb$_{1.5}$Te$_3$. (Zheng, Y. et al.: Mechanically robust BiSbTe alloys with superior thermoelectric performance: A case study of stable hierarchical nanostructured thermoelectric materials. *Advanced Energy Materials*. 2015. 5Copyright Wiley-VCH Verlag GmbH & Co. KGaA. Reproduced with permission.)

and MS-PAS specimens. Structural hierarchy and presence of nanometer-sized precipitates resulted in a sixfold increase in flexural strength and an eightfold increase in compressive strength. The stress–strain curve recorded during toughness testing shows that MS-PAS specimens have 30% higher K_{IC} than that of ZM ingots. The authors attributed the nonlinear behavior exhibited by the ZM ingots to crack deflection along the weakly bonded Te(1)–Te(1) layers. The authors also observed toughening mechanisms, such as crack deflection, bridging, and pullout, operative in MS-PAS specimens, due to the structural hierarchy in the microstructure, but could not explain why the MS-PAS specimens failed in a brittle manner.

The readers must note here that the MS-PAS technique also significantly reduced lattice thermal conductivity by as much as 50% and resulted in a peak ZT of 1.22 at 340 K.

14.1.4.3 Effects of Processing Parameters on the Properties of $Ca_3Co_4O_9$ TE Oxide Ceramics

In this study, the authors optimized the TE and mechanical properties of $Ca_3Co_4O_9$ TE oxide ceramics by varying the processing parameters for SPS.[143] This study assessed the effects of uniaxial pressure and sintering temperature on the microstructure and, hence, the mechanical and transport properties. Research shows that the application of external pressure, as well as higher sintering temperatures, increases the sintering driving force and, consequently, yields faster densification.[152–154] According to the presented case study, as the external pressure increased from 0 to 50 MPa, the bulk density rose from 60% to 99.6% of the theoretical density. This also caused a marginal increase in grain size and led to slight texturing, although not as much as induced by HP. Nevertheless, all the aforementioned factors affect the mechanical and electrical transport properties. This study did not measure any thermal transport properties.

The authors used instrumented indentation and three-point bend tests for obtaining the hardness, elastic modulus, flexural strength, and K_{IC}. All these properties monotonically increased with increasing pressure and with increasing sintering temperatures. As the pressure rose to 50 MPa, the hardness increased by a factor of 30, the modulus increased by a factor of 8, the flexural strength went up by a factor of 15, and K_{IC} increased by a factor of 7. The actual numbers have been included in Table 14.1. This drastic improvement in mechanical properties stemmed primarily from the considerable change in bulk density and secondarily, from texturing. Figure 14.9 shows pole figures and SEM micrographs of conventionally sintered and spark plasma-sintered $Ca_3Co_4O_9$ specimens. In addition, external pressure and higher sintering temperatures reduced the electrical resistivity of the specimens. The latter also increased the Seebeck coefficient. Correspondingly, the power factor of the TE material quadrupled from 50 to ~200 μW/m K^2 even at room temperatures.

While the effects of processing on microstructure and properties are manifested, microstructure engineering in TE materials, in general, involves additional complications, as explained in the introduction of this text.

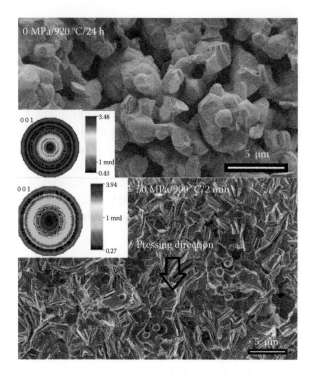

FIGURE 14.9 SEM micrographs and pole figures of (*top*) conventionally sintered Ca347 and (*bottom*) spark plasma sintered Ca349 at 900°C for 2 minutes. (Reprinted from *Ceramics International*, 40, Kenfaui, D. et al., Chateigner, D., and Noudem, J. G. Mechanical properties of $Ca_3Co_4O_9$ bulk oxides intended to be used in thermoelectric generators, 10237–10246, Copyright [2014], with permission from Elsevier.)

There exists a mature understanding of the effects of key microstructural features on individual TE transport properties in TE materials. While researchers have begun identifying appropriate microstructural features necessary for simultaneously optimizing mechanical and TE behavior, the case studies presented earlier for TE materials illustrate the inchoate nature of the field. Cues on microstructure–mechanical property correlations can be taken from the structural ceramics community to guide the optimization of processing techniques for producing TE materials.

One approach toward the improved reliability of TE materials involves the employment of processing techniques that reduce the intrinsic or microstructural flaw size and the surface flaw size. This includes the use of nanograined starting powders and techniques such as advanced sintering that result in higher densification. In addition to improving the mechanical properties, this approach enhances electrical and thermal transport properties.[155–158]

14.1.4.3.1 Microstructure "Design"

Another approach involves the engineering or "design" of the microstructure, for example, the interaction of secondary-phase particles with the advancing crack. Such restraining mechanisms, commonly known as toughening mechanisms, reduce the stress field and, hence, the driving force at the crack tip, thus

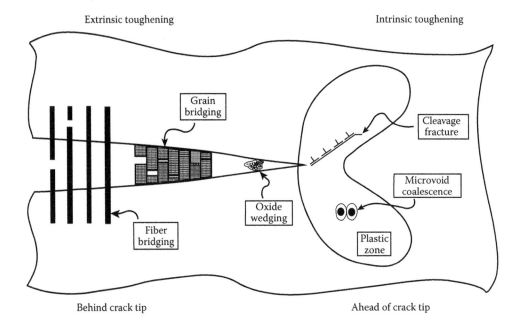

Extrinsic toughening   Intrinsic toughening

Grain bridging

Cleavage fracture

Microvoid coalescence

Oxide wedging

Fiber bridging

Plastic zone

Behind crack tip   Ahead of crack tip

FIGURE 14.10 Alternate classification of toughening mechanisms. (With kind permission from Springer Science+Business Media: *International Journal of Fracture*, Mechanisms of fatigue-crack propagation in ductile and brittle solids, 100, 1999, 55–83, Ritchie, R. O.)

escalating the energy required for crack propagation.[159–161] Based on whether the crack modifies the microstructure, these toughening mechanisms can be grouped into two classes. The first class encompasses noncumulative toughening mechanisms, such as crack deflection, crack bowing, and crack branching, where the crack tip consistently interacts with the microstructure. Such mechanisms remain unaffected by crack length. An alternate classification of toughening mechanisms is shown in Figure 14.10.[162] Crack deflection[163,164] produces a nonplanar crack, which could arise from either the residual strain in the composite or weak matrix/second-phase interfaces; in the latter case, the crack tip encounters an inhomogeneity, such as a second-phase particle, causing deflection along or across the weaker interfaces. This explains why a polycrystalline material exhibits a much higher K_{IC}, almost double, than its single crystal counterpart[165]: either grain boundaries serve as weak interfaces and cracks deflect along them or crystal anisotropy results in crack deflections when traversing through adjacent grains. Faber and Evans[166] reported crack deflection toughening in alumina-fluxed SiC and SiC/ZrO$_2$ composite. Lange[167] proposed that second-phase particles dispersed in a brittle matrix could pin a crack advancing in a plane and cause the crack to bow out between the pinning points. This, however, requires the second-phase particles in the brittle matrix to exhibit higher resistance to fracture than the matrix phase. Lange[168] attributed the toughening in borosilicate glass containing dispersed alumina particles, also observed by Hasselman and Fulrath[169] in 1966, to crack bowing.[170,171] Crack branching occurs when the stresses at the crack tip reach such high levels that the energy released during crack propagation exceeds the surface energy requirements for the propagation of two cracks. Crack branching and crack deflection collectively result in the relatively high strength and fracture toughness of silicon nitride.[172] The other class incorporates the cumulative mechanisms, such as microcrack toughening, transformation toughening, and crack bridging, which depend on change in crack length. The energy required for creating new surfaces varies with the crack length, and therefore, the resistance to crack propagation varies. In a ceramic matrix with a dispersed reinforcing secondary phase, for instance, whiskers, fibers, platelets, etc., as the crack front advances through the microstructure, energy is expended in debonding or pulling out these whiskers. Additionally, unbroken ligaments may remain in the zone behind the crack tip (wake zone). These unbroken ligaments exert a traction force to try to keep the crack faces together. This bridging zone shares the stress path and lowers the stresses at the crack tip. This mechanism, known as crack face bridging, contributes to toughening in all fiber-reinforced[173–176] and whisker-reinforced[177–180] composites with the appropriate interface and elastic relationship. The concept of microcrack toughening[181–185] originates from studies on zirconia-toughened ceramics. In this mechanism, the high stresses in the crack tip vicinity nucleate stable grain boundary microcracks, thereby redistributing the crack tip stress state. Transformation toughening,[186–188] also a

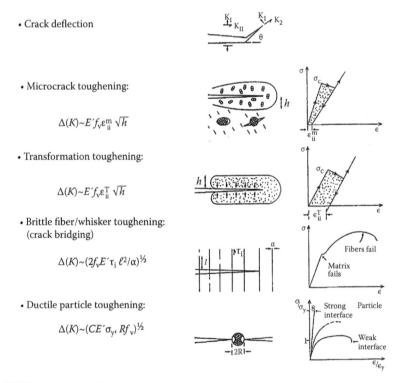

- Crack deflection

- Microcrack toughening:

$$\Delta(K) \sim E' f_v \varepsilon_{ii}^m \sqrt{h}$$

- Transformation toughening:

$$\Delta(K) \sim E' f_v \varepsilon_{ii}^T \sqrt{h}$$

- Brittle fiber/whisker toughening:
  (crack bridging)

$$\Delta(K) \sim (2 f_v E' \tau_i \ell^2/\alpha)^{\frac{1}{2}}$$

- Ductile particle toughening:

$$\Delta(K) \sim (C E' \sigma_y, R f_v)^{\frac{1}{2}}$$

FIGURE 14.11 Schematic illustration of various toughening mechanisms. (With kind permission from Springer Science+Business Media: *International Journal of Fracture*, Mechanisms of fatigue-crack propagation in ductile and brittle solids, 100, 1999, 55–83, Ritchie, R. O.)

wake zone mechanism, only occurs in materials where the stress-induced phase transition near the crack tip creates a volume change and applies a closing force on the crack-wake faces as the crack tip advances into transformation zone. This reduces the localized stress intensity factor at the crack tip, thereby shielding the crack tip. By engineering the microstructure, researchers have introduced one or more of these toughening mechanisms in numerous ceramic systems[189–208] and developed substantially tougher ceramic systems. Figure 14.11 schematically illustrates various toughening mechanisms.

The aforementioned case studies provide an overview of how variation in the microstructure critically controls the fracture toughness, which is observed as fracture strength for components and devices of TE materials. Different processing techniques can lead to drastically different microstructures such as in the case of ZM and MS-PAS techniques. The alignment of weak Te–Te layers in ZM leads to an incredibly brittle microstructure and low fracture strength, whereas MS-PAS creates a more varied microstructure that utilizes multiple toughening mechanisms to enhance fracture toughness. The variation of processing parameters to ensure full densification also significantly contributes to bulk scale properties. The elimination of porosity increases both mechanical and TE properties across the board. Finally, material processing can introduce flaws significantly larger in size than the microstructural dimension that significantly limit fracture strength. It is important to note that the microstructural influence on hardness has not been thoroughly discussed here. In certain cases, such as comparing the relative density of a given material, indentation measurements can reasonably predict fracture strength and fracture toughness. However, in general, hardness does not provide a good indicator for the mechanical properties of a material, especially in the application of TEs. Comparisons between different materials offer even less useful information, as evidenced by Figure 14.4.

14.1.5 Summary

Even with the development of high ZT materials, efficient TE conversion cannot be realized if cracks form or the material fails. Equally important are the TE transport properties; high conductivity requires a low crack population. While hardness may serve as a quick screening tool, it does not characterize the strength and resistance to fracture in materials. Toughness studies indicate that the commonly used TE materials are highly brittle in nature. Thermal stress analysis shows that TE devices fail under cyclic thermal loading, and the failure usually occurs at the interface between the hot side electrode and the TE leg. However, no detailed studies have attempted to characterize the strength at the interfaces. Future efforts should be devoted to not only the improvement of ZT and mechanical properties of TE materials, but also the mechanical characterization of interfaces. Fabrication of efficient and durable TE devices requires stronger interfaces in addition to stronger TE materials.

14.1.6 Acknowledgment

This work was partially supported by Solid State Solar Thermal Energy Conversion Center (S³TEC), an Energy Frontier Research Center funded by the US Department of Energy, Office of Science, Office of Basic Energy Science, under award number DE-SC0001299.

14.2 Mechanical Properties of Thermoelectric Materials Based on Mg₂Si

Yasuo Kogo and Tsutomu Iida

Since TE modules comprise various materials with different material properties, thermal stress induced during processing and/or in operation is one of the prime concerns for the reliability of the modules. Because of this, understanding the mechanical properties of the TE materials is of great importance for the modules, in particular when they are utilized in automotive applications in which the materials are subjected to thermomechanically severe environments.

Bearing this in mind, the mechanical properties of TE materials based on silicides are summarized in this section along with measurement techniques. Special attention is paid to the strength and fracture toughness, and the importance of these properties for the reliability of the TE modules is emphasized.

Various parts, such as TE chips, electrodes, and insulators, are bonded to each other in the module structure. TE modules consist of electrodes to extract power from the chips, and insulators to make contact between the modules and heat sources and heat sinks. These are metals, such as Cu or Ni, and ceramics, such as alumina, respectively. Assuming the modules have a traditional π-type structure, TE chips made of two different materials are used. All these materials are bonded and interconnected to fabricate the TE modules.

Thermal stresses are induced during cooling from the bonding temperature, since the two different materials bonded together have different material properties, especially the CTE. For bonding electrodes and/or insulators

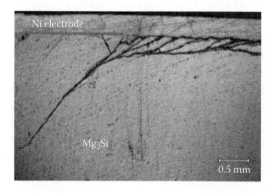

FIGURE 14.12 Thermal stress fracture of interface and TE material near the electrode.

to TE chips, silver solder or silver paste is used as a bonding agent. To obtain sufficient bonding strength, high-temperature processing is usually employed. After the bonding process, the bonded parts are cooled down to room temperature. Because of the different CTE of these materials, thermal stresses are induced at and around the bonded interface between the two different materials due to the difference in the amount of contraction during cooling.

Figure 14.12 shows the fractured interface of a TE material and a Ni electrode due to thermal stress. This structure was made by a single-step sintering process using plasma-assisted sintering (PAS). In this process, Mg_2Si powder was compacted and sintered together with Ni powder in a single step at a temperature of around 800°C, then the compact was cooled to room temperature. During the cooling process, cracks beginning at or near the interface deeply penetrated into the Mg_2Si chip because of the thermal stress induced by the difference in CTE between the Mg_2Si and the Ni. As shown here, the mechanical properties, as well as the TE ones, are critically important for highly reliable TE modules.

In order to design and fabricate TE modules, the mechanical properties of the constituent materials need to be well understood. To estimate the thermal stress induced in the modules, finite-element analysis (FEA) is often used. Input data for the analysis are the elastic properties, such as Young's modulus and Poisson's ratio, and the CTE. To enable the transient temperature gradient in the TE module to be considered, the thermal conductivity and the heat capacity are also required. For a more accurate numerical analysis, the temperature dependence of each property should be input.

One of the promising applications for TE materials based on silicides is an automotive TE generator. In current gasoline engine systems, more than 70% of the energy of the fuel is lost as waste heat in the exhaust gases and to the engine coolant. Recovering energy from the waste heat of vehicles is quite important because it increases the fuel efficiency and reduces the CO_2 emission. Since silicide materials have some significant advantages, such as their excellent TE properties, their abundance, which makes them cost effective, their nontoxicity, and their light weight, they are very promising candidates for TE generators for automobiles.

The durability of the TE modules in thermally and mechanically severe environments is one of the crucial issues in automotive applications. Automotive parts are exposed to severe vibration. For example, it is estimated that this is 10G for exhaust systems and 30G near the engine. In addition, the temperature of the exhaust gas steeply increases when an automobile is rapidly accelerated. When such temperature change is repeatedly applied to TE modules, it causes fatigue fractures induced by thermal stress.

Because of this, as well as understanding the TE properties, understanding the mechanical properties is far more important in automotive applications than in other ones. Even in less severe vibrational environments such as in industrial furnaces, thermal stress, as mentioned earlier, is one of the critical issues for TE modules. In addition to this, vibrational mechanical loading is applied in automotive applications, so the mechanical properties need to be well understood.

With this in mind, the mechanical properties of TE materials based on silicides are summarized in this review. Typical silicides used as TE materials are n-type Mg_2Si and p-type $MnSi$, which are expected to become available for practical use. The mechanical properties of these materials are the main focus of this review.

One of the fundamental properties for designing structures is Young's modulus. To estimate the induced stresses and/or strains in complex-shaped structures made of various materials, FEA is one of the most powerful tools and is widely used in the engineering field. Of the input data for the material in FEA, Young's modulus is the most fundamental and important property. This property represents the response of the material to applied stress, showing how much the material is deformed with stress. On the other hand, it shows how much stress is induced when the material is deformed.

An easy way to measure Young's modulus is by the ultrasonic method. The velocities of the longitudinal and transverse waves in the specimens are measured using an ultrasonic pulser receiver. Then, Young's modulus E can be calculated with the following equation:

TABLE 14.2 Mechanical Properties of Silicides Used as TE Material

Material	Young's Modulus (E GPa)	Hardness (H_v GPa)	Bending Strength σ_f(MPa)	Compression Strength (σ_f MPa)	Fracture Toughness (K_c MPa·m$^{1/2}$)	Ref.
MnSi	137		146.3	1140		Kim et al.[1]
MnSi$_{1.73}$	160	11.85	178	1083	1.63	Chere et al.[1]
	182					
Mg$_2$Si	115.6					Rogl et al.[3]
	113.5					Hu et al.[4]
	110.9					Jian et al.[5]
	76	4.2			0.88	Zhao et al.[6]
		4.38				Yang and Stabler[7]
	~117	4.8–5.6			0.7–1.3	He et al.[8]
		4.38			1.46	Ren et al.[9]
	105		57	430	1.2	Ma et al.[10]
		5.6			1.3	Davidge[11]
Mg$_2$Si$_{1-x}$Sn$_x$ (x = 0.4 to 0.6)	83 ± 25	3.54	79	492	0.99	Chere et al.[2]

Source: Gromnitskaya, E. L. Institute for High Pressure Physics, Troisk, unpublished, 2007; with kind permission from Springer Science+Business Media: *Journal of Electronic Materials,* 43, 2014, 6, Yakiv, G. et al.; Zhang, C. et al., *Journal of Physics D,* 42, 125403, 2009, Institute of Physics.

$$E = \frac{3\rho V_s^2 \left[V_1^2 - (4/3)V_s^2 \right]}{V_1^2 - V_s^2}, \tag{14.1}$$

where ρ denotes the density of the specimen, and V_1 and V_s are the sound velocities of the longitudinal and transverse waves, respectively.

The Young's moduli of the silicides used as TE materials are listed in Table 14.2.[209–219] The Young's modulus of MnSi is about 260 GPa, and that of Mg_2Si is about 110 GPa. Higher Young's modulus is advantageous to stabilize the shape of the structure. It should be noted, however, that thermal stresses induced in hot structures are proportional to $E\alpha\Delta T$. Because of this, high Young's modulus is not necessarily preferable for materials used in environments with high temperatures or high temperature gradients.

The TE material has to have sufficient strength. TE modules are exposed to thermal stresses induced during processing and in operation. In addition, mechanical stress is also induced during operation in vibrational environments. Such stresses in the structures can be estimated by FEA, and the structure is resilient when the materials used possess sufficient strength compared with the induced stresses. In other words, structural reliability depends to a large extent on the strength of the constituent materials.

The mechanical strength of TE materials is often evaluated by a bending test. There are various test methods for the mechanical strength such as tensile, bending, and compressive tests. In the case of brittle materials such as TE ones, a three- or four-point bending test is often employed. This is because the shape of the specimen is quite simple and the specimen can be set to fixtures without any difficulty. For a four-point bending test, the strength is calculated from the following equation:

$$\sigma = \frac{3P(L - l)}{2bh^2}, \tag{14.2}$$

where P is the maximum load, L and l are the outer and the inner span lengths, and b and h are the width and thickness of the specimen, respectively.

The compressive strength is also important for TE modules. TE modules are usually attached to heat sources using pressure, because the heat transfer is enhanced by pressing the TE module onto the heat source. Higher pressure results in better heat transfer. However, TE materials are subjected to continuous compressive loadings because of this. The bending and compressive strengths of MnSi and Mg_2Si are listed in Table 14.2. As shown in the table, MnSi has bending and compressive strengths of about 150 and 1100 MPa, respectively. Compared with MnSi, Mg_2Si has a considerably lower bending strength of 44 MPa and a compressive one of 440 MPa. As shown here, improving the mechanical properties of Mg_2Si is a key issue in order to realize TE generators made with silicides.

The fracture toughness is another important property as well as the strength. When we talk about the mechanical properties, strength is often focused on. For structural materials, however, fracture toughness or simply toughness is also quite an important property. A simple example for the difference in fracture toughness is the difference between glass and commercially available pure aluminum. Some glasses and aluminum have similar Young's modulus and strength. Even though the strength is similar, there are clear differences between them in their response to scratches or cracks. If a crack is introduced on a glass surface, the glass easily fractures into pieces with far lower stress than the strength. On the other hand, there is no effect on an aluminum specimen even if the same crack is introduced on the surface. Such a difference is due to the difference in fracture toughness between the glass and aluminum.

Fracture toughness is the measure of crack insensitivity, and low fracture toughness represents the brittleness of the material. As shown in Figure 14.13, even with similar Young's modulus and strength, glass fractures with a small deformation; aluminum shows large deformation before the final fracture. This means that the aluminum requires much higher mechanical energy than glass. This difference is the origin of the difference in fracture toughness. Since the strength of brittle materials, such as glass, dramatically decreases as a result of a small scratch on the surface, tougher materials with high fracture toughness are necessary to make highly reliable structures.

A simple way of evaluating fracture toughness is by indentation fracture method. In this method, a Vickers-type diamond indenter is used to indent the surface of a mirror-polished specimen to initiate cracks from the corners of the impression. The average crack length and impression size are measured by an optical microscope, and the fracture toughness is calculated from the following equation:

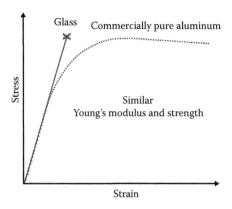

FIGURE 14.13 Stress–strain curves of glass and aluminum with similar Young's modulus and strength.

$$K_c = 0.018 \left(\frac{E}{HV} \right)^{1/2} \left(\frac{P}{c^{3/2}} \right) = 0.026 \frac{E^{1/2} P^{1/2} a}{c^{3/2}}, \qquad (14.3)$$

where E and P denote the Young's modulus of the specimen and the applied load, respectively, HV is the Vickers hardness, c is half of the average crack length, and a is half of the average diagonal length of the impression.

The fracture toughness of the silicides used as TE materials are listed in Table 14.3. In this table, the fracture toughness of other TE materials used in the midtemperature range and those of some engineering ceramics are also listed for reference. Generally speaking, the fracture toughness of TE materials used in the midtemperature range is similar to or lower than that of glass. In this class of TE materials, Mg_2Si shows relatively high fracture toughness. However, it is still insufficient compared with those of engineering ceramics, which are used as structural components in some automotive structures. Because of this, attempts to enhance the fracture toughness of Mg_2Si have been made.

One of the ways to enhance the fracture toughness is the incorporation of a metallic binder.[212] Up to 2 vol% Cu powder was mixed with Mg_2Si powder, and this mixture was sintered by PAS. The addition of Cu is expected to improve the grain boundary bonding. Figure 14.14 shows the indented surfaces of Mg_2Si (a) without the binder and (b) with 1.5 vol% of the Cu binder. Cracks have been arrested by the Cu particles at the grain boundaries, and the average crack length has been reduced.

TABLE 14.3 Fracture Toughness of TE Materials

Midtemperature TE Materials					Engineering Ceramics			
PbTe	PbSe	CoSb$_3$	Si$_{0.8}$Ge$_{0.2}$	Mg$_2$Si	Glass	SiC	Al$_2$O$_3$	Si$_3$N$_4$
0.35	0.8	0.8	1.0	–1.2	0.7	2–5	3–4	5–7

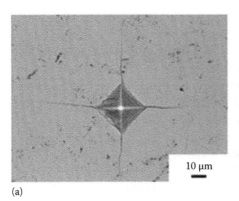
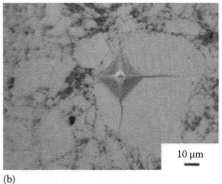

(a)   (b)

FIGURE 14.14 Indented surfaces of Mg_2Si (a) without Cu binder and (b) with 1.5 vol% Cu binder.

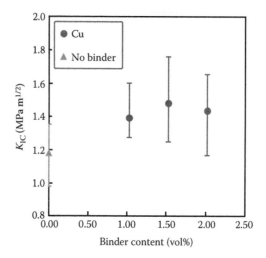

FIGURE 14.15 Effect of adding Cu on the fracture toughness of Mg_2Si.

Figure 14.15 shows the effect of adding Cu on the fracture toughness of Mg_2Si. As shown in the figure, the fracture toughness of Mg_2Si increases with the amount of Cu binder and reaches a maximum of 1.5 MPa $m^{1/2}$ with 1.5 vol% Cu binder, although the scatter in the data is rather large and needs to be minimized. It was confirmed that the TE properties remain unchanged even after adding 1.5 vol% Cu powder.

Another possible way of enhancing the fracture toughness is reducing the grain size of the Mg_2Si. Grain refinement of ceramic materials is known for its effectiveness in enhancing the strength and toughness. This is because the intrinsic crack size tends to be similar to the size of the grains. By reducing the intrinsic crack size, the strength of the ceramic is increased. Increasing the grain boundaries is expected to help improve the fracture toughness. Since similar effects are expected in the case of Mg_2Si, the mechanical and TE properties of Mg_2Si with various average grain size were evaluated.[12]

Figure 14.16 shows three-point bending test results of Mg_2Si with various average grain size (after sintering). Although the three-point bending strength is 50 MPa with an average grain size of 60 μm, it increases to 100 MPa with an average grain size of 10 μm. These results indicate that reducing the grain size is effective in enhancing the strength of Mg_2Si, although further reduction results in a decrease in strength.

A similar trend is observed in the fracture toughness of Mg_2Si. The fracture toughness increases from 1.1 MPa $m^{1/2}$ with an average grain size of 60 μm to 1.7 MPa $m^{1/2}$ with an average grain size of 5 μm.

Figure 14.17 shows the relationship between the TE properties (ZT) and the average grain size of Mg_2Si. With average grain sizes of 60 and 15 μm, Mg_2Si shows the same or slightly higher ZT, although further reduction results in a decrease in ZT. X-ray diffraction and elemental analysis by EDS revealed that the formation of MgO is a major reason for the degradation in the ZT value.

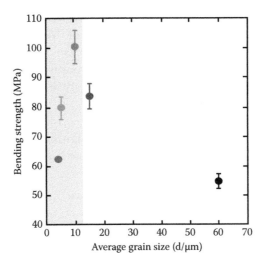

FIGURE 14.16 Bending strengths of Mg_2Si with different average grain sizes.

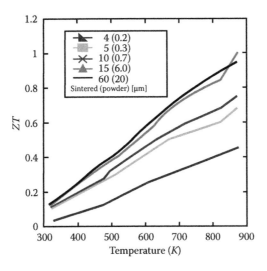

FIGURE 14.17 Effect of average grain size on the TE properties of Mg_2Si.

These results show that reducing the average grain size of Mg_2Si to 15 μm improves the increased the bending strength to 83 MPa and the fracture toughness to 1.5 MPa $m^{1/2}$ without degradation of the TE properties. This suggests that we can expect to make further improvements in the mechanical properties without the degradation of the TE properties, if the formation of MgO can be suppressed.

In future developments, the incorporation of a second phase might be the most effective way of enhancing the fracture toughness of Mg_2Si. Although Mg_2Si possesses sufficiently good TE properties, better mechanical properties are required from the structural viewpoint of the material. To further improve the mechanical properties of Mg_2Si, approaches used for engineering ceramics are helpful. The incorporation of ceramic particles to reinforce Mg_2Si is one of the most effective ways to enhance its toughness. Combinations of two or more mechanisms are also expected as a way of getting better properties.

For the development of silicides used as TE materials, improving the mechanical properties of Mg_2Si without significant degradation of its TE properties is the biggest challenge in the near future.

References

1. Kim, S. I. et al. Dense dislocation arrays embedded in grain boundaries for high-performance bulk thermoelectrics. *Science* **348**, 109–114 (2015).
2. Chere, E. K. et al. Enhancement of thermoelectric performance in n-type $PbTe_{1-y}Se_y$ by doping Cr and tuning Te:Se ratio. *Nano Energy* **13**, 355–367 (2015).
3. Rogl, G. et al. In-doped multifilled n-type skutterudites with ZT = 1.8. *Acta Mater.* **95**, 201–211 (2015).
4. Hu, X. et al. Power generation from nanostructured PbTe-based thermoelectrics: Comprehensive development from materials to modules. *Energy Environ. Sci.* **9**, 517–529 (2016).
5. Jian, Z. et al. Significant band engineering effect of YbTe for high performance thermoelectric PbTe. *J. Mater. Chem. C* **3**, 12410–12417 (2015).
6. Zhao, L. et al. Superior intrinsic thermoelectric performance with zT of 1.8 in single-crystal and melt-quenched highly dense $Cu_{2-x}Se$ bulks. *Sci. Rep.* **5**, 7671 (2015).
7. Yang, J., and Stabler, F. R. Automotive applications of thermoelectric materials. *J. Electron. Mater.* **38**, 1245–1251 (2009).
8. He, R. et al. Studies on mechanical properties of thermoelectric materials by nanoindentation. *Phys. Status Solidi A* **212**, 2191–2195 (2015).
9. Ren, F. et al. Mechanical characterization of PbTe-based thermoelectric materials. *Symposium U—Thermoelectric Power Generation* **1044**, 1044–U04-04 (2007).
10. Ma, J. M., Firdosy, S. A., Kaner, R. B., Fleurial, J.-P., and Ravi, V. A. Hardness and fracture toughness of thermoelectric $La_{3-x}Te_4$. *J. Mater. Sci.* **49**, 1150–1156 (2014).
11. Davidge, R. W. *Mechanical Behaviour of Ceramics.* (Cambridge University Press, Cambridge, 1979).
12. Brook, R. J. *Concise Encyclopedia of Advanced Ceramic Materials.* (Elsevier, Amsterdam, 2012).
13. Munz, D., and Fett, T. *Ceramics: Mechanical Properties, Failure Behaviour, Materials Selection.* (Springer Science and Business Media, Berlin, 2013).
14. Wachtman, J. B., Matthewson, M. J., and Cannon, W. R. *Mechanical Properties of Ceramics Revised Edition.* (John Wiley & Sons, Hoboken, NJ, 2009).
15. McPherson, J. W. *Reliability Physics and Engineering.* (Springer, New York, 2010).
16. Ravi, V. et al. Thermal expansion studies of selected high-temperature thermoelectric materials. *J. Electron. Mater.* **38**, 1433–1442 (2009).
17. Hsueh, C. H. Thermal stresses in elastic multilayer systems. *Thin Solid Films* **418**, 182–188 (2002).

18. Shackelford, J. F., and Groza, J. R. *Materials Processing Handbook*. (CRC Press, Boca Raton, FL, 2007).
19. Kingery, W. D. Factors affecting thermal stress resistance of ceramic materials. *J. Am. Ceram. Soc.* **38**, 3–15 (1955).
20. Hasselman, D. P. H. Elastic energy at fracture and surface energy as design criteria for thermal shock. *J. Am. Ceram. Soc.* **46**, 535–540 (1963).
21. Hasselman, D. P. H. Unified theory of thermal shock fracture initiation and crack propagation in brittle ceramics. *J. Am. Ceram. Soc.* **52**, 600–604 (1969).
22. Gałązka, K. et al. Phase formation, stability, and oxidation in (Ti, Zr, Hf) NiSn half-Heusler compounds. *Phys. Status Solidi A* **211**, 1259–1266 (2014).
23. Tsujii, N. et al. Phase stability and chemical composition dependence of the thermoelectric properties of the type-I clathrate $Ba_8Al_xSi4_{6-x}$ ($8 \leq x \leq 15$). *J. Solid State Chem.* **184**, 1293–1303 (2011).
24. Pedersen, B. L., Yin, H., Birkedal, H., Nygren, M., and Iversen, B. B. Cd Substitution in $M_xZn_{4-x}Sb_3$: Effect on thermal stability, crystal structure, phase transitions, and thermoelectric performance. *Chem. Mater.* **22**, 2375–2383 (2010).
25. Yang, J., and Caillat, T. Thermoelectric materials for space and automotive power deneration. *MRS Bull.* **31**, 224–229 (2006).
26. Mozharivskyj, Y., Pecharsky, A. O., Bud'ko, S., and Miller, G. J. A Promising thermoelectric material: Zn_4Sb_3 or $Zn_{6-8}Sb_5$, its composition, structure, stability, and polymorphs: Structure and stability of $Zn_{1-8}Sb$. *Chem. Mater.* **16**, 1580–1589 (2004).
27. Al-Merbati, A. S., Yilbas, B. S., and Sahin, A. Z. Thermodynamics and thermal stress analysis of thermoelectric power generator: Influence of pin geometry on device performance. *Appl. Therm. Eng.* **50**, 683–692 (2013).
28. Erturun, U., Erermis, K., and Mossi, K. Effect of various leg geometries on thermo-mechanical and power generation performance of thermoelectric devices. *Appl. Therm. Eng.* **73**, 128–141 (2014).
29. Suhir, E., and Shakouri, A. Assembly bonded at the ends: Could thinner and longer legs result in a lower thermal stress in a thermoelectric module design? *J. Appl. Mech.* **79**, 061010–061010 (2012).
30. Suhir, E., and Shakouri, A. Predicted thermal stress in a multileg thermoelectric module (TEM) design. *J. Appl. Mech.* **80**, 021012/1–021012/11 (2013).
31. Ziabari, A., Suhir, E., and Shakouri, A. Minimizing thermally induced interfacial shearing stress in a thermoelectric module with low fractional area coverage. *Microelectron. J.* **45**, 547–553 (2014).
32. Jia, X., and Gao, Y. Estimation of thermoelectric and mechanical performances of segmented thermoelectric generators under optimal operating conditions. *Appl. Therm. Eng.* **73**, 335–342 (2014).
33. Tachibana, M., and Fang, J. 2011 Chinese Materials Conference: An estimation of thermal stress of thermoelectric devices in the temperature cycling test. *Procedia Eng.* **27**, 177–185 (2012).
34. Wu, Y. et al. Numerical simulations on the temperature gradient and thermal stress of a thermoelectric power generator. *Energy Convers. Manag.* **88**, 915–927 (2014).
35. Clin, T., Turenne, S., Vasilevskiy, D., and Masut, R. A. Numerical simulation of the thermomechanical behavior of extruded bismuth telluride alloy module. *J. Electron. Mater.* **38**, 994–1001 (2009).
36. Turenne, S., Clin, T., Vasilevskiy, D., and Masut, R. A. Finite element thermomechanical modeling of large area thermoelectric generators based on bismuth telluride alloys. *J. Electron. Mater.* **39**, 1926–1933 (2010).

37. Jinushi, T., Okahara, M., Ishijima, Z., Shikata, H., and Kambe, M. Development of the high performance thermoelectric modules for high temperature heat sources. *Mater. Sci. Forum* **534**, 1521–1524 (2007).

38. Nemoto, T., Iida, T., Sato, J., Suda, H., and Takanashi, Y. Improvement in the durability and heat conduction of uni-leg thermoelectric modules using n-type Mg$_2$Si legs. *J. Electron. Mater.* **43**, 1890–1895 (2014).

39. Kambe, M., Jinushi, T., and Ishijima, Z. Encapsulated thermoelectric modules and compliant pads for advanced thermoelectric systems. *J. Electron. Mater.* **39**, 1418–1421 (2010).

40. Morschel, M., and Bastian, G. Interferometric analysis of thermomechanical deformations in thermoelectric generators. *J. Electron. Mater.* **42**, 1669–1675 (2013).

41. Salzgeber, K., Prenninger, P., Grytsiv, A., Rogl, P., and Bauer, E. Skutterudites: Thermoelectric materials for automotive applications? *J. Electron. Mater.* **39**, 2074–2078 (2010).

42. Hori, Y., Kusano, D., Ito, T., and Izumi, K. Analysis on thermo-mechanical stress of thermoelectric module. *1999. Eighteenth International Conference on Thermoelectrics*, 328–331 (1999).

43. Picard, M., Turenne, S., Vasilevskiy, D., and Masut, R. A. Numerical simulation of performance and thermomechanical behavior of thermoelectric modules with segmented bismuth-telluride-based legs. *J. Electron. Mater.* **42**, 2343–2349 (2013).

44. Lin, W. P., Wesolowski, D. E., and Lee, C. C. Barrier/bonding layers on bismuth telluride (Bi$_2$Te$_3$) for high temperature thermoelectric modules. *J. Mater. Sci. Mater. Electron.* **22**, 1313–1320 (2011).

45. Zhao, D., Geng, H., and Chen, L. Microstructure contact studies for skutterudite thermoelectric devices. *Int. J. Appl. Ceram. Technol.* **9**, 733–741 (2012).

46. Hatzikraniotis, E., Zorbas, K. T., Samaras, I., Kyratsi, T., and Paraskevopoulos, K. M. Efficiency study of a commercial thermoelectric power generator (TEG) under thermal cycling. *J. Electron. Mater.* **39**, 2112–2116 (2010).

47. Park, W., Barako, M. T., Marconnet, A. M., Asheghi, M., and Goodson, K. E. Effect of thermal cycling on commercial thermoelectric modules. *2012 13th IEEE Intersociety Conference on Thermal and Thermomechanical Phenomena in Electronic Systems (ITherm)*, 107–112 (2012).

48. Barako, M. T., Park, W., Marconnet, A. M., Asheghi, M., and Goodson, K. E. A reliability study with infrared imaging of thermoelectric modules under thermal cycling. *2012 13th IEEE Intersociety Conference on Thermal and Thermomechanical Phenomena in Electronic Systems (ITherm)*, 86–92 (2012).

49. Barako, M. T., Park, W., Marconnet, A. M., Asheghi, M., and Goodson, K. E. Thermal cycling, mechanical degradation, and the effective figure of merit of a thermoelectric module. *J. Electron. Mater.* **42**, 372–381 (2013).

50. Dikwa, J., Ateba, P. O., Quetel-Weben, S., and Ndjaka, J.-M. B. Influence of thermal cycles on thermoelectric uni-leg modules made of Ca$_3$Co$_4$O$_9$ /Ca$_{0.95}$Sm$_{0.05}$MnO$_3$ oxides. *J. Taibah Univ. Sci.* **8**, 385–393 (2014).

51. Pelletier, R., Turenne, S., Moreau, A., Vasilevskiy, D., and Masut, R. A. Evolution of mechanical stresses in extruded (Bi$_{1-x}$Sb$_x$)$_2$(Te$_{1-y}$Se$_y$)$_3$ thermoelectric alloys subjected to thermal shocks present in module fabrication processes. *ICT 2007: 26th International Conference on Thermoelectrics*, 49–54 (2007).

52. Mises, R. V. Mechanik der festen Körper im plastisch-deformablen Zustand. *Nachrichten Von Ges. Wiss. Zu Gött. Math.-Phys. Kl.* **1913**, 582–592 (1913).

53. Hill, R. *The Mathematical Theory of Plasticity*. Vol. 11. (Oxford University Press, Oxford, 1998).

54. Ford, H., and Alexander, J. M. *Advanced Mechanics of Materials*. (Longmans, London, 1963).

55. Kaibe, H. et al. Development of thermoelectric generating stacked modules aiming for 15% of conversion efficiency. *ICT 2005. 24th International Conference on Thermoelectrics, 2005*, 242–247 (2005).

56. Li, Y., Yang, X. Q., Zhai, P. C., and Zhang, Q. J. Thermal stress simulation and optimum design of $CoSb_3/Bi_2Te_3$ thermoelectric unicouples with graded Interlayers. *AIP Conf. Proc.* **973**, 297–302 (2008).

57. Sakamoto, T. et al. Stress analysis and output power measurement of an n-Mg_2Si thermoelectric power generator with an unconventional structure. *J. Electron. Mater.* **43**, 1620–1629 (2014).

58. Wereszczak, A. A., and Case, E. D. *Mechanical Response of Thermoelectric Materials*. (Oak Ridge National Laboratory, Oak Ridge, TN, 2015).

59. Chen, G., Mu, Y., Zhai, P., Li, G., and Zhang, Q. An investigation on the coupled thermal–mechanical–electrical response of automobile thermoelectric materials and devices. *J. Electron. Mater.* **42**, 1762–1770 (2013).

60. Bishop, R. F., Hill, R., and Mott, N. F. The theory of indentation and hardness tests. *Proc. Phys. Soc.* **57**, 147 (1945).

61. Bückle, H. Progress in micro-indentation hardness testing. *Metall. Rev.* **4**, 49–100 (1959).

62. Tabor, D. Indentation hardness: Fifty years on a personal view. *Philos. Mag. A* **74**, 1207–1212 (1996).

63. Hill, R., Storakers, B., and Zdunek, A. B. A Theoretical study of the Brinell hardness test. *Proc. R. Soc. Lond. Math. Phys. Eng. Sci.* **423**, 301–330 (1989).

64. Hudson, J. A. *Rock Testing and Site Characterization: Comprehensive Rock Engineering: Principles, Practice and Projects*. **3**, 107 (1993). (Elsevier, 2014).

65. Aydin, A. in *The ISRM Suggested Methods for Rock Characterization, Testing and Monitoring: 2007–2014* (ed.: Ulusay, R.), 25–33 (Springer International Publishing, New York, 2015).

66. Viles, H., Goudie, A., Grab, S., and Lalley, J. The use of the Schmidt Hammer and Equotip for rock hardness assessment in geomorphology and heritage science: A comparative analysis. *Earth Surf. Process. Landf.* **36**, 320–333 (2011).

67. Benedicks, C. A., and Christiansen, V. I. Investigations on the Herbert pendulum hardness tester. *J. Iron Steel Inst.* 219–248 (1924).

68. Matsubara, M., and Sakamoto, K. Improved Herbert Hardness Tester. *Exp. Tech.* **36**, 73–76 (2012).

69. Tabor, D. The hardness of solids. *Rev. Phys. Technol.* **1**, 145 (1970).

70. Tabor, D. The physical meaning of indentation and scratch hardness. *Br. J. Appl. Phys.* **7**, 159 (1956).

71. Tabor, D. A simple theory of static and dynamic hardness. *Proc. R. Soc. Lond. Math. Phys. Eng. Sci.* **192**, 247–274 (1948).

72. Fleck, N. A., Muller, G. M., Ashby, M. F., and Hutchinson, J. W. Strain gradient plasticity: Theory and experiment. *Acta Metall. Mater.* **42**, 475–487 (1994).

73. Fleck, N. A., and Hutchinson, J. W. Strain gradient plasticity. *Adv. Appl. Mech.* 296–361 (1997).

74. Gao, H., Huang, Y., Nix, W. D., and Hutchinson, J. W. Mechanism-based strain gradient plasticity—I. Theory. *J. Mech. Phys. Solids* **47**, 1239–1263 (1999).

75. Hull, D., and Bacon, D. J. *Introductions to Dislocations*. (Butterworth–Heinemann, Oxford, 2011).

76. Taylor, G. I. The mechanism of plastic deformation of crystals: Part I. Theoretical. *Proc. R. Soc. Lond. Ser. Contain. Pap. Math. Phys. Character* **145**, 362–387 (1934).

77. Friedel, J. *Dislocations: International Series of Monographs on Solid State Physics.* **3**, (Elsevier, Amsterdam, 2013).

78. Peierls, R. The size of a dislocation. *Proc. Phys. Soc.* **52**, 34 (1940).

79. Suzuki, T., and Takeuchi, S. Correlation of Peierls-Nabarro stress with crystal structure. *Rev. Phys. Appl. Paris.* **23**, 685 (1988).

80. Nabarro, F. R. N., and Hirth, J. P. *Dislocations in Solids.* **12**, (Elsevier, Amsterdam, 2004).

81. Courtney, T. H. *Mechanical Behavior of Materials.* (Waveband Press, Long Grove, IL, 2005).

82. Nieh, T. G., and Wadsworth, J. Hall-Petch relation in nanocrystalline solids. *Scr. Metall. Mater.* **25**, 955–958 (1991).

83. Chokshi, A. H., Rosen, A., Karch, J., and Gleiter, H. On the validity of the hall-petch relationship in nanocrystalline materials. *Scr. Metall.* **23**, 1679–1683 (1989).

84. Tvergaard, V., and Hutchinson, J. W. Microcracking in ceramics induced by thermal expansion or elastic anisotropy. *J. Am. Ceram. Soc.* **71**, 157–166 (1988).

85. Hertzberg, R. W. *Deformation and Fracture Mechanics of Engineering Materials.* (Wiley, Hoboken, NJ, 1989).

86. Broek, D. *Elementary engineering fracture mechanics.* (Springer Science and Business Media, Berlin, 2012).

87. Anderson, T. L., and Anderson, T. L. *Fracture Mechanics: Fundamentals and Applications.* (CRC Press, Boca, Raton, FL, 2005).

88. Janssen, M., Zuidema, J., and Wanhill, R. *Fracture Mechanics.* (CRC Press, Boca Raton, FL, 2004).

89. ASTM: PS070. Standard test methods for the determination of fracture toughness of advanced ceramics at ambient temperature. In *Annual Book of ASTM Standards* (ASTM International, West Conshohocken, PA, 1997).

90. Swab, J. J., Tice, J., Wereszczak, A. A., and Kraft, R. H. Fracture toughness of advanced structural ceramics: Applying ASTM C1421. *J. Am. Ceram. Soc.* **98**, 607–615 (2015).

91. Palmqvist, S. Method att bestamma segheten hos sproda material, sarskilt hardmetaller. *Jernkontorets Ann.* **141**, 300–307 (1957).

92. Evans, A. G., and Charles, E. A. Fracture toughness determinations by indentation. *J. Am. Ceram. Soc.* **59**, 371–372 (1976).

93. Nose, T., and Fujii, T. Evaluation of fracture toughness for ceramic materials by a single-edge-precracked-beam method. *J. Am. Ceram. Soc.* **71**, 328–333 (1988).

94. Sakai, M., and Bradt, R. C. Fracture toughness testing of brittle materials. *Int. Mater. Rev.* **38**, 53–78 (1993).

95. Quinn, G. D., and Bradt, R. C. On the Vickers indentation fracture toughness test. *J. Am. Ceram. Soc.* **90**, 673–680 (2007).

96. Chantikul, P., Anstis, G. R., Lawn, B. R., and Marshall, D. B. A critical evaluation of indentation techniques for measuring fracture toughness: II. Strength method. *J. Am. Ceram. Soc.* **64**, 539–543 (1981).

97. Anstis, G. R., Chantikul, P., Lawn, B. R., and Marshall, D. B. A critical evaluation of indentation techniques for measuring fracture toughness: I. Direct crack measurements. *J. Am. Ceram. Soc.* **64**, 533–538 (1981).

98. Niihara, K., Morena, R., and Hasselman, D. P. H. Indentation fracture toughness of brittle materials for Palmqvist cracks. *Fract. Mech. Ceram.* **5**, 97 (1981).

99. Niihara, K. A fracture mechanics analysis of indentation-induced Palmqvist crack in ceramics. *J. Mater. Sci. Lett.* **2**, 221–223 (1983).

100. Miyoshi, T., Sagawa, N., and Sassa, T. Study on fracture toughness evaluation for structural ceramics. *Rans. Jap. Soc. Mech. Eng.* **41**, 2487–2489 (1985).

101. Li, Z., Ghosh, A., Kobayashi, A. S., and Bradt, R. C. Indentation fracture toughness of sintered silicon carbide in the Palmqvist crack regime. *J. Am. Ceram. Soc.* **72**, 904–911 (1989).

102. Ponton, C. B., and Rawlings, R. D. Vickers indentation fracture toughness test: Part 1. Review of literature and formulation of standardised indentation toughness equations. *Mater. Sci. Technol.* **5**, 865–872 (1989).

103. Ponton, C. B., and Rawlings, R. D. Vickers indentation fracture toughness test: Part 2. Application and critical evaluation of standardised indentation toughness equations. *Mater. Sci. Technol.* **5**, 961–976 (1989).

104. Ghosh, A., Kobayashi, A. S., Li, Z., Henager, C. H. J., and Bradt, R. C. Vickers microindentation toughness of a sintered SiC in the median-crack regime. (1991). Technical Report.

105. Gahlawat, S. et al. Elastic constants determined by nanoindentation for p-type thermoelectric half-Heusler. *J. Appl. Phys.* **116**, 083516 (2014).

106. Rogl, G. et al. Mechanical properties of half-Heusler alloys. *Acta Mater.* **107**, 178–195 (2016).

107. Germond, J. D. *Structural Characterization and Thermoelectric Performance of ZrNiSn Half-Heusler Compound Synthesized by Mechanical Alloying.* (University of New Orleans, New Orleans, LA, 2010).

108. Fan, X. *Mechanical Characterization of Hydroxyapatite, Thermoelectric Materials and Doped Ceria.* (Michigan State University, East Lansing, MI, 2013).

109. Krez, J. Thermoelectric Properties in Phase-Separated Half-Heusler Materials. (Johannes Gutenberg-Universität, Mainz, 2014).

110. Kawaharada, Y., Kurosaki, K., Muta, H., Uno, M. and Yamanaka, S. Thermophysical properties of $NiZrSn_{1-x}Sb_x$ half-Heusler compounds. *J. Alloys Compd.* **381**, 9–11 (2004).

111. O'Connor, C. J. *Nanostructured Composite Materials for High Temperature Thermoelectric Energy Conversion.* (University of New Orleans, New Orleans, LA, 2012).

112. Verges, M. A. et al. Indentation testing of bulk $Zr_{0.5}Hf_{0.5}Co_{1-x}Ir_xSb_{0.99}Sn_{0.01}$ half-Heusler alloys. *Symposium DD—Thermoelectric Materials 2010-Growth, Properties, Novel Characterization Methods and Applications* **1267**, 1267–DD05–23 (2010).

113. Verges, M. A. et al. Young's modulus and hardness of $Zr_{0.5}Hf_{0.5}Ni_xPd_{1-x}$ $Sn_{0.99}Sb_{0.01}$ half-Heusler compounds. *Science* **3**, 659–666 (2011).

114. Sekimoto, T., Kurosaki, K., Muta, H., and Yamanaka, S. Thermoelectric properties of (Ti, Zr, Hf)CoSb type half-Heusler compounds. *Mater. Trans.* **46**, 1481–1484 (2005).

115. Sekimoto, T., Kurosaki, K., Muta, H., and Yamanaka, S. Thermoelectric and thermophysical properties of $ErPd_X$ (X = Sb and Bi) half-Heusler compounds. *J. Appl. Phys.* **99**, (2006).

116. Sekimoto, T., Kurosaki, K., Muta, H., and Yamanaka, S. Thermoelectric properties of half-Heusler type LaPdBi and GdPdBi. *Mater. Trans.* **48**, 2079–2082 (2007).

117. Zhao, L. D., Zhang, B. P., Li, J. F., Zhang, H. L., and Liu, W. S. Enhanced thermoelectric and mechanical properties in textured n-type Bi_2Te_3 prepared by spark plasma sintering. *Solid State Sci.* **10**, 651–658 (2008).

118. Hayashi, T., Sekine, M., Suzuki, J., Horio, Y., and Takizawa, H. Thermoelectric and mechanical properties of angular extruded $Bi_{0.4}Sb_{1.6}Te_3$ compounds. *Mater. Trans.* **48**, 2724–2728 (2007).

119. Seo, J., Cho, D., Park, K., and Lee, C. Fabrication and thermoelectric properties of p-type $Bi_{0.5}Sb_{1.5}Te_3$ compounds by ingot extrusion. *Mater. Res. Bull.* **35**, 2157–2163 (2000).

120. Zhao, L. D. et al. Thermoelectric and mechanical properties of nano-SiC-dispersed Bi_2Te_3 fabricated by mechanical alloying and spark plasma sintering. *J. Alloys Compd.* **455**, 259–264 (2008).

121. Bhuiyan, M. H., Kim, T. S., Koo, J. M., and Hong, S. J. Microstructural behavior of the heat treated n-type 95% Bi_2Te_3–5%Bi_2Se_3 gas atomized thermoelectric powders. *J. Alloys Compd.* **509**, 1722–1728 (2011).

122. Seo, J., Lee, D., Lee, C., and Park, K. Microstructure, mechanical properties and thermoelectric properties of p-type Te-doped $Bi_{0.5}Sb_{1.5}Te_3$ compounds fabricated by hot extrusion. *J. Mater. Sci. Lett.* **16**, 1153–1156 (1997).

123. Liu, D. W., Li, J. F., Chen, C., and Zhang, B. P. Effects of SiC nanodispersion on the thermoelectric properties of p-type and n-type Bi_2Te_3-based alloys. *J. Electron. Mater.* **40**, 992–998 (2011).

124. Bomshtein, N., Spiridonov, G., Dashevsky, Z., and Gelbstien, Y. Thermoelectric, structural, and mechanical properties of spark-plasma-sintered submicro- and microstructured p-type $Bi_{0.5}Sb_{1.5}Te_3$. *J. Electron. Mater.* **41**, 1546–1553 (2012).

125. Zheng, Y. et al. High-temperature mechanical and thermoelectric properties of p-type $Bi_{0.5}Sb_{1.5}Te_3$ commercial zone melting ingots. *J. Electron. Mater.* **43**, 2017–2022 (2014).

126. Ren, F., Wang, H., Menchhofer, P. A., and Kiggans, J. O. Thermoelectric and mechanical properties of multi-walled carbon nanotube doped $Bi_{0.4}Sb_{1.6}Te_3$ thermoelectric material. *Appl. Phys. Lett.* **103**, (2013).

127. Zheng, Y. et al. Mechanically robust BiSbTe alloys with superior thermoelectric performance: A case study of stable hierarchical nanostructured thermoelectric materials. *Adv. Energy Mater.* **5**, (2015).

128. Kallel, A. C., Roux, G., and Martin, C. L. Thermoelectric and mechanical properties of a hot pressed nanostructured n-type $Si_{80}Ge_{20}$ alloy. *Mater. Sci. Eng. A* **564**, 65–70 (2013).

129. Bathula, S. et al. Microstructure and mechanical properties of thermoelectric nanostructured n-type silicon-germanium alloys synthesized employing spark plasma sintering. *Appl. Phys. Lett.* **105**, (2014).

130. Ni, J. E. et al. Room temperature Young's modulus, shear modulus, Poisson's ratio and hardness of PbTe–PbS thermoelectric materials. *Mater. Sci. Eng. B* **170**, 58–66 (2010).

131. Schmidt, R. D., Case, E. D., Zhao, L. D., and Kanatzidis, M. G. Mechanical properties of low-cost, earth-abundant chalcogenide thermoelectric materials, PbSe and PbS, with additions of 0–4% CdS or ZnS. *J. Mater. Sci.* **50**, 1770–1782 (2015).

132. Vasilevskiy, D., Masut, R. A., and Turenne, S. Thermoelectric and mechanical properties of novel hot-extruded PbTe n-type material. *J. Electron. Mater.* **41**, 1057–1061 (2012).

133. Ravi, V. et al. Mechanical properties of thermoelectric skutterudites. *AIP Conf. Proc.* **969**, 656–662 (2008).

134. Wan, S., Huang, X., Qiu, P., Bai, S., and Chen, L. The effect of short carbon fibers on the thermoelectric and mechanical properties of p-type $CeFe_4Sb_{12}$ skutterudite composites. *Mater. Des.* **67**, 379–384 (2015).

135. Salvador, J. R. et al. Thermoelectric and mechanical properties of melt spun and spark plasma sintered n-type Yb- and Ba-filled skutterudites. *Mater. Sci. Eng. B* **178**, 1087–1096 (2013).

136. Eilertsen, J., Subramanian, M. A., and Kruzic, J. J. Fracture toughness of Co_4Sb_{12} and $In_{0.1}Co_4Sb_{12}$ thermoelectric skutterudites evaluated by three methods. *J. Alloys Compd.* **552**, 492–498 (2013).

137. Duan, B. et al. Enhanced thermoelectric and mechanical properties of Te-substituted skutterudite via nano-TiN dispersion. *Scr. Mater.* **67**, 372–375 (2012).

138. Rogl, G. et al. New p- and n-type skutterudites with $ZT > 1$ and nearly identical thermal expansion and mechanical properties. *Acta Mater.* **61**, 4066–4079 (2013).

139. Schmidt, R. D. et al. Room temperature Young's modulus, shear modulus, and Poisson's ratio of $Ce_{0.9}Fe_{3.5}Co_{0.5}Sb_{12}$ and $Co_{0.95}Pd_{0.05}Te_{0.05}Sb_3$ skutterudite materials. *J. Alloys Compd.* **504**, 303–309 (2010).

140. Salvador, J. R., Yang, J., Shi, X., Wang, H., and Wereszczak, A. A. Transport and mechanical property evaluation of $(AgSbTe)_{1-x}(GeTe)_x$ (x = 0.80, 0.82, 0.85, 0.87, 0.90). *J. Solid State Chem.* **182**, 2088–2095 (2009).

141. Tyagi, K. et al. Crystal structure and mechanical properties of spark plasma sintered Cu_2Se: An efficient photovoltaic and thermoelectric material. *Solid State Commun.* **207**, 21–25 (2015).

142. Kenfaui, D., Chateigner, D., Gomina, M., and Noudem, J. G. Texture, mechanical and thermoelectric properties of $Ca_3Co_4O_9$ ceramics. *J. Alloys Compd.* **490**, 472–479 (2010).

143. Kenfaui, D., Gomina, M., Chateigner, D., and Noudem, J. G. Mechanical properties of $Ca_3Co_4O_9$ bulk oxides intended to be used in thermoelectric generators. *Ceram. Int.* **40**, 10237–10246 (2014).

144. Schmidt, R. D., Case, E. D., Lehr, G. J., and Morelli, D. T. Room temperature mechanical properties of polycrystalline $YbAl_3$, a promising low temperature thermoelectric material. *Intermetallics* **35**, 15–24 (2013).

145. White, K. W., Yu, F., and Fang, Y. in *Handbook of Ceramic Composites* (ed.: Bansal, N. P.) 251–275 (Springer, New York, 2005).

146. Gahlawat, S., He, R., Chen, S., Ren, Z. F. and White, K. W. Unpublished work.

147. Kamat, S., Su, X., Ballarini, R. and Heuer, A. H. Structural basis for the fracture toughness of the shell of the conch Strombus gigas. *Nature* **405**, 1036–1040 (2000).

148. Meyers, M. A. et al. Structural biological composites: An overview. *JOM* **58**, 35–41 (2006).

149. Tan, T. et al. Mechanical properties of functionally graded hierarchical bamboo structures. *Acta Biomater.* **7**, 3796–3803 (2011).

150. Lakes, R. Materials with structural hierarchy. *Nature* **361**, 511–515 (1993).

151. Sun, J., and Bhushan, B. Hierarchical structure and mechanical properties of nacre: A review. *RSC Adv.* **2**, 7617–7632 (2012).

152. German, R. *Sintering: From Empirical Observations to Scientific Principles.* (Butterworth–Heinemann, Oxford, 2014).

153. Atkinson, H. V., and Davies, S. Fundamental aspects of hot isostatic pressing: An overview. *Metall. Mater. Trans. A* **31**, 2981–3000 (2000).

154. Munir, Z. A., Anselmi-Tamburini, U., and Ohyanagi, M. The effect of electric field and pressure on the synthesis and consolidation of materials: A review of the spark plasma sintering method. *J. Mater. Sci.* **41**, 763–777 (2006).

155. Rowe, D. M., Shukla, V. S., and Savvides, N. Phonon scattering at grain boundaries in heavily doped fine-grained silicon-germanium alloys. *Nature* **290**, 765–766 (1981).
156. Toprak, M. S. et al. The impact of nanostructuring on the thermal conductivity of thermoelectric CoSb₃. *Adv. Funct. Mater.* **14**, 1189–1196 (2004).
157. Wang, X. W. et al. Enhanced thermoelectric figure of merit in nanostructured n-type silicon germanium bulk alloy. *Appl. Phys. Lett.* **93**, (2008).
158. Lan, Y., Minnich, A. J., Chen, G., and Ren, Z. Enhancement of thermoelectric gigure-of-merit by a bulk nanostructuring approach. *Adv. Funct. Mater.* **20**, 357–376 (2010).
159. Evans, A. G. High toughness ceramics. *Mater. Sci. Eng. A* **105**, 65–75 (1988).
160. Becher, P. F. Microstructural design of toughened ceramics. *J. Am. Ceram. Soc.* **74**, 255–269 (1991).
161. Steinbrech, R. W. Toughening mechanisms for ceramic materials. *J. Eur. Ceram. Soc.* **10**, 131–142 (1992).
162. Ritchie, R. O. Mechanisms of fatigue-crack propagation in ductile and brittle solids. *Int. J. Fract.* **100**, 55–83 (1999).
163. Faber, K. T., and Evans, A. G. Crack deflection processes—I. Theory. *Acta Metall.* **31**, 565–576 (1983).
164. Faber, K. T., and Evans, A. G. Crack deflection processes—II. Experiment. *Acta Metall.* **31**, 577–584 (1983).
165. Wiederhorn, S. M. Brittle fracture and toughening mechanisms in ceramics. *Annu. Rev. Mater. Sci.* **14**, 373–403 (1984).
166. Faber, K. T., and Evans, A. G. Intergranular crack-deflection toughening in silicon carbide. *J. Am. Ceram. Soc.* **66**, C-94–C-95 (1983).
167. Lange, F. F. The interaction of a crack front with a second-phase dispersion. *Philos. Mag.* **22**, 0983–0992 (1970).
168. Lange, F. F. Fracture energy and strength behavior of a sodium borosilicate glass-Al₂O₃ composite system. *J. Am. Ceram. Soc.* **54**, 614–620 (1971).
169. Hasselman, D. P. H., and Fulrath, R. M. Proposed fracture theory of a dispersion-strengthened glass matrix. *J. Am. Ceram. Soc.* **49**, 68–72 (1966).
170. Evans, A. G. The strength of brittle materials containing second phase dispersions. *Philos. Mag.* **26**, 1327–1344 (1972).
171. Green, D. J., Nicholson, P. S., and Embury, J. D. Fracture of a brittle particulate composite. *J. Mater. Sci.* **14**, 1413–1420 (1979).
172. Schnitzeler, J., Göring, F., and Ziegler, G. An experimental method for characterizing crack propagation in dense silicon nitride. *Proceedings of the Second European Ceramic Society Conference (ECerS '91) held on 11–14 September 1991 in Augsburg, FRG.* (1991).
173. Marshall, D. B., Cox, B. N., and Evans, A. G. The mechanics of matrix cracking in brittle-matrix fiber composites. *Acta Metall.* **33**, 2013–2021 (1985).
174. Budiansky, B., Hutchinson, J. W., and Evans, A. G. Matrix fracture in fiber-reinforced ceramics. *J. Mech. Phys. Solids* **34**, 167–189 (1986).
175. Evans, A. G., and Marshall, D. B. The mechanical performance of fiber reinforced ceramic matrix composites. *Symposium G—High-Temperature/High-Performance Composites* **120**, 213 (1988).
176. Evans, A. G., and Marshall, D. B. Overview no. 85: The mechanical behavior of ceramic matrix composites. *Acta Metall.* **37**, 2567–2583 (1989).
177. Evans, A. G., Rühle, M., Daigleish, B. J., and Thouless, M. D. On prevalent whisker toughening mechanisms in ceramics. *Symposium E—Advances in Structural Ceramics* **78**, 259 (1986).

178. Becher, P. F., Tiegs, T. N., Ogle, J. C., and Warwick, W. H. in *Fracture Mechanics of Ceramics: Volume 7 Composites, Impact, Statistics, and High-Temperature Phenomena* (eds: Bradt, R. C., Evans, A. G., Hasselman, D. P. H., and Lange, F. F.) 61–73 (Springer, New York, 1986).

179. Becher, P. F., Hsueh, C. H., Angelini, P., and Tiegs, T. N. Toughening behavior in whisker-reinforced ceramic matrix composites. *J. Am. Ceram. Soc.* **71**, 1050–1061 (1988).

180. Bengisu, M., Inal, O. T., and Tosyali, O. On whisker toughening in ceramic materials. *Acta Metall. Mater.* **39**, 2509–2517 (1991).

181. Hutchinson, J. W. Crack tip shielding by micro-cracking in brittle solids. *Acta Metall.* **35**, 1605–1619 (1987).

182. Evans, A. G., and Faber, K. T. Crack-growth resistance of microcracking brittle materials. *J. Am. Ceram. Soc.* **67**, 255–260 (1984).

183. Fu, Y. *Mechanics of Microcrack Toughening in Ceramics.* (Lawrence Livermore National Laboratory, Livermore, CA, 1983).

184. Faber, K. T., Iwagoshi, T., and Ghosh, A. Toughening by stress-induced microcracking in two-phase ceramics. *J. Am. Ceram. Soc.* **71**, C-399–C-401 (1988).

185. Shum, D. K. M., and Hutchinson, J. W. On toughening by microcracks. *Mech. Mater.* **9**, 83–91 (1990).

186. Rose, L. R. F. The mechanics of transformation toughening. *Proc. R. Soc. Lond. Math. Phys. Eng. Sci.* **412**, 169–197 (1987).

187. McMeeking, R. M., and Evans, A. G. Mechanics of transformation-toughening in brittle materials. *J. Am. Ceram. Soc.* **65**, 242–246 (1982).

188. Evans, A. G., and Heuer, A. H. Review—Transformation toughening in ceramics: Martensitic transformations in crack-tip stress fields. *J. Am. Ceram. Soc.* **63**, 241–248 (1980).

189. Carter, D. H., and Hurley, G. F. Crack deflection as a toughening mechanism in SiC-whisker-reinforced $MoSi_2$. *J. Am. Ceram. Soc.* **70**, C-79–C-81 (1987).

190. Kim, D.-H., and Kim, C. H. Toughening behavior of silicon carbide with additions of yttria and alumina. *J. Am. Ceram. Soc.* **73**, 1431–1434 (1990).

191. Xia, Z. et al. Direct observation of toughening mechanisms in carbon nanotube ceramic matrix composites. *Acta Mater.* **52**, 931–944 (2004).

192. Becher, P. F., and Wei, G. C. Toughening behavior in SiC-whisker-reinforced alumina. *J. Am. Ceram. Soc.* **67**, C-267–C-269 (1984).

193. Yu, F., and White, K. W. Relationship between microstructure and mechanical performance of a 70% silicon nitride–30% barium aluminum silicate self-reinforced ceramic composite. *J. Am. Ceram. Soc.* **84**, 5–12 (2001).

194. Silvestroni, L., Sciti, D., Melandri, C., and Guicciardi, S. Toughened ZrB_2-based ceramics through SiC whisker or SiC chopped fiber additions. *J. Eur. Ceram. Soc.* **30**, 2155–2164 (2010).

195. White, K. W., and Guazzone, L. Elevated-temperature toughening mechanisms in a SiC/Al_2O_3 composite. *J. Am. Ceram. Soc.* **74**, 2280–2285 (1991).

196. Rühle, M., Evans, A. G., McMeeking, R. M., Charalambides, P. G., and Hutchinson, J. W. Microcrack toughening in alumina/zirconia. *Acta Metall.* **35**, 2701–2710 (1987).

197. Sigl, L. S., and Kleebe, H.-J. Microcracking in B_4C-TiB_2 composites. *J. Am. Ceram. Soc.* **78**, 2374–2380 (1995).

198. Rühle, M., Claussen, N., and Heuer, A. H. Transformation and microcrack toughening as complementary processes in ZrO_2-toughened Al_2O_3. *J. Am. Ceram. Soc.* **69**, 195–197 (1986).

199. Hannink, R. H. J., Kelly, P. M., and Muddle, B. C. Transformation toughening in zirconia-containing ceramics. *J. Am. Ceram. Soc.* **83**, 461–487 (2000).

200. Heuer, A. H. Transformation toughening in ZrO_2-containing ceramics. *J. Am. Ceram. Soc.* **70**, 689–698 (1987).

201. Porter, D. L., Evans, A. G., and Heuer, A. H. Transformation-toughening in partially-stabilized zirconia (PSZ). *Acta Metall.* **27**, 1649–1654 (1979).

202. Clarke, D. R., and Schwartz, B. Transformation toughening of glass ceramics. *J. Mater. Res.* **2**, 801–804 (1987).

203. Soboyejo, W., Brooks, D., Chen, L.-C., and Lederich, R. Transformation toughening and fracture behavior of molybdenum disilicide composites reinforced with partially stabilized zirconia. *J. Am. Ceram. Soc.* **78**, 1481–1488 (1995).

204. Kirchner, H. P., Gruver, R. M., Swain, M. V., and Garvie, R. C. Crack branching in transformation-toughened zirconia. *J. Am. Ceram. Soc.* **64**, 529–533 (1981).

205. Zhang, H., Wang, D., Chen, S., and Liu, X. Toughening of $MoSi_2$ doped by La_2O_3 particles. *Mater. Sci. Eng. A* **345**, 118–121 (2003).

206. Wang, J., and Stevens, R. Zirconia-toughened alumina (ZTA) ceramics. *J. Mater. Sci.* **24**, 3421–3440 (1989).

207. Hay, J. C., and White, K. W. Grain-bridging mechanisms in monolithic alumina and spinel. *J. Am. Ceram. Soc.* **76**, 1849–1854 (1993).

208. Claussen, N., and Petzow, G. in *Tailoring Multiphase and Composite Ceramics* (eds.: Tressler, R. E., Messing, G. L., Pantano, C. G., and Newnham, R. E.) 649–662 (Springer, New York, 1986).

209. Gromnitskaya, E. L. Institute for High Pressure Physics, Troisk, unpublished, (2007).

210. Yakiv, G. et al. Physical, mechanical, and structural properties of highly efficient nanostructured n- and p-silicides for practical thermoelectric applications. *J. Electron. Mater.* **43**, 6 (2014).

211. Zhang, C. et al. First-principle study of typical precipitates in creep resistant magnesium alloys. *J. Phys. D* **42**, 125403 (2009).

212. Masashi I. et al. Mechanical properties of Mg_2Si with metallic binders. *Jpn. J. Appl. Phys.* **54**,. 07JC03 (2015).

213. Tani, J. et al. Lattice dynamics of Mg_2Si and Mg_2Ge compounds from first-principles calculations. *Comput. Mater. Sci.* **42**, 531 (2008).

214. Whitten, W. B. et al. Elastic constants and lattice vibration frequencies of Mg_2Si. *J. Phys.Chem. Solids* **26**, 49 (1965).

215. Milekhine, V. et al. Mechanical properties of $FeSi$ (ε), $FeSi_2$ (ζa) and Mg_2Si. *Intermetallics* **10**, 743 (2002).

216. Li, G. H. et al. Magnesium silicide intermetallic alloys. *Metall. Trans. A* **24A**, 2383 (1993).

217. Schmidt, R. D. et al. Room-temperature mechanical properties and slow crack growth behavior of Mg_2Si thermoelectric materials. *J. Electron Mater.* **41**, 1210–1216 (2012).

218. Xiong, W. et al. *Met. Soc. China* **16**, 987–991 (2006).

219. Nakamura, T. et al. *J. Alloys Compds.* (To be submitted).

PART II

THERMOELECTRIC CONTACTS

CHAPTER 15

Contact for Bi$_2$Te$_3$-Based Thermoelectric Leg

Weishu Liu, Yucheng Lan, and Zhifeng Ren

Contents

15.1 Introduction

TE devices share a very simple configuration with metal interconnects (named as metal conductors, conductive metals, and internal electrodes), serially joining the p-type and n-type legs between two ceramic substrates through diffusive bonding or soldering. However, a reliable contact between the metal interconnect and the TE leg requires low contact resistance, strong bonding strength, and good thermal stability.[1] As a result, it is a really tough aim. Here, the contact issue of TE module involves not just one interface, but regions containing metallization layer, bonding materials, and barrier layer in some cases. Contact resistance involves all the contributions from the interfaces, thin layer between electrodes, and TE elements. In the case of soldering bond, the contact resistance mainly comes from the interface between the metallization layer and TE elements. Hereafter, the contact issue that we will discuss is focused on the interface between the metallization layer and the TE elements if without specific clarification.

The definition of the contact resistivity (or specific contact resistance) ρ_c, in the unit of microohms multiply square centimeter, is therefore an equivalent quality, which can be measured by two common methods. Figure 15.1a and b[2,3] shows a scanning probe measuring the voltage varying across the contact interface when a direct current (I) is applied. Contact resistivity ρ_c is derived by multiplying the voltage jump (V_{jump}) by the area of the cross section (A), i.e., $\rho_c = A \cdot V_{jump}/I$.[2,4] The accuracy of ρ_c is limited by the distance between the nearest step and the probe size. In contrast, the Cox–Strack technique is better for the case that the metal layer is very thin.[3,5] Usually, different sized metal pads (e.g., 50–200 μm) are made on the TE material, then the total electrical resistance (R_T) is measured for various pad sizes (d) and then plotted against the reciprocal pad size, as shown in

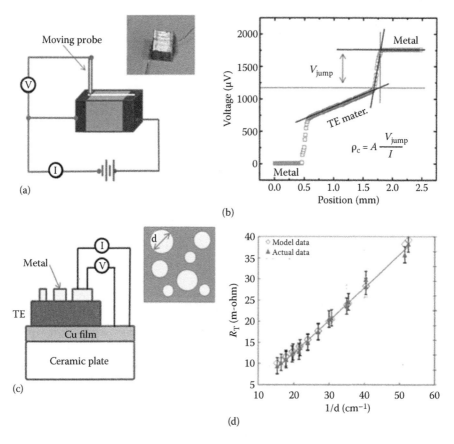

(a)

(b)

(c)

(d)

FIGURE 15.1 Contact resistivity measurement. (a) Schematic picture of scanning probe method for contact resistance; (b) scanning voltage as a function of the position. (Liu, W. S. et al., *Journal of Materials Chemistry A*, 1, 13093–13100, 2013. Reproduced by permission of The Royal Society of Chemistry.) (c) Schematic picture of Cox and Strack method; (d) R_T measured for different metal pads as a function of the reversed pad size. (Redrawn from Gupta, R. P., *Electrochemical and Solid-State Letters*, 12, H302–H304, 2009.)

Figure 15.1c and d. Finally, a ρ_c-containing formula is used to fit the cure of R_T versus *1/d* as follows[5]:

$$R_T = \left(\frac{4\rho_c}{\pi}\right)\frac{1}{d^2} + \frac{\rho_{TE}}{\pi}\arctan\left(\frac{4t}{d}\right)\frac{1}{d} + R_0, \qquad (15.1)$$

where ρ_{TE} is the electrical resistivity of TE material, t is the thickness of metal film, R_0 is the back side electrical resistance between TE material and back electrode film (Cu film shown in Figure 15.1c). There is another method being used for contact resistivity measurement.[6] However, the ways mentioned earlier are more popular.[7–9]

Bi$_2$Te$_3$-based TE devices were commercially available for the cooling purposes or thermal management for a long time. Conventionally, a nickel layer is firstly deposited on the Bi$_2$Te$_3$-based leg and soldered with the Cu interconnect. However, some thermal stability issues and bonding strength problems have arisen when Bi$_2$Te$_3$-based TE devices were tested for TE power generation purpose. Conventionally, the contact design of these devices is an industrial technique secret, not openly released. In this chapter, the authors shine some light on the contact designs and challenges for the Bi$_2$Te$_3$-based TE leg in the both cooling purpose and power generation case.

15.2 Contact for Cooling Purpose

15.2.1 Early Works

In Bi$_2$Te$_3$-based devices, copper is usually preferred as the metal interconnect between the n-type and p-type TE legs (or elements) because of its low electrical resistivity (1.67×10^{-8} Ω m). Generally, there are at least four methods being used to join the Cu interconnect and Bi$_2$Te$_3$ leg, including brazing, welding, soldering, and diffusive bonding. In the early reports, the loss of TE cooling efficiency due to the contact resistance could be as high as 40%. Haba[10] reported that the loss could be reduced down to 10% after adapting a finely rough and fluxed Bi$_2$Te$_3$ surface before soldering with Cu electrode. In Haba's route, the Bi$_2$Te$_3$ surface was firstly vapor blasted by very fine suspended pumices. The saturated zinc chloride solution in methyl alcohol was then used as the flux for n-type Bi$_2$Te$_3$ leg, while the saturated lithium chloride solution in methyl alcohol was adapted for the p-type Bi$_2$Te$_3$ leg. The fluxed surface was important for the wetting in the later soldering process by using a tin–antimony–bismuth solder (Sn: 47.5%; Sb: 2.5%; Bi: 50%). It was found that the optimum soldering temperature was in the range of 266–274°C. Then, the Cu electrode was fluxed by zinc chloride or ammonium chloride and then soldered with a typical tin–lead solder (Sn: 60%; Pb: 40%). Finally, the Bi$_2$Te$_3$ leg and Cu electrode were joined at 230°C with fast cooling for strong bonding. The diffusion of Sn into an n-type Bi$_2$Te$_3$ could cause a high electrical resistance. Instead of using tin–antimony–bismuth solder, Rosi and Bernoff[11] used an antimony–silver solder (composition: Sb: 48%,

Ag: 52%; working temperature: 490–520°C) for Bi_2Te_3 legs and another antimony–silver solder (composition: Sb: 45%, Ag 55%; working temperature: 500°C) for the copper electrode, finally reaching a junction resistance as low as 2×10^{-4} Ω in a single p–n pair Bi_2Te_3 junction. However, no leg dimension was released in their patent, and we cannot estimate their contact resistivity. The diffusion of Ag into an n-type Bi_2Te_3 leg could also generate detrimental effect to the TE performance. Liao et al.[12] uncovered that some SnTe or $Pb_{1-x}Sn_xTe$ compounds were formed when the SnPb (composition: Sn: 63%, Pb: 37%) and SnAgCu (composition: Sn: 95.5%, Ag: 4%, Cu: 0.5%) solders were used to directly join the p-type and n-type bismuth telluride legs with copper interconnects. It was found that the contact resistivity, within a range of 10–100 μΩ cm^2, highly depended on both the thickness and composition of the interfacial compounds. In addition to the regular solder, Hill[13] patented a method using bismuth thin shims or wafer as the bonding material to join the Bi_2Te_3 legs and the metal interconnect and reach a quite thermally stable contact interface with a contact resistivity of 50 μΩ cm^2. These early solutions joining the TE legs and metal interconnect were very simple and easy to implement. In current opinion, these early solutions might be imperfect, but they are enough to move novel TE solid-state refrigerators from being a laboratory concept into being pioneer industry protocol products in 50 years.

15.2.2 Effect of Surface Roughness

Later, a nickel barrier layer or metalized layer was adapted in Bi_2Te_3-based refrigeration devices to reduce the negative impact from the solder.[4,14] Lan et al.[15] confirmed the barrier effect of nickel which successfully prevents the diffusion of the Sn from solder into both the p-type $(Bi, Sb)_2Te_3$ and n-type $Bi_2(Te, Se)_3$ TE elements by using the EDS spectrum of TEM. If without Ni layer, Sn from a regular solder process could diffuse into 4 μm of the Bi_2Te_3 elements in a soldering process with a typical thermal history of 200–230°C for 1–2 min. Some of the typical contact resistivity of Ni/Bi_2Te_3 interface could be found in Mengali and Seiler's[4] work, which was measured by using a scanning probe method. For the plating Ni, the contact resistivity was 1.7–2.6 μΩ cm^2 in p-type leg, while it was 7–10 μΩ cm^2 in the n-type leg. Recently, Gupta et al.[3] applied the Cox–Strack technique to measure the ρ_c of Ni/Bi_2Te_3, and it was found to be approximately 5 μΩ cm^2. The advantage of nickel-metalized layer is that it speeds up the industrial acceptance of TE solid-state refrigerators.

The nickel layer can be made through both the chemical route (i.e., electrodeless plating or electrochemical deposition) and physical route (i.e., sputtering or spray). However, the bonding strength (or adhesive strength) between the nickel layer and the Bi_2Te_3 surface was very weak, especially for the one made by chemical route. A finely rough Bi_2Te_3 surface could increase the surface area and hence improve the bonding strength at the Ni/Bi_2Te_3 interface. Weitzman[14] reported that an etched Bi_2Te_3 surface had bonding strength as high as 7.2–10.6 MPa for p-type leg, while the bonding strength was 9.7–12.2 MPa for n-type leg, which was higher than that of a sandblasted surface (1.9–3.1 MPa for p-type leg; 3.1–6.7 MPa for n-type leg). The improved bonding strength was related by the anchor effect of the rough surface. Weitzman's etching process was conducted in a solution (nitric acid/hydrofluoric acid/glacial acetic acid = 3:2:2) for 15 s for the p-type Bi$_2$Te$_3$ legs, while for the n-type Bi$_2$Te$_3$ legs, it was performed in another solution (nitric acid/hydrochloric acid = 1:1) for 135 s to remove ~100 μm materials. After the etching, the p-type legs were rinsed with water, while the n-type legs were required to be cleaned in a solution of glacial acetic acid. The final roughening surface was observable with the naked eye. However, there was no more detailed roughness information available in Weitzman's patent. The real connection between bonding strength and the surface roughness was unclear. Additionally, other solutions were used for the Bi$_2$Te$_3$-based material etching, as summarized in Table 15.1.[16–20] For example, in contrast to the strong HNO$_3$- or HCl-containing solutions, a gentler methanol solution of bromine was reported for a controllable etching for Bi$_2$Te$_3$.[17]

TABLE 15.1 Chemical Etching Solutions for Bismuth Telluride

	Solution	Etching Rate	Rinsing and Cleaning Solution	Ref.
1	3 HNO$_3$ 2 HF 2 glacial CH$_3$COOH	15 s removes 100μm p-type Bi$_2$Te$_3$	Water	Hill[13]
2	1 HNO$_3$ 1 HF	135 s removes 100 μm n-type Bi$_2$Te$_3$	Glacial acetic acid solution	Weitzman[14]
3	2 HNO$_3$ 1 HCl 6 H$_2$O	1–2 min creates hexagonal pits in Bi$_2$Te$_3$	DI water and dried on filter paper	Teramoto and Takayanagi[16]
4	10 mL HNO$_3$ 10 mL HCl 40 mL H$_2$O/1 g iodine	1–2 min creates triangular pits in Bi$_2$Te$_3$	Ethanol	Teramoto and Takayanagi[16]
5	0.6% bromine in methanol	30 s to 2 min creates triangular pits in Bi$_2$Te$_3$	N/A	Sagar and Faust Jr.[17]
6	3 HCl 1 HNO$_3$ 2 H$_2$O	0.2 μm/min in Bi$_2$Te$_3$	N/A	Morgan and Taylor[18]
7	1 HCl 1 H$_2$O$_2$ 2 H$_2$O	0.7 μm/min in Bi$_2$Te$_3$	N/A	Diliberto et al.[19]
8	3 HNO$_3$ 6 Citric acid 1 H$_2$O	12 s creates triangular pits in Bi$_2$Te$_3$	N/A	Patel et al.[20]

15.2.3 Effect of Surface Cleanness

Besides the roughness of the TE elements surface, the careful surface cleaning before sputtering or plating the metallization layer is another important factor to determine the interface performance, especially the contact resistivity. Plasma cleaning before the sputtering was helpful in removing the surface oxides and hence in reducing contact resistivity from 10^{-4} to 10^{-6} Ω cm^2 in the (Bi, Sb)$_2$(Te, Se)$_3$ system.[9] Lin et al.[21] reported a chemical cleaning process, including ammonia hydroxide, hydrogen peroxide, and deionized (DI) water at 70°C for more than 15 min, finally with 2% dilute HF immersion for possible oxide layer removal. Iyore[22] adapted a more elaborate cleaning recipe for p-type Bi$_2$Te$_3$ ingot wafer, with roughness of 1 μm (peak-to-peak) and 150 nm (root-mean-square [rms]), before sputtering Ni and electrodeless plating Ni. The cleaning process had three steps: (1) continuous stream of acetone cleaning, followed by isopropyl alcohol and DI water, and then blowing dry with nitrogen; (2) chemical cleaning with a solution of H$_2$O:H$_2$O$_2$:HCl = 2:1:1 for 30 s; and (3) chemical cleaning with a solution of 30% ammonium hydroxide in water for 1 min. The sputtering of Ni was conducted under a power density of approximately 10 W/cm^2 for 2 h while maintaining a substrate temperature of 200°C. Finally, 2 μm electrodeless plating Ni was done in Marlow Industries under an 85°C bath temperature. It was found that the cleaning route of steps 2 and 3 could reduce the contact resistance from 8 to 3 μΩ cm^2. Furthermore, in Iyore's[22] thesis, he also suggested that the contact resistivity would be further reduced to less than 1 μΩ cm^2 by fine–polishing the surface to have a roughness of 40 nm (peak-to-peak) and 3 nm (rms), with a cost of weaker bonding strength, however.[22]

Feng et al.[23] reported a polishing strategy that fabricated a surface of Bi$_2$Te$_3$-based nanostructured materials with a 1.9 rms roughness. The polishing route involved electrolysis, mechanical polishing, chemical mechanical polishing, and postcleaning. Firstly, the electrolytic etching was implemented under 1.5 V for 30 s in a solution of 10% NaOH and 5% tartaric acid. The roughness was reduced from 133 to 52 nm (rms). Secondly, mechanical polishing was conducted on a Buehler trident cloth pad with an upper soft cloth surface and lower hard polyurethane support. The slurry was 5 wt% 0.05 μm alumina suspensions. The polishing down force was 1.1 psi with a counterclockwise plate rotation speed of 110 revolutions per minute. The roughness was then reduced from 52 to 3.2 nm (rms). Thirdly, chemical–mechanical polishing (CMP) employed slurry with 1% hydrogen peroxide, 0.1% benzotriazole, and 5 wt% 0.05 μm colloidal silica at a pH value of 5 adjusted by nitric acid. The roughness was finally minimized down to 1.9 nm (rms). Fourthly, the post-CMP cleaning was carried out in an ethanol solution of 0.5% bromine firstly to remove the thin oxide on Bi$_2$Te$_3$-based alloys. After the bromine–ethanol treatment, a 5% additive (product name: Clean-100, Wako Pure Chemical Industry Ltd) was chosen to improve the wettability to dissolve the organic components and promote liftoff of the particles from the substrate. After 1 min of immersing in the bromine–ethanol solution at room temperature and then a 5 min immersion in Clean-100, the contact angle of water on the Bi_2Te_3 surface was significantly reduced from 79.7° to 53.6°, as shown in Figure 15.2a and b. The improved wettability was a direct reflection of the cleaning state of the surface, which consequently affected the uniformity of the lateral sputtering nickel film. Figure 15.2c and d shows that the surface with smaller water contact angle has relatively smaller roughness. It suggested that the metal atoms could bond easier on a cleaner substrate surface and hence resulted into a Frank–van der Merwe growth rather than a Volmer–Weber growth according to a wetting or nonwetting of liquid controlling principle for the metal film growth. Feng et al.[24] also suggested that the post-CPM cleaning was necessary to remove the hydrophobic thin oxide layer after CMP because the oxide layer functions poorly as a nucleation site for metallization. The wettability of the Bi_2Te_3 surface could be further improved by immersing in a 1% MPS (3-mercaptopropyl trimethoxysilane)

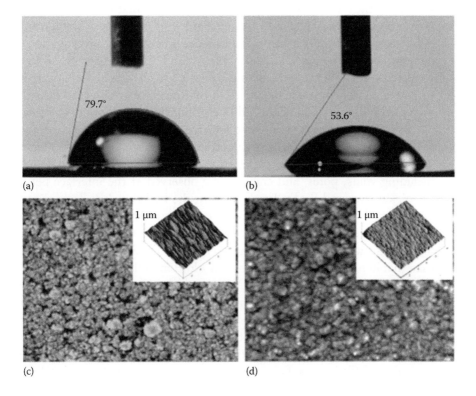

FIGURE 15.2 Effect of surface cleaning of Bi_2Te_3 substrate on the sputtering nickel. Water drop contact angle of nanostructured $Bi_2Te_{3-y}Se_y$ alloys (a) before and (b) after post-CMP cleaning. SEM images of 100 nm Ni sputtered on the surface of nanostructured $Bi_2Te_{3-y}Se_y$ alloys (c) without and (d) with post-CMP cleaning. The insets of (c) and (d) are their corresponding atomic force microscopy (AFM) images. The AFM scan size is 10 μm; the scan rate is 1 Hz. The z axis is 1 μm/div for height and x axis is 2 μm/div. The roughness is 41.59 nm for the inset of (c) and 13.51 nm for the inset of (d). (Reprinted from *Electrochimica Acta*, 56, Feng, H. P. et al., Studies on surface preparation and smoothness of nanostructured Bi_2Te_3–based alloys by electrochemical and mechanical methods, 3079–3084, Copyright [2011], with permission from Elsevier.)

for 50 min and hence finally improving the uniformity of the electroplating Ni. The final contact resistivity was as low as 0.3–1.1 $\mu\Omega$ cm^2 for p-type Bi_2Te_3 leg, while it was 0.68–1.2 $\mu\Omega$ cm^2 for n-type type Bi_2Te_3 leg.

Although both Iyore's and Feng's strategies are real reasonable for a super clean surface before the nickel sputtering or plating, both of them are complex and not yet widely adapted by the TE industry. Currently, the widely adapted route for the preparation of Ni contact layer on Bi_2Te_3 usually includes only three steps: (1) Bi_2Te_3 roughing by sand blowing or chemical etching, (2) chemical cleaning, and (3) electrochemically depositing or electrodeless plating Ni.

15.3 Contact for Power Generation

15.3.1 Early Efforts

In the commercial Bi_2Te_3-based devices, a 3–10 μm nickel layer is usually adapted as the metallization layer. Recently, the thermal stability of the interface between Ni layer and Bi_2Te_3 elements got attentions. Hatzikraniotis et al.[25] conducted a long-term performance and stability investigation of a commercial Bi_2Te_3-based module (Melcor HT9-3-25), with the cold side fixed at 24°C while the hot side was subjected to a thermal cycle from 30°C to 200°C. A continuously decreased output power was observed due to the increased resistance and decreased open circuit voltage. Park et al.[26] carried out a similar thermal cycling test for another commercial Bi_2Te_3 module (Ferrotec 9500/127/060). The peak output power displayed an 11% reduction after 6000 repeated thermal cycles from 30 to 160°C. Mechanical damages, including voids, pores, and cracks, were observed at the interface between the nickel layer and Bi_2Te_3 leg. Generally, the traditional Bi_2Te_3 devices are not yet ready for power generation due to the interface issue between the metalized layer and TE elements. It is known that at the same thermal history the diffusion of the Ni in the n-type $Bi_2(Te, Se)_3$ leg is much serious than that in p-type $(Bi, Sb)_2Te_3$ leg.[15] In addition to the diffusion, some NiTe intermetallic compounds, such as $NiTe_2$, was also observed at the interface of Ni/Bi_2Te_3.[27] Iyore et al.[22,28] investigated the thermal stability of the interface between sputtering nickel and n-type bismuth tellurium selenide ingot by systematical annealing at different temperatures. They found that the active temperature for the diffusion reaction was as low as 100°C. The temperature-dependent diffusion coefficients of nickel in the n-type bismuth tellurium selenide ingot were obtained, i.e., $D(100°C) = 5 \times 10^{-20}$ m^2/s, $D(150°C) = 7 \times 10^{-19}$ m^2/s, and $D(200°C) = 6 \times 10^{-18}$ m^2/s. The temperature-independent component was deduced as $D_0 = 2 \times 10^{-10}$ m^2/s, and the activated energy were calculated as 0.7 eV. The diffusion of the Ni in the Bi_2Te_3-based materials is less active than Cu,[29] Ag,[30] and Au,[31] e.g., D at room temperature of Cu along the cleave plane of Bi_2Te_3 single is ~10^{-6} m^2/s. When the operation temperature of the Bi_2Te_3-based devices is lower than 100°C, the diffusion of Ni at the interface is negligible, and hence, the interface Ni/Bi_2Te_3 is thermally stable. However, for power generation, the devices were required to operate at at least 200°C or even a higher temperature to fulfill the maximized output power or efficiency.[32] The interface stability, therefore, becomes a new technique problem.

15.3.2 Challenge

For the real power generation, the bonding strength between the nickel layer and the TE elements becomes even more critical than the contact resistance. As mentioned before, increasing the roughness of the Bi$_2$Te$_3$ elements surface before Ni plating does not only increase contact area but also add some anchor points. The bonding strength of the state-of-the-art Bi$_2$Te$_3$-based leg in the commercial devices is around 10 MPa, which is good enough for most of the refrigeration applications near room temperature but not enough for power generation. The bonding strength issue also embarrassed researchers working on flat-panel solar TE power generators.[2,24,32] Slight efficiency deterioration was observed when the power generator worked under a steady model with T_h = 220°C and T_c = 20°C because nickel diffusion at the contact interface caused an increased contact resistance.[24] By adding a thin Ti layer before the sputtering nickel, or coating with MPS (3-mercaptopropyltrimethoxysilane) before plating nickel, nickel diffusion could be reduced with a slightly reduced bonding strength, but not prevented. However, when it worked under the thermal cycle case, the weak bonding strength turned into a short slab. A large decrease in the efficiency was observed after the device was subjected to a thermal cycle between 220°C and 20°C. It was found that some microcracks generated at the Ni/Bi$_2$Te$_3$ interface due to the high thermal stress over the bonding strength. In order to get higher bonding strength at the Ni/Bi$_2$Te$_3$ interface, a diffusion bonding was used to join the Ni and Bi$_2$Te$_3$ by directly hot pressing the Ni powder layer with Bi$_2$Te$_3$-based bulk. A significantly improved bonding strength (>20 MPa) was obtained in both n-type Ni/Bi$_2$Te$_{2.7}$Se$_{0.3}$ and p-type Ni/Bi$_{0.4}$Sb$_{1.6}$Te$_{0.4}$. However, a larger contact resistivity was observed in Ni/Bi$_2$Te$_{2.7}$Se$_{0.3}$ (~230 µΩ cm^2) rather than in Ni/Bi$_{0.4}$Sb$_{1.6}$Te$_{0.4}$ (~1 µΩ cm^2), which therefore turned into a new challenge.

15.3.3 Understanding

The puzzle was related to the interfacial diffusion and reaction. Figure 15.3 shows the microstructure of the contact region (cross section) between the Ni metallization layer and TE elements in both the p-type leg Ni/Bi$_{0.4}$Sb$_{1.6}$Te$_3$/Ni and n-type leg Ni/Bi$_2$Te$_{2.7}$Se$_{0.3}$/Ni with a diffusion bonding temperature range from 400 to 500°C.[2] A clear interface reaction layer (IRL) was identified in both the p-type and n-type legs, which appeared as a gray layer between the Ni (dark region) and TE (light region). With the increased HP temperature from 400°C to 500°C, the thickness of the IRL also increased from ~3 to 13–16 µm for p-type leg and from ~4 to 17–21 µm for n-type leg, confirming the activated energy-controlled Ni diffusion mechanism. Furthermore, Figure 15.4 shows the atomic composition profile crossing from the nickel side to TE material side by selected-area (~2 × 20 µm^2) SEM-EDS in both p-type leg Ni/Bi$_{0.4}$Sb$_{1.6}$Te$_3$/Ni and n-type leg

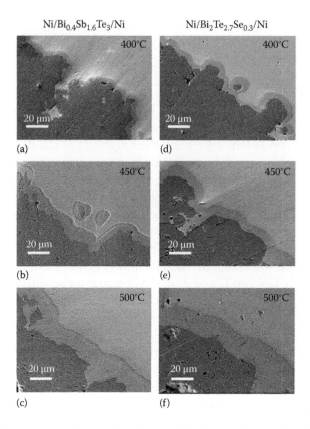

FIGURE 15.3 Microstructure of contact interface made by directly hot pressing the Ni powders and TE powders together. (a–c) $Ni/Bi_{0.4}Sb_{1.6}Te_3/Ni$, hot pressed at 400°C, 450°C, and 500°C, respectively. (d–f) $Ni/Bi_2Te_{2.7}Se_{0.3}/Ni$, hot pressed at 400°C, 450°C, and 500°C, respectively. The dark region is nickel element, and the light region is TE element. (Liu, W. S. et al., *Journal of Materials Chemistry A*, 1, 13093–13100, 2013. Reproduced by permission of The Royal Society of Chemistry.)

$Ni/Bi_2Te_{2.7}Se_{0.3}/Ni$.[2] According to the concentration profile, it seemed that a thicker Ni_3Te_2 and a thinner $NiTe$ were formed at the $Ni/Bi_{0.4}Sb_{1.6}Te_3$ interface, while a thinner Ni_3Te_2 and a thicker $NiTe$ were formed at the $Ni/Bi_2Te_{2.7}Se_{0.3}$ interface. A more careful analysis suggested that the phase of IRL in p-type $Ni/Bi_{0.4}Sb_{1.6}Te_3/Ni$ leg was closer to the Ni_2SbTe (PDF #01-089-7153), which has a similar hexagonal crystalline structure ($P6_3$/mmc, No. 194) with $NiTe$, but with Sb and Te sharing the same atomic site in the corresponding crystal structure. According to the chemical composition of the IRL of $Ni/Bi_{0.4}Sb_{1.6}Te_3$ interface as shown in Figure 15.4a, the impurity should be $NiTe_{1-x}Sb_x$. In other words, both Te and Sb from $Bi_{0.4}Sb_{1.6}Te_3$ reacted with Ni at the $Ni/Bi_{0.4}Sb_{1.6}Te_3$ interface, explaining no obvious Te deficiency in $Bi_{0.4}Sb_{1.6}Te_3$ shown in Figure 15.4a, but significant Te deficiency in $Bi_2Te_{2.7}Se_{0.3}$ shown in Figure 15.4b.

Additionally, it was noted that some "abnormally" high concentration of Ni (2–7%) even 10–20 μm away from the IRL was found in the SEM-EDS compositional scanning. A more careful TEM study was therefore carried out to investigate the diffusion of the Ni. Figure 15.5 shows the typical TEM

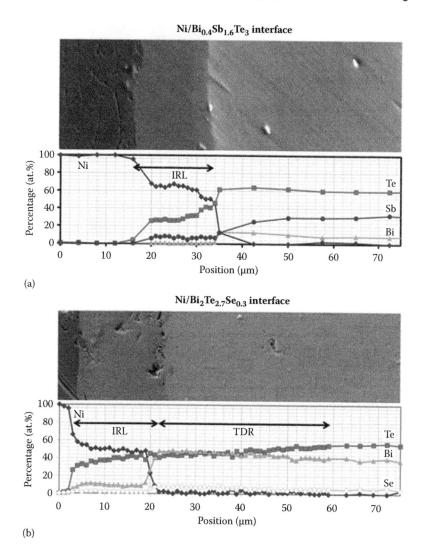

FIGURE 15.4 Comparison of composition profile between (a) Ni/Bi$_{0.4}$Sb$_{1.6}$Te$_3$ interface and (b) Ni/Bi$_2$Te$_{2.7}$Se$_{0.3}$ interface obtained from a selected-area SEM-EDS. IRL, interface reaction layer; TDR, Te-deficient region. (Liu, W. S. et al., *Journal of Materials Chemistry A*, 1, 13093–13100, 2013. Reproduced by permission of The Royal Society of Chemistry.)

images and EDS compositions of selected points at the Ni/Bi$_2$Te$_{2.7}$Se$_{0.3}$ interface.[2] One special feature was the sharp tips near the Ni/Bi$_2$Te$_{2.7}$Se$_{0.3}$ interface in the TE material side. The TEM-EDS suggested that most of these sharp tips are composed of abnormally high concentrations of Ni (>50 at.%), which is summarized in Figure 15.5a. A near pure Ni needle was found, as shown in Figure 15.5d, which could be a direct evidence of the fast diffusion of Ni along the Bi$_2$Te$_{2.7}$Se$_{0.3}$ grain boundary. A schematic model for the diffusion of Ni into Bi$_2$Te$_{2.7}$Se$_{0.3}$ is shown in Figure 15.5e. Similar fast diffusion along grain boundary was also reported at the Cu/Bi$_2$Te$_3$ interface.[33] The effect of special Ni needle is positive for the high bonding strength.

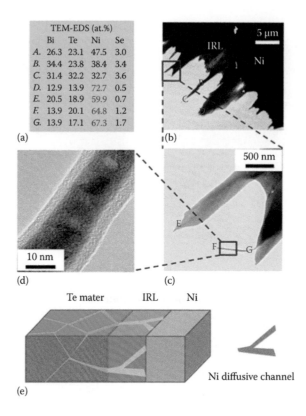

TEM-EDS (at.%)			
Bi	Te	Ni	Se
A. 26.3	23.1	47.5	3.0
B. 34.4	23.8	38.4	3.4
C. 31.4	32.2	32.7	3.6
D. 12.9	13.9	72.7	0.5
E. 20.5	18.9	59.9	0.7
F. 13.9	20.1	64.8	1.2
G. 13.9	17.1	67.3	1.7

(a) (b) (c) (d) (e)

Te mater IRL Ni

Ni diffusive channel

FIGURE 15.5 (a) Composition of selected interesting points (*A–G*) by TEM-EDS; (b–d) typical TEM image of the $Ni/Bi_2Te_{2.7}Se_{0.3}$ interface; (e) a schematic model for the fast diffusion of Ni along the grain boundary. (Liu, W. S. et al., *Journal of Materials Chemistry A*, 1, 13093–13100, 2013. Reproduced by permission of The Royal Society of Chemistry.)

The next question would be about the real cause of the high contact resistance at the $Ni/Bi_2Te_{2.7}Se_{0.3}$ interface. The first guess was related to the new products due the interface reaction. The possible $NiSe$, $NiTe$, Ni_2Te_3, Ni_3Te_2, and $(Bi_{1-x}Ni_x)_2(Te_{2.7}Se_{0.3})_{3-\delta}$ were also synthesized by BM and HP. It was found that all the $NiSe$, $NiTe$, Ni_2Te_3, and Ni_3Te_2 compounds have a semi-metal behavior with electrical resistivity of 0.5–0.7 $\mu\Omega$ m and Seebeck coefficient of –5 to –15 $\mu V/K$, which were clearly not a reason for the high contact resistance. In contrast, some $(Bi_{1-x}Ni_x)_2(Te_{2.7}Se_{0.3})_{3-\delta}$ materials showed a p-type behavior with a very high electric resistivity (> 200 $\mu\Omega$ m), when $x > 0.02$ and $-0.02 < \delta < 0.04$. It was clear that the high contact resistance in n-type $Ni/Bi_2Te_{2.7}Se_{0.3}/Ni$ leg was caused by this highly resistive p-type composition region due to Ni getting into the X site under the chalcogen-deficiency case,[2] as shown in Figure 15.6. Regarding the diffusion of Ni into TE materials, the V_x due to chalcogen transferring toward the reaction interface could be another channel besides the grain boundary channel, finally forming some p-type defect Ni_X. Although the formation of Ni_X neutralized the negatively charged carrier and raised the electrical resistivity in n-type legs, it increased the positively charged carrier and reduced electrical

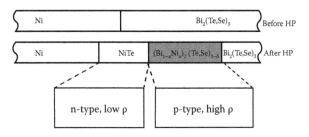

FIGURE 15.6 The schematic figure of the interface reaction between the Ni and Bi$_2$(Te,Se)$_3$, in which the region of (Bi$_{1-x}$Ni$_x$)$_2$(Te,Se)$_{3-\delta}$ with Ni dopant and chalcogen deficiency is responsible for the high contact resistivity.

resistivity in p-type legs. This was why we see the high contact resistance only in n-type legs, not in p-type legs. It is worth pointing out that the doping behavior of Ni in Bi$_2$Te$_{2.7}$Se$_{0.3}$ when Cu is present can be different. Lukas et al.[34] reported that 2% Ni could slightly increase the carrier concentration of Cu$_{0.01}$Bi$_2$Te$_{2.7}$Se$_{0.3}$ from 4.29 × 10^{19} cm^{-3} to 5.27 × 10^{19} cm^{-3}. When HP was similarly used to bond the Ni power layers on the Cu$_{0.01}$Bi$_2$Te$_{2.7}$Se$_{0.3}$ elements, the contact resistance was unbelievably as low as 1μΩ cm^2, but it was true. However, when this Ni/Cu$_{0.01}$Bi$_2$Te$_{2.7}$Se$_{0.3}$/Ni leg was subjected to a temperature gradient, the electrical resistivity of the Cu$_{0.01}$Bi$_2$Te$_{2.7}$Se$_{0.3}$ elements changed along the heat flux direction; the hot side had much higher electric resistivity than the cold side due to the movement of the mobile Cu ions. We have tried many strategies to stabilize the Cu, but failed. The authors would like to repeat Goldsmid's[35] early warning 30 years ago: "One should, of course, avoid plating with copper, silver, or gold, because of the ease with which these metals diffuse through bismuth telluride."

15.3.4 Solution

It was suggested that the key to avoid the high contact resistance was to prevent the formation of the p-type region, i.e., (Bi$_{1-x}$Ni$_x$)$_2$(Te$_{2.7}$Se$_{0.3}$)$_{3-\delta}$, within the n-type Bi$_2$Te$_{2.7}$Se$_{0.3}$ leg. Two Ni–Te compounds were used to form a sandwich structure, i.e., Ni/Ni$_3$Te$_2$/Ni$_2$Te$_3$/Bi$_2$Te$_{2.7}$Se$_{0.3}$/Ni$_2$Te$_3$/Ni$_3$Te$_2$/Ni, and hot pressed at 500°C, in which Ni, Ni$_3$Te$_2$, and Ni$_2$Te$_3$ were powders and Bi$_2$Te$_{2.7}$Se$_{0.3}$ was a hot pressed puck. It was found that contact resistance was reduced to 4 μΩ cm^2, which was 50 times lower than the one without Ni$_2$Te$_3$ and Ni$_3$Te$_2$ barrier layers. However, the bonding strength of the contact was also reduced down to ~11 MPa. In order to balance the contact resistance and the bonding strength, it is necessary to allow some degree interface reaction without causing the formation of a p-type (Bi$_{1-x}$Ni$_x$)$_2$(Te$_{2.7}$Se$_{0.3}$)$_{3-\delta}$. According to this guideline, 90% Bi$_2$Te$_{2.7}$Se$_{0.3}$–10% NiSe$_2$ powders (BL-1) and Bi$_2$Te$_{2.7}$Se$_{0.3}$–1% SbI$_3$ powders (BL-2) were investigated as the barrier layer between Ni and Bi$_2$Te$_{2.7}$Se$_{0.3}$ in the n-type leg.[2] Figure 15.7 shows voltage scanning measurement for the contact resistance measurement of the n-type Bi$_2$Te$_{2.7}$Se$_{0.3}$ legs with BL-1 and BL-2 as the barrier layer, respectively, by hot pressing at 425°C. The n-type Bi$_2$Te$_{2.7}$Se$_{0.3}$ leg with BL-1 barrier

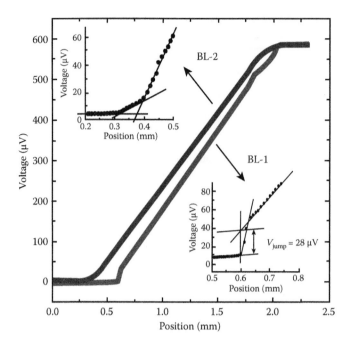

FIGURE 15.7 Contact resistance measurement by the scanning probe for two n-type legs with barrier layers: Ni/BL-1/Bi$_2$Te$_{2.7}$Se$_{0.3}$/BL-1/Ni and Ni/BL-2/Bi$_2$Te$_{2.7}$Se$_{0.3}$/BL-2/Ni, which is made by hot pressing at 425°C. BL-1: 90% Bi$_2$Te$_{2.7}$Se$_{0.3}$ + 10% NiSe$_2$; BL-2: 1% SbI$_3$-doped Bi$_2$Te$_{2.7}$Se$_{0.3}$. (Liu, W. S. et al., *Journal of Materials Chemistry A*, 1, 13093–13100, 2013. Reproduced by permission of The Royal Society of Chemistry.)

layer showed a balanced bonding strength (16 MPa) and contact resistance (9 μΩ cm²). On the other hand, the one with BL-2 barrier layer showed even negligible contact resistivity (<1 μΩ cm²) and comparable bonding strength (16 MPa). The efficiency measurement (T_h = 220°C and T_c = 20°C) of a device with the n-type leg with BL-2 barrier layer suggested a quite stable contact interface for over 150 h without notable efficiency degradation, which was in contrast to the obvious efficiency degradation in the one using the n-type leg with sputtering Ni as the metallization layer without barrier layer. In addition to the use of a barrier layer with a high carrier concentration, the use of an ion implantation-modified surface could be another promising way to raise the surface carrier concentration. Talor et al.[9] reported that the surface carrier concentration could increase 10–50 times than the regular TE performance-optimized state for the (Bi, Sb)$_2$(Te, Se)$_3$ (× 10¹⁹ cm⁻³) by using iodine ion for the n-type element and silver ion for the p-type one. Implantation made the contact resistivity decrease from 1.7 × 10⁻⁶ to 4.5 × 10⁻⁷ Ω cm² for n-type Bi$_2$Te$_3$ and from 7.7 × 10⁻⁷ to 2.7 × 10⁻⁷ Ω cm² for p-type Bi$_2$Te$_3$. However, the cost of the ion implantation may be too high for waste heat harvest purpose.

Besides the usage of barrier layer for Ni, seeking other new metals or alloys for the metalized layer was another alternative way for the contact design of Bi$_2$Te$_3$-based legs. Xiong et al.[36] conducted a first principle calculation to compare the interface stability of the Ni/Bi_2Te_3 and Co/Bi_2Te_3. They suggested that the Co/Bi_2Te_3 interface should be more thermally stable. Later, Gupta et al.[37] experimentally confirmed that the Co/Bi_2Te_3 interface could be thermally stable up to 200°C while Ni/Bi_2Te_3 was thermally stable at lower than 100°C.[3] Patel et al.[20] suggested that if only thermal stability is considered, the interface of Ti/Bi_2Te_3 could be thermally stable up to 250°C. However, if we desired a strong bonding, we have to allow some solid diffusion or reaction. When the temperature is high enough, the reaction at the Co/Bi_2Te_3 interface could cause similar contact problem. We tried directly hot pressing Co and $Fe_{0.85}Cr_{0.15}$ powders on the n-type $Bi_2Te_{2.7}Se_{0.3}$ puck at 500°C. It was found that the $Co/Bi_2Te_{2.7}Se_{0.3}$ interface had high contact resistivity over 200 $\mu\Omega$ cm², while the $Fe_{0.85}Cr_{0.15}/Bi_2Te_{2.7}Se_{0.3}$ interface had a low contact resistivity of 22 $\mu\Omega$ cm². Li et al.[7] reported that pure Sb was used as the metallization layer for the p-type $Bi_{0.5}Sb_{1.5}Te_3$ by diffusion bonding through SPS at 420°C, which showed a contact resistivity of 3 $\mu\Omega$ cm². The thermal stability of the $Sb/Bi_{0.5}Sb_{1.5}Te_3$ was investigated by annealing at 300°C for 30 days. After annealing, a 50 μm diffusion layer was formed and resulted in the contact resistivity rising to 8 $\mu\Omega$ cm².

Recently, we released a new metallization alloy NiFeInS for the n-type $Bi_2Te_{2.7}Se_{0.3}$ elements, which could have much better balance for the low contact resistivity, high bonding strength, and good thermal stability.[38] Figure 15.8 shows the voltage scanning measurement for the contact resistivity of $NiFeInS/Bi_2Te_{2.7}Se_{0.3}$ joined at 425°C. It showed a contact resistivity of 2–4 $\mu\Omega$ cm². A thermal cycle was used to test the thermal stability and the bonding strength, with each cycle involving heating from 25°C to 250°C at 300°C/hour, holding the temperature at 250°C for 1 hour, cooling down from 250°C to 25°C at 500°C/hour, and holding the temperature

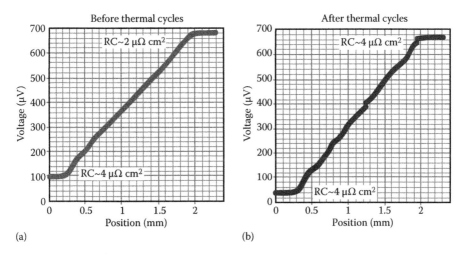

FIGURE 15.8 Comparison of probe scanning measurement for the n-type $Bi_2Te_{2.7}Se_{0.3}$ leg with NiFeInS metal contact layer (a) before and (b) after 12 thermal cycles from 25°C to 250°C. (From Ren, Z. F., and Liu, W. S., US Patent No. 62/150,686, 2015.)

at 25°C for 1 hour. After 12 thermal cycles, there were no notable changes in the contact resistivity for the interface of NiFeInS/$Bi_2Te_{2.7}Se_{0.3}$. In contrast, the interface with pure Ni as the metallization layer showed an increase in the contact resistance at both sides, i.e., from 2 to 6 $\mu\Omega$ cm² for one side and from 14 to 64 $\mu\Omega$ cm² for another side after being subjected to the same thermal cycles. The new $Bi_2Te_{2.7}Se_{0.3}$ leg with NiFeInS alloys as metallization layer was successfully applied in the second-generation solar TE generator.[39]

Additionally, thermal spray was another way to join the metallization layer on the TE elements surface for strong bonding. Weitzman[14] mentioned earlier that nickel as well other metals could be joined onto the Bi_2Te_3-based TE elements by high-speed spray of powers.[14] Zhang et al.[40] reported a high bonding strength of 18.9 MPa for sprayed Ni (65.4 μm thick) on p-type Bi_2Te_3 ingot and 24 MPa for a sprayed Ni (24.4 μm thick) on N-type Bi_2Te_3 ingot. In their process, sand blowing with zirconium corundum, under the conditions of a sand blowing pressure of 0.1–0.2 MPa and a blowing distance of 150–200 mm and angle of 45–80°, was firstly used to remove the surface contaminations. The plasma spray was conducted under the conditions of 500–700 Å current, 5–7 L/min hydrogen flow, 40–60 L/min argon flow, gun speed of 800–2000 mm/s, nickel powder (–45 + 15 mesh) feeding rate of 80–120 g/min, and a spraying distance of 150–180 mm. However, no contact resistivity was released in their patent. Li et al.[41,42] used similar route to spray Mo metallization layer on Bi_2Te_3 materials. The Mo/Bi_2Te_3 interface shows good thermal stability under 300°C annealing but high contact resistance of 80 $\mu\Omega$ cm². However, some microcracks were observed after it was subjected to a thermal shock test from 300 to 0°C, suggesting a high thermal stress raised from the CTE mismatching. The thermal spray method was also used to directly assemble the Bi_2Te_3 device, including making the Ni/Al metallization layer (50 μm) and Zn interconnect (0.6 mm).[43] This Bi_2Te_3 device showed promise to be more thermally stable at temperatures higher than 200°C.

15.4 Outlook

In this chapter, we have reviewed the main technique issue in Bi_2Te_3-based TE devices, i.e., the contact design, which involves the choice of metal interconnect and joint method. For the widely adapted soldering process, the interface between metallization layer and TE elements is a bottleneck. For the cooling application, low contact resistance is the focus for a good interface, which is mainly determined by the oxidized layer, surface cleanness, and surface roughness. However, for power generation, the contact interface requires not only a low contact resistivity, but also a high bonding strength and good thermal stability. It is a real challenge. Ni and Co are good for the p-type BiSbTe elements, but not good for the n-type. Recently, a new alloy of NiFeInS shows promise to solve the problem met in the n-type BiTeSe elements. Aside from the conventional diffusion bonding by HP, a thermal spraying method was recently accepted by the industry as a new way to form the metallization layer and the metal interconnects.

References

1. W. S. Liu, Q. Jie, H. S. Kim, and Z. F. Ren, Current progress and future challenge in thermoelectric power generation from materials to devices, *Acta Mater.* 87, 357–376, 2015.
2. W. S. Liu, H. Z. Wang, L. J. Wang, X. W. Wang, G. Joshi, G. Chen, and Z. F. Ren, Understanding of the contact of nanostructured thermoelectric n-type Bi$_2$Te$_{2.7}$Se$_{0.3}$ legs for power generation applications, *J. Mater. Chem. A* 1, 13093–13100, 2013.
3. R. P. Gupta, J. B. White, O. D. Iyore, U. Chakrabarti, H. N. Alshareef, and B. E. Gnade, Determination of contact resistivity by Cox and Strack method for metal contact to bulk bismuth antimony telluride. *Electrochem. Solid-State Lett.* 12, H302–H304, 2009.
4. O. J. Mengali and M. R. Seiler, Contact resistance studies on thermoelectric materials, *Adv. Energy Conversion* 2, 59–68, 1962.
5. R. H. Cox and H. Strack, Ohmic contact for GaAs devices, *Solid-State Electron.* 10, 1213–1214, 1967.
6. R. P. Gupta, R. Mccarty, and J. Sharp, Practical contact resistance measurement method for bulk Bi$_2$Te$_3$ thermoelectric devices, *J. Electron. Mater.* 43, 1068–1612, 2014.
7. F. Li, X. Y. Huang, W. Jiang, and L. D. Chen, Interface microstructure and performance of Sb contacts in bismuth telluride-based thermoelectric elements, *J. Electron. Mater.* 42, 1219–1224, 2013.
8. H. S. Kim, K. Kikuchi, T. Itoh, T. Iida, and M. Taya, Design of segmented thermoelectric generator based on cost-effective and light-weight thermoelectric alloys, *Mater. Sci. Eng. B* 185, 45–52, 2014.
9. P. J. Talor, J. R. Maddux, G. Meissner, R. Venkatasubramanian, G. Bulman, J. Pierce, R. Gupta et al., Controlled improvement in specific contact resistivity for thermoelectric materials by ion implantation, *Appl. Phys. Lett.* 103, 043902, 2013.
10. V. Haba, Method and materials for obtaining low contact resistance bond to bismuth telluride, US Patent No. 3017693, 1962; US Patent No. 3079455, 1963.
11. F. D. Rosi and R. A. Bernoff, Method and materials for obtaining low resistance bonds to thermoelectric bodies, US Patent No. 3037064, 1962.
12. C. N. Liao, C. H. Lee, and W. J. Chen, Effect of interfacial compound formation on contact resistivity of soldered junction between bismuth telluride-based thermoelements and copper, *Electrochem. Solid-State Lett.* 10, P23–P25, 2007.
13. L. R. Hill, Method of the boding bismuth-containing bodies, US Patent No. 3110100, 1963.
14. L. H. Weitzman, Etching bismuth telluride, US Patent No. 3338765, 1967.
15. Y. C. Lan, D. Z. Wang, G. Chen, and Z. F. Ren, Diffusion of nickel and tin in p-type (Bi,Sb)$_2$Te$_3$ and n-type Bi$_2$(Te,Se)$_3$ thermoelectric materials, *Appl. Phys. Lett.* 92, 101910, 2008.

16. I. Teramoto and S. Takayanagi, Dislocation etch pits on bismuth telluride crystals, *J. Appl. Phys.* 32, 119–120, 1961.
17. A. Sagar and J. W. Faust Jr., Dislocation studies in Bi_2Te_3 by etch-pit technique, *J. Appl. Phys.* 38, 482–490, 1967.
18. B. Morgan and P. Taylor. *Patterning of Bi_2Te_3 Polycrystalline Thin-Films on Silicon*. Army Research Laboratory, Adelphi, MD, No. ARL-TR-4351, 2008.
19. S. Diliberto, S. Michel, C. Boulanger, J. M. Lecuire, M. Jagle, S. Drost, and H. Bottner, A technology for devices prototyping based on electrodeposited thermoelectric V-VI layers, in *22nd International Conference on Thermoelectrics*, pp. 661–664, 2003.
20. P. Patel, S. M. Vyas, V. Patel, S. Thakor, M. P. Jani, and G. R. Pandya, Crystal growth and etching of Bi_2Te_3 alloys, *Int. J. Adv. Electron. Comp. Eng.* 2, 217–220, 2012.
21. W. P. Lin, D. E. Wesolowski, and C. C. Lee, Barrier/bonding layers on bismuth telluride for high temperature thermoelectric modules, *J. Mater. Sci.: Mater. Electron.* 22, 1313–1320, 2011.
22. O. D. Iyore, *Interface Characterization of Contacts to Bulk Bismuth Telluride Alloys*, Master's Thesis, University of Texas at Dallas, Richardson, TX, UMI No. 1470835, 2009.
23. H. P. Feng, B. Yu, S. Chen, K. Collins, C. He, Z. F. Ren, and G. Chen, Studies on surface preparation and smoothness of nanostructured Bi_2Te_3-based alloys by electrochemical and mechanical methods, *Electrochim. Acta* 56, 3079–3084, 2011.
24. S. P. Feng, Y. H. Chang, J. Yang, B. Poudel, B. Yu, Z. F. Ren, and G. Chen, Reliable contact fabrication on nanostructured Bi_2Te_3-based thermoelectric materials, *Phys. Chem. Chem. Phys.* 15, 6757–6762, 2013.
25. E. Hatzikraniotis, K. T. Zorbas, I. Samaras, Th. Kyratsi, and K. M. Paraskevopoulos, Efficiency study of a commercial thermoelectric power generator (TEG) under thermal cycling, *J. Electron. Mater.* 39, 2112–2116, 2010.
26. W. Park, M. T. Barako, A. M. Marconnet, M. Asheghi, and K. E. Goodson, Effect of thermal cycling on commercial thermoelectric modules, in *13th Intersociety Conference on Thermal and Thermomechanical Phenomena in Electronic Systems*, pp. 107–112, 2012.
27. T. Kacsich, E. Kolawa, J. P. Fleurial, T. Caillat, and M. A. Nicolet, Film of Ni-7 at% V, Pd, Pt and Ta-Si-N as diffusion barrier for copper on Bi_2Te_3 *J. Phys. D: Appl. Phys.* 31, 2406–2411, 1998.
28. O. D. Iyore, T. H. Lee, R. P. Gupta, J. B. White, H. N. Alshareef, M. J. Kim, and B. E. Gnade, Interface characterization of nickel contact to bulk bismuth telluride selenide, *Surf. Interface Analysis* 41, 440–444, 2009.
29. R. O. Carlson, Anisotropic diffusion of the copper into bismuth telluride, *J. Phys. Chem. Solids* 13, 65–70, 1960.
30. H. P. Dibbs and J. R. Tremblay, Thermal diffusion of silver in single crystal bismuth telluride, *J. Appl. Phys.* 39, 2976–2977, 1968.
31. J. D. Keys and H. M. Dutton, Solid solubility of gold in single-crystal bismuth telluride, *J. Appl. Phys.* 34, 1830–1831, 1963.
32. D. Kraemer, B. Poudel, H. P. Feng, J. C. Caylor, B. Yu, X. Yan, Y. Ma et al., High-performance flat-panel solar thermoelectric generators with high thermal concentration, *Nat. Mater.* 10, 532–538, 2011.
33. S. N. Zhevnenko, D. V. Vaganov, and E. I. Gershman, Rapid penetration of bismuth from solid Bi_2Te_3 along grain boundary in Cu and Cu based alloys, *J. Mater. Sci.* 46, 4248–4253, 2011.

34. K. C. Lukas, W. S. Liu, Z. F. Ren, and C. P. Opeil, Transport properties of Ni, Co, Fe, Mn doped Cu$_{0.01}$Bi$_2$Te$_{2.7}$Se$_{0.3}$ for thermoelectric device applications, *J. Appl. Phys.* 112, 054509/1–054509/5, 2012.

35. H. J. Goldsmid, *Electronic Refrigeneration*, Pion Limited, London, 1986.

36. K. Xiong, W. C. Wang, H. N. Alshareef, R. P. Gupta, J. B. White, B. E. Gnade, and K. Cho, Electronic structures and stability of Ni/Bi$_2$Te$_3$ and Co/Bi$_2$Te$_3$ interface, *J. Phys. D: Appl. Phys.* 43, 115303–115308, 2010.

37. R. P. Gupta, X. Xiong, J. B. White, K. Cho, H. N. Alshareef, and B. E. Gnade, Low resistance ohmic contacts to Bi$_2$Te$_3$ using Ni and Co metalization, *J. Electrochem. Soc.* 157, H666–H670, 2010.

38. Z. F. Ren and W. S. Liu, Contacts for Bi$_2$Te$_3$-based materials and methods of manufacture, US Patent No. 62/150,686, 2015.

39. D. Kraemer, Q. Jie, K. McEnaney, F. Cao, W. S. Liu, L. A. Weinstein, J. Loomis, Z. F. Ren, and G. Chen, Concentrating solar thermoelectric generators with a peak efficiency of 7.4%, *Nat. Energy* 1, 16153, 2016

40. J. F. Zhang, K. Yang, C. M. Deng, C. G. Deng, M. Liu, M. J. Dai, and K. S. Zhou, Method for spray thick nickel coating on surface of semiconductor, Chinese Patent: CN 104357784 A, 2014.

41. X. Y. Li, L. D. Chen, X. G. Xia, Y. S. Tang, S. Y. Tao, X. G. Xia, Y. Q. Wu, and R. P. Lu, Bismuth-telluride-based thermoelectric electrification device and manufacturing method thereof, Chinese Patent: CN 101409324 A, 2009.

42. X. Y. Li, L. D. Chen, X. G. Xia, Y. S. Tang, S. Y. Tao, X. G. Xia, Y. Q. Wu, and R. P. Lu, Bismuth telluride based thermoelectric generation device, Chinese Patent: CN. 201408783Y, 2010.

43. J. H. Zheng, Heat-resistant stable bismuth telluride-based thermoelectric semiconductor generator and preparation method thereof, Chinese Patent: CN 101847685 B, 2011.

CHAPTER 16

Contacts for Skutterudites

Qing Jie and Zhifeng Ren

Contents

16.1 Introduction

Filled SKUs show promising physical and mechanical properties for power generation by using the intermediate temperature heat sources such as automobiles. Therefore, quite a few groups around the world are studying the TE power generation modules by using filled SKU material, including the Jet Propulsion Laboratory in the United States, GM Global Research & Development in the United States, National Institute of Advanced Industrial Science and Technology in Japan, Shanghai Institute of Ceramics in China, and Fraunhofer Institute for Physical Measurement Techniques in Germany. García-Cañadas et al. [1] tried to use $Zn_{78}Al_{22}$ solder bar to directly bond filled SKU legs to alumina plates with attached Cu stripes to form the module. The average specific contact resistivity is as high as 477 $\mu\Omega\cdot cm^2$. After just seven thermal cycles, the internal resistance of the module increased from 0.7 Ω to higher than 8 Ω [1]. Their results show that directly brazing the filled SKU legs to the conducting strips is not a good choice because the elements in the brazing alloy diffuse into and react with SKUs. As shown in Figure 16.1, a good electrical contact layer serving as both electrode and diffusion barrier is necessary. Due to the low electrical resistivity of filled SKU material, good contact material with low specific contact resistivity is critical for obtaining high-performance TE modules.

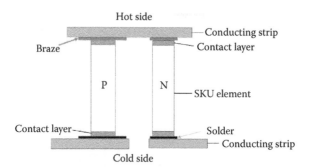

FIGURE 16.1 Drawing of p-/n-type element structure of the SKU module.

It is extremely tricky to choose the right contact layer material, because ideally, it should meet all the eight criteria listed in the following:

1. Matched CTE with the TE elements
2. High electrical conductivity
3. High thermal conductivity
4. Possibility to be made very thin to reduce total electrical and thermal resistances
5. Low contact resistivity at the interface between the contact layer and the TE layer
6. Stability at the working high temperature
7. Ability to form strong mechanical bonds with the TE layer
8. Higher yield strength than the solder at its operating temperature

Usually, the CTE matching should be the first consideration when we are choosing contact materials, because the thermal stress caused by CTE mismatching could cause cracks in TE materials and make it very difficult to obtain good legs for module making. For filled SKU modules, the n-type materials are usually $M_xCo_4Sb_{12}$ (where M could be a combination of multiple filler elements), with a CTE of about $9–12 \times 10^{-6}$ K^{-1} from room temperature to 600°C. The p-type materials are usually $N_yFe_{4-z}Co_zSb_{12}$ (N could be a combination of multiple filler elements, and z < 2), with a CTE of about $12–15 \times 10^{-6}$ K^{-1} from room temperature to 600°C. The different CTEs require different electrode materials for n- and p-type filled SKU legs.

16.2 Multiple-Layer Contact

Multiple-layer contact consists of a top electrode layer and one or more intermediate layers between the electrode and TE materials. The intermediate layer has been widely used to make TE legs. It is very useful in three cases: (1) When there is a big gap between the CTEs of the electrode and TE materials, intermediate layers with an intermediate CTE can effectively reduce the thermal stress caused by the CTE mismatch. (2) When the electrode quickly reacts with TE material, an intermediate layer can be inserted as a diffusion barrier to slow down or prevent the reaction. (3) When the electrode does not bond to TE material well, an intermediate layer can be used to bond them together. In some cases, the intermediate layer serves two or all three purposes. The following are some examples.

Ni is a good candidate as electrode material for TE power generation application, due to its high oxidization resistance and good wettability at soldering or brazing. However, the CTE of Ni is about 13×10^{-6} K^{-1}, which is bigger than that of n-type filled SKU. Another problem for Ni is the diffusion and reaction between Ni and filled SKUs. To use Ni as the electrode for filled SKU, a buffer layer is needed between them to reduce thermal stress as well as act as a diffusion barrier. Fan et al. [2] used Mo–Ti/Mo–Cu as the buffer layer to reduce the thermal stress between $Yb_{0.3}Co_4Sb_{12}$ and Ni electrode caused by the CTE mismatch. The contact resistivity remained below 9 $\mu\Omega$ cm^2 after 12 days of aging at 550°C, indicating that the interfacial reaction is slow [2].

Mo and Mo–Cu alloys are another candidate of contact material for filled SKU. They have high electrical and thermal conductivity. It is very easy to tune the CTE of Mo–Cu alloys by tuning the ratio of these two elements. However, it is very hard to directly bond Mo/Mo–Cu ally onto filled SKUs. Ti has a thermal expansion coefficient very close to $CoSb_3$, 8.6×10^{-6} K^{-1}. It was first introduced as an interlayer between Mo and $CoSb_3$ by Fan et al. [3] in 2004. $CoSb_3$ was effectively bonded to Mo through the insertion of the Ti barrier layer. Zhao et al. [4] used Ti foil as a buffer layer to join $CoSb_3$ and Mo–Cu alloy contact with an effective contact resistivity of 20–30 $\mu\Omega$ cm^2 [4–7]. It is worth pointing out that this contact resistivity number was calculated using the total voltage change across the Ti layer. So it includes the contact resistances on two interfaces ($CoSb_3$/Ti and Ti/electrode) and the resistance of Ti layer, as shown in Figure 16.2. The interface stability has been investigated in detail for this system. The results showed that an intermediate layer of $TiSb_2$ and TiSb formed at the Ti–$CoSb_3$ interface, and a composition gradient alloy layer was found at the Ti–Mo–Cu interface. It was also found that the thickness of the intermetallic compound layer increases, and hence, the shear strength of the $CoSb_3$/electrode joint quickly dropped with thermal aging. The specific contact resistivity increased from 20 to 28 $\mu\Omega$ cm^2 after aging at 550°C for 20 days. Later on, Zhao et al. [8] used W–Cu alloy to replace Mo–Cu as the electrode material for $CoSb_3$. Since Ti is still used as a buffer layer, the effective contact resistivity is similar and the reaction between Ti and $CoSb_3$ is still a problem [8]. To reduce the rate of interfacial reaction between Ti and n-type filled SKU, $Yb_{0.6}Co_4Sb_{12}$, small amounts of Al particles were introduced into the Ti-dominated diffusion barrier [9]. The Ti–Al alloy layer maintains high reaction activity during sintering. Experimental results indicated that Ti_3Al and TiAl alloys grew on the surface of Ti particles, forming the Ti/Ti_3Al/TiAl core–shell structure after sintering or during the operations at high temperatures. These compounds suppress the activity between Ti–Al alloy layer and n-type

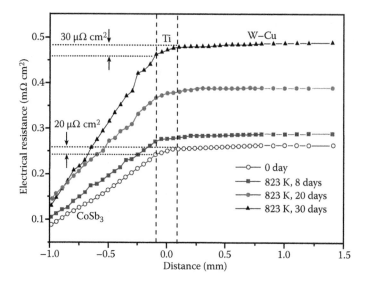

FIGURE 16.2 Electrical contact resistivity of the $CoSb_3/Ti/W_{80}Cu_{20}$ TE element. (Reprinted from *Journal of Alloys and Compounds*, 517, Zhao, D. G. et al., Fabrication and reliability evaluation of $CoSb_3$/W–Cu thermoelectric element, 198–203, Copyright (2012), with permission from Elsevier.)

filled SKU after sintering or during high-temperature service, as shown in Figure 16.3. Furthermore, it does not raise the interfacial thermal/electrical resistance, while it maintains a good bonding strength. The contact resistivity was initially around 10 $\mu\Omega$ cm^2 and remained below 12 $\mu\Omega$ cm^2 after 16 days of aging at 600°C.

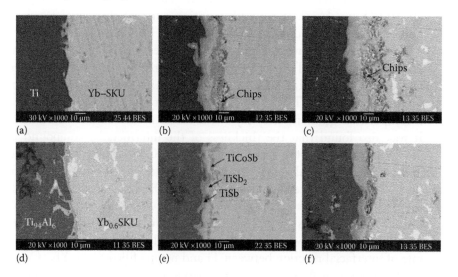

FIGURE 16.3 Evolution of the Ti/Yb–SKU (skutterudite) interfacial microstructure aged at 600°C for different times: (a) as sintered; (b) 8 days; (c) 16 days. Evolution of the $Ti_{94}Al_6$/Yb–SKU (skutterudite) interfacial microstructure aged at 600°C for different times: (d) as sintered; (e) 8 days; (f) 16 days. (Reprinted from *Journal of Alloys and Compounds*, 610, Gu, M. et al., Microstructural evolution of the interfacial layer in the Ti–Al/$Yb_{0.6}Co_4Sb_{12}$ thermoelectric joints at high temperature, 665–670, Copyright (2014), with permission from Elsevier.)

Instead of a barrier layer, Ti has also been used as electrode for filled SKU. In 2001, Caillat et al. [10] tried Ti as the electrode by adding metal powder on the top of the SKU legs during HP. The initial specific contact resistivity of Ti on both n- and p-type filled SKUs is lower than 5 $\mu\Omega$ cm^2. To improve the stability of the contact interface, Fleurial et al. [11] added a Zr layer as diffusion barrier between Ti and filled SKU. The specific contact resistivity between Zr and SKU was reported to be 19 $\mu\Omega$ cm^2 [11]. The aging tests were performed at 500°C for 2 months or 600°C for 2 weeks. The thickness of the intermetallic compound layer formed at Zr/CoSb$_3$ interface only increased a little after aging test, which indicates the high stability of the interface. The interface stability should come from the refractory interface compound ZrSb$_2$, which effectively prevents further diffusion of the Sb through Zr layer.

16.3 Single-Layer Contact

Although using intermediate barrier layer between electrode and filled SKU effectively reduced the thermal stress and improved the chemical stability of the contact, this method made the preparation of the TE legs more complicated and is adding the extra resistance into the total leg resistance through contact making. Ideally, a single layer of contact layer acting as electrode as well as diffusion barrier is better than multiple layers. Several other groups have tried this method. Wojciechowski et al. [12] prepared Ni, Mo, and Cr$_{80}$Si$_{20}$ layers by magnetron sputtering method on the surface of CoSb$_3$ as the contact layer. The thickness of the layers are 1–2.5 μm. Since the contact layers are very thin, the thermal stress from CTE mismatch is not a concern. On the other hand, such thin contact layers are not capable of acting as diffusion barrier. After brazing the legs onto Cu, the diffusion of the brazing alloy into the TE material leads to significant degradation of the TE properties of the leg. Salvador et al. [13,14] deposited thin Mo layer via arc spraying method as a single-layer electrode as well as diffusion barrier on both n- and p-type filled SKUs. The specific contact resistivity is estimated to be 30 $\mu\Omega$ cm^2 [13,14].

If Ti is used as electrode for p-type filled SKU, no barrier layer is need. Gu et al. [15] investigate the interfacial stability between Ti and a series of p-type filled SKUs. The jointed samples were aged at 550°C under vacuum for different durations. The initial specific contact resistivity was around 3 $\mu\Omega$ cm^2. After 30 days of aging, the contact resistivity still remained below 6 $\mu\Omega$ cm^2 [15]. On the control side, in case of the Ti/CoSb$_3$ joints, the contact resistivity rapidly rose from the initial 5 to 16 $\mu\Omega$ cm^2 after only 1 day of aging.

Guo et al. [16] reported using arc-melted Co–Ni–family metals as electrode as well as diffusion barrier in filled SKU module. As shown in Figure 16.4, by tuning the composition of the electrode composition, they can match their CTEs with the n- and p-type filled SKUs specifically. Although no specific contact resistivity data were reported for these legs, the interface was reported to be very stable after a long time of thermal aging and cycling

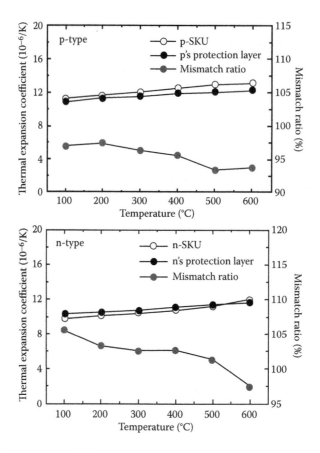

FIGURE 16.4 Changes of the CTE and percentage mismatch with temperature for the SKU material and its electrode (Co–Ni–family metal); (*top*), p-type; (*bottom*), n-type. (With kind permission from Springer Science+Business Media: *Journal of Electronic Materials*, Development of skutterudite thermoelectric materials and modules, 41, 2012, 1036–1042, Guo, J. Q. et al.)

test [17]. The aging test was performed with a hot-side temperature of 600°C and cold-side temperature of 80°C. After aging for 8000 h, the maximum electrical power output did not change. For the thermal cycling, the hot-side temperature was periodically changed from 200 to 600°C while the cold side was always kept at 40°C. The maximum power output decreased by 9.5% after 450 cycles, and the module's electrical resistance increased by about 5%.

Cobalt silicide is another type of contact material for filled SKU. Cobalt silicides have high electrical and thermal conductivities. These compounds are also very stable at the working temperature of filled SKU because of their high melting point (above 1300°C). Muto et al. [18] chose $CoSi_2$ as the contact material for the n-type and Co_2Si for the p-type filled SKU for CTE matching. As shown in Figure 16.5, the specific contact resistivities at the interface for both types are around 1–2 $\mu\Omega$ cm², which is about one magnitude lower than the numbers reported by other groups [19]. However, due to the high melting point of $CoSi_2$ and Co_2Si, a complicated procedure has to be used to produce crack-free leg samples. The contact pellets have to

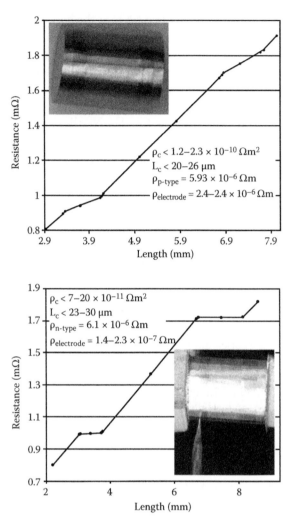

The figure shows two graphs. The top graph plots Resistance (mΩ) versus Length (mm):

$\rho_c < 1.2\text{–}2.3 \times 10^{-10}\ \Omega m^2$
$L_c < 20\text{–}26\ \mu m$
$\rho_{p\text{-type}} = 5.93 \times 10^{-6}\ \Omega m$
$\rho_{electrode} = 2.4\text{–}2.4 \times 10^{-6}\ \Omega m$

The bottom graph plots Resistance (mΩ) versus Length (mm):

$\rho_c < 7\text{–}20 \times 10^{-11}\ \Omega m^2$
$L_c < 23\text{–}30\ \mu m$
$\rho_{n\text{-type}} = 6.1 \times 10^{-6}\ \Omega m$
$\rho_{electrode} = 1.4\text{–}2.3 \times 10^{-7}\ \Omega m$

FIGURE 16.5 Resistance versus length of filled SKU device sample. The electrical contact resistivity, contact region length, TE bulk resistivity, and electrode bulk resistivity are given. (*Top*), p-type; (*bottom*), n-type. (From Muto, A., *Thermoelectric Device Characterization and Solar Thermoelectric System Modeling*, Doctoral dissertation, Massachusetts Institute of Technology, Cambridge, MA, 2011.)

be sintered at 1200°C first. After polishing and cleaning, the pellets will be inserted into the same hot press die and embedded in the filled SKU powder for the second hot pressing at 700–800°C, which served as the sintering of filled SKU layer as well as the bonding. Furthermore, the electrical and thermal conductivity of Co_2Si is relatively low compared with other materials reported as the electrode for p-type filled SKU.

16.4 Summary

Choosing the contact/electrode material for filled SKUs is a very challenging work due to the high reactivity of Sb. Traditional contact/electrode materials, such as Ni and Ti, react with filled SKU at the bonding or working temperature of the device and form intermetallic compounds at the interface. The reaction leads to higher contact resistance, which degrades the device performance and in the end the failure of the bonding and device. Several metal or alloys, such as Ti–Al alloy and Zr, have been used as the diffusion barrier layer to solve this problem. However, the total contact resistance is still high. Several novel materials, such as Co–Fe–Ni alloy and cobalt silicides, have been tried as the single-layer contact bonded to filled SKU and show promising contact quality. Long-time stability and interface reaction investigation are yet to be done.

Besides the specific contact resistivity, the bonding strength and long-time stability tests are also very important for the reliability of the filled SKU power generation devices. These tests have been performed differently by different groups. Standards for testing are needed for the comparison of the results from different groups and will benefit the growth of this field.

References

1. García-Cañadas, J., Powell, A. V., Kaltzoglou, A., Vaqueiro, P., and Min, G. (2013). Fabrication and evaluation of a skutterudite-based thermoelectric module for high-temperature applications, *Journal of Electronic Materials*, 42, 1369–1374.
2. Fan, X. C., Gu, M., Shi, X., Chen, L. D., Bai, S. Q., and Nunna, R. (2015). Fabrication and reliability evaluation of $Yb_{0.3}Co_4Sb_{12}$/Mo–Ti/Mo–Cu/Ni thermoelectric joints, *Ceramics International*, 41, 7590–7595.
3. Fan, J. F., Chen, L. D., Bai, S. Q., and Shi, X. (2004). Joining of Mo to $CoSb_3$ by spark plasma sintering by inserting a Ti interlayer, *Materials Letters*, 58, 3876–3878.
4. Zhao, D. G., Li, X. Y., He, L., Jiang, W., and Chen, L. D. (2009). Interfacial evolution behavior and reliability evaluation of $CoSb_3$/Ti/Mo–Cu thermoelectric joints during accelerated thermal aging, *Journal of Alloys and Compounds*, 477, 425–431.
5. Zhao, D. G., Li, X. Y., He, L., Jiang, W., and Chen, L. D. (2009). High temperature reliability evaluation of $CoSb_3$/electrode thermoelectric joints, *Intermetallics*, 17, 136–141.
6. Zhao, D. G., Geng, H. R., and Chen, L. D. (2012). Microstructure contact studies for skutterudite thermoelectric devices, *International Journal of Applied Ceramic Technology*, 9, 733–741.
7. Zhao, D. G., Li, X. Y., Jiang, W., and Chen, L. D. (2009). Fabrication of $CoSb_3$/MoCu thermoelectric joint by one-step SPS and evaluation, *Journal of Inorganic Materials*, 24, 545–548.
8. Zhao, D. G., Geng, H. R., and Teng, X. Y. (2012). Fabrication and reliability evaluation of $CoSb_3$/W–Cu thermoelectric element, *Journal of Alloys and Compounds*, 517, 198–203.
9. Gu, M., Xia, X. G., Li, X. Y., Huang, X. Y., and Chen, L. D. (2014). Microstructural evolution of the interfacial layer in the Ti–Al/$Yb_{0.6}Co_4Sb_{12}$ thermoelectric joints at high temperature, *Journal of Alloys and Compounds*, 610, 665–670.

10. Caillat, T., Fleurial, J. P., Snyder, G. J., and Borshchevsky, A. (2001). Development of high efficiency segmented thermoelectric unicouples, *Proceedings of 20th Int. Conf. on Thermoelectrics*, 282–285.
11. Fleurial, J. P., Caillat, T., and Chi, S. C. (2012). *Electrical Contacts for Skutterudite Thermoelectric Materials*, US Patent US 2012/0006376 A1.
12. Wojciechowski, K. T., Zybala, R., and Mania, R. (2011). High temperature $CoSb_3$–Cu junctions, *Microelectronics Reliability*, 51, 1198–1202.
13. Salvador, J. R., Cho, J. Y., Ye, Z., Moczygemba, J. E., Thompson, A. J., Sharp, J. W., König, J. D. et al. (2013). Thermal to electrical energy conversion of skutterudite-based thermoelectric modules, *Journal of Electronic Materials*, 42, 7, 1389–1399.
14. Salvador, J. R., Cho, J. Y., Ye, Z., Moczygemba, J. E., Thompson, A. J., Sharp, J. W., König, J. D. et al. (2014). Conversion efficiency of skutterudite-based thermoelectric modules, *Physical Chemistry Chemical Physics*, 16, 12510–12520.
15. Gu, M., Xia, X. G., Huang, X. Y., Bai, S. Q., Li, X. Y., and Chen, L. D. (2016). Study on the interfacial stability of p-type $Ti/Ce_yFe_xCo_{4-x}Sb_{12}$ thermoelectric joints at high temperature, *Journal of Alloys and Compounds*, 671, 238–244.
16. Guo, J. Q., Geng, H. Y., Ochi, T., Suzuki, S., Kikuchi, M., Yamaguchi, Y., and Ito, S. (2012). Development of skutterudite thermoelectric materials and modules, *Journal of Electronic Materials*, 41, 1036–1042.
17. Ochi, T., Nie, G., Suzuki, S., Kikuchi, M., Ito, S., and Guo, J. Q. (2014). Power-generation performance and durability of a skutterudite thermoelectric generator, *Journal of Electronic Materials*, 43, 2344–2347.
18. Muto, A., Yang, J., Poudel, B., Ren, Z. F., and Chen, G. (2013). Skutterudite unicouple characterization for energy harvesting applications, *Advanced Energy Materials*, 3, 245–251.
19. Muto, A. (2011). *Thermoelectric Device Characterization and Solar Thermoelectric System Modeling* (Doctoral dissertation, Massachusetts Institute of Technology, Cambridge, MA).

CHAPTER 17

Contacts for PbTe

Cheng-Chieh Li and C. Robert Kao

Contents

17.1 Introduction

In recent years, subjects on green energy production have attracted much attention worldwide. It is urgent to cut down on energy consumption and to reduce carbon footprint. In 2010, the International Energy Agency announced a blue map, declaring that the global CO_2 emissions should be reduced to 14 billion tons before 2050. Thus, environment-friendly TE materials have begun to gain attention due to the capability of power regeneration and its negligible carbon emission so as to transform industrial and automotive waste heat into renewable clean energy.

The usage of TEs could be traced back to the 1950s (Fritt, 1960; Heikes and Ure, 1961; Yu et al., 1970). During that time, enhanced Na-doped PbTe TE materials were embedded on the wall of nuclear reactor, called radioisotope TE generators (RTGs), for the earliest National Aeronautics and Space Administration–Jet Propulsion Laboratory (NASA-JPL) missions. The function of TE was designed to generate electrical power by converting the heat released from the nuclear decay of radioactive isotopes (typically plutonium-238). Table 17.1 lists the history of PbTe-based RTGs for NASA-JPL space exploration (Fleurial et al., 2011). In those missions, the heavy doped PbTe material could have a figure of merit ZT as high as ~1.4, and the conversion efficiency could reach 5.1–6.2%. Yet for the higher temperature spectrum, NASA-JPL indicated

TABLE 17.1 **PbTe-Based RTGs Used for Space Exploration Missions**

Mission	RTG Name	Launch Year	Mission Length (Years)
Transit 4A	SNAP-3B7	1961	15
Transit 4B	SNAP-3B8	1962	9
Apollo 12	SNAP-27 RTG	1969	8
Pioneer 10	SNAP-19 RTG	1972	34
Triad-01-1X	SNAP-9A	1972	15
Pioneer 11	SNAP-19 RTG	1973	35
Viking 1	SNAP-19 RTG	1975	4
Viking 2	SNAP-19 RTG	1975	6
MSL	MMRTG	2011	3

Source: Fleurial, J. et al., *Thermoelectrics: From Space Power Systems to Terrestrial Waste Heat Recovery Applications*, 2011 Thermoelectrics Applications Workshop, San Diego, CA, 2011.

that the power output of Si-Ge-based RTGs on the Voyager spacecraft dropped by 22% in over 14 years, which represents a 1.6% degradation in power output per year. One of the major degrading factors was due to the increase in the interfacial contact resistance (Yang and Caillat, 2006). This result implies that a suitable contact material is needed for high-temperature operations.

17.2 Fabrication Process of PbTe Modules

Figure 17.1 shows the typical fabrication process of TE modules. The commercially available ingot was first metallized by a diffusion barrier layer and

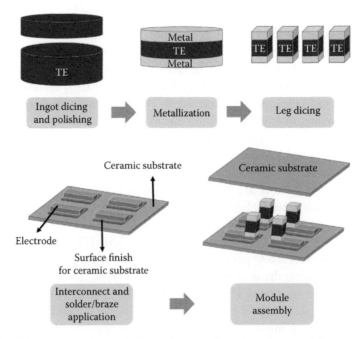

FIGURE 17.1 Schematic showing typical steps for manufacturing of TE module.

then wire sawed into small thermolegs. Finally, all these thermolegs were soldered or brazed onto the ceramic substrate, usually Al_2O_3, and interconnected to the counterpart ceramic substrate by soldering or brazing again. Cheng et al. (2013) demonstrated the assembly of PbTe modules by using a Ni diffusion barrier layer and Al–Si alloys as the brazing filler. But unfortunately, the assembly did not get good bonding result due to severe interfacial reaction. Hence, to choose an appropriate brazing filler is a tough challenge to this date.

Among all the soldering-related techniques, solid–liquid interdiffusion (SLID) technique has become more popular in recent years. The details of this technique are described in Section 17.4.2.

Conventionally, commercially available TE modules are based on planar structures. It is of interests that the idea of "ring-shaped" TE legs have been carried out for tubular TE modules for automotive waste heat recovery. In this design, ring-shaped legs of p- and n-type TE materials are alternately arranged along the axis of the tube with insulating rings in between to prevent electrical shortcut, as shown in Figure 17.2 (Schmitz et al., 2013). The tubular type of module may be advantageous compared with planar designs when the heat source and sink are fluids or gases and has much less mechanical stresses due to thermal expansion. However, the sintering of rings and tubes are more difficult compared with standard specimens. Thus, the process parameters of sintering will have to be adjusted many times in order to obtain a perfect specimen.

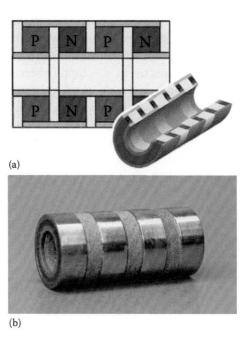

(a)

(b)

FIGURE 17.2 (a) Section drawing of a tubular TE module. (b) Tubular PbTe demonstration module consisting of four p- and n-legs connected with nickel bridges. (With kind permission from Springer Science+Business Media: *Journal of Electronic Materials*, Preparation of ring-shaped thermoelectric legs from PbTe powders for tubular thermoelectric modules, 42, 2013, 1702–1706, Schmitz, A. et al.)

17.3 Effect of Contact Materials on Figures of Merit

The performance of a TE module is conventionally gauged by the conversion efficiency and power output, while the cooling performance is gauged by the coefficient of performance (COP). The theoretical values could be obtained by using a theory developed by Ioffe (1957). However, the contributions of thermal and electrical contact were neglected in this theoretical method. Hence, a more complete method which takes the thermal and electrical contact of the modules into account has been developed over the past few years, providing a better accuracy for modeling TE modules.

Considering the existence of thermal and electrical contacts, the output voltage V and current I are modified by (Rowe and Min, 1996)

$$V = \frac{N\alpha(T_h - T_c)}{1 + 2rl_c/l}, \tag{17.1}$$

$$I = \frac{A\alpha(T_h - T_c)}{2\rho(n+l)(1 + 2rl_c/l)}, \tag{17.2}$$

where N is the number of thermocouples in a module, α is the Seebeck coefficient of the TE material, and T_h and T_c are temperatures at the hot and cold sides of the module, respectively. $r = \lambda/\lambda_c$ (λ is the thermal conductivity of TE materials, and λ_c is the thermal conductivity of contact materials). l_c is the thickness of contact layer. A and l are the cross-sectional area and the length of TE material, respectively. $n = 2\rho_c/\rho$ (where ρ is the electrical resistivity of TE materials, and ρ_c is the electrical resistivity of contact materials). n and r are usually referred to as electrical and thermal contact parameters, respectively. For commercially available Peltier modules, appropriate n and r values are usually around 0.1 mm and 0.2, respectively.

From the modified output voltage V and current I, the power output P and conversation efficiency Φ of a TE module could be derived as (Rowe and Min, 1996)

$$P = \frac{\alpha^2}{2\rho} \frac{AN(T_h - T_c)^2}{(n+l)(1 + 2rl_c/l)^2}, \tag{17.3}$$

$$\Phi = \frac{\left[(T_h - T_c)/T_h\right]}{(1 + 2rl_c/l)^2 \left\{2 - (1/2)\left[(T_h - T_c)/T_h\right] + (4/ZT_h)\left[(1+n)/(1 + 2rl_c)\right]\right\}}, \tag{17.4}$$

where $Z = \alpha^2/(\rho\lambda)$.

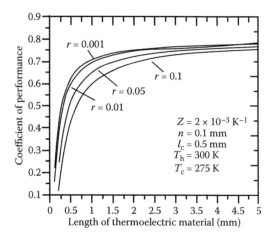

FIGURE 17.3 COP as a function of length of TE material for r = 0.001–0.1.

On the other hand, the COP could be also derived when operated in the Peltier mode (Rowe and Min, 1996):

$$\text{COP} = \frac{l}{l+2rl_c}\left(\frac{T_c}{T_h-T_c}\frac{\left[1+Z\bar{T}l/(n+l)\right]^{1/2}-T_h/T_c}{\left[1+Z\bar{T}l/(n+l)\right]^{1/2}+1}-\frac{rl_c}{l}\right), \quad (17.5)$$

where $\bar{T}$ is the mean temperature across the TE module ($T_h + T_c/2$). Figure 17.3 shows the relation of COP with TE material length (a typical condition for a Peltier module: ΔT = 25 K with T_h at 300 K). For commercially available modules, the values are $n \sim 0.1$ mm, $r \sim 0.1$, $l_c \sim 0.5$ mm, and $Z = 2 \times 10^{-3}$ K^{-1}. A dramatic decrease in COP can be observed when the length of the TE material is below 1 mm. Hence, the effect of contact resistances on the COP becomes significant when the length of the TE material is relatively short. It is important to reduce the contact resistances if short length of TE material is employed.

17.4 Literature Studies of Contact Materials for PbTe

The main reliability concerns of TE modules are the thermomechanical stresses from CTE mismatches, the intermetallic growth at the interface, and the sublimation of TE material. According to these failure mechanisms, the tendency to failures increases as temperature becomes higher. The following contents will deal with these aspects of the concerns.

17.4.1 Thermomechanical Stresses from Coefficient of Thermal Expansion Mismatches

With regard to CTE mismatch, an illustration of the thermal stress distribution is shown in Figure 17.4 (Clin et al., 2009). This simulation shows that the temperature difference of 100°C could severely deform the module

FIGURE 17.4 Von Mises stress distribution in a module subjected to a temperature difference of 100°C while keeping unconstrained surfaces (cold side on the bottom). The top alumina substrate is included in the simulation but is not shown. The red color represents higher calculated von Mises stress values. (With kind permission from Springer Science+Business Media: *Journal of Electronic Materials*, Numerical simulation of the thermomechanical behavior of extruded bismuth telluride alloy module, 38, 2009, 994, Clin, T. H. et al.)

(both surfaces unconstrained; the cold side is at the bottom of the module). In addition, the relative displacement of the center of the substrate with respect to the corners is on the order of 2 µm in the z direction perpendicular to the substrate face. One could observe that the stress in the module is mainly concentrated at the interface between the thermolegs and the conductors, suggesting that the reliability of TE module is highly dependent on the strength of the contacts. Figure 17.5 shows that the higher stress values would be located in the top or bottom corners of the legs if the stress during assembly was transferred to the bending of thermolegs (Clin et al., 2009). When the temperature difference between the hot and cold sides increases, the level of stress would increase accordingly. Thus, the effect of CTE mismatch would be more pronounced when the temperature difference increases.

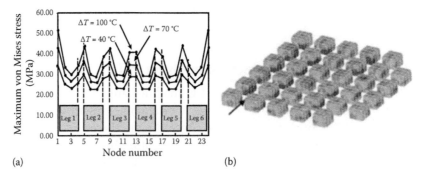

(a)   (b)

FIGURE 17.5 (a) Stress distribution along a line passing through the corners of the legs located in the direction of the arrow, as shown in the (b) module representation. The stress values are given for the three ΔT values simulated, for the hot side of the legs in a free-standing module. (With kind permission from Springer Science+Business Media: *Journal of Electronic Materials*, Numerical simulation of the thermomechanical behavior of extruded bismuth telluride alloy module, 38, 2009, 994, Clin, T. H. et al.)

17.4.2 Intermetallic Growth at the Interface

High-efficiency TE materials have been extensively developed over the past decade (Snyder and Toberer, 2008). Some materials which have *ZT* nearly equal to 2 can convert up to 20% of Carnot efficiency when incorporated into a device (Biswas et al., 2012). Even though significant progress has been achieved in the development of high-efficiency TE compounds, the development of TE modules to accommodate these materials has not been as rapid. The optimized service temperature for n-type and p-type PbTe-based TE materials falls in the range from 200°C to 600°C. During high-temperature operations, not only the severe diffusion but also the brittleness of the chemical reaction products could lead to the degradation of TE performance and reliability problems. Moreover, various techniques have been used to reduce the contact resistance, such as compression loading at the hot junction together with soft soldering of the cold junctions. Even with these mechanically complicated techniques, the contact resistance still excessively contributes to the total resistance of the modules. Consequently, appropriate bonding materials and fabrication methods should be developed for PbTe TE modules that allow reliable long-term operation.

The use of a bonding layer, inserted between the TE component and the electrode, is needed for lowering the electrical and thermal resistance. For instance, PbTe was first bonded to pure iron electrodes at 858°C in early 1960s (Weinstein and Mlavsky, 1962). Figure 17.6 shows that the bond has a contact resistance of less than 10 $\mu\Omega$ cm^2, and is mechanically stronger than the lead telluride itself (Weinstein and Mlavsky, 1962). It can be also seen that the overall resistance of the doubly bonded thermolegs is less than that of the unbonded one over 300°C. According to the vertical section of the Fe–Pb–Te ternary phase diagram (Rustamov and Abilov, 1987), the Fe-rich (α) phase coexists with PbTe, and the solubility limit of iron in PbTe is less than 3 at.% below 600°C. This thermodynamic assessment shows the possibility to preserve the integrity of whole modules when using Fe as a bonding

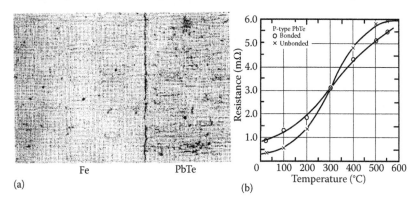

FIGURE 17.6 (a) Electromicrograph of Fe–PbTe (I$_2$-doped) bond. (b) Resistance of bonded and unbonded p-type PbTe as a function of temperature. (Reprinted with permission from Weinstein, M., and Mlavsky, A. I., *Review of Scientific Instruments*, 33, 1119, 1962. Copyright 1962, American Institute of Physics.)

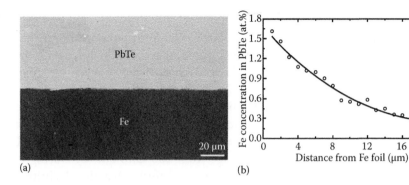

(a)

(b)

FIGURE 17.7 (a) Micrograph of the interface of Fe/PbTe joints bonded at 800°C for 60 min. (b) EPMA result of Fe concentration in PbTe from the interface between Fe foil and PbTe. (With kind permission from Springer Science+Business Media: *Journal of Materials Science*, Bonding and high-temperature reliability of NiFeMo alloy/n-type PbTe joints for thermoelectric module applications, 50, 2015, Xia, H. et al.)

material. Another empirical work supported the possibility of this idea (Xia et al., 2015). It can be observed that no reaction product formed at the interface as shown in Figure 17.7 (Xia et al., 2015), and Fe atoms only diffused approximately 20 μm into the PbTe during the bonding process. However, the iron bond is very sensitive and susceptible to oxidation during fabrication and operation. Thus, iron is still not a practical contact material for massive production unless the oxidation problem is resolved.

Nickel was designed as a bonding layer or a diffusion barrier layer over the past few years. Orihashi et al. (1998) found that there was no abrupt change of potential voltage at the interface between Ni and n-PbTe. However, a dramatic potential change was found at the interface between Ni and p-$Pb_{0.5}Sn_{0.5}$Te due to the chemical reaction product. The formation of Ni–Te compound, which deforms energy barrier or carrier compensation across the interface in the Ni/p-$Pb_{0.5}Sn_{0.5}$Te joint, may increase the interface resistance. One further study has been carried out for examining the reaction products between Ni and n-type PbTe (Xia et al., 2014b). Figure 17.8 shows the micrographs of two different positions at the interface of as-bonded Ni/PbTe joints (Xia et al., 2014b). At site A, one could observe that pure Pb has penetrated into the grain boundaries of the Ni foil, and some $Ni_{3\pm x}Te_2$ formed along the Ni grain boundaries and the Ni/PbTe interface. Figure 17.8c and d shows that a $Ni_5Pb_2Te_3$ phase (region 1) formed at the interface of Ni/PbTe joints. It can also be observed that long cracks formed in and around the ternary phase, indicating that the CTE mismatch between the ternary phase and the substrate is huge. A eutectic structure (region 2) can also be seen between the ternary phase and Ni. The formation process of this eutectic microstructure is assumed by the following steps: (1) Ni atoms diffused into PbTe to form a liquid layer composed of Pb, Te, and Ni at the Ni/PbTe interface during sintering; (2) large $Ni_5Pb_2Te_3$ particles formed from the liquid layer as the temperature decreased; and (3) when the temperature eventually reached the eutectic temperature, the eutectic structure formed. This implies that more

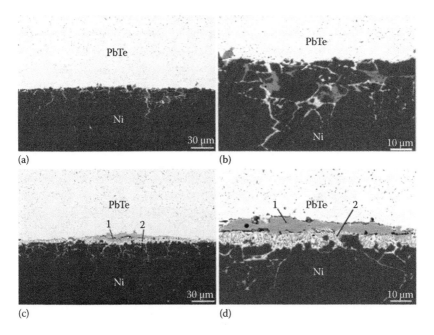

FIGURE 17.8 SEM micrographs of the interface of Ni/PbTe joints bonded at 650°C for 120 min at various positions: (a, b) site A and (c, d) site B. (With kind permission from Springer Science+Business Media: *Journal of Materials Science*, Bonding and interfacial reaction between Ni foil and n-type PbTe thermoelectric materials for thermoelectric module applications, 49, 2014, 1716, Xia, H. et al.)

and more nickel telluride will form and be accompanied by many voids and cracks during device operation. Hence, an alternative metallization layer to replace nickel is a must for a long-term operation.

To improve the wetting ability of Fe, a 50% PbTe and 50% Fe layer has been inserted between Ag electrodes and n-PbTe (Singh et al., 2008). No interaction between PbTe and Fe was observed, which in agreement with the phase diagram of PbTe–Fe. Apart from this, NiFeMo alloy has also been used as another solution of pure Ni or Fe (Xia et al., 2015). Figure 17.9 shows that the NiFeMo alloy foil bonds very well to n-type PbTe at 700°C with various bonding times (Xia et al., 2015). No porous layer can be found at the interface. However, some PbTe, or pure Pb, diffuse into the NiFeMo grains and grain boundaries, similar to the Ni case. With increasing bonding time, the materials could penetrate the NiFeMo alloy foil deeper and deeper along its grain boundaries. Figure 17.10 shows the high-temperature storage results that aged at 600°C for 240 h (Xia et al., 2015). The microstructure is similar to the as-bonded ones. But interestingly, the Seebeck coefficient of bulk PbTe with NiFeMo contacts is similar to that without NiFeMo contacts, indicating that the bond does not evidently affect the TE properties. However, the continued formation and penetration of eutectic liquid NiFeMo–PbTe and liquid Pb along the NiFeMo grain boundaries will eventually lead to contact degradation. Hence, if a long lifetime is required for NiFeMo/PbTe joints, the hot-side operation temperature was recommended to be not above 500°C.

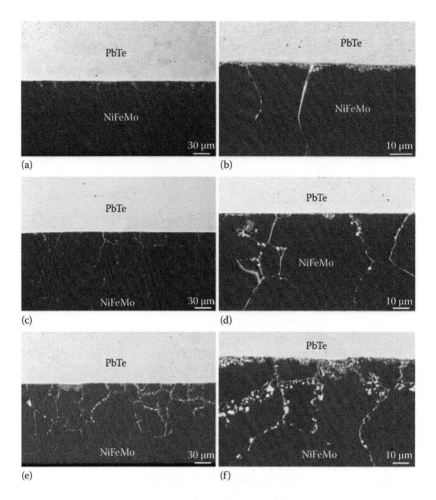

FIGURE 17.9 Electron micrograph of the interface of NiFeMo/PbTe joints bonded at 700°C for (a, b) 60, (c, d) 150, and (e, f) 300 min. (With kind permission from Springer Science+Business Media: *Journal of Materials Science*, Bonding and high-temperature reliability of NiFeMo alloy/n-type PbTe joints for thermoelectric module applications, 50, 2015, Xia, H. et al.)

In the early history of NASA, many systematic bonding tests were performed for PbTe. The selection and screening of potential braze and shoe materials for lead telluride are listed in Table 17.2 (Eiss, 1966). The bonding results show that many braze materials could be bonded into PbTe and form a bond that is metallurgically sound. However, in most cases, such bonded joints are difficult to survive or adequately perform for extended time under TE generator operating conditions. Generally speaking, the cause of failure is the diffusion of braze material into the TE material. This diffusion may affect Seebeck coefficient, electrical resistance, and so on. Especially, p-PbTe is more susceptible to degradation for this reason than n-PbTe. But among all the selected candidates, SnTe was found to be the most promising braze material, which shows the smallest deleterious effect on the TE properties of PbTe.

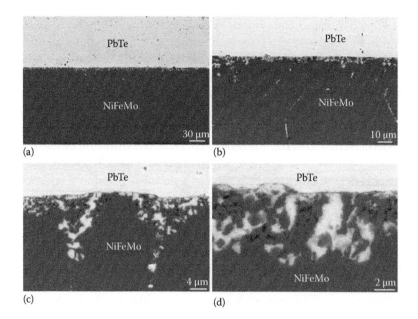

(a)   (b)
(c)   (d)

FIGURE 17.10 Electron micrograph of the interface of NiFeMo/PbTe joints bonded at 700°C for 60 min after accelerated aging at 600°C for 240 h. (With kind permission from Springer Science+Business Media: *Journal of Materials Science*, Bonding and high-temperature reliability of NiFeMo alloy/n-type PbTe joints for thermoelectric module applications, 50, 2015, Xia, H. et al.)

TABLE 17.2 Wettability of PbTe with Different Braze Materials

Braze Material	Temperature (°C)	Wetting Results
Sn	270	Good adhesion
Bi	300	Good adhesion
Se	233	Good adhesion
Sb	700	Good adhesion
In	192	No bond
Cu	657	Sample melted
SnTe	860	Good adhesion
Bi_2Te_3	648	Good adhesion with some pores in Bi_2Te_3 adjacent to interface
InSb	525	Good adhesion with some cracks in PbTe
CdSb	612	Braze separated from n-PbTe before mounting; p-PbTe sample had two intermediate phases
InTe	747	Good adhesion with some pores and cracks in InTe
Sb_2Te_3	670	Good adhesion with pores in p-PbTe adjacent to interface; signs of cracking or separation in n-PbTe interface
AuZn	869	No bond
$Ag-_{44}Sb$	666	Extensive penetration into PbTe, and phase form in interface
$Sb-_{23.5}Cu$	673	Good adhesion
$Sb-_{20.1}Zn$	649	Two phases form in interface
$Sb-_{30}Bi$	665	Good adhesion with few cracks in PbTe

Source: Eiss, A. L., *Thermoelectric Bonding Study*, NASA, Washington, DC, 1966.

For PbTe-based compounds, a new n-type nanostructured bulk material $AgPb_mSbTe_{2+m}$ (or LAST for lead–antimony–silver–tellurium) and a p-type $Ag(PbSn)_mSbTe_{2+m}$ (or LASTT for lead–antimony–silver–tin–tellurium) could reach $ZT > 2$ within the temperature range of 300–600°C. One systematic study has been carried out to find appropriate electrodes and bonding materials for hot-side contacts, as listed in Table 17.3 (Matchanov et al., 2011). The excellent low contact resistance bonds could be obtained using SS316 layers. However, for other materials, the diffusion of the Cu into the LAST or LASTT material could be observed when $T_{hot} > 550$ K, degrading the TE performance. SnTe served as a promising bonding layer after aging test for more than 8 h, similar to NASA experiments mentioned before, shown in Figure 17.11 (Matchanov et al., 2011). The resistance of the LAST and LASTT modules with different metal electrodes are listed in Table 17.4 (Matchanov et al., 2011). There is a sharp increase for nickel and for iron electrodes from 3 to 5 and from 20 to 50 mΩ after thermal cycling, respectively. The increase of the resistance was associated with cracks formed in the adjacent area of TE materials near contact. This might be due to the CTE mismatch between the TE material and electrode.

According to the vertical section of the AgPbTe ternary phase diagram, Ag could react with PbTe to form Ag_2Te intermetallic compound (Grieb et al., 1998). One recent work has pointed out the good TE properties of Ag_2Te itself, and furthermore, Ag also enhances the performance of PbTe (Pei et al., 2011). Moreover, the CTE of Ag is very close to that of PbTe, reducing

TABLE 17.3 Results of Diffusion Bonding for LAST and LASTT

Contact Material	Temperature (°C)	Time (h)	Result
Mn/FeMnC	720	6	Good bond with high resistance
Mn/Si/Mn/Si/ SS316	750	4	Good bond with high resistance
SnTe/FeMnC	800	1	Good bond
SnTe/SS316/Cu	650	1	Good bond
SnTe/Mo/Brass	720	4	Poor bond
Mo/Sb/Mo/Cu	650	16	Good bond with low resistance; Cu diffusion
Mo/SnTe/Mo/Cu	650	16	Good bond with low resistance; Cu diffusion
Mo/PbSn/Mo/Cu	650	16	Good bond with low resistance; Cu diffusion
Mo/In/SS316	760	4	Good bond with high resistance
Mo/Sb/SS316	770	4	Good bond with high resistance
Mn/Sb/SS316	810	4	Good bond with high resistance

Source: Applied Solar Energy, Investigation of the hot side contacts to nanostructured LASTt (T) material: Part I, 47, 2011, 90, Matchanov, N. A. et al. With kind permission from Springer Science+Business Media.

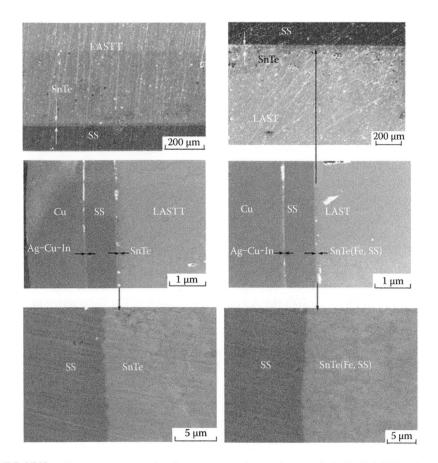

FIGURE 17.11 Electron micrograph of various interfaces of electrode/SnTe/LASTT after high-temperature tests for more than 8 h. (With kind permission from Springer Science+Business Media: *Applied Solar Energy*, Investigation of the hot side contacts to nanostructured LASTt(T) material: Part I, 47, 2011, 90, Matchanov, N. A. et al.)

TABLE 17.4 Electrical Resistances of LAST and LASTT Modules with Various Electrodes

Electrode Material	Resistance (mΩ)
Ni coated Cu	4.3
Cu (SS316 coated)	5.5–5.6
Ni	3.5–6.49
SS316	4.3–6
Ag/PB616/SS316	5.5–10
Cu/PB616/SS316	5.5–10

Source: Applied Solar Energy, Investigation of the hot side contacts to nanostructured LASTt(T) material: Part I, 47, 2011, 90, Matchanov, N. A. et al. With kind permission from Springer Science+Business Media.

the potential reliability issues arising from the CTE mismatch. Hence, Ag was chosen as one of the bonding material for PbTe. The microstructure evolution of the Ag/PbTe/Ag at 400°C is shown in Figure 17.12 (Li et al., 2015a). A nearly perfect Ag/PbTe interface without cracks is observed even after 1000 h of aging. But more and more Ag_2Te particles appear and uniformly distribute within PbTe. The amount of Ag_2Te increases following parabolic kinetics, which suggests that the formation of this phase is a diffusion-limited process. In contrast, the Ag foils vigorously reacted with PbTe to form a large scale of Ag_2Te and voids after annealing at 550°C, and both the upper and the lower Ag foils have been consumed (Figure 17.13) (Li et al., 2015a). So the usage of Ag should be $T_{hot} \leq 400°C$.

Similar to Ag, Cu was also selected as a bonding material for both PbTe due to its close CTE as well as its well-known electrode properties. Figure 17.14 shows the microstructure evolution of the Cu/PbTe/Cu at 400°C (Li et al., 2015b). The interface is uniform and free of cracks even after 1000 h of aging. But some Cu_2Te phase can be detected within the PbTe matrix, as Ag_2Te in PbTe. Unlike the Ag case, the growth of Cu_2Te is much faster than

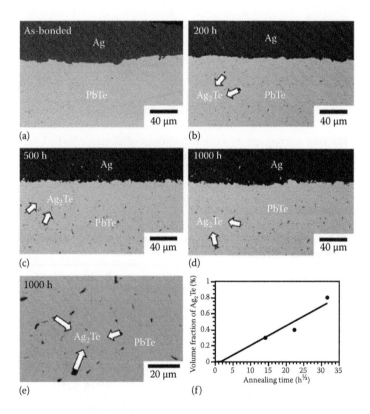

FIGURE 17.12 Interfaces of Ag/PbTe/Ag after aging at 400°C for (a) 0, (b) 200, (c) 500, and (d) 1000 h. (e) Zoom-in view for the PbTe region of (d). (f) Volume fraction of Ag_2Te as a function of the square root of the annealing time. (Reprinted from *Energy Conversion and Management*, 98, Li, C. C. et al., Silver as a highly effective bonding layer for lead telluride thermoelectric modules assembled by rapid hot-pressing, 134–137, Copyright [2015], with permission from Elsevier.)

FIGURE 17.13 Micrographs of Ag/PbTe/Ag after aging at 550°C for 50 h. (a) Low-magnification view showing that the Ag foil had disappeared. (b) Zoom-in view showing the coexistence of the reactants and the reaction products. (Reprinted from *Energy Conversion and Management*, 98, Li, C. C. et al., Silver as a highly effective bonding layer for lead telluride thermoelectric modules assembled by rapid hot-pressing, 134–137, Copyright [2015], with permission from Elsevier.)

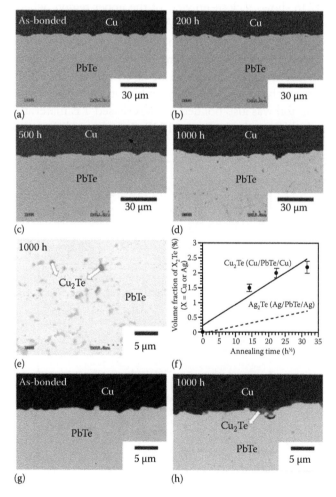

FIGURE 17.14 Interfaces of Cu/PbTe/Cu after aging at 400°C for (a) 0, (b) 200, (c) 500, and (d) 1000 h. (e) Zoom-in view of the PbTe matrix of (d). (f) Volume fraction of X_2Te (X = Cu or Ag) as a function of the square root of the aging time. (g) and (h) are the zoom-in views of (a) and (d), respectively. (Li, C. C. et al., *Journal of Materials Chemistry C*, 3, 10590, 2015. Reproduced by permission of The Royal Society of Chemistry.)

that of Ag_2Te in $Ag/PbTe/Ag$ under the same heat treatment. Hence, one should use thicker Cu foil to avoid the exhaustion of Cu.

It is well known that $(Pb, Sn)Te$ compounds serve as promising p-type TE materials. Figure 17.15 shows microstructure observation of the $Cu/Pb_{0.6}Sn_{0.4}Te/Cu$ after bonding (Li et al., 2015b). As one can see, both the upper and the lower thick Cu foils were nearly exhausted and Cu atoms could diffuse more than 1.4 mm depth along the grain boundaries of $Pb_{0.6}Sn_{0.4}Te$. Moreover, one could observe that $Pb_{0.6}Sn_{0.4}Te$ was surrounded by $PbTe$, Cu_2Te, and Cu_3Sn. The growth of Cu_3Sn and Cu_2Te at grain boundaries may result in the depletion of Sn and Te from $Pb_{0.6}Sn_{0.4}Te$ adjacent to the boundaries, then causing $Pb_{0.6}Sn_{0.4}Te$ to transform into the $PbTe$ phase. On the other hand, for the case of as-bonded $Ag/Pb_{0.6}Sn_{0.4}Te/Ag$, both the upper and the lower Ag foils are totally consumed and reacted along the grain boundaries of $Pb_{0.6}Sn_{0.4}Te$. Figure 17.16 shows a close-up view of the upper region, where $Pb_{0.6}Sn_{0.4}Te$ is surrounded by $PbTe$, Ag_2Te, and Ag_4Sn (Li et al., 2015b). In short, both Cu and Ag bonding layers face the same difficulty in maintaining a stable interface with $Pb_{0.6}Sn_{0.4}Te$ due to the severe Cu–Sn and Ag–Sn reactions. Thus, one should choose an appropriate bonding layer, which would not vigorously react with Sn, to reach the goal of long-term high-temperature operation.

Pure Nb foil was also bonded to PbTe-based TE materials as the hot-side contact via HP method. Figure 17.17 shows two different positions at the

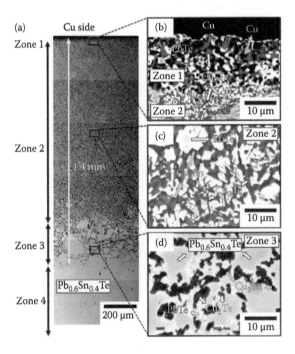

FIGURE 17.15 (a) Series electron overlay images of $Cu/Pb_{0.6}Sn_{0.4}Te/Cu$ across from Cu toward inner $Pb_{0.6}Sn_{0.4}Te$ after bonding. Zoom-in views of the regions of (b) zone 1 and zone 2, (c) zone 2, and (d) zone 3. (Li, C. C. et al., *Journal of Materials Chemistry C*, 3, 10590, 2015. Reproduced by permission of The Royal Society of Chemistry.)

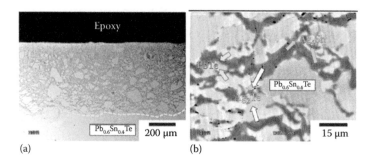

(a)   (b)

FIGURE 17.16 Electron micrographs of the as-bonded $Ag/Pb_{0.6}Sn_{0.4}Te/Ag$ sample. (a) Low-magnification view showing that the Ag foil was totally consumed. (b) Zoom-in view showing the reaction zone above the yellow curve outlined in a. (Li, C. C. et al., *Journal of Materials Chemistry C*, 3, 10590, 2015. Reproduced by permission of The Royal Society of Chemistry.)

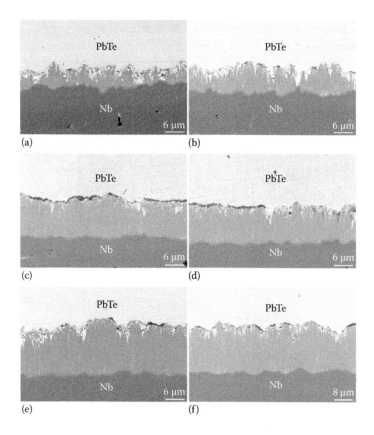

(a)   (b)

(c)   (d)

(e)   (f)

FIGURE 17.17 Electron micrographs of the interface of Nb/PbTe joints bonded at 700°C for (a, b) 60 (c, d) 150, and (e, f) 300 min. (With kind permission from Springer Science+Business Media: *Journal of Electronic Materials*, Interfacial reaction between Nb foil and n-type PbTe thermoelectric materials during thermoelectric contact fabrication, 43, 2014, 4064, Xia, H. et al.)

interface of the Nb/PbTe joints bonded at 700°C (Xia et al., 2014a). One can observe that a needlelike phase, Nb_3Te_4, forms between the Nb foil and n-type PbTe. At some positions of the Nb/PbTe interface, a thin black layer, pure Nb, can be also seen between the needlelike phase and PbTe. Fracture surface analysis can provide valuable insight into the failure mechanism of joints. Figure 17.18a and b shows the fracture surfaces of the Nb/PbTe joint bonded at 700°C on the PbTe side and on the Nb side, respectively (Xia et al., 2014a). These results indicated the Nb/PbTe joint mainly fractured at the interface between Nb and Nb_3Te_4 and within the PbTe matrix, as illustrated in Figure 17.18c. The weak adhesion between Nb and Nb_3Te_4 might be due to the uniform orientation of the needlelike Nb_3Te_4 grains.

High-temperature joining processes, such as brazing, have massive diffusion degradation and can retain residual thermal stress that can potentially reduce reliability. Consequently, many novel bonding materials and fabrication methods are proposed for TE modules that allow reliable long-term operation. SLID bonding technique is recently introduced for fabricating TE modules. The main principle of this technique is making low-melting temperature metal react with high melting-temperature metal to form a stable and high melting-temperature intermetallic compound, as illustrated in Figure 17.19. The SLID bonding of PbTe into Au-metallized Al_2O_3 has been successfully conducted by using Ag/In/Ag system (Li et al., 2015c). According to the Ag–In phase diagram (Baren and Massalski, 1990), once the temperature goes beyond 166°C, the molten In would react with Ag to form Ag_2In grains and eventually solidify into the room temperature as

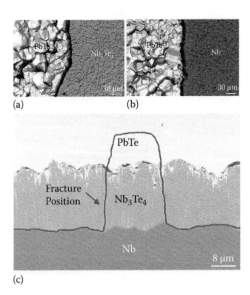

(a)   (b)

(c)

FIGURE 17.18 Electron micrographs of the fracture surfaces of the Nb/PbTe joint bonded at 700°C for 300 min: (a) on the PbTe side, (b) on the Nb side, and (c) schematic of the fracture position. (With kind permission from Springer Science+Business Media: *Journal of Electronic Materials*, Interfacial reaction between Nb foil and n-type PbTe thermoelectric materials during thermoelectric contact fabrication, 43, 2014, 4064, Xia, H. et al.)

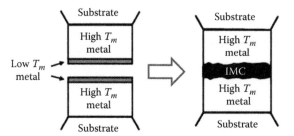

FIGURE 17.19 Schematic of the principle of SLID technique.

Ag$_2$In and Ag-rich layers, which could theoretically sustain the temperature of up to 660°C. The as-bonded joint also shows that the wetting zone could reach more than 90%. But interestingly, after aging for 1000 h at 400°C, the Ag$_2$In phase and Au layer were transformed into many small ternary Ag–In–Au compounds within the Ag-rich matrix. Although the phase transformation did occur during high–temperature storage, the joint still remained stable and mechanically sound.

One modified Ag/In SLID process has been proposed to increase the maximal shear strengths of 10.9–13.2 MPa through the enhancement by precoating 1 μm layer of Sn between the (Pb, Sn)Te TE material and Ni barrier layer, shown in Figure 17.20 (Chuang et al., 2014). The micrographs in Figure 17.21 show that the Ni$_3$Sn$_4$ and Cu$_{11}$In$_7$ intermetallic compounds grew thicker after high-temperature storage at 400°C for 200 h, while the Ag$_3$In intermetallics layer remained almost constant (Chuang et al., 2014). The shear strengths of these high-temperature storage (Pb, Sn)Te/Cu joints dropped to values of 10.5–10.8 MPa, as shown in Figure 17.20. In Figure 17.22, it can be seen that the fracture path occurred in the Ni$_3$Sn$_4$ intermetallics between the (Pb,Sn)Te material and Ni barrier layer (Chuang et al., 2014).

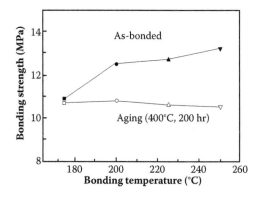

FIGURE 17.20 Bonding strengths of the (Pb,Sn)Te/Cu joints with In interlayers bonded at various temperatures for 30 min by using the modified SLID process and after high-temperature storage at 400°C for 200 h. (Reprinted from *Journal of Alloys and Compounds*, 613, Chuang, T. H. et al., Improvement of bonding strength of a (Pb,Sn)Te–Cu contact manufactured in a low temperature SLID-bonding process, 46–54, Copyright [2014], with permission from Elsevier.)

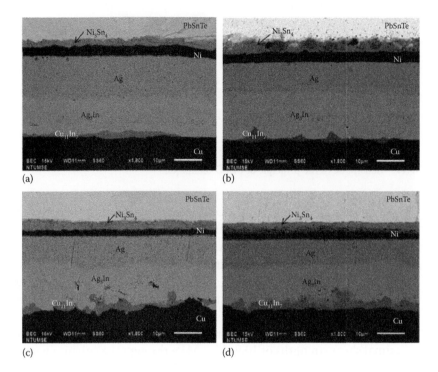

FIGURE 17.21 Morphology of the intermetallic compounds formed at the interfaces of the (Pb, Sn) Te/Cu joints with In interlayers bonded with the modified SLID process at various temperatures for 30 min and high-temperature storage at 400°C for 200 h: (a) 175°C, (b) 200°C, (c) 225°C, and (d) 250°C. (Reprinted from *Journal of Alloys and Compounds*, 613, Chuang, T. H. et al., Improvement of bonding strength of a (Pb,Sn)Te–Cu contact manufactured in a low temperature SLID-bonding process, 46–54, Copyright [2014], with permission from Elsevier.)

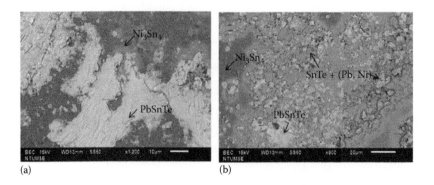

FIGURE 17.22 Typical fractography of high-temperature stored (Pb, Sn)Te/Cu joints with In interlayers bonded with the modified SLID process after shear tests: (a) TE material side and (b) electrode side. (Reprinted from *Journal of Alloys and Compounds*, 613, Chuang, T. H. et al., Improvement of bonding strength of a (Pb,Sn)Te–Cu contact manufactured in a low temperature SLID-bonding process, 46–54, Copyright [2014], with permission from Elsevier.)

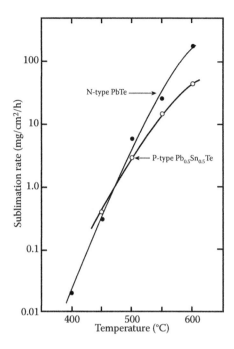

FIGURE 17.23 Rate of sublimation of (cold pressed and sintered) PbTe and $Pb_{0.5}Sn_{0.5}Te$ in vacuum. (Reprinted from *Advanced Energy Conversion*, 6, Bates, H. E., and Weinstein, M., Sublimation rates *In vacuo* of PbTe and $Pb_{0.5}Sn_{0.5}Te$ thermoelements, 177–180, Copyright [1966], with permission from Elsevier.)

17.4.3 Sublimation of Thermoelectric Material

Sublimation is one of the primary degradation mechanisms for TE power generation devices. The sublimation of the TE materials near the hot junction of the couples can result in a cross-sectional reduction of the thermoleg to increase the resistance, which in turn will decrease the power output of the TE module. If substantial sublimation occurs over time, this could lead to mechanical failure of the junction between the TE material and the metallization at the couple hot junction. The rate of sublimation in vacuum as a function of temperature for cold-pressed and sintered PbTe and $Pb_{0.5}Sn_{0.5}Te$ is shown in Figure 17.23 (Bates and Weinstein, 1966). The rates for PbTe varied from 0.02 to 175 mg/cm²/h, and for $Pb_{0.5}Sn_{0.5}Te$, from 0.0001 mg/cm²/h at 450°C to 45 mg/cm²/h at 600°C.

17.5 Future Prospect and Challenges

Despite many new materials having been discovered over the past decades, the development of PbTe for waste heat recovery applications remains extremely challenging. The major challenges lie in the areas of materials, processes, and integration. In order to obtain higher conversion efficiency or COP, one needs high-ZT TE materials and suitable contact materials that cover a broad temperature range. In addition, the uncertainty in material,

TABLE 17.5 Treatment Conditions and Detected Phases

Thermoelectric Material	Temperature (°C)	Time (h)	Oxides
PbTe	250	5	Traces of α-PbTeO$_3$ (orthorhombic), PbTeO$_3$ (tetragonal)
	350	5	α-PbTeO$_3$ (orthorhombic), PbTeO$_3$ (tetragonal)
	500	11	α-PbTeO$_3$ (orthorhombic)
	580	4	β-PbTeO$_3$, PbTeO$_3$ (tetragonal)
Pb$_{0.8}$Sn$_{0.2}$Te	400	30	SnO$_2$
	500	30	SnO$_2$, PbTeO$_3$(tetragonal), Pb$_2$SnO$_4$(tetragonal), PbSnO$_3$(cubic, pyrochlore type structure), α-PbO$_2$

Source: Berchenko, N. N. et al.: Oxidation of PbSnTe solid solutions and of PbSn alloys. *Surface and Interface Analysis.* 2006. 38. 518. Copyright Wiley-VCH Verlag GmbH & Co. KGaA. Reproduced with permission.

module, and subsystem costs, and in the original equipment manufacturer market size, is also a major factor inhibiting this technology development. There is still room for both performance enhancement and cost reduction.

In short, the suitable contact materials should have no massive solubility and severe reaction to TE materials and could retard the diffusion of elements from joints, electrodes, and substrates. In addition, the electrical and thermal resistances have to be low, so does the CTE mismatch. Moreover, high-temperature oxidation of TE material and contact material could also lead to the degradation of power output and failure. At high oxygen partial pressures and low temperatures, PbTeO$_3$ layer or Pb$_2$TeO$_4$/PbTeO$_3$ multilayers were found on the PbTe surface. When the temperature was higher than 819 K, PbTeO$_3$ would melt (Chen et al., 1997). The result of another oxidation work of PbTe and Pb$_{0.8}$Sn$_{0.2}$Te has also been listed in Table 17.5 (Berchenko et al., 2006). Hence, the hermetic sealing or ceramic coatings should be used in order to prevent oxidation.

References

Baren, M. R., and Massalski, T. B., *Binary Alloy Phase Diagrams* (Materials Park, OH: ASM International, 1990).

Bates, H. E., and Weinstein, M., Sublimation rates *In vacuo* of PbTe and Pb$_{0.5}$Sn$_{0.5}$Te thermoelements, *Advanced Energy Conversion* 6 (1966): 177.

Berchenko, N. N., Nikiforov, A. Y., and Fadeyev, S. V., Oxidation of PbSnTe solid solutions and of PbSn alloys, *Surface and Interface Analysis* 38 (2006): 518.

Biswas, K., He, J., Blum, I. D., Wu, C. I., Hogan, T. P., Seidman, D. N., Dravid, V. P., and Kanatzidis, M. G., High-performance bulk thermoelectrics with all-scale hierarchical architectures, *Nature* 489 (2012): 414.

Chen, L., Goto, T., Tu, R., and Hirai, T., *High-temperature oxidation behavior of PbTe and oxidation-resistive glass coating* (paper presented at the Proceedings of ICT 97. XVII International Conference on Thermoelectrics, Dresden, August 26–29, 1997).

Cheng, C. P., Ke, Y. S., Dai, M. J., Liu, C. K., Liao, L. L., and Cheng, C. H., *High-Temperature Failure Analysis of Thermoelectric PbTe Module Using Al-Si Bonding* (paper presented at the 2015 IEEE Electronic Components and Technology Conference, Tainan, October 7–9, 2013).

Chuang, T. H., Yeh, W. T., Chuang, C. H., and Hwang, J. D., Improvement of bonding strength of a (Pb,Sn)Te–Cu contact manufactured in a low temperature SLID-bonding process, *Journal of Alloys and Compounds* 613 (2014): 46.

Clin, T. H., Turenne, S., Vasilevskiy, D., and Masut, R. A., Numerical simulation of the thermomechanical behavior of extruded bismuth telluride alloy module, *Journal of Electronic Materials* 38 (2009): 994.

Eiss, A. L., *Thermoelectric Bonding Study* (Washington, DC: NASA, 1966).

Fleurial, J., Caillat, T., Nesmith, B. J., Ewell, R. C., Woerner, D. F., Carr, G. C., and JonesJet, L. E., *Thermoelectrics: From Space Power Systems to Terrestrial Waste Heat Recovery Applications* (slides presented at the 2011 Thermoelectrics Applications Workshop, San Diego, CA, January 3–6, 2011).

Fritts, R. W., *Thermoelectric Materials and Devices* (New York: Reinhold, 1960), 143–162.

Grieb, B., Lugscheider, E., and Wilden, J., Silver-lead-tellurium, ternary alloys, *VCH* 194 (1988): 465.

Heikes, R. R., and Ure, R. W., *Thermoelectricity: Science and Engineering* (New York: Interscience, 1961), 405–442.

Ioffe, A. F., *Semiconductor Thermoelements, and Thermoelectric Cooling* (London: Infosearch, 1957).

Li, C. C., Drymiotis, F., Liao, L. L., Dai, M. J., Liu, C. K., Chen, C. L., Chen, Y. Y., Kao, C. R., and Snyder, G. J., Silver as a highly effective bonding layer for lead telluride thermoelectric modules assembled by rapid hot-pressing, *Energy Conversion and Management* 98 (2015a): 134.

Li, C. C., Drymiotis, F., Liao, L. L., Hung, H. T., Ke, J. H., Liu, C. K., Kao, C. R., and Snyder, G. J., Interfacial reactions between PbTe-based thermoelectric materials and Cu and Ag bonding materials, *Journal of Materials Chemistry C* 3 (2015b): 10590.

Li, C. C., Hsu, S. J., Lee, C. C., Liao, L. L., Dai, M. J., Liu, C. K., Zhu, Z. X., Yang, H. W., Ke, J. H., Kao, C. R., and Snyder, G. J., Development of interconnection materials for Bi_2Te_3 and PbTe thermoelectric module by using SLID technique (paper presented at the 2015 IEEE Electronic Components and Technology Conference, San Diego, California, May 26–29, 2015c).

Matchanov, N. A., Farhan, M., D'Angelo, J., Timm, E. J., Hogan, T. P., Schock, H., Case, E. D., and Kanatzidis, M. G., Investigation of the hot side contacts to nanostructured LASTt (T) material: Part I, *Applied Solar Energy*, 47 (2011): 90.

Orihashi, M., Noda, Y., Chen, L., Kang, Y., Moro, K., and Hirai, T., *Ni/n-PbTe and Ni/p-$Pb_{0.5}Sn_{0.5}Te$ Joining by Plasma Activated Sintering* (paper presented at the Proceedings ICT 98. XVII International Conference on Thermoelectrics, Nagoya, May 24–28, 1998).

Pei, Y., Lensch-Falk, J., Toberer, E. S., Medlin, D. L., and Snyder, G. J., High thermoelectric performance in PbTe due to large nanoscale Ag_2Te precipitates and La doping, *Advanced Functional Materials* 21 (2011): 241.

Rowe, D. M., and Min, G., Design theory of thermoelectric modules for electrical power generation, *IEE Proceedings—Science, Measurement and Technology* 143 (1996): 351.

Rustamov, P. G., and Abilov, C. I., Phase diagrams of PbTe-Fe system, *Russian Journal of Inorganic Chemistry* 32 (1987): 1016.

Schmitz, A., Stiewe, C., and Müller, E., Preparation of ring-shaped thermoelectric legs from PbTe powders for tubular thermoelectric modules, *Journal of Electronic Materials*, 42 (2013): 1702.

Singh, A., Bhattacharya, S., Thinaharan, C., Aswal, D. K., Gupta, S. K., Yakhmi, J. V., and Bhanumurthy, K., Development of low resistance electrical contacts for thermoelectric devices based on n-type PbTe and p-type TAGS-85 $((AgSbTe_2)_{0.15}(GeTe)_{0.85})$, *Journal of Physics D: Applied Physics*, 42 (2008): 015502.

Snyder, G. J., and Toberer, E. S., Complex thermoelectric materials, *Nature Materials* 7 (2008): 105.

Weinstein, M., and Mlavsky, A. I., Bonding of lead telluride to pure iron electrodes, *Review of Scientific Instruments*, 33 (1962): 1119.

Xia, H., Chen, C. L., Drymiotis, F., Wu, A., Chen, Y. Y., and Snyder, G. J., Interfacial reaction between Nb foil and n-type PbTe thermoelectric materials during thermoelectric contact fabrication, *Journal of Electronic Materials* 43 (2014a): 4064.

Xia, H., Drymiotis, F., Chen, C. L., Wu, A., and Snyder, G. J., Bonding and interfacial reaction between Ni foil and n-type PbTe thermoelectric materials for thermoelectric module applications, *Journal of Materials Science* 49 (2014b): 1716.

Xia, H., Drymiotis, F., Chen, C. L., Wu, A., Chen, Y. Y., and Snyder, G. J., Bonding and high-temperature reliability of NiFeMo alloy/n-type PbTe joints for thermoelectric module applications, *Journal of Materials Science* 50 (2015): 2700.

Yang, J., and Caillat, T., Thermoelectric materials for space and automotive power generation, *MRS Bulletin* 31 (2006): 224.

Yu, I., Ravich, B. A. E., and Smirnov, I. A., *Semiconducting Lead Chalcogenides* (New York: Plenum Press, 1970).

PART III

THERMOELECTRIC MODULES AND SYSTEMS

CHAPTER 18

Bismuth Telluride Modules

Pham Hoang Ngan, Nini Pryds, and Li Han

Contents

18.1 Introduction

BiTe modules are the pioneers of TE modules that are available on the market and being sold annually in large quantities. At first sight, the modules were "vuot troi" for their usage in cooling applications. The range of interest has then been vastly broadening into power generation. For the use of these materials as generators, BiTe modules have the unique advantage of being most efficient for applications within the temperature regime where the hot side can reach a maximum temperature of up to 250°C [1]. This is a temperature range in which not many other heat-harvesting techniques are effective. As a matter of fact, in the temperature range from room temperature to 250°C, the TE performance of BiTe materials is still leading compared to those of other TE materials after so many years of materials development [2,3]. The TE figure of merit ZT of bismuth telluride had been above 1 at room temperature ever since the discovery of the material [4], and it is still showing increasing ZT values with advanced knowledge on material science [1,5,6], thus promising further boosting of the module's performance.

Therefore, further development of BiTe module is still highly attractive in today's need for effective energy usage. This goal comes along with critical requirement to improve electrical and thermal contact at interfaces between TE materials and other parts of the module optimize design for TE modules to perform their best in different application cases.

In order to meet the application requirements, the dimensions of BiTe modules varied from millimeters to submicrometers. Millimeter-thick BiTe modules or bulk modules are used as coolers or generators in applications such as cooking stoves [7], boilers, and marine applications [8] and in solar thermal systems [9]. Micrometer-thick or thin film BiTe modules are needed for applications in microelectronic systems, such as computer chips and optical laser, where the module plays a significant role as coolers and temperature controllers [10–12]. Even weak heat released from human body can also be effectively used by BiTe thin films and flexible BiTe modules to power certain devices such as healthcare chips and sensors [13] and wristwatches [13,14], usually known as wearable devices [13,14].

This chapter focuses on the methods for the construction of BiTe modules. The status of both commercial and noncommercial BiTe modules regarding their specifications and performances is given. Finally, the remaining challenges and future potentials for BiTe modules are discussed.

18.2 Construction of a BiTe Module

In general, the construction of a TE module is the key step to realize the desired theoretical performance of a module. The construction of a TE module is a complex procedure starting from material synthesis, TE element fabrication, and assembly of TE element with electrodes and substrates. In this chapter, the assembly of TE element with electrodes is the focus, because the quality of the bonding might be the determining factor of the final performance and practical use of the module. In general, a good joint requires various criteria: good adhesion between parts, low parasitic contact resistances in order to minimize parasitic electrical and thermal loss, and stability during operation [15–17]. Advanced processes for fabricating good contacts are henceforth required.

The contact fabrication process includes aspects relating to the selection of method to join TE material with electrode, selection of suitable material for the joining, consideration of a diffusion barrier between TE and electrode materials, and optimization of all the component materials so that they are physically and chemically stable together. In addition, fabrication process can be varied with the scale and dimensions required for TE module [2,6]. As the thickness of the TE module is being scaled down, i.e., in the case of thin film thermoelectric generators (TEGs), processes to build these connections changed from assembly of individual legs to thin film deposition-based procedures [2,6,18].

In the following, the processes to fabricate BiTe modules, bulk, thin film, and flexible BiTe modules are discussed.

18.2.1 Bulk Module

A typical process to manufacture a bulk BiTe module is illustrated in Figure 18.1. A typical fabrication process of a bulk TE module includes three basic steps: preparation of TE legs, connection of the TE legs to the electrodes, and attachment of electrodes to substrate. These electrodes could be attached to a firm, often insulating, substrate either before or after being joined with TE element. Finally, the module fabrication is completed by making a connection with the outer load/circuit. Additionally, the module fabrication could be finished by encapsulating the module into a protective material or structure to prevent it from external interaction, enhance mechanical endurance, and furthermore reduce the degradation of TE material during operation at high temperatures and reduce possible oxidation at the joining surfaces [19,20].

18.2.1.1 Preparation of TE Material

The fabrication process of a BiTe module varied based on the form of the starting TE material. Usually, BiTe materials are prepared in bulk form before being joined to the electrode. The material synthesis methods can be conventional Bridgman method [21–23], ZM [24], ingot growing [25], or mechanical alloying from pure component materials [26]. Usually, one intrinsic feature of BiTe materials is low mechanical strength and strong anisotropic TE properties because of the weak Te–Te van der Waals bonds along the c axis of BiTe crystal structure, whereas the Te–Bi and Bi–Te bonds along the ab plane are ionic covalent [27–29]. This means that during

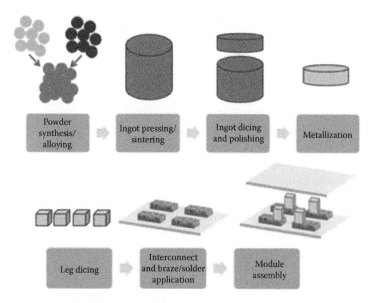

FIGURE 18.1 Typical process for manufacturing bulk BiTe modules. (Reprinted from *Sustainable Materials and Technologies*, 1–2, LeBlanc, S., Thermoelectric generators: Linking material properties and systems engineering for waste heat recovery applications, 26–35, Copyright [2014], with permission from Elsevier.)

preparation, care must be taken to prevent TE element from being damaged and select the directions with better TE performance during assembling [2]. Recently, another processing step is added to mitigate the anisotropic disadvantages: bulk BiTe batches are powderized by high-energy BM and then consolidated again by HP [26] or SPS [19]. In that way, the obtained BiTe has enhanced mechanical strength and less anisotropy.

After the synthesis, bulk BiTe is then cut into TE legs with predetermined dimensions according to the design of the module [30]. The leg surfaces are then often polished and cleaned to remove all contaminations that might disrupt the joining to the electrode, thereby reducing the quality of the obtained contact [31–33].

18.2.1.2 Diffusion Barrier

It has been widely acknowledged that operation under hot-side temperature near 200°C causes interdiffusion and/or reaction between BiTe and electrode materials and, as a consequence, negatively affects the performance and stability of module [34–36]. The interdiffusion could degrade the TE material and introduce brittleness and crack regions to the TE legs during operation and should therefore be diminished [2,37–41]. The deposition of a buffer layer to prevent the interdifussion is so far the principal solution to the problem [2,37–41]. A buffer layer with thickness of up to ~100 μm is usually deposited on BiTe surface, or in some cases, it is also deposited onto the surface of the electrode [41,42]. This buffer layer is preferred to be as thin as possible because parasitic electrical and thermal losses usually exist [2,15,43,44].

There has been a great deal of effort to select appropriate material for the buffer layer. Nickel has been conventionally used as a diffusion barrier [45,46], yet there was evidence showing that the material does not completely prevent the undesired diffusion [45,46]. Nickel itself also diffuses into BiTe at temperatures as low as 200°C, forming a NiTe layer which is mechanically weak that it might cause the module to deconstruct [47–49]. Furthermore, the Ni layer is reported to reduce the adhesiveness between BiTe and electrode material [41], where cracks that appeared resulted in bond strength of only about 2.1 MPa, and satisfactory bond improvement has not been achieved so far [47].

Recently, Lin et al. [34] investigated Pd, Ni/Au, Ag, and Ti/Au as diffusion barrier layers for BiTe with Ag electrode and found that Ti/Au was the best one. The diffusion barrier failed only after aging at 250°C for 50 h, when the crack between the Ti and Au layers started to form [34].

Au and Mo diffusion barrier has also been tested by other researchers [29], and positive results were obtained, showing that Au almost did not diffuse into BiTe. In another report, Kacsich et al. [46] found that 100 nm thick $Ta_{40}Si_{14}N_{46}$ films were shown to be able to prevent the interdifussion after 50 h at 250°C and 1 h at 350°C annealing. Another candidate is chromium, which is usually employed as a metallization layer for ceramic substrates and was also to be capable of mitigating the diffusion problem [39]. TE output test with hot-side temperature cycled from 250°C to 50°C has been conducted on a BiTe leg whose contact was built with a thin layer of chromium (thickness: ~200 μm) and has shown to be stable after over 50 cycles, which indicated that chromium can effectively prevent the diffusion of tin-based solder into BiTe [39]. Regardless, up to now, there is still not yet an ideal barrier layer for BiTe modules [50].

18.2.1.3 Selection of Electrode and Substrate Materials

Suitable electrode materials for TE module in general should have high electrical conductivity. Moreover, electrode materials should be able to maintain chemical stability (diminished interdifussion) and create low contact resistance with BiTe material. Copper is popularly utilized as electrode material for BiTe modules because of its high electrical conductivity of 1.7×10^{-8} Ω m at room temperature [51] and reasonable cost compared with other high-conductivity metals such as gold and silver. Copper electrodes can be in the form of strips or a layer deposited on an electrically insulating substrate. The thickness of the electrode varies with expected maximum output current of module in a disproportional manner, for example, a 0.4 mm thick copper electrode is expected for a 6 A module [52].

In many fabrication processes, electrodes were patterned onto substrate before joining with TE materials [52]. This step can be done by electron beam physical deposition [53], screen printing [54], and sputtering [31]. However, as previously mentioned, Cu is highly diffusive into BiTe at the working temperature of 250°C, [6,38,45,48]; therefore, alternatives such as Ag [53,55], Sb [50], Ti, Pd, and Au have been tried [34]. The result showed that the usage of Ag, Sb, and Pd did not deter the appearance of interdiffusion. The Ti/Au layer deposited onto BiTe material, although performed promisingly as interdifusion barrier, however failed in long-term stability test.

In terms of producing low contact resistance with BiTe, an investigation conducted by Vizel et al. [38] on different electrode materials, including Cu, Ag, Ni, and Fe, has shown that Cu had the advantage of producing the lowest contact resistance with BiTe material. Moreover, Yamashita and Tomiyoshi [56] have done an exclusive investigation on the effects of metal electrodes, including Ag, Al, Zn, Mo, Cu, Au, Ni, and Pt, on the TE power of BiTe compounds. They suggested that the selection of metal electrode affects the Seebeck coefficient, electrical resistivity, and thermopower at the contact interface and, therefore, the overall efficiency. For n-type BiTe samples, Au and Ag electrode materials resulted in TE power output that is 5–7% higher than that of Cu electrode. For p-type BiTe samples, the enhancement of Au and Ag electrode is 13.2–9.1%.

Selection of electrode material is also dependent on applications; for example, for applications in flexible devices, polyimide electrodes are favorable thanks to its high flexibility, low cost, and abundancy [57].

The substrate of the TE module in general and the BiTe modules in particular have the function of transferring heat throughout the modules and therefore requires high-thermal conductivity materials. Most of the current

TE modules use alumina substrate because it guarantees high thermal conductivity (18 W/m K at RT) for heat transfer through TE module and electrical insulation between module and heat source [52]. The thickness of the substrate usually ranges from 3 to 5 mm [58]. Besides alumina, alumina nitride and barium oxide have also been used as substrate materials, yet their prices are higher than that of alumina [52,55]. In miniature TE modules, the substrate material is usually silicon [27]. The common drawback of current substrate materials is the significant mismatch of their thermal expansion coefficient, which is ~16 parts per million (ppm)/°C for BiTe material [27,29], and ~5 ppm/°C for alumina [27,29].

After the selection of appropriate substrate and electrode materials, electrodes are attached to substrates by different deposition methods, for instance, electron beam deposition [53], electroplating [34] and sputtering [59]. These processes provide strong adhesion between electrode and substrate [34,53,59].

18.2.1.4 Joining of BiTe Element and Electrode

After the selection of electrode and substrate materials, the joining between the BiTe material and electrodes is conducted. This is the key step as the quality of the joint determines the final TE output of TE modules. Different methods can be used, and they are classifiable into soldering- and nonsoldering-based methods.

18.2.1.4.1 Soldering-Based Joining

Soldering-based methods require the usage of a third material, called solder alloy, to join TE and electrodes [43,60]. The material can be in the form of powder [53], foil, or paste [33,55], placed in between TE element and electrode. The joined structure is then heated and pressed so that the solder alloy melt or become soft, then cooled to firmly join the structure. The solder material should have a lower melting point than the softening temperature of BiTe (350°C) [3,38] to prevent the TE material from being degraded (by sublimation of volatile components Sb and Te). On the other hand, the melting temperature of the solder alloy should be higher than the working temperature of TE module to prevent the failure of the module during operation.

Due to being able to moderately address temperature requirements, Sn-based solder alloys have been widely used [41,48]. Namely, they are Sn–Ag, Sn–Ag–Cu, Sn–Cu, Sn–Ag–Ni, Bi–Sn–Sb, Au–Sn [52,55,61], and Bi–Sn [26,52]. However, the unavoidable drawback of using solder alloys is the introduction of contamination and possible formation of undesired phases that could harm the TE performance of the whole module [43]. These phases may act as another type of doping to TE materials and therefore degrade its intrinsic performance. For example, it was found that Sn rapidly diffuses into BiTe material, acts as dopant, or reacts with the element of the BiTe material, and even forms a Sn-rich region near the TE material, equivalently being considered as a resistor connected across the elements [40,61]. Buist and Roman [62] reported that at 138°C, there were excessively thick reaction layers (they, however, did not identify the composition of this layer) formed between solder Bi–Sn-based and TE material, degrading the reliability of the TE joint. In some cases where temperature at the hot side of module was accidently raised beyond 300°C, these types of Bi–Sn-based solders could melt and cause the module to break [52]. The hot-side temperature limit for BiTe module with Bi–Sn solder was suggested to be even less: 80°C due to low melting point of 138°C of the solder [63].

Pb–Sb solder was tried as an alternative to Sn-based solders [64], yet there was vigorous diffusion of Pb and reaction of with BiTe, causing the formation of a region with poor TE performance [63]. Furthermore, on the perspective of environmental sustainability, lead (Pb) is toxic and should be avoided [55].

In order to avoid the introduction of undesired elements from solder alloy and/or from a diffusion barrier, alloys of Bi and Sb [38,45,50] were studied as possible solder materials. These alloys contain about 50–99 wt.% Bi and 50–1 wt.% Sb. The modules using the 95:5 joining solder were tested at temperatures of 165–200°C and show no failures after 1000 temperature cycles. Further analysis by Vizel et al. [38] showed that the Bi in the solder might react with the electrode material (Cu, Ag, Ni, or Fe) and form a Bi-rich region bounded by Sb-based phases. This formation is not expectable because intermetallic compounds in general possess mechanically brittle feature, thus weakening the whole TE couples, especially during operation.

With the purpose of increasing the temperature limit for BiTe module, Ag powders were used as joining material using a pressure-assisted sintering process to join BiTe strength of 11 ± 1 N/mm^2 bonding without observable interdiffusion between the materials. The effect of Ag onto the stability of BiTe module, however, might still need an element with Ag electrode [53,55]. Strong bonds were obtained, associated with shear strength of 11 ± 1 Nmm^{-2} bonding without observable interdiffusion between the materials. The effect of Ag onto stability of BiTe module however might still need further research as different result was observed in another work [34]. The diffusion of Ag into BiTe was severely observed at 250°C, which again calls for the possibility of including a diffusion barrier.

18.2.1.4.2 Nonsoldering-Based Joining

In these methods, the synthesis of BiTe legs simultaneously occurs with the joining process. The hot side of BiTe modules are limited by the melting point of solder layer [53], and undesired interphases between solder alloy and TE material could result in high contact resistance and degradation of TE module's performance. Therefore, other techniques to directly join BiTe to electrodes are increasingly being developed and investigated.

For example, BiTe powders were hot pressed into an electrode at ~300–400°C to obtain BiTe legs which are directly connected to the electrode [65]. This process appeared to be promising because of better adhesion compared with the utilization of solder alloys [65]. A similar way is using hot pressed Ni powder onto BiTe powders at 400°C to form a bond by the interface reaction of Ni and BiTe, which is claimed to be suitable for BiTe in the form of nanopowder because it enhances the bonding strength between Ni electrode and BiTe material [47].

The solder-free process to assemble TE pairs of legs into a module could be a complex issue that requires various strategies and joining procedures [18]. For example, the solution of making contact for bulk crystalline BiTe materials cannot be utilized for nanostructured BiTe materials, because it can otherwise result in weak bonding strength of lower than 10 MPa, contact instability and, therefore, large contact resistance [47]. With the usage of appropriate joining process, i.e., direct hot pressing BiTe powder to Ni powder, the bonding strength was improved to 20 MPa. The method even resulted in low specific contact resistance below 1 $\mu\Omega$ cm^2 for p-type BiTe, even though for n-type BiTe, the reduction of area-specific contact resistance was moderate: ~210 $\mu\Omega$ cm^2 [47].

Undergoing mechanical stress during operation is unavoidable for TE modules, thus necessitating the requirement to enhance the strength and durability of modules [20,66]. In bulk TE modules, space between TE elements can be filled by an insulator such as epoxy or polyimide with the purpose of improving the hardness of the whole module, protecting TE material from possible degradation during operation, and allowing the usage of TE module as a monolithic structure [1,67].

18.2.2 Thin Film Module

Thin film TE modules refer to modules with thickness in the range of 5–100 μm [57,68–70], which are applied in low power applications where heat sources are at a small scale such as several to tens of microwatts [14,71]. Thin film TE modules are fabricated by techniques that are commonly used in microelectronics [70,72,73]. These processes include electrochemical deposition [73,74], molecular beam epitaxy [75], sputtering [76], metal–organic CVD [77], physical vapor deposition [78], CVD, flash evaporation [79], thermal coevaporation [80], and vacuum arc evaporation [57]. Typically, layers of the same type of the electrodes materials are initially deposited onto desired substrate with predetermined pattern, followed by the deposition of a diffusion barrier layer on top of the electrode. Afterward, films of p- or n-type material are subsequently deposited and patterned one by one. The device is then completed with another diffusion barrier, electrode layer, and the upper substrate layer on top of TE material [37]. These methods often require long time periods to prepare the starting materials or relatively complicated and expensive equipments. New processes to fabricate thin film BiTe module is therefore being used.

The main drawback of the conventional manufacturing technologies to fabricate thin film TE module is its high cost and long-time processing [33,72]. Although the recently proposed thermal spray fabricating process has a lower cost, the method suffers certain technological problems [72]. For example, the evaporating process might result in nonstoichiometric composition of the final BiTe material due to the very different vapor pressures of the elemental substances Bi, Sb, and Te [72]. In addition, inhomogeneity between the surface and inner layer might occur, especially in thermal evaporation and in sputtering [72]. One possible solution to overcome this problem is to use thermal coevaporation or cosputtering [76], where the deposition rate of each element is independently controlled and an optimal composition can be achieved.

Another promising fabrication technology of thin film BiTe module is screen printing, which allows large quantity production. [69,81]. The technology includes five steps: powder synthesis, material characterization, formulation, screen printing, and heat treatment [69,81]. BiTe p- and n-type materials in the form of powder are dispersed in solvent-containing binder at an appropriate concentration to form "ink." The ink is then deposited onto glass and ceramic substrates using a screen printer. A similar step is repeated for electrode deposition. Then, the layered films are dried to remove the solvent and binder, followed by a sintering process at 350°C for 30 min under nitrogen atmosphere. Resulted BiTe films have a thickness of 15 um and a density of 70–75%.

Thermal spray and laser machining are recent, promising methods to fabricate BiTe module thanks to the robustness and scalability in material processing enabled by laser machining [33]. The powder of BiTe materials was fed into a jet and was melted then accelerated toward substrate surface, where they quickly solidified and formed a dense coating layer, which was patterned using a motion-controlled laser beam. The method has the advantage of being simple, although there are still problems that need to be mitigated such as cracks developing on the contact between the insulating layer and the bottom metal substrate [33]. These problems limited the yield of functioning TE couples to 26 out of 40 obtained couples after fabrication.

Conventionally, the substrates of modules are highly thermally conductive and electrically insulating materials such as silicon carbide (SiC), aluminium nitride (AlN), boron nitride (BN), or beryllium oxide (BeO) [73]. For applications relating to microelectronic systems, substrate materials such as SiO_2 and polyimide are used for the simplification of manufacturing process [57]. Substrate materials can be metallized by the deposition of an active metal layer such as Ti or Cr [73], Cu [57], Al [80], and Au, Mo [67]. It is possible that during operation, the diffusion of electrode material into BiTe will occur [6,45,46]. The diffusion in principle can be controlled with a layer of diffusion barrier similar to that in bulk TE module fabrication. The diffusion might spontaneously cease due to the solubility limit of the materials, for example, at 250°C, the solubility limit of Au in n-type BiTe is ~30 ppm, whereas it is 200 ppm in p-type BiTe [82].

18.3 Status of BiTe Module

Overall pictures of the developing progress of BiTe modules on noncommercial and commercial levels are given in Tables 18.1 and 18.2, respectively. In these tables, power density P_d is calculated as output power P_{out} per TE element area, and efficiency η is the ratio between output power and heat that the TE module utilizes [1,83].

TABLE 18.1 Noncommercial Modules

Year [Author]	Materials	Temperature Difference ΔT (°C)	Dimension (μm)	Performance per 1 Couple P_d (W/cm²), P_{out} (W), η (%)	Substrate Material, Joining Method (between BiTe and Electrode)
1999 [71]; bulk	p-$(Bi_{0.25}Sb_{0.75})_2Te_3$ n-$Bi_2(Se_{0.1}Te_{0.9})_3$	$\Delta T = 10°C$	20 × 2.5 μm² 1200 μm thick	3 W/cm²	Electrode: kapton foil, ceramic substrate DC-magnetron sputtering
1999 [14]; bulk	p-$Bi_{0.4}Sb_{1.6}Te_3$ n-Bi_2Te_3	$\Delta T = 61°C$	80 × 80 μm² 600 μm thick[a]	4 W/cm²	Sintering Si wafer, Ni solder
2007 [83]; bulk, ring structure	Bi_2Te_3	$\Delta T = 80°C$	2000 μm Thick[a]	0.003 W—four ring prototype	Cu
2014 [13]; bulk		$\Delta T = 10°C$	71 μm² 150 μm thick	2.39×10^{-6} W/cm²	Electrode: Ti (10 nm)/Au (300 nm)
1999 [68]; thin film	p-$Bi_{0.5}Sb_{1.5}Te_3$ n-$Bi_2Te_{2.4}Se_{0.6}$	$\Delta T = 1°C$	670 × 40 μm² 2000 μm thick[a]	0.26×10^{-6} W/cm²	Thin glass, flash evaporation method using patterning Diffusion barrier: 300 nm-thick Al Bonding material: polymer Interconnect: Al
2001 [84]; thin film		$\Delta T = 70°C$	5.2 μm thick	(Not mentioned)	Superlattice thin film device
2002–2003 [85]; thin film	Sb_2Te_3 Bi_2Te_3	ΔT not mentioned	20 μm thick Diameter: 60 μm	40×10^{-6} W/cm²	Si substrate, +SiO_2 layer, Ni interconnects, ECD, and photolithograpy

(Continued)

TABLE 18.1 (CONTINUED) Noncommercial Modules

Year [Author]	Materials	Temperature Difference ΔT (°C)	Dimension (μm)	Performance per 1 Couple P_d (W/cm²), P_{out} (W), η (%)	Substrate Material, Joining Method (between BiTe and Electrode)
2004 [86]; thin film	$(Bi,Sb)_2Te_3$ Bi_2Te_3	$\Delta T = 5°C$	1400×600 μm²ᵃ 20 μm thick	0.6×10^{-6} W/mm²	Si wafer, interconnect, solder, by sputtering
2009 [87]; thin film and flexible	p- and n-type $Bi_{2+x}Te_{3-x}$	$\Delta T = 48.4°C$	126 μm thick	123.6×10^{-6} W/cm²	Au interconnects Device fabricated by electrochemical deposition Substrate is a polymer mold
2012 [88]; thin film	p-Bi_2Te_3/Sb_2Te_3 superlattice n-Bi_2Te_3/$Bi_2Te_{0.85}Se_{0.15}$ superlattice	$\Delta T = 100°C$	2000×2300 μm² 700 μm thickᵃ	3.17 W/cm²	Thin film superlattice TE device Substrate: GaAs Method: metallorganic CVD
2013 [89]; thin film	n-type Bi_2Te_3 p-type Sb_2Te_3	$\Delta T = 60°C$		0.73×10^{-6} W η = 1.5%	TE films deposited by sputtering Deposited on soda-lime glass substrates
2014 [90] (Flex TEG)	p-$(Bi_{0.15}Sb_{0.85})_2Te_3$	$\Delta T = 34°C$	2600×7600 μm² 200 μm thick	1.1×10^{-3} W/cm²	Substrate: polyethylene teraphthalate Adhesion layer: $CoSb_3$
2015 [91] (Flex TEG)	p-$Bi_{0.5}Sb_{1.5}Te_3$ n-Bi_2Te_3	$\Delta T = 20°C$	3500×600 μm² 100 μm thick	2.8 W/cm²	Polyimide Silver epoxy
2015 [92]	p- and n-type BiTe-based	$\Delta T = 190°C$	3200×1700 μm²ᵃ	4.38 W/cm² η = 3.2%	Cu electrode, solder, polymer film as insulting substrate

ᵃ Thickness of the whole module.

TABLE 18.2 Commercial Modules

Year [Author]	Materials	Temperature Condition	Dimensions, Scale	Performance: P_d (W/cm^2), P_{out} (W), η (%)
1999 Citizen Watch Co. Ltd. 1999 [93]	p-type BiSbTe n-type BiTe	$\Delta T = 1°C$	$7.5 \times 7.5 \times 1.5$ mm^3	24.9×10^{-6} W/cm^2
2006 Yamaha Co. (flexible module) [93]	BiTe-based materials	$\Delta T = 100°C$	8×14 mm^2	0.5 W/cm^2
2009 KELK Ltd. Komatsu [94]	BiTe-based materials	$\Delta T = 250°C$ (280–50°C)	$50 \times 50 \times 4.2$ mm^3	1 W/cm^2, $\eta = 7.2\%$
2011 Panasonic [95]	1-leg module $Bi_{0.5}Sb_{1.5}Te_3$	$\Delta T = 85°C$ (95–10°C)		2.5 W/pipe
HZ20 Hi-z.com [96]	BiTe-based materials	$T_c = 30°C$ $T_h = 230°C$ Max $T_h = 250°C$	$75 \times 75 \times 71$ mm^3	19 W 0.34 W/cm^2
HZ14 [96]	BiTe-based materials	$T_c = 30°C$ $T_h = 230°C$ Max $T_h = 250°C$	$62.7 \times 62.7 \times 49$ mm^3	14 W 0.36 W/cm^2
G1-1.4-219-1.14 Tellurex.com [97]	BiTe-based materials	$T_c = 50°C$ $T_h = 150°C$ Max $T_h = 175°C$	$54 \times 54 \times 219$ mm^3	5.7 W 0.2 W/cm^2

(Continued)

TABLE 18.2 (CONTINUED) Commercial Modules

Year [Author]	Materials	Temperature Condition	Dimensions, Scale	Performance: P_d (W/cm^2), P_{out} (W), η (%)
TEG-12610-5.1 Tecteg.com [98]	BiTe-based materials	$T_c = 30°C$ $T_h = 300°C$ Max $T_h = 330°C$	40 × 40 × 126 mm^3	7.1 W 0.44 W/cm^2
TEG-127-230-32e Thermalforce.de [99]	BiTe-based materials	$T_c = 20°C$ $T_h = 200°C$ Max $T_h = 230°C$	40 × 40 × 127 mm^3	13.3 W 0.83 W/cm^2
TEG-12656-0.6 Thermonamic [100]	BiTe-based materials	$T_c = 30°C$ $T_h = 300°C$ Max $T_h = 330°C$	56 × 56 × 126 mm^3	16.2 W 0.52 W/cm^2
TEG-12656-0.8 Thermonamic [100]	BiTe-based materials	$T_c = 30°C$ $T_h = 300°C$	56 × 56 × 126 mm^3	14.5 W 0.46 W/cm^2
TMH400302055 Wise life technology [101]	BiTe-based materials	$\Delta T = 40°C$	4 × 4 × 0.64 mm^3	13 W 0.2 W/cm^2
NEXTREME [102]	BiTe-based materials	$T_c = 100°C$ $T_h = 150°C$	5 × 4 × 0.7 mm^3	0.87 W/cm^2 η = 1.29%
GM250-127-28-10 TEG [103]	BiTe-based materials	$\Delta T = 220°C$	62 × 62 × 4 mm^3	0.74 W/cm^2

With the recent advances in the development of BiTe modules, the output power density has been increasing from microwatts per square centimeter to several watts per square centimeter and reached high flexibility in sizes from micrometers to millimeters.

BiTe modules with high maximum output power of 3–4 W/cm^2 with temperature gradients of 10°C and 61°C have been fabricated by Kishi et al. [14] and Stark and Stordeur [71] in 1999. The techniques used to fabricate these modules also enable the construction of thin film BiTe modules. In 2007, a BiTe module with new structure optimized for exhaust tubes was developed and reached the power output of 0.003 W under a temperature difference of 80°C, although the dimensions of these modules were not provided [83]. There were numerous companies invested in the development of BiTe modules such as Yamaha [93], Komatsu [94], Panasonic [95], Hi-Z [96], Tellurex [97], Tecteg [98], and Nextreme [83], to name a few (Table 18.2). Prototypes in laboratories usually have higher output performance compared with products in the market, whose top output power is in the range of 0.2–1 W/cm^2 (Table 18.2). In consequence, attempts to raise the TE outputs of products to approach those of modules of laboratories are still under way.

The development of thin film and flexible BiTe modules is also flourishing with the application potentials. Their TE power output meets the range of ~0.2 to ~100 μW/cm^2 (Table 18.1). Exceptionally, under a temperature gradient of 100°C, thin film superlattice BiTe module developed by Venkatasubramanian et al. [88] could reach the power output density of 3.17 W/cm^2. Flexible BiTe module constructed by Madan et al. [91] could also deliver a power density of 2.8 W/cm^2 with working temperature gradient of 20°C.

18.4 Remaining Challenges and Future Potential

Progress to improve BiTe modules from the performance aspect to fabrication technology still attracts research effort, although the module has a long history of development.

BiTe modules possess unique features that make them advantageous:

- Until the present day, they are the most effective method to make use of heat in the temperature range below 250°C.
- With a long history of development, fabrication techniques of the module have been matured, and facility for large-scale fabrication is being developed for a high level of cost-effectiveness and high throughput [12,13,71,87].
- Along with the maturity of module fabrication techniques, numerous designs to effectively apply BiTe module into various application setups have been elaborated and tested. Henceforth, the readiness and adaptability of BiTe module into practical application is of high level [12,84,89].
- BiTe modules do not require sources of high heat flux to generate power. Any available heat source would be sufficient for the device to operate and produce microwatt to milliwatt power [5].

Besides the strong advantages, there are still issues need to be solved:

- The TE properties of BiTe materials are strongly anisotropic. The electrical conductivity and thermal conductivity along the c axis are lower than those along the ab plane. The Seebeck coefficient along the c axis is, on the other hand, higher than that the ab plane [4,104]. Therefore, the figure of merit of BiTe material is higher in one direction than that along the other direction. This feature complicates the fabrication process of BiTe module should be taken carefully to select the optimum direction of BiTe material (Figure 18.2).
- BiTe material is mechanically weak due to the van der Waals bonds in its crystal structure, allowing easy cleavage along the ab plane [2]; therefore, it requires careful handling to eliminate cracks and damages which consequently degrade the final performance of module [71].
- BiTe modules are prone to interdiffusion problem that happens at the interface between BiTe and electrode materials. Due to the high electrical conductivity, Cu has been normally used as electrical collector in BiTe modules. However, severe diffusion of Cu into BiTe resulting in the performance degradation of BiTe is commonly known.
- Performance stability of BiTe modules remains to be a problem even with commercial ones. For example, long-term output test of commercial BiTe module Melcor HT 9-3-25, with cold side fixed at 24°C and hot-side temperature cycled from 30°C to 200°C, has been conducted by Hatzikraniotis et al. [105], showing a continuous reduction

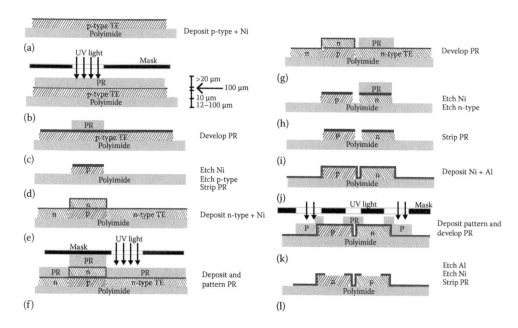

FIGURE 18.2 Fabrication process of a TE module. (With kind permission from Springer Science+Business Media: *Journal of Electronic Materials*, Fabrication of thermoelectric devices by applying microsystems technology, 39, 2010, 1516–1521, Goncalves, L. M. et al.)

of 3% of initial open circuit voltage (1.6 V) and 14% of initial power output (2.2 W). Test on another commercial BiTe module (Ferrotech 9500/127/060) reveals an output power reduction of 11% after 6000 times of cycling at the hot side temperature from 30°C to 160°C [106].

18.5 Summary

BiTe modules provide a significant and unique role as both coolers and power generators in a wide range of applications from micro- to macroscale. Despite the fact that BiTe modules have already been available on the market, there is still a demand for an advanced module fabrication process allowing the preservation of material TE performance at module scale and module durability. The remaining issue to thoroughly mitigate is still the diffusions between BiTe material and electrode. Moreover, a new approach to alleviate the challenge of thermal expansion coefficient mismatch is a crucial step to boost the level of reliability for the long-term operation of BiTe modules.

References

1. Rowe, D. M., *Thermoelectrics Handbook: Macro to Nano*. 2006: Taylor & Francis, Boca Raton, FL.
2. Liu, W. et al., Current progress and future challenges in thermoelectric power generation: From materials to devices. *Acta Materialia*, 2015. **87**: pp. 357–376.
3. Scherrer. H. and S. Scherrer, Bismuth telluride alloys for waste energy harvesting and cooling applications, in *Modules, Systems, and Applications in Thermoelectrics* (Edited by Rowe D. M.) (CRC Press 2012). pp. 6-1–6-18.
4. Goldsmid, H. J., Bismuth-the TE material for the future, in *2006 International Conference on Thermoelectrics*. 2006: Institute of Electrical and Electronics Engineers (IEEE), Piscataway, NJ.
5. Riffat, S. B., and X. Ma, Thermoelectrics: A review of present and potential applications. *Applied Thermal Engineering*, 2003. **23**(8): pp. 913–935.
6. LeBlanc, S., Thermoelectric generators: Linking material properties and systems engineering for waste heat recovery applications. *Sustainable Materials and Technologies*, 2014. **1–2**: pp. 26–35.
7. Gao, H. B. et al., Development of stove-powered thermoelectric generators: A review. *Applied Thermal Engineering*, 2016. **96**: pp. 297–310.
8. Barma, M. C. et al., Estimation of thermoelectric power generation by recovering waste heat from biomass fired thermal oil heater. *Energy Conversion and Management*, 2015. **98**: pp. 303–313.
9. Kraemer, D. et al., High-performance flat-panel solar thermoelectric generators with high thermal concentration. *Nature Materials*, 2011. **10**: pp. 532–538.
10. Snyder, G. J., M. Soto, R. Alley, D. Koester, and B. Conner, Hot spot cooling using embedded thermoelectric coolers, in *22nd IEEE Semi-Therm Symposium*. 2006: IEEE, Piscataway, NJ.
11. Zheng, X. F. et al., A review of thermoelectrics research—Recent developments and potentials for sustainable and renewable energy applications. *Renewable and Sustainable Energy Reviews*, 2014. **32**: pp. 486–503.

12. Chou, S. K. et al., Development of micro power generators—A review. *Applied Energy*, 2011. **88**(1): pp. 1–16.
13. Jung, K. K., and J. S. Ko, Thermoelectric generator based on a bismuth-telluride alloy fabricated by addition of ethylene glycol. *Current Applied Physics*, 2014. **14**(12): pp. 1788–1793.
14. Kishi, M. et al., Micro thermoelectric modules and their application to wrist-watches as an energy source. In *18th International Conference on Thermoelectrics*. 1999. p. 301.
15. Bjørk, R., The universal influence of contact resistance on the efficiency of a thermoelectric generator. *Journal of Electronic Materials*, 2015. **44**(8): pp. 2869–2876.
16. Annapragada, S. R. et al., Determination of electrical contact resistivity in thermoelectric modules (TEMs) from module-level measurements. *IEEE Transactions on Components, Packaging and Manufacturing Technology*, 2012. **2**(4): pp. 668–676.
17. Arai, K. et al., Improvement of electrical contact between TE material and Ni electrode interfaces by application of a buffer layer. *Journal of Electronic Materials*, 2012. **41**(6): pp. 1771–1777.
18. Zebarjadi, M. et al., Perspectives on thermoelectrics: From fundamentals to device applications. *Energy and Environmental Science*, 2012. **5**(1): pp. 5147–5162.
19. Brostow, W. D. et al., Bismuth telluride-based thermoelectric materials: Coatings as protection against thermal cycling effects. *Journal of Materials Research*, 2012. **27**(22): p. 2930.
20. Kambe, M., T. Jinushi, and Z. Ishijima, Encapsulated thermoelectric modules for advanced thermoelectric systems. *Journal of Electronic Materials*, 2013. **43**(6): pp. 1959–1965.
21. Custódio, M. C. C., and A. C. Hernandez, Tellurium-rich phase in n-type bismuth telluride crystals grown by the Bridgman technique. *Journal of Crystal Growth*, 1999. **205**(4): pp. 523–530.
22. Yamashita, O., and S. Sugihara, High-performance bismuth-telluride compounds with highly stable thermoelectric figure of merit. *Journal of Materials Science*, 2005. **40**(24): pp. 6439–6444.
23. Yamashita, O., S. Tomiyoshi, and K. Makita, Bismuth telluride compounds with high thermoelectric figures of merit. *Journal of Applied Physics*, 2003. **93**(1): pp. 386–374.
24. Hyun, D. B., T. S. Oh, J. S. Hwang, J. D. Shim, and N. V. Kolomoets, Electrical and thermoelectric properties of $90\%Bi_2Te_3\text{-}5\%Sb_2Te_3\text{-}5\%Sb_2Se_3$ single crystals doped with SbI_3. *Scripta Materialia*, 1999. **40**(1): p. 49.
25. Ha, H. P., D. B. Hyun, J. Y. Byun, Y. J. Oh, and E. P. Yoon, Enhancement of the yield of high-quality ingots in the zone-melting growth of p-type bismuth telluride alloys. *Journal of Materials Science*, 2002. **37**(21): pp. 4691–4696.
26. Kumpeerapun, T. et al., Performance of low-cost thermoelectric modules fabricated from hot pressing and cold pressing materials, in *Int. Conf. Thermoelectr. ICT, Proc.*, 2006: pp. 136–140.
27. Rowe, D. M., *CRC Handbook of Thermoelectrics*. New York. 16 (1995), 1251–1256.
28. Oh, T. S., D. Bin Hyun, and N. V. Kolomoets, Thermoelectric properties of the hot-pressed $(Bi,Sb)_2(Te,Se)_3$ alloys. *Scripta Materialia*, 2000. **42**(2000): pp. 849–854.

29. Wesolowski, D. E. et al., Development of a Bi_2Te_3-based thermoelectric generator with high-aspect ratio, free-standing legs. *Journal of Materials Research*, 2012. **27**(08): pp. 1149–1156.

30. Cheng, F. et al., Performance prediction and test of a Bi_2Te_3-based thermoelectric module for waste heat recovery. *Journal of Thermal Analysis and Calorimetry*, 2014. **118**(3): pp. 1781–1788.

31. Iyore, O. D. et al., Interface characterization of nickel contacts to bulk bismuth tellurium selenide. *Surface and Interface Analysis*, 2009. **41**(5): pp. 440–444.

32. Feng, S.-P. et al., Reliable contact fabrication on nanostructured Bi_2Te_3-based thermoelectric materials. *Physical Chemistry Chemical Physics*, 2013. **15**(18): p. 6757.

33. Tewolde, M. et al., Thermoelectric device fabrication using thermal spray and laser micromachining. *Journal of Thermal Spray Technology*, 2015. **25**(3): pp. 431–440.

34. Lin, W. P., D. E. Wesolowski, and C. C. Lee, Barrier/bonding layers on bismuth telluride (Bi_2Te_3) for high temperature thermoelectric modules. *Journal of Materials Science: Materials in Electronics*, 2011. **22**(9): pp. 1313–1320.

35. Thimont, Y. et al., Design of apparatus for Ni/Mg_2Si and $Ni/MnSi_{1.75}$ contact resistance determination for thermoelectric legs. *Journal of Electronic Materials*, 2014. **43**(6): pp. 2023–2028.

36. Xia, H. et al., Bonding and interfacial reaction between Ni foil and n-type PbTe thermoelectric materials for thermoelectric module applications. *Journal of Materials Science*, 2014. **49**(4): pp. 1716–1723.

37. Shi, X., Barrier/bonding layers on bismuth telluride (Bi_2Te_3) for high temperature thermoelectric modules. *Journal of Materials Science: Materials in Electronics*, 2011. **22**(2011): pp. 1313–1320.

38. Vizel, R. et al., Bonding of Bi_2Te_3-based thermoelectric legs to metallic contacts using $Bi_{0.82}Sb_{0.18}$ alloy. *Journal of Electronic Materials*, 2015. **45**(3): pp. 1296–1300.

39. Van Nong, N., T. H. Le, L. Han, H. N. Pham, and N. Pryds, Characterization of the contact between Bi_2Te_3-based materials and lead-free solder alloy under thermal cycling. In *Book of Abstract–34th Annual International Conference on Thermoelectrics (ICT 2015) and 13th European conference on Thermoelectrics (ECT 2015)*. 2015. 15C.5. Dresden.

40. Lan, Y. C. et al., Diffusion of nickel and tin in p-type $(Bi,Sb)_2Te_3$ and n-type $Bi_2(Te,Se)_3$ thermoelectric materials. *Applied Physics Letters*, 2008. **92**(10): p. 101910.

41. Yang, C. L. et al., Diffusion soldering of $Bi_{0.5}Sb_{1.5}Te_3$ thermoelectric material with Cu electrode. *Journal of Materials Engineering and Performance*, 2013. **22**(7): pp. 2029–2037.

42. Aswal, D. K., R. Basu, and A. Singh, Key issues in development of thermoelectric power generators: High figure-of-merit materials and their highly conducting interfaces with metallic interconnects. *Energy Conversion and Management*, 2016. **114**: pp. 50–67.

43. Ngan, P. H. et al., On the challenges of reducing contact resistances in thermoelectric generators based on half-Heusler alloys. *Journal of Electronic Materials*. **45**(1): pp. 594–601.

44. Hung, L. T. et al., Segmented thermoelectric oxide-based module for high-temperature waste heat harvesting. *Energy Technology*, 2015. **3**(11): pp. 1143–1151.

45. Yahatz, M., and J. Harper, *Fabrication of Thermoelectric Modules and Solder for Such Fabrication*. 1998: Melcor Corporation, Lawrence, USA.

46. Kacsich, T., E. Kolawa, J. P. Fleurial, T. Caillat, and M.-A. Nicolet, Films of Ni–7 at% V, Pd, Pt and Ta–Si–N as diffusion barriers for copper on Bi_2Te_3. *Journal of Physics D: Applied Physics*, 1998. **31**: pp. 2406–2411.

47. Liu, W. et al., Understanding of the contact of nanostructured thermoelectric n-type $Bi_2Te_{2.7}Se_{0.3}$ legs for power generation applications. *Journal of Materials Chemistry A*, 2013. **1**(42): pp. 13093.

48. Lin, T. Y., C. N. Liao, and A. T. Wu, Evaluation of diffusion barrier between lead-free solder systems and thermoelectric materials. *Journal of Electronic Materials*, 2011. **41**(1): pp. 153–158.

49. Mengali, O. J., and M. R. Seiler, Contact resistance studies on thermoelectric materials. *Advanced Energy Conversion*, 1962. **2**: pp. 59–68.

50. Li, F. et al., Interface microstructure and performance of Sb contacts in bismuth telluride-based thermoelectric elements. *Journal of Electronic Materials*, 2013. **42**(6): pp. 1219–1224.

51. Chambers, R. G., Electrical conductivity of metals. In *Electronics in Metals and Semiconductors*. 1990: Springer Netherlands, Dordrecht. pp. 120–132.

52. Sharp, J., and J. Bierschenk, The prevalence of standard large modules in thermoelectric applications. *Journal of Electronic Materials*, 2014. **44**(6): pp. 1763–1767.

53. Stranz, A., A. Waag, and E. Peiner, Investigation of thermoelectric parameters of Bi_2Te_3: TEGs assembled using pressure-assisted silver powder sintering-based joining technology. *Journal of Electronic Materials*, 2015. **44**(6): pp. 2055–2060.

54. Lee, H.-B. et al., Thin-film thermoelectric module for power generator applications using a screen-printing method. *Journal of Electronic Materials*, 2011. **40**(5): pp. 615–619.

55. Kähler, J. et al., Thermoelectric coolers with sintered silver interconnects. *Journal of Electronic Materials*, 2014. **43**(6): pp. 2397–2404.

56. Yamashita, O., H. Odahara, and S. Tomiyoshi, Effect of metal electrode on thermoelectric power in bismuth telluride compounds. *Journal of Material Science*, 2004. **39**: pp. 5653–5658.

57. Kato, K. et al., Fabrication of a flexible bismuth telluride power generation module using microporous polyimide films as substrates. *Journal of Electronic Materials*, 2013. **43**(6): pp. 1733–1739.

58. Messele, Y., E. Yilma, and R. Nasser, Design of a single stage thermoelectric power generator module with specific application on the automotive industry. *Advances in Intelligent Systems and Computing*, 2015. **334**: pp. 215–231.

59. Kishi, M. et al., Fabrication of a miniature thermoelectric module with elements composed of sintered Bi-Te compounds. in *16th International Conference on Thermoelectrics*. 1997.

60. Jacobson, D. M., and G. Humpston, *Principles of Soldering*. 2004: ASM International, Materials Park, OH.

61. Haba, V., *Method and Materials for Obtaining Low Resistance Bonds to Bismuth Telluride*. US patent: US3079455 A (1963). RCA Corp, New York.

62. Buist, R. J., and S. J. Roman, Development of a burst voltage measurement system for high-resolution contact resistance tests of thermoelectric heterojunctions. In *Eighteenth International Conference on Thermoelectrics*. 1999: IEEE, Piscataway, NJ.

63. Yahatz, M., and J. Harper, *Fabrication of Thermoelectric Modules and Solder for Such Fabrication*, US Patent. 1996: Melcor Corporation.

64. Fleurial, J.-P. et al., Thick-Film TE Microdevices, in *International Conference on Thermoelectrics, ICT, Proceedings*. 1999. pp. 294–300.

65. Poudel, B. et al., *Methods of Fabricating Thermoelectric Elements*. 2014, Google Patents.
66. Nelson, J. L., M. D. Gilley, and D. A. Johnson, *Encapsulated Thermoelectric Heat Pump and Method of Manufacture*. 1990, Google Patents.
67. Snyder, G. J. et al., Testing of milliwatt power source components. In *2002. Proceedings ICT'02: Twenty-First International Conference on Thermoelectrics*. 2002: IEEE, Piscataway, NJ. pp. 463–470.
68. Kim, I.-H., (Bi,Sb)$_2$(Te,Se)$_3$-based thin film thermoelectric generators. *Materials Letters*, 2000. **43**: pp. 221–224.
69. Navone, C. et al., Optimization and fabrication of a thick printed thermoelectric device. *Journal of Electronic Materials*, 2011. **40**(5): pp. 789–793.
70. Snyder, G. J. et al., Thermoelectric microdevice fabricated by a MEMS-like electrochemical process. *Nature Materials*, 2003. **2**(8): pp. 528–31.
71. Stark, I., and M. Stordeur, New micro thermoelectric devices based on bismuth telluride-type thin solid films. In *International Conference on Thermoelectrics*. 2000: IEEE, Piscataway, NJ. pp. 465–472.
72. Goncalves, L. M., P. Alpuim, and J. H. Correia, Fabrication of thermoelectric devices by applying microsystems technology. *Journal of Electronic Materials*, 2010. **39**(9): pp. 1516–1521.
73. Pleurial, J.-P. et al., Microfabricated thermoelectric power-generation devices. US patent: US6787691B2 (2004).
74. Lin, J. R. et al., Thermoelectric microdevice fabrication process and evaluation at the Jet Propulsion Laboratory (JPL). *Thermoelectrics, 2002. Proceedings ICT '02. Twenty-First International Conference on* (2002). pp. 535–539.
75. Harman, T. C., P. J. Taylor, M. P. Walsh, and B. E. LaForge, Quantum dot superlattice thermoelectric materials and devices. *Science*, 2002. **297**: pp. 2229–2232.
76. Kim, D.-H. et al., Effect of deposition temperature on the structural and thermoelectric properties of bismuth telluride thin films grown by co-sputtering. *Thin Solid Films*, 2006. **510**(1–2): pp. 148–153.
77. Kwon, S.-D. et al., Fabrication of bismuth telluride-based alloy thin film thermoelectric devices grown by metal organic chemical vapor deposition. *Journal of Electronic Materials*, 2009. **38**(7): pp. 920–924.
78. Makala, R. S., K. Jagannadham, and B. C. Sales, Pulsed laser deposition of Bi$_2$Te$_3$-based thermoelectric thin films. *Journal of Applied Physics*, 2003. **94**(6): p. 3907.
79. Takashiri, M. et al., Fabrication and characterization of bismuth-telluride-based alloy thin film thermoelectric generators by flash evaporation method. *Sensors and Actuators A: Physical*, 2007. **138**(2): pp. 329–334.
80. Carmo, J. P., L. M. Gonçalves, and J. H. Correia, Micro and nanodevices for thermoelectric converters, in *Scanning Probe Microscopy in Nanoscience and Nanotechnology 2* (Edited by Bharat Bhushan), 2011. pp. 791–812.
81. Dimitriadou, I. A. et al., Feasibility study on screen printing as a fabrication technique for low-cost thermoelectric devices, in *Proceedings of the 11th European Conference on Thermoelectrics*, 2014. pp. 177–182.
82. Shaughnessy, M. C. et al., Energetics and diffusion of gold in bismuth telluride-based thermoelectric compounds. *Journal of Applied Physics*, 2014. **115**(6): p. 063705.
83. Min, G., and D. M. Rowe, Ring-structured thermoelectric module. *Semiconductor Science and Technology*, 2007. **22**(8): pp. 880–883.

84. Venkatasubramanian, R., E. Siivola, T. Colpitts, and B. O'Quinn, Thin-film thermoelectric devices with high room-temperature figure of merit. *Nature*, 2001. **413**: pp. 597–601.

85. Lim, J. R. et al., Thermoelectric microdevice fabrication process and evaluation at Jet Propulsion Laboratory (JPL). In *21st International Conference on Thermoelectrics (2002)*. 2002: IEEE, Piscataway, NJ.

86. Bottner, H. et al., New thermoelectric components using microsystem technologies. *Journal of Microelectromechanical Systems*, 2004. **13**(3): pp. 414–420.

87. Glatz, W. et al., Bi_2Te_3 based flexible micro thermoelectric generator with optimized design. *Journal of Microelectromechanical Systems*, 2009. **18**(3): pp. 763–772.

88. Venkatasubramanian, R. et al., Thin-film superlattice thermoelectric devices for energy harvesting and thermal management. In *Thermoelectrics and its Energy Harvesting: Modules, Systems, and Applications in Thermoelectrics*, D. M. Rowe, Editor. 2012: CRC Press, Boca Raton, FL.

89. Fan, P. et al., The high performance of a thin film thermoelectric generator with heat flow running parallel to film surface. *Applied Physics Letters*, 2013. **102**(3): p. 033904.

90. Baba, S. et al., Formation and characterization of polyethylene terephthalate-based $(Bi_{0.15}Sb_{0.85})_2Te_3$ thermoelectric modules with $CoSb_3$ adhesion layer by aerosol deposition. *Journal of Alloys and Compounds*, 2014. **589**: pp. 56–60.

91. Madan, D. et al., Printed flexible thermoelectric generators for use on low levels of waste heat. *Applied Energy*, 2015. **156**: pp. 587–592.

92. Hu, X. et al., Power generation evaluated on a bismuth telluride unicouple module. *Journal of Electronic Materials*, 2015. **44**(6): pp. 1785–1790.

93. Martin-Gonzalez, M. et al., The state of the art on thermoelectric devices in Japan. *Materials Today: Proceedings*. 2015.

94. Komatsu, Komatsu to launch sales of the world's highest efficiency thermoelectric generation modules developed in-house. http://www.komatsu.com /CompanyInfo/press/2009012714011528411.html.

95. http://news.panasonic.com/global/press/data/en110630-4/en110630-4.html.

96. Hi-Z, http://www.hi-z.com/.

97. Tellurex, http://www.tellurex.com/.

98. Tecteg, http://www.tecteg.com/.

99. Thermalforce.de, http://www.thermalforce.de/.

100. Thermonamic, http://www.thermonamic.com/.

101. Hsu, C.-T. et al., Experiments and simulations on low-temperature waste heat harvesting system by thermoelectric power generators. *Applied Energy*, 2011. **88**(4): pp. 1291–1297.

102. Tom Schneider, R. A., D. Koester, and S. Lee, *Thin Film Thermoelectric Power Generation*. 2007: Nextreme Thermal Solution, Durham, NC.

103. http://docs-asia.electrocomponents.com/webdocs/1110/0900766b81110d9d.pdf.

104. Yan, X. et al., Experimental studies on anisotropic thermoelectric properties and structures of n-type $Bi_2Te_{2.7}Se_{0.3}$. *Nano Letters*, 2010. **10**: pp. 3373–8.

105. Hatzikraniotis, E. et al., Efficiency study of a commercial thermoelectric power generator (TEG) under thermal cycling. *Journal of Electronic Materials*, 2009. **39**(9): pp. 2112–2116.

106. Park, W., M. T. Barako, A. M. Marconnet, M. Asheghi, and K. E. Goodson, Effect of thermal cycling on commercial thermoelectric modules. In *13th IEEE Intersociety Conference on Thermal and Thermomechanical Phenomena in Electronic Systems (ITherm)*. 2012: IEEE, Piscataway, NJ.

CHAPTER 19

Half-Heusler Modules

Giri Raj Joshi, Yucheng Lan, and Zhifeng Ren

Contents

19.1 Introduction of Half-Heusler Materials

19.1.1 Literature Review

Half-Heusler (HH) is a kind of intermetallic compound with excellent TE properties. As discussed in Chapter 8 ("Half-Heuslers for High Temperatures"), various HHs have been discovered and characterized, such as MCoSb [1–13] and MNiSn-based compounds (where M = Ti, Zr, Hf) [14–26], as well as (Ti,Zr)CoSn [13] and (Nb,Ti)FeSb [27]. These HH compounds crystallize in the cubic MgAgAs-type structure with the space group of $F\bar{4}3m$, consisting of four interpenetrating face-centered-cubic sublattices. The crystallographic sites (0, 0, 0) and (1/4, 1/4, 1/4) are occupied by two different transition metal elements, the (1/2, 1/2, 1/2) site is occupied by sp elements such as Sn, Sb, or Bi, whereas site (3/4, 3/4, 3/4) is vacant. The (3/4, 3/4, 3/4) vacant site can be easily doped to induce mass fluctuation at atomic level in the crystals to scatter phonons, reducing thermal conductivity and exhibiting glass-like thermal conductivities. Additionally, each filled sublattice can be independently tuned through substitutions to optimize the HH materials, decreasing thermal conductivity. At the same time, the narrow bandgap semiconductors have 18 valence electrons and are excellent electron crystals, resulting in excellent electrical conductivity and large Seebeck coefficient. For instance, the Seebeck coefficients of TiNiSn, ZrNiSn, and HfNiSn HHs are in the range of 200–400 uV/K at room temperature [16]. Therefore HH compounds have attracted considerable attentions since the 1990s as a promising TE material [14–16,28,29], working around 700 K and above.

TE properties of HH materials have been improved by doping, substitutions, and alloying based on their unique crystallographic structures, significantly decreasing the thermal conductivity of these compounds while simultaneously maintaining their high power factors, leading to *ZT* values in the range of 1 [10,30–34]. The *ZT* value is comparable to other TE materials, such as Bi_2Te_3. Nanostructuring has been employed further to decrease the thermal conductivity of the HH materials by increasing phonon scattering by grain boundaries [12]. *ZT* is further enhanced.

Additionally, the HH compounds are thermally stable at high temperatures, with good mechanical properties besides excellent TE properties. Therefore, HH compounds have been widely applied in moderate-temperature TE devices for power generation. Physical properties of nanostructured HH materials have been reviewed [35–39].

Up to now, HH compounds have been fabricated into modules/devices because of their excellent properties. Table 19.1 summarizes some TE devices fabricated from the materials. Their maximum power density at a given temperature gradient, dimensions, and the number of legs of TE generators are listed.

There are three kinds of devices listed in Table 19.1. Unileg modules are the simplest structures. TE HH unicouples are fabricated from both p-type and n-type HH legs, such as n-type materials (Zr,Hf)NiSn and p-type materials (Zr,Hf)CoSb [40] and n-type ZrNiSn-based alloys and p-type FeNbSb HH compounds [41]. Hybrid TE modules are composed of an n- or p-type HH element, while another p- or n-type element is not composed of HH compounds. These kinds of modules can be fabricated into multicouple devices, such as power generation devices fabricated from p- and n-type nanostructured HH compounds [42]. The details of these devices are discussed in the chapter.

In Table 19.1, the device maximum efficiency η can be expressed by

$$\eta = \frac{\sqrt{1 + Z(T_h + T_c)/2} - 1}{\sqrt{1 + Z(T_h + T_c)/2} + T_c/T_h} \eta_{\text{Carnot}}, \tag{19.1}$$

where $\eta_{\text{Carnot}} = (T_h - T_c)/T_h$, $Z = S^2\sigma/\kappa$, S is the Seebeck coefficient, σ is the electrical conductivity, and κ is the thermal conductivity. T_h is the temperature of hot sides, and T_c is the temperature of cold sides. $\Delta T = T_h - T_c$. The highest device efficiency of 10% was achieved [43]. The built modules exhibited a maximum power density of 3.2 W/cm^2.

TABLE 19.1 Various Modules Fabricated from Half-Heusler Compounds

Type	HH Materials	Leg Number and Dimensions (mm)	ΔT and T_h(K)	Power Density (mW/cm²)	Power Density (mW/cm³)	Device Efficiency (%)	Ref.
Unileg	n-type $[Ti_{0.33}Zr_{0.33}Hf_{0.33}]NiSn$ $(ZT = 0.45)$	4, 2 × 2 × 4	565, 868	275	700	n/a	Populoh et al. [44]
p–n couple	p-type $(Hf_{0.3}Zr_{0.7})Co(Sn_{0.3}Sb_{0.7})$ nanocomposite $(ZT = 0.8)$/n-type $(Hf_{0.6}Zr_{0.4})Ni(Sn_{0.995}Sb_{0.005})$ nanocomposite $(ZT = 1.05)$	a p–n pair	657, 970	n/a	n/a	8.7	Poon et al. [42]
Unicouple	p-type HH $(ZT = 0.5)$/ n-type HH $(ZT = 0.9)$	a p–n pair, 5 × 5 × 7	422, 743	n/a	n/a	4.0	Yamamoto and Nagase [45]
Generator	p-type $Hf_{0.5}Zr_{0.5}CoSn_{0.2}Sb_{0.8}$ polycrystals/n-type $Ti_{0.6}Hf_{0.4}NiSn$ polycrystals	a single pair, 5 × 5 × 8	550, 873	n/a	n/a	5.0	Ngan et al. [46]
Generator	p-type $Hf_{0.5}Zr_{0.5}CoSb_{0.8}Sn_{0.2}$ nanocomposites/n-type $Hf_{0.75}Zr_{0.25}NiSn_{0.99}Sb_{0.01}$ nanocomposites	a single pair, 2.5 × 2.5 × 2	678, n/a	8.9	n/a	8.9	Joshi and Poudel [47]
Hybrid single couple	p-type TAGS/n-type $Hf_{0.6}Zr_{0.4}NiSn_{0.98}Sb_{0.02}$	a single pair, 1 × 1 × 2	523, 832	n/a	n/a	9.2	Bulman and Cook [43]
Hybrid module	p-type TAGS/n-type $Hf_{0.6}Zr_{0.4}NiSn_{0.98}Sb_{0.02}$	49-couples, 1 × 1 × 2	497, 873	n/a	n/a	10	Bulman and Cook [43]
Prototype module	p-type $FeNb_{1-x}Hf_xSb$ $(ZT = 1.5)$/n-type $FeNb_{1-y}Zr_ySb$	8 couple, n/a	655, 1000	2.2	n/a	6.2	Fu et al. [41]

19.1.2 Module Fabrication Challenges

In order to commercialize HH devices, several issues should be considered during module fabrication.

- *Thermal stability*: Long-term stability and reproducibility are very important for automotive and industrial waste heat recovery applications of TE modules/devices. In order to decrease the thermal conductivity of TE materials, doping and nanostructuring are usually employed. The produced TE materials are usually thermally unstable. Recently, the effect of thermal cycling upon the TE performance of p-type HH materials was investigated [48]. The long-term stability of the HH modules were examined under thermal shock tests [40]. It was reported that there is no degradation of either the TE materials or the soldered contacts after 1000 thermal shock cycles exposed to air (hot-side temperature T_h: 550°C; cold-side temperature T_c: 20°C; maximum heating rate: 20 K/s). The n-type $Ti_{0.3}Zr_{0.35}Hf_{0.35}NiSn$ and p-type $Ti_{0.26}Sc_{0.04}Zr_{0.35}Hf_{0.35}NiSn$ HH materials were also stable even after 500 cycles (1700 h) in a temperature range from 373 to 873 K [49]. The figure of merit ZT for both n- and p-type compounds does not significantly change. Therefore, HH materials are promising candidates for medium-temperature TE modules.
- *Cost*: HH usually contains the expensive element Hf, increasing the cost of HH devices. In order to lower the cost of HH modules, HH materials with reduced Hf content were investigated [13]. Furthermore, a kind of hafnium-free p-type TE compound with good ZT values was recently achieved [50]. The newly discovered HH compounds would lower the price of HH devices.
- *Contact and others*: Contact metallization on HH materials is challenging due to their application at higher temperatures (100–700°C). During applications, a continuous thermal cycling from hot-side to cold-side temperatures creates an enormous stress at the junction of HH and metal layers due to their dissimilar physical and mechanical properties, and ultimately the devices reach failure modes. It is important to identify a process of metallization for the materials having very low contact resistance and that are mechanically strong, which can sustain the long-term thermal cycling operations.

19.2 Half-Heusler Modules

19.2.1 Contact Development

Contact layers of metal are generally developed on TE materials to facilitate bonding between TE and metal header/electrical connection, and the basic requirement for a good contact on TE materials is to develop a strong bonding and very low contact resistance between the metal layer and TE materials. There are two methods reported to make an electrical connection on HH materials [40,41,43,47–51].

TABLE 19.2   Different Metallization Techniques and Corresponding Properties Published in Literature

Metallization Techniques	Metal Layer	Contact Resistance ($\times 10^{-6}$ Ω cm^2)	Bonding Strength in MPa	Ref.
Brazing	Copper	5	75	Joshi et al. [51]
Hot pressed	Incusil/silver	90	n/a	Ngan et al. [46]
Hot pressed	Titanium	1	50	Joshi and Poudel [47]

First common method is the brazing of HH with a metal header at high temperatures (>700°C). This technique provides a strong bonding as well as very low contact resistance between HH and metal header and can sustain the thermal stress due to long-term thermal cycling during applications. However, no reports have been published for long-term operation of more than 1000 thermal cycles, which is essential for the reliability of the modules.

The other method is the hot pressing of metal powder/foil with HH materials at high temperatures, which also provides a strong bonding and very low contact resistance between TE and metal layer. Due to strong bonding, this contact can sustain long-term thermal cycling operations. However, the mass production of such contact will be challenging.

The reported contact resistance and bonding strength from the metallization techniques mentioned earlier are summarized in Table 19.2.

19.2.2 Half-Heusler Module Performance

A typical TE module consists of a series/parallel connection of p–n leg pairs.

In general, such TE modules are characterized by measuring power output/power density, efficiency, and their stability under long-term operation. For a module to be used as a power generator, it should be placed in between a heat source and a heat sink. If ΔT is the temperature difference across the module due to heat source temperature (T_h) and heat sink temperatures (T_c), then the voltage generated due to temperature gradient ($\Delta T = T_h - T_c$) is

$$V = \alpha \Delta T, \tag{19.2}$$

where α is the Seebeck coefficient of the module. If R_L is the load resistance connected with the module to complete the circuit, and R_i is the internal resistance of the module, then the electrical current owing through the load resistance/module due to thermal gradient is

$$I = V/(R_L + R_i) = \alpha \Delta T/(R_L + R_i), \tag{19.3}$$

and the power output from the module is given by

$$P = I^2 R_L = \alpha^2 \Delta T^2 R_L/(R_L + R_i)^2. \tag{19.4}$$

If we neglect the second-order effects, the maximum output power P_{max} of a TE module is achieved if the load resistance R_L equals the internal resistance R_i, as given by

$$P = I^2 R_L = \alpha^2 \Delta T^2 / (4R_i). \tag{19.5}$$

In power generation mode, α^2/R_i represents the TE figure of merit (Z) of the module. From Equation 19.5, the power output of the module is directly proportional to the figure of merit of the module and ΔT across the module.

The efficiency (η) of a TE module is defined as the ratio of power output (P) across load resistance R_L to the heat flow through the module (Q):

$$\eta = P/Q = I^2 R_L / Q = [\alpha \Delta T / (R_L + R_i)]^2 R_L / Q. \tag{19.6}$$

Since the module performance also depends on the interconnects between different pairs of legs, only modules with more than five pairs of legs are discussed in this section to better represent the real application scenarios. In literature, two types of HH materials are used to fabricate HH modules: one with bulk materials and the other one is nanocomposite.

- *HH modules using bulk materials*: Bartholomé et al. reported 1.6 cm × 1.6 cm HH module using bulk materials which is prepared using vacuum induction melting process, followed by long-term annealing [40]. The module consists of 14 legs of the same size. The leg size is 2.5 × 2.5 × 3 mm^3, and they are connected with metal header using brazing technique. Table 19.3 summarizes the detail performance of the bulk HH modules while operating between temperatures of 20°C and 547°C.

  The power density from this module is low due to lower average ZT values across the operation temperatures and can be improved by improving the ZT of the materials. However, these modules are tested for long-term operation only up to 150 thermal cycles, which might not be enough to understand the reliability of the modules especially due to the lower mechanical strength of the bulk materials.

- *HH modules using bulk nanomaterials*: HH modules using bulk nanomaterials are reported with very high power outputs [41,51]. Table 19.4

TABLE 19.3 HH Module Performance Using Bulk TE Materials

Materials	Average ZT	Module Size	P_{max}(W)	Power Density (W/cm^2)	Predicted Efficiency (%)
n-type: $Zr_{0.4}Hf_{0.6}NiSn_{0.98}Sb_{0.02}$; p-type: $Zr_{0.5}Hf_{0.5}CoSb_{0.8}Sn_{0.2}$	0.44	1.6 cm × 1.6 cm	2.8	1.1	5

TABLE 19.4 HH Module Performance Using Nanostructured Bulk TE Materials

Materials	Average ZT	ΔT	Leg Size	Module Size	P_{max} (W)	Power Density (W/cm²)	Efficiency (%)	Ref.
n-type: ZrNiSn-based; p-type: FeNb$_{1-x}$Hf$_x$Sb	0.8	655	4 mm × 4 mm × 10 mm	2 cm × 2 cm	8.9	2.2	6.2	Fu et al. [41]
n-type: Hf$_{0.25}$Zr$_{0.25}$Ti$_{0.25}$NiSn$_{0.99}$Sb$_{0.01}$; p-type: Hf$_{0.5}$Zr$_{0.5}$CoSb$_{0.8}$Sn$_{0.2}$	0.65	500	1.8 mm × 1.8 mm × 2 mm	2.65 cm × 2.65 cm	15.5	2.2	4.5	Joshi et al. [51]

presents a list of high-performance HH modules using nanostructured materials. Most of these modules are fabricated by brazing HH legs with copper-bonded ceramic substrates [52].

Out of the modules presented in Tables 19.3 and 19.4, the HH module presented by Joshi et al. [51] was manufactured in commercial scale. Figure 19.1 presents the typical performance curves of such commercial HH module fabricated using nanostructured TE materials [51]. This module contains 60 p- and 60 n-type legs with dimension of $1.8 \times 1.8 \times 2$ mm^3.

A maximum power of 15 W was produced (power density of 2.2 W/cm^2) with a temperature difference of 500°C and cold-side temperature of 100°C. These modules have a very high power density compared to the other competitive TE materials, which makes the HH material system as a favorable choice to be used in mid-high-temperature applications. Furthermore, these modules show stable module properties after long-term operations of up to 1000 thermal cycles and 1000 h of soak test (Figure 19.2). The stable long-term performance is due to the excellent mechanical integrity of the nanostructured HH materials [51].

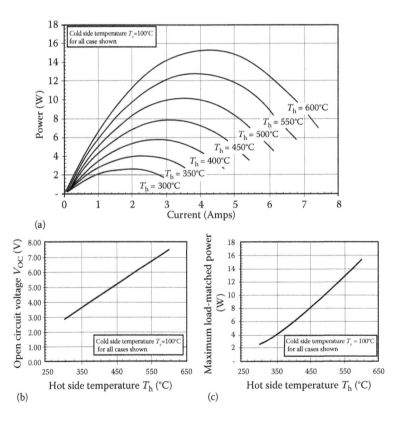

(a)

(b)

(c)

FIGURE 19.1 (a) Power vs. current curves, (b) open circuit voltage vs. hot-side temperature, and (c) minimum load-matched power vs. hot-side temperature of 60-couple HH modules with nanostructured materials. Cold side temperature T_c is 100°C for all cases shown.

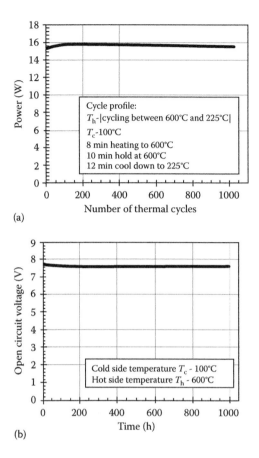

(a)

(b)

FIGURE 19.2 Long-term (a) thermal cycling test and (b) soak test of nanostructured HH module.

19.3 Half-Heusler Module Applications

HH modules are best suited for mid-high-temperature application (200–700°C) such as waste heat recovery from the exhaust of vehicles and industrial waste heat due to their very high power density (>2 W/cm^2) and thermal and mechanical stability. Reports of successful application of HH modules to produce up to 1 kW of power have been published in literature [53,54].

Nanostructured HH modules [51] have been used to generate power from vehicle exhaust heat (>550°C) [53] and residential water boilers, exploiting the temperature of burning natural gas (>1000°C) [54].

Figure 19.3 presents the average hot- and cold-side temperatures of a TEG fabricated using nanostructured HH modules for (1) waste heat recovery of vehicles with respect to inlet exhaust temperature and (2) TEG system power and voltage output with respect to current. A maximum power output of 1 kW (power density of 1.4 W/cm^2) was produced with temperature difference of 370°C [53].

Table 19.5 summarizes the reported results from the different applications of nanostructured HH modules in TEG.

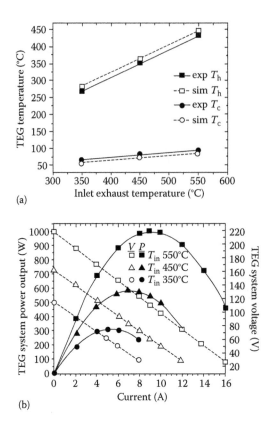

(a)

(b)

FIGURE 19.3 (a) Average hot (T_h)- and cold (T_c)-side temperatures of TEG vs. inlet exhaust temperature (T_{in}) and (b) TEG power and voltage output vs. current.

TABLE 19.5 Results of Application of Nanostructured HH Modules

Application	Module Hot Side	Module Cold Side	No. of Module Used	Total Electric Power Generated (W)	Efficiency (%)	Ref.
Vehicle exhaust	450°C	80°C	400	1140	n/a	Zhang et al. [53]
Residential boilers (Micro CHP)	600°C	100°C	16	186	4	Zhang et al. [54]

Source: Joshi, G. et al., High temperature half-Heusler modules for power generation applications. *International Conference of Thermoelectrics*, Dresden, Germany, 2015.

19.4 Summary

HH modules are one of the reliable TE modules with very high power density (>2 W/cm^2) and efficiency (>4%) for mid-high-temperature applications (200–700°C). Bulk and nanostructured HH materials have been used to fabricate such modules. The modules with nanostructured materials are reported to be mechanically strong for module integrity, which provides thermally and mechanically stable modules for long-term operation. The high power density of the HH modules is due to good *ZT* of the materials, very low contact resistance of the device, and excellent mechanical properties. Contact resistance as low as 5 μΩ cm^2 and bonding strength higher than 75 MPa have been reported. This contact resistance is very low and strength is very high compared to other competitive TE materials for mid-temperature applications. These properties allow the HH modules to be best suitable for applications below 700°C.

References

1. R. Kuentzler, R. Clad, G. Schmerber, and Y. Dossmann, Gap at the Fermi level and magnetism in RMSn ternary compounds (R = Ti, Zr, Hf and M = Fe, Co, Ni). *Journal of Magnetism and Magnetic Materials*, 1992, 104, 1976–1978.

2. J. Tobola, J. Pierre, S. Kaprzyk, R. V. Skolozdra, and M. A. Kouacou, Crossover from semiconductor to magnetic metal in semi-Heusler phases as a function of valence electron concentration. *Journal of Physics: Condensed Matter*, 1998, 10, 1013.

3. Y. Xia, S. Bhattacharya, V. Ponnambalam, A. L. Pope, S. J. Poon, and T. M. Tritt, Thermoelectric properties of semimetallic (Zr, Hf)CoSb half-Heusler phases. *Journal of Applied Physics*, 2000, 88, 1952–1955.

4. Y. Xia, V. Ponnambalam, S. Bhattacharya, A. L. Pope, S. J. Poon, and T. M. Tritt, Electrical transport properties of TiCoSb half-Heusler phases that exhibit high resistivity. *Journal of Physics: Condensed Matter*, 2001, 13, 77.

5. M. Zhou, L. Chen, W. Zhang, and C. Feng, Disorder scattering effect on the high-temperature lattice thermal conductivity of TiCoSb-based half-Heusler compounds. *Journal of Applied Physics*, 2005, 98, 013708.

6. M. Zhou, C. Feng, L. Chen, and X. Huang, Effects of partial substitution of Co by Ni on the high-temperature thermoelectric properties of TiCoSb-based half-Heusler compounds. *Journal of Alloys Compounds*, 2005, 391, 194–197.

7. T. Wu, W. Jiang, X. Li, Y. Zhou, and L. Chen, Thermoelectric properties of *p*-type Fe-doped TiCoSb half-Heusler compounds. *Journal of Applied Physics*, 2007, 102, 103705.

8. M. Zhou, L. Chen, C. Feng, D. Wang, and J.-F. Li, Moderate-temperature thermoelectric properties of TiCoSb-based half-Heusler compounds Ti$_{1-x}$Ta$_x$ CoSb. *Journal of Applied Physics*, 2007, 101, 113714.

9. W. J. Xie, X. F. Tang, and Q. J. Zhang, Fast preparation and thermal transport property of TiCoSb-based half-Heusler compounds. *Chinese Physics*, 2007, 16, 3549–3552.

10. W. Xie, Q. Jin, and X. Tang, The preparation and thermoelectric properties of $Ti_{0.5}Zr_{0.25}Hf_{0.25}Co_{1-x}Ni_xSb$ half-Heusler compounds. *Journal of Applied Physics*, 2008, 103, 043711.

11. P. Qiu, X. Huang, X. Chen, and L. Chen, Enhanced thermoelectric performance by the combination of alloying and doping in TiCoSb-based half-Heusler compounds. *Journal of Applied Physics*, 2009, 106, 103703.

12. X. Yan, G. Joshi, W. Liu, Y. Lan, H. Wang, S. Lee, J. W. Simonson et al., Enhanced thermoelectric figure of merit of *p*-type half-Heuslers. *Nano Letters*, 2011, 11, 556–560.

13. R. He, H. S. Kim, Y. Lan, D. Wang, S. Chen, and Z. Ren, Investigating the thermoelectric properties of *p*-type half-Heusler $Hf_x(ZrTi)_{1-x}CoSb_{0.8}Sn_{0.2}$ by reducing Hf concentration for power generation. *RSC Advances*, 2014, 4, 64711–64716.

14. F. G. Aliev, V. V. Kozyrkov, V. V. Moshchalkov, R. V. Scolozdra, and K. Durczewski, Narrow band in the intermetallic compounds MNiSn (*M* = Ti, Zr, Hf) *Zeitschrift fur Physik B: Condensed Matter*, 1990, 80, 353–357.

15. H. Hohl, A. Ramirez, W. Kaefer, K. Fess, C. Thurner, C. Kloc, and E. Bucher, A new class of materials with promising thermoelectric properties: MNiSn (*M* = Ti, Zr, Hf). *Symposium Q: Thermoelectric Materials–New Directions and Approaches*, 1997, p. 109.

16. C. Uher, J. Yang, S. Hu, D. T. Morelli, and G. P. Meisner, Transport properties of pure and doped MNiSn (*M* = Zr, Hf). *Physical Review B: Condensed Matter and Materials Physics*, 1999, 59, 8615–8621.

17. S. Bhattacharya, A. L. Pope, R. T. Littleton, T. M. Tritt, V. Ponnambalam, Y. Xia, and S. J. Poon, Effect of Sb doping on the thermoelectric properties of Ti-based half-Heusler compounds, $TiNiSn_{1-x}Sb_x$. *Applied Physics Letters*, 2000, 77, 2476–2478.

18. Q. Shen, L. Chen, T. Goto, T. Hirai, J. Yang, G. P. Meisner, and C. Uher, Effects of partial substitution of Ni by Pd on the thermoelectric properties of ZrNiSn-based half-Heusler compounds. *Applied Physics Letters*, 2001, 79, 4165–4167.

19. S. Katsuyama, H. Matsushima, and M. Ito, Effect of substitution for Ni by Co and/or Cu on the thermoelectric properties of half-Heusler ZrNiSn. *Journal of Alloys Compounds*, 2004, 385, 232–237.

20. H. Muta, T. Yamaguchi, K. Kurosaki, and S. Yamanaka, Thermoelectric properties of ZrNiSn based half-Heusler compounds. *ICT 2005. 24th International Conference on Thermoelectrics, 2005*, 2005, pp. 351–354.

21. H. Muta, T. Kanemitsu, K. Kurosaki, and S. Yamanaka, Substitution effect on thermoelectric properties of ZrNiSn based half-Heusler compounds. *Materials Transactions*, 2006, 47, 1453–1457.

22. S. Katsuyama, R. Matsuo, and M. Ito, Thermoelectric properties of half-Heusler alloys $Zr_{1-x}Y_xNiSn_{1-y}Sb_y$. *Journal of Alloys Compounds*, 2007, 428, 262–267.

23. C. Yu, T. J. Zhu, R.-Z. Shi, Y. Zhang, X.-B. Zhao, and J. He, High-performance half-Heusler thermoelectric materials $Hf_{1-x}Zr_xNiSn_{1-y}Sb_y$ prepared by levitation melting and spark plasma sintering. *Acta Materialia*, 2009, 57, 2757–2764.

24. V. A. Romaka, D. Fruchart, V. V. Romaka, E. K. Hlil, Y. V. Stadnyk, Y. K. Gorelenko, and L. G. Akselrud, Features of the structural, electrokinetic, and magnetic properties of the heavily doped ZrNiSn semiconductor: Dy acceptor impurity. *Semiconductors*, 2009, 43, 7–13.

25. P. Qiu, J. Yang, X. Huang, X. Chen, and L. Chen, Effect of antisite defects on band structure and thermoelectric performance of ZrNiSn half-Heusler alloys. *Applied Physics Letters*, 2010, 96, 152105.

26. T. J. Zhu, K. Xiao, C. Yu, J. J. Shen, S. H. Yang, A. J. Zhou, X. B. Zhao, and J. He, Effects of yttrium doping on the thermoelectric properties of $Hf_{0.6}Zr_{0.4}NiSn_{0.98}Sb_{0.02}$ half-Heusler alloys. *Journal of Applied Physics*, 2010, 108, 044903.

27. G. Joshi, R. He, M. Engber, G. Samsonidze, T. Pantha, E. Dahal, K. Dahal et al., NbFeSb-based *p*-type half-Heuslers for power generation applications. *Energy and Environmental Science*, 2014, 7, 4070–4076.

28. G. A. Slack, New Materials and Performance Limits for Thermoelectric Cooling. In *New Materials and Performance Limits for Thermoelectric Cooling*, ed. D. M. Rowe, CRC Press, Boca Raton, FL, 1995.

29. T. M. Tritt, M. L. Wilson, A. L. Johnson, S. L. Gault, and R. Stroud, Potential of quasicrystals and quasicrystal approximants for new and improved thermoelectric materials. *Proceedings ICT'97. XVI International Conference on Thermoelectrics, 1997*, Dresden, 1997, pp. 454–458.

30. S. Sakurada and N. Shutoh, Effect of Ti substitution on the thermoelectric properties of (Zr,Hf)NiSn half-Heusler compounds. *Applied Physics Letters*, 2005, 86, 082105.

31. S. R. Culp, S. J. Poon, N. Hickman, T. M. Tritt, and J. Blumm, Effect of substitutions on the thermoelectric figure of merit of half-Heusler phases at 800°C. *Applied Physics Letters*, 2006, 88, 042106.

32. S. R. Culp, J. W. Simonson, S. J. Poon, V. Ponnambalam, J. Edwards, and T. M. Tritt, (Zr,Hf)Co(Sb,Sn) half-Heusler phases as high-temperature 700°C *p*-type thermoelectric materials. *Applied Physics Letters*, 2008, 93, 022105.

33. W. Xie, J. He, S. Zhu, X. Su, S. Wang, T. Holgate, J. Graff et al., Simultaneously optimizing the independent thermoelectric properties in (Ti,Zr,Hf)(Co,Ni)Sb alloy by *in situ* forming InSb nanoinclusions. *Acta Materialia*, 2010, 58, 4705–4713.

34. S. Populoh, M. Aguirre, O. Brunko, K. Galazka, Y. Lu, and A. Weidenka, High figure of merit in (Ti,Zr,Hf)NiSn half-Heusler alloys. *Scripta Materialia*, 2012, 66, 1073–1076.

35. G. S. Nolas, J. Poon, and M. Kanatzidis, Recent developments in bulk thermoelectric materials. *MRS Bulletin*, 2006, 31, 199–205.

36. Y. Lan, A. J. Minnich, G. Chen, and Z. Ren, Enhancement of thermoelectric figure-of-merit by a bulk nanostructuring approach. *Advanced Functional Materials*, 2010, 20, 357–376.

37. W. Liu, X. Yan, G. Chen, and Z. Ren, Recent advances in thermoelectric nanocomposites. *Nano Energy*, 2012, 1, 42–56.

38. W. Xie, A. Weidenkaff, X. Tang, Q. Zhang, J. Poon, and T. M. Tritt, Recent advances in nanostructured thermoelectric half-Heusler compounds. *Nanomaterials*, 2012, 2, 379.

39. S. Chen and Z. Ren, Recent progress of half-Heusler for moderate temperature thermoelectric applications. *Materials Today*, 2013, 16, 387–395.

40. K. Bartholomé, B. Balke, D. Zuckermann, M. Köhne, M. Müller, K. Tarantik, and J. K. König, Thermoelectric modules based on half-Heusler materials produced in large quantities. *Journal of Electronic Materials*, 2014, 43, 1775–1781.

41. C. Fu, S. Bai, Y. Liu, Y. Tang, L. Chen, X. Zhao, and T. Zhu, Realizing high figure of merit in heavy-band *p*-type half-Heusler thermoelectric materials. *Nature Communications*, 2015, 6, 1.

42. S. J. Poon, D. Wu, S. Zhu, W. Xie, T. M. Tritt, P. Thomas, and R. Venkatasubramanian, Half-Heusler phases and nanocomposites as emerging high-ZT thermoelectric materials. *Journal of Materials Research*, 2011, 26, 2795–2802.

43. G. Bulman and B. Cook, High-efficiency energy harvesting using TAGS-85/half-Heusler thermoelectric devices. *Proceedings of SPIE: Energy Harvesting and Storage: Materials, Devices, and Applications V*, 2014, p. 911507.

44. S. Populoh, O. Brunko, K. Galazka, W. Xie, and A. Weidenka, Half-Heusler (TiZrHf)NiSn unileg module with high powder density. *Materials*, 2013, 6, 1326–1332.

45. X. Hu, A. Yamamoto, and K. Nagase, Characterization of half-Heusler unicouple for thermoelectric conversion. *Journal of Applied Physics*, 2015, 117, 225102.

46. P. H. Ngan, N. Van Nong, L. T. Hung, B. Balke, L. Han, E. M. J. Hedegaard, S. Linderoth, and N. Pryds, On the challenges of reducing contact resistances in thermoelectric generators based on half-Heusler alloys. *Journal of Electronic Materials*, 2016, 45, 594–601.

47. G. Joshi and B. Poudel, Efficient and robust thermoelectric power generation device using hot-pressed metal contacts on nanostructured half-Heusler alloys. *Journal of Electronic Materials*, 2016, 45, 1–5.

48. E. Rausch, B. Balke, S. Ouardi, and C. Felser, Long-term stability of (Ti/Zr/Hf)CoSb$_{1-x}$Sn$_x$ thermoelectric p-type half-Heusler compounds upon thermal cycling. *Energy Technology*, 2015, 3, 1217–1224.

49. J. Krez, B. Balke, S. Ouardi, S. Selle, T. Hoche, C. Felser, W. Hermes, and M. Schwind, Long-term stability of phase-separated half-Heusler compounds. *Physical Chemistry Chemical Physics*, 2015, 17, 29854–29858.

50. C. Fu, T. Zhu, Y. Liu, H. Xie and X. Zhao, Band engineering of high performance p-type FeNbSb based half-Heusler thermoelectric materials for figure of merit zT > 1. *Energy and Environmental Science*, 2015, 8, 216–220.

51. G. Joshi, P. Banjade, J. Keane, S. Parera, B. Poudel, and C. Ballinger, High temperature half-Heusler modules for power generation applications. *International Conference of Thermoelectrics*, Dresden, 2015.

52. X. Wang and J. D'Angelo, Thermoelectric Module and Method of Making Same. US Patent Application: US20140360549, 2014.

53. Y. Zhang, M. Cleary, X. Wang, N. Kempf, L. Schoensee, J. Yang, G. Joshi, and L. Meda, High-temperature and high-power-density nanostructured thermoelectric generator for automotive waste heat recovery. *Energy Conversion and Management*, 2015, 105, 946–950.

54. Y. Zhang, X. Wang, M. Cleary, L. Schoensee, N. Kempf, and J. Richardson, High-performance nanostructured thermoelectric generators for micro combined heat and power systems. *Applied Thermal Engineering*, 2016, 96, 83–87.

CHAPTER **20**

Silicide Modules
Practical Issues in Developing Mg₂Si with Good Stability for Generating Power from Waste Heat Sources

Tsutomu Iida, Yasuo Kogo, Atsuo Yasumori,
Keishi Nishio, and Naomi Hirayama

Contents

20.1 Introduction

Magnesium silicide (Mg_2Si) in the antifluorite structure has been suggested as a TE material for future generations of thermoelectric generators (TEGs) operating at mid-high-temperatures (600–900 K). Mg_2Si has the benefit of being environmentally benign and has sufficient performance to enable it to improve the energy efficiency of combustion systems, such as automotive engines, boilers, and industrial furnaces. In contrast to other mid-high-temperature-range TE materials, simple Mg_2Si and Mg_2Si-based alloys have the merits of being nontoxic and sustainable, and moreover, the cost of production is low. Since TEGs are devices with relatively high mass, a significant amount of material is needed for practical production processes; thus, the relative abundance of its constituent elements is a key feature of Mg_2Si, and this will help to broaden its application in the field of TE devices and encourage the development of waste heat recovery systems. The use of Mg_2Si as a possible TE material was suggested many years ago [1–14]. It has the benefit of a narrow bandgap

[1–3,9–14] and is expected to operate as a TE material at mid-temperatures, and furthermore, it is known that impurity doping can enhance its TE properties [4–8,12–29]. In order to maximize the TE properties of Mg_2Si-related TE materials, theoretical calculations have also been made [6,19,30–37].

For practical TE applications, the tuning of the electrical conductivity of the base material is an important issue, and in this regard, doping Mg_2Si is possible simply by incorporating impurities. However, theoretical studies using calculations based on first principles have shown that interstitial Mg defects have stable states under thermodynamically stable conditions, exhibiting donor-like behavior [30]. This feature better promotes n-type material and is a critical factor when doping to produce p-type Mg_2Si. Regarding this, many experimental results have been reported. A quantitative analysis of magnesium silicides with various compositions (i.e., $Mg_{2-x}Si$ with various values of x) using XRD revealed that the Mg ions at interstitial 4B sites were stable even in Mg-deficient material, and the probability of occupation at an interstitial site was 0.5% and independent of the Mg–Si composition [38]. The introduction of interstitial Mg ions in Mg_2Si by means of irradiation with fast neutrons had an effect on the electrical conductivity, and subsequent annealing at approximately >623 K changed the electronic transition, which is associated with a significant variation in the carrier concentration. The electronic transition observed after a postirradiation heat treatment was seen to correlate with the relaxation process of the interstitial Mg ions. The interstitial Mg ions and associated Mg vacancies are defects introduced by the radiation; thus, the relaxation and recovery processes observed may include radiation damage; however, the possibility of variations in the interstitial Mg sites can be seen in the experimental results [39].

One of the most common methods for fabricating Mg_2Si TE power generators starts with the synthesis of metallic Mg and Si using well-defined sintering processes [40,41]. Durability is an important issue for TE devices, and one problem when operating Mg_2Si TE devices in air at mid-temperatures is the oxidation of the Mg_2Si, which is usually fatal for TE devices. This is predominantly the oxidation of residual metallic Mg in the synthesized matrix. This metallic Mg can easily react to form MgO, eroding the Mg_2Si, giving rise to the degradation of the electrical and mechanical properties, and finally inducing the decomposition of the Mg_2Si. The metallic Mg can be successfully reduced by using a well-configured all-molten synthesis method. Using this method, the residual Mg can be completely eradicated, and good durability even in air at an operating temperature of ~900 K has been achieved [8,42]. Starting with nanosized Mg_2Si particles is desirable for TE devices since this helps reduce the thermal conductivity; however, even with the use of a durable Mg_2Si source material, there are many difficulties in forming nanosized (10–100 nm) particles of Mg_2Si without oxidation using the common planetary BM method.

Significant enhancement of the PF by applying pressure to doped Mg_2Si has been observed [43]. The pressure applied was 15 GPa, providing ~1% shrinkage of the lattice, and it was shown that the PF increased by a factor of 4 to ~8 × 10^{-3} W/m K^2, brought about by a 10-fold increase in electrical conductivity. This implies that incorporating dopants with large ionic radii can provide local lattice compression, resulting in increased TE PFs.

n-Type Mg$_2$Si with a good PF and a sufficiently high dimensionless figure of merit *ZT* has been demonstrated, while the value of *ZT* for p-type material was found to be insufficient. This correlates with the formation of native donor-like interstitial defects under equilibrium conditions. For p-type doping of Mg$_2$Si, there are several reports of Ag being used in substitutional sites [44]; however, sufficient electrical activation was not realized. Instead of using Ag, the group 2 element Sr may work as a functional acceptor. Material starting with a mixture of Mg$_2$Si and Mg$_2$Sr, and then sintered using a PAS process, has been made [45]. The conduction type of the synthesized Mg$_2$Si$_{1-x}$Sr$_x$ matrix varied as a function of the composition, changing from n-type to p-type with increasing *x*. Practical p-type characteristics were obtained with *x* > 0.3. The latest achievements regarding Mg$_2$Si$_{1-x}$Sr$_x$ are increased output with higher PFs and *ZT* values. For a possible Mg$_2$Si-based TEG, Mg$_2$Si can be used for the n-type leg, while there is a choice of p-type legs with TE properties that coordinate with n-type Mg$_2$Si. As things stand, suggested p-type materials include the newly exploited tetrahedrite compounds [45–50], higher manganese silicide [51–58], and SrMgSi [59]. On the other hand, unileg structures using solely n-Mg$_2$Si, which avoid troublesome issues such as the different thermal expansion characteristics and different TE properties, are possible.

In this chapter, we examine the following features with a view of implementing them in Mg$_2$Si TE devices: (1) impurity doping in order to enhance the TE properties of Mg$_2$Si, mainly for n-type material; (2) appropriate electrodes with sufficient low contact resistance; (3) problems due to degradation, the durability of Mg$_2$Si, and the formation of a stable passivated surface; and (4) possible etchants for Mg$_2$Si to clean and stabilize the surface for later processing.

20.2 Impurity Doping (n-Type)

An important aspect of Mg$_2$Si is its capability of being doped in order to modify its electrical conductivity, thermal conductivity, and durability at elevated operating temperatures. As has been examined and mentioned earlier, Mg$_2$Si commonly exhibits n-type conductivity due to donor-like interstitial sites of Mg formed when in it is in thermodynamic equilibrium. For n-type conductivity, the group 13 element Al and group 15 elements Bi and Sb are well-known n-type dopants for Mg$_2$Si [4,7,8,12,60]. In the case of Bi doping, *ZT* > 1 has been obtained, while the doped matrix rapidly deteriorated at elevated temperatures. Al-doped Mg$_2$Si is easy to process but gives only a moderate *ZT* value of ~0.7. The current prominent dopant for Mg$_2$Si in terms of obtaining a good PF and low thermal conductivity is Sb, even though it is not a completely toxic-free material and Sb-doped Mg$_2$Si is difficult to sinter. The Seebeck coefficient, electrical resistivity, and thermal conductivity of polycrystalline $Mg_2Si_{1-x}Sb_x$ with $0 \leq x \leq 0.37$ has been investigated, where it was found that doping with Sb decreases the thermal conductivity. On the other hand, increasing the amount of Sb brings about the formation of Mg vacancies, especially for highly doped specimens [60].

In view of the material development, increasing the values of such properties as the dimensionless figure of merit (ZT) and the PF at given operating temperatures is unquestionably important. There has been much work published regarding the common doping characteristics. Subsequent endeavors have been toward implementing a TEG setup for practical TE applications. Using our recently developed Mg_2Si TE chip fabrication process, which includes the synthesis of the materials, doping, sintering, and surface treatments, we produced n-type Mg_2Si devices with high power generation densities and low degradation, even in air at elevated temperatures. These characteristics of n-type Mg_2Si are now ready for translation to practical TEGs. Here, we examine Sb, the most promising n-type impurity for Mg_2Si and look for ways to transfer this to a practical production process.

Sb-doped Mg_2Si is known to be the most stable form of n-type Mg_2Si even at elevated temperatures and provides material with adequate TE properties with sufficiently low resistivity; therefore, we focus on Sb as the best possible dopant for realizing durable mass-produced TE chips. Actually, Sb is the principal n-type dopant in Si technology for heavily doped n-type substrates on the order of $\sim 10^{21}$ cm^{-3}, which are used in high-power device applications. Typical n-type dopants at substitutional Si sites in Mg_2Si can be compared with those dopants in Si technology. The solid solubility of Sb in Mg_2Si is reportedly $\sim 10^{21}$ cm^{-3}, but the typical electrical activation is less than $\sim 10^{20}$ cm^{-3}, which is rather low compared with that in Si [60,61].

Currently, process-reliable and thermally durable Mg_2Si source material is mainly fabricated using a melt synthesis method. In order to obtain better durability at operating temperatures of ~ 900 K in an air ambient, complete reduction of the residual metallic Mg during the melt synthesis is needed. Better TE properties at temperatures of >800 K are obtained using sintered pellets of Mg_2Si using an SPS method at ~ 1400 K and starting with all-molten synthesized polycrystalline Mg_2Si, which has previously been pulverized to <50 μm. A sintering process using a conventional hot press has been used to compact Mg_2Si powder, while the typically required longer process times and higher process temperatures than those used in the SPS may give rise to deterioration mainly due to partial oxidization of the Mg_2Si. Regarding the SPS of Sb-doped Mg_2Si, the sintering and reproducibility has proven to be difficult in terms of scalability if the doping content is beyond 3 at%.

To improve the sintering procedure for Sb-doped Mg_2Si, the inclusion of a metallic binder, such as Ni, Cu, Zn, and Al, during sintering improves the reproducibility and scalability of the sintering process [62]. Pellets of up to ~ 50 μm in diameter have been formed, with improved Young's modulus, bending strength, Vickers hardness, and fracture toughness [63].

Regarding the increase in the value of *ZT* by impurity doping, electrically active n-type dopant atoms at substitutional Si sites can increase the electrical conductivity, while variations in local lattice constants can reduce the thermal conductivity if the ionic radii of the dopant atom is larger than that of Si; the importance of which can be seen in

$$ZT = \frac{S^2 \sigma}{\kappa} T, \qquad (20.1)$$

where S is the Seebeck coefficient, σ is the electrical conductivity, and κ is the thermal conductivity. Since the thermal conductivity characteristics of Mg_2Si are neither attractive nor competitive compared with other mid-temperature TE materials, such as lead telluride, SKU, and HH, it makes sense to dope Mg_2Si with Sb. Not only the large ionic radii of Sb but also the thermodynamic stability and increased phonon scattering in Mg_2Si with higher concentrations of Sb make this a viable proposition [12,64]. Increasing the dopant concentration of Sb can effectively lower the thermal conductivity, while it brings about an increase in electrical conductivity because of the electrical activation of Sb at substitutional Si sites, implying that a large current density can be obtained. Increasing the value of *ZT* is seen as being rather good from the materials science point of view. However, TE chips with low output voltages and large currents make designing a practical TEG difficult. A possible approach for this issue is additional doping with an isoelectric impurity, with which the thermal conductivity under normal conditions can be varied, while the electrical conductivity remains unaffected.

One possible isoelectric impurity for Sb-doped Mg_2Si is the group 12 element Zn, which is expected to predominantly be a substituent atom in place of Mg. Figure 20.1 shows the results of the temperature-dependent PF and thermal conductivity of Zn-doped Mg_2Si. The Zn dopant was incorporated during a melt synthesis process at 1378 K, and the resultant polycrystalline Mg_2Si was pulverized to powder with a size of 25–50 μm, then placed into a graphite die, and sintered by a PAS technique using an ELENIX Ed-PAS-III-Es. The sintering was performed at a pressure of 60 MPa in an Ar (0.06 MPa) atmosphere, and the sintering temperature and time (typically ~1153 K for 15 min) were varied depending on the type of impurity. The temperature-dependent TE properties were measured using an ADVANCE-RIKO ZEM3 to determine the Seebeck coefficient (S) and the electrical conductivity (σ), and an ADVANCE-RIKO TC-1200H to determine the thermal conductivity (κ) over a temperature.

It is clear that the PF observed for the Zn-doped samples exhibits no significant concentration dependence and the values are comparable to those of the undoped one. Moreover, the incorporation of Zn has led to a decrease in thermal conductivity in the lower temperature range from room temperature up to ~600 K, which further decreases with increasing Zn. At temperatures

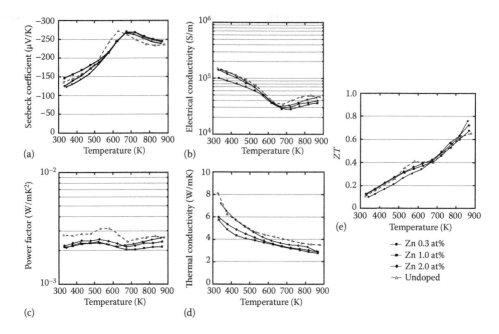

FIGURE 20.1 (a) Temperature-dependent Seebeck coefficient, (b) electrical conductivity, (c) calculated PF, (d) thermal conductivity, and (e) ZT value of specimens doped only with Zn compared with undoped and Sb-doped ones. The concentrations of Zn are either 0.3, 1.0, or 2.0 at%.

above ~700 K, the observed thermal conductivity is comparable to that of the Sb-doped specimen. The results of the calculated ZT values of the Zn-doped specimens for various Zn concentrations are also shown in Figure 20.1. It can be seen that there is no notable correlation between the observed ZT values and the amount of Zn in the Mg_2Si. Reducing the thermal conductivity while maintaining the electrical conductivity can be useful for tuning the different TE properties, i.e., independently varying the numerator and denominator in Equation 20.1. A clear use of Zn doping with regard to realizing a practical TEG design is its ability to preserve current generation while lowering the thermal conductivity.

The effect of codoping with Sb and the isoelectric impurity zinc (Zn) on the TE properties was examined. We expect (1) low thermal conductivity and high electrical activation due to the donor impurity of Sb and (2) a decrease in the thermal conductivity while maintaining the electrical conductivity due to the addition of Zn. The Sb and Zn dopants were incorporated during a melt synthesis process at 1378 K. The concentrations of the Sb and Zn were varied from 0.5 to 1.0 at%, respectively. The codoped specimens typically had no cracks and a relative density of 98% or more, indicating that the sinterability was improved and the reproducibility was better than that of simple Sb-doped samples. The incorporation of Zn along with Sb promotes the reproducibility of the sintering process for even higher Sb-doping concentrations, up to 1.0 at%, while doping with Sb alone in our experiment remained as unsuccessful as ever.

In order to understand the morphological microstructure and the elemental distribution, the polished surfaces of a sample codoped with Sb (1.0 at%) and Zn (0.5 at%) were examined using a field-emission scanning electron microscope (FE-SEM) JEOL JSM-7500F, and the elemental contents at given points were analyzed by FE-SEM EDX measurements. Elemental Zn could barely be detected in the Mg_2Si grains, although there were traces of it in the grain boundaries. So it might be that some of the Zn had precipitated in a metallic form, while we could also see that the Zn was not widely distributed around the clearly precipitated portions. The FE-SEM EDX measurements revealed that definite Mg_2Si grains and some other substances appeared at the grain boundaries. Codoping with Sb and Zn dramatically improved the sintering reproducibility with yields of up to 100%, while the basic virtues of durability and low resistivity associated with Sb-doped Mg_2Si were retained.

The results of the temperature-dependent Seebeck coefficient, electrical conductivity, and calculated PF of sintered samples doped with Sb and Zn are plotted in Figure 20.2, over the temperature range from 350 to 873 K. For comparison, the results of a specimen doped with only Sb, with an Sb concentration of 0.5 at%, are also shown. Basically, the samples that are codoped with Sb and Zn display n-type conductivity, since the observed Seebeck coefficient is negative for all samples over the whole measured temperature range. Recent valence electron density distributions determined from the structure factors of the synchrotron X-ray diffraction patterns using the maximum

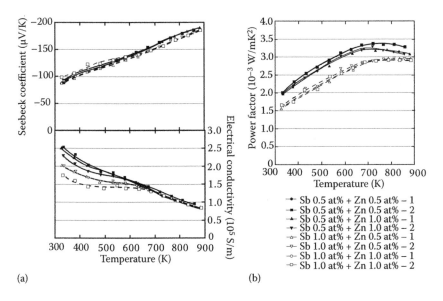

(a)   (b)

FIGURE 20.2 (a) Temperature-dependent Seebeck coefficient and electrical conductivity and (b) calculated PF of sintered samples doped with Sb and Zn. The concentrations of Sb and Zn are either 0.5 or 1.0 at%.

entropy method suggest that Sb occupies substitutional Si sites and Zn occupies substitutional Mg sites [65]. Thus, the concept of codoping with Sb and Zn makes sense. For identical concentrations of Sb, the specimens codoped with Sb and Zn display similar variations of the Seebeck coefficient and electrical conductivity with those solely doped with Sb at elevated temperatures, while slight decreases at lower temperatures (less than ~500 K) can clearly be observed for all the codoped specimens. In the case of Sb and Zn doping, the maximum decrease in electrical conductivity is when the concentrations of Sb and Zn are both 1.0 at%. In this study, the slight decrease in electrical conductivity was observed in measurements that were repeatedly performed; however, a detailed analysis was not done in this case. The Zn precipitation observed in the FE-SEM EDX measurements at a concentration of 0.5 at% implies that 1.0 at% Zn may give rise to greater precipitation at the grain boundaries and the formation of ZnO after heat cycle operations at elevated temperatures.

As shown in Figure 20.2, the values of the calculated PFs for the codoped specimens with Sb concentrations of 0.5 at% are ~3.4 × 10^{-3} W/m K^2 for temperatures beyond 650 K. These are slightly higher values than those obtained for the sample solely doped with Sb. Our initial attempts at codoping with 0.5 at% Sb and 0.5 at% Zn gave PFs of 1.0–2.5 × 10^{-3} W/m K^2; however, by modifying the various sintering process parameters, such as the compaction temperature, the pressure, the sequencing, and the pre- and postprocessing, we were able to increase this to ~3.4 × 10^{-3} W/m K^2.

The variations in thermal conductivity of the samples codoped with Sb and Zn are shown in Figure 20.3 for temperatures ranging from room temperature to 873 K. The observed thermal conductivity values for codoped specimens with 0.5 at% Sb monotonically decrease with increasing sample temperature of up to ~873 K, where the dominant contribution is typically from an increase in the phonon scattering contribution. Increasing the Sb doping concentration forces down the thermal conductivity values, especially

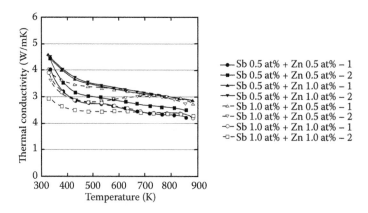

FIGURE 20.3 Measured thermal conductivity of samples codoped with Sb and Zn in the range from room temperature to 873 K.

in the lower temperature range (less than ~600 K). For specimens codoped with Sb and Zn at 1.0 and 0.5 at%, respectively, lower thermal conductivity values than those of material doped with only Sb were observed over the whole measured temperature range, and the observed values start to level out at ~2.5 W/m K above ~650 K, finally reaching 2.1 W/m K above 850 K.

Incorporating Zn leads to better thermal conductivity values compared to adding Sb only. It has previously been observed that increasing the Sb content from 0.5 to 1.0 at% can contribute to a decrease in thermal conductivity, but this behavior was not evident in the samples codoped with Sb and Zn. It may be speculated that the metallic precipitates of excess Zn at the grain boundaries might reduce phonon scattering, increasing the heat conduction. The lowest values of thermal conductivity were obtained for specimens doped with Sb:Zn = 0.5: 0.5 at% and 1.0:1.0 at%. In order to better understand the mechanism through which codoping with Sb and Zn reduces the thermal conductivity, a more systematic investigation with more variations in the Sb and Zn contents is underway.

Figure 20.4 shows the evaluated values of ZT with respect to temperature for samples doped with only Sb and for the codoped specimens. The ZT values of the codoped samples with Sb:Zn = 0.5:0.5 at% and 1.0:1.0 at% exhibit higher values than those of the Sb-doped sample without Zn. The maximum value of 1.26 at 873 K is for the codoped specimen with 0.5 at% Sb and 0.5 at% Zn. Specimens with other combinations of doping concentration have ZT values from 1.1 to 1.2, where it is worth pointing out that the ZT values achieved here are for specimens prepared using mass-produced source materials and production-type work flows.

As has already been mentioned, the combination of Zn with Sb can provide two practical benefits at specific operating temperatures: (1) reduction of the thermal conductivity, thus increasing the value of ZT, and (2) no significant increase in electrical conductivity or effect on the PF characteristics due to the presence of Zn. Tuning the TE properties of the material by reducing

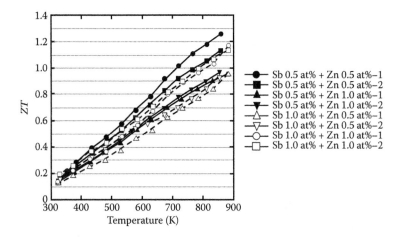

FIGURE 20.4 Dimensionless figure of merit (ZT) of Mg$_2$Si samples codoped with Sb and Zn.

the electrical conductivity while maintaining the value of the PF will have an impact on practical TEG design; that is, the electrical power generated from the TEG can be obtained with a reduced current component. Using the ANSYS code and the TE properties of the codoped sample with Sb:Zn = 0.5:0.5 at%, we calculated the power generated by $5 \times 5 \times 5$ mm^3 and $3 \times 3 \times 7$ mm^3 TE chips to be 3.6 W/cm^2 and 2.7 W/cm^2, respectively.

20.3 Electrode

Mg_2Si is a typical semiconductor material, and drawing out the generated electricity from a TE chip requires a well-matched electrode material with a sufficiently low ohmic contact and well-defined interfacial electrical characteristics. One possible method of forming an electrode on Mg_2Si is by the so-called monobloc sintering method, which is a simultaneous sintering process by means of PAS, which forms an electrode during the sintering of the Mg_2Si matrix. To form a Ni electrode, which is one of the most prevalent electrodes for mid-temperature TE materials, Ni powder (99.99%, 1–3 μm) is placed above and below the Mg_2Si powder in the graphite sintering crucible and then sintered with appropriate sintering conditions to form sufficient TE properties for the Mg_2Si. In this case, the range of temperature for monobloc sintering is from 1093 K to 1163 K, since this is the temperature range for forming distinct Mg_2Si.

Figure 20.5a shows optical microscopy images at the Ni/Mg_2Si interface for Sb (0.5 at%)-doped and Al (1.0 at%)-doped Mg_2Si specimens. A good coherent interface between the Mg_2Si and Ni was formed, and the contact resistances observed were 5.20×10^{-10} Ω m^2 and 3.62×10^{-9} Ω m^2 for Sb-doped and Al-doped specimens, respectively. The obtained value of the contact resistance for the Sb-doped specimen is sufficiently low and a kind of benchmark value for n-type Mg_2Si. At the Ni/Mg_2Si interface, intermediate layers were observed for both the Sb- and Al-doped specimens, which we denote as the η-phase and the ω-phases. To understand these two intermediate layers, high resolution cross-sectional transmission electron microscope (XTEM) observations were performed using a JEOL JEM-4010 at 400 keV. Figure 20.5b shows the measured XTEM image for the Ni/Mg_2Si interface of an Al-doped Mg_2Si sample [66]. From analyses of the diffraction patterns from selected areas of the Ni/Mg_2Si interface, the compositions of the η- and ω-phases were revealed to be the intermetallic compounds $Mg_6Si_7Ni_{16}$ (Pearson symbol: cF116; prototype: $Mn_{23}Th_6$) and $Mg_{33}Si_{37}Ni_{30}$ (hP55; Ag_7Te_4), respectively. These are shown in Figure 20.6. Additionally, another phase of $MgNi_2$ (hP24; $MgNi_2$) was discovered between the Ni electrode, and the η-phase. These intermediate layers were formed during the monobloc sintering process and were found to be rather stable. Subsequent aging tests on several specimens at 873 K in air for 1000 h indicated no notable broadening due to Ni diffusion. To understand the thermodynamic stability of the η- and ω-phases, a first-principles calculation was made using the CASTEP code, and this indicates that the ω-phase seems to be a stable phase and the η-phase seems to be an unstable one [67]. As shown in Figure 20.5b, however, there may be an interblended portion between the

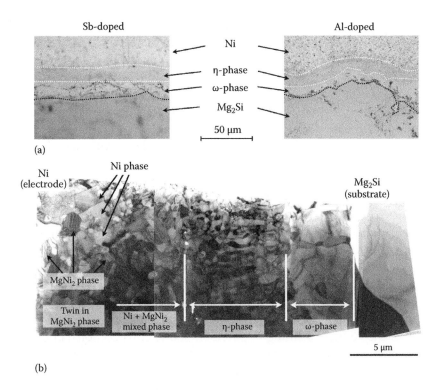

(a)

(b)

FIGURE 20.5 Observations of the interfaces between the Ni electrodes and the Mg_2Si matrix; (a) optical microscope images, (b) cross-sectional image of the sample made using a high-resolution transmission electron microscope. The sample was cut from a sintered pellet.

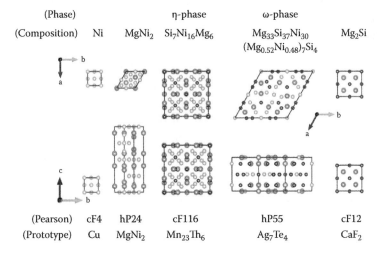

FIGURE 20.6 Compounds at the Ni/Mg_2Si interface determined by the analyses of the diffraction patterns of selected areas shown in the high-resolution transmission electron microscope images.

η-phase and ω-phase; thus, the thermal stability of the whole intermediate layer needs to be investigated in a more systematic manner.

Ni is one of the most common electrode materials for mid-temperature TE materials. On the other hand, there has not been a widespread testing of other possible electrode materials. Therefore, we did extensive tests to determine the interfacial characteristics of several metallic silicide electrodes, such as $CoSi_2$ + Ni ($CoSi_2$:Ni = 1:1), $CrSi_2$ + Ni ($CrSi_2$:Ni = 1:1), and NiSi, which can all be formed within the constraints of the monobloc sintering method. The results obtained are summarized in Table 20.1. For Al-doped specimens, the contact resistance depends on the type of electrode used, and the lowest value obtained was on the order of $\sim 10^{-10}$ Ω m^2 when $CoSi_2$ + Ni was used. Sb-doped Mg_2Si exhibited no dependence with regard to the type of electrode material at all; nevertheless, the values were sufficiently low at $\sim 10^{-10}$ Ω m^2. After aging at 873 K for 1000 h in air, there was no conspicuous deterioration in the Al-doped specimens with the various electrodes, whereas the reproducibility of the Sb-doped ones deteriorated, which was perceived to be mainly due to imperfections of sample compaction introducing voids and microcracks at the electrode/Mg_2Si interface.

In order to realize a simple and scalable process for forming the electrodes, a metallic paste printing method using Ni paste containing submicrometer-sized Ni powder was used to deposit electrodes on a compacted Sb-doped Mg_2Si specimen. The Mg_2Si pellet was subsequently annealed at ~ 900 K after the Ni paste printing. Figure 20.7 shows cross-sectional optical microscope images of the Ni paste/Mg_2Si interface, indicating that a less uneven interface structure was obtained compared to that obtained in the monobloc-sintered sample. There is no clear intermediate structure such as the η-phase and ω-phase that appear after monobloc sintering, which is mainly due to the lower process temperature used for the paste printing method. For the printing method, the powder in the Ni paste was smaller in size than the Mg_2Si powder (one to ten micrometers) in order to make better mechanical and electrical connection. The measured contact resistance was 1.5 × 10^{-9} Ω m^2, which is three times larger than that for the monobloc-sintered Sb-doped Mg_2Si; however, this is low enough to be used in a practical device fabrication process. If we increase the heat treatment temperature for the paste-printed specimen up to ~ 1100 K, then some layers with different contrast appear at the interface in the microscope images. The SEM-EDX measurements at the Ni paste/Mg_2Si interface showed no notable interdiffusion of the constituent elements of Ni and Mg_2Si during the formation process

TABLE 20.1 Measured Contact Resistance for Al-/Sb-Doped Mg_2Si with Ni, $CoSi_2$ + Ni, $CrSi_2$ + Ni, and NiSi Electrodes

	Al 1.0 at.% Doped Mg_2Si				Sb 0.5 at.% Doped Mg_2Si			
	Ni	$CoSi_2$ + Ni	$CrSi_2$ + Ni	NiSi	Ni	$CoSi_2$ + Ni	$CrSi_2$ + Ni	NiSi
Resistivity of Mg_2Si ($\times 10^{-6}$ Ω m)	9.87	9.34	10.28	9.45	4.02	3.89	4.15	3.86
Contact resistance ($\times 10^{-9}$ Ω m^2)	3.62	0.72	1.02	2.76	0.52	0.37	0.57	0.69

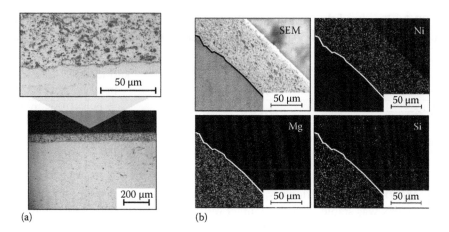

(a)   (b)

FIGURE 20.7   Cross-sectional (a) optical microscope images and (b) SEM-EDX composition analysis for the Ni paste/Mg$_2$Si interface.

as is shown in Figure 20.7; nevertheless, the interface characteristics at the operating temperature after aging and heat cycling need to be assessed in order to verify their practical capability.

The results of a preliminary atmospheric aging test at 873 K for 1000 h are summarized in Figure 20.8 and Table 20.2. The surfaces of the tested TE chips and the electrode interface were seen to be durable during the

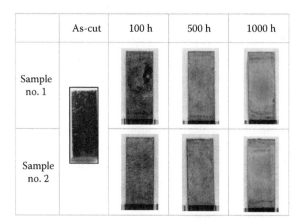

FIGURE 20.8   Appearance of TE chips after aging at 873 K for 1000 h.

TABLE 20.2   Results of the Variation in Contact Resistance of a Ni Electrode Prepared by the Paste Printing Method on Mg$_2$Si Codoped with Sb and Zn

Aging Duration	0 h	100 h	500 h	1000 h
Mg$_2$Si resistivity	3.36	3.97	3.75	3.85
($\times 10^{-6}$ Ω m)	3.60	3.96	4.04	3.69
Contact resistance	1.52	1.03	1.41	1.55
($\times 10^{-9}$ Ω m^2)	1.56	1.47	1.68	1.58

Note: Aging tests were performed at 873 K for 1000 h.

TABLE 20.3 Results of the Variation in Contact Resistance of a Ni Electrode Prepared by the Paste Printing Method on Mg$_2$Si as Compared with the Monobloc Sintering Method

	P_{max} [mW]	$P_{density}$ [W/cm^2]	Efficiency [%]	Internal Resistance [mΩ]	Contact Resistance [×10^{-9} Ω m^2]
Monobloc sintering	110	1.20	3.0	7.0	0.264
Screen printing	157	1.62	3.5	5.6	1.52

experiments. Additionally, the measured contact resistance of the Mg$_2$Si matrix was stable, and the contact resistance showed no notable change even after 1000 h of exposure. These results regarding the paste printing process are still at a preliminary stage; thus, the optimization of the whole process, including the paste content, the printing procedure, and the annealing time and temperature, needs to be done. In view of the reproducibility, some of the samples showed cracks at the Mg$_2$Si/Ni electrode. As we have seen in Section 20.3, a number of different types of electrode with low contact resistance can be used on the Sb-doped Mg$_2$Si matrix, whereas the results for the Al-doped Mg$_2$Si matrix are more variable. Thus, a similar experiment using the paste printing process on Al-doped specimens will throw more light on this topic.

The power generated by single TE Mg$_2$Si chips with Ni electrodes formed using the monobloc and paste printing methods was measured. The size of the specimens was 3 × 3 × 7 mm^3 in both cases. The power curve measurement was performed using ADVANCE-RIKO Mini-PEM apparatus, and the results are summarized in Table 20.3. The total power generation performance is well balanced for the sample with the electrode formed by the paste printing method. Briefly, the internal resistance of the monobloc-sintered sample is slightly higher than that of the paste printed one. This is mainly because of the sintering flexibility, i.e., the process conditions for the monobloc sintering process have to be compromised to form both the Mg$_2$Si matrix and the Ni electrode, so the process temperature is not optimum for the formation of the Mg$_2$Si, while the best sintering conditions for Mg$_2$Si can be applied when the paste printing method is used, thereby resulting in the different internal resistances.

20.4 Typical Modes of Degradation and Passivation

Although Mg$_2$Si has a long history as a TE material, the technological breakthrough that has enabled its industrialization is the containment of oxidation at the operating temperature in air. Basically, this has been brought about by eliminating residual Mg during the fabrication process [44,64]. However, other oxidation processes have been discovered, which are associated with processes such as the synthesis, the sintering, the electrode formation, and the TE chip manufacture. In order to obtain practical TEGs with reasonable durability using Mg$_2$Si, further investigation regarding deterioration needs to be done in order to quickly move forward toward production. In principle, if we can fabricate pure enough Mg_2Si, then the dominant degradation process of Mg_2Si is associated with the oxidation of the constituent elements, Mg and Si, as opposed to the direct decomposition of the matrix. The most stable oxide in the Ellingham diagram of the constituent elements is MgO; thus, preventing this from forming at elevated temperatures is a key issue.

Figure 20.9 shows undoped Mg_2Si with no residual metallic Mg, where the Mg_2Si wafer ($3 \times 3 \times 1$ mm³) was placed in a chamber and heated up to 873 K for 1000 h in an air/Ar (forming gas) ambient. The surface was optically examined using a LaserTech VL2000DX laser microscope during heating, and the sample was subsequently analyzed by XRD using 40 KeV Cu kα x-rays in a RIGAKU Ultima IV and by FE-SEM EDX with a JEOL JSM-7500F. After 1000 h of aging, bold grain boundaries and flecks on the grains appeared, resulting in the growth of a diffraction peak in the XRD spectrum due to MgO with increasing aging time. Precise local composition analysis using the FE-SEM EDX revealed the formation of MgO along the grain boundaries, when the duration of heat treatment increased. Note that the sintered Mg_2Si initially includes a tiny amount of MgO at the grain boundaries and in voids; then as the sample temperature increases, the Mg_2Si changes to MgO in areas adjacent to the initial MgO and then spreads to the surrounding areas. The oxidation of Mg_2Si to form MgO can be suppressed by heat treatment in an Ar or Ar/H_2 (forming gas) ambient, but the inclusion of MgO in the Mg_2Si caused significant deterioration at the grain boundaries, even under inert gas conditions, giving rise to degradation of the TE properties.

If one can successfully form good Mg_2Si with less oxide contamination and tightly held grains, surface passivation to eliminate the exposure to oxygen may be helpful. A possible tough passivation is a glass coating with a similar CTE to that of Mg_2Si and which shows no interdiffusion of the constituent elements. The measured CTE of Mg_2Si is 16.0×10^{-6} K⁻¹ (523–873 K). An appropriate glass composition is $15Na_2O\text{-}15K_2O\text{-}10CaO\text{-}10Al_2O_3\text{-}50SiO_2$

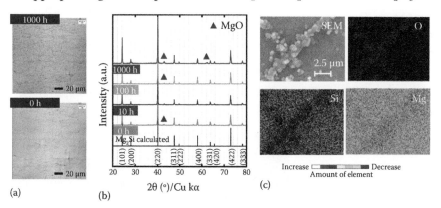

(a) (b) (c)

FIGURE 20.9 Oxidation of the surface of Sb- and Zn-doped Mg_2Si at an elevated temperature for up to 1000 h. The morphology of the surface was examined using (a) a laser microscope, then locally analyzed using (c) a field-emission-type SEM-EDX. (b) The XRD results reveal an increase in MgO with increasing aging time.

(mol%), which, using Appen's experimental formula, has a similar CTE to Mg_2Si [67]. The determination of the CTE was carried out by means of thermomechanical analysis (TMA) apparatus. Figure 20.10 shows the thermal expansion curves for Mg_2Si and the annealed $15Na_2O$-$15K_2O$-$10CaO$-$10Al_2O_3$-$50SiO_2$ glass as a function of temperature, which shows the curves to be well matched. The glass was ground into fine particles, and the Mg_2Si was coated with this by a simple room-temperature dipping method, then it was sintered at 973 K for 60 min under vacuum. The estimated CTE of the glass was 15.1×10^{-6} K^{-1}, and the measured CTE value was 14.6×10^{-6} K^{-1} (523–873 K). The softening temperature is 869 K. Sintering of the glass causes crystallization. A bulk glass comprising kalsilite ($KAlSiO_4$), calcium silicate (Ca_2SiO_4), calcium aluminate ($Ca_3Al_2O_6$), and a residual glassy phase subject to the same heat treatment conditions gave CTE values close to those of Mg_2Si as shown in Figure 20.10. The crystallized glass coating has some voids; thus, process optimization and densification to achieve a homogeneous coating is under way. Figure 20.11 shows the results of interdiffusion analysis

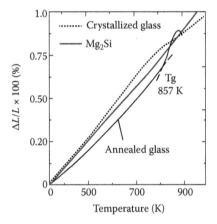

FIGURE 20.10 Results of the measured CTEs for annealed and crystallized glasses and the Mg_2Si matrix using TMA.

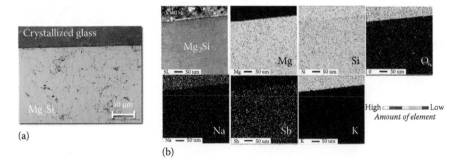

FIGURE 20.11 (a) Cross-sectional optical microscope image and (b) EPMA composition mapping images of the interface between the glass coating and Mg_2Si. The EPMA analysis was performed for elements of Mg, Si, O, Na, Sb, and K in the wavelength dispersion mode.

measured at the passivation/Mg$_2$Si interface by EPMA with a JEOL JXA-8900 after aging at 873 K for 300 h in air. As is shown in the EPMA mapping results, no notable diffusion of the constituent elements occurred at the interface.

20.5 Chemical Etching

Chemical etching of Mg$_2$Si is one of the important process technologies needed to push forward toward industrial production. This is needed to remove the surface-native oxide and unexpected contaminants and to stabilize and coordinate the surface conditions for subsequent processes, such as the formation of electrodes and passivation. However, Mg$_2$Si is highly reactive with water; thus, it is difficult to etch the surface and condition it for subsequent processes. Basically, since the current dominant method for fabricating Mg$_2$Si is the melt process [44,64], and the residual metallic Mg can be removed, etching experiments are also possible; however, there have been almost no reports. Here, we show some preliminary experimental results for etching processes for Mg$_2$Si using acid and alkaline solutions.

Figure 20.12 shows the results of surface etching of Mg$_2$Si codoped with Sb (0.5 at%) and Zn (0.5 at%) for both acidic and alkaline etchants. For Mg$_2$Si, etching is performed by corrosion of the metallic constituent element

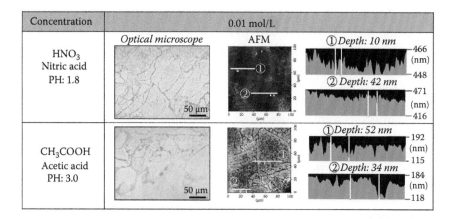

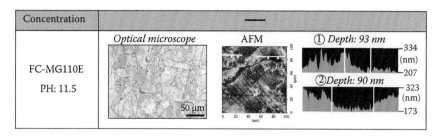

FIGURE 20.12 Surface morphology of Mg$_2$Si samples etched in HNO$_3$, CH$_3$COOH, and the NaOH-based etchant, FCMG110.

Mg. The reaction during etching in an acidic solution is as follows: Mg → $Mg^{2+}(aq) + 2e^-/2H^+ + 2e^- → H_2$, which together give $Mg + 2H^+ → Mg^{2+}(aq) + H_2$ [68]. Nitric acid (HNO_3), acetic acid (CH_3COOH), and hydrochloric acid (HCl) were used as the basic components. The morphological stability of the surface, the etching speed, the correlation between the etching direction and the crystal orientation, and the influence at the grain boundaries were evaluated. The acidic etchants and conditions used were as follows: (1) HNO_3: 0.01 mol/L, pH = 1.8, at 313 K for 1 min and (2) CH_3COOH: 0.01 mol/L, pH = 3.0, at 313 K for 1 min. Samples were subsequently cleaned in acetone in an ultrasonic bath. The original surfaces of the samples were polished with an abrasive consisting of 0.5 μm diamond paste. The etched surface morphology and etch depth were examined using a Nomarski contrast optical microscope and a Hitachi SI-AF01 atomic force microscope. In both cases, the surfaces obtained were stable after etching, and the etch depths were 480 and 890 nm for HNO_3 and CH_3COOH, respectively. The CH_3COOH etchant had a larger reaction rate at the grain boundaries than the HNO_3 etchant.

Etching with an alkaline etchant was performed using potassium hydroxide (KOH) and the sodium hydroxide (NaOH)-based commercial Mg–Al alloy cleaner FCMG110 (Nihon Parkerizing), which is used as an industrial degreasing agent for Mg–Al alloys. Under alkaline etching conditions, the reaction during etching from the potential pH diagram is presumed to be as follows: $Mg → Mg^{2+}(aq) + 2e^-/1/2O_2 + H_2O + 2e^- → 2OH^-$, which together give $Mg + H_2O + 1/2O_2 → Mg(OH)_2(s)$ [69]. The resultant surface after etching with FCMG110 with pH = 11.5, at 313 K for 1 min, had sharp and traceable characteristics with a shallow etching depth of ~160 nm as compared with the acidic etchants. X-ray photoelectron spectroscopy was used to measure the shift in Mg1s, Si2p, and O1s in the spectrum as a function of depth. This showed that unstable native oxides, hydrocarbons, and other contaminants were removed, and a stable mixture of MgO and SiO_2 with an abrupt interface between the oxide layer and Mg_2Si matrix was formed. In view of the surface stability and less thick oxide layer, FCMG110 seems a good prospect at the moment.

20.6 Summary

There is an increasing demand for integrated waste heat recovery systems in industrial furnaces and automobiles, for which there are many new and interesting TE materials and other new materials in various stages of development. Simple Mg_2Si is one possible TE material, since it possesses some notable features such as its light weight, abundance, and freedom from toxicity. Recent development results have already prepared it to follow on from Pb–Te as a TE device in the mid-temperature range. Although the latest ZT values for n-type Mg_2Si are moderate (~1.2), the current PF is ~3.4 × 10^{-3} W/m K^2, which shows that we can extract a lot of power from a TE chip. However, the technological steps required to realize practical TEG devices that can be put into production need to be developed, as does their integration into various kinds of heat source. This requires reliable TEGs. One problem for designing TEGs using Mg_2Si is the lack of an efficient p-type conductivity material; however, combination with other good p-type materials is a pragmatic solution. We know that there is a long way to go before we have a material of acceptable durability that can be used in production. However, the peripheral technologies, such as the formation of low contact resistance electrodes, durable electrode material, the degradation tolerance, passivation coatings, relevant soldering for midtemperature operation, thermal flow control, thermal impedance matching, electrical impedance matching, DC–DC convertor matching, and so on also need to be developed. In this chapter, we have focused not only on the basic TE material characteristics but also on engineering aspects such as degradation, passivation, and etching. In order to provide for the many potential applications for TEGs, the peripheral and neighboring technology development for each promising TE material needs to be extensive, but also sustainable. This can also promote the development of new or modified exciting TE phenomena and materials.

References

1. R. G. Morris, R. D. Redin, and G. C. Danielson, *Phys. Rev.* **109**, 1909 (1958).
2. P. Koenig, D. W. Lynch, and G. C. Danielson, *J. Phys. Chem. Solids* **20**, 122 (1961).
3. A. Stella, A. D. Brotherrs, H. Hopkings, and D. W. Lynch, *Phys. Status Solidi* **23**, 697 (1967).
4. J. Tani and H. Kido, *Physica B* **223**, 364 (2005).
5. V. K. Zaitsev, M. I. Fedorov, E. A. Gurieva, I. S. Eremin, P. P. Konstantinov, A. Yu. Samunin, and M. V. Vedernikov, *Phys. Rev. B* **74**, 045207 (2006).
6. J. Tani and H. Kido, *Intermetallics* **15**, 1202 (2007).
7. J. Tani and H. Kido, *J. Alloys Compd.* **466**, 335 (2008).
8. M. Akasaka, T. Iida, A. Matsumoto, K. Yamanaka, Y. Takanashi, T. Imai, and N. Hamada, *J. Appl. Phys.* **104**, 013703 (2008).
9. P. Boulet, M. J. Verstraete, J.-P. Crocombette, M. Briki, and M.-C. Record, *Comput. Mater. Sci.* **50**, 847 (2011).
10. P. Pandit and S. P. Sanyal, *Ind. J. Pure Appl. Phys.* **49**, 692 (2011).
11. W. Liu, X. Tan, K. Yin, H. Liu, X. Tang, J. Shi, Q. Zhang, and C. Uher, *Phys. Rev. Lett.* **108**, 166601 (2012).
12. T. Sakamoto, T. Iida, N. Fukushima, Y. Honda, M. Tada, Y. Taguchi, Y. Mito, H. Taguchi, and Y. Takanashi, *Thin Solid Films* **519**, 8528 (2011).
13. S. Battiston, S. Fiameni, M. Saleemi, S. Boldrini, A. Famengo, F. Agresti, M. Stingaciu, M. S. Toprak, M. Fabrizio, and S. Barison, *J. Electron. Mater.* **42**, 1956 (2013).
14. M. Ioannou, G. Polymeris, E. Hatzikraniotis, A. U. Khan, K. M. Paraskevopoulos, and Th. Kyratsi, *J. Electron. Mater.* **42**, 1827 (2013).
15. T. Kajitani, M. Kubouchi, S. Kikuchi, K. Hayashi, T. Ueno, Y. Miyazaki, and K. Yubuta, *J. Electron. Mater.* **42**, 1855 (2013).
16. S. Fiameni, A. Famengo, S. Boldrini, S. Battiston, M. Saleemi, M. Stingaciu, M. Jhonsson, S. Barison, and M. Fabrizio, *J. Electron. Mater.* **42**, 2062 (2013).

17. S. Fiameni, S. Battiston, S. Boldrini, A. Famengo, F. Agresti, S. Barison, and M. Fabrizio, *J. Solid State Chem.* **193**, 142 (2012).
18. M. Ioannou, K. Chrissafis, E. Pavlidou, F. Gascoin, and Th. Kyratsi, *J. Solid State Chem.* **197**, 172 (2013).
19. Y. Imai, A. Watanabe, and M. Mukaida, *J. Alloys Compd.* **358**, 257 (2003).
20. W. Xiong, X. Qin, and L. Wang, *J. Mater. Sci. Technol.* **23**, 595 (2007).
21. L. Wang, X. Y. Qin, W. Xiong, and X. G. Zhu, *Mater. Sci. Eng. A* **459**, 216 (2007).
22. S. K. Bux, M. T. Yeung, E. S. Toberer, G. J. Snyder, R. B. Kaner, and J.-P. Fleurial, *J. Mater. Chem.* **21**, 12259 (2011).
23. Q. S. Meng, W. H. Fan, R. X. Chen, and Z. A. Munir, *J. Alloys Compd.* **509**, 7922 (2011).
24. M. Saleemi, M. S. Toprak, S. Fiameni, S. Boldrini, S. Battiston, A. Famengo, M. Stingaciu, M. Johnsson, and M. Muhammed, *J. Mater. Sci.* **48**, 1940 (2013).
25. D. Cederkrantz, N. Farahi, K. A. Borup, B. B. Iversen, M. Nygren, and A. E. C. Palmqvist, *J. Appl. Phys.* **111**, 023701 (2012).
26. S. Nakamura, Y. Mori, and K. Takarabe, *Phys. Status Solidi C* **10**, 1145 (2013).
27. S. Muthiah, J. Pulikkotil, A. K. Srivastava, A. Kumar, B. D. Pathak, A. Dhar, and R. C. Budhani, *Appl. Phys. Lett.* **103**, 053901 (2013).
28. X. Hu, D. Mayson, and M. R. Barnett, *J. Alloys Compd.* **589**, 485 (2014).
29. J. Zhao, Z. Liu, J. Reid, K. Takarabe, T. Iida, B. Wang, Y. Uwatoko, and J. S. Tse, *J. Mater. Chem.* **A3**, 19774 (2015).
30. A. Kato, T. Yagi, and N. Fukusako, *J. Phys. Condens. Matter* **21**, 205801 (2009).
31. P. Zwolenski, J. Tobola, and S. Kaprzyk, *J. Electron. Mater.* **40**, 889 (2011).
32. H. Wang, W. Chu, and H. Jin, *Comput. Mater. Sci.* **60**, 224 (2012).
33. N. Satyala and D. Vashaee, *J. Electron. Mater.* **41**, 1785 (2012).
34. J. J. Pulikkotil, D. J. Singh, S. Auluck, M. Saravanan, D. K. Misra, A. Dhar, and R. C. Budhani, *Phys. Rev. B* **86**, 155204 (2012).
35. N. Hirayama, T. Iida, S. Morioka, M. Sakamoto, K. Nishio, Y. Kogo, Y. Takanashi, and N. Hamada, *Jpn. J. Appl. Phys.* **54**, 7S2 07JC05-1 (2015).
36. N. Hirayama, T. Iida, H. Funashima, S. Morioka, M. Sakamoto, K. Nishio, Y. Kogo, Y. Takanashi, and N. Hamada, *J. Electron. Mater.* **44**, 1656 (2014).
37. N. Hirayama, T. Iida, S. Morioka, M. Sakamoto, K. Nishio, Y. Kogo, Y. Takanashi, and N. Hamada, *J. Mater. Res.* **30**, 2564 (2015).
38. M. Kubouchi, K. Hayashi, and Y. Miyazaki, *J. Alloys Compd.* **617**, 389 (2014).
39. A. E. Karkin, V. I. Voronin, N. V. Morozova, S. V. Ovsyannikov, K. Takarabe, Y. Mori, S. Nakamura, and V. V. Shchennikov, *J. Phys. Chem. C* **120**, 9692 (2016).
40. T. Kajikawa, K. Shida, K. Shiraishi, M. Omori, and T. Hirai, *Proc. 16th Intl. Conf. on Thermoelectrics 1997, IEEE*, 275 (1997).
41. T. Kajikawa, K. Shida, K. Shiraishi, T. Ito, M. Omori, and T. Hirai, *Proc. 17th Intl. Conf. on Thermoelectrics 1998, IEEE*, 362 (1998).
42. M. Akasaka, T. Iida, J. Soga, N. Kato, T. Sakuma, Y. Higuchi, and Y. Takanashi, *J. Crystal Growth* **304**, 196 (2007).
43. N. V. Morozova, S. V. Ovsyannikov, I. V. Korobeinikov, A. E. Karkin, K. Takarabe, Y. Mori, S. Nakamura, and V. V. Shchennikov, *J. Appl. Phys.* **115**, 213705 (2014).
44. T. Sakamoto, T. Iida, A. Matsumoto, Y. Honda, T. Nemoto, J. Sato, T. Nakajima, H. Taguchi, and Y. Takanashi, *J. Electr. Mater.* **39**, 1708 (2010).

45. K. Suekuni, K. Tsuruta, T. Ariga, and M. Koyano, *Appl. Phys. Express* **5**, 051201 (2012).
46. K. Suekuni, K. Tsuruta, M. Kunii, H. Nishiate, E. Nishibori, S. Maki, M. Ohta, A. Yamamoto, and M. Koyano, *J. Appl. Phys.* **113**, 043712 (2013).
47. K. Suekuni, Y. Tomizawa, T. Ozaki, and M. Koyano, *J. Appl. Phys.* **115**, 143702 (2014).
48. X. Lu and D. T. Morelli, *J. Electron. Mater.* **43**, 1983 (2013).
49. X. Lu and D. T. Morelli, Y. Xia, and V. Ozolins, *Chem. Mater.* **27**, 408 (2015).
50. Alphabet Energy PowerCard, https://www.alphabetenergy.com/product/power card/ (2016).
51. I. Nishida, *J. Mater. Sci.* **7**, 435 (1972).
52. I. Kawasumi, M. Sakata, I. Nishida, and K. Masumoto, *J. Mater. Sci.* **16**, 355 (1981).
53. E. Groβ, M. Riffel, and U. Stöhrer, *J. Mater. Res.* **10**, 34 (1995).
54. M. I. Fedorov, V. K. Zaitsev, F. Yu. Solomkin, and M. V. Vedernikov, *Tech. Phys. Lett.* **23**, 602 (1997).
55. M. Umemoto, Z. G. Liu, R. Omatsuzawa, and K. Tsuchiya, *Mater. Sci. Forum* **918**, 343 (2000).
56. T. Itoh and M. Yamada, *J. Electron. Mater.* **38**, 925 (2009).
57. A. Famengo, S. Boldrini, S. Battiston, S. Fiameni, A. Miozzo, M. Fabrizio, and S. Barison, *Proc. the 11th European Conf. on Thermoelectrics*, 89 (2014).
58. T. Kajitani, T. Ueno, Y. Miyazaki, K. Hayashi, T. Fujiwara, R. Ihara, T. Nakamura, and M. Takakura, *J. Electron. Mater.* **43**, 1993 (2013).
59. T. Kajitani, M. Kubouchi, S. Kikuchi, K. Hayashi, T. Ueno, Y. Miyazaki, and K. Yubuta, *J. Electr. Mater.* **42**, 1855 (2013).
60. G. S. Nolas, D. Wang, and M. Beekman, *Phys. Rev. B* **76**, 235204 (2007).
61. K. Arai, A. Sasaki, Y. Kimori, M. Iida, T. Nakamura, Y. Yamaguchi, K. Fujimoto, R. Tamura, T. Iida, and K. Nishio, *Mater. Sci. Eng. B* **195**, 45 (2015).
62. Y. Hayatsu, T. Iida, T. Sakamoto, S. Kurosaki, K. Nishio, Y. Kogo, and Y. Takanashi, *J. Solid State Chem.* **193**, 161 (2012).
63. M. Ishikawa, T. Nakamura, S. Hirata, T. Iida, K. Nishio, and Y. Kogo, *Jpn. J. Appl. Phys.* **54**, 07JC03 (2015).
64. T. Sakamoto, T. Iida, S. Kurosaki, K. Yano, H. Taguchi, K. Nishio, and Y. Takanashi, *J. Electr. Mater.* **40**, 629 (2011).
65. J. Zhao, J. Reid, T. Iida, K. Takarabe, M. Wu, and J. S. Tse, *J. Alloys Compd.* **681**, 66 (2016).
66. H. Sugawara, submitted to *J. Alloys Compd.*
67. A. A. Appen, *Khimiya Stekla (Glass Chemistry, in Russian)*, 2nd Edition, Leningrad, Khimiya Publishing House, 1974, pp. 350.
68. G. L. Makar and J. Kruger, *Int. Mater. Rev.* **38**, 138 (1993).
69. Y. Imai, submitted to *J. Alloys Compd.*

CHAPTER **21**

Oxide Modules

Le Thanh Hung, Ngo Van Nong, and Nini Pryds

Contents

21.1 Introduction

TE oxides have been considered as promising materials due to their non-toxicity, low cost, and chemical stability at high temperatures [1–3]. Studied results show great potential for applications in TEG at high temperature and have thus drawn much attention over the years. This chapter targets to summarize the research and development of exploring the usage of TE oxide-based materials for high-temperature TEGs. The performance of oxide materials and devices in the first part of this chapter will be considered under the ideal theoretical condition of no parasitic losses and disregard other factors such as mechanical properties, thermal expansion coefficient, and chemical stability.

In general, the conversion efficiency of a TEG is governed by device figure of merit ZT:

$$ZT = \frac{S^2 T}{KR},\tag{21.1}$$

where S, K, and R are the total Seebeck coefficient, the total thermal conductance, and the total resistance of the module, respectively.

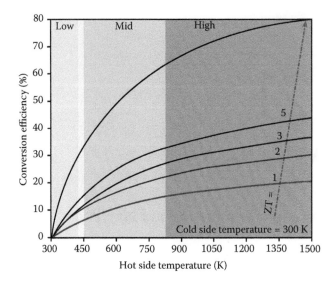

FIGURE 21.1 Calculated conversion heat–electricity efficiency as a function of temperature and device figure of merit.

The maximum conversion efficiency of TE module is often given as

$$\eta_{max} = \frac{T_h - T_c}{T_h} \frac{\sqrt{1+Z\bar{T}}-1}{\sqrt{1+Z\bar{T}}+(T_c/T_h)}. \quad (21.2)$$

The first term of Equation 21.2 is known as the Carnot efficiency, which is the upper limit conversion efficiency at infinite ZT, as shown in Figure 21.1. The second term is the contribution by the intrinsic properties of the TE materials, i.e., Seebeck coefficient and electrical and thermal conductivities. It is clearly seen that the maximum conversion efficiency of module increases by increasing the hot side temperature as a result of the contribution from the first term in Equation 21.2, i.e., Carnot efficiency. For high-temperature TEG application in air, devices fabricated from intermetallic compounds often show oxidation, sublimation, and volatility processes in the high-temperature range [4,5]. Therefore, encapsulation is needed to protect the devices. By considering the reasons mentioned earlier, the high-temperature oxide TE module is of great interest in high-temperature heat–electricity conversion, e.g., waste heat recovery from steel industrial and thermal power plants.

21.2 Calculated Efficiency of All Oxide Materials and Modules

There have been many studies over the years on the development of high-performance TE oxide materials. Figure 21.2a and b shows a selection of the state-of-the-art material figure of merit ZT ($Z = S^2/\rho\kappa$), for p-type and n-type oxide materials. Oxide materials exhibit widely their peak ZT values ranging from 0.1 at 1173 K for NiO [6], NdCoO$_3$ [7], and LaSrCuO$_4$ [8], 0.2 for Ca$_3$Co$_2$O$_6$ [9], and 0.65 for Ca$_3$Co$_4$O$_9$ nanocomposite [10] at 1173 K and to

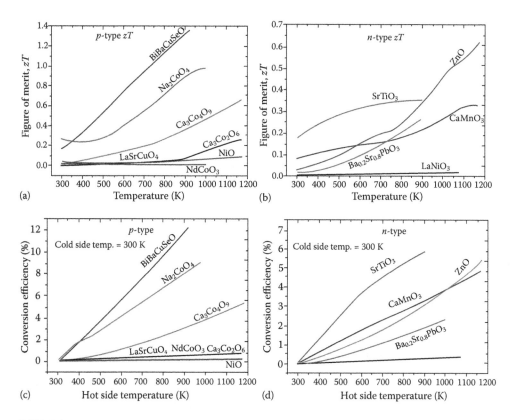

FIGURE 21.2 Material figure of merit ZT of state of-the-art oxide material (a) p-type and (b) n-type. The calculated maximum efficiency of single (c) p-type and (d) n-type TE oxide elements as a function of hot-side temperature.

the value of close to 1 at 1000 K for Na_2CoO_4 [11] and over 1 at 900 K for BiCuSeO [12]. For n-type oxide materials, the $LaNiO_3$ [13,14] attains the lowest magnitude of ZT < 0.05 in the temperature range of 300–1100 K. Perovskites $BaSrPbO_3$ [15], $CaMnO_3$ [16], and $SrTiO_3$ [17] obtain a medium ZT value of around 0.3 at 1173 K, while dual doped ZnO [18] presents the highest peak ZT value of 0.65 at 1173 K. Since most oxide materials (except BiCuSeO) have a small ZT value in the low and mid-temperature ranges of 300–700 K, TEG-based oxide has gained much interest mainly for high-temperature waste heat harvesting applications, e.g., steel or cement industrials.

To estimate the efficiency of oxide materials under the actual working condition, we have used numerical modeling [1,19–21].

The efficiency of a single material can be expressed as [1,20,21]

$$\eta = 1 - \frac{S_c T_c + \dfrac{1}{u_c}}{S_h T_h + \dfrac{1}{u_h}}, \qquad (21.3)$$

where S_c, T_c, u_c, S_h, T_h, and u_h are the Seebeck coefficient, temperature, and reduced current density at the cold- (subscript by c) and hot-side (subscript by h) temperatures, respectively. The relative current density (reduced current density) $u = J/\kappa\nabla T$ is defined as the ratio of the electrical current density (J) to the conduction heat flux (κ) [1,20,21]. In this calculation, the cold-side temperature is fixed at 300 K while the temperature of the hot side is varied. It should be noted that neither heat losses nor electrical losses are included in these calculations. Therefore, the obtained values can be considered as the upper limit performance achievable from the intrinsic TE properties. Figure 21.2c and d shows the calculated efficiency as a function of the hot-side temperature (cold side fixed at 300 K) for p-type and n-type oxide materials as a single element. For the conversion efficiency of all p-type curves in Figure 21.2c, the order of efficiency values of a single material has a similar tendency with ZT curves, as shown in Figure 21.2a. With the smallest ZT values, NiO, NdCoO$_3$, and LaSrCuO$_4$ have a maximum efficiency of less than 1% in the whole temperature of 300 and 1173 K. Although the ZT value of Ca$_3$Co$_2$O$_6$ rapidly increases with increasing temperature (Figure 21.2a), its conversion efficiency is still lower than 1% due to low-performance contribution in the whole temperature range. With the maximum conversion efficiency of 5.4%, the Ca$_3$Co$_4$O$_9$ is the most the promising p-type oxide material for temperature region of above 1000 K, while below 1000 K, Na$_2$CoO$_4$ and BiCuSeO oxyselenides exhibit conversion efficiencies of 10% and 12%, respectively. As for the n-type, LaNiO$_3$ shows an efficiency value which is lower than 0.1%, while BaSrPbO$_3$ shows a maximum value of around 2%. Interestingly, although the peak ZT of SrTiO$_3$ exhibits lower value, its maximum conversion efficiency is higher than that of dual doped ZnO; it is due to the conversion efficiency results from the accumulated performance of the whole working temperature range. The ZT value of CaMnO$_3$ is higher than that of the dual doped ZnO at temperatures below 600 K, while it is significantly lower in the temperature region of above 600 K, as shown in Figure 21.2b. However, the efficiency value of CaMnO$_3$ is higher than the efficiency of dual doped ZnO in the whole temperature range from 300 to 1000 K. In fact, the total conversion efficiency can be calculated as $\eta = \eta_c\eta_r$, where $\eta_c = (T_h - T_c)/T_c$ is the Carnot efficiency, and $\eta_r = [1 - u (S/z)]/[1 + (1/uST)]$ is the reduced efficiency [21]. Mathematically, η_r maximizes when $u = \left(\sqrt{1 + zT} - 1\right)/ST = s$, which is called the compatibility factor [21,22]. In the case where $s = u$, the material can obtain the maximum achievable total conversion efficiency. The smaller the value of ($s - u$), the larger achievable total conversion efficiency is obtained. Therefore, s and u values are important parameters and are needed to be considered for designing high-efficiency materials [1,23,24].

By connecting the p-type and n-type oxide elements, the efficiency of a unicouple can be computed as

$$\eta_{\text{unicouple}} = 1 - \frac{S_{c,p}T_{c,p} + 1/u_{c,p} - S_{c,n}T_{c,n} - 1/u_{c,n}}{S_{h,p}T_{h,p} + 1/u_{h,p} - S_{h,n}T_{h,n} - 1/u_{h,n}}, \tag{21.4}$$

where the subscripts n and p symbolize the n- and p-type legs. The calculated efficiency using Equation 21.4 for some promising unicouple-based oxide TE materials are summarized in Table 21.1. The hot-side temperature is defined by

TABLE 21.1 Calculated Maximum Conversion Efficiency of Unicouples with Various Maximum Hot-Side Temperatures and the Combination of p–n Oxide Leg

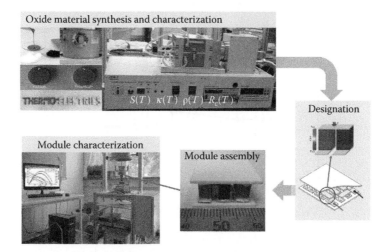	$Ca_3Co_4O_9$ (T_{hot}/1173 K) (%)	$Ca_3Co_2O_6$ (T_{hot}/1173 K) (%)	Na_2CoO_4 (T_{hot}/1000 K) (%)	BiCuSeO (T_{hot}/900 K) (%)
ZnO (T_{hot}/1173 K)	5.3	1.4	6.1	5.2
$CaMnO_3$ (T_{hot}/1173 K)	5.0	1.7	5.7	5.1
$Ba_{0.2}Sr_{0.8}PbO_3$ (T_{hot}/1000 K)	2.9	0.9	3.1	3.6
$SrTiO_3$ (T_{hot}/900 K)	4.1	0.89	7.0	7.9

the limited stability of the comprising TE materials, e.g., 900 K for any unicouple made by p-BiCuSeO and n-SrTiO₃. It can be seen that the combination of p- and n-type oxides resulted in different conversion efficiencies; the unicouples with $Ca_3Co_2O_6$ show the smallest values ranging from 0.9% to 1.7%; in contrast, the unicouple of p-BiCuSeO and n-SrTiO₃ exhibits the highest conversion efficiency of up to 7.9%. For the highest hot-side temperature of 1173 K, the p-type $Ca_3Co_4O_9$ combines with either n-type ZnO or CaMnO₃, showing a 5% efficiency, which is comparable with that of a current commercially available TE module based on Bi-Te materials. These results suggest that oxide TE materials have a great potential for high-temperature TEG for waste heat harvesting.

21.3 Oxide Module Construction and Characterization

Figure 21.3 shows the process steps for a typical module fabrication and characterization. The p- and n-type material processing first optimizes their

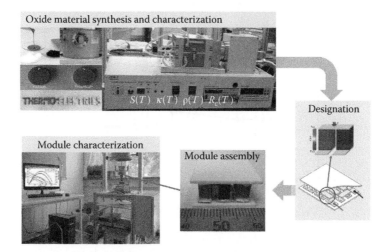

FIGURE 21.3 Schematic of the whole TE oxide module construction process: from material synthesis and characterization, modeling design, to build up and test module.

TE properties. The solid-state reaction is the preferred synthesis method for fabricating the oxide TE powder materials. For densification, hot press or SPS techniques are often used [19,25–28]. Following the synthesis and the densification, the TE properties of the materials are measured as a function of the temperature and fed into a numerical model, where the optimal output performance of the module is then calculated for a given optimized geometry of the legs and number of p–n couples [25]. The p- and n-type legs were then cut into the dimensions suggested by the modeling results. The unicouple contains two TE materials, n- and p-type, coupled together with a metal electrode. Silver (Ag) is the most used material for electrodes in module assembly [8,25–45] since it has high electrical and thermal conductivities and can resist high temperature in air.

The joining of Ag electrodes with TE legs is often done by the brazing method with filler materials which can be either pure Ag [25,28,35], Pt paste [29], Ag paste [31,33,38–45] or a mixture of Ag and oxide powder [30,32,37] depending on the materials in use. Joining by diffusion between oxide materials and Ag electrodes was also suggested, as described by Souma et al. [26,27]. This technique can tightly join two materials, be faster than brazing joining method, and does not require a filler metal. One of the crucial demands from the joining contacts is low electrical resistance and stability. A low resistance and stable contact are required since they ensure high performance and long-term stability of the TE module. Figure 21.4 shows typical experimental results of contact resistances of single p-type $Ca_3Co_4O_9$, single-segmented p-type $Ca_3Co_4O_9$/HH alloy, and n-type-doped ZnO oxide with Ag electrode using brazing joining method [19,25]. There is no sign of cracks or air gap observed at the interfaces, which results in low contact resistance

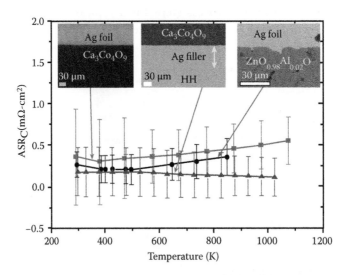

FIGURE 21.4 Picture of interface and contact resistance of the Ag with p-type and n-type oxide materials. (Hung, L. T. et al.: Segmented thermoelectric oxide-based module for high-temperature waste heat harvesting. *Energy Technology*. 2015. 3. 1143–1151. Copyright Wiley-VCH Verlag GmbH & Co. KGaA. Reproduced with permission.)

in the range of 100–500 $\mu\Omega$ cm^2 [19,25]. The entire module with the electrical insulation substrate is finally assembled by hot press method. It is crucial that each step in Figure 21.3 should be properly done, and any failure in one of the steps must be carefully checked or come back to optimize the previous steps.

21.4 All Oxide Thermoelectric Modules

Table 21.2 is a summary of TE oxide modules reported in the literature over the last 16 years. The size of those modules is varied largely from a TEG module comprising of a single unicouple up to 140 unicouples. Although the ZT value of 0.93 at 960 K of the p-type $NaCo_2O_4$ is very high, only a few oxide modules have been studied using $NaCo_2O_4$ [26,27,36] as p-type leg. The p-type $Ca_3Co_4O_9$ is mostly used because this material is highly stable at a high temperature in air [46]. For n-type, $CaMnO_3$ or doped ZnO are dominated materials since the higher ZT n-type $SrTiO_3$ is not stable in air at an elevated temperature [2]. As shown in Table 21.2, the power generation characteristics have been conducted under various conditions of the hot- and cold-side temperatures, resulting in differences in open circuit voltage and maximum output power. To give a comparison of studied oxide modules, the power density is often used [25,39,45]. However, one should be careful when using the value of power density, which might also lead to an unreasonable conclusion, because the power density depends on the length and the number of the TE legs. A more clear explanation can be described in the following.

The performance of module comprising p–n couples can be expressed in terms of the output voltage and current values. The open circuit voltage V_{OC} can be directly calculated from the definition of the Seebeck coefficient as [47]

$$V_{OC} = n \int_{T_c}^{T_h} \{S_p(T) - S_n(T)\} dT, \qquad (21.5)$$

where n is the number of p–n couples, S_p and S_n are the Seebeck coefficients of p- and n-type legs, respectively. Moreover, the value of the electric current I can be calculated from Ohm's law as

$$I = \frac{V_{OC}}{R_{int} + R_{Load}}, \qquad (21.6)$$

where R_{int} is the total of the internal resistances of the module that is contributed by the sum of the resistances of the p–n legs, and R_{Load} is the external resistive load. The internal resistance can be described as $R_{int} = R_{legs} + R_C$,

TABLE 21.2 Power Generation Characteristics of Oxide-Based TE Modules Reported in the Literature

Ref.	Year	Materials	No. of p-n Couples	Jointing Technique	T_{hot} (K)	ΔT (K)	V_{OC} (V)	V_{OC}/ Couples (V)	P_{max} (mW)	Leg Size (mm)	Power Density (mW/cm²)	Efficiency (%) Cal	Mea
Shina et al. [28]	2001	p-$Li_{0.025}Ni_{0.975}O$ n-$Ba_{0.2}Sr_{0.8}PbO_3$	1	Cosintering, cold side Ag paste	978	552	0.12	0.12	7.91	3 × 4 × 14	32.9	0.64	–
Shina et al. [28]	2001	p-$Li_{0.025}Ni_{0.975}O$ n-$Ba_{0.2}Sr_{0.8}PbO_3$	4	Cosintering, cold side Ag paste	1164	539	0.4	0.1	34.4	3 × 4 × 14	35.8	–	–
Matsubara et al. [29]	2001	p-$Ca_{2.75}Gd_{0.25}Co_4O_9$ n-$Ca_{0.92}La_{0.08}MnO_3$	8	Pt paste	1046	390	0.98	0.12	63.5	3 × 3 × 25	44.1	1.1	–
Funahashi et al. [30]	2004	p-$Ca_{2.7}Bi_{0.3}Co_4O_9$ n-$La_{0.9}Bi_{0.1}NiO_3$	1	Ag paste, 6 wt.% oxide powder	1073	500	0.1	0.1	94	3.7 × 4–4.53 × 4.7	317–280ᵃ	–	–
Reddy et al. [31]	2005	p-$Ca_3Co_4O_9$ n-$Ca_{0.95}Sm_{0.05}MnO_3$	2	Ag paste	1025	925	0.4	0.2	31.5–24.5	4 × 4 × 5	98.4	–	–
Funahashi et al. [32]	2006	p-$Ca_{2.7}Bi_{0.3}Co_4O_9$ n-$La_{0.9}Bi_{0.1}NiO_3$	140	Ag paste, 6 wt.% oxide	1072	551	4.5	0.03	150	1.3 × 1.3 × 5	31.7	1.3	–
Souma et al. [26]	2006	p-$NaCo_2O_4$ n-$Zn_{0.98}Al_{0.02}O$	12	Diffusion welding	839	462	0.8	0.06	58	3 × 4 × 10	20.1	–	–
Urata et al. [33]	2006	p-$Ca_{2.7}Bi_{0.3}Co_4O_9$ n-$CaMn_{0.98}Mo_{0.02}O_3$	8	Ag paste	897	565	1	0.12	170	5 × 5 × 4.5	42.5	–	–
Funahashi et al. [34]	2007	p-$Ca_{2.7}Bi_{0.3}Co_4O_9$ n-$La_{0.9}Bi_{0.1}NiO_3$	1	Ag paste	1073	500	0.1	0.1	177	3.7 × 4–4.53 × 4.7	528–598ᵃ	–	–
Urata et al. [35]	2007	p-$Ca_{2.7}Bi_{0.3}Co_4O_9$ n-$CaMn_{0.98}Mo_{0.02}O_3$	8	Ag	1273	975	0.7	0.09	340	5 × 5 × 4.5	85	–	–
Souma et al. [27]	2008	p-$NaCo_2O_4$ n-$Zn_{0.98}Al_{0.02}O$	12	Diffusion welding	934	455	0.8	0.067	52.5	3 × 4 × 10	18.2	–	–
Park et al. [36]	2009	p-$Na(Co_{0.95}Ni_{0.05})_2O_4$ n-$Zn_{0.99}Sn_{0.01}O$	1	Ag paste	923	422	0.14	0.14	0.027	7 × 7 × 17	0.03	–	–
Tomeš et al. [8]	2010	p-$La_{1.98}Sr_{0.02}CuO_4$ n-$CaMn_{0.98}Mo_{0.02}O_3$	2	Ag paste	941	622	0.464	0.23	88.8	4.5 × 4.5 × 5	109.6	–	0.073

(Continued)

TABLE 21.2 (CONTINUED) Power Generation Characteristics of Oxide-Based TE Modules Reported in the Literature

Ref.	Year	Materials	No. of p-n Couples	Jointing Technique	T_{hot} (K)	ΔT (K)	V_{OC} (V)	V_{OC}/Couples (V)	P_{max} (mW)	Leg Size (mm)	Power Density (mW/cm²)	Efficiency (%) Cal	Mea
Tomeš et al. [37]	2010	p-$GdCo_{0.95}Ni_{0.05}O_3$ n-$CaMn_{0.98}Mo_{0.02}O_3$	2	Ag/CuO paste	800	500	0.34	0.17	40	4 × 4 × 5	62.5	–	–
Choi et al. [38]	2011	p-$Ca_3Co_4O_9$ n-$(ZnO)_7$-In_2O_3	44	Ag paste	1100	673	1.8	0.04	423	15 × 15 × 27	2.1	–	–
Lim et al. [39]	2011	p-$Ca_3Co_4O_9$ n-$Ca_{0.9}Nd_{0.1}MnO_3$	1	Ag paste	1175	727	0.19	0.19	95	8.5 × 6 × 10	93.2	–	–
Inagoya et al. [40]	2011	p-$Nd_{0.095}Ca_{0.005}CoO_3$ n-$LaCo_{0.99}Mn_{0.01}O_3$	10	Ag paste	704	399	1	0.1	44	23 mm² × 20	9.6	–	–
Han et al. [41]	2011	p-$Ca_3Co_{3.8}Ag_{0.2}O_9$ n-$Ca_{0.98}Sm_{0.02}O_3$	2	Ag paste	873	523	0.33	0.16	36.8	3 × 6 × 6	51.1	–	–
Funahashi [42]	2011	p-$Ca_{2.7}Bi_{0.3}Co_4O_9$ n-$Ca_{0.9}Yb_{0.1}MnO_3$	108	Ag paste	873	400	10.5	0.097	12000	7 × 3.5 × 5	226.7[b]	–	–
Park and Lee [43]	2013	p-$Ca_{2.76}Cu_{0.24}Co_4O_9$ n-$Ca_{0.8}Dy_{0.2}MnO_3$	4	Ag paste	937	321	0.28	0.07	31	7 × 9 × 25	6.15	–	–
Mele et al. [44]	2014	p-$Ca_3Co_4O_9$ n-$Zn_{0.98}Al_{0.02}O$	6	Ag paste	773	248	0.12	0.02	2.26	4 × 4 × 10	1.2	–	–
Saucke et al. [45]	2015	p-$Ca_3Co_{3.9}O_{9.3}$ n-$CaMn_{0.97}W_{0.03}O_3$	2	Ag paste	1051	727	0.36	0.18	200.4	40.77 mm² × 4	491[c]	1.08	1.08
Hung et al. [25]	2015	p-$Ca_3Co_4O_9$ n-$Zn_{0.98}Al_{0.02}O$	4	Ag	1173	700	0.67	0.17	256	4 × 4 × 8	200	0.72	0.37
Hung et al. [25]	2015	p-HH/$Ca_3Co_4O_9$ n-$Zn_{0.98}Al_{0.02}O$	4	Ag	1173	700	0.76	0.19	829	4 × 4 × 8	650	1.8	1.16

[a] The reported data from the same author is somewhat inconsistent. In addition, the theoretical calculation under ideal conditions, i.e., no losses shows the maximum power density of 200 mW/cm². Therefore, experimental needs to be confirmed.
[b] The value is calculated based on the experimental data in the article.
[c] The length of the legs is computed from the ratio of volume and area.

and R_{Load}, where R_{legs} is the sum of the resistances of the p–n legs without metal electrodes, and R_C is the contact resistance contribution from all interfacial contact between electrodes and the legs. The output power as a function of electric current and the resistive load is given by

$$P = I^2 R_{\text{Load}} = V_{\text{OC}}^2 \left[\frac{R_{\text{Load}}}{(R_{\text{int}} + R_{\text{Load}})^2} \right]. \tag{21.7}$$

The maximum output power is obtained when the external load resistance is equal to the internal resistance ($R_{\text{int}} = R_{\text{Load}}$):

$$P_{\text{Max}} = \frac{V_{\text{OC}}^2}{4R_{\text{int}}} = \frac{V_{\text{OC}}^2}{4R_C + 4R_{\text{legs}}} = \frac{V_{\text{OC}}^2}{4R_C + 4n \left[\rho_p \left(l_p / A_p \right) + \rho_n \left(l_n / A_n \right) \right]}, \tag{21.8}$$

where ρ_p, ρ_n, A_p, A_n, l_p, and l_n are values of the electrical resistivity, cross-sectional area, and length of the p-type (subscript p) and n-type (subscript by n), respectively.

It is clearly seen from Equations 21.5 and 21.8 that the maximum output power is proportional to the total number of the p–n couples and Seebeck coefficient, but its magnitude is inversely proportional to the length of p-type and n-type legs, i.e., the shorter length will provide higher output power.

The power density is defined as the ratio of P_{max} divided by the total area times the number of legs $n(A_p + A_n)$:

$$P_{\text{density}} = \frac{P_{\text{Max}}}{n(A_p + A_n)}. \tag{21.9}$$

Equations 21.8 and 21.9 indicate that the power density increases by decreasing the length of the TE legs as result of increasing P_{max}. As presented in Table 21.2, the highest value of power density falls on the modules with a short leg length such as those in the studies by Funahashi [42] and Saucke et al. In contrast, the module constructed from a longer length has a low value of powder density, e.g., from the studies by Souma et al. [26] and Matsubara et al. [29].

To compare the performance of modules, the conversion efficiency η, which is the ratio of electric power to the total heat input, is a more appropriate way to use [47]. η can be defined as:

$$\eta = \frac{P}{Q_h} = \frac{P}{K(T_h - T_c) + V_{\text{OC}} T_h I - 0.5 I^2 R_{\text{legs}}}. \tag{21.10}$$

Equations 21.7 and 21.10 imply that the efficiency value of the devices is independent from the length of TE legs. As can be seen from Table 21.2, there

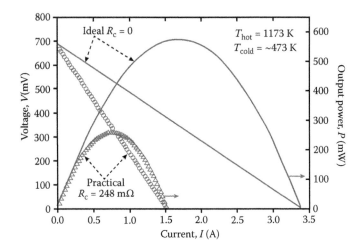

FIGURE 21.5 Experimental (*open circle and triangle*) and ideal conditions (*solid line*) of voltages and output power for the 4 p–n couples comprising from n-type $Zn_{0.98}Al_{0.02}$ and p-type $Ca_3Co_4O_9$. (Hung, L. T. et al.: Segmented thermoelectric oxide-based module for high-temperature waste heat harvesting. *Energy Technology*. 2015. 3. 1143–1151. Copyright Wiley-VCH Verlag GmbH & Co. KGaA. Reproduced with permission.)

is only a few research reported on the value of efficiency making it difficult to collect the data. This may be because the measurement of the heat flux on the hot side of the TEG is complicated especially in the high-temperature range. The maximum obtained efficiency value of all studied data is currently less than 2%, and this value is far from theoretical calculation values, as listed in Table 21.1. One of the main sources, which lead to low conversion efficiency value, is the contact resistances, i.e., high electrical and thermal contact resistances at the interface of oxide materials and metal electrodes. In the experiment, the contact resistance of oxide–metal is in the range of 25–400 $\mu\Omega$ cm^2 for Ag and $Ca_3Co_4O_9$ [19,25,45], and it is in the range of 100–500 $\mu\Omega$ cm^2 for Ag and doped ZnO [25]. Figure 21.5 gives an example of how contact resistance influences the output performance of the oxide module. An interfacial contact resistance of 248 mΩ leads to the suppression of the maximum output power from 575 mW for the ideal theoretical condition to 256 mW for the practically measured condition. Therefore, the study of interfacial contact resistance is a critical area of research to improve the performance of any TEG. In summary, although considerable attempts have been made to produce highly efficient TE oxide modules, the performance of those reported oxide TEGs is still low due to the low performance of oxide materials in the low-mid temperature and high interfacial oxide–metal contact resistances.

21.5 Segmented Oxide-Based Modules

As aforementioned, one of the main drawbacks of reported oxide TEG is the low performance of oxide materials in the temperature range of 300–700 K. In this context, there are two possible solutions where oxide TEG can be combined with other materials: either in cascaded or segmented generators. In a cascade generator, the oxide TEG is stacked with other high-performance TEG working at a mid-low temperature that is normally made of alloys. By this way, the performance of the module can be improved with about twice compared to that of single oxide-based module [42]. However, this type of module often requires an additional electronic device to collect the maximum output of two electric circuits in each single stage. The electrical wires are needed for the connection, and hence, a large amount of heat losses exists in electrical connection wires either heat conductance loss due to the low electrical resistivity or Joule heating loss due to the high electrical resistance [48]. In contrast, a segmented TEG requires only a single electric circuit. In a segmented TE module, the p-type and/or n-type legs are designed by a segmentation of different materials with their highest value of *ZT*. One of the criteria to ensure an improvement in efficiency is that the difference in compatibility factor $\left(s = \sqrt{(1 + zT)}/ST\right)$ values of the selected materials have to be within a factor of two [1,20,22,49]. According to Hung et al. [50], the high conversion efficiency of 12.2% can be achieved in a unicouple of segmented p-type $Ca_3Co_4O_9$ and n-type dual doped ZnO with state-of-the art TE materials of, e.g., BiTe, PbTe, and HH alloys. Figure 21.6 shows the result of a segmented oxide module using n-type $Zn_{0.98}Al_{0.02}O$ and segmented p-type half-Heusler-$Ca_3Co_4O_9$. Its performance was analyzed and compared to nonsegmented module with similar dimension under the same testing conditions. The obtained output power of those modules under the same measurement conditions clearly indicates that segmentation is an effective way to boot up the performance of oxide module. An efficiency of 1.16% is achieved for segmented module, which is three times higher than that of nonsegmented one, as shown in Table 21.2. Furthermore, the conversion

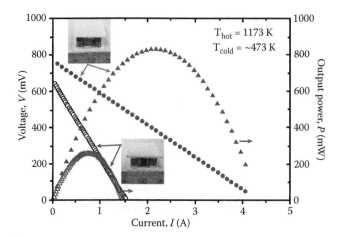

FIGURE 21.6 Power generation characteristics of 4 p–n couples oxide module using n-type $Zn_{0.98}Al_{0.02}O$ and p-type $Ca_3Co_4O_9$ and 4 p–n couples segmented oxide module using n-type $Zn_{0.98}Al_{0.02}$ and segmented p-type half-Heusler-$Ca_3Co_4O_9$. (Hung, L. T. et al.: Segmented thermoelectric oxide-based module for high-temperature waste heat harvesting. *Energy Technology*. 2015. 3. 1143–1151. Copyright Wiley-VCH Verlag GmbH & Co. KGaA. Reproduced with permission.)

efficiency of 5% has also been achieved in single p-type segmented leg of misfit-layered cobaltite and HH alloy [19]. In both cases, the segmented legs are highly stable at a high temperature in air.

21.6 Outlook and Challenges

Oxide TE modules have been rapidly developed in the last 16 years with many attempts to improve the conversion efficiency of TEG modules for high-temperature waste heat harvesting. Some of the efforts have been successfully realized, and promising devices are now commercially available [51]. The theoretical calculation has pointed out that TEG with the efficiency of 7.9% could be possibly obtained in oxide materials if parasitic losses can be minimized.

There is a big gap existing between the performance of materials and performance of devices. Although many studies have been conducted to improve the performance of oxide materials, the maximum conversion efficiency of current oxide module is still lower than 2%. In fact, TE properties of oxide materials are studied as a function of temperature, thermal cycling, or heat treatment duration in which all parasitic losses were disregarded. In practice, oxide TE legs in module operate under a large temperature gradient and generating power conditions with the contribution of electrical and thermal losses. Thus, studying the performance of oxide material under working conditions would help reduce the currently existing gap of materials and devices.

We have also clearly pointed out that one of the main sources leading to the decrease in the performance of oxide TEG is the contact resistances. Understanding the contact resistance requires more study to theoretically clarify the origin of the resistance and to not only make a good electrical connection but also have durability from thermal cycling and large temperature span. Also, studying metal–oxide interfacial contact resistances under module level could be an important topic.

The combination of taking the advantages of oxides and other intermetallic alloys in segmented TEG has been realized as one of the most efficient ways to produce high efficiency. The stable and high conversion efficiency of 5% was achieved on segmented p-type-based oxide materials. The future study, therefore, should target segmented n-type oxide materials.

References

1. Hung, L. T., Van Nong, N., Linderoth, S., and Pryds, N. Segmentation of low-cost high efficiency oxide-based thermoelectric materials. *Phys Status Solidi* 2015;212:767–74.
2. Koumoto, K., Wang, Y., Zhang, R., Kosuga, A., and Funahashi, R. Oxide thermoelectric materials: A nanostructuring approach. *Annu Rev Mater Res* 2010;40:363–94.
3. Koumoto, K. Oxide Thermoelectrics. *Thermoelectric Handbook*. CRC Press, Boca Raton, FL; 2005;1–35.
4. Yang, J., and Caillat, T. Thermoelectric materials for space and automotive power generation. *MRS Bull* 2006;31:224–9.

5. Kambe, M., Jinushi, T., and Ishijima, Z. Encapsulated thermoelectric modules and compliant pads for advanced thermoelectric systems. *J Electron Mater* 2010;39:1418–21.

6. Shin, W., and Murayama, N. Li-doped nickel oxide as a thermoelectric material. *Jpn J Appl Phys, Part 2 Lett* 1999;38:1336–8.

7. Moon, J.-W., Seo, W.-S., Okabe, H., Okawa, T., and Koumoto, K. Ca-doped $RCoO_3$ (R = Gd, Sm, Nd, Pr) as thermoelectric materials. *J Mater Chem* 2000;10:2007–9.

8. Tomeš, P., Trottmann, M., Suter, C., Aguirre, M. H., Steinfeld, A., Haueter, P. et al. Thermoelectric oxide modules (TOMs) for the direct conversion of simulated solar radiation into electrical energy. *Materials (Basel)* 2010;3: 2801–14.

9. Nong, N. V., and Ohtaki, M. Thermoelectric properties and local electronic structure of rare earth-doped $Ca_3Co_2O_6$. *2006 25th International Conference on Thermoelectrics* 2006:62–5.

10. Van Nong, N., Pryds, N., Linderoth, S., and Ohtaki, M. Enhancement of the thermoelectric performance of p-type layered oxide $Ca_3Co_4O_{(9+\delta)}$ through heavy doping and metallic nanoinclusions. *Adv Mater* 2011;23:2484–90.

11. Ito, M., and Furumoto, D. Microstructure and thermoelectric properties of $Na_xCo_2O_4$/Ag composite synthesized by the polymerized complex method. *J Alloys Compd* 2008;450:517–20.

12. Sui, J., Li, J., He, J., Pei, Y.-L., Berardan, D., Wu, H. et al. Texturation boosts the thermoelectric performance of BiCuSeO oxyselenides. *Energy Environ Sci* 2013;6:2916.

13. Funahashi, R. Mikami, M. Urata, S. Kouuchi, T. Mizuno, K. and Chong, K. Thermoelectric properties of Ni-based oxides. *Mat Res Soc Symp Proc* 2004;793:1–11.

14. Hsiao, C.-L., Chang, W.-C., and Qi, X. Sol-gel synthesis and characterisation of nanostructured $LaNiO_{3-x}$ for thermoelectric applications. *Sci Adv Mater* 2014;6:1406–11.

15. Yasukawa, M., and Murayama, N. A promising oxide material for high-temperature thermoelectric energy conversion: $Ba_{1-x}Sr_xPbO_3$ solid solution system. *Mater Sci Eng B* 1998;54:64–9.

16. Bocher, L., Aguirre, M. H., Logvinovich, D., Shkabko, A., Robert, R., Trottmann, M. et al. $CaMn_{1-x}Nb_xO_3$ ($x \le 0.08$) perovskite-type phases as promising new high-temperature n-type thermoelectric materials. *Inorg Chem* 2008;47:8077–85.

17. Wang, N., He, H., Ba, Y., Wan, C., and Koumoto, K. Thermoelectric properties of Nb-doped $SrTiO_3$ ceramics enhanced by potassium titanate nanowires addition. *J Ceram Soc Japan* 2010;118:1098–101.

18. Ohtaki, M., Araki, K., and Yamamoto, K. High thermoelectric performance of dually doped ZnO ceramics. *J Electron Mater* 2009;38:1234–8.

19. Hung, L. T., Van Nong, N., Snyder, G. J., Viet, M. H., Balke, B., Han, L. et al. High performance p-type segmented leg of misfit-layered cobaltite and half-Heusler alloy. *Energy Convers Manag* 2015;99:20–7.

20. Ngan, P. H., Christensen, D. V., Snyder, G. J., Hung, L. T., Linderoth, S., Nong, N. V. et al. Towards high efficiency segmented thermoelectric unicouples. *Phys Status Solidi* 2014;211:9–17.

21. Snyder, G. J. Thermoelectric power generation. Thermoelectrics Handbook. CRC Press, Boca Raton, FL; 2005:1–26.

22. Snyder, G., and Ursell, T. Thermoelectric efficiency and compatibility. *Phys Rev Lett* 2003;91:148301.
23. Müller, E., Zabrocki, K., Goupil, C., Snyder, G.J. and Seifert, W. *Functionally graded thermoelectric generator and cooler elements.* Chapter 4 in Thermoelectrics and its Energy Harvesting, Vol 1, edited by D. M. Rowe. CRC Press, 2012.
24. Seifert, W., Pluschke, V., Goupil, C., Zabrocki, K., Müller, E., and Snyder, G. J. Maximum performance in self-compatible thermoelectric elements. *J Mater Res* 2011;26:1933–9.
25. Hung, L. T., Van Nong, N., Han, L., Bjørk, R., Ngan, P. H., Holgate, T. C. et al. Segmented thermoelectric oxide-based module for high-temperature waste heat harvesting. *Energy Technol* 2015;3:1143–51.
26. Souma, T., Ohtaki, M., Shigeno, M., Ohba, Y., Nakamura, N., and Shimozaki, T. Fabrication and power generation characteristics of p-$NaCo_2O_4$ n-ZnO oxide thermoelectric modules. *Proceedings of ICT* 2006;603–6.
27. Souma, T., Ohtaki, M., Ohnishi, K., Shigeno, M., Ohba, Y., and Shimozaki, T. Power generation characteristics of oxide thermoelectric modules incorporating nanostructured ZnO sintered materials. *2007 26th International Conference on Thermoelectrics* 2007:38–41.
28. Shina, W., Murayamaa, N., Ikedab, K., and Sagob, S. Thermoelectric power generation using Li-doped NiO and $(Ba, Sr)PbO_3$ module. *J Power Sources* 2001;103:80–5.
29. Matsubara, I., Funahashi, R., Takeuchi, T., Sodeoka, S., Shimizu, T., and Ueno, K. Fabrication of an all-oxide thermoelectric power generator. *Appl Phys Lett* 2001;78:3627.
30. Funahashi, R., Urata, S., Mizuno, K., Kouuchi, T., and Mikami, M. $Ca_{2.7}Bi_{0.3}Co_4O_9/La_{0.9}Bi_{0.1}NiO_3$ thermoelectric devices with high output power density. *Appl Phys Lett* 2004;85:1036.
31. Reddy, E. S., Noudem, J. G., Hebert, S., and Goupil, C. Fabrication and properties of four-leg oxide thermoelectric modules. *J Phys D Appl Phys* 2005;38:3751–5.
32. Funahashi, R., Mikami, M., Mihara, T., Urata, S., and Ando, N. A portable thermoelectric-power-generating module composed of oxide devices. *J Appl Phys* 2006;99:66117.
33. Urata, S., Funahashi, R., and Mihara, T. Power generation of p-type $Ca_3Co_4O_9$/n-type $CaMnO_3$ module. *2006 25th International Conference on Thermoelectrics* 2006:501–4.
34. Funahashi, R., and Urata, S. Fabrication and application of an oxide thermoelectric system. *Int J Appl Ceram Technol* 2007;4:297–307.
35. Urata, S., Funahashi, R., Mihara, T., Urata, R., Mihara, T. S., and Funahashi, R. Power generation of p-type $Ca_3Co_4O_9$/n-type $CaMnO_3$ module. *2006 25th International Conference on Thermoelectrics* 2006:501–4.
36. Park, K., Choi, J. W., and Lee, C. W. Characteristics of thermoelectric power modules based on p-type $Na(Co_{0.95}Ni_{0.05})_2O_4$ and n-type $Zn_{0.99}Sn_{0.01}O$. *J Alloys Compd* 2009;486:785–9.
37. Tomeš, P., Robert, R., Trottmann, M., Bocher, L., Aguirre, M. H., Bitschi, A. et al. Synthesis and characterization of new ceramic thermoelectrics implemented in a thermoelectric oxide module. *J Electron Mater* 2010;39:1696–703.
38. Choi, S.-M., Lee, K.-H., Lim, C.-H., and Seo, W.-S. Oxide-based thermoelectric power generation module using p-type $Ca_3Co_4O_9$ and n-type $(ZnO)_7In_2O_3$ legs. *Energy Convers Manag* 2011;52:335–9.

39. Lim, C.-H., Choi, S.-M., Seo, W.-S., and Park, H.-H. A power-generation test for oxide-based thermoelectric modules using p-type $Ca_3Co_4O_9$ and n-Type $Ca_{0.9}Nd_{0.1}MnO_3$ legs. *J Electron Mater* 2011;41:1247–55.

40. Inagoya, A., Sawaki, D., Horiuchi, Y., Urata, S., Funahashi, R., and Terasaki, I. Thermoelectric module made of perovskite cobalt oxides with large thermopower. *J Appl Phys* 2011;110:123712.

41. Han, L., Jiang, Y., Li, S., Su, H., Lan, X., Qin, K. et al. High temperature thermoelectric properties and energy transfer devices of $Ca_3Co_{4-x}Ag_xO_9$ and $Ca_{1-y}Sm_yMnO_3$. *J Alloys Compd* 2011;509:8970.

42. Funahashi, R. Waste heat recovery using thermoelectric oxide materials. *Sci Adv Mater* 2011;3:682–6.

43. Park, K., and Lee, G. W. Fabrication and thermoelectric power of π-shaped $Ca_3Co_4O_9$/$CaMnO_3$ modules for renewable energy conversion. *Energy* 2013; 60:87–93.

44. Mele, P., Kamei, H., Yasumune, H., Matsumoto, K., and Miyazaki, K. Development of thermoelectric module based on dense $Ca_3Co_4O_9$ and $Zn_{0.98}Al_{0.02}O$ legs. *Met Mater Int* 2014;20:389–97.

45. Saucke, G., Populoh, S., Thiel, P., Xie, W., Funahashi, R., and Weidenkaff, A. Compatibility approach for the improvement of oxide thermoelectric converters for industrial heat recovery applications. *J Appl Phys* 2015;118:1–8.

46. Van Nong, N., Pryds, N., Linderoth, S., and Ohtaki, M. Enhancement of the thermoelectric performance of p-type layered oxide $Ca_3Co_4O_{9+\delta}$ through heavy doping and metallic nanoinclusions. *Adv Mater* 2011;23:2484–90.

47. Cobble, M. Calculations of generator performance. *CRC Handbook of Thermoelectrics*, vol. 2, CRC Press, Boca Raton, FL; 1995.

48. Heikes, R. R., and Ure, R. W. Thermoelectricity: Science and engineering. Interscience Publishers, Genva; 1961.

49. Caillat, T., Fleurial, J.-P., Snyder, G. J., and Borshchevsky, A. Development of high efficiency segmented thermoelectric *unicouples. Proceedings of ICT2001: 20the International Conference on Thermoelectrics.* (Cat. No.01TH8589):282–5.

50. Hung, L. T., Van Nong, N., Linderoth, S., and Pryds, N. Segmentation of low-cost high efficiency oxide-based thermoelectric materials. *Phys Status Solidi Appl Mater Sci* 2015;212:767–74.

51. Tecteg Power Generator. http://tecteg.com/.

CHAPTER 22

Solar Thermoelectric Power Generators

Yucheng Lan and Zhifeng Ren

Contents

22.1 Introduction

Thermoelectric generators (TEGs) can directly convert heat into electricity. Based on heat sources, TEGs can be cataloged as geothermal TEGs [1–3] (geothermal heat as energy sources), automotive TEGs [4–6] (engine or catalytic converter waste heat as energy sources), radioisotope TEGs [7] (radioisotope heat as energy sources), and others [8–15]. This chapter focuses on solar thermoelectric generators (STEGs), which harvest solar energy into electricity.

The sun is the most plentiful energy source for the earth. The total solar energy absorbed by earth's atmosphere, oceans, and lands is approximately 3,850,000 exajoules (1 EJ = 10^{18} J) per year while human energy consumption was 539 EJ in 2010. In other words, all the solar energy received from the sun in 1 h can satisfy the whole world's demand for 1 year. Therefore, solar-derived electricity represents a vast and abundant renewable energy resource.

Solar energy has been harvested through photovoltaic techniques and thermal routes. Solar TE methods are one kind of thermal routes to generate electricity from solar energy using TEGs through the Seebeck effect.

STEGs have been designed since the end of the nineteenth century and recently gained research attentions due to improvements in TE materials. The first documented STEG design can be dated back to 1888 [16,17]. Solar rays were concentrated by a lens upon one side of TE elements that were coated with absorbent materials. Adjustable tracking devices were subsequently added to the designed STEGs [18,19].

The first experimental results for STEGs were published in 1913 [20]. An asphalt layer was employed as a solar absorber and copper–constantan thermocouples as TEGs. The hot-side temperature of the devices was 100°C. The device efficiency was very low, 0.008%.

The first effective STEG was reported in 1954 [21], using flat-plate solar collectors in combination with $ZnSb/Bi_{91}Sb_9$ TE alloys. The devices demonstrated 0.6% efficiency, which increased to 3.4% when a 50-fold concentrating lens was added.

Later, many theoretical results were reported while few experimental results were published. These early experimental works showed a low device efficiency (usually <1%), primarily due to low figure of merit ZT (<0.4) of TE modules, low solar concentration, and low hot-side temperatures.

A significant improvement, 4.6% device efficiency, was experimentally demonstrated in 2011 [22]. The demonstrated flat-panel STEGs consisted of selective absorbers to absorb solar energy, vacuum enclosures to minimize conductive and convective losses, nanostructured TE modules with high ZT to generate electricity, and thermal concentration designs to increase hot-side temperatures. Five years later, the same group increased the STEG efficiency to 7.9% [23] by using a concentrated optical/thermal design and nanostructured TE modules. Both the STEGs do not require tracking system and can effectively convert solar energy into electricity.

Some historical experimental results of STEGs are listed in Table 22.1. Here the device efficiency η_{STEG} of STEGs is defined as the ratio of the output electricity $W_{electricity}$ to the input solar energy W_{solar}:

$$\eta_{STEG} = W_{electricity}/W_{solar}. \tag{22.1}$$

Figure 22.1 shows the historical experimental efficiency of STEGs.

A typical STEG usually consists of four subsystems [24]: a solar collection subsystem (collector), a solar absorption subsystem (absorber), a TEG (TE module), and an auxiliary management subsystem (heat sink, insulator, vacuum enclosure, etc.). Solar collectors can guide incident solar radiation upon solar absorbers. Solar absorbers convert the guided solar energy (electromagnetic energy) to heat (thermal energy). TEGs convert heat into electricity through the Seebeck effect.

TABLE 22.1 Summary of STEG Experimental Results

Year	η_{STEG}	Optical/Thermal Subsystems	TE Materials (p-/n-Type)	T_c(°C)	ΔT(°C)	$C_{optical}$	$C_{thermal}$	Ref.
1913, 1922	0.008%	Flat-plate window glasses/asphaltum layer	copper-constantan	RT	70[a]	1	1	Coblentz et al. [20,25]
1954	0.068%	Flat-plate glasses/blackened copper plate	chromel-constantan	20	60–80	1	1	Telkes [21]
1954	0.63%	Flat-plate glasses/blackened copper plate	$ZnSb/Bi_{91}Sb_9$	20	60–80	1	1	Telkes [21]
1954	3.35%	Fresnel-lens/blackened copper plate	$ZnSb/Bi_{91}Sb_9$	20	240	50	1	Telkes [21]
1980	<0.5%	Semi-parabolic concentrator/matt-black aluminum flat-plate collector	Bi_2Te_3	30	120	n/a	1	Goldsmid et al. [26]
2010	3%	Solar concentrator and Fresnel lens/selective surface coating	Bi_2Te_3	RT	~150	66	1	Amatya and Ram [27]
2011	2.9%	Parabolic dish concentrator/Cu plates	Bi_2Te_3	RT	109	n/a	n/a	Fan et al. [28]
2011	4.6%	Flat glass/wavelength-selective solar absorber	Bi_2Te_3	20	200	1	299	Kraemer et al. [22]
2011	5.2%	Lens/wavelength-selective solar absorber	Bi_2Te_3	20	200	1.5	196	Kraemer et al. [22]
2016	7.4%	Lens/wavelength-selective solar absorber	Bi_2Te_3 and SKU	20	400	50	5.4	Kraemer et al. [23]
2016	9.6%	Lens/wavelength-selective solar absorber	Bi_2Te_3 and SKU	20	550	200	1.4	Kraemer et al. [23]

Note: η_{STEG}, STEG efficiency; $C_{optical}$, optical concentration ratio; $C_{thermal}$, thermal concentration ratio; T_h, TE hot-side temperature; T_c, TE cold-side temperature; $\Delta T = T_h - T_c$; RT, room temperature.

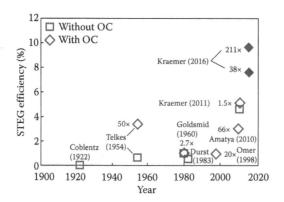

FIGURE 22.1 Historical overview of experimental efficiency demonstrations of STEGs with (*diamond symbol*) and without (*square symbol*) optical concentration (OC). STEG efficiencies are based on the solar radiation flux incident on the solar absorber, excluding optical losses from concentrating optics: Amatya and Ram [27], Coblentz [25], Durst et al. [29], Goldsmid [30], Kraemer et al. [22,23], Omer and Infield [31], and Telkes [21]. (Reprinted by permission from Kraemer, D. et al., *Nature Energy*, 1, 16153, 2016. Copyright 2011, Macmillan Publishers Ltd.)

The overall efficiency of a STEG device η_{STEG} depends on the optical collection efficiency $\eta_{optical}$, thermal absorption efficiency $\eta_{thermal}$, TE efficiency η_{TEG}, and auxiliary efficiency $\eta_{auxiliary}$. The device efficiency (η_{STEG}) of a STEG is usually expressed as the product of these efficiencies [23]:

$$\eta_{STEG} = \eta_{optical} \times \eta_{thermal} \times \eta_{TEG} \times \eta_{auxiliary}. \tag{22.2}$$

Here, the optical collection efficiency $\eta_{optical}$ is the transmission efficiency of incident solar radiation onto solar absorbers through solar collectors. The optical efficiency $\eta_{optical}$ can be determined by optical losses including the loss of the diffused light and possible glass transmission affecting the solar radiation flux incident on solar absorbers. The thermal absorption efficiency $\eta_{thermal}$ is the conversion efficiency of solar irradiation energy to thermal energy by solar absorbers. The absorption efficiency $\eta_{thermal}$ is the measurement of how efficiently the solar irradiation incident on solar absorbers is converted into heat. The TE efficiency η_{TEG} is the conversion efficiency of thermal energy to electricity by TEGs. The auxiliary efficiency $\eta_{auxiliary}$ accounts for possible system parasitic losses such as electricity consumption for pumping and cooling.

According to Equation 22.2, any innovative designs or techniques, such as an evacuated enclosure to prevent air convection and conduction losses to increase the auxiliary efficiency $\eta_{auxiliary}$, can improve the device efficiency η_{STEG} of STEGs.

22.1.1 Optical Subsystem: Solar Collectors/Concentrators

Figure 22.2 shows four kinds of optical systems used in typical STEGs. Such types of solar energy collectors have also been used for photovoltaic cells and other thermal methods to harvest solar energy [32,33].

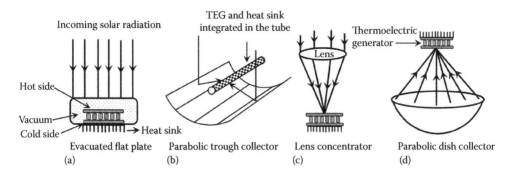

FIGURE 22.2 Different optical concentration systems. (a) Without optical concentrator. (b) A parabolic mirror as an optical concentrator. (c) An optical lens as an optical concentrator. (d) A parabolic disk as an optical concentrator. (Sundarraj, P. et al., *RSC Advances*, 4, 46860–46874, 2014. Reproduced by permission of The Royal Society of Chemistry.)

The simplest optical subsystem is a transparent flat plate, as shown in Figure 22.2a. It is a nonfocusing plate and can be a window glass or low-reflection coated glass [21]. Solar irradiation can transmit through the flat plates with a transmittance of 80–96% at normal incidence. Correspondingly, its optical efficiency $\eta_{optical}$ is the transmittance, and the optical concentration ratio $C_{optical}$ is 1 for the optical subsystem.

Solar irradiation energy can also be concentrated using optical lenses (Figure 22.2c) and reflectors (Figure 22.2b and d) to produce higher temperatures at foci. These optical lenses and reflectors are termed concentrators. Solar absorbers and TE modules are placed at the foci of the concentrators. Large temperature differences can be created across TE modules.

The optical concentration ratio of concentrators $C_{optical}$ is defined as the ratio of the optical concentrator area to the absorber area upon which the solar light is concentrated. The concentration ratio $C_{optical}$ is 10–200 depending on solar concentrators. Depending on the dimensional size of TE modules, an optical concentration $C_{optical}$ of 40–200 is usually needed to create an appreciable temperature difference of 200°C. However, an optical concentration by a factor of 40–200 usually requires tracking, incurring an additional cost to the system, which is unattractive, given a low efficiency.

Considering 1 sun to be approximately 1 kW/m², the flux after concentration is given by the solar flux multiplied by the optical concentration ratio $C_{optical}$.

The optical concentrator efficiency $\eta_{optical}$ is the ratio of solar flux that reaches thermal absorbers to the total solar flux incident upon optical concentrators. Typical optical efficiencies are 40–90% depending on concentrators and solar tracking systems [32].

22.1.2 Thermal Absorption Subsystem: Solar Absorbers

There are three main kinds of possible thermal absorption constructions, as shown in Figure 22.3 [34].

The first are flat-plate absorbers, as shown in Figure 22.3a, with TEGs attached to it. The incident solar irradiation is reflected by absorbers, causing

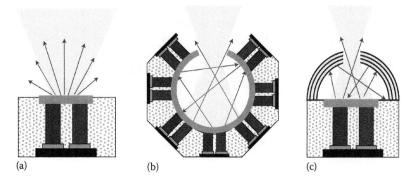

FIGURE 22.3 STEG designs that have been either modeled or experimentally demonstrated. (a) Flat-plate absorber with no cavity to limit radiative losses. (b) Cavity lined with many TEG modules. (c) Cavity with nonuniform walls. Medium-dark blocks: TE legs. Lighter-dark blocks: solar absorbers. Shadow regions: solar irradiation. (With kind permission from Springer Science+Business Media: *Journal of Electronic Materials*, High-temperature high-efficiency solar thermoelectric generators, 43, 2014, 2348–2355, Baranowski, L. L. et al.)

solar energy losses. Therefore, the quality of the flat-plate absorbers, absorptance, and emittance dominates the thermal absorption efficiency $\eta_{thermal}$.

The second kind of absorption construction uses a blackbody-type trap, with a narrow opening at the focus as shown in Figure 22.3b, to diminish reradiation losses from the target. In the second design, hot sides of TE modules are in contact with the outer wall of the blackbody trap.

Figure 22.3c shows the third kind of design. A flat-plate absorber is enclosed by a nonuniform wall. The focused solar beam passes through the narrow opening of the cavity onto a flat-plate absorber. The reradiation can be reflected back by the inner wall to reduce solar energy losses.

In these optical absorption constructions, flat-plate absorbers are critical to the efficiency $\eta_{thermal}$ of the subsystems. Blackened metal plates were employed as flat-plate absorbers during the early development of STEGs, such as aluminum plates coated with matte black paint [26], silver foils painted with graphite [21], and copper plates [28]. The low absorptance of these blackened metal plates degraded the device efficiency of STEGs. In order to improve thermal absorption efficiency $\eta_{thermal}$, solar-selective absorbers have been developed. Al–SiO–Al–SiO-coated Al foils were used for solar absorbers that have long-term stability up to 600°C [35,36]. Its absorptivity $\alpha > 0.8$ and its emissivity $\varepsilon < 0.08$ for the solar spectrum, giving an α/ε ratio of 10. The coating had an average 2% reflection from 0.3 to 1.3 μm, with reflection at 2 μm of 60% or better and reflection beyond 2.5 μm of 90% or better [36]. Temperature-dependent absorptance of this kind of absorbers is shown as the curve *a)* in Figure 22.4a. Alternating six-layer Al$_2$O$_3$–Mo Honeywell coating was also developed as high-temperature selective solar absorbers [37]. It was stable to 1200 K. Its solar absorptance was 85–91% and its emittance was 6–16% at 540°C. The curve *b)* in Figure 22.4a [38] shows its temperature-dependent absorption and emittance. Other high-temperature absorbers consisting of alternating insulating layer/metal layer/

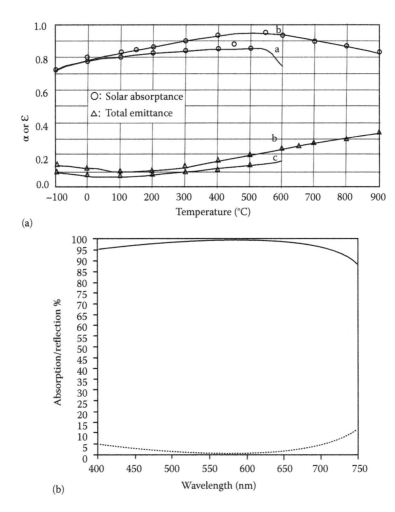

(a)

(b)

FIGURE 22.4 (a) Absorption (α) and emittance (ε) of Al–SiO–Al–SiO on Al coating (curves marked *a*) and alternating six-layer Al$_2$O$_3$–Mo coating on Mo (curves marked *b*) as a function of temperature. (Reprinted from *Advanced Energy Conversion*, 7, Fuschillo, N., and Gibson, R., Germanium-silicon, lead telluride, and bismuth telluride alloy solar thermoelectric generators for Venus and Mercury probes, Copyright [1967], with permission from Elsevier.) (b) Absorption (solid line) and reflection (dashed line) spectra of 20 nm Au/SiO$_2$ nanocomposite deposited on a 100 nm gold film with 25 nm SiO$_2$ spacer layer measured at an angle of incidence of 6°. (Hedayati, M. K. et al.: Design of a perfect black absorber at visible frequencies using plasmonic metamaterials. *Advanced Materials.* 2011, 23, 5410–5414. Copyright Wiley-VCH Verlag GmbH & Co. KGaA. Reproduced with permission.)

insulating layer were also developed that exhibited solar absorptivities above 0.8 and thermal emissivities below 0.15 [37,39–41].

Metal multilayer coatings are a kind of solar wavelength-selective absorbers operating at 500°C [42]. A deposited nine-layer metal coating can achieve an absorptance of 0.937 and emittance of 0.24 at room temperature and 0.34 at 500°C. Figure 22.4b shows the optical properties of plasmonic metamaterials that were employed as perfect black absorbers at visible frequencies [38]. Optically thick metallic Au films were deposited on glass substrates, followed by a thin dielectric SiO$_2$ layer acting as the spacer layer. Then, a thin

(20 nm) nearly percolated film of Au/SiO$_2$ nanocomposite was deposited on the top. Such kind of films almost absorbed 100% visible light [38].

An ideal solar-selective absorber should have unity absorptance in the short wavelength range of the solar spectrum with a step transition to zero absorptance in the infrared region to prevent the emittance of blackbody radiation. The transition should occur at a cut-off wavelength that depends on both the temperature of the absorber surface and optical concentration. With the development of solar selective-absorbers, the thermal absorption efficiency $\eta_{thermal}$ is close to 100%.

22.1.3 Thermoelectric Subsystem: TEGs

STEG system efficiency η_{STEG} depends on TE efficiency η_{TEG} besides optical efficiency $\eta_{optical}$ and thermal efficiency $\eta_{thermal}$. At present, the TE efficiency η_{TEG} is the bottleneck to limit the application of STEGs.

TE devices are usually consisted of n- and p-type TE legs connected electrically in series and thermally in parallel. TEGs utilize the Seebeck effect to generate electricity when one side of TEGs is maintained at a higher temperature compared to the other side. The efficiency of an ideal TE device η_{TEG} can be written as a function of temperature and the figure of merit ZT:

$$\eta_{TEG} = \frac{T_h - T_c}{T_h} \frac{\sqrt{1+(ZT)_M} - 1}{\sqrt{1+(ZT)_M + T_c/T_h}}, \tag{22.3}$$

where T_c is the cold-side temperature, T_h is the hot-side temperature, and ZT is the effective figure of merit of TE legs between T_c and T_h. ZT has been greatly enhanced in the past decades through nanostructuring, as discussed in Chapters 2 through 10.

TE nanocomposites with high ZT have been fabricated into various TE devices [43–46]. These TE devices can be utilized in STEGs. Table 22.2 summarizes some typical TE nanocomposites that can be used for STEGs. The ZT values of these nanostructured bulk materials were significantly improved to around 1.0 from ~0.5. According to Equation 22.3, the TE efficiency η_{TEG} should be increased if these TE nanocomposites were integrated into STEGs.

Figure 22.5 [47] shows the figures of merit of some advanced TE nanocomposite materials. ZT peak values of the TE nanocomposites are higher than 1.0. Depending on temperature-dependent TE properties, different types of TE materials can be chosen from room temperature to 1000°C for STEGs. For instance, bismuth telluride alloys (peak ZT occurs around 100°C) can be used in room-temperature STEGs working from 30°C to 200°C [22,33]. PbTe/PbSe alloys, skutterudite, and half-Heusler compounds can be utilized in medium-temperature STEGs working from 200°C to 500°C [33]. SiGe alloys can operate at high temperature, around 1000°C [7,33]. These nanostructured bulks should play an important role in improving the device efficiency of STEG systems.

TABLE 22.2   Some Typical TE Nanocomposites for Solar TE Energy Conversion

TE Material	Type	ZT_{max}	Reference
Bi_2Te_3			
$Bi_2Te_{2.7}Se_{0.3}$	n	1.04	Yan et al. [48]
$Bi_{0.4}Sb_{1.6}Te_3$	p	1.80	Fan et al. [49]
PbTe			
$Ag_{0.53}Pb_{18}Sb_{1.2}Te_{20}$	n	1.70	Cook et al. [50]
$Pb_{0.98}Na_{0.02}Te_{0.85}Se_{0.15}$	p	1.80	Pei et al. [51]
Skutterudite			
$Yb_{0.2}Co_4Sb_{12.3}$	n	1.30	Li et al. [52]
Half-Heusler			
$Hf_{0.5}Zr_{0.25}Ti_{0.25}NiSn_{0.99}Sb_{0.01}$	n	1	Joshi et al. [54]
$Hf_{0.8}Ti_{0.2}CoSb_{0.8}Sn_{0.2}$	p	1	Yan et al. [54]
SiGe alloys			
$Si_{80}Ge_{20}$	n	1.30	Raag et al. [55]
$Si_{80}Ge_{20}$	p	0.95	Poudel et al. [56]

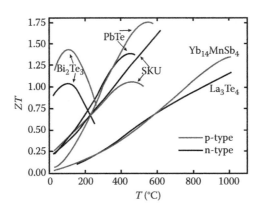

FIGURE 22.5 *ZT* of some TE nanocomposites. SKU: skutterudite. (Baranowski, L. L. et al., *Energy and Environmental Science*, 5, 9055–9067, 2012. Reproduced by permission of The Royal Society of Chemistry.)

22.1.4 Auxiliary Subsystem: Heat Sinks and Others

A key challenge in solar TE power conversion is to create a significant temperature difference across TE devices. A critical approach is by using heat sinks to the cold sides of TE devices. According to Equation 22.3, the TE efficiency η_{TEG} can be increased when a large temperature difference can be created between the ends of TE devices, enhancing the overall efficiency η_{STEG} of STEG devices.

22.1.5 Applications

STEGs have been attractive for deep-space probe missions in near-sun orbits [35,36,57,58] because of their resistance to high radiation intensity. STEGs could advantageously replace the standard power sources such as the solar (cells) panel or the radioisotope TEGs in space. For example, silicon–germanium alloys were manufactured into flat-plate solar TEGs [59] as a nonpropulsive power source for near-sun missions at a distance of 0.25 AU from the sun [55]. The designed generators can produce approximately 150 W of electrical power at a load voltage of about 28 V, operating at a nominal temperature of 1500°F [59].

Unfortunately, the device efficiency of STEGs is still low and cannot compete with solar cells and hot pipes yet. More works are being carried out to increase its efficiency.

22.2 Flat-Panel Nonconcentrated STEGs

Flat-panel nonconcentrated STEGs are the simplest design to harvest solar energy in all STEGs. The first experimental STEGs were designed as this structure in 1913 [20]. Black asphalt was deposited onto copper foils as solar receivers to absorb solar radiation [20,25]. Copper–constantan thermocouples were employed as TE devices, soldered to the copper foil receivers. The TE thermocouples were connected in series and sealed into an insulating base, from which the TE thermocouples were cooled by ambient water. The thermocouples were then covered by air-spaced glass panes to diminish heat losses from the hot receiver. The flat-glasses also worked as optical collectors, $C_{optical}$ = 1. A temperature difference of 17°C was created between the hot sides and cold sides of the TE devices. The device efficiency η_{STEG}, the ratio of the electricity output to the total incident solar energy, was about 0.008% of the first STEGs.

In 1954, Telkes [21] used a similar design to investigate flat-panel collector STEGs. Figure 22.6 shows the STEG devices. Copper sheets were blackened with ball milled graphite to absorb solar radiation, used as solar receivers of the STEGs. p-Type ZnSb alloys (with 35:65 wt%) and n-type $Bi_{91}Sb_9$ alloys were employed as p-type and n-type TE elements, respectively, soldered to the copper sheet receivers. The solar receivers were then located behind glass panes with a transmission of 83%, $\eta_{optical}$ ~ 83%. The cold side of the TEGs (Figure 22.6b) was maintained at a fixed temperature (about

(a)   (b)

FIGURE 22.6 Telkes's STEG consisting of 25 TE junctions. (a) Front and (b) back views. (Reprinted with permission from Telkes, M., *Journal of Applied Physics*, 25, 765–777, 1954. Copyright 1954, American Institute of Physics.)

20°C). A temperature difference of 60–80°C was created between the cold side and the receivers when two glass panels were framed, while 100–130°C was achieved when a four-glass pane frame was used to reflect solar irradiation. The transparent glass panes diminished heat losses through limiting air convection and increasing reflection/absorption of solar re-irradiation from the receivers.

The highest device efficiency of 0.63% was achieved under a temperature difference of 60–80°C, as shown in Figure 22.7. An efficiency of 1.05% was expected under a temperature difference of 100–130°C. Fifty percent of the incident solar energy may be collected at a temperature difference of 63°C when using two glass panes. The efficiency could be improved if four low-reflection panels were used to increases the temperature difference to 100°C. Most heat was lost through air convection in Telkes's devices, leading to a low optical/thermal efficiency and a low overall efficiency of the STEGs.

Goldsmid et al. [26] employed 31 Bi_2Te_3 modules as TEGs of an STEG in 1980. An absorber was an aluminum plate with 0.5 × 0.5 m^2 of cross section coated with matte black paint (Figure 22.8). The absorber was attached to the TE modules via an aluminum block. The heat sink consisting of aluminum fin arrays was attached to the TE modules through an aluminum block that was supported by an aluminum back plate. The TE modules had a cross section of about 54 × 54 mm^2 and a thickness of about 6 mm. The *ZT* of the TE modules was about 0.72 at 300 K. The front of the absorber was covered with two plates of low-iron glass, each 2 mm in thickness. The STEG was tested under solar irradiation with a heat input of 105 W. A temperature difference of 70 K was expected across the TE modules. An STEG efficiency η_{STEG} was less than 0.5%.

Flat-plate solar TEGs were also tested in a vacuum. Fuschillo et al. [36] employed very thin coated aluminum plates as absorbers, which gave an absorptivity of >0.8 and an emissivity of <0.08. Nickel disks were soldered to

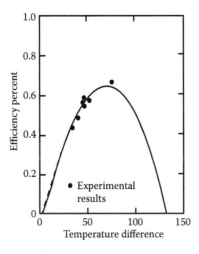

FIGURE 22.7 Efficiency of Telkes's STEG, with solar energy on clear days. (Reprinted with permission from Telkes, M., *Journal of Applied Physics*, 25, 765–777, 1954. Copyright 1954, American Institute of Physics.)

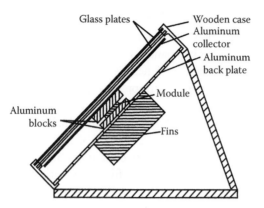

FIGURE 22.8 Schematic section through the STEG with a flat-plate collector. (Reprinted with permission from Goldsmid, H. J. et al., Solar thermoelectric generation using bismuth telluride alloys, *Solar Energy*, 24, 435–440, 1980. Copyright 1980, Elsevier.)

the aluminum absorbers at the TE leg locations, and four TE legs (pairs of n- and p-type legs) were soldered to the nickel disks. The TE legs were made of nonoriented, dense pressed bismuth telluride alloy materials. The figure of merit ZT is ~0.67 in the range of 350–520 K. The TE legs were of the size $0.1245 \times 0.1245 \times 0.25$ cm³. Then the STEGs were tested in a vacuum of better than 10^{-4} torr. The upper plate of the STEGs was maintained at 245°C utilizing 250 W radiant heating lamps radiating onto the absorber plates through glass windows in the vacuum manifold. Another side of TE legs was maintained at 70°C by conduction to the metal manifolds supporting the STEGs. An efficiency η_{STEG} of 4.37% was achieved in the flat-plate STEGs.

Typical experimental parameters of some flat-panel STEGs are listed in Table 22.1. The flat-plate nonconcentrated solar generators were limited to moderate temperature ranges. Therefore, this type of STEGs cannot utilize the increasing efficiency of thermocouples at higher temperatures.

The flat-panel STEGs were applied for terrestrial and space applications [25,35,36].

In summary, flat-plate STEGs can only generate moderate temperature differences around 100°C, and the TE efficiency is low. Overall efficiency η_{STEG} is not higher than 1%.

22.3 Thermal Concentration STEGs

In order to increase temperature differences over TE modules, the thermal concentration has been employed in STEGs since 2011. Figure 22.9 shows a typical structure of a STEG cell with thermal concentration. TE legs with the small cross-sectional area were attached to a absorber with a large surface area. Highly solar-absorbing surfaces (solar absorbers) can convert the solar radiation into heat and thermally concentrate it onto the TE elements through lateral heat conduction within the highly thermally conductive absorber substrate. This method of concentrating heat

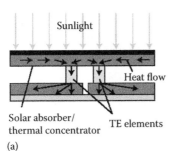

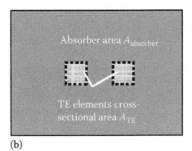

Sunlight

Heat flow

Solar absorber/
thermal concentrator   TE elements

(a)

Absorber area A_{absorber}

TE elements cross-
sectional area A_{TE}

(b)

FIGURE 22.9   STEGs with thermal concentration. (a) Cross-sectional and (b) plan views. (Reprinted by permission from Kraemer, D. et al., *Nature Materials*, 10, 532–538, 2011. Copyright 2011, Macmillan Publishers Ltd.)

by conduction has been used in various solar thermal systems [22,33]. The thermal concentration ratio C_{thermal} is defined by the ratio of the area of the absorber A_{absorber} and the cross-sectional area of the TE elements A_{TE}:

$$C_{\text{thermal}} = \frac{A_{\text{absorber}}}{A_{\text{TE}}}. \qquad (22.4)$$

The absorbers can be circular [23] or rectangular [22], as shown in Figure 22.10.

Figure 22.11 shows a thermal concentration STEG. The STEG consisted of a pair of n-/p-type TE elements, a flat-panel solar-selective absorber that also acts as a thermal concentrator, and two electrodes that serve as heat spreaders and radiation shields. The solar absorber was a multilayer thin-film spectrally selective surface on a copper substrate. The selective surface had a specified solar absorptance of 94.4% and a thermal emittance of 5% at 100°C. The TE elements were made from nanostructured Bi_2Te_3 alloys with their properties given in previous publications [43,56,60]. The dimensions of

Solar absorber

Segmented
legs

Copper electrodes

T-type thermocouples

(a)

Solar absorber

11 mm

Segmented TE legs

(b)

FIGURE 22.10   STEGs with thermal concentration. (a) Rectangular absorber C_{thermal} = 1.4. (b) Circular absorber C_{thermal} = 5.4. (Reprinted by permission from Kraemer, D. et al., *Nature Energy*, 1, 16153, 2016. Copyright 2011, Macmillan Publishers Ltd.)

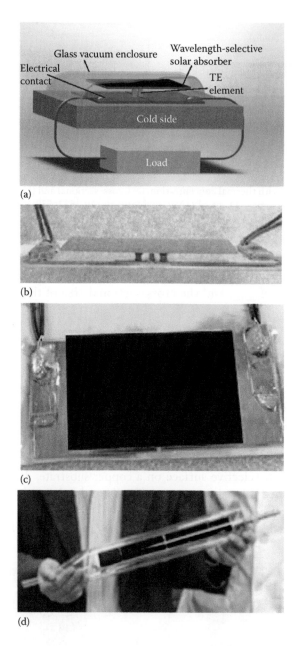

(a)

(b)

(c)

(d)

FIGURE 22.11 Thermal concentration STEG cell. (a) Structure of an STEG cell. (b) Side view and (c) top view of the device. (d) Prototype sealed in a vacuum tube. (Reprinted by permission from Kraemer, D. et al., *Nature Materials*, 10, 532–538, 2011. Copyright 2011, Macmillan Publishers Ltd.)

a typical p-type TE element were $1.35 \times 1.35 \times 1.65$ mm^3, and the n-type elements were of similar dimensions. The effective ZT of TE materials at optimal operation conditions was 1.03, whereas that of the best conventional materials was approximately 0.89. The TE elements were electrically connected in series and sandwiched between the solar absorber plate and the heat sink (copper plates). The copper plates were mounted onto a ceramic plate for electrical insulation and put in good thermal contact with a temperature-controlled sample holder. The copper substrate of the solar absorber serves as a heat concentrator by laterally conducting heat to the TE elements, whereas the bottom copper plates serve as electrodes, heat spreaders, and radiation shields that reduce radiation losses from the rear side of the selective absorber. The device was surrounded by a glass enclosure (Figure 22.11a A solar simulator with an AM1.5G filter was used as the light source. The STEG efficiency $\eta = IV/(q_i A a)$ was calculated from the voltage V, the current I, the absorber area $A_{absorber}$, and the incident solar radiation flux q_i. This efficiency included the glass transmission loss and the solar absorber loss as well as possible parasitic electrical and thermal losses of the system. The transmittance of the glass enclosure is measured to be 94%.

Figure 22.12 shows a typical performance curve of STEGs under illumination intensities corresponding to AM1.5G (1 kW/m²) with optical concentration $C_{optical} = 1$. The corresponding thermal concentration of the STEGs was 299. The peak efficiency was 4.6% at AM1.5G conditions when the cold side was maintained at 20°C. The efficiency was in agreement with theoretical predictions [22].

Figure 22.13 shows *I–V* and *I–P* curves of the STEGs. The voltage–current relation is v theoretically linear.

Figure 22.14 plots the measured peak efficiency as a function of the thermal concentration. There was an optimal thermal concentration for a given TE element geometry and solar intensity. If the thermal concentration was too low, the optothermal efficiency $\eta_{optical-thermal}$ is high because the solar absorber is at a lower temperature and the radiation loss is low; however, the TE device efficiency η_{TE} is low because of the smaller temperature difference. If the thermal concentration is too high, the maximum temperature reached is too high, and, thus $\eta_{optical-thermal}$ is too low. From the experimental results, the optimal thermal concentrations for the chosen dimensions of the Bi_2Te_3 elements was approximately 299 (AM1.5G, 1 kW/m²) and 196 (1.5 kW/m²).

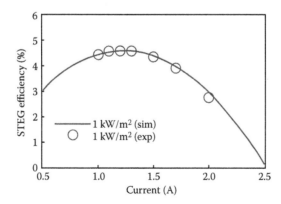

FIGURE 22.12 Efficiency as a function of the cell current at incident solar radiation fluxes of 1 kW/m². Open circles are experimental data; lines are modeling results with thermal concentration $C_{thermal} = 299$. For all experimental data, the cold side was maintained at 20°C. (Reprinted by permission from Kraemer, D. et al., *Nature Materials*, 10, 532–538, 2011. Copyright 2011, Macmillan Publishers Ltd.)

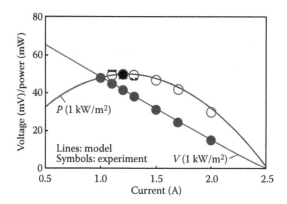

FIGURE 22.13 *I–V* and *I–P* characteristics of the STEGs at incident solar radiation flux of 1 kW/m². Open/filled circles are experimental data; lines are modeling results with thermal concentrations $C_{thermal}$ = 299. $C_{optical}$ = 1. For all experimental data, the cold side was maintained at 20°C. (Reprinted by permission from Kraemer, D. et al., *Nature Materials*, 10, 532–538, 2011. Copyright 2011, Macmillan Publishers Ltd.)

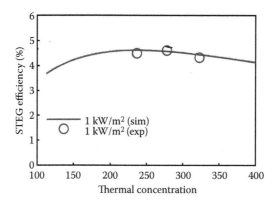

FIGURE 22.14 STEG efficiency as a function of thermal concentration at incident solar radiation flux of 1 kW/m². $C_{optical}$ = 1. For all experimental data, the cold side of the devices was maintained at 20°C. (Reprinted by permission from Kraemer, D. et al., *Nature Materials*, 10, 532–538, 2011. Copyright 2011, Macmillan Publishers Ltd.)

More than 100 similar devices were tested and routinely achieved an efficiency in the range 4.3–4.6% at AM1.5G conditions [22]. This efficiency was seven times higher than that reported by Telkes [21]. The experimental efficiency was close to theoretical values. It was predicted that the device efficiency for the Bi_2Te_3-based flat-plate thermal concentration STEG was 5% at AM1.5G conditions and with a cold junction temperature of 25°C [61].

Several factors can enable efficiency improvements: (1) nanostructured materials with high ZT, (2) a selective surface, and (3) high thermal concentration and operation in an evacuated environment. More experimental results showed that the device efficiency also depends on solar radiation intensity and cold-side temperature. A maximum efficiency of 8–10% was predicted for STEGs with ZT = 1.5–2.0 and with a moderate optical concentration (less than 3). The efficiency approaches 14% with a $ZT = 2$ and 10 times optical concentration [22].

22.4 Optical Concentration STEGs

The conception of optical concentration was proposed in 1888 [16,17]. In the first documented STEGs, the solar rays were concentrated upon solar receivers.

In 1954, Telkes [21] used a condenser lens of 8.6 cm diameter to concentrate solar irradiation. The optical concentration $C_{optical}$ was 50×. A temperature difference of 247°C was generated across TE elements made of a p-type ZnSb alloy and an n-type Bi-based alloy. The device efficiency was increased to 3.35% when the lens was employed as a concentrator.

Goldsmid et al. [26] used semiparabolic solar concentrators to concentrate solar radiation, as shown in Figure 22.15. The concentrated solar energy was collected by a blackened aluminum collector. The heat sink was maintained at a temperature of 30°C. A temperature difference of 120°C was achieved across TEGs. However, the maximum overall efficiency of the system was no more than about 0.5% because three-quarters of solar energy was lost directly to the surroundings (although the top of the collector was covered by an acrylic layer to reduce the convection losses).

Lertsatitthanakorn et al. [62] reported a solar parabolic concentrator to collect solar energy, as shown in Figure 22.16. The parabolic concentrator was constructed from eight parabolic segments of steel, and the concave face of the concentrator was lined with a highly reflective aluminum foil (85% reflective). The absorber and TE modules were located at the focus. The size of the absorber was quite small compared with the reflector area, resulting in a high optical concentration.

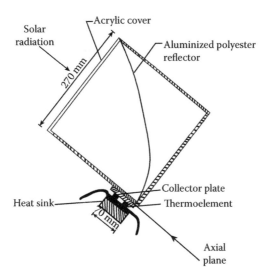

FIGURE 22.15 Section through generating thermocouple and semiparabolic concentrator. (Reprinted from Goldsmid, H. J. et al., Solar thermoelectric generation using bismuth telluride alloys, *Solar Energy*, 24, 435–440, 1980. Copyright 1980, with permission from Elsevier.)

FIGURE 22.16 Photograph of a solar parabolic concentrator coupled to a TE module. (Reprinted from Lertsatitthanakorn, C. et al., Electricity generation from a solar parabolic concentrator coupled to a thermoelectric module, *Energy Procedia*, 52, 150–158 2014. Copyright 2014, with permission from Elsevier.)

Fan et al. [28] also used a parabolic dish concentrator to collect solar radiation and reflected onto a Bi_2Te_3 TEG installed close to the focal point of the parabolic disk. High temperatures were achieved on the bottom face of the TEG. The other face was cooled by forced convection of water using a liquid-cooled heat sink to maintain a large temperature difference across the TEG. A conversion efficiency of 2.9% was reported. However, a low efficiency of 0.78% was reported by other research groups for a similar parabolic dish system [63].

Amatya and Ram [27] combined a parabolic reflection concentrator and a Fresnel lens to concentrate solar energy to 66× suns onto the fixed focal spot, as shown in Figure 22.17. The hot side of a TEG located at this spot was heated up under the concentrated sunlight. A system efficiency of 3% was achieved for a commercial Bi_2Te_3 module with material $ZT \sim 0.9$. A conversion efficiency of 5.6% was expected at 120× suns when n-type $ErAs:(InGaAs)_{1-x}(InAlAs)x$ and p-type $(AgSbTe)x(PbSnTe)_{1-x}$ were employed as TEGs.

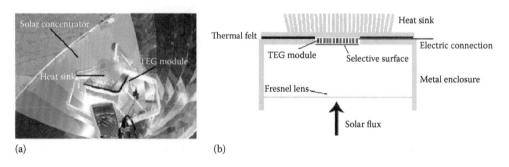

FIGURE 22.17 (a) STEG, showing the concentrator and the heat sink. (b) Schematic of the enclosure holding the TEG. A Fresnel lens is used to increase the flux concentration. (With kind permission from Springer Science+Business Media: *Journal of Electronic Materials*, Solar thermoelectric generator for micropower applications, 39, 2010, 1735–1740, Amatya, R., and Ram, R. J.)

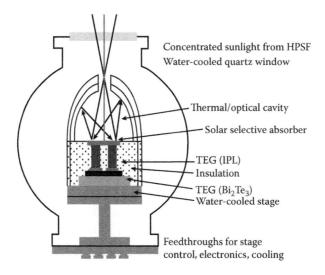

Concentrated sunlight from HPSF
Water-cooled quartz window

Thermal/optical cavity

Solar selective absorber

TEG (IPL)
Insulation
TEG (Bi₂Te₃)
Water-cooled stage

Feedthroughs for stage
control, electronics, cooling

FIGURE 22.18 Schematic of a cavity STEG. (Reprinted from Olsen, M. L. et al., A high-temperature, high-efficiency solar thermoelectric generator prototype, *Energy Procedia*, 49, 1460–1469 2014. Copyright 2014, with permission from Elsevier.)

Similar to the combination of a parabolic reflector and a Fresnel lens, the incident sunlight can also be first concentrated by Fresnel lenses and then focused onto solar collectors/TEGs by reflective flat mirrors [64]. It was predicted that the highest efficiencies of 9.8%, 13.5%, and 14.1% for Bi_2Te_3, SKU, and LAST alloys, respectively, were possible.

Figure 22.18 shows a design of cavity concentration. The thermally insulating cavity allowed a concentrated sunlight to enter through an aperture but further limit radiative losses. The optical concentration system produced a high optical concentration (200×–300× suns) at the absorber surface to achieve a sunlight-generated $T_h = 1000°C$ [42]. A device efficiency of 15% was predicted when n-type SKU and La_3Te_4 and p-type SKU and $Yb_{14}MnSb_{11}$ were employed for TE modules.

Suter et al. [65] simulated a TE solar cavity receiver STEG containing an array of TE modules (p-type $La_{1.98}Sr_{0.02}CuO_4$ and n-type $CaMn_{0.98}Nb_{0.02}O_3$) under 600× suns. The solar-to-electricity efficiency was predicted to be 11%.

Theory modeling indicated that optical concentration STEGs can have efficiencies exceeding 10% based on a geometric optical concentration ratio of 45× at 600°C using SKU and bismuth telluride materials [66].

22.5 Optical/Thermal Coconcentration STEGs

The TE efficiency depends on temperature differences over TE devices, as shown by Equation 22.3. Therefore, optical concentration or thermal concentration has been employed in STEG devices to increase temperature difference beside heat sinks, as discussed in the previous sections. In order to effectively generate higher hot-side temperatures, optical/thermal coconcentration has been used in recent STEG devices [22,23,26,67].

Figure 22.19a shows the concept of STEGs with a coconcentration system. A wavelength-selective surface of area A_s absorbs the solar radiation and hence raises its temperature. The selective surface is coupled to the hot side of a pair of p-type and n-type TE elements that generate electricity. Solar radiation is directed toward the selective surface, with optical concentration $C_{optical}$. The selective surface area is equal to the focal area of the optical system, and the optical concentration ratio $C_{optical}$ is defined as the ratio of the cross-sectional area of the aperture of the optical system divided by the frontal area of the selective surface. As incoming solar radiation is absorbed by the selective surface, heat is conducted along the selective surface to the TE elements. This process is called thermal concentration, and the thermal concentration ratio $C_{thermal}$ is defined as the area of the selective surface divided by the cross-sectional area of the TEG. Numerical results suggested [67] that an STEG system efficiency larger than 5% can be achieved with a hot-side operational temperature between 150°C and 250°C, as shown in Figure 22.19b.

The co-concentration concept was experimentally approved in flat-panel STEG devices with high thermal concentration [22]. Kraemer et al. [22] employed a system shown in Figure 22.11. Solar irradiation was optically concentrated from 1 to 1.5 kW/m² ($C_{optical}$ = 1.5). Nanostructured Bi$_2$Te$_3$ alloys were employed as p- and n-type TE elements. Figure 22.20 shows the typical performance curves of STEGs under illumination intensities with optical concentration $C_{optical}$ = 1.5 and optimal thermal concentration of $C_{thermal}$ = 196. A STEG efficiency of 5.2% was achieved [22], slightly higher than that without optical concentration (4.6%), using 1.5 optical

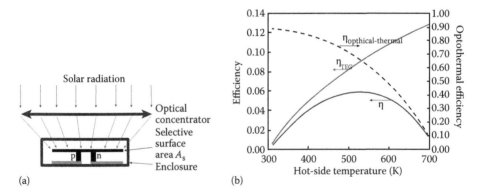

(a)   (b)

FIGURE 22.19 (a) Unit cell of an STEG. Solar radiation may go through an optical concentrator with a concentration ratio $C_{optical}$, transmit through a glass enclosure with a transmittance, and be absorbed by a wavelength-selective surface with a solar absorptance and effective thermal emittance. The absorbed heat conducts through TE elements and is dissipated to the environment at the cold side. Thermal concentration $C_{thermal}$ is the ratio of the selective surface area divided by the total cross-sectional area of the p-type and the n-type TE elements. (b) In a STEG cell, optothermal efficiency ($\eta_{optical-thermal}$) decreases while the TEG efficiency increases with increasing the hot-side temperature, leading to an optimal operational temperature that maximizes the system efficiency (ZT = 1). (Reprinted with permission from Chen, G., *Journal of Applied Physics*, 109, 104908, 2011. Copyright 2011, American Institute of Physics.)

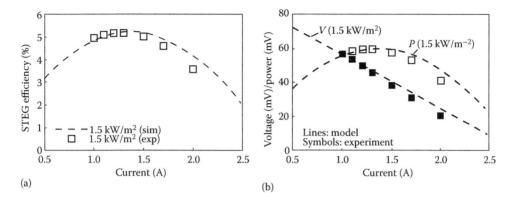

(a)   (b)

FIGURE 22.20 STEG cell performance characteristics. Typical STEG cell characteristics at incident solar radiation flux of 1.5 kW/m². Open/filled squares are experimental data; lines are modeling results with thermal concentrations C_{thermal} = 196 and optical concentration C_{optical} = 1.5. (a) Efficiency as a function of the cell current. (b) I–V and I–P characteristics of the STEGs. For all experimental data, the cold side was maintained at 20°C. (Reprinted by permission from Kraemer, D. et al., *Nature Materials*, 10, 532–538, 2011. Copyright 2011, Macmillan Publishers Ltd.)

concentration of sunlight in combination with thermal concentration C_{thermal} = 196 of solar energy by heat conduction.

Five years later after the first co-concentraction STEGs [22], Kraemer et al. [23] optimized optical concentration and employed a pair of segmented p- and n-type TE elements to improve STEG efficiency. Figure 22.21 shows the design and an experimental prototype. The copper plate was brazed to

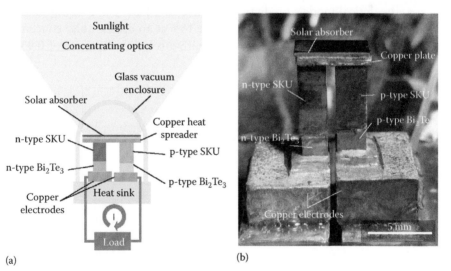

(a)   (b)

FIGURE 22.21 (a) STEG based on a pair of segmented p-/n-type TE legs consisting of doped bismuth telluride (Bi₂Te₃) and SKU materials. The segmented legs are sandwiched between a high-temperature spectrally selective solar absorber and a heat sink with copper electrodes and surrounded by a glass vacuum enclosure. Concentrating optics focuses the incident sunlight onto the solar absorber. (b) STEG cell optimized for high optical concentration (C_{optical} = 200) with a geometric thermal concentration ratio C_{thermal} of 1.4. (Reprinted by permission from Kraemer, D. et al., *Nature Energy*, 1, 16153, 2016. Copyright 2016, Macmillan Publishers Ltd.)

the SKU ends as the hot junction electrically connecting the n-/p-type segmented legs in series. The typical dimensions of the p-type segmented leg including contact pads were approximately $3 \times 3 \times 9$ mm^3, with the bismuth telluride and SKU sections being 1.75 and 5.9 mm long, respectively. The n-type segmented leg was of similar dimensions. The cold-junction copper electrodes for each of the two bismuth telluride legs were soldered onto individual TE coolers, which were attached to a liquid-cooled stage. The segmented TE legs consisting of bismuth telluride and SKU materials were surrounded by tight copper radiation shields to reduce radiation heat loss. The assembly was tested inside a vacuum chamber with vacuum levels of 10^{-3} to 10^{-2} Pa under concentrated solar radiation from a solar simulator.

Figure 22.22 shows the STEG efficiency of an STEG optimized for high optical concentration and low thermal concentration ratio. The combined concentration ratio is 280 ($C_{optical} \times C_{thermal} = 200 \times 1.4$). The STEGs operated with an absorber temperature of approximately 560–580 °C. The TEG efficiency η_{TEG} is 11%, and the absorber efficiency $\eta_{absorber}$ is 90% under optimized conditions. A peak efficiency of 9.6% was reported at a normal solar irradiance of 211 kW/m^2.

Figure 22.23 shows typical STEG efficiency characteristics at high thermal concentration ratio ($C_{thermal} = 5.4$) and low optical concentration ratio ($C_{optical} \sim 50$) with a high-temperature spectrally selective solar absorber. The peak STEG efficiency was 7.6% due to large thermal radiation heat loss of the solar absorber. The absorber operating temperature at the peak STEG performance was approximately 420°C.

Theoretical models predicted [23] that an STEG with an efficient device ZT of 1 can reach an efficiency of over 16% at an optical solar concentration of 1000×, based on a solar absorber with a solar absorptance of 0.93 and a solar absorber temperate of 1000°C.

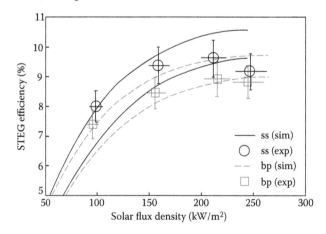

FIGURE 22.22 STEG efficiency as a function of incident solar flux density. Error bars are the result of uncertainty in incident radiation power measurement. Thermal concentration $C_{thermal} = 1.4$ and optical concentration $C_{optical} = 200$. *ss*, high-temperature spectrally selective solar absorbers; *bp*, black paint solar absorbers. (Reprinted by permission from Kraemer D. et al., *Nature Energy*, 1, 16153, 2016. Copyright 2016, Macmillan publishers Ltd.)

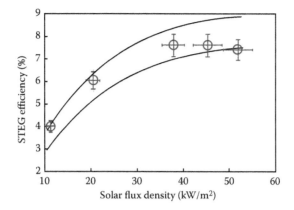

FIGURE 22.23 STEG efficiency as a function of incident solar flux density. Error bars are the result of uncertainty in incident radiation power measurement. Thermal concentration $C_{thermal}$ = 5.4 and optical concentration $C_{optical}$ = 50. Open circles are experimental data; lines are modeling results of upper and lower bounds. (Reprinted by permission from Kraemer D. et al., *Nature Energy* 1, 16153, 2016. Copyright 2016, Macmillan Publishers Ltd.)

22.6 Solar Hybrid TEGs

22.6.1 Thermal TEGs

Tubular solar collectors were used to concentrate solar heat and pump TEGs [29] in the 1980s. TE modules were subjected to a temperature difference of 100°C within the temperature range of 0–200°C, converting heat to electrical energy with an overall efficiency of about 0.75%.

Zhang et al. [68] recently developed a thermal hybrid system where a TEG module was placed at one end of an evacuated tube of solar water heater, as shown in Figure 22.24. The thermal efficiency of this system was about 47%, and the electrical efficiency was only 1.06% with a peak output power of 40 W under daily solar exposure. The electrical efficiency of this

FIGURE 22.24 Pilot TEG cogenerator composed of two evacuated tubular solar collectors, each with 18 tubes. Each tube contains a TE module. (Reprinted from Zhang, M. et al., Efficient, low-cost solar thermoelectric cogenerators comprising evacuated tubular solar collectors and thermoelectric modules, *Applied Energy*, 109, 51–59, 2013. Copyright 2013, with permission from Elsevier.)

system was reduced mainly due to narrow temperature differences ΔT over TE modules and ZT value of about 0.59 of the used TEG modules. Higher device efficiency was expected under higher solar flux and higher ZT.

Besides experiments, some theoretical researches have been reported on thermal hybrid TEG systems [69,70]. An electrical efficiency about 1% was expected.

Up to now, the electrical efficiency of hot pipe/TE hybrid systems is low, about less than a few percentages [71]. It was believed that the hybrid systems could offer small, mobile, transportable, and off-grid power and heating systems, being commercially attractive once TE modules are inexpensive.

22.6.2 Photovoltaic TEGs

Solar photovoltaic TE (PV-TEG) hybrid technology was proposed to utilize the entire solar spectrum to improve conversion efficiency. Figure 22.25 shows a typical PV-TEG hybrid system [72]. TEGs were sandwiched between silicon PV modules and a heat extractor. Generated heat under illumination in each PV cell passed through individual TEG that thermally contacted with PV cells from one side, and with the heat extractor from the other side, generating electricity through the Seebeck effect. The solar radiation can be concentrated or nonconcentrated onto PV cells. A TEG efficiency of about 5% was reported in such a PV-TEG system consisted of a parabolic mirror to achieve a temperature difference of $\Delta T = 150°C$ across Bi_2Te_3 TEG modules and crystalline Si solar PV modules with conversion efficiency of around 14% [72,73].

Wang et al. [74] developed a PV-TEG system consisted of a series-connected dye-sensitized solar cell (DSSC), a solar selective absorber (SSA), and a TEG. The whole purpose was to utilize both the high- and low-energy photon for energy conversion with the help of DSSC and SSA-TEG configuration. The overall conversion efficiency of the PV-TEG system was 13.8%.

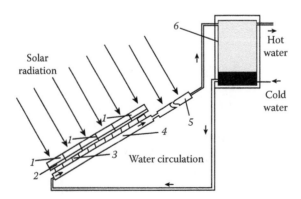

FIGURE 22.25 Scheme of hybrid PV/TEG system. *1*, solar cell; *2*, back electrode of the cell; *3*, TEG; *4*, heat extractor; *5*, plane collector; *6*, thermal tank. (Reprinted from Chávez-Urbiola, E. A. et al., Solar hybrid systems with thermoelectric generators, *Solar Energy*, 86, 369–378, 2012. Copyright 2012, with permission from Elsevier.)

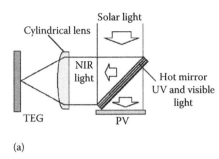

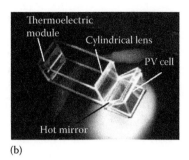

(a)   (b)

FIGURE 22.26 (a) Schematic illustrations of the hybrid PV-TEG module. (b) Photograph of the PV-TE hybrid generator. (Mizoshiri, M. et al., *Japanese Journal of Applied Physics*, 51, 2012. Institute of Physics.)

Zhang et al. [75] developed the first polymer-based PV-TEG system for power production, using a P3HT/IC60BA for making PV cells. This system was able to produce 9–11 mW/cm² power density when the temperature difference was about 5–9 °C, achieving much higher power of the hybrid system.

Mizoshiri et al. [76] developed another type of PV-TEG hybrid system, as shown in Figure 22.26. A hot mirror was employed to separate sunlight into ultraviolet (UV) to visible solar light for PV and near-infrared (NIR) light for TEG module. A cylindrical lens was used to concentrate the NIR light on the TE module. A p-type $Bi_{0.5}Sb_{1.5}Te_3$/n-type $Bi_2Te_{2.7}Se_{0.3}$ TE thin film TEG was used in the system with a temperature difference of about 20°C. It was found that hybridization had led to an improvement of about 1.3% compared to the PV panel alone.

Most of PV-TEG hybrid systems are in the initial stages of research [24], paving the way to simultaneously harvest electricity from solar light and solar heat to get a ful utilization of solar energy. These hybrid systems are promising and challenging. These types of hybrid devices may represent the future of the solar energy conversion, if the cost and system efficiency can be restricted in competitive ranges.

22.6.3 PCM-STEGs

Solar TE devices can generate electricity at daytime while not at night. In order to continuously generate electricity in the absence of solar energy, phase-change materials (PCMs) have been attached to STEGs [77–79]. STEGs can produce electricity from the stored heat energy in PCM without solar irradiation.

Up to now, a large number of PCMs are available in the temperature range of 5–190°C [80,81] with a heat of fusion that closed to water's (333–334 kJ/kg) (such as erythritol with 339 kJ/kg at 118°C, 38.5% MgCl + 61.5% NaCl with 328 kJ/kg at 435°C), even higher (such as NaCl with 492 kJ/kg at 800°C, KF with 452 kJ/kg at 857°C, $NaCO_3$–$BaCO_3$/MgO with 415 kJ/kg at 500–850°C), covering pure inorganic and organic materials as well as mixtures. Some PCMs, such as NaCl and KF, can store three times more heat per unit volume than water. Depending on the working temperature range of STEGs, suitable PCMs can be chosen to store heat at daytime and release energy to STEGs at night.

Figure 22.27 shows a basic concept of hybrid PCM-STEGs [77]. The PCM is placed at back of STEGs where it is not irradiated by the sun and insulated from the external environment except a contact face with STEGs. At daytime, solar radiation powers the STEGs. At the same time, the heat flow passes through STEGs to heat PCMs, which initially are solid. The temperature of PCM rises over the whole day until it reaches the melting point of PCM. At night, the outside face of STEGs becomes cold (Figure 22.27b) while PCM maintains the temperature at around its melting point. Therefore, the PCM works as a heat source at night, and the heat flows from PCM to TEGs. Again, STEGs can generate electricity during night.

Figure 22.28a shows another kind of hybrid PCM-STEG system. The hybrid STEG was composed of a dome-shaped polydimethylsiloxane (PDMS) lens to concentrate the solar radiation, n-octadecane PCM, a wavelength-selective solar absorber, TEG modules, and a cooling part. The PCM was placed between the optical concentrator and the STEG. PCM absorbed/stored solar energy at daytime beside concentrated solar radiation. Solar energy was first concentrated by the dome-shaped PDMS lens and then changed the PCM state from solid to liquid by absorbing the concentrated energy as heat. The solar energy transmitted through PCM was collected by the solar absorber and converted to heat. The converted heat was further converted to electricity by TEG modules. The temperature difference over the TEG modules is shown in Figure 22.28b. When the solar irradiation was removed (marked as "light off"), the stored energy was released via the phase change of the PCM from liquid to solid. A temperature difference was produced by the released heat from the PCM, under the condition without solar energy, as shown in Figure 22.28b. Different from the hybrid PCM-STEGs shown in Figure 22.27, the direction of the heat flux emitted from the PCM here was the same as that in the daytime. Figure 22.28c compares the output power and energy of the PCM-STEG and traditional STEGs. The PCM contributed 34 mJ to the total produced electricity.

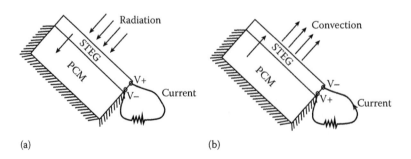

(a)   (b)

FIGURE 22.27 Hybrid PCM-STEGs. (a) Solar radiation on TEG elements and PCM during daytime. (b) PCM works as the heat source at night to power STEGs. (Reprinted from *Sensors and Actuators A: Physical*, 163, Agbossou, A. et al., Solar micro-energy harvesting based on thermoelectric and latent heat effects. part I: Theoretical analysis, 277–283, Copyright 2010, with permission from Elsevier.)

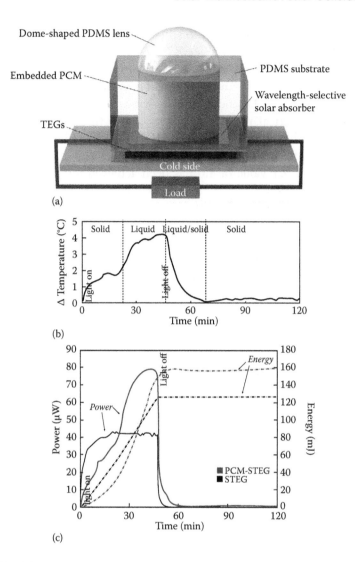

FIGURE 22.28 (a) Structure of a hybrid PCM-STEG. (b) Temperature difference of the hybrid STEG. (c) Comparison of the output power and energy of the hybrid STEG and traditional STEG. (Reprinted by permission from Kim, M.-S. et al., *Scientific Reports*, 6, 27913, 2016. Copyright 2016, Macmillan Publishers Ltd.)

Jo et al. [82] also reported the delayed output power of a PCM-STEG system after a heat source was removed from the hybrid TEG.

Hybrid PCM-STEG systems were also attached near aircraft engines to power autonomous health-monitoring sensors [83] and wireless sensor nodes [84].

22.7 Conclusions and Outlook on STEGs

The concept of STEGs was proposed 100 years ago, and the overall efficiency of STEG devices is slowly developed for decades. The highest device efficiency is only 9.6% [23] up to now because of low-ZT TE materials and heat losses. Luckily, many novel TE materials with high ZT have been recently developed [43,85–90] to efficiently convert heat into electricity, and other TE materials with higher ZT are expected. Solar-selective absorbers have been fabricated [91] to collect solar irradiation without energy losses. All these new materials and design will provide a significant opportunity to efficiently harvest solar energy shortly soon.

Some theoretical modeling predicted [47] that an STEG consisted of the state-of-the-art TE materials ($ZT \sim 1$) could achieve 15.9% generator efficiency under an incident flux of 100 kW/m^2 and a hot-side temperature of 1000°C. A ZT value of 1.0 has been achieved for some state-of-the-art TE nanocomposites, such as SiGe nanocomposites ($ZT = 1.3$ at 900°C for n-type [92] and $ZT = 0.95$ at 800°C for p-type [93]). More theoretical studies claimed that the STEG efficiency of TE materials with $ZT > 2$ may achieve an efficiency of 16–20% in the intermediate temperature range (300–600°C) [94], 24% near 1000°C [23], and even 30.6% at 1500°C [47]. All these predicted efficiencies will be experimentally verified once new TE materials with $ZT \geq 2$ are available. Therefore, STEGs will be competitive with concentrated solar power plants and attract more and more attentions in the next decades.

References

1. X. Niu, J. Yu, and S. Wang, Experimental study on low-temperature waste heat thermoelectric generator, *Journal of Power Sources*, vol. 188, no. 2, pp. 621–626, 2009.
2. R. Ahiska and H. Mamur, Design and implementation of a new portable thermoelectric generator for low geothermal temperatures, *IET Renewable Power Generation*, vol. 7, no. 6, pp. 700–706, 2013.
3. R. Ahiska and S. Dişlitaş, Microcontroller based thermoelectric generator application, *Gazi University Journal of Science*, vol. 19, no. 2, pp. 135–141, 2006.
4. C.-T. Hsu, D.-J. Yao, K.-J. Ye, and B. Yu, Renewable energy of waste heat recovery system for automobiles, *Journal of Renewable Sustainable Energy*, vol. 2, p. 013105, 2010.
5. J. G. Haidar and J. I. Ghojel, Waste heat recovery from the exhaust of low-power diesel engine using thermoelectric generators, *Proceedings of XX International Conference on Thermoelectrics*, pp. 413–418, 2001.
6. M. A. Karri, E. F. Thacher, and B. T. Helenbrook, Exhaust energy conversion by thermoelectric generator: Two case studies, *Energy Conversion and Management*, vol. 52, no. 3, pp. 1596–1611, 2011.
7. M. S. El-Genk, H. H. Saber, and T. Caillat, Efficient segmented thermoelectric uni-couples for space power applications, *Energy Conversion and Management*, vol. 44, no. 11, pp. 1755–1772, 2003.
8. S. Riffat and X. Ma, Thermoelectrics: A review of present and potential applications, *Applied Thermal Engineering*, vol. 23, no. 8, pp. 913–935, 2003.
9. S. LeBlanc, Thermoelectric generators: Linking material properties and systems engineering for waste heat recovery applications, *Sustainable Materials and Technologies*, vols. 1–2, pp. 26–35, 2014.
10. T. Torfs, V. Leonov, R. F. Yazicioglu, P. Merken, C. Van Hoof, R. J. Vullers, and B. Gyselinckx, Wearable autonomous wireless electro-encephalography system fully powered by human body heat, *Sensors, 2008 IEEE*, pp. 1269–1272, 2008.

11. V. Leonov and R. J. M. Vullers, Wearable electronics self-powered by using human body heat: The state of the art and the perspective, *Journal of Renewable Sustainable Energy*, vol. 1, p. 062701, 2009.

12. L. Francioso, C. De Pascali, I. Farella, C. Martucci, P. Cretì, P. Siciliano, and A. Perrone, Flexible thermoelectric generator for ambient assisted living wearable biometric sensors, *Journal of Power Sources*, vol. 196, no. 6, pp. 3239–3243, 2011.

13. R. Y. Nuwayhid, A. Shihadeh, and N. Ghaddar, Development and testing of a domestic woodstove thermoelectric generator with natural convection cooling, *Energy Conversion and Management*, vol. 46, nos. 9–10, pp. 1631–1643, 2005.

14. D. Champier, J. P. Bédécarrats, T. Kousksou, M. Rivaletto, F. Strub, and P. Pignolet, Study of a TE (thermoelectric) generator incorporated in a multifunction wood stove, *Energy*, vol. 36, no. 3, pp. 1518–1526, 2011.

15. H. B. Gao, G. H. Huang, H. J. Li, Z. G. Qu, and Y. J. Zhang, Development of stove-powered thermoelectric generators: A review, *Applied Thermal Engineering*, vol. 96, pp. 297–310, 2016.

16. E. Weston, Apparatus for utilizing solar radiant energy, US Patent, US389124A, 1888.

17. E. Weston, Art of utilizing solar radiant energy, US Patent, US389125A, 1888.

18. M. L. Severy, Apparatus for generating electricity by solar heat, US Patent, US527379A, 1894.

19. M. L. Severy, Apparatus for mounting and operating thermopiles, US Patent, US527377A, 1894.

20. W. W. Coblentz, Thermal generator, US Patent, US1077219A, 1913.

21. M. Telkes, Solar thermoelectric generators, *Journal of Applied Physics*, vol. 25, no. 6, pp. 765–777, 1954.

22. D. Kraemer, B. Poudel, H.-P. Feng, J. C. Caylor, B. Yu, X. Yan, Y. Ma et al., High-performance flat-panel solar thermoelectric generators with high thermal concentration, *Nature Materials*, vol. 10, no. 7, pp. 532–538, 2011.

23. D. Kraemer, Q. Jie, K. McEnaney, F. Cao, W. Liu, L. A. Weinstein, J. Loomis, Z. Ren, and G. Chen, Concentrating solar thermoelectric generators with a peak efficiency of 7.4%, *Nature Energy*, vol. 1, p. 16153, 2016.

24. P. Sundarraj, D. Maity, S. S. Roy, and R. A. Taylor, Recent advances in thermoelectric materials and solar thermoelectric generators—A critical review, *RSC Advances*, vol. 4, no. 87, pp. 46860–46874, 2014.

25. W. W. Coblentz, Harnessing heat from the sun, *Scientific American*, vol. 127, no. 5, p. 324, 1922.

26. H. J. Goldsmid, J. E. Giutronich, and M. M. Kaila, Solar thermoelectric generation using bismuth telluride alloys, *Solar Energy*, vol. 24, no. 5, pp. 435–440, 1980.

27. R. Amatya and R. J. Ram, Solar thermoelectric generator for micropower applications, *Journal of Electronic Materials*, vol. 39, no. 9, pp. 1735–1740, 2010.

28. H. Fan, R. Singh, and A. Akbarzadeh, Electric power generation from thermoelectric cells using a solar dish concentrator, *Journal of Electronic Materials*, vol. 40, no. 5, pp. 1311–1320, 2011.

29. T. Durst, L. B. Harris, and H. J. Goldsmid, Studies of a thermoelectric generator operating from tubular solar collectors, *Solar Energy*, vol. 3, no. 4, pp. 421–425, 1983.

30. H. J. Goldsmid, *Applications of Thermoelectricity*. Pergamon Press Ltd., Methuen, 1960.

31. S. A. Omer and D. G. Infield, Design optimization of thermoelectric devices for solar power generation, *Solar Energy Materials and Solar Cells*, vol. 53, nos. 1–2, pp. 67–82, 1998.

32. S. A. Kalogirou, Solar thermal collectors and applications, *Progress in Energy and Combustion Science*, vol. 30, no. 3, pp. 231–295, 2004.

33. D. Mills, Advances in solar thermal electricity technology, *Solar Energy*, vol. 76, no. 1–3, pp. 19–31, 2004.

34. L. L. Baranowski, E. L. Warren, and E. S. Toberer, High-temperature high-efficiency solar thermoelectric generators, *Journal of Electronic Materials*, vol. 43, no. 6, pp. 2348–2355, 2014.

35. N. Fuschillo and R. Gibson, Germanium-silicon, lead telluride, and bismuth telluride alloy solar thermoelectric generators for Venus and Mercury probes, *Advanced Energy Conversion*, vol. 7, no. 1, pp. 43–52, 1967.

36. N. Fuschillo, R. Gibson, F. K. Eggleston, and J. Epstein, Flat plate solar thermo-electric generator for near-earth orbits, *Advanced Energy Conversion*, vol. 6, no. 2, pp. 103–125, 1966.

37. R. N. Schmidt and K. C. Park, High-temperature space-stable selective solar absorber coatings, *Applied Optics*, vol. 4, no. 8, pp. 917–925, 1965.

38. M. K. Hedayati, M. Javaherirahim, B. Mozooni, R. Abdelaziz, A. Tavassolizadeh, V. S. K. Chakravadhanula, V. Zaporojtchenko, T. Strunkus, F. Faupel, and M. Elbahri, Design of a perfect black absorber at visible frequencies using plasmonic metamaterials, *Advanced Materials*, vol. 23, no. 45, pp. 5410–5414, 2011.

39. N. P. Sergeant, O. Pincon, M. Agrawal, and P. Peumans, Design of wide-angle solar-selective absorbers using aperiodic metal-dielectric stacks, *Optics Express*, vol. 17, no. 25, pp. 22800–22812, 2009.

40. H. Sai, H. Yugami, Y. Kanamori, and K. Hane, Solar selective absorbers based on two-dimensional W surface gratings with submicron periods for high-temperature photothermal conversion, *Solar Energy Materials and Solar Cells*, vol. 79, no. 1, pp. 35–49, 2003.

41. F. Cao, D. Kraemer, T. Sun, Y. Lan, G. Chen, and Z. Ren, Enhanced thermal stability of W-Ni-Al$_2$O$_3$ cermet-based spectrally selective solar absorbers with tungsten infrared reflectors, *Advanced Energy Materials*, vol. 5, no. 2, p. 1401042, 2015.

42. M. L. Olsen, E. L. Warren, P. A. Parilla, E. S. Toberer, C. E. Kennedy, G. J. Snyder, S. A. Firdosy et al., A high-temperature, high-efficiency solar thermoelectric generator prototype, *Energy Procedia*, vol. 49, pp. 1460–1469, 2014.

43. G. J. Snyder and E. S. Toberer, Complex thermoelectric materials, *Nature Materials*, vol. 7, no. 2, pp. 105–114, 2008.

44. A. J. Minnich, M. S. Dresselhaus, Z. F. Ren, and G. Chen, Bulk nanostructured thermoelectric materials: Current research and future prospects, *Energy and Environmental Science*, vol. 2, no. 5, pp. 466–479, 2009.

45. C. J. Vineis, A. Shakouri, A. Majumdar, and M. G. Kanatzidis, Nanostructured thermoelectrics: Big efficiency gains from small features, *Advanced Materials*, vol. 22, no. 36, pp. 3970–3980, 2010.

46. S. K. Bux, J.-P. Fleurial, and R. B. Kaner, Nanostructured materials for thermoelectric applications, *Chemical Communications*, vol. 46, no. 44, pp. 8311–8324, 2010.

47. L. L. Baranowski, G. J. Snyder, and E. S. Toberer, Concentrated solar thermo-electric generators, *Energy and Environmental Science*, vol. 5, no. 10, pp. 9055–9067, 2012.

48. X. Yan, B. Poudel, Y. Ma, W. S. Liu, G. Joshi, H. Wang, Y. Lan, D. Wang, G. Chen, and Z. F. Ren, Experimental studies on anisotropic thermoelectric properties and structures of n-type $Bi_2Te_{2.7}Se_{0.3}$, *Nano Letters*, vol. 10, no. 9, pp. 3373–3378, 2010.

49. S. Fan, J. Zhao, J. Guo, Q. Yan, J. Ma, and H. H. Hng, p-Type $Bi_{0.4}Sb_{1.6}Te_3Bi_{0.4}Sb_{1.6}Te_3$ nanocomposites with enhanced figure of merit, *Applied Physics Letters*, vol. 96, p. 182104, 2010.

50. B. A. Cook, M. J. Kramer, J. L. Harringa, M.-K. Han, D.-Y. Chung, and M. G. Kanatzidis, Analysis of nanostructuring in high figure-of-merit $Ag_{1-x}Pb_mSbTe_{2+m}$ thermoelectric materials, *Advanced Functional Materials*, vol. 19, no. 8, pp. 1254–1259, 2009.

51. Y. Pei, X. Shi, A. LaLonde, H. Wang, L. Chen, and G. J. Snyder, Convergence of electronic bands for high performance bulk thermoelectrics, *Nature*, vol. 473, no. 7345, pp. 66–69, 2011.

52. H. Li, X. Tang, Q. Zhang, and C. Uher, Rapid preparation method of bulk nanostructured $Yb_{0.3}Co_4Sb_{12+y}$ compounds and their improved thermoelectric performance, *Applied Physics Letters*, vol. 93, no. 25, p. 252109, 2008.

53. G. Joshi, T. Dahal, S. Chen, H. Wang, J. Shiomi, G. Chen, and Z. Ren, Enhancement of thermoelectric figure-of-merit at low temperatures by titanium substitution for hafnium in n-type half-Heuslers $Hf_{0.75-x}Ti_xZr_{0.25}NiSn_{0.99}Sb_{0.01}$, *Nano Energy*, vol. 2, no. 1, pp. 82–87, 2013.

54. X. Yan, W. Liu, H. Wang, S. Chen, J. Shiomi, K. Esfarjani, H. Wang, D. Wang, G. Chen, and Z. Ren, Stronger phonon scattering by larger differences in atomic mass and size in p-type half-Heuslers $Hf_{1-x}Ti_xCoSb_{0.8}Sn_{0.2}$, *Energy and Environmental Science*, vol. 5, no. 6, pp. 7543–7548, 2012.

55. V. Raag, L. Hankins, and M. Swerdling, Design concepts of solar thermoelectric generators in space applications, in *International Conference on Thermoelectric Energy Conversion, Proceedings*, pp. 60–65, Institute of Electrical and Electronics Engineers, Inc., Piscataway, NJ, 1978.

56. B. Poudel, Q. Hao, Y. Ma, Y. Lan, A. Minnich, B. Yu, X. Yan et al., High thermoelectric performance of nanostructured bismuth antimony telluride bulk alloys, *Science*, vol. 320, no. 5876, pp. 634–638, 2008.

57. A. J. Krause and J. L. McCabria, *Solar Thermoelectric Generator System Concept and Feasibility Study*. Westinghouse Electric Corp., Pittsburgh, PA, 1962.

58. H. Scherrer, L. Vikhor, B. Lenoir, A. Dauscher, and P. Poinas, Solar thermolectric generator based on skutterudites, *Journal of Power Sources*, vol. 115, no. 1, pp. 141–148, 2003.

59. V. Raag and R. E. Berlin, A silicon-germanium solar thermoelectric generator, *Energy Conversion*, vol. 8, pp. 161–168, 1968.

60. Y. Ma, Q. Hao, B. Poudel, Y. Lan, B. Yu, D. Wang, G. Chen, and Z. Ren, Enhanced thermoelectric figure-of-merit in p-type nanostructured bismuth antimony tellurium alloys made from elemental chunks, *Nano Letters*, vol. 8, no. 8, pp. 2580–2584, 2008.

61. D. Kraemer, K. McEnaney, M. Chiesa, and G. Chen, Modeling and optimization of solar thermoelectric generators for terrestrial applications, *Solar Energy*, vol. 86, no. 5, pp. 1338–1350, 2012.

62. C. Lertsatitthanakorn, J. Jamradloedluk, and M. Rungsiyopas, Electricity generation from a solar parabolic concentrator coupled to a thermoelectric module, *Energy Procedia*, vol. 52, pp. 150–158, 2014.

63. M. Eswaramoorthy, S. Shanmugam, and A. Veerappan, Experimental study on solar parabolic dish thermoelectric generator, *International Journal of Energy Engineering*, vol. 3, no. 3, p. 62, 2013.

64. P. Li, L. Cai, P. Zhai, X. Tang, Q. Zhang, and M. Niino, Design of a concentration solar thermoelectric generator, *Journal of Electronic Materials*, vol. 39, no. 9, pp. 1522–1530, 2010.

65. C. Suter, P. Tomeš, A. Weidenkaff, and A. Steinfeld, A solar cavity-receiver packed with an array of thermoelectric converter modules, *Solar Energy*, vol. 85, no. 7, pp. 1511–1518, 2011.

66. K. McEnaney, D. Kraemer, Z. Ren, and G. Chen, Modeling of concentrating solar thermoelectric generators, *Journal of Applied Physics*, vol. 110, no. 7, p. 074502, 2011.

67. G. Chen, Theoretical efficiency of solar thermoelectric energy generators, *Journal of Applied Physics*, vol. 109, no. 10, p. 104908, 2011.

68. M. Zhang, L. Miao, Y. P. Kang, S. Tanemura, C. A. Fisher, G. Xu, C. X. Li, and G. Z. Fan, Efficient, low-cost solar thermoelectric cogenerators comprising evacuated tubular solar collectors and thermoelectric modules, *Applied Energy*, vol. 109, pp. 51–59, 2013.

69. W. He, Y. Su, S. Riffat, J. Hou, and J. Ji, Parametrical analysis of the design and performance of a solar heat pipe thermoelectric generator unit, *Applied Energy*, vol. 88, no. 12, pp. 5083–5089, 2011.

70. W. He, Y. Su, Y. Wang, S. Riffat, and J. Ji, A study on incorporation of thermoelectric modules with evacuated-tube heat-pipe solar collectors, *Renewable Energy*, vol. 37, no. 1, pp. 142–149, 2012.

71. K. S. Ong, Review of solar, heat pipe and thermoelectric hybrid systems for power generation and heating, *International Journal of Low-Carbon Technologies*, vol. 0, pp. 1–6, 2015.

72. E. A. Chávez-Urbiola, Y. V. Vorobiev, and L. P. Bulat, Solar hybrid systems with thermoelectric generators, *Solar Energy*, vol. 86, no. 1, pp. 369–378, 2012.

73. E. A. Chávez-Urbiola and Y. Vorobiev, Investigation of solar hybrid electric/thermal system with radiation concentrator and thermoelectric generator, *International Journal of Photoenergy*, vol. 2013, p. 704087, 2013.

74. N. Wang, L. Han, H. He, N.-H. Park, and K. Koumoto, A novel high-performance photovoltaic-thermoelectric hybrid device, *Energy and Environmental Science*, vol. 4, no. 9, pp. 3676–3679, 2011.

75. Y. Zhang, J. Fang, C. He, H. Yan, Z. Wei, and Y. Li, Integrated energy-harvesting system by combining the advantages of polymer solar cells and thermoelectric devices, *The Journal of Physical Chemistry C*, vol. 117, no. 47, pp. 24685–24691, 2013.

76. M. Mizoshiri, M. Mikami, and K. Ozaki, Thermal–photovoltaic hybrid solar generator using thin-film thermoelectric modules, *Japanese Journal of Applied Physics*, vol. 51, p. 06FL07, 2012.

77. A. Agbossou, Q. Zhang, G. Sebald, and D. Guyomar, Solar micro-energy harvesting based on thermoelectric and latent heat effects. part I: Theoretical analysis, *Sensors and Actuators A: Physical*, vol. 163, no. 1, pp. 277–283, 2010.

78. Q. Zhang, A. Agbossou, Z. Feng, and M. Cosnier, Solar micro-energy harvesting based on thermoelectric and latent heat effects. part II: Experimental analysis, *Sensors and Actuators A: Physical*, vol. 163, no. 1, pp. 284–290, 2010.

79. M.-S. Kim, M.-K. Kim, S.-E. Jo, C. Joo, and Y.-J. Kim, Refraction-assisted solar thermoelectric generator based on phase-change lens, *Scientific Reports*, vol. 6, p. 27913, 2016.
80. B. Zalba, J. M. Marín, L. F. Cabeza, and H. Mehling, Review on thermal energy storage with phase change: Materials, heat transfer analysis and applications, *Applied Thermal Engineering*, vol. 23, no. 3, pp. 251–283, 2003.
81. V. V. Tyagi and D. Buddhi, PCM thermal storage in buildings: A state of art, *Renewable and Sustainable Energy Reviews*, vol. 11, no. 6, pp. 1146–1166, 2007.
82. S.-E. Jo, M.-S. Kim, M.-K. Kim, and Y.-J. Kim, Power generation of a thermo-electric generator with phase change materials, *Smart Materials and Structures*, vol. 22, no. 11, p. 115008, 2013.
83. A. Elefsiniotis, T. Becker, and U. Schmid, Thermoelectric energy harvesting using phase change materials (PCMs) in high temperature environments in aircraft, *Journal of Electronic Materials*, vol. 43, no. 6, pp. 1809–1814, 2014.
84. D. Samson, M. Kluge, T. Becker, and U. Schmid, Wireless sensor node powered by aircraft specific thermoelectric energy harvesting, *Sensors and Actuators A: Physical*, vol. 172, no. 1, pp. 240–244, 2011.
85. J. R. Sootsman, D. Y. Chung, and M. G. Kanatzidis, New and old concepts in thermoelectric materials, *Angewandte Chemie International Edition*, vol. 48, no. 46, pp. 8616–8639, 2009.
86. Y. Lan, A. J. Minnich, G. Chen, and Z. Ren, Enhancement of thermoelectric figure-of-merit by a bulk nanostructuring approach, *Advanced Functional Materials*, vol. 20, no. 3, pp. 357–376, 2010.
87. J.-F. Li, W.-S. Liu, L.-D. Zhao, and M. Zhou, High-performance nanostructured thermoelectric materials, *NPG Asia Mater*, vol. 2, pp. 152–158, 2010.
88. T. M. Tritt, Thermoelectric phenomena, materials, and applications, *Annual Review of Materials Research*, vol. 41, no. 1, pp. 433–448, 2011.
89. W. Liu, X. Yan, G. Chen, and Z. Ren, Recent advances in thermoelectric nanocomposites, *Nano Energy*, vol. 1, no. 1, pp. 42–56, 2012.
90. S. Chen and Z. Ren, Recent progress of half-Heusler for moderate temperature thermoelectric applications, *Materials Today*, vol. 16, no. 10, pp. 387–395, 2013.
91. F. Cao, K. McEnaney, G. Chen, and Z. Ren, A review of cermet-based spectrally selective solar absorbers, *Energy and Environmental Science*, vol. 7, no. 5, pp. 1615–1627, 2014.
92. X. W. Wang, H. Lee, Y. C. Lan, G. H. Zhu, G. Joshi, D. Z. Wang, J. Yang et al., Enhanced thermoelectric figure of merit in nanostructured n-type silicon germanium bulk alloy, *Applied Physics Letters*, vol. 93, no. 19, p. 193121, 2008.
93. G. Joshi, H. Lee, Y. Lan, X. Wang, G. Zhu, D. Wang, R. W. Gould et al., Enhanced thermoelectric figure-of-merit in nanostructured p-type silicon germanium bulk alloys, *Nano Letters*, vol. 8, no. 12, pp. 4670–4674, 2008.
94. K. Biswas, J. He, I. D. Blum, C.-I. Wu, T. P. Hogan, D. N. Seidman, V. P. Dravid, and M. G. Kanatzidis, High-performance bulk thermoelectrics with all-scale hierarchical architectures, *Nature*, vol. 489, no. 7416, pp. 414–418, 2012.

Index

Page numbers followed by f and t indicate figures and tables, respectively.

by CPI Group (UK) Ltd, Croydon, CR0 4YY

01/11/2024

782603-0018